DYNAMICAL PROCESSES IN CONDENSED MATTER

ADVANCES IN CHEMICAL PHYSICS

VOLUME LXIII

DYNAMICAL PROCESSES IN CONDENSED MATTER

Edited by

MYRON W. EVANS

University College of North Wales

ADVANCES IN CHEMICAL PHYSICS
VOLUME LXIII

Series editors

Ilya Prigogine

University of Brussels
Brussels, Belgium
and
University of Texas
Austin, Texas

Stuart A. Rice

Department of Chemistry
and
The James Franck Institute
University of Chicago
Chicago, Illinois

AN INTERSCIENCE® PUBLICATION

JOHN WILEY & SONS

New York · Chichester · Brisbane · Toronto · Singapore

An Interscience® Publication

Library of Congress Cataloging in Publication Data:

Main entry under title:

Dynamical processes in condensed matter.

  (Advances in chemical physics, ISSN 0065-2385; v. 63)
  "An Interscience publication."
  Includes indexes.
  1. Condensed matter—Addresses, essays, lectures.
2. Chemistry, Physical and theoretical—Addresses,
essays, lectures.   I. Evans, Myron W. (Myron Wyn),
1950–   II. Series.

QD453.A27 vol. 63 [QC173.4.C65] 541 s [530.4] 85-5328
ISBN 0-471-80778-8

Printed in the United States of America

10 9 8 7 6 5 4 3 2 1

INTRODUCTION

Few of us can any longer keep up with the flood of scientific literature, even in specialized subfields. Any attempt to do more, and be broadly educated with respect to a large domain of science, has the appearance of tilting at windmills. Yet the synthesis of ideas drawn from different subjects into new, powerful, general concepts is as valuable as ever, and the desire to remain educated persists in all scientists. This series, *Advances in Chemical Physics*, is devoted to helping the reader obtain general information about a wide variety of topics in chemical physics, which field we interpret very broadly. Our intent is to have experts present comprehensive analyses of subjects of interest and to encourage the expression of individual points of view. We hope that this approach to the presentation of an overview of a subject with both stimulate new research and serve as a personalized learning text for beginners in a field.

ILYA PRIGOGINE
STUART A. RICE

PREFACE

Volume LXII of this series was concerned with the axioms developed recently by Professor Grigolini, Professor Pastori, and their co-workers for the solution of the differential equations that govern (for example) the evolution of conditional probability density functions. Numerical algorithms were devised and applied to problems that depend on equations of this type. It was emphasized that one particular class of equations may have applications in several different areas, and that the techniques developed in pursuit of a solution of one type of problem may be of considerable utility in a different discipline. In this way, a large number of seemingly unrelated observations may be explained in terms of a much smaller number of axioms. This is a feature of chemical physics and related disciplines, and it is natural, therefore, to follow Volume LXII, devoted mainly to axioms and applications, with a second volume describing experimental, numerical, and related analytical methods and observations.

It may seem that the only thing common to the articles that follow is that they deal with condensed matter. A symmetry of purpose in what follows may not be found easily without reference to Volume LXII, whose ideas find application, in some way or another, to the subjects reviewed individually in this volume. Each article in this volume is complete in itself, but is also intended to summarize a field of work in which the axioms of Volume LXII have been or could be of use.

The first article, by M. Andretta, R. Serra, and G. Zanarini, industrial specialists of TEMA S.p.A., Bologna, Italy, in association with K. Pendergast, Cross Keys College, is on the nature of soliton diffusion in activated polymers. This class of polymers is already being made the basis for a new high-technology industry following the discovery that their conductivities may range over many orders of magnitude when they are doped with the appropriate ionic or similar material. This clearly written description offers a wide-ranging introductory review.

There follows a useful account by W. T. Coffey of the development and applications of the Kramers equations. Methods akin to those of Volume LXII are applied therein to solve these equations numerically.

The review by L. A. Dissado, R. Nigmatullin, and R. M. Hill develops the relationship between a progressive loss of memory and structure relaxation in various condensed media, for which a universal character of the frequency dependence of properties such as dielectric loss has been established over a multidecade range. This review provides some interesting insights into the methods of Volume LXII from a different viewpoint.

The article by G. J. Evans illustrates (with photographs) his recent discovery of the inverse of the important Costa–Ribeiro thermodielectric effect, that is, the ability of electric fields to bring about liquid-to-solid phase changes of technological importance. The theory of such phase changes may well be based in the future on the Kramers equations with external fields, described in Volume LXII. It is already possible in this context to follow analytically another important effect discovered recently and described here by G. J. Evans—that of ultrafine spectral detail in the far-infrared spectrum of liquid water. The analogous work of Bloembergen and Purcell has been reviewed already in the *Advances in Chemical Physics* series.

The review by M. W. Evans and G. J. Evans deals with the formidable array of spectroscopic techniques now available for investigating stochastic processes in condensed matter, and attempts to draw (with the aid of computer simulations) a pattern of conclusions from the results for some liquids, such as dichloromethane, that have been studied intensively using these techniques. This body of results presents an interesting opportunity for testing the self-consistency of the experimental data using the analytical methods of Volume LXII.

The most important development in the study of "pure" molecular dynamics in the last decade has been that of computer simulation. The rapid increase in the power and speed of computers, the introduction of array processing, and the improved comprehensibility of software have made this technique indispensable for the study of, for example, spectral bandshapes in the molecular liquid state. It has been referred to and used extensively in Volume LXII and is again used in the review article by Evans and Evans. It is therefore appropriate that an article be devoted to a description of the many new simulation techniques that have been developed. The article by D. Fincham and D. M. Heyes describes in detail the software of the SERC CCP5 group and others, and includes actual examples of FORTRAN programs for methods such as distributed array processing, constant-pressure simulation, field-effect simulation with the difference method, and the linked-cell method for handling samples of up to 10,000 or even more polyatomic molecules. This article should be of particular interest to non-specialists, who might be encouraged to use the software and documentation available free on request from the SERC's Daresbury Laboratory, near Warrington, U.K.

The nature of vibrational relaxation in molecular liquids is potentially of great interest for the determination and prediction of chemical reactions in solution. The article by M. F. Herman and E. Kluk is therefore timely in its description of vibrational relaxation, and may be cross-referenced to the analytical description, using reduced model theory, in Volume LXII, and to the numerical methods described by Fincham and Heyes in this volume.

Some aspects of the Kramers equation, and of the methods devised by Grigolini for its solution, are mentioned by F. Marchesoni. This article may be cross-referenced to the one by M. W. Evans and G. J. Evans, in which an attempt is made to use a variety of resources for studying selected samples.

Some of the most subtle and beautiful observational results in research on molecular condensed matter have emerged from the study of liquid crystals. J. K. Moscicki reviews some of his recent theoretical and experimental work in this field, carried out at the University College of Wales, Aberystwyth, U.K.; Laboratoire du CNRS Maurice Letoit, France; and Cornell University, U.S.A. The richness and variety of the dynamical behavior of rigid rodlike polymer liquid crystals, revealed through the Kerr effect, and of the equilibrium phase diagrams of these polymers bear ample witness to the interest these results might have for future analysis with a variety of methods.

In his article, W. Schröer describes and extends the results of his Cambridge University Ph.D. thesis on the structural properties of molecular liquids. These are exemplified by the various generalizations he has made of the Kirkwood–Fröhlich theory and of the Clausius–Mosotti function in order to account for the medium-range order imposed by polyatomic molecules in the liquid state. These analytical results are particularly important for interpretation of the numerical indications given by computer simulations. The structure and dynamics of the molecular liquid state are inextricably interwoven, and Schröer's contribution provides an invaluable complement to the dynamical considerations that make up most of Volumes LXII and LXIII.

The final article, by J. K. Vij and F. Hufnagel, focuses on the technology of contemporary microwave and far-infrared spectroscopy, exemplified by Fourier-transform interferometry and klystron and laser spectroscopy. The availability of powerful dedicated computers has yielded a considerable advance in interferometer capability. Completely automated instruments are now available with ranges of 2–24,000 cm^{-1} at resolutions of 0.01 cm^{-1}. At this limit the experimental behavior is often at its most interesting, and therefore fertile ground for application of the ideas of Volume LXII can be found in these data, particularly at high resolution.

M. W. EVANS

Bangor, Wales
March 1985

CONTRIBUTORS TO VOLUME LXIII

M. ANDRETTA, TEMA S.p.A. (ENI Group), Bologna, Italy

WILLIAM COFFEY, Department of Microelectronics and Electrical Engineering, Trinity College, Dublin, Ireland

L. A. DISSADO, Department of Physics, Chelsea College, University of London, London, United Kingdom

GARETH J. EVANS, Department of Chemistry, University College of Wales, Aberystwyth, Dyfed, Wales

M. W. EVANS, Department of Applied Mathematics and Computing, University College of North Wales, Bangor, Gwynedd, Wales

D. FINCHAM, DAP Support Unit, Queen Mary College, University of London, London, United Kingdom

M. F. HERMAN, Department of Chemistry, Tulane University, New Orleans, Louisiana

D. M. HEYES, Department of Chemistry, Royal Holloway College, University of London, Surrey, United Kingdom

R. M. HILL, Department of Physics, Chelsea College, University of London, London, United Kingdom

F. HUFNAGEL, Microwave Laboratory, Institute of Physics, Johannes Gutenberg University, Mainz, Federal Republic of Germany

E. KLUK, Department of Chemistry, Tulane University, New Orleans, Louisiana

F. MARCHESONI, Dipartimento di Fisica, Università di Perugia, Perugia, Italy

J. K. MOSCICKI, Institute of Physics, Jagellonian University, Cracow, Poland

R. NIGMATULLIN, Department of Chemistry, Kazan State University, Kazan, U.S.S.R.

K. PENDERGAST, Department of Earth and Natural Science, Cross Keys College, Cross Keys, South Wales

WOLFFRAM SCHRÖER, Department of Chemistry, University of Bremen, Bremen, Federal Republic of Germany

R. SERRA, TEMA S.p.A. (ENI Group), Bologna, Italy

J. K. VIJ, Department of Microelectronics and Electrical Engineering, Trinity College, Dublin, Ireland

G. ZANARINI, Dipartimento di Fisica, Universita di Bologna, Bologna, Italy

CONTENTS

TRANSPORT PROPERTIES AND SOLITON MODELS OF POLYACETYLENE

M. ANDRETTA

TEMA S.p.A. (ENI Group), Fiera District, Bologna, Italy

R. SERRA

TEMA S.p.A. (ENI Group), Fiera District, Bologna, Italy

G. ZANARINI

Dipartimento di Fisica, Università di Bologna, Bologna, Italy

K. PENDERGAST

*Department of Earth and Natural Science, Cross Keys College,
Cross Keys, Gwent, South Wales*

CONTENTS

1

I. INTRODUCTION

Over the last 10 years interest has been increasing in the possible applications of conducting polymers. In principle, it is possible to combine the electrical properties of metals and semiconductors with the mechanical characteristics and workability of polymers. Using such an approach, the problems of weight, mechanical fragility, cost, and so on could be drastically reduced and new applications found.

Polymeric materials have already found vast application as dielectrics. The first attempt to provide polymers with significant electrical conductivity involved mixing metallic powders with the polymer. In 1973[1] metallic conductivity was observed in polymeric crystals of nitrogen and sulfur [$(SN)_x$]. Although this is the only example of a polymer possessing intrinsically metallic properties, even more important from an applicative standpoint was the practical demonstration that the electrical conductivity of a polymeric material could be changed by impurity-doping analogous to that used in traditional semiconductors. This effect of doping was particularly evident in polyacetylene, $(CH)_x$, first synthesized by Natta et al.[2] more than 20 years ago.

Interest in this material has grown only recently, after Shirakawa et al. succeeded in synthesizing it in the form of a film[3] and after the discovery by the group at Pennsylvania University that dopants such as AsF_5, alkali metals, and halogens drastically changed its electrical properties.[4,5] In fact, the electrical conductivity can vary over 12 orders of magnitude as a function of dopant concentration, passing from typical insulator values through the semiconductor range to almost metallic conductivity.

Analogous effects of doping with AsF_5 and alkali metals have been observed in other polymers such as poly(p-phenylene) and polypyrrole.[1] Polyacetylene (PAC), however, is the most important conducting polymer, as is shown by the large number of references to $(CH)_x$ in the scientific and patent literature. In this regard we mention only the application of lithium iodide-doped PAC films in lightweight rechargeable batteries and the realization of low-cost (but limited efficiency) solar cells.[1] Potential applications include the use of PAC in thermoelectric generators, piezoelectric converters, high-tension screening, and electrophotography.

The central importance of PAC among electrically active polymers is linked to another aspect, which has motivated the choices made in writing this review: alongside the experimental work, there has been a notable increase in attempts at a theoretical understanding of the peculiar properties of these polymers. At first, the theoretical work aimed at comparing the properties of polymeric conductors with those of traditional semiconductors, taking advantage of the developments taking place in the field of amorphous semiconductors. These theoretical developments, in turn, were the fruit of a generalized effort to reduce costs in the integrated circuit industry, which has succeeded in finding some applications in which noncrystalline materials can be used.

The attempt to base the theory of electrical conductivity in polymers on that for amorphous semiconductors has, however, had only partial success. As will be shown later, such a theoretical approach cannot constitute a unified framework within which to interpret the increasing amount of experimental data (not only electrical, but also magnetic and optical).

This situation is analogous to that which characterized the field of traditional semiconductors up to the 1940s. Then, Shockley made the decisive step forward for both semiconductor theory and technology by choosing to work with a simple material, crystalline germanium, the most directly accessible theoretically.[6] The choice of PAC now as the material on which to concentrate experimental and theoretical effort is analogous to the choice of crystalline germanium then.

The development of soliton models, although limited so far to the schematization of a single PAC (or even of a polyene) chain, has led to a unified theoretical framework capable not only of interpreting experimental data, but also of suggesting new experiments. If, however, we examine the ample literature on PAC, the impression we receive is somewhat vague. The urgency of publishing new scientific results, the contrasts between different schools, and the industrial influences in play lead to a confused and fragmentary picture.

The first part of this review applies the main soliton models to the structure of PAC, pointing out the analogies and differences between the different theoretical approaches and briefly comparing the experimental data relating to magnetic and optical properties. The electrical properties are discussed in more detail, forming the second part of the review. The experimental data and theoretical hypotheses are presented in order of increasing dopant concentration, thus allowing the identification of the regions in which the electrical properties show different characteristics and the illustration of the dominant conduction mechanism for each region. Finally, in the appendixes, we discuss the relationship between the continuous and the discrete structural models in more detail.

The literature on the subject is large and ever-growing. This review will therefore be not a complete account of the existing literature, but rather a choice of arguments integrated in a unified picture of the unusual properties of this fascinating material.

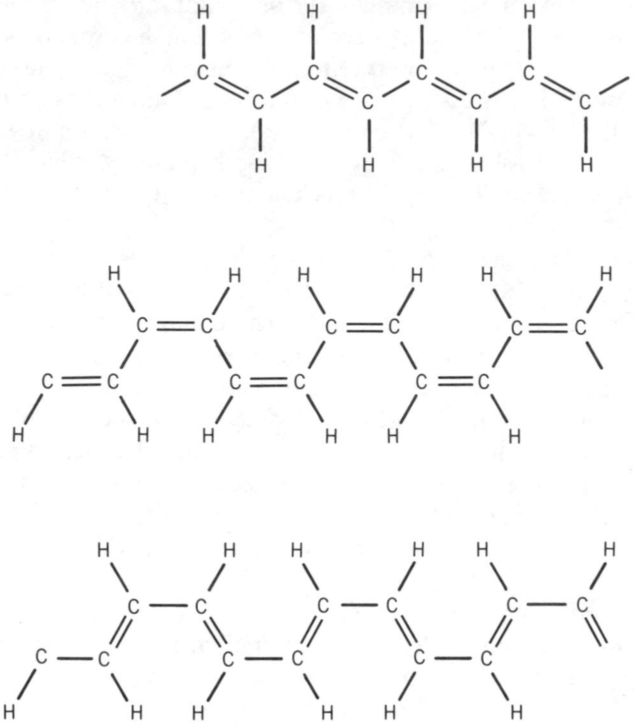

trans–trans–oid

cis–trans–oid

trans–cis–oid

Figure 1. Isomers of polyacetylene.

II. SOLITON MODELS OF POLYACETYLENE

A. General Characteristics of Polyacetylene

There are three isomers of PAC [$(CH)_x$], the structures of which are shown in Fig. 1. The most thermodynamically stable isomer is the *trans*-transoid (abbreviated as trans), which is also the most interesting as regards electrical properties, as well as the most studied.

It is possible to control the *cis/trans* ratio of a PAC film by varying the temperature of polymerization or by thermal isomerization of a predominantly *cis* sample synthesized at low temperature (−78°C).[7] It is also thought that impurity doping and the addition of oxygen lead to *cis*–*trans* isomerization.[8,9]

From here onward, we shall refer only to *trans*-$(CH)_x$ unless otherwise stated.

In order that the reader may better understand the nature of the various electronic bonds in PAC, we will briefly examine the types of orbitals involved. In PAC three of the four valence electrons of each carbon atom give rise to sp^2 hybrid orbitals, which form σ-bonds with both the hydrogen atom and the two neighboring carbon atoms. The wavefunction of the remaining p electron is involved in π-bonding between adjacent carbon atoms.

One would therefore suppose, at first sight, that *trans*-PAC has a resonant structure giving rise to C—C bonds of equal length. Experimental data, however, exclude this hypothesis[10] and indicate alternation of short (1.36 Å) with long (1.43 Å) bonds. Moreover, theoretical considerations suggest that the formation of a structure with alternate double bonds ("dimerized structure"[‡]), as opposed to a uniform structure, is energetically favored.

The substance of Peierls' argument for this is as follows:[12] the nondimerized structure would correspond to a metallic type of material, since there would be no gap between the π and π^* levels. The introduction of a spatially periodic potential, associated with small changes in the atomic positions with respect to the nondimerized structure, creates a gap at the Fermi level, thus lowering the energy of the ground state. Therefore, there is a reduction in the electronic energy of the chain that is greatest when the distortion has a wavevector equal to twice the Fermi wavevector, K_f, which is defined as a function of the Fermi energy E_f by

$$\frac{\hbar^2 K_f^2}{2m_e} = E_f \qquad (1)$$

where m_e is the mass of the electron and $\hbar$ is the reduced Planck constant.

In one-dimensional materials the gain in electronic energy due to Peierls' distortion is always greater than the increase in lattice energy.[3] It is thus expected that the ground state is the distorted dimerized one.

The alternating nature of the bonds, which creates a gap at the Fermi level, drastically changes the electrical properties of the material. Thus the nondimerized structure should show a metallic behavior, whereas in this case we clearly have semiconductor characteristics. In fact, the existence of a gap of about 1.4 eV has been observed experimentally in PAC.[4]

PAC, from a model point of view, seems therefore analogous to a traditional disordered semiconductor. The disorder can be attributed both to imperfect crystallization and to structural changes caused by the inclusion of possible impurities.

Anticipating a subject that will be examined more thoroughly in Section III, we note that this hypothetical model is also supported by the existence

---

[‡] The use of the term "dimerized structure" to denote the presence of alternate short and long bonds is now commonplace in the PAC literature.[11] This clearly has no correspondence to the term "dimer" as used in organic chemistry.

of a semiconductor-to-metal transition at dopant levels of a few percent, which was initially interpreted as an Anderson transition, typical of disordered materials.[10] It is therefore possible to describe some of the electrical properties of PAC without referring to unusual concepts such as the topological "soliton."

However, more complete understanding of the electrical properties of doped PAC, especially at low dopant levels, is possible within the scope of the soliton model. Moreover, this model also accounts for the magnetic and optical properties of the material, thus providing a unified picture of its electromagnetic behavior.

At this point, therefore, we will take the opportunity of introducing the fundamental concepts of the soliton theories of PAC and comparing them with the principle magnetic and optical experimental data. In the next section we will discuss in more detail the mechanisms proposed to explain PAC's electrical conduction.

The concept at the root of the soliton models of PAC is linked to its magnetic behavior.[15,16] *Trans*-$(CH)_x$ shows an intense electron spin resonance (ESR) signal, which demonstrates the existence of unpaired spins, at a level of about 1 for every 3000 carbon atoms. It is possible to identify the π-electron system as the origin of the signal. The existence of unpaired spins is also shown by the presence of a Curie-type component in the paramagnetic susceptibility,[4] which is associated with the presence of magnetic dipoles that can orient themselves freely in an external field, and therefore with the presence of singly occupied levels.[2]‡ The amplitude of the ESR signal suggests that the unpaired spins in PAC have considerable mobility down to a temperature of 9 K. On doping, for example, with AsF_5, there is a drastic reduction in the ESR signal and in the Curie susceptibility, accompanied by a spectacular increase in the electrical conductivity.

Goldberg and coworkers having noted that the presence of free and mobile spins was associated with low conductivity, concluded that the ESR signal had its origin in neutral defects present in the chain.[17] They formulated the hypothesis that these defects were "domain walls" that separated different portions of the polymeric chain, characterized by changes of phase in the alternation of the bonds (see Fig. 2). By analogy with ferromagnetic materials, they supposed that the domain walls of PAC would be delocalized, extending over about 10 lattice constants. The development of this hypothesis by Goldberg et al.[17] has led to the above-mentioned "soliton theory" of PAC.

In reality, several different approaches of this type exist: the soliton hypothesis was independently proposed by M. J. Rice[18] and by Su, Schrieffer, and Heeger[11] (SSH) in March 1979. Rice's description is of a more phenom-

---

‡We will return to this subject in Section II.F [Eq. (27)].

Figure 2. A domain wall separating two portions of *trans*-(CH)$_x$ by means of a change in phase in the alternation of the bonds.

enological character than that of SSH, who start from a detailed microscopic model of the PAC chain. In August of the same year, Takayama, Lin-Liu, and Maki[19] (TLM) proposed a model of *trans*-(CH)$_x$ that may be considered the continuous limit of the SSH model.[‡]

In the following sections, we will review these three models, which give the same qualitative, and often the same quantitative, results, and will discuss the relationships between them.

B. Rice's Model

To understand Rice's hypotheses on PAC, a short discussion of the concept of a charge density wave (CDW) is necessary. We have seen in the previous section that a hypothetical one dimensional conductor is, in reality, subject to the phenomenon known as Peierls' distortion. Thus the atoms in the lattice will occupy positions different from those of the reference configuration, which, in PAC, corresponds to the nondimerized structure in which all the C—C bonds have the same length. This periodic lattice distortion (PLD) is necessarily accompanied by a periodic distortion in the electron density distribution of the system (i.e., the CDW), as was demonstrated by Fröhlich.[21]

The PLD and the CDW together form the "condensate,"[3] the "position" of which may be identified by means of a parameter ϕ. Let us approximate the discrete chain by means of a continuous chain. Both $u(x)$, which specifies the displacement of an atom from its reference position, and $\delta n(x)$, which describes the variation in the macroscopic density of the conduction electrons with respect to the metallic state, would show sinusoidal trends of the type $\sin(qx + \phi)$.

In a translationally invariant system such as that considered by Fröhlich, the condensate energy E_c is independent of ϕ and it can therefore find itself in a state of uniform motion. Fröhlich proposed this mechanism as a possible explanation of the phenomenon of superconductivity, before the formulation of the now generally accepted Bardeen–Cooper–Schrieffer (BCS) theory.[22]

[‡] In the same period, S. A. Brazovskii independently formulated a continuous model of PAC to describe soliton excitations.[20] We will not discuss this in detail here, as an analysis of this type would not introduce any new fundamental aspects and would only complicate the description.

In recent years, with the experimental realization of almost one-dimensional metals, Lee, Rice, and Anderson[23] have noted that the translational invariance that Fröhlich assumed could be violated by several factors, the main ones being as follows:

1. The presence of lattice defects (particularly charged impurities).
2. Interactions between chains (the presence of CDWs on neighboring chains).
3. The fact that the lattice is discrete and not continuous, so that values of ϕ exist that optimize the energy of interaction between conduction electrons and ions.

The energy of the condensate, E_c, thus depends on the phase, and this gives rise to a bias F_R proportional to $\partial E_c / \partial \phi$. Thus there is a preferential position, ϕ_0, characterized by $F_R(\phi_0) = 0$. The condensate is thus "pinned."

The distortion of the electron density distribution in PAC has a wavelength equal to $2a$, where a is the lattice constant for the nondimerized reference state (see Fig. 2). Under these conditions the effects of pinning associated with the discontinuity of the lattice are very large, and it is unreasonable to think that the collective translational motion of the CDW, proposed by Fröhlich, could significantly contribute to the electrical conductivity.

Later, Rice et al.[24] demonstrated that even a pinned condensate could show dc conductivity due to the presence of nonlinear excitations (solitons), also known as "ϕ particles." These are solutions of the nonlinear equations of motion of the phase of the condensate, and take the form of solitary waves moving with constant velocity. They can be physically interpreted as transition zones separating parts of the chain where the value of the phase is constant.

These mobile solitons contribute to the conductivity with a term proportional to

$$\exp\left(-\frac{E_s}{k_B T}\right)$$

where E_s is the energy of formation of the soliton, T the absolute temperature, and k_B the Boltzmann constant (see also Section III).

Let us now deal with the specific case of PAC.

We will treat the σ-electrons of $(CH)_x$ using the adiabatic approximation and focus our attention on the π-electrons, of which, in the neutral material, there is one for every CH group.

The theory of Peierls leads to the conclusion that the ground state of the chain will be characterized by a periodic distortion with a wavevector equal to $2K_f$, where K_f, defined by Eq. (1), is related to the number of electrons per site, v, by the relationship[25]

$$K_f = \frac{\pi}{2a} v$$

In the neutral material, $v = 1$ and thus the CDW, as we have already noted, has a wavelength equal to $2a$ and is commensurate with the lattice.

Let us now consider the changes in the electronic structure associated with the addition or subtraction of electrons from PAC. First, let us study the addition of a single electron, which, as can be seen from the above equation, alters the wavevector of the CDW so that it is no longer commensurate with the lattice. This would cost a large amount of energy. On the other hand, it may be expected that the minimum-energy configuration consists of a large localized distortion (soliton) in a small portion of the chain, which thus maintains the commensurability between the CDW and the lattice for most of its length.

The added electron goes into an electronic level, created by the soliton distortion, at the center of the Peierls gap associated with the structure having $v = 1$.[18]

The above case constitutes an example of "frustration,"[26] a concept which is central to the understanding of nonlinear systems. The basic idea is that numerous physical systems do not always resolve the competition between opposing tendencies by giving rise to spatially homogeneous compromise solutions. In some cases, they give indeed rise to strong localized distortions that separate essentially unperturbed areas.

Let us now examine Rice's hypotheses quantitatively. We will, as usual, take the nondimerized structure as our reference, whereas the ground state of a neutral PAC chain actually presents alternating long and short bonds.

Let u_n be the projection along the axis of the chain of the nth CH group with respect to the reference position. Let us also call phase A and phase B the two possible structures of the ground state, which are chemically and energetically identical and distinguishable only by different phases in the alternation of the single and double bonds (see Fig. 3). Phase A and phase B are characterized by the relationships

$$u_n = (-1)^n u_0 \quad \text{(phase A)}$$
$$u_n = -(-1)^n u_0 \quad \text{(phase B)}$$

(2)

We also define $u(x)$ as the continuous limit of the expression $(-1)^n u_n$.

The addition of the electron will cause the formation of a soliton (phase kink) linking two different parts of the chain, one in phase A and the other

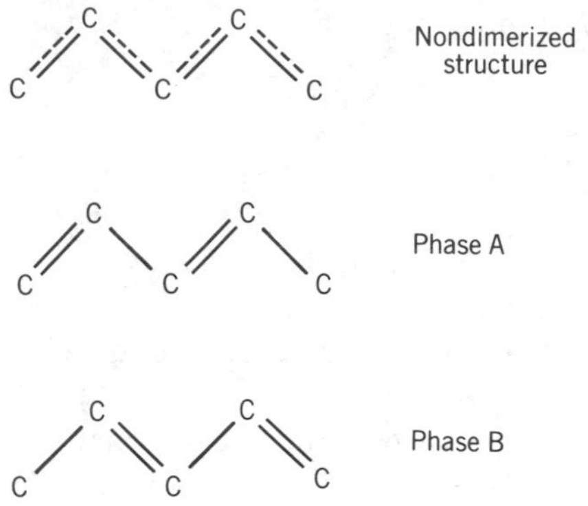

Nondimerized
structure

Phase A

Phase B

Figure 3

in phase B. Generally, in the case of a static soliton, the function $u(x)$ will take the form

$$u(x) = u_0 f\left(\frac{x - x_0}{L}\right) \tag{3}$$

where u_0 is the amplitude of the distortion in the ground state [see Eq. (2)], $f(y)$ is an odd function of its argument, which rapidly tends toward ± 1 when $|y| > 1$, x_0 is the center of mass of the soliton, and L is a measure of its half-width. The expression for $f(y)$ depends on the specific model chosen for PAC. Rice adopts a Hamiltonian of the type

$$H = \int dx \left[\frac{1}{2}D\left(\frac{\partial u}{\partial t}\right)^2 - \frac{1}{2}C\left(\frac{\partial u}{\partial x}\right)^2 + V(u)\right] \tag{4}$$

with D and C as positive constants. The first term is the kinetic energy of the ionic motion, while the second describes the coupling between closest-neighbor ions in the elastic approximation. The "on site" potential $V(u)$ describes the positive (or negative) attraction toward the reference position and is chosen in a phenomenological manner.

In our case, it is required that the nondimerized configuration ($u = 0$) does not correspond to a position of stable equilibrium, owing to the Peierls distortion. It is also required that two possible stable equilibrium states exist that are energetically degenerate and correspond to phases A and B of Fig. 3. The simplest potential of this type, adopted by Rice, is

$$V(u) = \frac{1}{2}Au^2 + \frac{1}{4}Bu^4 \tag{5}$$

with $A < 0$ and $B > 0$ (see Fig. 4).[‡]

[‡] The potential of Eq. (5) (the Landau–Ginzburg potential) is familiar in other fields of physics[27,28] and has been employed in the description of phase transitions of other kinds associated with the change of sign of the parameter A (see Fig. 4).

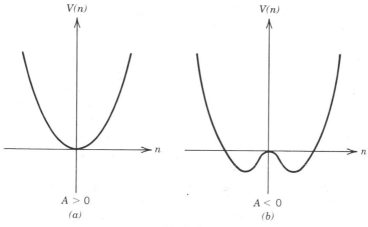

Figure 4

Krumhansl and Schrieffer[29] and, independently, Varma[30] demonstrated that the equation of motion associated with the Hamiltonian defined by Eqs. (4) and (5) allows a solution of the type

$$u(x,t) = u_0 \tanh\left(\frac{x - vt}{L}\right)$$

which describes a soliton with uniform motion conserving its original shape (L and v constant). The corresponding static solution, of the type defined by Eq. (3), is

$$u(x,t) = u_0 \tanh\left(\frac{x - x_0}{L}\right) \tag{6}$$

This describes a stationary "phase kink," with its center in x_0, in a chain of infinite length subject to the boundary conditions

$$\lim_{x \to +\infty} u(x,t) = u_0 \quad \text{and} \quad \lim_{x \to -\infty} u(x,t) = -u_0 \tag{7}$$

General considerations that we will not go into here allow us to associate this phase kink with an energetic level at the halfway point of the Peierls gap.[31] There are three possible cases. First, this level may be occupied by a single electron; the soliton is therefore neutral, and may be considered an excited state of the pure material. There are cases in which the ground state (the perfectly dimerized chain) cannot be realized because of topological constraints, for example, when there is an odd number of CH groups between two carbon atoms that are sp^3 hybridized instead of sp^2 hybridized. The cause of such sp^3 hybridization may be, for example, the presence of cross-linking between the different polymeric chains.

In this case, the configuration in which two portions of the chain having different phases are separated by a kink (see Fig. 2) is the most stable energetically. The ground state of a finite chain of PAC having an odd number of carbon atoms also presents an alternation of phase (with a related soliton), if one takes into account the boundary effects that favor termination with a double bond.[31-33] This leads us to the conclusion that undoped PAC should have a neutral soliton concentration of the order of about one soliton per chain.

Because the soliton energy level is occupied by a single electron (free spin), it will give rise to an ESR signal and a Curie paramagnetism, whereas a neutral soliton will not contribute to the electrical conductivity. The agreement between these predictions and the experimental results noted in the previous section is evident.

The soliton model is also capable of suggesting a convincing doping mechanism for PAC. If the energy level associated with the soliton is doubly occupied (the case we will consider here to clarify our ideas) or empty, then the soliton will be charged.

A donor impurity will give up its electron to the polymeric chain directly in the conduction band or in the soliton level, depending on which of the two alternatives is energetically favored. By measuring the energy from the Fermi level, Rice estimated that the energy of formation of a charged soliton is

$$E \approx \frac{4\Delta}{3\pi} \tag{8}$$

where 2Δ is the Peierls gap.[‡] The doping mechanism that leads to the formation of charged solitons is thus preferred with respect to direct promotion into the conduction band. A doubly occupied soliton will not give rise to an ESR signal or to Curie magnetism, which is consistent with the observations of Goldberg et al.[7] cited above.

We will defer discussion of the mechanisms of electric conduction in PAC until the Section III. Here we will limit ourselves to noting that a charged soliton generated by a charged impurity is presumably bound to the latter

[‡] The reasoning, in a nutshell, is as follows: It can be demonstrated that the ratio between the half-width L of the soliton and the lattice constant a is equal to $W/2\Delta$, where W is the amplitude of the π-band (~ 10 eV). Thus $2L \approx 10a$. It can also be shown that the Peierls condensation energy per site, ε_c, is equal to $\Delta^2/\pi W$. Thus, Eq. (8) can be deduced from the amplitude of the soliton and the energy per site.

by a Coulomb interaction energy much greater than the thermal energy $k_B T$ at room temperature.[18] Thus charged solitons generated in this way cannot be charge carriers in the material.

Rice and Mele[34] extended the original approach to take into account important physical factors that had been ignored. In the initial work, the formation energy of a charged soliton, $S^{\pm}$, was the same as that necessary for the creation of a neutral soliton, S^0, since the Coulomb repulsion between the two electrons (or the two vacancies in a soliton without electrons) occupying the soliton level had been ignored. Rice and Mele took account of this by using an internal potential energy term for the charged soliton of the type

$$V(L) = \left(A + \frac{U_0}{\varepsilon} \right) \frac{1}{L} + BL$$

where A, U_0, and B are positive constants and ε is the dielectric constant. The corresponding internal potential energy term for the neutral soliton does not have the term containing U_0.

Rice and Mele also took into account in an approximate way the stabilizing effect of the Coulomb interaction between the impurity and charged solitons by considering the dopant molecule as a point charge and assuming a uniform charge distribution for the soliton.

The basic results were as follows: A charged soliton far away from impurities has an energy very close to that of ionization in the conduction band. The crudeness of the model does not allow us to determine which of the two alternatives is more convenient. The situation changes drastically, however, if one takes into account the soliton–impurity interaction (at a separation of 2 Å). The energy of formation of the charged soliton is then only slightly higher (by $\sim 10\%$) than that of a neutral soliton. Therefore, even this more detailed analysis confirms the main predictions of the initial model.

Before concluding this discussion of Rice's hypothesis, we note that the use of a phenomenological Hamiltonian, such as that defined by Eqs. (5) and (6), appears at this point to be completely arbitrary. The ability of the model to describe the unusual phenomena in neutral and doped PAC is quite striking, but any attempt at a more complete treatment must, in some way, begin with a microscopic model. As we shall see, the theory of Su, Schrieffer, and Heeger[11] fulfills this requirement and also allows us to understand the qualitative success of Rice's simpler approach.

C. The Su, Schrieffer, and Heeger Model

The difference between the bonding and antibonding energies of the σ-electrons of PAC is large ($\simeq 10$ eV) with respect to phonon and soliton energies ($\simeq 0.5$ eV). This allowed Su, Schrieffer, and Heeger (SSH) to treat the electrons using the adiabatic approximation. The Hamiltonian of the model is the sum of three terms:

$$H = H_K + H_{PI} + H_{EI} \tag{9}$$

The first term is the kinetic energy

$$H_K = \frac{1}{2} \sum_n M\dot{u}_n^2 \tag{10}$$

where M is the mass of the CH group and we sum over all the CH groups of the chain.

There is then a term describing elastic interaction between closest-neighbors of the type

$$H_{PI} = \frac{1}{2} \sum_n K(u_{n+1} - u_n)^2 \tag{11}$$

Is is also assumed that the π-electrons can be treated using the strong bonding approximation

$$H_{EI} = -\sum_{n,s} t_{n+1,n}\left(c_{n+1,s}^\dagger c_{n,s} + c_{n,s}c_{n+1,s}^\dagger\right) \tag{12}$$

The meaning of the creation and annihilation operators is as follows: $c_{n,s}^\dagger(c_{n,s})$ creates (or destroys) an electron with spin s at the nth CH group; and $t_{n+1,n}$ is the hopping integral, which describes the variation in energy associated with the jumping of an electron between the nth and the $(n+1)$th carbon atom. It is only by means of this latter term that the coupling between the electronic and phononic degrees of freedom is introduced into the model. It is in fact assumed that it is possible to expand the hopping integral about the value t_0 that it takes in the nondimerized reference state by using the linear expression

$$t_{n+1,n} = t_0 - \alpha(u_{n+1} - u_n) \tag{13}$$

where α is an electron–phonon coupling constant.

The model lacks explicit references to the interactions between electrons, which can be partly taken into account by using "screened" values of the parameters t_0 and α. In any case, this approach would have to be abandoned when the electron–electron repulsion is very strong.

Ignoring the kinetic term, SSH first calculated the energy of the nondimerized (metallic) state, the band structure of which is shown in Fig. 5.

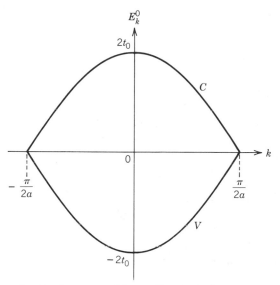

Figure 5.   Band structure of the nondimerized state of *trans*-PAC.

They then calculated the energy of the ground state of the chain, which corresponds to the perfectly dimerized structure $u_n = (-1)^n \bar{u}$, with $\bar{u}$ constant. Diagonalizing the Hamiltonian (without the kinetic term) yielded

$$H_d = \sum_{K,s} E_K \left(n^c_{K,s} - n^v_{K,s} \right) + 2NK\bar{u}^2$$

where N is the total number of carbon atoms and $n^c_{K,s}(n^v_{K,s})$ is the occupancy operator, which has as eigenvalues the numbers of electrons having momentum k and spin s in the conduction (or valence) band. E_K is given by

$$E_K = \sqrt{\left(2t_0 \cos ka \right)^2 + \left(4\alpha\bar{u} \sin ka \right)^2}$$

The energy of the ground state $[n^v_{K,s} = 1, \ n^c_{K,s} = 0 \ \forall(k,s)]$ is

$$E_0(\bar{u}) = -2\sum_K E_K + 2NK\bar{u}^2$$

By substituting an integral for the summation, it is possible to estimate $E_0(\bar{u})$ using the expression

$$E_0(\bar{u}) \approx -\frac{4Nt_0}{\pi} E(1-Z^2) + \frac{NKt_0^2 Z^2}{2\alpha^2}, \quad Z = \frac{2\alpha\bar{u}}{t_0}$$

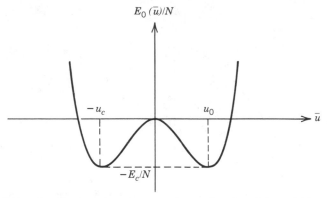

Figure 6.   Condensation energy per site as a function of the variable u.

where $E(1 - Z^2)$ is the elliptic integral. In PAC the gap is estimated as 2Δ $= 1.4$ eV, the width of the bands as $W = 10$ eV, and the elastic constant as $K = 21$ eV Å^{-2}. Using these values, the condensation energy per site, $E_0(\bar{u})/N$, shows the behavior illustrated in Fig. 6. The ground state is characterized by a value of $\bar{u} = u_0$ equal to 4×10^{-2} Å, and the height of the energy barrier is 1.5×10^{-2} eV.

Comparing Fig. 6 with Fig. 4b, which represents the behavior of the "on site" potential of Rice's model [Eq. (6)], we can understand the "microscopic" reasons why Rice's description gives qualitatively correct results. In fact, the form adopted for the on-site potential simulates the behavior of the energy as a function of the dimerization parameter $\bar{u}$, and therefore simulates, at least qualitatively, the interaction between the vibrational and the electronic degrees of freedom of the chain.

Having studied the ground state, SSH then considered an excited configuration, comprising a phase kink, of the type

$$u_n = (-1)^n u_0, \qquad\qquad n \le -m$$

$$u_n = -(-1)^n u_0 \tanh\left(\frac{n}{L}\right), \qquad -m < n < m \qquad (14)$$

$$u_n = -(-1)^n u_0, \qquad\qquad n \ge m$$

The choice of the hyperbolic tangent is not motivated, as in Rice's model, by the existence of exact solutions of this type. It is to be taken only as a trial function having the desired characteristics. Su, Schrieffer, and Heeger state that the use of other, analogous trial functions does not lead to very different results.

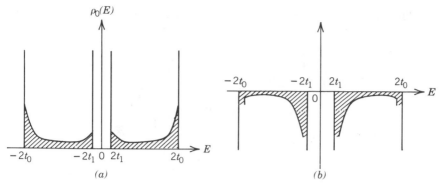

Figure 7.   (a) Density of states of phase A or B of PAC. (b) Density of states in the presence of a soliton.

A soliton formation energy, E_s, of about 0.4 eV, corresponding to a gap of 1.4 eV, is obtained, along with a value of L equal to 7. The soliton is thus spread over about 14 atoms, and the pinning effects due to the discrete nature of the lattice are small. The variation in energy associated with the movement of the center of the soliton between two adjacent lattice positions is about 2×10^{-3} eV, which presumably gives the defect a high mobility even at low temperatures (down to about 30 K, as estimated by SSH).

The variation in the density of states $\Delta\rho(E)$ associated with the presence of the domain wall is illustrated in Fig. 7. The presence of a level localized at the center of the gap must be reconciled with the conservation of the total number of electronic states of the π and π^* bands:

$$\int_{-\infty}^{\infty} \Delta\rho(E)\, dE = 0$$

From this relationship and from the observation that $\Delta\rho(E)$ is an even function, it follows that the valence and the conduction bands each lack one-half of a state with respect to the perfectly dimerized state. Since the valence band is completely filled, the extra electron occupies the halfway level of the gap; the soliton is therefore neutral, with a spin of $\frac{1}{2}$. The addition or removal of an electron gives rise to a charged kink, with zero spin. As we can see, the situation is exactly analogous to that described by Rice.

Also in this case, if we ignore the Coulomb interaction between coupled electrons in the soliton level, the formation energies E_s of a charged and of a neutral kink are the same, being given by

$$E_s \approx \tfrac{3}{5}\Delta$$

Thus, SSH also sustain the hypothesis that the charge-transfer mechanism after doping consists of the formation of charged solitons, as opposed to direct promotion into the conduction band.

We have so far considered only the static case; a soliton moving uniformly with velocity v may be described by the equation

$$u_n(t) = (-1)^n u_0 \tanh\left(\frac{na - vt}{L}\right) \tag{15}$$

The translational mass M_s of the soliton is defined by the relationship

$$\tfrac{1}{2}M_s v^2 = \sum_n \tfrac{1}{2}M\dot{u}_n^2 \tag{16}$$

and turns out to be equal to six electronic masses. This suggests that quantum effects are important, and that the use of models in which the soliton is treated as a classical particle demands great caution.

D. The Takayama, Lin-Liu, and Maki Model

The Takayama, Lin-Liu, and Maki (TLM) model[19] may be considered as the continuous limit of the SSH theory discussed in the previous section. Its importance is that it makes possible an analytical treatment that can give all the static solutions.[35]

The SSH Hamiltonian in the continuous limit becomes (see Appendix B)

$$H = \frac{K}{8\alpha^2} \int \frac{dx}{a} \Delta^2(x, t) + \frac{M}{32a^2} \int \frac{dx}{a} \left(\frac{\partial \Delta(x, t)}{\partial t}\right)^2$$
$$+ \int \frac{dx}{a} \psi^\dagger \left[-iv_F\sigma_3 \frac{\partial}{\partial x} + \Delta(x, t)\sigma_1\right]\psi \tag{17}$$

where $\Delta(x, t)$ is linked to the continuous limit $u(x, t)$ of the quantity $(-1)^n u_n$ (see Section II.B) by the relationship

$$\Delta(x, t) = 4\alpha u(x, t) \tag{18}$$

(M, K, a, and α have the same meanings as in the preceding sections.) $\Delta(x, t)$ is proportional to the local extent of Peierls' distortion: the proportionality constant is chosen so that, in the case of perfect dimerization, it is equal to the half-width, Δ, of the gap. The meanings of the other symbols in Eq. (17) are as follows: v_F is the Fermi velocity, equal to $2t_0a$, where t_0 is the hopping integral of the nondimerized state [Eq. (13)], whereas ψ is the spinor (two-component wavefunction) that describes the electronic field:

$$\psi(x, t) = \begin{pmatrix} w(x, t) \\ v(x, t) \end{pmatrix} \tag{19}$$

Finally, the σ_i are Pauli matrices.[36]

Because we are looking for static solutions, for the moment we will ignore the kinetic term. One can demonstrate that the Hamiltonian (17) can be associated with the eigenvalue equation

$$
\begin{pmatrix} -i\nu_F \dfrac{\partial}{\partial x} & \Delta(x) \\[2ex] \Delta(x) & i\nu_F \dfrac{\partial}{\partial x} \end{pmatrix} \begin{pmatrix} w_n \\[1ex] \upsilon_n \end{pmatrix} = \varepsilon_n \begin{pmatrix} w_n \\[1ex] \upsilon_n \end{pmatrix}
\tag{20}
$$

and with the self-consistent equation

$$
\Delta(x) = -\frac{4\alpha a^2}{K} \sum_{n,s} \upsilon_n^*(x) w_n(x)
\tag{21}
$$

In the mean field approximation, the energy of the chain is

$$
\Gamma_{MF} \quad \sum_{n,s} \upsilon_n \quad + \quad \frac{K}{8\alpha^2} \int \Delta^2(x)\, dx
$$

where the summation is extended over all the energetic levels below the Fermi level (at $T = 0$ K).

One can calculate the energy of a soliton of the type

$$
\Delta(x) = \tanh\left(\frac{x}{L}\right)
\tag{22}
$$

The energy is optimized for

$$
L = \frac{\hbar \nu_F}{\Delta}
\tag{23}
$$

where Δ is the half-width of the gap and is equal to

$$
E_s = \frac{2\Delta}{\pi}
\tag{24}
$$

which is in good agreement with the results of SSH and slightly lower than the value obtained by Rice [Eq. (8)].

For a soliton moving with a constant velocity v, the energy of formation E_s also turns out to be a function of v. Whereas SSH ignored this contribution, TLM were able to calculate it analytically and to demonstrate that it is insignificant in PAC.

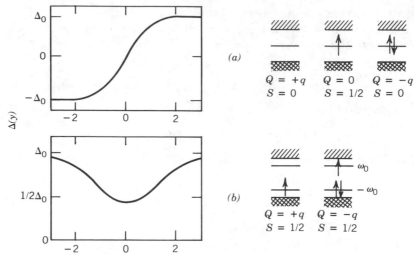

Figure 8. Intrinsic defects of *trans*-PAC with their relative electronic levels: (*a*) soliton; (*b*) polaron. [The abscissa represents the spatial coordinate x, and the ordinate the variable $\Delta(y)$.]

It has been shown[35] that the TLM equations [Eqs. (20) and (21)] can take solutions of the polaronic type, which can be written

$$\Delta(x) = \Delta - \frac{v_F}{L}\left[\tanh\left(\frac{x+x_0}{L}\right) - \tanh\left(\frac{x-x_0}{L}\right)\right] \qquad (25)$$

where L is a parameter defining the shape of the polaron. The soliton [Eq. (22)] links two portions of the chain characterized by different phases of bond alternation, whereas the polaron [Eq. (25)] links two portions having the same phase (Fig. 8).

Thus, the latter does not enjoy the topological stability properties of the kinks, but is an excited state of the chain that can freely decay to the dimerized ground state. Two energy levels are associated with a polaronic defect. These are placed symmetrically about the gap, with energies $\varepsilon = +\omega_0$, where

$$\omega_0 = \sqrt{\Delta^2 - \frac{v_F^2}{L^2}}$$

The relationship for determining the parameter L is

$$\tanh\left(\frac{2u_0}{L}\right) = \frac{v_F}{L\Delta}$$

The neutral polaron has two electrons in the lower level and none in the higher one; for the reasons indicated above it is unstable and relaxes to the ground state. The structure having one or three electrons, however, is stable and requires an energy of formation of about 0.9Δ, whereas that containing four electrons gives rise to a kink–antikink pair that separates.

Another type of static solution of the TLM model is a series of regularly spaced kinks and antikinks (soliton crystal). This solution can be particularly important in the study of the semiconductor–metal transition in *trans*-PAC.

Bredas et al.[37] carried out *ab initio* calculations (based on the Hartree–Fock SCF–MNDO method) on "short" (40 carbon atoms) PAC chains showing that in the absence of dopant, there is a difference in bond length of 0.105 Å between the alternate single and double bonds. They also predicted the existence of solitons having the following properties: The neutral soliton may be fitted satisfactorily using a tanh function with half-width $L = 3$ and an electronic wavefunction covering about 11 carbon atoms. The positively charged soliton extends over about 11 CH groups and has an electronic wavefunction covering about 17 carbon atoms. The negatively charged soliton is much more compact ($L \approx 2$) and is not very well represented by a tanh function. In this case the associated CDW is practically identical to that of the positively charged soliton.

The calculations of Bredas et al. also predict the existence of polarons that, as in the case of solitons, are narrower if negatively charged.

Thus, the *ab initio* calculations qualitatively confirm the results of the more simplified models that use analytical methods. The quantitative discrepancies are due to the fact that Bredas et al. explicitly took into account more specific physical effects, such as the polarization of the π-orbitals.

The results of Bredas and co-workers have been confirmed by Zerbi and Zannoni.[38] Their dynamic calculations and the experimental infrared and Raman spectra they obtained show that in doped PAC defects exist that are localized in a symmetrical manner with respect to the dopant molecules. According to Zerbi and Zannoni, these defects extend over about three double bonds, whereas the associated electronic wavefunctions are much more diffuse.

E.   Some Observations on Doping Mechanisms

Polarons, which were introduced at the end of the previous section, are also nonlinear excitations of the soliton type, but do not invert the phase and may be considered as bonded kink–antikink pairs.

The study of polarons is of particular importance in the analysis of doping mechanisms in PAC.[39] If we imagine that an impurity (e.g., a donor) introduced into the material gives up its electron to an already existing soliton level (a topological kink stabilized by cross-links or edge or other effects), the situation is straightforward. It is possible, however, for dopant levels to be much higher than the concentration of native solitons. Previous considerations, together with the apparent absence of contributions to the electrical conductivity by the extended-state bands, suggest that the impurities themselves can generate further solitonic excitations (kinks or polarons) that play the role of electron acceptors in the charge-transfer process. The introduction of a kink in a chain that is initially all phase A or all phase B leads to a phase change in a large portion of the chain; associated with this change is an activation energy proportional to the length of the same portion of the chain. It is thus plausible that the introduction of the impurity generates a localized effect that does not significantly alter the topology of the chain at large distances from the impurity. This suggests the hypothesis that the doping takes place by the formation of charged polarons rather than solitons.[‡]

We have seen that polarons of the type described by Eq. (25) constitute exact solutions of the TLM equations. Analogous studies have been carried out using the SSH model[42-44] to consider polarons of the type

$$u_n = (-1)^n u_0 \tanh\left(\frac{n - n_0/2}{m}\right)\tanh\left(\frac{n + n_0/2}{m}\right) \tag{26}$$

The results obtained are substantially in agreement with those of the continuous model: a polaron (a state linked to a kink–antikink pair) can exist with a net charge of $+q$, but the addition of a further electron or vacancy destabilizes the system, which relaxes to a configuration having two separate charged kinks in a period that can be estimated, with the help of computer simulation, to be on the order of 10^{-15}.

In conclusion, it is reasonable to suppose that in cases of donor impurity doping, the charge transfer takes place by the formation of polarons with charges of $-q$. The addition of another impurity near the charged polaron thus gives rise to a doubly occupied polaron, which is unstable and generates, in turn, a pair of charged kinks. Different kinks in the same chain then form a soliton crystal, which is associated with a band at the center of the Peierls gap. In the initial doping phase there will also be transfer of electrons from the impurity to the native soliton levels of the material.

[‡] This hypothesis is also supported by numerical calculations by several different authors that demonstrate that in the presence of impurities, the polaron state has a lower energy than the soliton state.[40,41]

F. Brief Comparison with Experimental Results

The interpretation of the experimental data on PAC is complicated by the disorder present in the material, which is due to both imperfect crystallization and the effect of the introduction of high concentration of impurities. It is particularly difficult to study the different properties as a function of y, the dopant concentration, because the active impurities that play a part in the charge transfer process constitute only a fraction (which is not constant) of the total quantity of dopant introduced into the material.[44] We therefore do not intend to give a complete review of the experimental results and their relationships with the theoretical models, except for those pertaining to the electrical conductivity, which will be dealt with in detail in the following section.

In this section we will only describe briefly the magnetic and optical properties of PAC within the context of the soliton model. In our opinion, as mentioned above, the importance of this model derives not so much from its explanation of a single experimental fact in such a manner as to exclude alternative explanations as from its ability to explain different phenomena and observations within a single framework and to formulate verifiable predictions and thus stimulate further experiments.

Undoped PAC has a low electrical conductivity, on the order of 10^{-5} ohm^{-1}cm^{-1}, and contains a free spin concentration, as estimated from ESR measurements, of about 1 per 3000 carbon atoms.[14] The width of the resonance indicates that these spins are mobile down to temperatures of only a few degrees Kelvin. It has also been possible to determine experimentally the ratio between the spin diffusion coefficient along the chain, $D_{\parallel}$, and that in the orthogonal direction, $D_{\perp}$.[45,46] Its value is on the order of 10^5–10^6, in agreement, at least qualitatively, with the soliton model. Maki[47] has also shown that the diffusion coefficient $D_{\parallel}$ calculated with the TLM Hamiltonian is in quantitative agreement with the experimental values.

If the doping takes place with the formation of charged solitons, this would be expected to be accompanied by a decrease in the number of free spins, N_c. This phenomenon was observed by Ikehata et al.[48] and is illustrated in Fig. 9.

The values of the variables N_c and χ_P (Pauli susceptibility) were obtained by magnetic susceptibility (χ) measurements at different temperatures by using the relationship

$$\chi(T) = \chi_P + \frac{N_c \mu_B^2}{k_B T} \qquad (27)$$

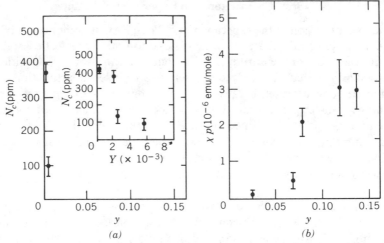

Figure 9.   (a) Free spin concentration as a function of y. (From Ikehata et al.[48]) (b) Pauli susceptibility as a function of y. (From Ikehata et al.[48])

where μ_B is the Bohr magneton. χ_P is independent of the temperature, and is the type of susceptibility expected in a metal.

Anticipating a subject to be covered in more detail in the next section, we briefly note the existence of a semiconductor–metal transition at dopant levels of several percent. This transition can be seen in Fig. 9b as a sudden rise in the Pauli susceptibility. The second term in the right-hand side of Eq. (27) is the Curie susceptibility, which is associated with the presence of magnetic dipoles that are free to orient themselves in an external field. The contribution of the singly occupied soliton levels is of this type, and Fig. 9a shows the predicted fall in N_c associated with the increase in impurity concentration.

We also note that, according to Ikehata et al.,[48] the critical impurity concentration that produces a metallic behavior of the magnetic susceptibility (~ 5%) is about one order of magnitude higher than that which gives rise to a metallic electrical conductivity. We will return to this observation later.

The necessity of invoking the soliton hypothesis to explain the magnetic properties of PAC has been refuted by the IBM group.[49-51] They maintain that the doped material is composed of metallic islands separated by dielectric zones. The formation of the metallic islands is due to the nonhomogeneous nature of the doping process, which gives rise to regions having high impurity concentrations alternating with regions of essentially neutral PAC. Increasing the macroscopic concentration of the dopant, y, increases the density of the metallic zones; the conduction takes place by the formation

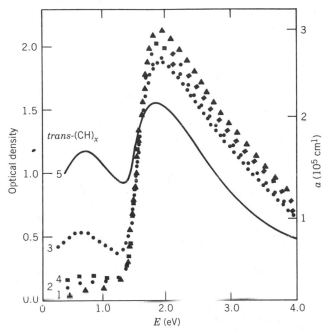

Figure 10.   Absorption spectrum of *trans*-PAC doped with AsF_5. 1, neutral PAC; 2, 0.01% AsF_5; 3, 0.1% AsF_5; 4, compensated with NH_3; 5, 0.5% AsF_5. (From Suzuki et al.[53])

of "percolating" paths between these zones. Without entering into the merits of this arguments, we note that there are discrepancies between the results of the IBM group and those obtained by other workers, especially those at the University of Pennsylvania (for a review see, e.g., refs. 14 and 52). Also, Ikehata et al.[48] have perfected a technique that allows much more homogeneous doping than is obtainable with conventional methods.

Important information on the structure of PAC has also been obtained from the absorption spectra. Let us first consider the region of the spectrum corresponding to the energy differences on the order of 1 eV. Figure 10 shows the data of Suzuki et al.[53] relating to neutral PAC and the same doped with AsF_5. Analogous results have been obtained with dopants such as I_2 (ref. 44) and Br_2 (ref. 54).

The strong absorption band at 1.4 eV (peaking at 2 eV), which is assigned to direct transitions between the valence and conduction bands, can be seen. A second peak can be clearly distinguished, the intensity of which is a growing function of y. The location of this peak is reasonably close to the value predicted by the soliton model, that is at the halfway point of the gap.

This is one of the more convincing pieces of experimental data in favor of the soliton theory, or rather the soliton–polaron theory, of PAC.[55] One could object that the presence of structural disorder would in any case give rise to the formation of energy levels within the gap.[56-58] The soliton model does, however, succeed in predicting a particular behavior of the density of states that is consistent with the optical absorption spectrum actually observed.[59] Also, in contrast to the more "traditional" theory, the soliton model's predictions agree with the experimentally observed fact[60] that the energy and the form of the peak at the midpoint of the gap at low dopant concentrations is independent of the relative percentage and type of dopant introduced. One must not, however, forget that generally an exact verification is rendered difficult by the lack of precision in the evaluation of the effective impurity concentration, by three-dimensional effects, and by lattice disorder.[59]

Another interesting region of the PAC spectrum is the infrared, although, as we shall see, interpretation of the data is somewhat confusing.

Following low-level doping, the appearance of two new vibrational modes was observed at 1370 and 900 cm^{-1}.[61] Mele and Rice,[62] after studying the phonon spectrum of doped PAC in the presence of charged solitons, claimed that they could account for the presence of these two vibrational modes. They also predicted the existence of a further vibrational mode in the infrared, somewhere between 300 and 500 cm^{-1}, due to the Coulomb interaction between the charged solitons and the impurities that generate them.[18,62]

Horovitz,[63] on the other hand, maintained that the infrared absorption frequencies and their intensity ratios are independent of the detailed charge configuration and therefore cannot be used to demonstrate the existence of solitons. Also, contrary to the assignment made by Mele and Rice,[62] he assigned the mode observed at 900 cm^{-1} to the CDW electrostatically bound to the charged impurity ("pinned mode").

A study of the absorption spectrum of PAC under particular experimental conditions confirmed the validity of Horovitz's hypothesis. As we shall see later, it is possible to produce a photocurrent in trans-$(CH)_x$, which is attributed to the movement of photogenerated charged solitons. Under these conditions, given the absence of dopant, the "pinned mode" should also be absent. In fact, the spectra of Vardeny et al.[64] do not show the characteristic mode at 900 cm^{-1}, confirming the thesis of Horovitz. Kivelson claims that the reason for the different results obtained by Mele and Rice was their use of an inadequate perturbative development, and certainly does not reflect any deficiency of the soliton model.[66]

Rice[65] has also studied the effects of putative dopant concentration ranges in which mobile charged solitons are present. That these may exist was suggested to him by the previously mentioned observation of Ikehata[48] that there is a large range of dopant concentration levels at which PAC shows a high electrical conductivity but the Pauli magnetic susceptibility is almost zero. Rice demonstrated that for a charged soliton moving with uniform velocity, the coupling between the translational motion and the internal vibrational degrees of freedom of the soliton generates a new band in the infrared, the observation of which would constitute important evidence for the existence of mobile charged kinks. It is difficult, however, to locate this band precisely; there are two very different estimates, 1280 and 780 cm^{-1}.[65] A more recent hypothesis about the "transition region" of Ikehata[48] suggests that the prevailing conduction mechanism is variable-range electronic hopping, typical of amorphous semiconductors.[67]

We shall see in more detail in the next section, which is dedicated to the mechanisms of electrical conductivity, that the contribution of mobile charged solitons to the conductivity is important for the photocurrent in neutral PAC and when PAC is doped with ionized impurities solvated in polar solutions.

In any case, it is evident that the interpretation of the infrared absorption spectrum of PAC is extremely complicated. Although soliton models are capable of reproducing fairly well the observed phenomena, it is probably impossible to use these data to exclude alternative explanations.

Comparison of the behavior of trans-(CH)$_x$, with which we have been concerned so far, with that of the cis isomer raises important issues concerning the soliton hypothesis. Although the chemistry is essentially the same for the two cases, they exhibit surprisingly different phenomena. This is particularly evident in photoconduction and luminescence experiments. We will therefore first take a brief look at the characteristics of cis-(CH)$_x$.

A topological kink-type soliton, which inverts the phase of the bond alternation, should separate a trans–cis-oid portion of the chain from a cis–trans-oid portion (see Fig. 1). The situation is qualitatively different in the trans isomer, where the soliton separates two "phases" that energetically are degenerate and in all aspects equivalent. The energy of the cis–trans-oid state, however, is greater than that of the trans–cis-oid. This may be represented by a double-well potential model analogous to those of Fig. 4 and 6, but with the potential wells at different levels (see Fig. 11).

Since one of the two structures is energetically favored, stable mobile kinks are not possible in cis-(CH)$_x$. There could exist, however, static kinks related to the termination of a chain with an impurity or other "pinning center." Electron spin resonance experiments carried out on samples containing the cis isomer showed the presence of spins that were free, but not mobile. These were attributed[4] to small zones of trans-(CH)$_x$ included in the predominantly cis chains.

Note that in cis-(CH)$_x$ the formation of polarons (kink–antikink bound pairs) is possible and that these are unstable unless stabilized by electron transfer. The energy difference between the two "phases" in this molecule

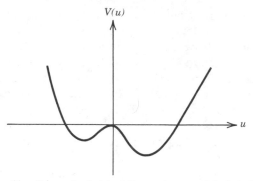

Figure 11.   Schematic sketch of the on-site potential of *cis*-PAC.

gives a confinement energy[35] that tends to keep the pair bound together and, in the case of charge transfer from the impurity, may also give rise to a polaron having a charge of $\pm 2q$ (this is, however, unstable in the *trans* isomer, where it relaxes into a kink–antikink pair with infinite separation).[‡]

It is also worth noting that the study of doping in *cis*-PAC is extremely difficult, because it is thought that the doping induces isomerization to the more stable *trans* form.[14] This hypothesis, however has recently become the subject of renewed debate.[68]

Nonetheless, it is possible to study the effect of the photogeneration of electron–vacancy pairs in both isomers.[69,70]

The main observations are as follows:[71-75] A light impulse induces the passage of a photocurrent in *trans*-$(CH)_x$, in which luminescence phenomena are not observed. In contrast, in *cis*-$(CH)_x$ there is luminescence but no photocurrent. Also, the threshold energy for photoconduction in *trans*-PAC is clearly less than the 2Δ gap between the valence and conduction bands, whereas it is fairly close to the value of Δ predicted by the soliton theories. Also, the photogeneration branching ratio, which represents the ratio between charged and neutral solitons photogenerated in *trans*-$(CH)_x$, is greater than 10^4.

These phenomena may be interpreted in the light of studies carried out using both the discrete SSH model and the continuous TLM one.[76-82] The results obtained show that in *trans*-PAC, two processes leading to the photogeneration of charges can take place. One is the direct production of a charged soliton–antisoliton pair, helped by quantum fluctuations in the ground state, with a threshold energy equal to the energy required to form

---

[‡]Naturally, it is possible to have a singly charged polaron in the *cis*, as well as the *trans*, isomer.

two charged solitons. The other is direct promotion of an electron from the valence to the conduction band. In the latter case, the generated electron–hole pair is unstable and decays, in 10^{-13}s, into a charged soliton–antisoliton pair.

If the separation of the pair, under the effect of an external field, is sufficiently rapid to prevent recombination, the absence of luminescence is perfectly comprehensible. Besides, as we know from studies on the doped material, the presence of charged solitons is associated with conduction phenomena, and therefore with the photocurrent.[‡]

The surprising difference in the behavior of the *cis* isomer can also be explained within the framework of the soliton model. In fact, the existence of a confinement energy due to the energy difference between the *cis – trans*-oid and the *trans – cis*-oid configurations does not allow the separation of the soliton–antisoliton pair in such a way as to prevent the radiative recombination. This explains the observed luminescence and the disappearance of the charged solitons associated with the absence of a photocurrent.

In conclusion, the predictions of the soliton theories are consistent with the magnetic and optical properties of PAC, albeit within certain limits set by the disorder in the material and the corresponding interpretative difficulties. In the next section, we shall see that the dependence of the electrical conductivity on the different parameters can also be understood within the soliton model.

III. ELECTRICAL CONDUCTION MECHANISMS IN POLYACETYLENE

A. Introduction

From the earliest experimental studies on polyacetylene (PAC)[62,83,84] arose the problem of explaining the results of measurement of its electrical properties. At low impurity levels (concentration of impurity introduced, $y \leq 1\%$) it shows a behavior that can be considered that of an amorphous semiconductor. At higher dopant levels, however, its electrical properties (such as conductivity and thermoelectric power) become ones typical of disordered metals. In any case, as discussed in the previous chapter, the normal theories on amorphous solids are not capable of giving an accurate and satisfactory explanation of all the experimental results, in particular the abnormal dependence of the Pauli and Curie magnetic susceptibilities on the concentration of charge carriers.[14]

[‡] These observations are sufficient for our present purposes. We defer the discussion of photoconduction mechanisms to Section III.

We have already shown how the hypothesis of soliton formation along the $(CH)_x$ chains can explain the various magnetic and optical properties. We will now examine in more detail the mechanisms of conduction in PAC, along with the theories proposed to explain the dependence of the conductivity on impurity concentration and temperature.

It is worthwhile remembering, however, that the soliton model of PAC is not yet universally accepted (see Section II). In particular, some workers do not agree that it is necessary to invoke the presence of solitons to explain the conduction mechanisms, even at low impurity levels ($y < 1\%$).[49]

According to these detractors, charge flow, at whatever dopant level, takes place by "hopping" between metal islands that form due to the non-uniformity of the impurity concentration within the polymeric material. The supporters of this conduction model maintain that the experimental evidence, especially the magnetic susceptibility and electron spin resonance (ESR) data, which generally favor the soliton hypothesis, derive from measurements made at the limits of sensitivity of the instrumentation, and do not conflict with a model of inhomogeneous doping and consequent formation of metallic islands.[49] However, as we shall see further on, more recent experimental results on the dependence, in only slightly doped samples, of conductivity on the temperature and the pressure,[85] as well as the photoconduction and luminescence measurements carried out on PAC,[69,70] support the existence of soliton-type defects along the $(CH)_x$ chains. These results are, in fact, in excellent agreement with the predictions of the soliton theories and could be explained in terms of alternative models only with great difficulty.

It is true, however, that at high dopant levels ($y > 1\%$) the soliton model predicts results analogous to those that may be deduced from the theories of amorphous solids. In fact, when the concentration of defects along the $(CH)_x$ chains is particularly high, the electrical properties of PAC are determined mainly by the structural disorder created by the doping, and can therefore be explained with models developed for the study of disordered metals.[10,67,86,87]

We will now analyze in more detail the various mechanisms proposed to explain the electrical properties of PAC at various dopant levels.

B. Conductivity of "Pure" and Lightly Doped Polyacetylene

As explained above, various experiments, especially those on the magnetic properties of "pure"[‡] and doped PAC, have revealed characteristics that would be difficult to understand without hypothesizing the presence of

---

[‡] It is worth noting that even with the most refined synthetic techniques, it is not possible to obtain a PAC sample with a concentration y of native impurities per carbon atom lower than about 10^{-4}. This value is much higher than that which can be obtained for materials in common usage in electronic devices (e.g., Si or Ge), which can be less than 10^{-13}.

Figure 12.   Schematic representation of conduction by thermally liberated charged solitons.

solitons (or, more precisely, domain walls or phase kinks) that are mobile and can link two portions of a polymeric chain having different phases of bond alternation.

Within the framework of the soliton model of PAC, it is assumed that the doping of this material causes a charge transfer from the impurities to the polymeric chains; this can take place both by transfer to neutral structural defects that already exist and by the creation of new charged solitons.[11]

The latter, unlike the neutral "phase kinks," could, by virtue of their motion, contribute directly to the transport of charge within PAC. This would give, for the case where the dc conductivity depended on the charged soli tons no longer bound to the ionized impurities that generated them, the equation

$$\sigma \propto \frac{q^2}{k_B T} D_c \exp\left(-\frac{E_b}{k_B T}\right) \tag{28}$$

where q is the electronic charge, k_B is the Boltzmann constant, D_c is the diffusion constant of a free charged soliton, and E_b is the energy of binding between a charged soliton and the impurity that generated it (see Fig. 12).

Su, Schrieffer, and Heeger[11] estimated the value of this binding energy by considering the Coulomb interaction, screened by the macroscopic dielectric constant $\varepsilon \simeq 10$, between a charged soliton and a punctiform ionized impurity placed at a distance d from the polymeric chain. On the basis of the experimental results, SSH assumed this distance to be equal to 2.4 Å, obtaining for the binding energy of a soliton $E_b \simeq 0.3$ eV (of the order of the thermal energy at 3500 K). It follows that at room temperature and at dopant levels that exclude any considerable overlap of the electronic wavefunctions of the various phase kinks, the direct contribution of the charged solitons generated by impurities to the electrical conductivity should be quite insignificant. It is worth underlining that the value of $E_b \simeq 0.3$ eV is very probably an underestimate, since the Coulomb screening at a distance on the order of 2 Å is very small.[‡]

[‡]The situation is different in the case where the impurity atoms are solvated. As noted by André et al.,[88] solvating the donor impurities (Na, Li) introduced into the PAC sample greatly increased (up to four orders of magnitude) the electrical conductivity, presumably as a result of the motion of charged solitons.

Another possible conduction mechanism in "pure" PAC, also based on the presence of free charged solitons, was proposed by Rice in 1979. In that paper he formulated the hypothesis that the formation of soliton–antisoliton pairs can take place due to thermal agitation, with a consequent charge transfer between the two localized energy states. In this case, PAC would contain thermally generated charged solitons not related to any impurity. Therefore the equation for electrical conductivity would have the form

$$\sigma \propto \frac{q^2}{k_B T} D_c \exp\left[-\left(\frac{E_s}{k_B T}\right)\right] \qquad (29)$$

where E_s now represents the formation energy of a soliton. Considering, however, that the calculations of SSH[11] and of Rice[18] both give $E_s \simeq 0.4$ eV, we can be fairly sure that the contribution to the transport of charge by thermally generated solitons will also be quite small.

A direct confirmation of the need to use other mechanisms to explain the transport of charge within PAC derives from an analysis of experimental data on the dependence of the conductivity on temperature.[4,67,85,89] Some examples are shown in Fig. 13 of the results obtained by Chiang et a.[4] on PAC samples containing different amounts of impurities. Generally, in all the experimental work, two common characteristics emerge. The first is a strong dependence of resistivity on the dopant concentration. It is in fact possible, by introducing a suitable amount of impurity, to vary the conductivity of PAC over 12 orders of magnitude, reaching a maximum value, for concentrations $y > 1\%$, at 300 K, on the order of 10^2 ohm^{-1}cm^{-1}.[4]

The experimental data also show that, even at dopant levels at which the resistivity of PAC takes on normal semiconductor values ($y < 1\%$), $\ln \sigma$ does not depend linearly on the reciprocal of the temperature. Generally, in fact, the conductivity σ of PAC shows less temperature dependence than would be expected for directly activated charge carriers [see Eqs. (28) and (29)].

These two characteristics of the transport properties in slightly doped PAC led to the idea, based on the earliest research carried out in this field, that the conductivity of this material was due to some mechanism involving "hopping" between localized states associated, in some way, with the charged impurities present within the samples.[10]

In this way, it was possible to explain the dependence of the electrical conductivity on both the dopant concentration and the temperature. In particular, the temperature dependence of σ in samples with $y < 1\%$ is indicative of the fact that in PAC there is no unique activation energy but rather a wide spectrum of energies, due to either dynamic disorder (in the Kivelson model) or static disorder (variable-range hopping).

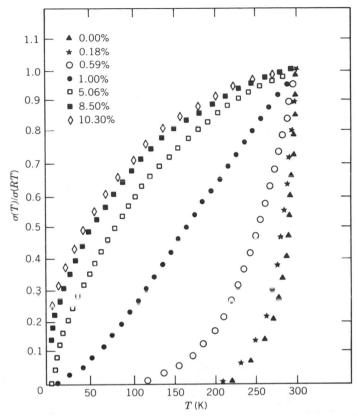

Figure 13.   Temperature dependence of the conductivity of PAC samples with different dopant levels. "$\sigma(RT)$" indicates the value of the conductivity at room temperature. (Reproduced from ref. 49.)

C.   The Kivelson Model

In a series of papers published since 1981[89-91] Kivelson proposed and developed a conduction model for lightly doped PAC ($y < 1\%$) that not only fits well into soliton theory, but also gives predictions that appear to be confirmed by experimental results.[85,92]

In this model, charge transport is effected by "hopping" between localized energy states, although the specific mechanism of the electronic transition has certain aspects different from those of the mechanism proposed for amorphous solids.[93] In fact, in PAC, it is assumed that the quantum leap of the electrons takes place from a doubly occupied electronic level, corresponding to a charged soliton bound to an impurity, to that of a neighbor-

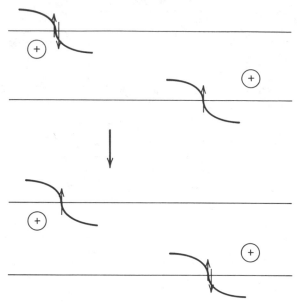

Figure 14.   Electronic hopping between soliton states.

ing, neutral soliton. If the latter "phase kink" does not find itself close to a charged impurity, such an electronic transition requires a quantity of energy ($\Delta E \simeq 0.3$ eV) equal to that required to liberate a charged soliton.

If, however, the neutral soliton, while moving along the polymeric chain, passes close to another impurity, the energy difference between the two electronic states may be reduced to only a few $k_B T$ (see Fig. 14).

Consequently, the fact that the localized levels involved in the electronic hopping are almost degenerate assures that the major contribution to the transport of charge within PAC derives from transitions between soliton states corresponding to impurities that are close neighbors. From this point of view, Kivelson's hypothesis is analogous to fixed-range hopping in disordered materials.[93] This also assumes that transitions take place mainly between states of neighboring impurities. In any case, the localized states in amorphous materials have a much wider energy spectrum, so that the average energy difference between close neighbors is much higher than in PAC. This means that at fairly low temperatures, the number of phonons that can give rise to hopping between adjacent electronic levels is considerably reduced. Thus, within the material, we have mainly quantum leaps between localized states at larger relative distances, because in this way the probability of finding almost degenerate impurity levels increases (variable-range hopping).

Given its importance and the amount of experimental verification that the Kivelson model has acquired over the past few years, we will now give a reasoned critical deduction of it, with the main aim of showing the reasons and the physical mechanisms on which it is based. The need for this is increased by the presence of some obscure points and inaccuracies in the original articles.

As is intuitively easy to understand, in a material where charge transport takes place by hopping between localized states, the conductivity σ depends on the effective rate of transition between the various electronic levels. (This will be covered in more detail later.) In the specific case of two localized states, i and j, this rate is given by[91]

$$\Gamma_{ij} = \overline{n_i(1 - n_j)\nu_{ij}} \tag{30}$$

where n_i is the probability that the ith soliton energy level is doubly occupied, ν_{ij} is the rate of transition of an electron from state i to state j (which is independent of their occupancies), and the overbar indicates an averaging operation, over the thermal distribution, of the quantities below it.

At thermal equilibrium this gives

$$\overline{n_i(1 - n_j)\nu_{ij}} = \bar{n}_i(1 - \bar{n}_j)\bar{\nu}_{ij}$$
$$= n_i^0(1 - n_j^0)\nu_{ij}^0 \tag{31}$$

where the superscript 0 indicates the equilibrium value ($n_i^0 = \{\exp[(E_i - E_F)/k_B T]+1\}^{-1}$, with E_F as the Fermi energy within the material).

Since we propose to study stationary states, we take the soliton distribution within the material at a fixed time t. This assumption allows us to calculate the transition rate ν_{ij} between the initial state i and the final state j, at a mutual distance of $\mathbf{R}_{ij} = (\mathbf{R}_j - \mathbf{R}_i)$, using Fermi's Golden Rule:

$$\nu_{ij}(\mathbf{R}_{ij}, E_i - E_j) = \frac{2\pi}{\hbar}|\langle i|H_e|j\rangle|^2\rho(E_j) \tag{32}$$

where $\rho(E_j)$ is the final soliton density of states, $|\langle i|H_e|j\rangle|$ represents the modulus of the ijth element in the perturbative matrix, and H_e is the approximate Hamiltonian of the electron–phonon coupling.

The equilibrium value, $\bar{\nu}_{ij}$, is found by averaging Eq. (32) over the thermal distributions $P_{ch}(E_i)$ and $P_n(E_j)$ of the charged initial and neutral final soliton states:

$$\nu_{ij}(\mathbf{R}_{ij}, E_i - E_j) = \nu_{ij}(\mathbf{R}_{ij}, T)$$
$$= \iint \nu_{ij}(R_{ij}, E_i - E_j)P_{ch}(E_i)P_n(E_j)\,dE_i\,dE_j \tag{33}$$

The exact evaluation of $\nu_{ij}(\mathbf{R}_{ij}, T)$ however, meets with considerable difficulties: first of all, the calculation of Eq. (33) requires knowledge of all the phonon wavefunctions in the presence of solitons. Thus only approximate calculations can be used;[62] this implies that both the formula for the perturbative Hamiltonian H_e, which describes the electron–phonon interaction and the form of the soliton density of states, cannot be calculated in a precise manner. Second, all the theoretical work developed so far has been based on one-dimensional PAC models (see previous section). This means that, for any treatment of the hopping mechanism, which, as we shall see later, occurs predominantly in a three-dimensional manner, one is obliged to introduce further approximations.[91] Given, therefore, the intrinsic difficulty of evaluating the average rate of transition, in his papers[89-91] Kivelson proposed an empirical form of Eq. (33). We will now investigate the reasonableness of taking this route.

It is usual in problems of this type to assume that the transition rate ν_{ij} $(\mathbf{R}_{ij}, E_i - E_j)$, expressed by Eq. (32), is proportional to the square of the overlap $S(\mathbf{R}_{ij})$ between the electronic wavefunctions. If the spatial distribution of the localized states is uniform, the integral that appears in Eq. (33) may be subdivided in the following manner:

$$\nu_{ij}(\mathbf{R}_{ij}, T) = S^2(\mathbf{R}_{ij}) \int\int g(E_i - E_j) P_{ch}(E_i) P_n(E_j) \, dE_i \, dE_j \qquad (34)$$

where $g(E_i - E_j)$ is the electron–phonon coupling function, which depends only on the energy difference between the levels being considered.

From this, it follows that the complicated thermal averaging over the initial and final states expressed by the integral of Eq. (34) is a function only of the absolute temperature T. For the average transition rate we can therefore write

$$\nu_{ij}(\mathbf{R}_{ij}, T) = S^2(R_{ij}) \gamma(T) \qquad (35)$$

Regarding an explicit evaluation of the terms appearing in Eq. (35), by analogy with hypotheses generally made in related problems,[93] Kivelson assumed

$$S(\mathbf{R}_{ij}) = \exp\left[\left(\frac{R_\parallel}{\xi_\parallel}\right)^2 + \left(\frac{R_\perp}{\xi_\perp}\right)^2\right]^{1/2} \qquad (36)$$

where $R_\parallel$ and $R_\perp$ are, respectively, the components of $\mathbf{R}_{ij}$ parallel to and perpendicular to the chain, and $\xi_\parallel$ and $\xi_\perp$ are the decay lengths of the electronic wavefunction in those same directions ($\xi_\parallel \simeq 4\xi_\perp$).

For $\gamma(T)$, Kivelson chose a fitting function of the type

$$\gamma(T) = \gamma_1 \left(\frac{T}{T_0} \right)^x \tag{37}$$

where the parameters γ_1, T_0, and x can be determined, for example, by a comparison with the experimental results on electrical conductivity. The use of an equation such as Eq. (37) is justified, according to Kivelson, by the necessity of reconciling the results of the approximate calculations, which show that the interaction between the soliton electrons and the lattice vibrations has a maximum corresponding to the center of the optical region of the phononic spectrum ($\hbar\omega \simeq 0.15$ eV), with the considerable amplitude of the band, which extends into the high-energy acoustic regions ($\hbar\omega \simeq 0.05$ eV).

In particular, this latter observation leads to the exclusion, for $\gamma(T)$, of any fitting function peaking strongly about a maximum, making an expression such as Eq. (37) more acceptable. All of these considerations, put together in a quantitatively more rigorous manner, but with mathematical aspects that go beyond the scope of this review, allowed Kivelson to use an approximate expression for $\gamma(T)$, valid for the temperature range around 300 K:

$$\gamma(T) \simeq \frac{10^{18}}{N} \left(\frac{T}{300} \right)^{10} \tag{38}$$

where N is the average number of carbon atoms in a PAC chain ($N \simeq 10^4$).

Taking into account Eqs. (35)–(38) we therefore obtain, for the average transition rate,

$$\nu_{ij}(\mathbf{R}_{ij}, T) = \frac{\gamma_0}{N} \left(\frac{T}{T_0} \right)^x \exp\left[(R_\parallel/\xi_\parallel)^2 + (R_\perp/\xi_\perp)^2 \right] \tag{39}$$

with $\gamma_0 \simeq 10^8$ Hz, $T_0 \simeq 300$ K, and $x \simeq 10$.

The latter expression then allows us to determine the effective transition rate at thermodynamic equilibrium [Eqs. (30) and (31)]:

$$\Gamma_{ji}^0 = \Gamma_{ij}^0 = n^0(1 - n^0)\nu_{ij}(\mathbf{R}_{ij}, T) \tag{40}$$

where $n^0 = y_{ch}/(y_n + y_{ch})$ is the fraction of doubly occupied soliton states and $(1 - n^0)$ represents the fraction, $y_n/(y_n + y_{ch})$, of singly occupied soliton states.

When an electric field $\mathscr{E}$ is introduced, the effective net rate, $\Gamma_{ij} - \Gamma_{ji}$, takes on a nonzero value proportional to the current within the material.

To determine the net charge flow from the ith to the jth electronic level, we follow the standard lines of reasoning developed within the theories for amorphous solids.[94,95] In the presence of an electric field, one must add a perturbative factor $q\mathscr{E} \cdot \mathbf{R}$ to the energy E_i of the ith localized state:

$$E_i \rightarrow E_i + q\mathscr{E} \cdot \mathbf{R}_i \tag{41}$$

One must also take into account the variation that the application of such a field brings about in the probability of the occupancy of the states. This may be done by substituting for the Fermi energy E_F (equal to the equilibrium value of the chemical potential of the material) a position-dependent electrochemical potential:

$$\mu_i = E_F + \delta\mu_i(\mathbf{R}_i) \tag{42}$$

(μ_i is a Fermi pseudopotential within the material).

If we assume that $q\mathscr{E} \cdot \mathbf{R}_i \ll k_B T$ and that all the perturbative terms are therefore small with respect to the thermodynamic equilibrium values, then, with series expansion, we have[93-95]

$$\Gamma_{ij} - \Gamma_{ji} = \frac{\Gamma_{ij}^0}{k_B T}\left(q\mathscr{E} \cdot \mathbf{R}_{ij} + \delta\mu_i - \delta\mu_j\right) \tag{43}$$

In this equation, the terms within the brackets represent the potential energy difference between the two points i and j due to the effect of the external applied field $\mathscr{E}$. Given the physical significance of $\Gamma_{ij} - \Gamma_{ji}$ we can therefore define the quantity

$$\begin{aligned} G_{ij} &= \frac{q^2 \Gamma_{ij}^0}{k_B T} = \frac{q^2 n^0 (1 - n^0) \nu_{ij}}{k_B T} \\ &= \frac{q^2 n_0 (1 - n_0) S^2(R_{ij}) \gamma(T)}{k_B T} \end{aligned} \tag{44}$$

as the conductance between the two points i and j.

To find the expression for the dc conductivity σ, we need only to find a relationship between G_{ij}, defined by Eq. (44), which has so-called local validity, and the macroscopic quantity σ.

As will become clear later, toward this end it is necessary to decide whether the transport of charge is predominantly mono- or multidimensional, that is, whether the electronic transitions take place mainly between soliton states in the same chain (intrachain hopping) or in different chains (interchain hopping).[‡]

First, let us try to find an intuitive criterion for discriminating between these two possibilities. From what has been shown above, it is clear that two main factors decide which type of hopping prevails. On the one hand, there is the considerable anisotropy of the decay lengths of the wavefunctions of the soliton electrons ($\xi_\parallel \simeq 4\xi_\perp$), which would imply the considerable predominance of transitions along the same chain (intrachain hopping). On the other hand, a significant role is played by the geometrical factor deriving from the spatial distribution of the impurities. If, in fact, the various active impurity centers are distributed throughout the material in a completely random and uniform manner, their average separation, along a single chain, is proportional to $(C_{imp}b^2)^{-1}$, where C_{imp} is the impurity concentration (per unit volume) and b is the average separation between the various polymeric chains ($b \simeq 4.4$ Å). However, if one considers the total volumetric distribution, the distance between these centers depends linearly on $(C_{imp})^{-1/3}$. At low dopant concentrations the above-mentioned "geometrical factor" therefore dominates over the anisotropy of the electronic wavefunctions, and the hopping must take place in all three directions.

Let us now try to better quantify this concept and to see whether or not the conductivity clearly has three-dimensional characteristics within the field of applicability of the Kivelson model.

Toward this end, let us consider the expression for the average transition rate in the cases of one- and three-dimensional fixed-range hopping. In the first case, by analogy with Eq. (39), we have

$$\nu_{1D} \propto \exp\left(\frac{-2R_1}{\xi_\parallel}\right) \tag{45}$$

where $R_1 \simeq (C_{imp}b^2)^{-1}$ is the average distance between closest-neighbor impurities along a single chain.

In the case of three-dimensional hopping, however, from Eq. (39) we obtain

$$\nu_{3D} \propto \exp\left(\frac{-2R_3}{\xi}\right) \tag{46}$$

[‡]We must be careful not to confuse the mono- or multidimensional character of the transport of charge within PAC with the anisotropy of the conductivity, which cannot be excluded even in the case of interchain hopping.

where R_3 represents the typical separation between impurities

$$R_3 = \left(\frac{4\pi}{3} C_{imp} \right)^{-1/3} \qquad (47)$$

and ξ is the average three-dimensional decay length,[89-91] defined as

$$\xi = \left(\xi_\parallel \xi_\perp^2 \right)^{1/3} \qquad (48)$$

Therefore, within the model outlined here, the transport of charge within PAC will be prevalently one- or three-dimensional according to whether, for the majority of transitions between localized states, ν_{1D} is greater or smaller than ν_{3D}.

An approximate estimate of the critical impurity concentration C_{imp}^c, for which, whatever the electronic level considered, $\nu_{1D} = \nu_{3D}$, may be directly obtained on the basis of Eqs. (46) and (47).

Equalizing the two exponents, we have

$$C_{imp}^c = \left(\frac{4\pi}{3} \right)^{1/2} \frac{\xi_\perp}{\xi_\parallel} \frac{1}{b^3} \qquad (49)$$

Substituting for the quantities appearing in Eq. (49) with their numerical values, we obtain $C_{imp}^c \simeq 6 \times 10^{-3} \text{Å}^{-3}$, which corresponds to a relative impurity concentration y of $\sim 45\%$.[‡]

It can thus be stated that in the whole range of applicability of the Kivelson model ($y < 1\%$), the transport of charge takes place by "quantum leaps" between soliton levels belonging mainly to different polymeric chains.

On the basis of the above conclusions, we can now combine Eq. (44), which calculates the average conductance between two localized states i and j, with the macroscopically measured dc conductivity. This is a typical percolative problem[91,96-98] for which analytical solutions have been found. With

---

[‡]A more rigorous percolation theory[91,96] allows one to deduce, with greater precision, that the fraction f_{1D} of electronic states involved in one-dimensional fixed-range hopping is given by

$$f_{1D} = \left(\frac{3}{\pi} \right)^{1/2} \frac{\xi_\parallel}{\xi_\perp} C_{imp} b^3 + O\left(C_{imp}^2 \right) \qquad (50)$$

Given their significance, Eqs. (49) and (50) may be compared by setting the latter equal to 1. The value of C_{imp}^c obtained in this way differs from that obtained from Eq. (49) by only a factor of 2.

a suitable choice of fitting parameters, these solutions are in good agreement with numerical simulations based on Monte Carlo methods.

In particular, the line Kivelson followed in his papers[89-91] is based on an approximate model developed by Butcher et al.[96,97] for the treatment of analogous problems in amorphous semiconductors.

The formula for the dc conductivity derived from this model is

$$\sigma = \frac{Aq^2\gamma(T)}{k_BT} \frac{\xi}{R_3^2} \frac{y_ny_{ch}}{(y_n+y_{ch})^2} \exp\left(\frac{-2BR_3}{\xi}\right) \qquad (51)$$

where $\Gamma(T)=(\partial_0/N)(T/T_0)^x$,

$\gamma_0 \simeq 10$ Hz
$T \simeq 300$ K
$x \simeq 10$
$R_3 = [(4\pi/3)C_{imp}]^{-1/3}$
$\xi = (\xi_{\parallel}\xi_{\perp}^2)^{1/3}$

and where $A = 0.45$ and $B = 1.39$ are two dimensionless constants whose values depend on the specific nature of the particular percolative problem (i.e., the number of directions in which conduction takes place).

D.   Comparison with Experimental Data and Discussion of the Kivelson Model

As stated previously, the Kivelson model for the transport of charge in lightly doped PAC ($y<1\%$) currently constitutes a suitable framework for the study of the electrical properties of this material. The model seems to fit both the experimental data and the structural theories of PAC better than any other model. Recently Croboczek and Summerfield[99] have applied a new theory, called the Extended Pair Approximation (EPA),[100] to the calculation of the dc and ac conductivities of PAC. Their approach, based on a representation in terms of equivalent circuits, gives frequency and temperature dependences of the conductivity that agree with experimental results better than those obtained with the Kivelson theory do. It is worth noting, however, that the model requires five fitting parameters to agree with the experimental data, and also does not enter into the merits of the various descriptions of the physical nature of the states within the gap. To describe the trend in the density of states, the authors used three Gaussian functions that, as they themselves state, could be produced not only by solitons but also by any other type of defect within the material.

The hypothesis that the conductivity of PAC is due to the hopping of electrons between almost degenerate soliton levels uniformly distributed throughout the active regions of the material allows the problem of the conductivity to be treated in a manner in many ways analogous to that used for the description of fixed-range hopping in amorphous semiconductors.

In particular, the factorization of the transition rate $\nu_{ij}(\mathbf{R}_{ij}, T)$ into two terms [Eq. (35)], one depending only on the distance between the states (and therefore on the impurity concentration) and the other only on the absolute temperature (Kivelson factorization), is possible because, at any temperature, the quantum leaps take place between closest-neighbor localized levels (see Section III.C).

There are several consequences of this separation. In the first place, the ac conductivity for high frequencies ω can be obtained, as in the case of noncrystalline semiconductors, by using the pair approximation[70,90,91] and is given by

$$\sigma_{AC}(\omega) \propto \omega \left[\ln\left(\frac{2}{\Gamma(T)} \right) \right]^4 \tag{52}$$

with $\Gamma(T) = n_0(1 - n_0)\gamma(T)$.

The fact that σ_{AC}'s dependence on the temperature is determined by exactly the same factor $\gamma(T)$ that appears in the formula for the dc conductivity is linked directly to the possibility of carrying out the Kivelson factorization.

Recent experimental measurements[92] carried out on slightly doped trans-$(CH)_x$ did, in fact, show a considerable variation in σ_{AC} with changes in temperature, in good agreement with the predictions of the model.

Another consequence of the separation carried out on the transition rate $\nu_{ij}(\mathbf{R}_{ij}, T)$ [Eq. (35)] derives from the predictions that can be made regarding the temperature dependence of the electrical conductivity at different pressures. As can be seen from Eq. (51), the dominant effect of pressure changes on the conductivity occurs via the variation of the average distance between the impurities R_3, which appears in the exponent of that expression, while all the other factors remain practically the same. The relative variation of the average distance R_3 between impurities after a pressure change $\Delta P = (P - P_0)$ may be written as

$$\frac{1}{R_3} \frac{\Delta R_3}{\Delta P} = \mu \frac{1}{V} \frac{\Delta V}{\Delta P} = -\mu K \tag{53}$$

where V and K are, respectively, the total volume and the compressibility of the PAC sample, and μ is a dimensionless parameter less than 1 introduced to take into account, in the relationship between the total volume change and the variations in the average distance between impurities, the essentially "spongy" nature of the PAC samples. (For isotropic homogeneous material, $\mu = \frac{1}{3}$.) From Eqs. (51) and (53), we obtain

$$\sigma(T, P) = \sigma_0(T)\exp\left[-\frac{2BR_3(P_0)}{\xi}(1 - \mu K\Delta P)\right] \qquad (54)$$

It then follows, according to the Kivelson model, that the ratio of the conductivity values at different pressures does not depend on the temperature.

Figure 15 shows the results of measurements carried out by Moses et al.[85] in 1981 on a PAC sample doped with 0.03% impurity.

Even though the data relating to the temperature dependence of the resistance at constant pressure seem to be consistent with both the Kivelson model and the variable-range hopping model (when suitable sets of fitting parameters are chosen),[10,93,95] the differences in the conductivity values at

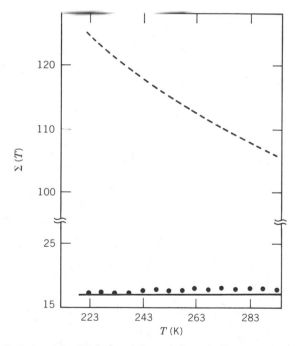

Figure 15.   Experimental values (dotted line) of the ratio $\Sigma(T) = \sigma(T, 8740 \text{ bar})/\sigma(T,0)$ compared with theoretical predictions. Continuous line represents predictions of Kivelson model; dashed line represents predictions based on variable-range hopping. (Data taken from Moses et al.[85]

different pressures can agree only with the hypothesis of quantum leaps between almost degenerate localized soliton states.[‡]

A further consequence of the Kivelson factorization is that, within the validity of the model, the form of the expression relating σ to the temperature should be independent of the impurity concentration. This prediction was already confirmed by the earliest experiments carried out on the electrical properties of PAC.[4]

In PAC, not only is there movement of charge, there is also diffusion of spin. As mentioned in the previous section, it was precisely the need to explain the existence, in the neutral material, of free and mobile spins, that gave birth to the hypothesis of topological solitons along the PAC chains.

It is thus clear that the spin-diffusion constant perpendicular to the chain, $D_\perp$, is linked to the electrical conductivity by the Einstein relationship

$$D_\perp = \frac{k_B T}{q^2 C_{imp}} \sigma_\perp \tag{55}$$

The diffusion constant $D_\parallel$ parallel to the chain, on the other hand, is determined mainly by the motion of neutral solitons.

It is, in fact, reasonable to assume that the movement of a neutral "phase kink" constitutes a spin-transfer mechanism that is much more efficient than "hopping" between localized electronic states.

[‡] In the case of variable-ranging hopping, in fact, the conductivity follows a law of the type $\sigma = \sigma_0 \exp[-(T_0/T)^{1/4}]$, where $T_0 = 16/[\xi^2 N(E_F) K_B]$ and $N(E_F)$ is the localized density of states near the Fermi level.[93,95] When the pressure increases, the localized density of states increases due to both the change in volume V and the change in the amplitude E of the gap, so that

$$\frac{\delta N(E_F)}{N(E_F)} \simeq -\left(\frac{\delta E_g}{E_g} + \frac{\delta V}{V}\right)$$

We therefore have

$$\sigma(T, P) = \sigma_0 \exp\left\{-\left(\frac{T_0}{T}\right)^{1/4}\left[1 - \tfrac{1}{4}(\varepsilon + K)P\right]\right\}$$

where $\varepsilon = \dfrac{\delta E_g}{E_g \delta P}$. It thus follows that with variable-range hopping it is not possible to factor $\sigma(t, P)$ into two separate parts, one depending only on T and the other only on P, it is for the fixed-range hopping proposed by Kivelson.

Consequently, we see that

$$D_\parallel \gg \frac{k_B T}{q^2 C_{\text{imp}}} \sigma_\parallel \tag{56}$$

Both of these predictions have recently been confirmed experimentally.[91] Within the Kivelson model, attempts have been made to explain the unusual photoelectric behavior of PAC[69,70] (see Section II).

In fact, as first observed by Su and Schrieffer[76] and later confirmed by other theoretical work,[101-103] an electron–vacancy pair generated by photoabsorption decays, over a period on the order of 10^{-13} s, into a charged soliton–antisoliton pair. It was thus assumed that in lightly doped trans-$(CH)_x$ photoconductivity is essentially due to this soliton production, which, by increasing the density of the localized levels, increases the possibility of hopping between electronic states.

However, several other workers[74] have formulated the hypothesis that the photocurrent is due directly to the movement of charged solitons produced by the decay of the electron–hole pairs, which, not being bound to any impurity, are free to move along the polymeric chain. The dynamics of optically generated charge carriers thus constitutes a field of study that needs to be clarified.

Even though the Kivelson model has recently received ample and satisfying experimental confirmation, it cannot be denied that the calculations on which it is based represent something of a simplification. For example, we have already seen the need to introduce an approximate form for the electron–phonon coupling constant. Besides, as Kivelson has clearly stressed,[91] his treatment completely ignores soliton–soliton interaction.‡ This problem has still not been resolved satisfactorily. A certain coupling energy should exist even between neutral solitons due to the coupling between the π-electrons of the contiguous PAC chains. This coupling brings about a configuration where chains have single bonds corresponding to double bonds in neighboring and parallel chains; this is a lower-energy configuration than that with paired double bonds. The presence of two solitons, either from the same chain or from contiguous ones, varies the relative phase of a certain portion of the two chains, thus producing a kind of soliton "confinement

---

‡ With particular regard to electron–electron interaction within the same doubly occupied soliton level, the effect of their mutual repulsion is implicitly considered, given that the treatment excludes quantum transitions that would transform two neutral solitons into two "phase kinks" of opposite charge.

On the other hand, if electron–electron coupling were important, the Kivelson model would require further revision, since it is based on the theoretical deductions of SSH, which assume that this type of interaction can be ignored.

energy." However, theoretical calculations carried out with the aim of evaluating this interaction energy do not show a great deal of agreement.

Baughman and Moss[104] treated the coupling in terms of Coulomb forces and obtained a confinement energy of about 2×10^{-4} eV for the C_2H_2 group. On the other hand, Baeriswyl and Maki,[105] generalizing the SSH Hamiltonian, suggested that the coupling was due to a term $t_\perp$ deriving from interchain hopping. A $t_\perp$ value of $\sim 7.5 \times 10^{-2}$ eV was obtained from the pressure dependence of the optical absorption spectra. This gave a confinement energy per site of about 3×10^{-3} eV, equal to a thermal energy of 35 K. This would imply that in the crystalline regions of the material at room temperature, the solitons should be confined to a relative distance of about 10–20 lattice constants. This result, if confirmed by further theoretical calculations, could have several consequences for the models described above. In any case, Schrieffer[106] has recently demonstrated that such a confinement effect should take place only over limited time intervals. Thus the confinement of solitons is an open problem and still requires substantial theoretical clarification.

Moreover, it must be added that, as mentioned above, most of the charged solitons are linked to impurities of opposite sign, and it is therefore reasonable to suppose that the Coulomb interaction between ionized topological defects belonging to different chains will be quite small.

The situation is probably different in the case of coupling between a charged and a neutral soliton. In fact, theoretical calculations yield a value of about 0.3 eV for the binding energy of a charged soliton–neutral antisoliton pair, that is, a polaron (see Section II.5). It is to be noted, however, that although this energy seems fairly high (on the order of the thermal energy at 3500 K), the effects that these polarons could have on the electrical properties of PAC have not yet been examined in detail. In any case, within the conductivity model outlined here, it is reasonable to suppose that the above formulae retain their validity, independently of the specific type of structural defect present within PAC, as long as the corresponding energy states are almost degenerate. In fact, apart from the specific form of the average transition rate, $\nu_{ij}(\mathbf{R}_{ij}, T)$ [see Eq. (35)], the only condition necessary for the Kivelson factorization is the existence of localized levels having energy differences of only a few $k_B T$, at whatever temperature.

Knowledge of the origin and nature of the electronic states is required only for the explicit evaluation of the terms into which the average transition rate has been factored, terms that, given the intrinsic difficulty of the calculations required for their determination, are approximate anyway because of the use of suitable fitting functions.

Also, in PAC the conductivity shows anisotropic characteristics due to the fact that the different decay lengths ($\xi_\parallel \simeq 4\xi_\perp$) cause the quantum leaps to take place prevalently in the direction parallel to the chain, and therefore $\sigma_{\parallel}$ > $\sigma_{\perp}$. The effect of this anisotropy, which should not be confused with the three-dimensional character of intersoliton hopping, is ignored in the derivation of Eq. (52) because the values of R_1, R_2, and R_3 are substituted with the average values of R and . The formula obtained therefore refers to the average value of the conductivity in PAC, $\sigma = \frac{1}{3}(\sigma_{\parallel} + 2\sigma_{\perp})$.

Notwithstanding all of these simplifications and approximations, the Kivelson model is the theoretical framework that so far allows the best matching of the experimental results with the hypotheses that have been developed to explain other properties of PAC. The most interesting aspect of the ideas outlined above is that we have taken into consideration, at any temperature T, only those quantum leaps between localized soliton states having energy differences of only a few $k_B T$. As a result of this, it is possible to factor the expression for the conductivity into two functions, one depending only on the average distance between impurities and the other only on the absolute temperature.

The electrical conductivity model proposed by Kivelson is, therefore, a particular type of fixed-range hopping, characterized both by dynamic disorder, deriving from the motion of neutral solitons along the various polymeric chains, and by structural disorder due to the presence of dopant atoms. As will be made clear in the next section, the latter plays a very important role in the dynamics of the semiconductor–metal transition observed in heavily doped PAC.

E. The Electrical Conductivity of Heavily Doped Polyacetylene

As would be expected, at dopant concentrations $y > 1\%$, the interaction between charged phase kinks generated by the impurities produces a "splitting" of the localized electronic levels within the gap. It thus follows that the Kivelson model, which is based on hopping between degenerate soliton states, can no longer be applied.

With respect to this point, A. J. Epstein et al.[67] proposed, for PAC with dopant concentrations in the range $y \simeq 1$–2%,[‡] a mechanism of variable-range hopping between levels in the soliton band. According to this hypothesis, the temperature dependence of the conductivity takes on the same form as for variable-range hopping in amorphous semiconductors:

$$\sigma(T) = \sigma_0 \exp\left[-\left(\frac{T_0}{T}\right)^{1/4}\right] \tag{57}$$

---

‡ The values given here are approximate. The exact values depend on the type of PAC with which one is dealing, on the dopant, and so on.

where σ_0 is a constant that depends on the electron–phonon coupling integral, $T_0 = 16/[\xi^2 N(E_F)k_B]$, and $N(E_F)$ is the localized density of states near the Fermi level.[93]

For a better understanding of the electrical characteristics of PAC at even higher dopant levels and, in particular, the exact form of the semiconductor–metal transition observed for impurity concentrations $y \sim 2\%$,[‡] we must examine the effect of varying the impurity level on the electronic structure of the polymer. Theoretical study on this particular point was first done in 1980, by Mele and Rice.[87] They considered, among other things, the effects of both the random arrangement of the impurity ions (and therefore, of the charged solitons bound to them) and the electrostatic interaction between the various polymeric chains on the distribution of the energy levels.

The results obtained are shown in Figs. 16–18. The first point to be noted, as previously mentioned, is that the width of the soliton band increases with increasing dopant concentration y (see Fig. 16). If the effects of both the Coulomb interaction with the randomly distributed impurity ions and the interactions between the various chains are ignored, there is a nonzero density of states near the Fermi level, and therefore metallic characteristics, only for impurity levels above 14% (see Fig. 16).[‡]

The presence of charged, randomly distributed dopant atoms and the interactions between the various chains further widen the band of localized levels, reducing the size of the forbidden energy gap.[107] In this way, even for $y \simeq 2\%$, there are localized states near the Fermi energy (see Figs. 17 and 18).

The studies carried out have thus shown the importance of the above effects of static structural disorder on the mechanism of the semiconductor–metal transition. In fact, it is precisely these localized states deriving from perturbations in the "ideal" structure of the soliton levels that considerably reduce the critical impurity concentration at which this transition takes place.

It is thus worth stressing how the above-mentioned variation in the electrical properties of PAC is different from that which provides the basis for the semiconductor–metal transition in amorphous materials.[93,95] In the latter case, the appearance of typically metallic characteristics is due to the

---

[‡] On this point it is to be noted that, taking as the average length of the charged solitons the distance that separates 14 carbon atoms along the PAC chain, one obtains for the critical concentration n_c at which the electronic wavefunctions of the various phase kinks begin to overlap a value of $\sim 7\%$.

It is worth stressing that all the theoretical values reported below should be treated with caution, as their determination is closely linked to the shapes and the dimensions of the various structural defects, for which it is possible to give only rough estimates within the models discussed here.

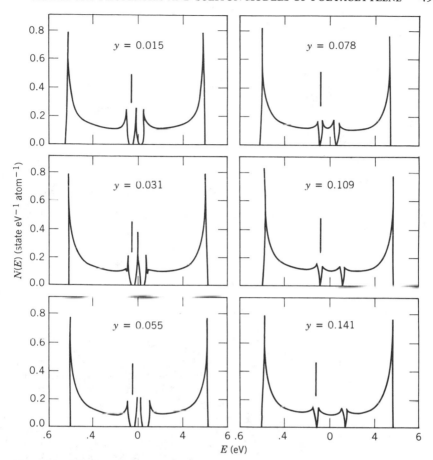

Figure 16.   The density of states of the π-electrons of $(CH)_x$ for the case in which the effects of "Coulomb pinning" between charged solitons and dopant ions randomly distributed throughout the PAC sample, as well as the interactions between the various chains, are ignored. The concentration y of acceptor impurities is given for each case; the vertical lines show the positions of the Fermi levels. (Reproduced from Mele and Rice.[87])

Coulomb screening exerted by the outer electrons of the impurity atoms, which, at high dopant concentrations, prevents the existence of bound external electronic states. In PAC, however, the presence of electronic levels at the Fermi energy is essentially brought about by the overlap of the soliton electronic wavefunctions.

From Fig. 17 it can also be seen that although the localized states near the Fermi level begin to come into existence at relative dopant levels of ~ 2%,

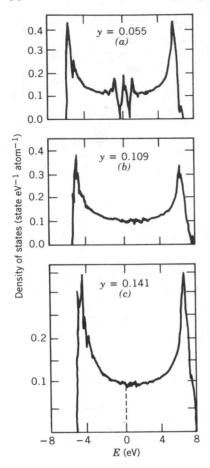

Figure 17. The density of states of the π-electrons of $(CH)_x$ doped with acceptor atoms for the case in which the Coulomb coupling between charged solitons and the randomly distributed impurity ions, as well as the interactions between the various chains, is taken into account. The concentration y of acceptor impurities is given for each case; the vertical lines show the positions of the Fermi levels. (Reproduced from Mele and Rice.[87])

the value of the order parameter u_0 remains appreciably different from zero until concentrations of more than 10% are reached. This indicates that even in the metallic-behavior region, significant portions of the polymeric chains conserve a structure having bond alternation.

Since the width of the Peierls gap between the bonding and antibonding bands of the π-electrons is proportional to the value of the order parameter u_0, the disappearance of this gap between the extended bands should occur only at concentrations greater than 10%. At these dopant levels, the higher extreme of the valence band reaches the lower extreme of the conduction band and the density of states near the Fermi level reaches its maximum value.

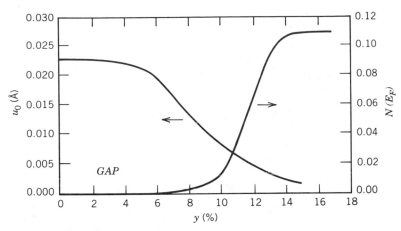

Figure 18.   The density of states $N(E_F)$ at the Fermi level and the order parameter u_0 as functions of the dopant concentration y, taking into account, as for Fig. 17, the effects of static structural disorder. (Reproduced from Mele and Rice.[87])

As proof of the precision of their experimental predictions, Mele and Rice[87] reported the measurements made by Epstein on the Pauli susceptibility of PAC as a function of the concentration of I_3^- ions present. The experimental data, in agreement with the theoretical estimates (see Fig. 18), show a saturation around values corresponding to ~ 0.11 state eV^{-1} atom^{-1} obtained for relative I_3^- percentages of $\sim 10\%$.

With regard to the conduction mechanism in "metallic" PAC, it is worth recalling that in this material the single polymeric chains are grouped together in fibers with diameters on the order of hundreds of angstroms and lengths about 10 times greater.[108] It follows that although at high dopant levels the single chains have a metallic-type resistivity, the total conductivity of the material is significantly influenced by the presence of interchain and interfiber contact. Understanding the charge-transport mechanisms in PAC with high relative percentages of impurity is also complicated by the fact that the doping does not take place in a completely uniform manner. This causes the formation within the polymer of large, heavily doped regions separated by regions of PAC having a clearly inferior conductivities. For these reasons various researchers[86,87,109] hold the opinion that for impurity concentrations greater than 2%, charge transport within PAC can be interpreted using a model of electron hopping between metallic "islands" separated by zones of relatively higher resistivity. This situation may be shown schematically as in Fig. 19.

Since the probability of an electron passing from one metallic region to another significantly depends on the distance between them, it is reasonable

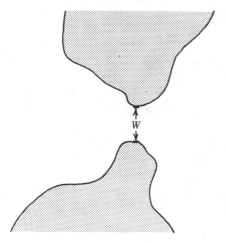

Figure 19.   Schematic representation of the region in which two conducting zones in PAC, represented by the shaded areas, are closest to each other. The heavier part of the boundary line represents the portion of the surface involved in the hopping.

to assume that the only areas involved in hopping are those parts of the metallic zones that are closest to each other (heavier part of the boundary line in Fig. 19).

From the point of view of the energetic structure of the material, the presence of dielectric regions separating zones that show a metallic type of conductivity may be described by introducing, in randomly distributed positions, potential barriers dividing the metallic portions of the material (see Fig. 20).

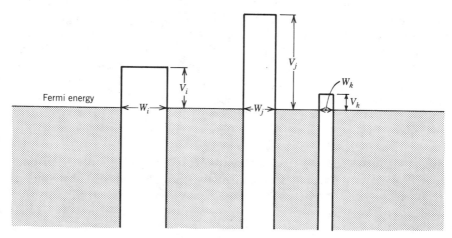

Figure 20

The electrons can thus pass from one metallic island to another either by tunneling or as a result of thermal agitation. Since the height, V_n, of the nth potential barrier (see Fig. 20) depends on the surface charge density at the points between which the electronic hopping takes place, Sheng[86] showed how thermal fluctuations in the concentration of charge carriers give rise to random variations in the values of V_n, which considerably influences the passage of electrons from one metallic zone to the next.

The formula for the temperature dependence of the conductivity is calculated by considering the contributions to the transport of charge within the material arising from the electrons passing the fluctuating potential barriers both by quantum tunneling and by thermal agitation.

Even though the physical idea at the basis of this model is fairly simple, the calculations necessary for the determination of the expression for the conductivity of heavily doped PAC are quite complicated, and the evaluation of several quantities can be carried out only by using numerical techniques. Besides, as could be expected from what has been said so far, the derivation of the conductivity formula for metallic PAC requires a knowledge of the shapes and the distribution of the potential barriers within the material. Since, however, it is not possible to determine these quantities *a priori*, one is obliged to introduce several fitting parameters into the expression.

In order not to lengthen this review with a mathematical treatment that would not improve the physical understanding of the model, we report directly the formula obtained by Sheng[86] on the basis of the considerations made above:

$$\sigma = \sigma_0 \exp\left[-\frac{T_1}{T}\varepsilon^2 - \frac{T_1}{T_0}\varphi(\varepsilon) \right] \qquad (58)$$

where σ_0 is a fairly complicated function of the temperature, which can, however, be considered as approximately constant; T_0 represents the critical temperature below which the conductivity becomes temperature independent because the barriers are then passed only by tunneling; T_1 is the temperature above which the average energy of the electrons is sufficient to allow their direct crossing of the barrier by thermal excitation; and ε is a parameter that depends on the value of the applied field. Finally, $\varphi(\varepsilon)$ is a function that depends on the shape that is assumed for the potential barriers within the material.

Although the different fitting parameters appearing in Eq. (58) make the formula quite general and applicable to a wide range of disordered materials, such as C–PVC compounds and heavily doped gallium arsenide, the agreement with experimental data on the temperature dependence of the conductivity of heavily doped PAC is very good (see Fig. 21).

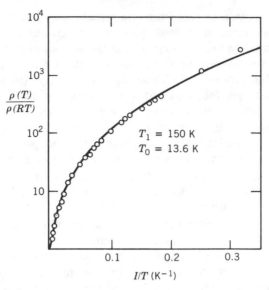

Figure 21.   Comparison of the experimental data of Chiang et al.[4] on the temperature dependence of the resistivity of heavily doped PAC with theoretical predictions based on the model involving hopping between metallic islands. $\rho(RT)$ is the resistivity value at room temperature. The continuous line was calculated using Eq. (58) and the values given in the figure. (Reproduced from Sheng.[86])

While recognizing the extreme caution required in the application of a formula containing several fitting parameters, such as Eq. (58), we believe that the excellent agreement with the experimental data confirms the validity of the physical hypotheses at the basis of the charge-transport model involving a hopping mechanism between metallic islands within heavily doped PAC.

IV.   CONCLUSIONS

To conclude this review of the physical properties of PAC and of the hypotheses that have been proposed for their interpretation, we can say that over the last few years, beneath a surface of apparent confusion, a possible unitary framework for understanding the properties of this material is beginning to take shape. The fundamental concept forming the basis of this unifying outlook is the soliton, a relatively new idea within solid state physics that has been previously applied in hydrodynamics. In only a few years the notion of a soliton has been developed as a unifying element in different areas of physics, so much so that it may be defined as a new *paradigm* in the study of nonlinear phenomena.[26] In the past, an approximate analytical solution of a nonlinear differential equation was searched for using perturbative techniques, employing the solutions of the corresponding linearized equation. The discovery that a wide range of equations allow exact solutions that take the form of solitary waves has given rise to a completely different approach that even allows description of sectors of the "solution space" previously inaccessible with perturbative techniques.

In the specific case of PAC, the origin of the soliton hypothesis is probably to be found in charge density wave (CDW) theory, by means of which it has already been shown that nonlinear excitations of the solitary-wave type can exist.

Rice's simple and ingenious model is not substantially different from the ϕ^4 model of the CDW,[24] and represents the first step toward an understanding of the role of solitons in determining the electromagnetic and optical properties of PAC.

The use of a Ginsburg–Landau type of approach such as that proposed by Rice was considered an oversimplification by Su, Schrieffer, and Heeger[14] (SSH), who proposed a microscopic model of a PAC chain (or rather a polyene chain, given that the hydrogen atoms are not explicitly included in the model). The SSH Hamiltonian is probably the best existing model of PAC from the point of view of the "readability" of the physical hypotheses. As mentioned above, such a model is not completely free of limitations. In particular, one would need to consider, besides the possible interactions between chains already noted, the effects of quantum fluctuations. Several theoretical studies on this matter,[110-112] however, have demonstrated that such fluctuations do not detract from the qualitative results of the simple SSH model. The possibility of applying the continuous limit has also been verified (TLM model). Also worth noting are the *ab initio* calculations on segments of PAC chains,[37] which have confirmed the validity of the soliton model, based on the strong bonding approximation. In particular, the doping mechanisms must be those described previously, even in the light of these new and more accurate calculations.

We cannot state that the soliton hypothesis is proven beyond any reasonable doubt. We have seen that the experimental situation is confusing, and that it is possible to describe a portion of the observed phenomenology by referring to more familiar concepts belonging to the theory of disordered semiconductors. However, we stress the success of the soliton approach in explaining and bringing together widely differing phenomena, including electrical, magnetic, and optical properties, both in the pure and in the doped material.

The use of simplified models and the limits of their validities can also be understood within the context of the soliton theory. It is also quite remarkable how the theory can explain the different behavior of the *cis* and *trans* isomers, which would otherwise be difficult to understand.

One interesting point raised by chemists regarding the approach described here is that the dopant concentrations reached are practically sufficient for the formation of a new material (e.g., PAC iodide) whose properties can no longer be interpreted as those of doped PAC. In our opinion, however, the soliton hypothesis may be considered a specific description of the chemical bond between impurities and chains, and of its influence on the band structure of the "compound." Once this point of view is accepted, the contrast between the two approaches disappears, or becomes simply a question of terminology.

We also note that in general, the presence of disorder in a semiconductor gives rise to localized states within the energy gap between the valence and conduction bands. Thus it is possible to reinterpret many of the experimental results examined simply by introducing a particular density of states together with suitable hypotheses on the mobility of the defects, without having to refer to the soliton concept. However, given that, in order to agree with experimental results, the new density of states must approach that given by the soliton model, this operation is unsatisfactory, since it compels us to postulate that which we could otherwise derive directly from more fundamental hypotheses. The situation would be different if other microscopic hypotheses could be found that were capable of justifying an analogous behavior of the density of states.

Two great benefits of a unifying model are that it can suggest experiments that would otherwise not even be taken into consideration and provide a key to the understanding of the abundant data already collected. As already mentioned, the soliton theory of PAC is a long way from being definitive. In fact, although a satisfactory static view of the material has been obtained, we know very little of the evolution of a chain containing more than one soliton and/or polaron. Studies on the behavior of this type of system are in progress, involving mainly computer simulations based on the SSH Hamiltonian. Other research, using Rice's model, is being carried out to obtain information on collisions between solitons. We consider the study of the dynamics of multisoliton systems to represent one of the main areas for future research on PAC. Other aspects are those of understanding the doping mechanisms and the soliton–polaron "interplay" in the material in the presence of charge transfer.

There is undoubtedly theoretical interest in the hypothesis that under certain conditions there exist mobile charged solitons. These seem to be present in the doped material when the ionized impurities are introduced from a polar solution.[88] Also, it is thought that mobile charged solitons are responsible for the photocurrent in pure PAC. Regarding the possibility of experimental verification of these hypotheses, we note that an analysis of the elec-. trical dependence of the electrical conductivity on different variables will probably not give decisive answers. In fact, mobile charged solitons may contribute to the transport of charge along a chain, whereas the macroscopic conductivity depends also on the transport of charge between chains. An important experiment has been suggested by Rice, according to whom the presence of mobile charged solitons should give rise to a new absorption in the infrared spectrum deriving from the coupling between the kink's translational motion and its vibrational degrees of freedom.

A promising area for theoretical research on PAC is the statistical mechanics of soliton systems. Explorations of this field are necessary if we are to pass from a microscopic description, usually limited to a single isolated chain, to the quantities that describe the macroscopic behavior of the material. Studies of the statistical mechanics of equilibria have demonstrated the possibility of treating, in an approximate manner, the elementary excitations in the chain by assigning to them probabilities that are functions of their energies.[46] It would thus be interesting to follow the statistical mechanics of these systems, focusing particular attention on nonequilibrium states and on the properties of soliton transport.[113]

We have already mentioned the difficulty of interpreting data relating to the dependence of different variables on the dopant concentration. Most studies refer to the macroscopic impurity concentration y, obtained directly or indirectly from the material–dopant mass balance. The quantity used in theoretical calculations, however, is the effective ion concentration (y_{eff}) of impurities that bond to the chain, giving rise to the phenomena described above. It is possible that dopant "molecules" that are not very active form within the polymeric matrix, so that the definitions of concentration do not coincide. It has recently been proposed[44,54] to estimate the fraction of active impurities by optical absorption measurements; samples doped with I_2 and Br_2 clearly show saturation phenomena such that above a certain level, further addition of dopant can no longer cause significant changes in y_{eff}. To clarify this point, it would be helpful to carry out a systematic collection of data (electrical, magnetic, and optical) as a function of y_{eff}, which in turn should be determined as a function of y from optical absorption or other measurements.

An analysis of the charge-transport mechanisms would also be greatly aided by experiments on the anisotropy of the electrical conductivity, as long as one could define the degree of orientation of the sample under examination, and not only the drafting (elongation) ratio, a variable that cannot be interpreted directly. Polyacetylene having good workability and very high drafting ratios, recently synthesized at ENI Laboratories (San Donato Milanese, Italy), could prove to be extremely useful for such studies.

Another class of experiments of great applicative importance consists of those relating to the kinetics of doping and to the kinetics of the variation of the properties associated with doping. A theoretical interpretation of these types of data, which we have not covered in this review, is generally more complicated than that for stationary situations, which are of more common interest. Nonetheless, this is a very important area for the development of future work on PAC and for the study of electrically active polymers in general.

Finally, given that possible technological applications depend on the possibility of discriminating between the different models (which is in any case of theoretical interest), more detailed experiments will be required to obtain a clear and unequivocal explanation of certain physical processes.

Important areas for future experimental work are the frequency dependence of the ac conductivity; infrared absorption mechanisms; and the processes of photogeneration, photoconductivity, and luminescence in *cis*- and *trans*-PAC.

APPENDIX A. THE TAKAYAMA, LIN-LIU, AND MAKI MODEL AS A CONTINUOUS LIMIT OF THE SU, SCHRIEFFER, AND HEEGER MODEL

In this appendix we examine the derivation of the model Hamiltonian of Takayama, Lin-Liu, and Maki (TLM), which may be considered a continuous limit of the discrete Hamiltonian of Su, Schrieffer, and Heeger (SSH). As the increasing amount of literature demonstrates, the possibility, offered by the continuous model, of being treated exactly in an analytical manner makes the TLM Hamiltonian a particularly important and promising instrument for studying both the static and the dynamic properties of polyacetylene (PAC). The techniques employed here are also useful for the study of other one-dimensional materials.

First, we will summarize the fundamental aspects of the SSH model. As a reference structure, let us take the nondimerized structure of *trans*-PAC (shown schematically in Fig. 3 in Section II.B).[‡]

[‡] We recall here that, according to both predictions of the various theoretical models (and *ab initio* calculations) and experimental measurements, the ground state of a neutral chain of PAC shows an alternation of bonds (perfect dimerization). The nondimerized structure referred to in the text is therefore only a useful basis for the construction of the model, and is not related to the actual position of the CH groups along a PAC chain.

If we let u_j be the projection, along the chain axis, of the change in position of the jth CH group with respect to the reference position (see Section II.B), then the Hamiltonian of a PAC chain containing N CH groups is[11,114]

$$H = -\sum_{s}\sum_{j=1}^{N} t_{j+1,j}\left[C_{j,s}^{\dagger}C_{j+1,s} + C_{j+1,s}^{\dagger}C_{j,s}\right] + \frac{1}{2}\sum_{j=1}^{N} K(u_{j+1} - u_j)^2$$

$$+ \frac{1}{2}\sum_{j=1}^{N} M\dot{u}_j^2 \tag{59}$$

where $t_{j+1,j}$ is the hopping integral, $C_{j,s}^{\dagger}$ and $C_{j,s}$ are, respectively, the creation and annihilation operators of a π-electron with spin s on the jth CH group, K is the effective elastic constant of the σ-electrons, and M is the total mass of a CH group.

In the case of nondimerization, ignoring the kinetic term, the SSH Hamiltonian reduces to only the term deriving from the energy of the electrons:

$$H_0 = -\sum_{j,s} t_0\left[C_{j+1,s}^{\dagger}C_{j,s} + C_{j,s}^{\dagger}C_{j+1,s}\right] \tag{60}$$

This latter equation may then be written in the spectral form

$$H_0 = -\sum_{j,s} t_0\left[{}_0|j+1\rangle_{ss}\langle j|_0 + {}_0|j\rangle_{ss}\langle j+1|_0\right] \tag{61}$$

where ${}_0|j\rangle_s$ denotes the Wannier function associated with the electrons of spin s at the point $X_j = ja$ (a is the lattice constant). The subscript zero indicates a reference to the nondimerized structure.

As mentioned earlier, the ground state of a neutral PAC chain has alternate single (C—C) and double (C=C) bonds (perfectly dimerized structure). This means that the basic lattice unit in PAC is C_2H_2.

Let us use the vector $R_j = (2j + \frac{1}{2})a$ for each unit cell consisting of two CH groups placed, respectively, in $X_j^1 = 2ja$ and $X_j^2 = (2j+1)a$. There are thus two "tight-binding" functions ${}_0|1j\rangle_s$ and ${}_0|2j\rangle_s$ associated with each lattice cell, so that Eq. (61) can be rewritten as

$$H_0 = -\sum_{j=1}^{N/2}\sum_{m=1}^{2}\sum_{s=\pm 1} t_0\left[{}_0|m, j+1\rangle_{ss}\langle j, m|_0 + {}_0|m, j\rangle_{ss}\langle j+1, m|_0\right]$$

$$\tag{62}$$

Changing to the representation in K space, the eigenstates of the Hamiltonian [Eq. (62)] are given by

$$_0|K_e\rangle_s = \sum_{j,m} a_e(K;j)_0|m,j\rangle_s \qquad a_e(K;j) = \frac{1}{\sqrt{N}} e^{iK_i a} \qquad (63)$$

$$_0|K_a\rangle_s = \sum_{j,m} a_a(K,j)_0|m,j\rangle_s \qquad a_a(K;j) = \frac{(-1)^j}{\sqrt{N}} e^{jK_j a} \qquad (64)$$

where $_0|K_e\rangle_s$ and $_0|K_a\rangle_s$ represent, respectively, the eigenfunctions of the bonding and antibonding π-electrons (having spin s and wavevector K).

Inverting the transformations [Eqs. (63) and (64)] gives the unperturbed Hamiltonian of the π-electrons in the K representation:

$$H_0^K = \sum_{K,s} 2t_0 \cos Ka \left[_0|K_a\rangle_{ss}\langle K_a|_0 - _0|K_e\rangle_{ss}\langle K_e|_0 \right]$$

$$= \sum_{K,s} 2t_0 \cos Ka \left[C_{K,s}^{a\dagger} C_{K,s}^a - C_{K,s}^e C_{K,s}^e \right] \qquad (65)$$

where $C_{K,s}^{e\dagger}, C_{K,s}^{a\dagger} (C_{K,s}^e, C_{K,s}^a)$ represent the creation (annihilation) operators of a bonding and an antibonding electron, respectively.

By defining the two-component state vector

$$\underline{_0|K\rangle_s} = \begin{pmatrix} _0|K_a\rangle_s \\ _0|K_e\rangle_s \end{pmatrix} \qquad (66)$$

we can rewrite Eq. (65) in the spinorial representation

$$H_0^K = \sum_{K,s} {}_s\langle K|_0 2t_0 \cos Ka \sigma_3 {}_0|K\rangle_s$$

$$= \sum_{K,s} {}_s\langle K|_0 \varepsilon(K) {}_0|K\rangle_s \qquad (67)$$

where $\sigma_3 = \begin{pmatrix} 1 & 0 \\ 0 & -1 \end{pmatrix}$ is the Pauli[‡] matrix and $\varepsilon(K) = 2t \cos Ka$ is the spectrum of the eigenvalues for the energy of the unperturbed state (nondimerized chain).

[‡]It is worth stressing that such a spinorial representation does not refer to the two components of the state vector that represent the two spin orientations, but to the bonding and antibonding electrons with the same wavevector K.

APPENDIX B.   PERTURBATIVE EXPANSION OF THE SSH HAMILTONIAN AND ITS CONTINUOUS LIMIT, THE TLM HAMILTONIAN

Let us now consider the case in which a perturbative term is added to the unperturbed SSH Hamiltonian of the π-electrons (see Appendix A) due to the "dimerization" of the lattice.[114]

If we again ignore the kinetic term, the Hamiltonian of our system becomes [see Eq. (59)]

$$H = -\sum_{j,s} t_{j+1,j}\left[C_{j,s}^{\dagger}C_{j+1,s} + C_{j+1,s}^{\dagger}C_{j,s}\right] + \frac{1}{2}\sum_{j}k(u_{j+1}-u_{j})^{2} \quad (68)$$

Since the variations in bond lengths u_j (~ 0.08 Å) are much smaller than the lattice constant a ($\simeq 1.2$ Å), we can expand the hopping integral $t_{j+1,j}$ about the nondimerized state, introducing the electron–phonon coupling constant α:

$$t_{j+1,j} = t_0 - \alpha(u_{j+1} - u_j) \quad (68a)$$

Equation (68) then becomes

$$H = -\sum_{j,s}\left[t_0 + \alpha(u_j - u_{j+1})\right]\left[C_{j,s}^{\dagger}C_{j+1,s} + C_{j+1,s}^{\dagger}C_{j,s}\right] + \tfrac{1}{2}K\sum_{j}(u_{j+1}-u_{j})^{2} \quad (69)$$

which can be rewritten as

$$H = H_0 + V$$

with

$$V = -\sum_{j,s}\alpha(u_j - u_{j+1})\left[C_{j,s}^{\dagger}C_{j+1,s} + C_{j+1,s}^{\dagger}C_{j,s}\right] + \frac{1}{2}\sum_{j}K(u_{j+1}-u_{j})^{2} \quad (70)$$

and H_0 as given by Eq. (60).

It can be demonstrated that the intensity of the perturbation on the spectrum of the eigenvalues of the unperturbed Hamiltonian H_0, due to the dimerization of the lattice increases when the modulus of the wavevector K increases.[11] Consequently, we will consider here only the effects of such a perturbation on the electrons with $|K| \cong K_F$, where K_F is the Fermi wavevector, defined as

$$K_F = \frac{\pi}{2a}$$

The perturbed eigenstates $|K\rangle_s$ can thus be expanded on the basis of the unperturbed Bloch functions with $K = \pm K_F$:[115]

$$|K_a\rangle_s = \sum_{j=1}^{N/2} \psi_{1Ks}(j) \left[\frac{1}{\sqrt{2}} \left(e^{i\pi j}{}_0 |2j\rangle_s + e^{i\pi(2j+1)/2}{}_0 |2j+1\rangle_s \right) \right]$$

$$|K_e\rangle_s = \sum_{j=1}^{N/2} \psi_{2Ks}(j) \left[\frac{i}{\sqrt{2}} \left(e^{-i\pi j}{}_0 |2j\rangle_s + e^{-i\pi(2j+1)/2}{}_0 |2j+1\rangle_s \right) \right] \quad (71)$$

We then define the vectors

$$\underline{|K\rangle}_s \equiv \begin{pmatrix} |K_1\rangle_s \\ |K_2\rangle_s \end{pmatrix} = \begin{pmatrix} |K_a\rangle_s \\ |K_e\rangle_s \end{pmatrix}; \qquad \psi_{K,s}(j) \equiv \begin{pmatrix} \psi_{1Ks}(j) \\ \psi_{2Ks}(j) \end{pmatrix}$$

so that Eq. (71) can be rewritten in the more compact form

$$|K_L\rangle_s = \sum_{j=1}^{N/2} \psi_{LKs}(j) \sum_{m:1}^{2} a(K_L; m_j)_0 |m; j\rangle_s, \qquad L = 1, 2 \quad (72)$$

Since $|K\rangle_s$ is an eigenstate of the perturbed Hamiltonian, this gives us[‡]

$$\underline{}_s\langle K| \left\{ \tfrac{1}{2} \left[(H_0 + V) + (H_0 + V)^\dagger \right] \right\} \underline{|K\rangle}_s = E(K) {}_s\langle K|K\rangle_s \quad (73)$$

Substituting in Eq. (73) the explicit form of $\underline{|K\rangle}_s$ given by Eq. (72), we obtain the following equation, which must be satisfied by the envelope functions $\psi_{LKs}(j)$:

$$\sum_{m,m'=1}^{2} \sum_{j,j'=1}^{N/2} \psi^\dagger_{iKs}(j) a^\dagger(K_i, m_j)$$

$$\cdot \left\{ \tfrac{1}{2} \left[(H_0 + V) + (H_0 + V)^\dagger \right] \right\} \cdot a(K_i, m_{j'}) \psi_{i'Ks}$$

$$= E(K) \delta_{i,i'} \sum_j |\psi_{iKs}(j)|^2 \quad (74)$$

[‡] In the following formulae the operator $H_0 + V$ is Hermitian.

We now make the further hypothesis that the perturbative potential V [see Eq. (70)] varies only slightly over the range of lengths in which the single Wannier functions $_0|m_j\rangle_s$ are significantly different from zero. Mathematically, this assumption can be expressed by putting

$$_s\langle m_j|_0V_0|m'_{j'}\rangle_s = \delta_{m,m'}\delta_{j,j'}V(m_j) \tag{75}$$

Also, using this hypothesis, the envelope functions $\psi_{mKs}(j)$ slowly vary with j; by introducing the continuous approximation in the spatial variable $X_j = ja$, we can therefore expand the term on the left-hand side of Eq. (74) in a Taylor series, to the first order, obtaining[59]

$$\sum_{j,s}\psi^\dagger_{K,s}\mathbf{h}\psi_{K,s}$$

$$\equiv \sum_{j,s}\psi^\dagger_{K,s}(j)\left[-i\frac{\partial[\varepsilon(K)]}{\partial K}\frac{\partial(\sigma_3)}{\partial X} + \tfrac{1}{2}(\Delta_{2j}+\Delta_{2j+1})(\sigma_1)\right]\psi_{Ks}(j)$$

$$+\frac{1}{2}\left(\frac{K}{4\alpha^2}\right)\sum_j(\Delta_j)^2 = E(K) \tag{76}$$

where $\sigma_1 = \begin{pmatrix} 0 & 1 \\ 1 & 0 \end{pmatrix}$ is the Pauli matrix.

The previous equation includes the order parameter Δ_j, defined as[‡]

$$\Delta_j = (-1)^j 2\alpha(u_j - u_{j+1}) \tag{77}$$

Using the linear approximation of the electronic dispersion at the extremities of the valence and conduction bands (Luttinger approximation)

$$\varepsilon(K) = 2t_0\cos\left[(K\pm K_F)a\right] = \pm 2t_0\sin Ka \simeq \pm 2t_0aK \equiv v_FK \tag{78}$$

where v_F is the Fermi velocity, Eq. (76) then becomes

$$H = \sum_{j,s}\psi^\dagger_{Ks}(j)\left\{-2t_0ai\frac{\partial}{\partial X}(\sigma_3) + \tfrac{1}{2}(\Delta_{2j}+\Delta_{2j+1})(\sigma_1)\right\}\psi_{Ks}(j)$$

$$+\frac{1}{2}\left(\frac{K}{4\alpha^2}\right)\sum_j(\Delta_j)^2 \tag{79}$$

[‡] This order parameter takes one of the two nonzero values $\Delta_j = \pm\Delta_0 = \pm 2\alpha u_0$ in the ground state in which the polymeric chain is perfectly dimerized.[11,59]

The continuous TLM Hamiltonian is obtained by substituting $\int_0^L dx/a$ for the summation over $j(\Sigma_j)$ in Eq. (79):

$$H(x) = \sum_s \int_0^L \frac{dx}{a} \psi_{Ks}^\dagger \left\{ -i\nu_F \frac{\partial}{\partial x}(\sigma_3) + \Delta(x)(\sigma_1) \right\} \psi_{Ks}(x)$$

$$+ \frac{1}{2}\left(\frac{K}{4\alpha^2}\right) \int_0^L \frac{dx}{a} \Delta^2(x) \equiv H_{MF} + H_r \qquad (80)$$

From Eqs. (76) and (80) the following Bogoliubov–de Gennes equations for $\psi_{1Ks}(x)$ and $\psi_{2Ks}(x)$ are obtained:[19,116]

$$E(K)\psi_{1Ks}(x) = -i\nu_F \frac{\partial}{\partial x}(\psi_{1Ks}(x)) + \Delta(x)\psi_{2Ks}(x)$$
$$E(K)\psi_{2Ks}(x) = i\nu_F \frac{\partial}{\partial x}(\psi_{2Ks}(x)) + \Delta(x)\psi_{1Ks}(x) \qquad (81)$$

To define completely and correctly the problem of the lattice structure of PAC, however, a further equation that allows the determination of $\Delta(x)$ is required. Toward this end we impose the further condition that $\Delta(x)$ takes the form that minimizes the total energy of the chain.

Thus, on carrying out the functional differentiation of the Hamiltonian [Eq. (80)] with respect to $\delta\Delta(x)$, we obtain, for $T = 0$ K, the following equation, which is self-consistent with Eqs. (81):

$$\Delta(x) = -2\left(\frac{4\alpha^2 a}{K}\right) \sum_{j,s} \psi_{2Ks}^*(x)\psi_{1Ks}(x) \qquad (82)$$

Acknowledgments

Thanks are due to TEMA S.p.A. for permission to publish this paper. Eng. P. Verrecchia and Dr. S. Serbassi strongly encouraged and supported our work. We gratefully acknowledge many discussions with Prof. L. D'Ilario, Professor B. H. Lavenda, Dr. G. Lugli, and Dr. G. Perego. Professor E. Tosatti read the manuscript carefully and suggested several improvements. We also acknowledge interesting discussions with Dr. S. Kivelson and Dr. M W. Evans.

We thank Dr. D. Jones for his careful English translation.

This work was partly supported by ENI Grant no. 3235/24/2/82.

References

1. K. J. Wynne and G. B. Street, *Ind. Eng. Chem. Prod. Res. Dev.*, **21**, 23 (1982).
2. G. Natta, G. Mazzanti and P. Corradini, *Lincei-Rend. Sc. Fis. Mat.*, **25**, 3 (1958).
3. T. Ito, H. Shirakawa and S. Ikeda, *J. Polym. Sci., Chem. Ed.*, **12**, 11 (1974).
4. C. Chiang et al., *Phys. Rev. Lett.*, **39**, 1098 (1977).
5. C. Chiang et al., *J. Am. Chem. Soc.*, **100**, 1013 (1978).

6. W. Shockley, *IEEE Trans. Electron Devices*, **ED-23**, 7, 597 (1976).

7. H. Shirakawa, T. Ito and S. Ikeda, *Makromol. Chem.*, **179**, 1565 (1978).

8. K. Shimamura, *Makromol. Chem. Rapid Commun.*, **3**, 269 (1983).

9. M. Aldissi, F. Schue, and M. Rolland, *Phys. Status Solidi A*, **69**, 733 (1982).

10. M. Andretta, *Prime Risultanze di Uno Studio Sulle Proprieta' Elettriche del Poliacetilene*, TEMA internal report, TEMA S.p.A., Bologna, Italy, 1982. See also references therein.

11. W. P. Su, J. R. Schrieffer, and A. J. Heeger, *Phys. Rev. B*, **22**, 2099 (1980); *Phys. Rev. Lett.*, **42**, 1698 (1979).

12. R. E. Peierls, *Quantum Theory of Solids*, Clarendon, London, 1955, p. 108.

13. M. J. Rice, in *Solitons in Condensed Matter Physics*, A. R. Bishop and T. Schneider, eds., Springer-Verlag, New York, 1978, p. 246.

14. A. J. Heeger and A. G. McDiarmid, in *Physics in One Dimension*, J. Bernasconi and T. Schneider, eds., Springer-Verlag, New York, 1981, p. 179.

15. M. Nechtschein et al., *Phys. Rev. Lett.*, **44**, 356 (1980).

16. M. Mehring et al., *Proceedings of the International Conference on Synthetic Low Dimensional Conductors and Superconductors*, suppl. to *J. Physique*, **Fasc. 6, Colloque n. 3** (*Conf. PCPC*), 217 (1983).

17. I. B. Goldberg et al., *J. Chem. Phys.*, **70**, 1132 (1979).

18. M. J. Rice, *Phys. Lett. A*, **71**, 152 (1979),

19. M. Takayama, Y. R. Lin-Liu, and K. Maki, *Phys. Rev. B*, **21**, 2388 (1980).

20. S. A. Brazovskii, *JETP Lett.*, **28**, 656 (1978); *JETP*, **51**, 342 (1980).

21. H. Fröhlich, *Proc. R. Soc. London A*, **223**, 296 (1954).

22. J. Bardeen, L. N. Cooper, and J. R. Schrieffer, *Phys. Rev.*, **106**, 162 (1957); *Phys. Rev.*, **108**, 1175 (1957).

23. P. A. Lee, T. M. Rice, and P. W. Anderson, *Solid State Commun.*, **14**, 703 (1974).

24. M. J. Rice et al., *Phys. Rev. Lett.*, **36**, 432 (1976).

25. G. A. Toombs, *Phys. Rep.*, **40**, 181 (1978).

26. A. R. Bishop, J. A. Krumhansl, and S. E. Trullinger, *Physica*, **1D**, 1 (1980).

27. H. Haken, *Synergetics*, 2nd enlarged ed., Springer-Verlag, New York, 1978, Chapter 6, p. 179.

28. L. D. Landau and E. M. Lifshitz, *Statistical Physics*, 3rd ed., Pergamon, Elmsford, New York, 1980, Chapter 14.

29. J. A. Krumhansl and J. R. Schrieffer, *Phys. Rev. B*, **11**, 3535 (1975).

30. C. M. Varma, *Phys. Rev. B*, **14**, 244 (1976).

31. D. Vanderbilt and E. J. Mele, *Phys. Rev. B*, **22**, 3939 (1980).

32. W. P. Su, *Solid State Commun.*, **35**, 899 (1980).

33. C. T. White, P. Brant, and M. L. Elert, *Conf. PCPC*, 443 (1983).

34. M. J. Rice and E. J. Mele, *Solid State Commun.*, **35**, 487 (1980).

35. A. R. Bishop, D. K. Campbell, and K. Fesser, *Mol. Cryst. Liq. Cryst.*, **77**, 253 (1981).

36. K. Gottfried, *Quantum Mechanics*, Vol. 1, W. A. Benjamin, 1966, Chapter 6, p. 275.

37. J. L. Bredas et al., *Conf. PCPC*, 373 (1983).

38. G. Zerbi and G. Zannoni, *Conf. PCPC*, 273 (1983).

39. K. Fesser, A. R. Bishop, and D. K. Campbell, *Phys. Rev. B*, **27** (8), 4804 (1983).

40. G. W. Brant and A. J. Glick, *Phys. Rev. B*, **26**, 5855 (1982); *J. Phys.*, **C15**, L391 (1982).

41. J. P. Albert and C. Jouanin, *Conf. PCPC*, 387 (1983); *Physica B + C*, **117/118**, no. 2, 620 (1983); *Solid State Commun.*, **47** (10), 825 (1983).

42. J. P. Albert, C. Jouanin, and P. Bernier, *Polymer Preprints*, **23**, 84 (1982).

43. J. L. Bredas, R. R. Chance, and R. Silbey, *Polymer Preprints*, **23**, 82 (1982).

44. J. Kanicki, E. Vander Donckt, and S. Boué, *J. Chem. Soc.*, *Faraday Trans.*, **2**, 77, 2157 (1981).

45. F. Devreux et al., in *Physics in One Dimension*, J. Bernasconi and T. Schneider, eds., Springer-Verlag, New York, 1981, p. 194.

46. M. Nechtschein et al., *Conf. PCPC*, 209 (1983).

47. K. Maki, *Phys. Rev. B*, **24**, 2181 (1982); *Phys. Rev. B*, **24**, 2187 (1982).

48. S. Ikehata et al., *Phys. Rev. Lett.*, **45**, 1123 (1980).

49. Y. Tomkiewicz et al., *Phys. Rev. Lett.*, **43**, 1532 (1979).

50. Y. Tomkiewicz et al., in *Physics in One Dimension*, J. Bernasconi and T. Schneider, eds., Springer-Verlag, New York, 1981, p. 214.

51. Y. Tomkiewicz et al., *Mol. Cryst. Liq. Cryst.*, **83**, 1049 (1982).

52. A. J. Heeger and A. G. McDiarmid, *Mol. Cryst. Liq. Cryst.*, **77**, 1 (1981).

53. N. Suzuki et al., *Phys. Rev. Lett.*, **45**, 1209 (1980).

54. J. Kanicki, S. Boué, and E. Vander Donckt, *Thin Solid Films*, **92**, 243 (1982).

55. S. Etemad et al., *Conf. PCPC*, 413 (1983).

56. D. Baeriswyl, in *Proceedings of the 3rd General Conference on Condensed Matter, Division EPS* (*Lausanne* 1983), to appear in *Helv. Phys. Acta*.

57. D. Baeriswyl et al., in *Electronic Properties of Polymers*, J. Mort and G. Pfister, eds., Wiley, New York, 1982.

58. A. Glick, *Phys. Rev. Lett.*, **49**, 804 (1982).

59. S. Kivelson et al., *Phys. Rev. B*, **25**, 4173 (1982).

60. A. Feldblum et al., *Phys. Rev. B*, **26**, 815 (1982).

61. C. R. Fincher et al., *Phys. Rev. B*, **19**, 4140 (1979).

62. E. J. Mele and M. J. Rice, *Phys. Rev. Lett.*, **45**, 926 (1980).

63. B. Horovitz, *Solid State Commun.*, **41**, 729 (1982).

64. J. F. Rabolt, T. C. Clarke, and G. B. Street, *J. Chem. Phys.*, **76**, 5781 (1982).

65. M. J. Rice, *Phys. Rev. B*, **24**, 3638 (1981).

66. S. Kivelson, personal communication.

67. A. J. Epstein et al., *Conf. PCPC*, 61 (1983); *Phys. Rev. Lett.*, **50**, 1866 (1983).

68. K. Menke et al., *Polymer Preprints*, **83**, 79 (1982).

69. S. Etemad et al., *Solid State Commun.*, **40**, 75 (1981).

70. L. Lauchlan et al., *Phys. Rev. B*, **24**, 3701 (1981).

71. Z. Vardeny et al., *Conf. PCPC*, 325 (1983).

72. J. Orenstein and G. L. Baker, *Phys. Rev. Lett.*, **49**, 1043 (1980).

73. C. V. Shank et al., *Phys. Rev. Lett.*, **49**, 1660 (1982).

74. J. D. Flood and A. J. Heeger, *Conf. PCPC*, 397 (1983) *Phys. Rev. B*, **28**, 2356 (1983).

75. G. B. Blanchet, C. R. Fincher, and A. J. Heeger, *Phys. Rev. Lett.*, in press.

76. W. P. Su and J. R. Schriefer, *Proc. Natl. Acad. Sci. USA*, **77**, 5626 (1980).

77. W. P. Su, *Mol. Cryst. Liq. Cryst.*, **77**, 265 (1981).

78. W. P. Su, S. Kivelson, and J. R. Schrieffer, in *Physics in One Dimension*, J. Bernasconi

and T. Schneider, eds., Springer-Verlag, New York, 1981, p. 201.

79. J. R. Schrieffer, R. Ball, and W. P. Su, *Conf. PCPC*, 429 (1983).
80. J. P. Setna and S. Kivelson, *Phys. Rev. B*, **26**, 3513 (1982).
81. Z.-B. Su and L. Yu, *Phys. Rev. B*, **27**, 5199 (1983).
82. A. R. Bishop et al., *Phys. Rev. Lett.*, **52**, 671 (1984).
83. C. R. Fincher et al., *Phys. Rev. B*, **20**, 1589 (1979).
84. Y. W. Park et al., *Solid State Commun.*, **29**, 747 (1979).
85. D. Moses et al., *Solid State Commun.*, **40**, 1007 (1981).
86. P. Sheng, *Phys. Rev. B*, **21**, 2180 (1980).
87. E. J. Mele and M. J. Rice, *Phys. Rev. B*, **23**, 5397 (1981).
88. J. J. André et al., *Conf. PCPC*, 199 (1983).
89. S. Kivelson, *Phys. Rev. Lett.*, **46**, 1344 (1981).
90. S. Kivelson, *Mol. Cryst. Liq. Cryst.*, **77**, 65 (1981).
91. S. Kivelson, *Phys. Rev. B*, **25**, 3798 (1982).
92. A. J. Epstein et al., *Polymer Preprints*, **83**, 88 (1982).
93. L. R. Friedman and D. P. Tunstall, The Metal–Non Metal Transition in Disordered Systems, in *Proceedings of the 19th Scottish Universities Summer School in Physics, St. Andrews*, 1978, Scottish Universities Summer School in Physics, 1978.
94. V. Ambegaokar et al., *Phys. Rev. B*, **4**, 2612 (1971)
95. O. Madelung, *Introduction to Solid State Theory*, Springer-Verlag, New York, 1978.
96. P. W. Butcher et al., *Philos. Mag.*, **36**, 19 (1977).
97. P. W. Butcher et al., in *Amorphous and Liquid Semiconductors*, W. Paul and W. Kastner, eds., North Holland, Amsterdam 1980.
98. M. Pollack, *J. Non-Cryst. Solids*, **11**, 1 (1972).
99. J. A. Croboczek and S. Summerfield, *Conf. PCPC*, 517 (1983).
100. S. Summerfield and P. N. Butcher, *J. Phys. C*, **15**, 7003 (1982).
101. S. A. Brazovskii and N. N. Kirova, *JETP Lett.*, **33**, 6 (1981).
102. D. K. Campbell and A. R. Bishop, *Phys. Rev. B*, **24**, 4859 (1981).
103. J. L. Bredas et al., *Mol. Cryst. Liq. Cryst.*, **77**, 319 (1981).
104. R. H. Baughman and G. Moss, *J. Chem. Phys.*, **77**, 6321 (1982).
105. D. Baeriswyl and K. Maki, *Phys. Rev. B*, **28**, 2068 (1983).
106. J. R. Schrieffer, to be published.
107. M. Kabbaj et al., *Conf. PCPC*, 463 (1983).
108. A. J. Epstein et al., *Polymer*, **23**, 1211 (1982).
109. M. Audenaert, G. Gusman, and R. Deltour, *Phys. Rev. B*, **24**, 7380 (1981).
110. M. Nakahara and K. Maki, *Phys. Rev. B*, **25**, 7789 (1982).
111. W. P. Su, *Solid State Commun.*, **42**, 497 (1982).
112. J. E. Hirsch and E. Fradkin, *Phys. Rev. Lett.*, **50**, 402 (1982).
113. M. A. Collins, "Solitons in Chemical Physics," in Advances in Chemical Physics, Vol. LIII, I. Prigogine and Stuart A. Rice, eds., Interscience, New York, 1983, pp. 225–339.
114. M. Andretta, R. Serra, and G. Zanarini *Modelli di Polimeri Elettricamente Attivi*, CLUEB, Bologna, 1983, p. 34.
115. S. Kivelson and C. D. Gellatt, Jr., *Phys. Rev. B*, **19**, 5160 (1979).
116. M. Cryot, *Rep. Prog. Phys.*, **36**, 103 (1973).

DEVELOPMENT AND APPLICATION
OF THE THEORY OF BROWNIAN MOTION

WILLIAM COFFEY

Department of Microelectronics and Electrical Engineering,
Trinity College, Dublin 2, Ireland[‡]
and
Division of Mechanics, Department of Physics,
The University of Athens, Panepistimiopolis,
Athens 621, Greece

CONTENTS

[‡] Permanent address.

I. INTRODUCTORY CONCEPTS

A. The Early Development of Statistical Mechanics

The development of statistical mechanics and the introduction of random processes into physics began in the 19th century, when physicists were attempting to show that heat in a medium is due to the random motion of the constituent molecules.

Maxwell, in 1859, considered gases as if they were made up of small rigid spheres distributed randomly but with uniform average density in a vessel.[1] In his model the molecules are supposed to have random velocities and to collide in a perfectly random fashion with each other and with the walls of the vessel. The process is also supposed to have lasted a long time, so that equilibrium conditions will have been attained. The position of a molecule is represented by Cartesian coordinates x, y, z and its velocity by coordinates u, v, w, so that

$$\frac{dx}{dt} = u$$

$$\frac{dy}{dt} = v$$

$$\frac{dz}{dt} = w$$

where t is time. Maxwell asked what the steady-state probability $f(u, v, w)$ $du\, dv\, dw$ was that the velocity components lie in small ranges between u and $u + du$, v and $v + dv$, and w and $w + dw$. His original argument, although not now regarded as completely satisfactory, is of interest both for its simplicity and for its historical importance.

B.   Maxwell's Formula for the Distribution of Velocities in a Gas

The derivation we give here is essentially that of Maxwell, with a few slight changes in nomenclature.

Let N be the whole number of particles. Let u, v, w be the components of the velocity of each particle in three rectangular directions, and let the number of particles for which u lies between u and $u + du$ be $Nf(u)\, du$, where $f(u)$ is a function of u to be determined.

The number of particles for which v lies between v and $v + dv$ will be $Nf(v)\, dv$, and the number for which w lies between w and $w + dw$ will be $Nf(w)\, dw$, where f always stands for the same function.

Now the existence of the velocity u does not in any way affect that of the velocities v or w, since these are all at right angles to each other and independent, so that the number of particles whose velocity lies between u and $u + du$, and also between v and $v + dv$ and also between w and $w + dw$ is

$$Nf(u)f(v)f(w)\, du\, dv\, dw$$

If we suppose the N particles to start from the origin at the same instant, then this will be the number in the element of volume $du\, dv\, dw$ after unit of time, and the number referred to unit of volume will be

$$Nf(u)f(v)f(w)$$

But the directions of the coordinates are perfectly arbitrary, and therefore this number must depend on the distance from the origin alone, that is

$$f(u)f(v)f(w) = \phi(u^2 + v^2 + w^2)$$

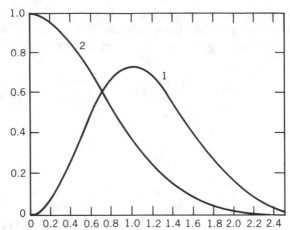

Figure 1.   Maxwellian distribution of velocities. Curve 1 is the grouping of the magnitudes of the velocities independently of their directions in space i.e., $2v^2e^{-v^2/2}$. Curve 2 is the magnitude of a single component i.e. e^{-v^2}.

Solving this functional equation, we find

$$f(u) = Ce^{Au^2}, \qquad \phi(u^2 + v^2 + w^2) = C^3 e^{A(u^2 + v^2 + w^2)}$$

This proof, though attractive because of its simplicity, is deemed unsatisfactory because it assumes the three velocity components to be independent. The distribution (Fig. 1) may, however, be justified from rigorous considerations.[2] The constant C^3 is a normalizing factor chosen so as to make the total probability unity:

$$C^3 = \left[\int_{-\infty}^{\infty} \int_{-\infty}^{\infty} \int_{-\infty}^{\infty} e^{A(u^2 + v^2 + w^2)} \, du \, dv \, dw \right]^{-1}$$

The constant A is $-m/2kT$ where m is the mass of a gas molecule. T is the absolute temperature, $k = R/N = 1.38 \times 10^{-23}$ J K^{-1}, R is the gas constant (8.314 J K mol^{-1}), N is Avogadro's number (6.023×10^{23} mol^{-1})

C.   Boltzmann's Generalization of Maxwell's Results

Boltzmann, in several papers beginning in 1868,[1] generalized Maxwell's results by supposing that the molecules are also subjected to a conservative field of force $V(x, y, z)$, so that the total energy of a molecule is*

$$E(x, y, z, u, v, w) = V(x, y, z) + \tfrac{1}{2}m(u^2 + v^2 + w^2)$$

*$V(x, y, z)$ is the potential energy corresponding to this force.

He then found that (C' is a constant)

$$f(x, y, z, u, v, w) = C' \exp\left[\frac{2A}{m} E(u, v, w, x, y, z)\right] \qquad (1.1)$$

Such a gas is said to have the Maxwell–Boltzmann probability distribution.

D. Boltzmann's H Theorem

The H theorem originally arose as part of Boltzmann's argument that a gas with an arbitrary initial probability distribution tends to approach the probability distribution given by Eq. (1.1) as time goes on. The theorem is intimately connected with the *Boltzmann equation* for the change with time of the probability distribution in a gas, which assumes that only collisions between two molecules are of any importance. The Boltzmann equation represents the first attempt to treat a gas not in a state of thermodynamic equilibrium.

By constructing the function

$$H = \int \int f \ln f \, d\mathbf{r} \, d\mathbf{v}$$

(the negative of the entropy), Boltzmann was able to show that

$$\frac{dH}{dt} \leq 0 \qquad (1.2)$$

in agreement with the conclusion from macroscopic thermodynamics that the entropy always appears to increase. In Eq. (1.2) in Boltzmann's original proof the time evolution of f was governed by the Boltzmann equation, while $\mathbf{r}$ and $\mathbf{v}$ denote the position and velocity of a molecule. The theorem is perfectly general, however, and does not depend *per se* on the Boltzmann equation.

E. Objections to the H Theorem

Loschmidt[3] pointed out in 1876 that the H theorem was incompatible with classical mechanics, since if a system governed by Newton's laws has at any instant the signs of all its velocities reversed, then that system should retrace its previous motion. This was apparently incompatible with the monotonic decrease of H predicted by Boltzmann.

Zermelo[3] argued in 1896 that the H theorem was incompatible with the Poincaré recurrence theorem, which requires that a conservative dynamical system return arbitrarily closely to any given initial state (for a gas, this means a reversion to any arbitrary initial distribution of temperature, however irregular). Boltzmann and later investigators successfully countered these arguments by recognizing the *probabilistic* nature of *H*. The very definition of *H* requires the introduction of a *mechanism* for the change of *H* with time.

F. The Equipartition Theorem

It follows from the Maxwell–Boltzmann distribution that the mean kinetic energy for each degree of freedom of the gas molecules is the same, that is

$$\tfrac{1}{2}m\langle u^2 \rangle = \tfrac{1}{2}m\langle v^2 \rangle = \tfrac{1}{2}m\langle w^2 \rangle = \tfrac{1}{2}kT$$

This property, which applies also when there are several different kinds of molecule in the gas, is known as the *equipartition of energy*. If we interpret the mean kinetic energy as temperature, the equipartition theorem implies that gases in contact reach a common temperature, in agreement with experiment.

G. Calculation of Averages in Classical Statistical Mechanics

Now that we have concluded our brief historical outline of the growth of classical statistical mechanics, it is convenient to summarize how averages of dynamical quantities may be calculated. Following Gibbs,[4] we will describe the state* of a given system in terms of coordinates $q_1, q_2, \ldots$ and conjugate momenta $p_1, p_2, \ldots$ (rather than velocities), the number of each being equal to the number of degrees of freedom of the system. Then, in the notation of Gibbs,[4] the Maxwell–Boltzmann law (i.e., the probability of finding the state of the system in the range $dq_1 \ldots dq_N, dp_1 \ldots dp_N$) is

$$f\,d\tau = Ce^{-\beta E(p_1 \cdots p_N, q_1 \cdots q_N)}\,dq_1\,dp_1 \ldots dq_N\,dp_N$$

E is the Hamiltonian or total energy of the system. As before, the coefficient *C* is chosen so that

$$\int f\,d\tau = 1$$

where the integration extends over all possible values of the variables. We suppose throughout that the system is in thermodynamic equilibrium at the

---

*See also section XV.

temperature T and $\beta = (kT)^{-1}$. Since there is a continual interchange of energy between the system and its surroundings, the total energy content E at a given temperature does not have a definite value. In calculating resulting averages for a system, it is convenient to introduce the partition function Z, defined by

$$Z = \int e^{-\beta E}\, dq_1\, dp_1 \ldots dq_N\, dp_N$$

The average value of any function of the p's and q's, $A(p_i, q_i)$, is then given by

$$\overline{A(p_i, q_i)} = \langle A(p_i, q_i) \rangle = \frac{1}{Z} \int A e^{-\beta E}\, d\tau$$

(We shall use overbars and angle brackets to denote averages; overbars will denote time averages, and angle brackets, ensemble averages.)

It is a fundamental tenet of statistical mechanics[1] that the time-average behavior of a system is equal to the ensemble-average behavior. In particular, the average total energy is given by

$$\langle E \rangle = \frac{1}{Z} \int E e^{-\beta E}\, d\tau = \frac{\partial}{\partial \beta} \ln Z$$

The average value of any coordinate is

$$\langle q_i \rangle = \frac{1}{Z} \int q_i e^{-\beta E}\, d\tau$$

Let us suppose that the coordinates are so chosen that the kinetic energy is expressed as a sum of squares of the momenta with constant coefficients, that is,

$$T(p_i) = \sum_i \frac{p_i^2}{2m_i}$$

and the Hamiltonian is thus

$$E = \sum_i \left(\frac{p_i^2}{2m_i} + V(q_i) \right)$$

Then in calculating Z, or the average of any function of the coordinates only, the integrals that occur are the *products* of integrals over the position and momentum coordinates, respectively. For example, in evaluating $\langle E \rangle$

$$\left\langle \frac{p_i^2}{2m_i} \right\rangle = \frac{\dfrac{1}{2m_i} \int p_i^2 e^{-\beta p_i^2/2m}\, dp_i \int \text{(over other coordinates)}}{\int e^{-\beta p_i^2/2m}\, dp_i \int \text{(over other coordinates)}}$$

Now

$$\int_{-\infty}^{\infty} e^{-ap^2}\, dp = \left(\frac{\pi}{a} \right)^{1/2},$$

$$\frac{\partial}{\partial a} \int_{-\infty}^{\infty} e^{-ap^2}\, dp = \int_{-\infty}^{\infty} p^2 e^{-ap^2}\, dp = \frac{1}{2a}\left(\frac{\pi}{a} \right)^{1/2}$$

The ratio is $1/2a$; hence the mean value of $p_i^2/2m$ is given by

$$\left\langle \frac{p_i^2}{2m} \right\rangle = \frac{1}{2m_i}\left(\frac{2m_i kT}{2} \right) = \tfrac{1}{2}kT$$

Thus we have the equipartition theorem described above: The mean value of the *kinetic energy* in any coordinates is $\tfrac{1}{2}kT$. The same is true of potential-energy terms, which are simply of the form $\tfrac{1}{2}K_i q_i^2$, assuming that q_i does not enter the Hamiltonian in any other form. If the coordinate q_i enters the Hamiltonian simply as (a harmonic oscillator)

$$\tfrac{1}{2}m_i p_i^2 + \tfrac{1}{2}K_i q_i^2$$

however, the particle cannot come into thermal equilibrium with the rest of the system; there must be some kind of interaction between it and the rest of the system. This is generally provided by some mechanism such as the collision of gas molecules with the particle which constitutes the oscillator. In general, if the *interaction energy depends solely on the coordinates*, not on the momenta, then the equipartition theorem holds.[5]

According to the theorem, associated with each degree of freedom is a mean kinetic energy $\tfrac{1}{2}kT$. One may then calculate the fluctuations due to this $\tfrac{1}{2}kT$ energy in perfect generality from the laws of statistical mechanics. This is possible because[6] the average energy of these random motions will be exactly the same for all systems at the same temperature (so long as each is in thermodynamic equilibrium with its surroundings), entirely independently of the natures of the systems and the mechanisms that produce them. The energy distribution will be a function of the particular system in question. Barnes and Silverman[6] show how the equipartition theorem may be used to set a natural limit to the ultimate sensitivity of all measuring devices.

H. Glossary of Statistical Terms

Before embarking on our account of the development of the theory of Brownian movement, we must give some elementary statistical definitions. Max Born, in his *Natural Philosophy of Cause and Chance*,[7] paints a delightful picture of the introduction of statistical techniques into physics:

> The new turn in physics was the introduction of atomistics and statistics. To follow up the history of atomistics into the remote past is not in the plan of this lecture. We can take it for granted that since the days of Democritos the hypothesis of matter being composed of ultimate and indivisible particles was familiar to every educated man. It was reviewed when the time was ripe. Lord Kelvin quotes frequently a Father Boscovich as one of the first to use atomistic considerations to solve physical problems; he lived in the eighteenth century, and there may have been others, of whom I know nothing, thinking on the same lines. The first systematic use of atomistics was made in chemistry, where it allowed the reduction of innumerable substances to a relatively small stock of elements. Physics followed considerably later because atomistics as such was of no great use without another fundamental idea, namely that the observable properties of matter are not intrinsic qualities of its smallest parts, but averages over distributions governed by the laws of chance.
>
> The theory of probability itself, which expresses these laws, is much older; it sprang not from the needs of natural science but from gambling and other, more or less disreputable, human activities.

(The early developments are given by Todhunter.[8])

The concept of that theory that we shall need is simply the random variable. This is a quantity that may take any of the values of a specified set with a specified relative frequency, or probability. The random variable could be a vector molecular property such as center-of-mass velocity, angular momentum, or dipole orientation; or a tensor, such as the polarizability. It is regarded as defined not only by a set of permissible values such as an ordinary mathematical variable has, but by an associated probability function expressing *how often* these values appear in the situation under discussion.

STOCHASTIC PROCESS. A stochastic process is a family of random variables $[X(t), t \in T]$, where t is some parameter, generally time, defined in a set T.

CORRELATION. In general, correlation is interdependence between quantitative or qualitative data. In a narrower sense, it is a relationship between measurable random variables.

AUTOCORRELATION FUNCTION. This function is the internal correlation between a number of series of observations (μ) in time or space (i.e., it measures the degree of correlation between the same random variables at different times).

CROSS-CORRELATION FUNCTION. A cross-correlation function measures the degree of correlation between different random variables at different times.

STATIONARY PROCESS OR STATIONARITY. A stationary process is one in which the point at which one starts the exercise of correlating a property X at $t = t_1$ with that same property at $t = t_2$ is of no consequence.

MARKOV PROCESS. A Markov process is one in which only the last state is relevant in determining future behavior.

FLUCTUATION. Fluctuation is the deviation from its average value of a random variable. The concept arises in physics because the statistical conception of matter in bulk implies that spontaneous deviations from equilibrium are possible. There are several different types of problem, some of them concerned with the deviations from the average or fluctuations found by repeated observations, as with Brownian motion of suspended particles.

CONDITIONAL PROBABILITY. This is the probability of an event given the occurrence of another event in the sample.[9]

MASTER EQUATION. Think of a particle moving in three dimensions on a lattice.[10] The probability of finding the particle at point $\mathbf{m}$ at a time t increases due to transitions from all other points $\mathbf{m}'$ to the point under consideration. It decreases due to transitions that cause it to leave $\mathbf{m}$ to go to any other point. Thus we have the equation for the probability $P(\mathbf{m}, t)$ of finding the particle at **m** at t (given that it was at **m′** at some earlier time):[10]

$$\dot{P}(\mathbf{m}, t) = \text{rate in} - \text{rate out}$$

This is the most elementary definition of the master equation.

TRANSITION PROBABILITY.   This is the "mechanism" of the jump process from point **m′** to point **m** in the previous definition. More precisely, it is the probability of a jump from point **m′** to point **m**.

WIENER–KHINCHIN THEOREM.[11,12]   This theorem states that the spectral density is the Fourier cosine transform of the autocorrelation function.

II. BROWNIAN MOTION

Brownian motion baffled investigators and excited enormous interest for many years, because it seemed inexplicable on the grounds of kinetic theory. Later, in the hands of Einstein and Perrin, it was to provide incontrovertible evidence for the existence of atoms and molecules. We begin by describing its discovery.

The first detailed account[13] of Brownian motion was given by the English botanist Robert Brown, who was interested in studying the plant life of the South Seas. Brown was concerned with studies of the fertilization process in plants, more precisely with the transport of pollen into the ovulum of a plant. He examined an aqueous suspension of pollen under a microscope and found that the pollen grains were in "rapid oscillatory motion." He then studied the pollens of several species, with similar results. Initially he thought that the movement was not only "vital" (in the sense of not being due to a physical cause), but peculiar to the male sexual cells of plants. He quickly disembarrassed himself of this explanation on observing that the motion was exhibited by grains of both *organic* and *inorganic* matter in suspension. We describe the evolution of Brown's reasoning in his own words:

> Having found as I believed a peculiar character in the motion of the par-
> ticles of pollen in water it occurred to me to appeal to this peculiarity as a test
> in certain cryptogamous plants, namely mosses and the genus *Equisetum* in
> which the existence of sex organs had not been universally admitted.... But I
> at the same time observed, that in bruising the ovula or seeds of *Equisetum*
> which at first happened accidentally, I so greatly increased the number of
> moving particles, that the source of the added quantity could not be doubted.
> I found also on bruising first the floral leaves of Mosses and then all other parts

of those plants, that I readily obtained similar particles not in equal quantity indeed, but equally in motion. My supposed test of the male organ was therefore necessarily abandoned.

He proceeds:

Reflecting on all the facts with which I had now become acquainted, I was disposed to believe that the minute particles or molecules of apparently uniform size, were in reality the supposed constituent or elementary molecules of organic bodies, first so considered by Buffon and Needham....

A.   Demonstration of the Motion for Inorganic Particles

Brown investigated whether the motion was limited to organic bodies:

A minute portion of silicified wood which exhibited the structure of Coniferae, was bruised and spherical particles or molecules in all respects like those so frequently mentioned were readily obtained from it; in such quantity, however, that the whole substance of the petrifaction seemed to be formed of them. From hence I inferred that these molecules were not limited to organic bodies, nor even to their products.

Later, he writes:

Rocks of all ages, including those in which organic remains have never been found yielded the molecules in abundance. Their existence was ascertained in each of the constituent minerals of granite a fragment of the Sphinx being one of the speciments observed.

B.   Enduring Character of the Motion

Brown described an experiment in which a drop of water of microscopic size, immersed in oil and containing just one particle, unceasingly exhibited the motion. These considerations led him to the theory, which he never stated as a conclusion,[13] that matter is composed of small particles, called by Brown "active molecules," that exhibit a rapid irregular motion having its origin in the particles themselves and not in the surrounding fluid.

III.   SUBSEQUENT 19TH-CENTURY ATTEMPTS TO EXPLAIN BROWNIAN MOVEMENT

Following Brown's work there were many years of speculation[6,15] as to the cause of the phenomenon before Einstein made conclusive mathematical predictions of a diffusive effect arising from the random thermal motions of particles in suspension. Most of the hypotheses advanced in the 19th century could be dismissed by considering Brown's experiment on the microscopic water drop enclosed in oil. According to Nelson,[13] the first investigator to express a notion close to the modern theory of Brownian movement (i.e., that the perpetual motion is caused by bombardment of the Brownian particle by the particles of the surrounding medium) was Chr. Weiner in 1863.

A. Gouy's Investigation

A very detailed experimental investigation was made by Gouy that greatly supported the kinetic-theory explanation. Gouy's conclusions may be summarized by the following seven points:[13]

1. The motion is very irregular, composed of translations and rotations, and the trajectory appears to have no tangent.
2. Two particles appear to move independently, even when they approach one another to within a distance less than their diameter.
3. The smaller the particles, the more active the motion.
4. The composition and density of the particles have no effect.
5. The less viscous the fluid, the more active the motion.
6. The higher the temperature, the more active the motion.
7. The motion never ceases.

Point 1 is of profound interest in view of the later work of N. Wiener, who proved that the sample points of the Brownian-motion trajectory are almost everywhere continuous, but nowhere differentiable. Despite these careful observations in favour of kinetic theory, however, several arguments always seemed to militate against it. We give below two of the most prominent.

B. An Early Attempt to Explain Brownian Motion in Terms of Collisions

Consider the *conservation* of momentum during an atomic collision with a macroscopic Brownian particle of mass M and velocity V. If the surrounding molecules each have mass m and velocity v, the velocity change Δv of the molecule on a *single* impact would be $(m/M)v$. If v is calculated from the kinetic-theory equation

$$\tfrac{1}{2}m\langle v^2 \rangle = \tfrac{3}{2}kT$$

and then the principle of conservation of momentum is applied, Δv for a 10^{-6}-m-diameter particle in water at room temperature (300 K) is $\sim 5 \times 10^{-8}$ m s^{-1}. The observed Brownian movement for this system, however, is *greater* than this by two orders of magnitude. Von Nägeli was aware of this discrepancy; however, he could not explain it in terms of collisions because he assumed that these would produce *zero net effect*. Thus he effectively calculated only the velocity change as a result of a single collision. His error lay in regarding the random collisions 'as occurring in regularly alternating directions that would keep bringing the target molecule back to its starting position. This assumption is invalid, because if N random collisions occur (see the random-walk problem),[11] the displacement (root-mean-square value) will be proportional to $\sqrt{N}$. Now if the time interval between successive observations of the particle is τ, N will be proportional to τ. Thus the root-mean-square value of the displacement is proportional to $\sqrt{\tau}$, and *not zero* as assumed by Von Nägeli.

C. Attempts to Explain Brownian Motion Using the Equipartition Theorem

Many investigators assumed (correctly) that the macroscopic Brownian particle could be treated simply as an enormous "atom" of mass M. This would also allow a test of the kinetic theory, because the law of equipartition of energy implied that the kinetic energy of translation of a Brownian particle and of a molecule should be equal. Thus the speed $\dot{s}$ of a Brownian particle should be given by that for a molecule:

$$\left\langle \frac{M\dot{s}^2}{2} \right\rangle = \tfrac{3}{2}kT \tag{3.1}$$

where $ds/dt = \dot{s}$. For the system described above, $\dot{s} \approx 2 \times 10^{-3}$ m s^{-1}, which is much greater than the visually observed value. The explanation is that the equipartition formula above holds only when the time between observations is of the order of the time between collisions. In practice we cannot make observations to such a fine degree.

To aid our argument let us consider $\langle \dot{s}^2 \rangle$ more closely. Suppose we observe for 3 min at 30-s intervals the motion of a Brownian particle and plot its two-dimensional "random walk." Its trajectory looks like that shown in Fig. 2a. Suppose the same random walk had been observed at intervals τ of 10 s. There would then be 3 times as many points, and the overall impression would be that the particle is moving $\sqrt{3}$ times as fast. During a time interval τ the particle undergoes millions of collisions, so as the time between observations is decreased still further, the apparent velocity continues to increase, and only when τ is of the order of the time between collisions will Eq. (3.1) hold true. We note that the trajectory that we will observe is in no sense the actual path of the particle cf. the following remarks of Fowler:[14] "We can never follow the details of the movement of the grain (Brownian

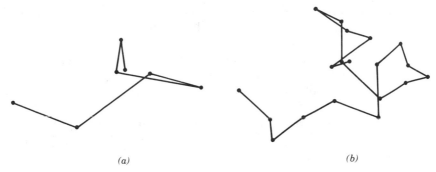

Figure 2.   Trajectory of a Brownian particle: (*a*) observed at intervals of 30 s and (*b*) at intervals of 10 s.

particle) which has a kink at every molecular collision—about 10^{21} a second in an ordinary liquid. What we observe in the way of displacements are of the nature of residual fluctuations about a mean value zero and have little direct connection with the actual detailed path of the grain, to our senses (pushed to their furthest in the form of the best cine camera taking pictures at 10^5 per sec) the details of the path are impossibly fine. It may be fairly compared in a crude way to the graph of a continuous function with no derivative." (We shall say more of this later.)

IV.   EINSTEIN'S EXPLANATION OF BROWNIAN MOVEMENT

It was left to Einstein in 1905 to explain Brownian movement essentially by combining* the elementary stochastic process known as the random walk with the Maxwell–Boltzmann distribution.[15] His ideas may be summarized thus: If a particle in a fluid without friction receives a blow due to a collision with a molecule, then the *velocity* of the particle changes. However, if the fluid is very viscous, the change in velocity is quickly dissipated and the net result of an impact is a change in the displacement of the particle. Thus Einstein assumed that the effect of collisions is to produce random jumps in the *position* of a Brownian particle; that is, the particle performs a kind of random walk. Taking the jumps in the walk as small, he obtained a partial differential equation for the probability distribution of the displacement in one dimension.[1] This equation is a diffusion equation similar to that for unsteady heat conduction. It is the simplest case of a class of equations that

---

*In the sense that the velocity distribution is in equilibrium but not that of the displacements.

have become known as the Fokker–Planck equations*. Einstein obtained its solution, from which he was able to show that the mean-square displacement of a Brownian particle should increase linearly with time. By using the fact that at equilibrium the Maxwellian distribution of velocities must hold, he was able to express the constants in the solution in terms of the temperature and the viscosity of the fluid. Einstein's formula for the mean-square displacement was verified experimentally by Perrin in 1908.[11] He obtained from Einstein's formula a value for Avogadro's number that agreed to within 19% of the accepted value. This provided powerful evidence for the molecular structure of matter.

A. The Random Walk

This problem seems first to have been treated, in somewhat disguised form, by Rayleigh[11,16] (1880, 1894), who wished to find the probability distribution function of a sum of n sinusoidal motions all having the same period and amplitude but a *random* distribution of phases. For $n \to \infty$, Rayleigh obtained a diffusion equation similar to that of Einstein. The first clear statement of the random walk problem seems to have been made by Karl Pearson in 1905:[11] "A man starts from a point O and walks ℓ yards in a straight line; he then turns through any angle whatever and walks another ℓ yards in a second straight line. He repeats this process n times. I require the probability that after these n stretches he is at a distance between r and $r + dr$ from his starting point zero."

Bachelier[17] (1900) made a mathematical model of the French Stock Exchange and obtained a diffusion equation similar to that of Einstein.[1] Later[1] (1911, 1912), he studied the related problem of the Gambler's Ruin, which is in effect a type of random walk problem. He showed that when the sequence of bets placed by the gambler is large, it is simpler to formulate a continuous model of the process. He was again led to a type of Fokker–Planck equation.

B. Calculations of the Mean-Square Displacement
of a Brownian Particle

We give here the calculation of the mean-square displacement of a Brownian grain as rendered by Langevin (1908).[11] This is easier for the beginner to understand than the Einstein method (which we shall describe later).

Langevin simply wrote down the equation of motion of the grain in suspension. He assumed that the forces acting on it could be divided into two

---

*Laplace[1] in 1812 obtained a partial differential equation similar to the Fokker—Planck equation in a disscussion of the mixing of balls drawn and replaced at random from two urns.

parts: (1) a systematic part $-\zeta\dot{x}(t)$ representing a dynamical friction experienced by the grain; and (2) a fluctuating part $F(t)$ that is characteristic of Brownian motion. His equation is thus

$$m\frac{d^2x(t)}{dt^2} + \zeta\frac{dx(t)}{dt} = F(t) \tag{4.1}$$

Regarding the friction term $-\zeta\dot{x}$, it is assumed that this is governed by Stokes's Law, which states that the frictional force decelerating a spherical particle of radius a and mass m is

$$\zeta\dot{x} = 6\pi\eta a\dot{x}$$

where η is the viscosity of the surrounding fluid. For the fluctuating part $F(t)$, the following assumptions are made:

1. $F(t)$ is independent of $\dot{x}$.
2. $F(t)$ varies *extremely rapidly* compared with the variations of $x(t)$.

The second assumption implies that time intervals of duration Δt exist such that during Δt the variations in $\dot{x}$ that are to be expected are very small indeed, whereas during the same time interval $F(t)$ may undergo several fluctuations. Alternatively, we may say that although $\dot{x}(t)$ and $\dot{x}(t+\Delta t)$ are expected to differ by a negligible amount, *no correlation* between $F(t)$ and $F(t+\Delta t)$ exists.

The Langevin equation is the very first example of a stochastic differential equation. In that equation $x(t)$ is a random variable; in other words, a variable that can only take on values with a certain probability. A very thorough account of the meaning of the Langevin equation was given by Doob (1942);[11] he showed that the Langevin equation should be interpreted not as a differential equation, but as an integral equation. These considerations, however, will not affect the elementary derivation of the formula for the mean-square displacement that we give here.

Let us multiply Eq. (4.1) by $x(t)$ so that it becomes

$$m\ddot{x}x = -\zeta\dot{x}x + F(t)x \tag{4.2}$$

Now

$$\dot{x}x = \frac{1}{2}\frac{d}{dt}(x^2) \tag{4.3}$$

and

$$\ddot{x}x = \frac{1}{2}\frac{d}{dt}\left[\frac{d(x^2)}{dt}\right] - \dot{x}^2 \tag{4.4}$$

Thus

$$\frac{m}{2}\frac{d}{dt}\left[\frac{d(x^2)}{dt}\right] - m\dot{x}^2 = \frac{-\zeta d(x^2)}{2\,dt} + Fx \qquad (4.5)$$

All the foregoing refer to one selected Brownian particle. Let us form Eq. (4.5) for each particle suspended in the fluid, and take the mean of all such equations for all particles:

$$\frac{m}{2}\frac{d}{dt}\left[\frac{\overline{d(x^2)}}{dt}\right] - m\overline{\dot{x}^2} = -\frac{\zeta}{2}\frac{\overline{d(x^2)}}{dt} + \overline{Fx} \qquad (4.6)$$

We now assume that $\overline{Fx}$ vanishes, because the force F varies in a completely irregular manner. In other words, the random force F and the displacement x are completely uncorrelated. (This can be proved rigorously.) We also assume that the equipartition theorem holds (i.e., the velocity process has reached its equilibrium value, which is given by the Maxwellian distribution), so that

$$\tfrac{1}{2}m\overline{\dot{x}^2} = \tfrac{1}{2}kT \qquad (4.7)$$

Thus Eq. (4.6) becomes

$$\frac{m}{2}\frac{d}{dt}\left(\frac{\overline{dx^2}}{dt}\right) + \frac{\zeta}{2}\frac{\overline{dx^2}}{dt} = kT \qquad (4.8)$$

Let us now write

$$\frac{\overline{dx^2}}{dt} = u \qquad (4.9)$$

and assume that the order of differentiation and averaging may be interchanged. Thus

$$\frac{m}{2}\frac{du}{dt} + \frac{\zeta}{2}u = kT \qquad (4.10)$$

The solution is

$$u = \frac{2kT}{\zeta} + C\exp\left(\frac{-\zeta t}{m}\right) \qquad (4.11)$$

where C is a constant of integration. Now if t is great compared with m/ζ, that is, if a small mass is combined with a large friction constant, the exponential term has no influence after the first extremely small time interval, and

$$u = \frac{\overline{dx^2}}{dt} = \frac{2kT}{\zeta} \tag{4.12}$$

(Our neglect of the term $\exp(-\zeta t/m)$ is tantamount to ignoring the effect of the inertia of the Brownian particle entirely; hence we say that inertial effects are excluded.)

If we integrate Eq. (4.12) from $t = 0$ to $t = \tau$, we get

$$\overline{x^2} - \overline{x_0^2} = \frac{2kT}{\zeta} \tau \tag{4.13}$$

If we now set $x_0 = 0$ when $t = 0$, and because of its small value write $\overline{(\Delta x)^2}$ instead of $\overline{x^2}$, then

$$\overline{(\Delta x)^2} = \frac{2kT}{\zeta} \tau \tag{4.14}$$

which is the formula of Einstein,[11] as derived by Langevin. $\overline{(\Delta x)^2}$ has the following meaning: We observe a Brownian grain at time 0 and at time τ. During the time interval $(0, \tau)$ it has undergone a displacement Δs, whose projection on the x-axis is Δx. The same grain is observed at later times 2τ, $3\tau, \ldots$, and Δx is determined for each interval. These values are squared and their mean value is calculated; hence $\overline{(\Delta x)^2}$. As we have already emphasized, these displacements that we observe are not in any sense the *detailed* path of the grain, nor is $\Delta x/\tau$ its velocity. What we observe is the *result* of millions of collisions.

C. Calculation of Avogadro's Number

According to Stokes's Law the friction coefficient ζ is

$$\zeta = 6\pi\eta a \tag{4.15}$$

where a is the radius of the Brownian particle.

If we combine this with Eq. (4.14) we get

$$\overline{(\Delta x)^2} = \frac{kT\tau}{3\pi\eta a} \tag{4.16}$$

or

$$\sqrt{\overline{(\Delta x)^2}} = \sqrt{\frac{RT\tau}{3\pi\eta aN}} \qquad (4.17)$$

where

$$N = \frac{R}{k} \qquad (4.18)$$

is the Avogadro number. If R, the gas constant, is known, all the other variables except N may be determined in a suitable experiment. Thus we may use Einstein's formula to estimate the Avogadro number. In 1908 Perrin computed the Avogadro number from observations of the Brownian movement, obtaining $N = 6.85 \times 10^{23}$ mol^{-1} (ref. 11). He also confirmed the relation between τ, η, and T predicted by the Einstein equation. For this work he was awarded the Nobel prize in 1926.

V.   DERIVATION OF THE FORMULA FOR THE MEAN-SQUARE DISPLACEMENT USING EINSTEIN'S METHOD

As we have indicated, Einstein derived the expression for the mean-square displacement of a Brownian particle by means of a diffusion equation. We now give this method.

Let us suppose there are f particles per unit volume between x and $x + dx$ at a time t (we shall confine ourselves to one dimension, since there is no significant loss of generality in doing so). After a time τ has elapsed, we consider a volume element of the same size at the point x'. Particles initially not in this volume element but near it have now entered it from *neighboring* elements. We suppose that the *probability* that a particle enters from an adjoining volume element is a function only of the distance $x' - x$ and the *difference* in time between two successive observations, τ. We denote this probability by $\phi(x' - x, \tau)$. This function also includes the case in which the particles were previously in the element x' if we set $x' - x = 0$. Since the particles must come from *some* volume element, the density at time τ is*

$$f(x', t + \tau) = \int_{-\infty}^{\infty} f(x, t)\phi(x' - x, \tau)\, dx \qquad (5.1)$$

This is a form of the master equation—specifically, it is the Chapman–

---

*ϕ is the transition probability or 'mechanism' of the process.

Kolmogorov equation. Let us now introduce the displacement

$$X = x - x'$$

and fix x' so that $dX = dx$. Then

$$f(x', t + \tau) = \int_{\infty}^{\infty} f(x' + X, t)\phi(X, \tau)\, dX \qquad (5.2)$$

The function $\phi(X, \tau)$ is *even*, since positive and negative displacements are equally likely (the Brownian particle has no memory); thus

$$\phi(X, \tau) = \phi(-X, \tau) \qquad (5.3)$$

We now suppose that τ is small and develop the left-hand side of Eq. (5.2) in powers of τ and the right-hand side in powers of X. Thus

$$f(x', t) + \tau \frac{\partial f}{\partial t} \cdots = \int_{-\infty}^{\infty} \left\{ f(x', t) + X \frac{\partial f}{\partial x} \right.$$

$$\left. + \frac{X^2}{2!} \frac{\partial^2 f}{\partial x^2} \cdots \right\} \phi(X, \tau)\, dX$$

$$= f(x', t) \int_{-\infty}^{\infty} \phi(X, \tau)\, dX + \frac{\partial f}{\partial x} \int_{-\infty}^{\infty} X\phi(X, \tau)\, dX$$

$$+ \frac{1}{2!} \frac{\partial^2 f}{\partial x^2} \int_{-\infty}^{\infty} X^2\phi(X, \tau)\, dX + \cdots \qquad (5.4)$$

Now since $\phi(X, \tau)$ is a probability density function and $\phi(-X, \tau) = \phi(X, \tau)$, we must have

$$\int_{-\infty}^{\infty} \phi(X, \tau)\, dX = 1 \qquad (5.5)$$

$$\int_{-\infty}^{\infty} X\phi(X, \tau)\, dX = 0 \qquad (5.6)$$

and

$$\int_{-\infty}^{\infty} X^2\phi(X, \tau)\, d\tau = \langle X^2 \rangle \qquad (5.7)$$

We also suppose that all higher-order terms such as $\langle X^4 \rangle$ on the right-hand side of Eq. (5.4) are all of order τ^2, so that that equation now becomes

$$\tau \frac{\partial f}{\partial t} = \frac{1}{2} \frac{\partial^2 f}{\partial x^2} \cdot \int_{-\infty}^{\infty} X^2 \phi(X, \tau) \, dX \tag{5.8}$$

The integral on the right in Eq. (5.8) represents the mean-square displacement $\langle X^2 \rangle$, since it is *the sum of the squares of the displacements each multiplied by the probability of its occurrence*. Suppose we observe the instantaneous positions of a Brownian grain at equal time intervals that are not too short. In this way we ensure that we may look upon the individual displacements as *independent*, on account of the random nature of the impacts. (If we make the observations too frequently, a given velocity of the particle will persist after the instant of observation and so affect the next observation.[18] This is what happens when the inertia of the Brownian particle is included in the theory. In practical observations on, for example, suspended pollen grains, it is nearly impossible to make observations at such small time intervals.) After a series of p observations the resulting displacement is

$$(X)^{(p)} = X_1 + X_2 + X_3 + \cdots + X_p \tag{5.9}$$

The square of the displacement is then

$$[X^{(p)}]^2 = (X_1 + X_2 + \cdots + X_p)^2$$
$$= \sum_1^p X_r^2 + 2 \sum_1^p \sum_1^p X_r X_s \tag{5.10}$$

If we now average over the displacements of a large number of particles, the double sum will vanish because of the *equal* probability of positive and negative displacements, and so

$$\langle [X^{(p)}]^2 \rangle = p \langle X^2 \rangle \tag{5.11}$$

This equation means that after a time $p\tau$ has elapsed, we observe a mean-square displacement $p \langle X^2 \rangle$, that is, p *times the value corresponding to the time* τ. Thus $\langle X^2 \rangle$ is proportional to the interval of observations, somewhat akin to a type of superposition principle.

A. The Diffusion Equation

Equation (5.8) is the diffusion equation, which is analogous to the unsteady-heat-conduction equation with the concentration f of Brownian grains taking the place of temperature. If the concentration of Brownian particles varies from point to point in the fluid, a stream of particles will tend to flow from the points of high concentration to those of low concentration. We set the *density* of this flow proportional to the slope of the concentration,[18] that is, to $-D$ grad f or $-D \, \partial f / \partial x$ in the one-dimensional case. The change in the number of particles within a small element $dx \, dy$ is due to the entry of particles from one side and their departure from the other. At the point x a current

$$\left(-D \frac{\partial f}{\partial x} \right)_x dy$$

enters, and at the point $x + dx$, the current

$$\left(-D \frac{\partial f}{\partial x} \right)_{x + dx} dy$$

leaves. The excess of the number entering over the number leaving is the rate at which the particles are diffusing into the small element, so that

$$\frac{\partial f}{\partial t} dx \, dy = D \frac{\partial^2 f}{\partial x^2} dx \, dy$$

Thus in the Einstein theory

$$D = \frac{\langle X^2 \rangle}{2\tau}$$

represents the diffusion coefficient of the particles.

B. Evaluation of D in Terms of Molecular Quantities

Let us suppose that the Brownian particles are in a field of force (e.g., the gravitational field of the Earth); then the Maxwell–Boltzmann distribution for the configuration of the particles must set in. This is

$$f = f_0 e^{-V/kT} \tag{5.12}$$

If the force is constant, as would be true when gravity acts on the Brownian particles, the potential energy V is

$$V = Fx \tag{5.13}$$

We may imagine the Maxwell–Boltzmann distribution to be set up as a result of the motion of the Brownian particles due to the *force together with a diffusion current* that seeks to satisfy Eq. (5.8). Now the velocity of a particle that is in equilibrium under the action of the applied force and viscosity is again given by Stokes's law:

$$v = \frac{F}{6\pi\eta a} \qquad (5.14)$$

where a is, as before, the radius of the particle, and η is the viscosity coefficient. The number of particles crossing unit area in unit time is then

$$j = \frac{fF}{6\pi\eta a} \qquad (5.15)$$

But in order to preserve equilibrium a diffusion current of equal strength flows in the opposite direction:

$$-D\frac{\partial f}{\partial x} = \frac{fF}{6\pi\eta a} \qquad (5.16)$$

This is the mathematical statement of the fact that the rate of diffusion under the concentration gradient must just balance the directed effect of the field of force. Now if

$$V = Fx \qquad (5.17)$$

we have from Eq. (5.12)

$$\frac{1}{f}\frac{\partial f}{\partial x} = -\frac{F}{kT} \qquad (5.18)$$

Thus with Eq. (5.16)

$$\frac{D}{kT} = \frac{1}{6\pi\eta a} \qquad (5.19)$$

or

$$\frac{\langle X^2 \rangle}{\tau} = \frac{kT}{3\pi\eta a} \qquad (5.20)$$

which is Einstein's formula. It is interesting to recall that he derived his formula without having observed Brownian movement, but predicted that such a movement should occur from the standpoint of the kinetic theory of matter.

We quote from his 1905 paper:[15]

> According to the molecular kinetic theory of heat, bodies of microscopically-visible size suspended in a liquid will perform movements of such magnitude that they can be easily observed in a microscope, on account of the molecular motions of heat. It is possible that the movements to be discussed here are identical with the so called 'Brownian molecular motion'; however, the information available to me regarding the latter is so lacking in precision, that I can form no judgement in the matter.
>
> If the movement discussed here can actually be observed (together with the laws relating it that one would expect to find), then classical thermodynamics can no longer be looked upon as applicable with precision* to bodies even of dimensions distinguishable in a microscope: an exact determination of actual atomic dimensions is then possible. On the other hand, had the prediction of the argument proved to be incorrect a weighty argument would be proved against the molecular-kinetic conception of heat.

And later:

> From the standpoint of the molecular kinetic theory of heat a dissolved molecule is differentiated from a suspended body *solely* by its dimensions and it is not apparent why a number of suspended particles should not produce the same osmotic pressure as the same number of molecules. We must assume that the suspended particles perform an irregular movement—even if a very slow one—in the liquid on account of the molecular movement of the liquid.

C. Solution of Eq. (5.8)

The solution of Eq. (5.8) is found by supposing that all the Brownian particles are initially placed near the point $x = 0$ at $t = 0$. This corresponds to finding the point-source solution (or Green's function) of the diffusion equation. The conditions to be imposed on the solution are

$$f(x,0) = \delta(x) \tag{5.21}$$

where $\delta(x)$ is the Dirac delta function, and

$$\int_{-\infty}^{\infty} f(x,t)\, dx = 1 \tag{5.22}$$

Equation (5.22) follows from the fact that f is a probability distribution. The

---

*See our later discussion of Boltzmann's H. Theorem.

solution subject to these conditions is*

$$f(x,t) = \frac{1}{\sqrt{4\pi Dt}} e^{-x^2/4Dt} \qquad (5.23)$$

and if $t = \tau$ (time between observations)

$$f(x,\tau) = \frac{1}{\sqrt{4\pi D\tau}} e^{-x^2/4D\tau} \qquad (5.24)$$

The law of distribution of displacements is therefore the Gaussian distribution.

D. Perrin's Verification of Einstein's Formula

Perrin obtained the following set of counts for the displacements of a grain of radius 2.1×10^{-5} cm at 30-s intervals.[11,14] Out of a number N of such observations, the number n of observed displacements between x_1 and x_2 should be

$$n = \frac{N}{\sqrt{2\pi}} \int_{x_1}^{x_2} \frac{1}{\sqrt{2D\tau}} e^{-x^2/4D\tau} dx$$

He found the following comparison between theory and experiment:

Observations and Calculations of the Distribution
of the Displacements of a Brownian Grain

Range of x ($\times 10^4$ cm)	1st Set		2nd Set		Total	
	Obs.	Calc.	Obs.	Calc.	Obs.	Calc.
0–3.4	82	91	86	84	168	175
3.4–6.8	66	70	65	63	131	133
6.8–10.2	46	39	31	36	77	75
10.2–17.0	27	23	23	21	50	44

E. Extension of Einstein's Formula to Rotational Motion

The arguments used to establish the formulae for the mean-square displacement may be extended to the rotational motion about a diameter of a particle in suspension. If θ is any coordinate (e.g., an angular one) and if $\langle \theta^2 \rangle = A^2$ is the mean-square displacement in time τ due to molecular agi-

---

* details in 9.

tation, then

$$\frac{\partial f}{\partial t} = D \frac{\partial^2 f}{\partial \theta^2} \left(D = \frac{A^2}{2\tau} \right)$$

The number of particles passing by diffusion across a given value θ of the coordinate in unit time is again

$$j = - D \frac{\partial f}{\partial \theta} \tag{5.25}$$

Suppose that an external torque of potential $V(\theta)$ acts on the body. At equilibrium we would then have

$$f = f_0 e^{-V(\theta)/kT}$$

For a particular particle under a torque λ, we have a steady angular velocity

$$\zeta \frac{d\theta}{dt} = \lambda$$

Then at equilibrium when $\lambda = - \partial V/\partial \theta$

$$D \frac{\partial f}{\partial \theta} = - \frac{f}{\zeta} \frac{\partial V}{\partial \theta}$$

But

$$\frac{1}{f} \frac{\partial f}{\partial \theta} = - \frac{\partial V}{\partial \theta} \frac{1}{kT}$$

and thus

$$D = \frac{kT}{\zeta} = \frac{1}{2} \frac{A^2}{\tau}$$

This is the generalization of Eq. (5.20). For a sphere rotating about a fixed axis the drag coefficient is

$$\zeta = 8\pi a^3 \eta \tag{5.26}$$

Therefore for rotational displacements

$$\left\langle \frac{\theta^2}{\tau} \right\rangle = \frac{A^2}{\tau} = \frac{kT}{4\pi a^3 \eta} \tag{5.27}$$

This equation was confirmed experimentally by Perrin.

F.   Limits of Validity of Einstein's Formula

Einstein's formula for the displacement cannot be applied for any arbitrarily small time. We give the original argument of Einstein that illustrates this (see ref. 15, p. 34). The *mean* rate of change of θ as a result of thermal agitation is

$$\frac{\sqrt{\langle \theta^2 \rangle}}{t} = \sqrt{\frac{2kT}{\zeta}} \frac{1}{\sqrt{t}} \tag{5.28}$$

This becomes infinitely great for indefinitely small intervals of time t. This is impossible, since each suspended particle would move with an infinitely great instantaneous velocity. The reason for this difficulty is that we have implicitly assumed in our development that events occurring *during* the interval of observation τ are *completely independent* of events in the time *immediately* preceding it. This is not true if τ is chosen small enough.

Einstein establishes a range of validity for his formula using the following argument. Suppose that the instantaneous rate of change of θ at an initial time t_0 is

$$\frac{d\theta}{dt_0} = \omega_0 \tag{5.29}$$

Let us further suppose that the angular velocity $\omega(t)$ at some later time t is not affected by the irregular thermal processes that occur in the time interval (t_0, t), but that the change in ω is determined *solely by the viscous drag* $\zeta\omega$. Then we have

$$I\frac{d\omega}{dt} = -\zeta\omega \tag{5.30}$$

I is defined by the condition that $I\omega^2/2$ must be the energy corresponding to the angular velocity $\omega(t)$. (The quantity I is termed μ by Einstein.) It is evidently the moment of inertia of the sphere about a diameter. By integration of Eq. (5.30) we get

$$\dot{\theta}(t) = \omega(t) = \omega_0 e^{-\zeta t/I}$$

This is negligible only when $t > I/\zeta$, that is, when the time interval between observations is large compared with I/ζ. If this condition is not satisfied, account must be taken of inertial effects. Einstein calculates that for bodies of 1-μm diameter and unit density in water at room temperature (300 K) the lower limit of applicability of the formula for $\langle \theta^2 \rangle$ is of the order of 10^{-7} s; this lower limit for the interval between observations increases in proportion to the square of the radius of the body. The same considerations hold for the translational as for the rotational motion of the grain. For practical purposes the inertial effects will only start to come into prominence when Brownian movement is used to model high-frequency relaxation processes such as dielectric relaxation and Kerr-effect relaxation.

G. Stochastic Processes

The initial equation (5.1) used by Einstein may be obtained from elementary properties of stochastic processes as follows:

A stochastic process is a family of random variables $[X(t), t \in T]$ where t is some parameter, generally the time, defined on a set T. It is convenient to decompose this set T into instants $t_1 < t_2 < t_3 \cdots < t_n < T$ and then to approximate the family of random variables $[X(t)]$ by $X(t_1)$, $X(t_2)$, and so on. We may describe the process by the following family of joint probability distributions: $f_1(x_1, t_1) \, dx_1$ is the probability of finding $X(t_1)$ in $(x_1, x_1 + dx_1)$; $f_2(x_1, t_1, x_2, t_2) \, dx_1 \, dx_2$ that of finding $X(t_1)$ in $(x_1, x_1 + dx_1)$ and $X(t_2)$ in $(x_2, x_2 + dx_2)$; and $f_3(x_1, t_1, x_2, t_2, x_3, t_3) \, dx_1 \, dx_2 \, dx_3$ that of finding $X(t_1)$ in $(x_1, x_1 + dx_1)$, $X(t_2)$ in $(x_2, x_2 + dx_2)$, and $X(t_3)$ in $(x_3, x_3 + dx_3)$; and so on to $f_n(x_1, t_1, x_2, t_2, \ldots, x_n, t_n)$ (Fig. 3).

The process is *stationary* when that probability distribution underlying the process during a given interval of time depends only on the length of that interval and not on when the interval began. That the underlying mechanism that causes random variables $[X(t)]$ to fluctuate does not change with the course of time is another way of saying this.

A *purely random process* is

$$f_1(x_1, t_1)$$
$$f_2(x_1, t_1, x_2, t_2) = f_1(x_1, t_1) f_1(x_2, t_2)$$
$$f_3(x_1, t_1, x_2, t_2, x_3, t_3) = f_1(x_1, t_1) f_1(x_2, t_2) f_1(x_3, t_3), \text{ etc.}$$

All the information is contained in the first probability density function, f_1.

The *Markov process* (see Section I.H.) strictly limits the dynamic possibilities, but is a more useful statistical concept in which all information about the process is contained in f_2. It is convenient at this point to introduce the *conditional probability density function* (cpdf).

If the probability that $X(t_n)$ is in $(x_n, x_n + dx_n)$ at time t_n depends only on $X(t_{n-1})$, x_n, and t_n, then

$$f_n(x_n, t_n | x_{n-1}, t_{n-1}, \ldots, x_1, t_1) = f_2(x_n, t_n | x_{n-1}, t_{n-1}).$$

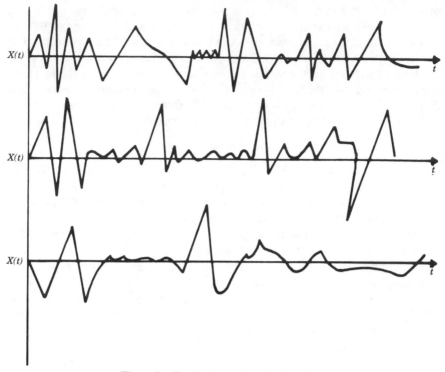

Figure 3.   Realizations of random functions.

This defines the Markov concept. A cpdf *depends* on other events, and is thus more in accord with the dynamical ideas governing, for example, atomic or molecular collisions.

Consider a set of instants $t_1 < t_2 < t_3$, where we suppose for the moment that t_1 and x_1 are fixed. Define $f_2(x_2, t_2 | x_1, t_1) \, dx_2$ as the probability that $X(t_2)$ is in $(x_2, x_2 + dx_2)$ given that $X(t_1)$ had the value x_1 at time t_1, and $f_3(x_3, t_3 | x_2, t_2; x_1, t_1) \, dx_3$ as the probability that $X(t_3)$ is in $(x_3, x_3 + dx_3)$ given that $X(t_2)$ has the value x_2 at time t_2 and that $X(t_1)$ has the value x_1 at time t_1.

If we multiply f_2 and f_3 and integrate with respect to x_2, all dependence on x_2 vanishes. The new probability density function depends only on x_1 at t_1:

$$f_3(x_3, t_3 | x_1, t_1) \, dx_3 = \int_{-\infty}^{\infty} f_2(x_2, t_2 | x_1, t_1) f_3(x_3, t_3 | x_2, t_2; x_1, t_1) \, dx_2 \, dx_3$$

or

$$f_3(x_3, t_3|x_1, t_1) = \int_{-\infty}^{\infty} f_2(x_2, t_2|x_1, t_1) f_3(x_3, t_3|x_2, t_2; x_1, t_1) \, dx_2$$

$$(5.31)$$

If we confine ourselves to Markov processes,

$$f_3(x_3, t_3|x_2, t_2; x_1, t_1) = f_2(x_3, t_3|x_2, t_2)$$

and

$$f_2(x_3, t_3|x_1, t_1) = \int_{-\infty}^{\infty} f_2(x_2, t_2|x_1, t_1) f_2(x_3, t_3|x_2, t_2) \, dx_2 \quad (5.32)$$

Equation (5.31) is termed the *Chapman–Kolmogorov equation* and Eq. (5.32) the *Smoluchowski equation*.* Eq. 5.1 is got from Eq. 5.32 by writing $x_3 = x'$, $x_2 = x$, $t_2 = t$, $t_3 = t + \tau$ so that (discounting the x_1, t_1)

$$f(x', t + \tau) = \int_{-\infty}^{\infty} f(x, t) f(x', t + \tau|x, t) \, dx \qquad (5.32a)$$

then assuming that the transition probability $(X = x - x')$

$$f(x', t + \tau|x, t) = f(x - x', \tau) = \phi(X, \tau)$$

VI.   APPLICATION TO ELECTRICAL SYSTEMS

Einstein also showed how his theory may be applied to conduction processes in a conductor. The charge carriers are regarded as charged Brownian particles; thus, if ζ is replaced by the electrical resistance of the conductor and the charge q replaces the displacement of the Brownian particle, then Einstein's formula gives, for the mean square charge that has flowed across a section of the conductor at time t,

$$\langle q^2 \rangle = \frac{2kT}{R} t \qquad (6.1)$$

De Haas-Lorentz, in her book *Die Brownsche Bewegung*, published in 1913 (which contains a very thorough account of the history of the phenomenon to that time), lists six electrical systems in which fluctuations are treated by means of the Brownian movement.[11] One of her examples is that of a conductor having self-inductance L and resistance R, and carrying a current i.

---

*Not to be confused with the partial differential equation of the same name to be discussed later. Conditional probabilities are written in modern notation opposite to that used in ref.[11].

The Langevin equation for this circuit is*

$$L = \frac{di(t)}{dt} + Ri = E(t)$$

If we let $\lambda = 1 - Rt/L$ and $x = \int E\,dt$, we get

$$\overline{i^2} = \frac{\overline{x^2}}{L^2} \frac{1}{(1-\lambda)^2} = \frac{\overline{x^2}}{2RLt}$$

If one degree of freedom is now assumed, $\frac{1}{2}L\overline{i^2}$ must be equal to $\frac{1}{2}kT$ by the equipartition theorem, and so

$$\overline{x^2} = 2RkTt$$

This shows the mean magnitude of the electromotive force that must exist in the conductor as a result of thermal fluctuations. The formula ignores effects of the inductance on the dynamical behavior. This is entirely analogous to the neglect of inertial effects in the theory of the Brownian motion of a particle. Equation (6.1), derived in 1913, is closely related to the work of Johnson (1925) and Nyquist (1928) (see Barnes and Silverman[6]).

VII.   BROWNIAN MOTION AS A NATURAL LIMIT TO ALL MEASURING PROCESSES

A very detailed account of this was given by Barnes and Silverman.[6] It is instructive to give part of their original account of the Brownian movement of a suspended mirror.

Let M be a very light mirror suspended upon a fine quartz fiber of torsion constant A. The motion of this system may be characterized by the one coordinate ϕ, which is the angle through which the system has rotated from its position of equilibrium. Since it is a motion of one degree of freedom, we may expect that the system will oscillate back and forth with a Brownian motion of such magnitude that

$$\tfrac{1}{2}A\overline{\phi^2} = \tfrac{1}{2}kT$$

An example will make clear the order of magnitude of these oscillations. At 18°C, $\frac{1}{2}kT = 2 \times 10^{-14}$ ergs. Assuming A to be 10^{-6} dyne cm rad^{-2}, as for a thin quartz fiber, we find $\overline{\delta\phi} = 2 \times 10^{-4}$ rad.

* We postulate an ensemble of LR circuits.

Gerlach (1927) allowed a beam of light to be reflected from such a mirror onto a distant scale and then made a study of the inherent zero unsteadiness of the system.[6] With a quartz fiber a few tenths of a micrometer in diameter and a few centimeters long, a mirror measuring 0.8×1.6 mm, and a scale at 1.5 m, he observed Brownian movements of several centimeters. A system with sufficiently thin fiber showed at 1 m a Brownian movement of over a meter. The mean-square deflection, $\overline{\phi^2}$, was successfully measured with an accuracy of 7%, and was found to agree very closely with the value predicted.

Kappler (1931), continuing this work, investigated the possibility of using the Brownian movement of such a suspended system to obtain a more accurate value of Avogadro's number.[6] A photographic record of the Brownian movement of a system, the equation $\frac{1}{2}A\overline{\phi^2} = \frac{1}{2}kT$, and measurements of A and $\overline{\phi^2}$ permit a value to be obtained for k, the Boltzmann constant. Having this, N is obtained from the relation

$$N = R/k$$

The constant of the fiber, A, was determined carefully to $\pm 0.2\%$. From 101 h of registrations at 287 K, $\overline{\phi^2}$ was found to be 4.178×10^{-6} radians^2 ($\pm 0.4\%$). From these values, N was found to be 6.059×10^{23}, which is probably accurate to $\pm 1\%$.

To determine whether the value of $\overline{\phi^2}$ had been made too large by some outside impulse, such as mechanical vibrations, Kappler derived curves at various pressures ranging from 1 atm to 10^{-4} mm Hg. The curves in Fig. 4 which were recorded at the two extremes of the pressure range, show clearly the pressure dependency of the form of the Brownian movement. It is quite striking that at low pressures the motion approaches the natural sinusoidal mode of oscillation of the system and tends to lose its random character. But for slight outside disturbances that make themselves felt only at the lowest pressures, all of the curves, in spite of the difference in their forms, yield for the Brownian movement of the system identical values of $\overline{\phi^2}$.

In a second paper, Kappler studied the influence exerted by these outside mechanical disturbances to see if the 1% error in the previous determination of N could be reduced. By studying Brownian movement for long periods of time with respect to its dependence on the original conditions (original velocities, etc.), one can determine what parts of $\overline{\phi^2}$ are due to spurious effects. In addition, one obtains the mean-square velocities, $\overline{u^2}$, and from them can make a quite independent determination of k, the Boltzmann constant, since $\frac{1}{2}m\overline{u^2} = \frac{1}{2}kT$.

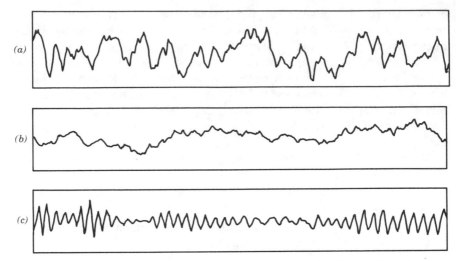

Figure 4.　Brownian fluctuation of a suspended mirror (from Kappler). (*a*) Restoring force 2.66×10^{-9} abs. units. $I = 6.1 \times 10^{-6}$. Camera distance 86.5 cm. Time 30 s, equivalent to 2 mm. Pressure 4×10^{-3} mm Hg. (*b*) Restoring force 9.428×10^{-9} abs. units. $I = 10^{-7}$. Camera distance 72.1 cm. Time 30 s, equivalent to 1 mm. Atmospheric pressure. (*c*) Same system as curve *b*, except that the pressure is 10^{-4} mm. (Taken from R. Barnes and S. Silverman, *Rev. Mod. Phys.* **6**, 162, 1934.)

Uhlenbeck and Goudsmit (1929) investigated theoretically the dependence of the form of Brownian movement on pressure.[6] They developed the displacement of the small mirror, for a time interval that was long compared with the free period of the system, into a Fourier series. The squares of the amplitudes of each Fourier component were found to be given by[6,14]

$$\overline{\phi_\kappa^2} = \frac{\pi^{1/2} m^{1/2} (8kT)^{3/2}}{\rho I T} \frac{1}{\pi kT (\omega^2 - \omega_k^2)^2 + 32 p^2 \rho^2 \omega_k^2}$$

where m is the mass of the gas molecules and ρ is the mass of the mirror per square centimeter.* They are, then, explicit functions of the pressure and the molecular weight of the surrounding gas molecules, as well as of the absolute temperature. The sums of these quantities, however, just as must be expected, are entirely independent of the pressure and the molecular weight, being functions only of T. Thus we get the result predicted from the equipartition theorem,

---

*I is the moment of inertia of the mirror about its axis of rotation.

equipartition theorem,

$$\tfrac{1}{2}A\overline{\phi^2} = \tfrac{1}{2}A \sum_{\kappa=0}^{\infty} \overline{\phi_\kappa^2} = \tfrac{1}{2}kT$$

The above discussion shows that an instrument such as a Nichols radiometer, which consists of a tube evacuated to about 0.01 mm Hg in which a pair of light vanes and a small mirror are suspended by a quartz fiber, must always have an unsteady zero. Experimentally, such instruments, in spite of all possible mechanical precautions, do show these residual deflections around their zeros.

For any suspended system with a fixed torsion constant A, there is, then, *a definite limit set by Brownian motion.*

VIII. SMOLUCHOWSKI

After Einstein, the next fundamental contribution to the theory of Brownian movement was made by the Polish physicist M. von Smoluchowski in 1906.[11] He considered the problem of the Brownian movement of a particle under the influence of an external force. He was able to show that if an external force $\kappa(x)$ acts on the assembly, then the distribution function satisfies the equation

$$\frac{\partial f(x,t)}{\partial t} = -\frac{\partial}{\partial x}\left[\frac{\kappa(x)}{\zeta}f(x,t)\right] + \frac{kT}{\zeta}\frac{\partial^2 f(x,t)}{\partial x^2} \qquad (8.1)$$

This equation is now generally known as the Smoluchowski equation. In particular, Smoluchowski gave a detailed account of the solution of Eq. (8.1) when the force $\kappa(x)$ is elastic, that is, when

$$\kappa(x) = -\gamma x$$

Equation (8.1) was subsequently applied by Debye[19] (1913) to explain very successfully the observed anomalous dispersion of polar liquids at radio frequencies.

The second fundamental contribution made by Smoluchowski was his theory of the fluctuations in concentration of a colloidal dispersion. In particular, he set forth the points of agreement and differences between the statistical and purely thermodynamic conceptions of natural processes, especially the apparent contradiction between the reversible laws of classical mechanics and irreversible thermodynamic processes. He drew the noteworthy conclusion that "a process appears irreversible (or reversible) according as whether the initial state is characterized by a long (or short) average time of recurrence compared to the time during which the system is under observation." This is of fundamental importance for the understanding of Boltzmann's H theorem.

IX. THE DEBYE THEORY OF DIELECTRIC RELAXATION

By 1913 it had been fairly well established that a decrease in relative permittivity with increasing frequency existed for certain kinds of molecules in which special atomic groups such as OH or NH_2 were present. Thus it seemed that the characteristic property of liquids with anomalous dispersion in the radio-frequency bands (including microwave and far-infrared frequencies) was due to the polar nature of the constituent molecules. Debye,[19] in a 1913 discussion of the origin of the observed anomalous dispersion, used Einstein's theory in an attempt to explain the dispersion.

A. Debye's Treatment

In his first model of the phenomenon, Debye considered an assembly of molecules, each carrying a permanent dipole moment, with each molecule compelled to rotate about an axis normal to itself. He supposed that the electrical interaction between each member of the assembly could be ignored, so that *on the average* all molecules of the assembly would behave in the same way. Thus it would suffice to consider the behavior of one molecule only. Hence the problem was reduced to considering the rotational Brownian movement in two dimensions of a dipole or rigid rotator subjected to an external, time-varying electric field **E**. The distribution function f used by Debye was defined as follows: $f(\theta, t) \, d\theta$ is the number of dipoles whose axes lie in an element of angle $d\theta$ on the circumference of a circle. Since the field **E** is time varying, f is a function both of the time and of the angle θ between the dipole and the field. It now follows that the Smoluchowski equation for the problem can be written simply by replacing the x-coordinate with the angular coordinate θ; thus we have the differential equation for f,

$$\frac{\partial f(\theta, t)}{\partial t} = \frac{\partial}{\partial \theta}\left(\frac{kT}{\zeta} \frac{\partial f(\theta, t)}{\partial \theta} + \frac{1}{\zeta} \frac{\partial V(\theta, t)}{\partial \theta} f(\theta, t) \right) \tag{9.1}$$

Equation (9.1) corresponds to the Langevin equation*

$$\zeta \dot{\theta}(t) + \frac{\partial V(\theta, t)}{\partial \theta} = \lambda(t) \tag{9.2}$$

---

*θ in 9.1 (or indeed the space variable in any probability density diffusion equation) must be regarded as an ensemble average of angles. In 9.2 on the other hand θ is a random variable. μ denotes the dipole moment in what follows.

where $\lambda(t)$ is the random torque arising from the Brownian movement of the surroundings. The torque $\lambda(t)$ has the same properties as F did in our discussion of the Langevin equation. We immediately see from Eq. (9.2) that the inertial term $I\ddot{\theta}$ arising from the finite mass of the dipole is ignored, so that the distribution functions calculated from Eq. (9.2) will be in error at high frequencies (short times).[20,21]

The potential energy due to the external field is (in the absence of induced dipole moments)

$$V = -\mu E \cos\theta = -\mathbf{\mu}\cdot\mathbf{E}$$

Thus Eq. (9.1) is reduced to

$$\frac{\partial f(\theta,t)}{\partial t} = \frac{\partial}{\partial\theta}\left(\frac{kT}{\zeta}\frac{\partial f(\theta,t)}{\partial\theta} + \frac{\mu E}{\zeta}f(\theta,t)\sin\theta\right) \tag{9.3}$$

Debye obtained two particular solutions for Eq. (9.3). The first type of solution he obtained was the so-called aftereffect solution, which may be described as follows. We suppose that the dielectric consisting of the assembly of noninteracting dipolar molecules has been influenced for a long time by a steady external field; thus on the average the axes of the dipoles are oriented mainly in the direction of the field. Let us now suddenly switch the field off; the axes of the dipoles then revert to their original, random orientations. This phenomenon of returning from the polarized state to the original chaotic state or vice versa is called dielectric relaxation. This effect may be studied by solving Eq. (9.3) for the following situation. Suppose that the constant field $\mathbf{E} = \mathbf{E}_0$ had been acting for a long time up to $t = 0$; thus for $t > 0$ the potential-energy term in Eq. (9.3) vanishes and that equation is reduced to

$$\frac{\partial f(\theta,t)}{\partial t} = \frac{kT}{\zeta}\frac{\partial^2 f(\theta,t)}{\partial\theta^2} \tag{9.4}$$

Furthermore, since $\mathbf{E}_0$ had been acting for a long time up to $t = 0$, the distribution at that time must be the Maxwell–Boltzmann distribution, so that

$$f(\theta,0) = A\exp\left(\frac{\mu E_0\cos\theta}{kT}\right) \tag{9.5}$$

To terms linear in the field strength this expression is approximated by

$$f(\theta,0) = A\left(1 + \frac{\mu E_0}{kT}\cos\theta\right) \tag{9.6}$$

where A is a given constant. This holds true if $\mu E_0 \ll kT$.

Equation (9.6) is the *linear* approximation to the distribution function. If the higher-order terms are included, we are taking account of the nonlinear behavior of the dielectric. This is discussed later, because it is irrelevant to the present case. The form of the initial condition now suggests that we seek a solution to Eq. (9.6) of the form

$$f(\theta, t) = A\left(1 + \frac{\mu E_0}{kT} g(t)\cos\theta\right) \tag{9.7}$$

On substituting this equation into Eq. (9.4) and solving the resulting ordinary differential equation for $g(t)$, which is simply

$$\frac{dg(t)}{dt} = -\frac{kT}{\zeta} g(t)$$

we find that $f(\theta, t)$ is given by

$$f = A\left(1 + \frac{\mu E_0}{kT} e^{-(kT/\zeta)t}\cos\theta\right)$$

We note that as $t \to \infty$, f is independent of θ, and the variable part of f decays with time. The time constant of the decay is

$$\tau_0 = \frac{\zeta}{kT} \tag{9.8}$$

which is called the Debye relaxation time. We now calculate the mean dipole moment, that is, the dipole moment averaged over the ensemble. We have

$$m_{E_0} = \mu\langle\cos\theta\rangle = \frac{\displaystyle\int_0^{2\pi}\mu f\cos\theta\,d\theta}{\displaystyle\int_0^{2\pi}f\,d\theta} = \frac{\mu^2 E_0}{2kT}e^{-t/\tau_0} \tag{9.9}$$

This is the mean component of the electric moment in the direction of the field. This component decays exponentially with the decay constant τ_0.

We now discuss the second problem posed by Debye, namely, the behavior of f when an alternating electric field is applied. We suppose that an alternating field has been applied to the dielectric for a long time, so the transient effect associated with the switching-on of the field may be neglected. If we express the field in complex notation as $E_m e^{i\omega t}$, then

$$\frac{\partial f(\theta, t)}{\partial t} = \frac{\partial}{\partial \theta}\left(\frac{kT}{\zeta} \frac{\partial f(\theta, t)}{\partial \theta} + \frac{\mu E_m}{\zeta} e^{i\omega t} \sin \theta \right) \tag{9.10}$$

In writing down Eq. (9.10) we assume that the linear approximation for f still holds; otherwise we cannot use the complex notation for the field. Debye next assumed that the solution of Eq. (9.10) may be written

$$f(\theta, t) = A\left(1 + B\frac{\mu E_m}{kT} e^{i\omega t} \cos \theta \right) \tag{9.11}$$

where B must be equal to unity for $t = 0$. On substituting this equation into Eq. (9.10) we readily find the value of B, and thus f, to be

$$f = A\left(1 + \frac{e^{i\omega t}}{1 + i\omega \tau_0} \frac{\mu E_m}{kT} \cos \theta \right) \tag{9.12}$$

With Eq. (9.9) the mean dipole moment is thus

$$\langle \mu \cos \theta \rangle = \frac{\mu^2}{2kT} \frac{E_m e^{i\omega t}}{1 + i\omega \tau_0} \tag{9.13}$$

or

$$\langle \mu \cos \theta \rangle = \frac{\mu^2}{2kT} \frac{E_m [(\cos \omega t + \omega \tau_0 \sin \omega t) + i(\sin \omega t - \omega \tau_0 \cos \omega t)]}{1 + \omega^2 \tau_0^2}$$

$$\tag{9.14}$$

Thus we see at once that there is a difference in phase between $\langle \mu \cos \theta \rangle$ and E. This phase difference persists if in place of $E_m e^{i\omega t}$ we take its real or imaginary parts $E_m \cos \omega t$ or $E_m \sin \omega t$. A second model in which the dipole is fixed in a sphere that is allowed to rotate in three dimensions was also given by Debye.[19] If inertial effects are discounted the only difference is that the relaxation time for the sphere model is *half* that of the disk model.

Debye made an estimate of the relaxation time τ_0 and the frequency bands in which anomalous dispersion should occur by assuming that Stokes's formula for the frictional torque acting on a rotating sphere, $\zeta = 8\pi\eta a^3$, applies to the dipole. Taking the case of water at room temperature where $\eta = 0.01$, and assuming that $a = 2 \times 10^{-8}$ cm, he found ζ to be 2×10^{-24}, which yields a relaxation time τ_0 of 0.25×10^{-10} s. (Note that the sphere formula

$\tau_0 = \zeta/2kT$ has been used rather than the ζ/kT of the disk model.) This relaxation time corresponds to a wavelength of the order of 1 cm, and because ζ varies as a^3, one would expect for different liquids a considerable range of wavelengths corresponding to microwave frequencies to be the characteristic region in which to expect an anomalous dispersion resulting from orientation. Note that Debye emphasized that the introduction of this numerical calculation is not necessary for the *formal* description of the relaxation effect. It is merely a device for estimating the frequency at which the relaxation should occur. In experimental work τ_0 is usually taken as a parameter.

B. Relation of the Debye Theory to Observed Quantities

If the phase angle in Eq. (9.14) is ϕ, that equation may be written

$$\langle \mu \cos \theta \rangle = \frac{\mu^2}{2kT} \frac{E_m}{\sqrt{1 + \omega^2 \tau_0^2}} e^{i(\omega t - \phi)} \qquad (9.15)$$

where

$$\tan \phi = \omega \tau_0$$

A *difference* in phase between field intensity and polarization is *always* accompanied by energy absorption. Due to the finite relaxation time, we will therefore encounter not only dispersion but *also* absorption. In general, the dispersion and absorption will be connected by the Kramers–Kronig dispersion relations.[9,21] It will be convenient for the discussion that follows to replace Eq. (9.15) with its three-dimensional equivalent, which is simply[9,12]

$$\langle \mu \cos \theta \rangle = \frac{\mu^2}{3kT} \frac{E_m}{\sqrt{1 + \omega^2 \tau_0^2}} e^{i(\omega t - \phi)} \qquad (9.16)$$

where we remember that

$$\tau_{\text{sp}} = \frac{\zeta}{2kT} = \tau_0 \qquad (9.17)$$

Assuming that no interaction takes place between the constituent molecules of a *macroscopic* sphere of dielectric, the permittivity of that sphere may be calculated from the equation

$$\frac{3\varepsilon_0(\varepsilon - 1)}{(\varepsilon + 2)} = n\alpha \qquad (9.18)$$

where n is the number of molecules per unit volume of the sphere, ε_0 is the permittivity of free space, and α is the polarizability per molecule. If we call α_∞ the polarizability at frequencies so high that the orientational mode of polarization has ceased to operate, we may write

$$n\alpha_\infty = \frac{3\varepsilon_0(\varepsilon_\infty - 1)}{(\varepsilon_\infty + 2)} \tag{9.19}$$

where ε_∞ is the permittivity at these frequencies. Equation (9.18) when combined with Eq. (9.16) now gives

$$\frac{\varepsilon(\omega) - 1}{\varepsilon(\omega) + 2} = \frac{n}{3\varepsilon_0}\left(\alpha_\infty + \frac{\mu^2}{3kT}\frac{1}{1 + i\omega\tau_0}\right) \tag{9.20}$$

It is convenient to write this equation in somewhat different form by multiplying both sides by the quotient of the molecular weight M and the density ρ. We then get for the left-hand side

$$\frac{\varepsilon(\omega) - 1}{\varepsilon(\omega) + 2}\frac{M}{\rho} \tag{9.21}$$

while for the right-hand side we get an expression involving nM/ρ which is equal to the Avogadro number. Thus Eq. (9.20) is rearranged to read

$$\frac{M}{\rho}\frac{\varepsilon(\omega) - 1}{\varepsilon(\omega) + 2} = P(\omega) = \frac{N}{3\varepsilon_0}\left(\alpha_\infty + \frac{\mu^2}{3kT}\frac{1}{1 + i\omega\tau_0}\right) \tag{9.22}$$

This equation may be solved for $\varepsilon(\omega)$ to give

$$\varepsilon(\omega) = \frac{1 + (2\rho/M)P(\omega)}{1 - (\rho/M)P(\omega)} \tag{9.23}$$

Following Debye, instead of characterizing the liquid by the two quantities α_∞ and $\mu^2/3kT$, we will use the two permittivities ε_∞ and ε_s defined by

$$\frac{\varepsilon_\infty - 1}{\varepsilon_\infty + 2}\frac{M}{\rho} = \frac{N}{3\varepsilon_0}\alpha_\infty \tag{9.24}$$

$$\frac{\varepsilon_s - 1}{\varepsilon_s + 2}\frac{M}{\rho} = \frac{N}{3\varepsilon_0}\left(\alpha_\infty + \frac{\mu^2}{3kT}\right) \tag{9.25}$$

Here ε_s is the static permittivity, that is, the permittivity for zero frequencies. We then find that

$$P(\omega) = \frac{M}{\rho} \left[\frac{\varepsilon_\infty - 1}{\varepsilon_\infty + 2} + \frac{1}{1 + i\omega\tau_0} \left(\frac{\varepsilon_s - 1}{\varepsilon_s + 2} - \frac{\varepsilon_\infty - 1}{\varepsilon_\infty + 2} \right) \right] \qquad (9.26)$$

and thus

$$\varepsilon(\omega) = \frac{\varepsilon_s/(\varepsilon_s + 2) + i\omega\tau_0\varepsilon_\infty/(\varepsilon_\infty + 2)}{1/(\varepsilon_s + 2) + i\omega\tau_0/(\varepsilon_\infty + 2)} \qquad (9.27)$$

Now

$$\varepsilon(\omega) = n^2(\omega)[1 - i\kappa(\omega)]^2$$

where n is the refractive index and κ is the absorption index. Both n and κ are real. We now introduce x via

$$x = \frac{\varepsilon_s + 2}{\varepsilon_\infty + 2} \omega\tau$$

By algebra,

$$n^2 = \frac{1}{2} \left[\sqrt{\frac{\varepsilon_s^2 + \varepsilon_\infty^2 x^2}{1 + x^2}} + \frac{\varepsilon_s + \varepsilon_\infty x^2}{1 + x^2} \right]$$

$$n^2\kappa^2 = \frac{1}{2} \left[\sqrt{\frac{\varepsilon_s^2 + \varepsilon_\infty^2 x^2}{1 + x^2}} - \frac{\varepsilon_s + \varepsilon_\infty x^2}{1 + x^2} \right]$$

These give values for n and $n\kappa$, respectively. In the interval $x = 0,\ldots, x = \infty$, the square of the refractive index ranges from $n^2 = \varepsilon_s$ to $n^2 = \varepsilon_\infty$; the product $n^2\kappa^2$ starts with 0 for $x = 0$, goes through a maximum, and comes back to 0 again for $x = \infty$. The same is true for the absorption index κ alone. The general behavior is illustrated in Fig. 5. It must again be noted that the theory does not include inertial effects, so it will be in error at high frequencies.

In practice the absorption is generally measured using a bridge method. The results are then given in terms of a phase angle Φ. Let us suppose that the potential difference across the plates of a capacitor is $Ve^{i\omega t}$. Then the permittivity $\varepsilon(\omega)$ has to be decomposed into a real part and an imaginary part:

$$\varepsilon(\omega) = \varepsilon'(\omega) - i\varepsilon''(\omega)$$

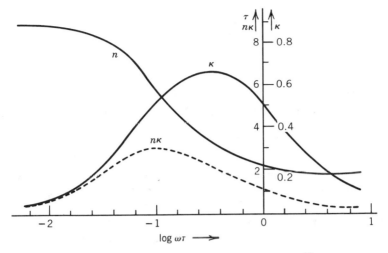

Figure 5.   Dispersion and absorption in polar liquids after Debye.[19]

The charge Q on one of the plates (omitting constant factors) will then be

$$Q = (\varepsilon' - i\varepsilon'')Ve^{i\omega t}$$

The phase angle Φ is then given by

$$\tan \Phi = \frac{\varepsilon''}{\varepsilon'}$$

where with the above equation

$$\varepsilon' = \varepsilon_\infty + \frac{\varepsilon_s - \varepsilon_\infty}{1 + x^2}$$

$$\varepsilon'' = \frac{\varepsilon_s - \varepsilon_\infty}{1 + x^2} x$$

$$\tan \Phi = \frac{\varepsilon_s - \varepsilon_\infty}{\varepsilon_s + \varepsilon_\infty x^2} x$$

This phase angle reaches its maximum at the same place in the spectrum where κ is a maximum, and thus

$$\tan \Phi_{max} = \frac{1}{2} \frac{\varepsilon_s - \varepsilon_\infty}{\sqrt{\varepsilon_s \varepsilon_\infty}}$$

At this frequency

$$\varepsilon' = \frac{2\varepsilon_s \varepsilon_\infty}{\varepsilon_\infty + \varepsilon_s}, \qquad \varepsilon'' = \frac{\varepsilon_s - \varepsilon_\infty}{\varepsilon_s + \varepsilon_\infty} \sqrt{\varepsilon_s \varepsilon_\infty}$$

The explanation of the microwave absorption must rank as one of the major triumphs of the theory of Brownian movement.

X.   DEVELOPMENTS SUBSEQUENT TO 1920

The next significant addition to the theory of Brownian movement was made in 1918 by L. S. Ornstein.[11,15] Using a method suggested by de Haas-Lorentz and Fürth,[15] he showed that if inertia was included, then the mean-square displacement of a Brownian particle was

$$\left\langle \left(x(t) - x_0 \right)^2 \right\rangle = \frac{2kTm}{\zeta^2} \left(\frac{\zeta t}{m} - 1 + e^{-\frac{\zeta t}{m}} \right) \qquad (10.1)$$

For times $t \gg m/\zeta$, this formula becomes Einstein's, that is,

$$\left\langle \left(x(t) - x_0 \right)^2 \right\rangle = \frac{2kTt}{\zeta}$$

whereas for times so short that the frictional force does not get a chance to operate

$$\left\langle \left(x(t) - x_0 \right)^2 \right\rangle = \frac{kTt^2}{m}$$

This is the formula one would expect if the molecule experienced no forces between collisions (the classical kinetic-theory result). For pollen particles and particles of microscopic size, however, Einstein's formula holds without correction. This will not be true when the theory is applied to molecular relaxation processes.

A detailed account of the derivation of Eq. (10.1) was given by Uhlenbeck and Ornstein (1930).[11] The starting point of their investigation was the Langevin equation, this time including the term

$$\frac{m \, dv(t)}{dt}$$

They then wrote down the formal solution of this equation, and calculated the mean-square value of the velocity of the Brownian particle from the formal solution using the properties of the random force of the Langevin equation. They next showed explicitly, by calculating the statistical moments of the solutions, that the Gaussian distribution must hold for v. This allowed them to write down the probability distribution of the velocities. By integrating the formal solution for v again, they obtained an expression for the position $x(t)$ of the Brownian particle. Again using the moment method, they were able to show that the probability distribution of the displacements is Gaussian. This immediately led to Eq. (10.1). (The details of the calculation as carried out by a more modern method are given in ref. 12, Chapter 4.)

In addition to treating the Brownian movement of the free particle, they also considered a harmonic oscillator and obtained expressions for the mean-square displacement, mean-square velocity, and so on (details in ref. 12, Chapter 4). Their starting point was again the Langevin equation,

$$\frac{m\,d^2x(t)}{dt^2} + \zeta\frac{dx(t)}{dt} + \gamma x = F(t)$$

where γx is the restoring force of the oscillator.

A. Mathematical Difficulties Associated with the Langevin Equation— Interpretation of That Equation as an Integral Equation by Doob

Doob (1942) pointed out a number of mathematical difficulties associated with the writing down of an equation such as the Langevin equation.[11] He was the first investigator to show how that equation should be interpreted. His approach also allowed great mathematical and conceptual simplifications. We illustrate this in the following discussion.

Ornstein and Uhlenbeck based their investigation on the *Langevin equation*

$$\frac{du(t)}{dt} = -\beta u(t) + A(t) \tag{10.2}$$

which is simply Newton's law of motion applied to a particle after dividing through by the mass. The first term on the right is due to the frictional resistance imposed on the particle by the medium surrounding the particle. The second term represents the *random forces* (molecular impacts); probability hypotheses are imposed on the $A(t)$, including relations between $A(t)$ and $u(t)$, in order to determine the $u(t)$ distribution. Unfortunately the $u(t)$ distribution has as we have seen, the property that the velocity function has *no time derivative*; thus the solution can hardly satisfy Eq. (10.2).

The question now poses itself: What exactly do we mean by Eq. (10.2)? To interpret the Langevin equation we write it as

$$du(t) = -\beta u(t)\, dt + dB(t) \tag{10.3}$$

and try to give these differentials a meaning. (We shall suppose that the $B(t)$ process is a differential process; that is, if, $t_1 < \cdots < t_n$, we suppose that $\{B(t_2) - B(t_1), \ldots, B(t_n) - B(t_{n-1})\}$ *are mutually independent* random variables. We shall also suppose *temporal homogeneity*, that is, *stationarity*, so that the distribution of $B(s+t) - B(s)$ is *independent* of the initial value s. We shall also suppose that

$$\langle B(s+t) - B(s) \rangle = 0$$

and that

$$\langle [B(s+t) - B(s)]^2 \rangle = c^2 t, \qquad t \geq 0$$

Let us integrate both sides of Eq. (10.3) after first multiplying both sides by a continuous function $f(t)$ of the time t. We then interpret Eq. (10.3) as meaning that for f, a continuous function of time, and for all t,

$$\int_{t=a}^{b} f(t)\, du(t) = -\beta \int_{a}^{b} f(t) u(t)\, dt + \int_{a}^{b} f(t)\, dB(t)$$

If $f(t) = 1$

$$u(b) - u(a) = -\beta \int_{a}^{b} u(t)\, dt + B(b) - B(a)$$

In the Ornstein–Uhlenbeck process, $u(t)$ exists and so does the displacement, so (writing $a = 0$, $b = t$)

$$B(t) - B(0) = u(t) - u(0) + \beta [X(t) - X(0)]$$

Let us now suppose that $b - a$ is long compared with the time between impacts, so that $B(b) - B(a)$ may be written as the series

$$B(b) - B(a) = \sum_{k=1}^{n} (B(t_k) - B(t_{k-1}))$$

where $t_0 = a < t_1 < t_2 \cdots < t_{n-1} < t_n = b$.

In the future when we write down a Langevin equation

$$\frac{du(t)}{dt} = -\beta u(t) + \frac{dB(t)}{dt}$$

we shall always interpret it as

$$u(b) - u(a) = -\beta \int_a^b u(t)\, dt + B(b) - B(a)$$

where

$$B(b) - B(a) = \sum_{k=1}^n \left[B(t_k) - B(t_{k-1}) \right]$$

Thus we see that the Langevin equation should always be interpreted as an integral equation. The treatment is readily extended to matrix Langevin equations, which occur in the analysis of higher-order systems.

B. The Wiener Process

The essence of Doob's investigation is contained in the stochastic process[11] known as the Wiener process (after Norbert Wiener). This is the mathematical idealization of Brownian motion and is described as follows:

Let $X(t)$ be a random variable that, for the sake of illustration only, is taken to denote the displacement after a time t of a particle undergoing Brownian motion, and let $X(0) = 0$. Further, let us consider a time interval (s, t) that is long compared with the time between impacts of the particles of the surrounding medium on the Brownian particle. We shall assume the following:

1. The displacement $X(t) - X(s)$ of the Brownian particle over (s, t) is the sum

    $$\sum_{k=1}^n \left[X(t_k) - X(t_{k-1}) \right]$$

    of the small displacements $X(t_k) - X(t_{k-1})$ of the Brownian particle caused by the impacts of the particles of the surrounding medium on the Brownian particle.

2. The probability distribution of $X(t_k)$ depends only on $X(t_{k-1})$ and not on $X(t_{k-2})$, $X(t_{k-3}), \ldots$, so that we have a Markov process. It follows that

    $$X(t_1) - X(s),\ X(t_2) - X(t_1), \ldots, X(t_n) - X(t_{n-1})$$

    are independent random variables. Formally, we say that the process has independent increments.

3. We have

    $$\left\langle X(t_1) - X(s) \right\rangle = \left\langle X(t_k) - X(t_{k-1}) \right\rangle = 0$$

Since we have assumed that $X(t) - X(s)$ is the sum of a large number of independent random variables, each having an arbitrary distribution, it follows from the central limit theorem[9] that the characteristic function $\phi(u)_{X(t)-X(s)}$ of the random variable $X(t) - X(s)$ is

$$\phi(u)_{X(t)-X(s)} = \exp\left\{-\tfrac{1}{2}u^2\left\langle\left[X(t)-X(s)\right]^2\right\rangle\right\}.$$

Formally, we say that a stochastic process consisting of a family of random variables $\{X(t), t \geq 0\}$ is a Wiener process if:[12]

1. $\{X(t), t \geq 0\}$ has stationary independent increments.
2. $X(t)$ is normally distributed for $t \geq 0$.
3. $\langle X(t) \rangle = 0$.
4. $X(0) = 0$.
5. $\left\langle\left[X(t)-X(s)\right]^2\right\rangle = c^2(t-s) = c^2|t-s|$.
6. $\langle X(s)X(t) \rangle = c^2\min(s,t) = c^2 s \wedge t.$*

We must now obtain an expression for $\langle[X(t)-X(s)]^2\rangle$. Let

$$\left\langle X^2(t) \right\rangle = f(t)$$

Then if t_1 and t_2 are distinct times, we have

$$f(t_1 + t_2) = \left\langle X^2(t_1+t_2) \right\rangle = \left\langle\left[X(t_1+t_2)-X(t_1)+X(t_1)-X(0)\right]^2\right\rangle$$

since

$$X(0) = 0$$

Multiplying out

$$f(t_1+t_2) = \left\langle\left[X(t_1+t_2)-X(t_1)\right]^2\right\rangle + \left\langle\left[X(t_1)-X(0)\right]^2\right\rangle$$

because

$$\left\langle\left[X(t_1+t_2)-X(t_1)\right]\left[X(t_1)-X(0)\right]\right\rangle$$
$$= \left\langle X(t_1+t_2)-X(t_1) \right\rangle\left\langle X(t_1)-X(0) \right\rangle = 0$$

since $X(t_1+t_2) - X(t_1)$ and $X(t_1) - X(0)$ are independent random vari-

---

* $\wedge$ denotes the minimum of s and t.

ables. By stationarity

$$f(t_1 + t_2) = \left\langle [X(t_2) - X(0)]^2 \right\rangle + \left\langle [X(t_1) - X(0)]^2 \right\rangle$$
$$= \left\langle X^2(t_2) \right\rangle + \left\langle X^2(t_1) \right\rangle$$

or

$$f(t_1 + t_2) = f(t_1) + f(t_2)$$

We let $t_2 = t_1$, $t_1 = -s$; then

$$f(t - s) = f(t) + f(-s)$$

The only function which will satisfy the equation is

$$f(t - s) = c^2(t - s)$$

where c is a constant to be determined.

Now, again by stationarity,

$$\left\langle [X(t) - X(s)]^2 \right\rangle = \left\langle [X(t - s) - X(0)]^2 \right\rangle = \left\langle X^2(t - s) \right\rangle$$

whence,

$$\left\langle [X(t) - X(s)]^2 \right\rangle = c^2(t - s) = c^2|t - s|$$

the modulus bars being inserted in order to ensure a positive variance, and

$$\phi(u)_{X(t) - X(s)} = \exp\left[-\tfrac{1}{2}u^2 c^2 |t - s| \right]$$

We now wish to evaluate $K(s, t)$, where

$$K(s, t) = \text{cov}[X(s), X(t)] = \left\langle [X(s) - \left\langle X(s) \right\rangle][X(t) - \left\langle X(t) \right\rangle] \right\rangle$$
$$= \left\langle X(s) X(t) \right\rangle$$

This may be written

$$K(s, t) = \left\langle X(s)[X(t) - X(s) + X(s)] \right\rangle$$
$$= c^2 \min(s, t) = c^2 s \wedge t$$

since

$$\left\langle X(s)[X(t) - X(s)] \right\rangle = 0$$

We shall now consider the differences $\xi(\Delta)$ of the Wiener process. In what follows the random variable $X(t)$ above is replaced by the symbol $B(t)$. We consider two overlapping time intervals $[t_1, t_2]$ and $[t_1', t_2']$ of length Δ and Δ' respectively:

$$\bullet - \bullet - \bullet - \bullet$$
$$t_1 \quad t_1' \quad t_2 \quad t_2'$$

with

$$t_1 < t_1' < t_2 < t_2'$$

We write

$$\xi(\Delta) = B(t_2) - B(t_1)$$
$$\xi(\Delta') = B(t_2') - B(t_1')$$

Multiplying out these equations, averaging, and making use of the covariance $K(s, t)$ we find that

$$\langle \xi(\Delta)\xi(\Delta') \rangle = c^2 |t_2 - t_1'| = c^2 |\Delta \cap \Delta'|$$

and the special case $\Delta = \Delta'$,

$$\langle \xi^2(\Delta) \rangle = c^2 |\Delta|$$

Note that we may interpret differentials such as $dB(t)$ in terms of ξ because

$$dB(t) = B(t + dt) - B(t) = \xi(dt)$$

and thus

$$\int_{-\infty}^{\infty} f(t)\, dB(t) = \int_{-\infty}^{\infty} f(t)\xi(dt)$$

We now wish to discuss the integral

$$\xi[f] = \int_{-\infty}^{\infty} f(t)\xi(dt)$$

termed a Wiener integral.[12] One may show that

$$\langle \xi^2[f] \rangle = c^2 \int_{-\infty}^{\infty} f^2(t)\, dt$$

and also that if $f(t)$ and $g(t)$ are continuous,[12]

$$\left\langle \int_{-\infty}^{\infty} f(t)\xi(dt) \int_{-\infty}^{\infty} g(t)\xi(dt) \right\rangle = c^2 \int_{-\infty}^{\infty} f(t)g(t)\, dt$$

or

$$\langle \xi[f]\xi[g] \rangle = c^2 \int_{-\infty}^{\infty} f(t)g(t)\, dt$$

where c^2 is a given constant. These are very useful formulas.

A full treatment of these topics from a mathematical point of view may be found in ref. 13. These results in functional integration appeared for the first time in the fundamental paper of Klein,[11,12] and were also discussed in some detail by Chandrasekhar (1943).[11]

C.   Klein's Discovery of the Diffusion Equation in Configuration–Velocity Space—the Kramers Equation

The results of Ornstein and Uhlenbeck were essentially derived by calculating statistical averages from the Langevin equation. They did not show how the probability distribution of the displacements for either the free particle or the oscillator could be calculated using the diffusion-equation method. To accomplish this one must write down the diffusion equation governing the behavior of the probability distribution in configuration–velocity space (which we shall term phase space). The solution of this equation (for given initial conditions) when integrated over the velocities will give the full configuration-space distribution. This was first achieved by Klein.[12] He showed that the phase-space distribution function

$$f(x, v, t|x(0), v(0), 0) = f(x, v, t) \tag{10.4}$$

for a Brownian particle moving along the x-axis under the influence of a potential $V(x)$ satisfies the equation

$$\frac{\partial f}{\partial t} + v\frac{\partial f}{\partial x} - \frac{1}{m}\frac{\partial V(x)}{\partial x}\frac{\partial f}{\partial v} = \frac{\zeta}{m}\frac{\partial}{\partial v}\left(vf + \frac{kT}{m}\frac{\partial f}{\partial v}\right) \tag{10.5}$$

This equation, although fully derived by Klein 19 years earlier, is often known as the Kramers equation after a 1940 paper of Kramers[11] that concerned the escape of particles over potential barriers as a result of shuttling action of Brownian motion. It is also called the Fokker–Planck–Kramers equation, the Kolmogorov equation, and the generalized Liouville equation.[13,20] In the course of this chapter it will be convenient for us to call it the Kramers equation. Equation (10.5) corresponds to the Langevin equation

$$m\frac{d^2x(t)}{dt^2} + \zeta\frac{dx}{dt} + \frac{\partial V}{\partial x} = F(t) \tag{10.6}$$

It thus includes inertial effects *exactly*, unlike the Einstein theory. Equation (10.6) may be solved with ease only in two cases: (1) that of the free Brownian particle; and (2) the case where $V = \frac{1}{2}\gamma x^2$ (the harmonic oscillator). As one would expect, the same is true of Eq. (10.5). One may show, by solving Eq. (10.5) and integrating the resulting expression for f over the velocities, that for the *free* Brownian particle

$$f(x,t) = \left(\frac{m\beta^2}{2\pi kT(2\beta t - 3 + 4e^{-\beta t} - e^{-2\beta t})}\right)^{1/2}$$

$$\times \exp\left\{-\frac{m\beta^2}{2kT}\frac{[x - x_0 - v_0(1 - e^{-\beta t})/\beta]^2}{2\beta t - 3 + 4e^{-\beta t} - e^{-2\beta t}}\right\} \tag{10.7}$$

where

$$\beta = \frac{\zeta}{m}$$

For $t \gg \beta^{-1}$ this equation becomes

$$f(x,t|x_0,0) = \left(\frac{m\beta}{4\pi kTt}\right)^{1/2}\exp\left[\frac{-m\beta(x - x_0)^2}{4kTt}\right] \tag{10.8}$$

which is the result of Einstein, whereas as $t \to 0$ it becomes the Dirac delta function

$$\delta(x - x_0)$$

Equation (10.7) is derived by first solving the Kramers equation when $V = 0$ and then integrating the resulting expression over the velocities to get the exact configuration-space distribution function. The word *exact* is used in the sense that inertial effects are *fully included* in the calculation. It is not in general possible to write down a diffusion equation that includes inertial effects exactly for the distribution function in configuration space *alone*.

D.   Relation of the Kramers Equation to the Smoluchowski Equation

Kramers[11] and Chandrasekhar[11] have shown how the Kramers equation may be related to the Smoluchowski equation. It is instructive to reproduce their argument here. We rewrite the Kramers equation as

$$\frac{\partial f}{\partial t} = \beta \left(\frac{\partial}{\partial v} - \frac{1}{\beta} \frac{\partial}{\partial x} \right) \left(fv + \frac{kT}{m} \frac{\partial f}{\partial v} - \frac{\kappa}{m\beta} f + \frac{kT}{m\beta} \frac{\partial f}{\partial x} \right)$$
$$+ \frac{\partial}{\partial x} \left(\frac{kT}{m\beta} \frac{\partial f}{\partial x} - \frac{\kappa}{m\beta} f \right) \tag{10.9}$$

where the force κ is given by

$$\kappa = - \frac{\partial V}{\partial x} \tag{10.10}$$

and

$$\beta = \frac{\zeta}{m} \tag{10.11}$$

Let us now integrate this equation along the straight line

$$x + \frac{v}{\beta} = \text{constant} = x_0$$

from $v = -\infty$ to $v = \infty$. We find that

$$\frac{\partial}{\partial t} \int_{x + v\beta^{-1} = x_0} f \, dv = \int_{x + v\beta^{-1} = x_0} \frac{\partial}{\partial x} \left(\frac{kT}{m\beta} \frac{\partial f}{\partial x} - \frac{\kappa}{m\beta} f \right) dv \tag{10.12}$$

We now suppose that $\kappa(x)$ does not change appreciably over distances of the order of $\sqrt{kT/m\beta^2}$. Then starting from an arbitrary initial distribution $W(x, v, 0)$ at time $t = 0$ we expect that a Maxwellian distribution of velocities will be established at all points after time intervals $\Delta t \gg \beta^{-1}$. If we are not interested in time intervals of the order of β^{-1}, we can write

$$f(x, v, t) = \left(\frac{m}{2\pi kT} \right)^{1/2} \exp\left(-\frac{mv^2}{2kT} \right) f_c(x, t) \tag{10.13}$$

Within these assumptions Eq. (10.12) becomes

$$\frac{\partial f_c(x, t)}{\partial t} = \frac{\partial}{\partial x} \left\{ \frac{kT}{m\beta} \frac{\partial f_c}{\partial x} - \frac{\kappa(x) f_c}{m\beta} \right\} \tag{10.14}$$

In writing down Eq. (10.14) we have supposed, following Chandrasekhar,[11] that in the domain of v from which the dominant contribution to the integral on the right-hand side of Eq. (10.12) arises (namely $|v| \leq (kT/m)^{1/2}$), the *variation* of x (which is of the order $|v|/\beta \approx (kT/m\beta^2)^{1/2}$) is *small* compared with the distances in configuration space, in which κ and f_c change appreciably. The diffusion equation is therefore

$$\frac{\partial f(x,t)}{\partial t} = \frac{\partial}{\partial x}\left(\frac{kT}{m\beta}\frac{\partial f}{\partial x} - \frac{\kappa}{m\beta}f\right) \tag{10.15}$$

(For convenience we have written $f_c = f$.) Equation (10.15) is the Smoluchowski equation. In our former notation it reads

$$\frac{\partial f(x,t)}{\partial t} = \frac{\partial}{\partial x}\left(\frac{kT}{\zeta}\frac{\partial f}{\partial x} + \frac{\partial V}{\partial x}\frac{f}{\zeta}\right) \tag{10.16}$$

Thus only when the conditions used to derive Eq. (10.16) apply, that is, when *inertial effects are ignored*, is there a simple differential equation for the distribution function in configuration space.

E. The Fokker–Planck Equation

Only in the special case of the *free* Brownian particle is it possible to write down a partial differential equation giving the evolution of the distribution function in velocity space *alone*. This is the Fokker–Planck equation,

$$\frac{\partial f}{\partial t} = \beta\frac{\partial}{\partial v}\left(fv + \frac{kT}{m}\frac{\partial f}{\partial v}\right) \tag{10.17}$$

where $f = f(v,t)$. Equation (10.17) corresponds to the Langevin equation

$$m\frac{dv}{dt} + m\beta v = F(t) \tag{10.18}$$

The fundamental solution of this is

$$f(v,t|v_0,0) = \left[\frac{m}{2\pi kT(1-e^{-2\beta t})}\right]^{1/2}$$

$$\times \exp\left[\frac{-m\left(v - v_0 e^{-\beta t}\right)^2}{2kT(1-e^{-2\beta t})}\right] \tag{10.19}$$

For time intervals much greater than β^{-1} this becomes the Maxwellian distribution

$$f(v,t|v_0,0) = \left(\frac{m}{2\pi kT}\right)^{1/2} \exp\left(\frac{-mv^2}{2kT}\right) \qquad (10.20)$$

which is *independent* of the initial value v_0 of v.

F. Summary of the Various Probability Density Diffusion Equations of the Theory

1.   The detailed derivation of the probability density diffusion equations described above will be given later. For the sake of generality we shall in what follows write the equations for motion in three dimensions. The position of a particle in phase space is denoted by the coordinates $\mathbf{r}$ and $\mathbf{v}$. In all cases the distribution function will be denoted by the letter f. The particular space in which f is defined will be indicated by writing equations such as $f = f(\mathbf{v}, t)$.

2.   We shall term the equation

$$\frac{\partial f}{\partial t} + \mathbf{v} \cdot \mathrm{grad}_r f - \frac{1}{m} \mathrm{grad}_v f \cdot \mathrm{grad}_r V = \beta\left(\mathrm{div}_v(\mathbf{v}f) + \frac{kT}{m}\nabla_v^2 f\right) \quad (10.21)$$

(where $f = f(\mathbf{r}, \mathbf{v}, t)$) the *Kramers equation*. All the information concerning the average behavior of a Brownian particle may be obtained by solving this equation. It corresponds to the Langevin equation

$$m\frac{d\mathbf{v}}{dt} + m\beta\mathbf{v}(t) + \mathrm{grad}_r V(\mathbf{r}) = \mathbf{F}(t) \qquad (10.22)$$

The Kramers equation is a partial differential equation in phase space. The configuration-space distribution function is gotten by integrating the phase-space distribution over the velocities.

3.   If the force $-\mathrm{grad}_r V(\mathbf{r})$ does not change appreciably over distances of the order of $(kT/m\beta^2)^{1/2}$ and if we are only interested in times $t \gg \beta^{-1}$ (that is, if inertial effects are not included), we may write down the equation giving the evolution of the distribution function in configuration space only. This is the *Smoluchowski equation* ($f = f(\mathbf{r}, t)$):

$$\frac{\partial f}{\partial t} = \mathrm{div}_r\left(\frac{kT}{m\beta}\mathrm{grad}_r f + \frac{f}{m\beta}\mathrm{grad}_r V\right) \qquad (10.23)$$

4. For a free Brownian particle only, one may write the diffusion equation in velocity space. It is the Fokker–Planck equation ($f = f(\mathbf{v}, t)$):

$$\frac{\partial f}{\partial t} = \beta \left(\text{div}_v(\mathbf{v}f) + \frac{kT}{m} \nabla_v^2 f \right) \tag{10.24}*$$

5. Closed-form solutions of the Kramers equation appear to be available only for the free particle and for the harmonic oscillator.

6. The Kramers equation is a special case of Boltzmann's equation in which the effect of collisions is represented by the term

$$\beta \left(\text{div}_v(\mathbf{v}f) + \frac{kT}{m} \nabla_v^2 f \right) \tag{10.25}$$

XI. SOLUTION OF THE KRAMERS EQUATION WHEN V IS NOT HARMONIC

The origins of this problem lie mainly in the 1940 paper of Kramers,[11] who, as we remarked earlier, was concerned with the passage of particles over potential barriers as a result of the shuttling action of the Brownian movement. Thus the following problem was studied [we use here the phraseology of Brinkman (1956)[12]]: "A particle moves in an external field of force but—in addition to this—is subject to the irregular forces of a surrounding medium in thermal equilibrium (Brownian movement). Originally the particle is caught in a potential hole but it may escape in the course of time by passing over a potential barrier.... Kramers calculated the probability of escape under the assumption of a stationary diffusion current and obtained the result that for such a situation the transition state method (as it is called) has a limited range of applicability." Thus the Kramers method of dealing with the problem assumes that

$$\frac{\partial f}{\partial t} = 0$$

in the Kramers equation. This is the so-called quasistationary approximation.

The first investigator to give a method of dealing with the problem when $\partial_t f \neq 0$, so that the full nonequilibrium distribution could be obtained, was Brinkman in 1956. The essence of Brinkman's method was to expand the

---

*If $\partial_t f$ in Eq. 10.24 is replaced by the hydrodynamical derivative $D_t f = \partial_t f + \mathbf{v} \cdot \nabla_\mathbf{r}$, the Kramers equation 10.21 takes on the form of Eq. 10.24 hence the reason for the generic term Fokker-Planck Equations.

velocity-dependent portion of the distribution function in the Kramers equation into a series of Hermite-like functions (more precisely, the Weber functions[12]). From this he obtained a hierarchy of partial differential–difference equations for the space–time-dependence of the distribution functions. In the case where the potential is periodic, both the potential and the spatial part of the equations may be expanded in Fourier series; this leads to a set of ordinary differential–difference equations the solution of which yields the time dependence of the distribution function. The solution of these equations may be reduced to that of

$$\dot{Y} = BY \tag{11.1}$$

where Y is an infinite vector of column vectors and B is an infinite square matrix whose elements are themselves matrices. Brinkman, although the originator of this method, was unable to solve these equations due to the unavailability of powerful digital computers at the time he was writing. These equations now have been solved numerically for several problems.[12] In the case where the potential vanishes (the free particle), the matrix elements of Eq. (11.1) may be found in closed form.

Brinkman's method is described in detail on page 136 of ref. 12. To guide the reader we very briefly sketch his procedure here.

The starting point is the Langevin equation (for rotation in 2 dimensions $m = I$, $x = \theta$)

$$m\frac{d^2x(t)}{dt^2} + \zeta\frac{dx(t)}{dt} + \frac{\partial V(x)}{\partial x} = F(t) \tag{11.2}$$

The Kramers equation corresponding to this is

$$\frac{\partial f}{\partial t} + \dot{x}\frac{\partial f}{\partial x} - \frac{1}{m}\frac{\partial V}{\partial x}\frac{\partial f}{\partial \dot{x}} = \frac{\zeta}{m}\frac{\partial}{\partial \dot{x}}\left(\dot{x}f + \frac{kT}{m}\frac{\partial f}{\partial \dot{x}}\right), \qquad \dot{x} = v \tag{11.3}$$

with the initial condition (for the fundamental solution)

$$f(x, \dot{x}, 0) = \delta(x - x_0)\delta(\dot{x} - \dot{x}_0) \tag{11.4}$$

We now assume that

$$f = \sum_{n=0}^{\infty} e^{-\dot{x}_1^2/4}D_n(\dot{x}_1)\sum_{p=-\infty}^{p=\infty} e^{ipx}A_p^n(t) \tag{11.5}$$

where

$$\dot{x}_1 = \dot{x}\sqrt{\frac{m}{kT}}$$

p and n are integers, and the D_n are the Weber functions.[12]

The coefficients $A_p^n(t)$ yield the time behavior of the system. We find by substituting Eq. (11.5) into Eq. (11.3) that the $A_p^n(t)$ satisfy (with $V(x) = -V_0 \cos x$)*

$$\frac{dA_p^n(t)}{dt} + n\frac{\zeta}{m}A_p^n(t) - \frac{iV_0}{2\sqrt{mkT}}\left[A_{p-1}^{n-1}(t) - A_{p+1}^{n-1}(t)\right]$$

$$+ ip\sqrt{\frac{kT}{m}}\left[(n+1)A_p^{n+1}(t) + A_p^{n-1}(t)\right] = 0 \qquad (11.6)$$

A related set of equations may be derived when the potential is not periodic. The difference is that we can no longer expand the solution in a Fourier series; instead, we must use some other set of orthogonal functions. The most suitable set will be the Hermite polynomials. Such an expansion will then yield a set of equations similar to Eq. (11.6). If the potential is not a simple cosine, but rather is just a periodic function, it can again be expanded in a Fourier series, and again a set of equations similar to (11.6) is obtained.

The set of differential–difference equations represents the equation of motion of an entry A_p^n in a column matrix $(A_p^n(t))_p$. Introducing Kronecker's delta, δ_{pq}, we may rewrite

$$\delta_{pq}\dot{A}_p^n(t) + \frac{n\zeta}{m}\delta_{pq}A_p^n(t) + (n+1)\sqrt{\frac{kT}{m}}\,ip\delta_{pq}A_p^{n+1}(t)$$

$$+ \left[ip\sqrt{\frac{kT}{m}}\delta_{pq} - \frac{iV_0}{2\sqrt{mkT}}(\delta_{p-1q} - \delta_{p+1q})\right]A_q^{n-1}(t) = 0 \quad (11.7)$$

Let us now introduce the matrices

$$\mathbf{I} = (1)_{p \times q} = (\delta_{pq})_{p \times q}$$

$$\mathbf{D} = (D)_{p \times q} = \left(\sqrt{\frac{kT}{m}}\,ip\delta_{pq}\right)$$

$$\hat{\mathbf{D}} = \left(ip\sqrt{\frac{kT}{m}}\delta_{pq} - \frac{iV_0}{2\sqrt{mkT}}(\delta_{p-1q} - \delta_{p+1q})\right)$$

*In many problems the amplitude of the potential may vary with time as in dielectric relaxation in an a.c. field, say $E_0 \cos \omega t$, then V_0 in Eq. 11.6 is replaced by $\mu E_0 \cos \omega t$ where μ is the dipole moment.

These equations yield the matrix equation for the $\mathbf{A}_n$

$$\mathbf{I}\frac{d\mathbf{A}_n}{dt} + n\frac{\zeta}{m}\mathbf{I}\mathbf{A}_n(t) + (n+1)\mathbf{D}\mathbf{I}\mathbf{A}_{n+1}(t) + \hat{\mathbf{D}}\mathbf{I}\mathbf{A}_{n-1}(t) = 0$$

We note that $\mathbf{I},\mathbf{D},\hat{\mathbf{D}}$ are $p \times q$ matrices, while $\mathbf{A}_n$ is a column vector of p rows. Further, $\mathbf{A}_r = 0$; $r < 0$. The formal solution is

$$\mathbf{I}\mathbf{A}_n(t) = \exp\left(\frac{-n\zeta}{m}t\right)\mathbf{I}\mathbf{A}_n(0)$$
$$+ \int_0^t \exp\left[-\frac{n\zeta}{m}(t-u)\right]\left[(n+1)\mathbf{D}\mathbf{A}_{n+1}(u) + \hat{\mathbf{D}}\mathbf{A}_{n-1}(u)\right]du$$

$$(11.8)$$

This set can in theory be solved by iteration using the Picard method; indeed, where $V_0 = 0$, that is, when there is no *explicit* dependence of $\mathbf{A}_n$ on $p-1$ and $p+1$, one may show that

$$\hat{A}_0^1(t) = \exp\left(\frac{-\zeta}{m}t\right), \qquad A_0^1(t) = \frac{A_0^1(t)}{A_1^0(0)}, \text{ etc.}$$

$$\hat{A}_1^0(t) = \exp\left[-\frac{kTm}{\zeta^2}\left(\frac{\zeta}{m}t - 1 + e^{-(\zeta/m)t}\right)\right]$$

which are the classical Ornstein–Uhlenbeck results. Such a course is not open to us here because the iteration procedure when $V_0 \neq 0$ rapidly becomes too complex to be carried out by hand. A better course, and one that gives very considerable insight into the nature of the $\mathbf{A}_n$, is to rearrange Eq. (11.8) to read

$$\begin{bmatrix} \mathbf{I}\dot{\mathbf{A}}_0 \\ \mathbf{I}\dot{\mathbf{A}}_1 \\ \vdots \\ \mathbf{I}\dot{\mathbf{A}}_{n-1} \\ \mathbf{I}\dot{\mathbf{A}}_n \end{bmatrix} = \begin{bmatrix} \mathbf{0} & -\mathbf{D} & \cdot & \cdot & \cdot & \cdot \\ -\hat{\mathbf{D}} & -\frac{\zeta}{m}\mathbf{I} & -2\mathbf{D} & \cdot & \cdot & \cdot \\ \vdots & \vdots & \vdots & \vdots & \vdots & \vdots \\ \cdot & \cdot & \cdot & \cdot & \cdot & \cdot \\ \cdot & \cdot & \cdot & \cdot & \cdot & \cdot \end{bmatrix} \begin{bmatrix} \mathbf{I}\mathbf{A}_0 \\ \mathbf{I}\mathbf{A}_1 \\ \vdots \\ \cdot \\ \cdot \end{bmatrix}$$

or in an obvious notation

$$\dot{\mathbf{Y}} = \mathbf{B}\mathbf{Y}$$

so that

$$\mathbf{Y}(t) = \mathbf{Y}_0 e^{\mathbf{B}t}$$

or

$$\mathbf{Y}(t) = \mathscr{L}^{-1}\{(s\mathbf{I} - \mathbf{B})^{-1}\}\mathbf{Y}_0$$

where $\mathscr{L}$ denotes the Laplace transform. $(s\mathbf{I} - \mathbf{B})^{-1}$ may be obtained to any degree of accuracy by successively limiting the size of the matrix $\mathbf{B}$ above to 2×2, 3×3, and so on. The solution is then gotten by solving the resulting set for $(s\mathbf{I} - \mathbf{B})^{-1}$. In particular, if we have the Maxwell–Boltzmann distribution for $\dot{x}_0$, then $\tilde{\mathbf{A}}_0(s)$ may be expressed as an infinite continued fraction involving the matrices $\mathbf{D}, \hat{\mathbf{D}}$, namely,

$$\tilde{\mathbf{A}}_0(s) = \cfrac{\mathbf{A}_0(0)}{s\mathbf{I} - \cfrac{\mathbf{D}\hat{\mathbf{D}}}{(s+\beta)\mathbf{I} - \cfrac{2\mathbf{D}\hat{\mathbf{D}}}{(s+2\beta)\mathbf{I} - \cfrac{3\mathbf{D}\hat{\mathbf{D}}}{(s+3\beta)\mathbf{I} -}}}}$$

where this function is *understood to mean*

$$\tilde{\mathbf{A}}_0(s) = \Big[s\mathbf{I} - \mathbf{D}\big[(s+\beta)\mathbf{I} - 2\mathbf{D}\big[(s+2\beta)\mathbf{I} - 3\mathbf{D}[(s+3\beta)\mathbf{I} - \cdots]^{-1}$$

$$\times \hat{\mathbf{D}}\big]^{-1}\hat{\mathbf{D}}\big]^{-1}\hat{\mathbf{D}}\Big]^{-1}\mathbf{A}_0(0), \qquad \beta = \frac{\zeta}{m} \qquad (11.9)$$

Before proceeding, we note that this continued fraction may be written in the compact form[22]

$$\tilde{\mathbf{A}}_0(s) = [s\mathbf{I} - \mathbf{K}_0(s)]^{-1}\mathbf{A}_0(0) \qquad (11.10)$$

where

$$\tilde{\mathbf{K}}_0(s) = \mathbf{D}\big[(s+\beta)\mathbf{I} - 2\mathbf{D}\big[(s+2\beta)\mathbf{I} - 3\mathbf{D}[(s+3\beta)\mathbf{I} - \cdots]^{-1}\hat{\mathbf{D}}\big]^{-1}\hat{\mathbf{D}}\big]^{-1}\hat{\mathbf{D}}$$

$$(11.11)$$

Equation (11.10) may be formally inverted into the time domain (using the convolution theorem) to read

$$\frac{d\mathbf{A}_0}{dt} = \int_0^t \mathbf{K}_0(t-\tau)\mathbf{A}_0(\tau)\,d\tau \qquad (11.12)$$

where $K_0(t - \tau)$ is called the *memory function*. This is yet another way (which has become very popular) of expressing the problem of solving the differential–difference Eqs. (11.7) associated with the Kramers equation. For large β, the memory function $\tilde{K}_0(s)$ may be expanded in powers of β^{-1} (assuming s is of the order of β). Up to the order β^{-5}, we have

$$\tilde{K}_0(s) = (s + \beta)^{-1}D\hat{D} + 2(s + \beta)^{-2}(s + 2\beta)^{-1}D^2\hat{D}^2$$
$$+ 6(s + \beta)^{-2}(s + 2\beta)^{-2}(s + 3\beta)^{-1}D^3\hat{D}^3$$
$$+ 4(s + \beta)^{-3}(s + 2\beta)^{-2}D(D\hat{D})^2\hat{D} + O(\beta^{-7}) \quad (11.13)$$

The inverse Laplace transform of this is then the expansion of the memory function:

$$K_0(t) = e^{-\beta t}D\hat{D} + \beta^{-2}(2\beta t e^{-\beta t} - 2e^{-\beta t} + 2e^{-2\beta t})D^2\hat{D}^2$$
$$+ \beta^{-4}\left\{ \left[3\beta t e^{-\beta t} - (15/2)e^{-\beta t} + 6\beta t e^{-2\beta t} + 6e^{-2\beta t} + \tfrac{3}{2}e^{-3\beta t} \right]D^3\hat{D}^3 \right.$$
$$+ \left[2(\beta t)^2 e^{-\beta t} - 8\beta t e^{-\beta t} + 12e^{-\beta t} \right.$$
$$\left. - 4\beta t e^{-2\beta t} - 12e^{-2\beta t} \right]D(D\hat{D})^2\hat{D} \right\} + O(\beta^{-6})$$

This method of solution will not, however, admit of resonance behavior, because it holds only for large β. A much more comprehensive treatment of the connection between memory functions and the Kramers equation is given by Risken.[22] We now return to Eq. (11.9). A detailed discussion of this continued fraction is given in ref. 12 and in the references quoted therein. The continued-fraction solution itself appears to have first been given by Risken and Vollmer.[12] In numerical calculations involving the continued fraction the matrices D and $\hat{D}$ are truncated with matrices of dimension $2Q + 1$ and the continued fraction is truncated after its Nth approximant. In the numerical treatment the numbers N and Q are determined in such a way that a further increase of N and Q does not change the final result with a given accuracy. The eigenvalues of the fraction are found using well-established numerical root-finding techniques. Once this procedure is completed, the inverse Laplace transform of the result then yields the time behavior of the configuration-space distribution function, from which any required ensemble average may be written down.[12]

The continued-fraction formulation of the problem is not confined to the special case where we have a Maxwellian initial distribution of the velocities. One may write down continued-fraction expressions for all the A_n for any *arbitrary* initial distribution (e.g., the delta-function distribution). The calculations, however, become very much more complicated, and unsuitable to an introductory article such as the present one. A preliminary discussion of the difficulties associated with various initial conditions is given by Coffey et al.[12] In all cases the striking feature of the inclusion of a potential-energy term in the Kramers equation is the prediction of a *denumerable* set of relaxation mechanisms for the time behavior of ensemble averages.

The complexity of the solution of the Brinkman equations when the initial velocity distribution is other than Maxwellian is well illustrated by the solution of Eq. (11.6). For an arbitrary initial distribution of velocities when $V_0 = 0$,

$$\tilde{A}_1^0(s) = i\sqrt{\frac{m}{kT}}\left[A_1^0(0)S_1^0(s) + A_1^1(0)S_1^0 S_1^1 + A_1^2(0)S_1^0(s)S_1^1(s)S_1^2(s)...\right]$$

where $S_p^n(s)$ is the continued fraction

$$\frac{-ip\sqrt{kT/m}}{s + n\beta + ip\sqrt{kT/m}\,S_p^{n+1}(s)}$$

For the Maxwellian initial distribution only the term $A_1^0(0)S_1^0(s)$ survives.

The solution of the Kramers equation where all external stimuli have been applied in the infinite past ($t = -\infty$) so that steady conditions have been attained by $t = 0$ is the *stationary* solution. This will consist of a time independent part $f_{st}(x, v)$ and (if the external forces are continually varying with time) a time dependent part $f_t(x, v, t)$. Thus

$$f(x, v, t) = f_{st}(x, v) + f_t(x, v, t)$$

Likewise the Kramers operator $\mathscr{L}$ where

$$\frac{\partial f}{\partial t} = \mathscr{L}f = \left[\mathscr{L}_K + \mathscr{L}_{ext}(t)\right]f \qquad (11.14)$$

has a time dependent and a time independent part.

$\mathscr{L}_{ext}(t)$ arises from the time varying external force. Since

$$\frac{\partial f_{st}}{\partial t} = \mathscr{L}_K f_{st} = 0$$

Eq. (11.14) becomes

$$\frac{\partial f_t}{\partial t} - \mathscr{L}_K f_t = \mathscr{L}_{ext} f_{st} + \mathscr{L}_{ext} f_t \qquad (11.15)$$

Now suppose that the applied time dependent force is small then the last term above is small. Neglecting this the formal solution of Eq. (11.15) is

$$f_t = \int_{-\infty}^{t} e^{\mathscr{L}_K(t-t')} \mathscr{L}_{ext}(t') f_{st}\, dt' \tag{11.16}$$

This is the *linear response* solution. It is of fundamental importance in relaxation theory. Note that this solution is independent of the initial conditions of the problem. Another solution of interest is where an external step force having been applied at $t = -\infty$ is switched off at $t = 0$ so that the interval of solution is now $0 < t < \infty$. The *linear* part of the response following the removal of the force is called the *aftereffect* solution. There is also the solution (again on $0 < t < \infty$) where the Brownian particles have a definite velocity at $t = 0$ (termed the *sharp* initial condition[22]). There the initial velocity distribution is $\delta(v - v_0)$. This is used for calculating velocity correlation functions. Finally there is the *fundamental* solution (Green's function) corresponding to a delta function initial distribution of *both* velocities and configurations. The solution for a Maxwellian initial distribution of velocities may be found by averaging over the Green's function. To illustrate the foregoing concepts we will consider Eq. (11.2) when in addition to the white noise force it is driven by a constant force F_0 and an alternating force $F_1 \cos \omega t$. (For the applications see Section X1, A below). The Brinkman equations are (assuming that x is periodic)

$$\frac{d\mathbf{A}_n(t)}{dt} + \frac{n\zeta}{m} \mathbf{A}_n(t) + (n+1)\mathbf{DA}_{n+1}(t) + \hat{\mathbf{D}}\mathbf{A}_{n-1}(t)$$

$$+ \frac{1}{\sqrt{mkT}} (F_0 \mathbf{I} + F_1 \mathbf{I} \cos \omega t) \mathbf{A}_{n-1}(t) = 0. \tag{11.17}$$

The operators $\mathbf{D}$ and $\hat{\mathbf{D}}$ are as in Eq. (11.8). The solution of most interest is the steady state one where $F_1 \cos \omega t$ has been applied at $t = -\infty$. This solution contains harmonics and subharmonics of the input force. It is difficult to obtain* because of the product term $(F_1 \mathbf{I} \cos \omega t) \mathbf{A}_{n-1}(t)$. The linear response however may be found by solving in accordance with Eq. (11.16), the set $(-\infty < t < \infty)$

$$\frac{d\mathbf{A}_n(t)}{dt} + n\frac{\zeta}{m} \mathbf{A}_n(t) + (n+1)\mathbf{DA}_{n+1}(t)$$

$$+ \left[\hat{\mathbf{D}} + \frac{F_0 \mathbf{I}}{\sqrt{mkT}} \right] \mathbf{A}_{n-1}(t) + \frac{(F_1 \cos \omega t) \mathbf{I} \mathbf{A}_{n-1}^{(0)}(0)}{\sqrt{mkT}} = 0 \tag{11.18}$$

$\mathbf{A}_{n-1}^{(0)}$ is the value of that quantity when $F_1 = 0$. Since steady conditions have

---

*The same considerations apply to dielectric relaxation under the influence of a strong field $E_0 \cos \omega t$. See footnote after Eq. (11.6).

been reached $A_{n-1}^{(0)}$ satisfies

$$\frac{n\zeta}{m}A_n^{(0)}(0)+(n+1)DA_{n+1}^{(0)}(0)+\hat{D}_1A_{n-1}^{(0)}(0)=0 \qquad (11.19)$$

$$\hat{D}_1=\hat{D}+\frac{F_0\mathbf{I}}{\sqrt{mkT}} \qquad (11.20)$$

The Laplace transform of Eq. (11.18) may be expressed as products of matrix continued fractions, thus allowing efficient numerical computation of the $\tilde{A}_n(s)$. The fundamental solution of (11.17) is also of interest. There we set $F_1=0$, have a delta function initial distribution of displacements and velocities, and solve the resulting equation: ($\mathbf{F}_0=F_0\mathbf{I}$ etc)

$$\frac{dA_n(t)}{dt}+\frac{n\zeta}{m}A_n(t)+(n+1)DA_{n+1}(t)+\hat{D}A_{n-1}(t)$$

$$+\frac{1}{\sqrt{mkT}}F_0A_{n-1}(t)=O \qquad (11.21)$$

The Laplace transform of $A_n(t)$ may be expressed as a series of products of matrix continued fractions. If the initial distribution of velocities is Maxwellian, the Laplace transform of Eq. (11.21) is obtained from Eq. (11.11) with $\hat{D}$ replaced by $\hat{D}_1$. The linear response to a step force applied at $t=0$ may also be found from Eq. (11.21) with $A_{n-1}(t)$ in the last term replaced (the solution interval being $-\infty<t<\infty$ so that equilibrium conditions have been attained by $t=0$) by $A_{n-1}^{(0)}(0)$. The superscript 0 denotes* the value of A_{n-1} when $F_0=0$. Risken and Vollmer[12] have calculated the frequency dependent mobility for this case and show that it has two peaks—one at zero frequency arising from the particles that go over the hills of the cosine potential, and the other at $\omega_0\simeq\sqrt{V_0}$ arising from the oscillations of the particles in the wells of the potential.

A. Physical Applications of the Theory of Brownian Movement in a Potential

The physical phenomena to which the theory has been applied are:

1. The current voltage characteristic of the Josephson tunneling junction.[23,24]
2. Dielectric and Kerr-effect relaxation of an assembly of dipolar molecules, including inertial effects and dipole–dipole interactions.[9]
3. The mobility of superionic conductors.[22]
4. Linewidths in nuclear magnetic resonance.[31]
5. Incoherent scattering of slow neutrons.[32]

---

* In this case.

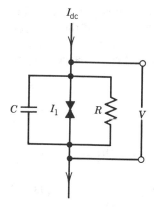

Figure 6.   Equivalent circuit of Josephson junction.

6.   Cycle slips in second-order phase-locked loops.[27]
7.   Quantum noise in ring laser gyros.[25,49,50]
8.   Thermalization of neutrons in a heavy gas moderator.[76]
9.   The photoelectromotive force in semiconductors.[26]
10. Escape of particles over potential barriers.[11,12,27-29]
11. The analytical evaluation of the line shape of single mode semiconductor lasers.[47]
12. Motion of single domain charge-density wave-systems.[48]
13. Light scattering from macromolecules.[39]

The Josephson tunneling junction is made up of two superconductors separated from each other by a thin layer of oxide.[22,23] We call[23] ψ_R and ψ_L pair wave functions for the right and left superconductor, respectively. The phase difference $\phi = \phi_L - \phi_R$ between these wave functions of the Cooper pairs is given by the Josephson equation[23]*

$$\dot{\phi} = \frac{2ev(t)}{\hbar} \tag{11.22}$$

where $v(t)$ is the potential difference across the junction; e is the charge on the electron; and $\hbar = h/2\pi$, where h is the Planck constant (Fig. 6). The junction is now modeled by a resistance R in parallel with a capacitance C across which is connected a dc current generator I_{dc} (representing the bias current applied to the junction). At the other end of the junction (across the resistance R) is connected a phase-dependent current generator, $I_1 \sin \phi$,

---

*In other words $E = \hbar \omega$

representing the Josephson supercurrent due to the Cooper pairs tunneling through the junction (I_1 is the maximum supercurrent). Since the junction operates at a temperature above absolute zero, there will be a noise current $i(t)$ superimposed on the bias current. The circuit thus behaves according to the equation.

$$I_{dc} + i(t) = C\frac{dv(t)}{dt} + Gv(t) + I_1\sin\phi(t) \tag{11.23}$$

where $G = R^{-1}$ and the white-noise current $i(t)$ satisfies the condition

$$\langle i(t_1)i(t_2)\rangle = \frac{2kT}{R}\delta(t_1 - t_2) \tag{11.24}$$

where t_1 and t_2 are distinct times. Using the Josephson equation (11.22), Eq. (11.23) may be cast into the form:

$$\left(\frac{\hbar}{2e}\right)^2 C\ddot{\phi} + \left(\frac{\hbar}{2e}\right)^2\frac{1}{R}\dot{\phi} + \frac{\hbar}{2e}I_1\sin\phi(t) = \frac{\hbar}{2e}(I_{dc} + i) \tag{11.25}$$

Equation (11.25) has the same form as that for a Brownian particle moving under the influence of both a cosine potential and a linear potential. The behavior of the system will be determined by the relative amplitudes of the constant and periodic terms. A more complete discussion of Eq. (11.25) is given in refs. 22 and 23, and in the papers quoted therein. The quantities of physical interest are the mean value of the voltage $\langle v(t)\rangle$ and the bias current I_{dc} as a function of $\langle v(t)\rangle$. In formulating this theory* of the Josephson junction, we have neglected the spatial variation of the phase ϕ. If this is included, the Langevin equation (11.25) becomes a partial differential equation closely related to the Sine–Gordon equation.[23]

We have already seen in Section IX how the mean dipole moment of an assembly of dipolar molecules may be calculated from the theory of Brownian movement via the equation

$$m_E = \mu\langle\cos\theta\rangle$$

The other important phenomenon in dielectrics is the Kerr effect, which is the occurrence of double refraction in a substance when it is placed in an electric field. A substance that exhibits the Kerr effect is ordinarily singly refracting, but becomes doubly refracting when an external electric field is applied. The substance behaves like a uniaxial crystal with the optic axis

---

*The theory is an example of diffusion of phase.

parallel to the electric field. The expression for the optical path difference is[30]

$$\Delta = (n_e - n_o)\ell = KE^2\ell\lambda \qquad (11.26)$$

where n_e is the extraordinary index, n_o the ordinary one, ℓ the path length in the medium, E the magnitude of the electric field, λ the wavelength *in vacuo*, and K the Kerr constant. The effect[30] is caused by either natural or induced optical anisotropy of the individual molecules of the medium and a lining up of the molecules by the applied electric field, and is particularly large for liquids made up of polar molecules with large anisotropies.[30] The *dynamic Kerr effect* is the variation of K with frequency. It is nonlinear in the field strength. In the rotating-sphere model of Debye, the terms in the square of the field strength are those proportional to $\langle P_2(\cos\theta)\rangle$ where P_2 is the second Legendre polynomial. Thus a theory of the dynamic Kerr effect* may be constructed using the Debye model and all its extensions.

The Debye model is again used in nuclear magnetic resonance (NMR). The essential idea is that an assembly of atomic nuclei that have been excited to some nonequilibrium spin distribution will return to thermal equilibrium by a mechanism that involves the coupling of the nuclear spins with their local molecular environment.[31] The rate of the return to equilibrium is characterized by a relaxation time. In spin-lattice relaxation[31] we distinguish between *intra*molecular interactions inside a molecule and *inter*molecular interactions among spins of different molecules. Inside a molecule, according to Abragam,[31] the variation of dipolar coupling of spins arises almost uniquely from the rotation of the molecule. In the interactions between molecules, their relative translations as well as their characteristic rotations must be considered. In the application of the Debye model to the first case, the usual polar angles θ and ϕ define the direction of the proton–proton axis and the probability density function is that of finding this axis in a given orientation at time t. The dipole–dipole coupling between two spins in the molecule may be expressed in terms of the spherical harmonics Y_2^m, where $m = 1, 2$.[31] Then with the aid of the probability function for orientation of spins as rendered by the Debye theory, the correlation functions of the spins may be calculated, and hence the spin-lattice relaxation time and the spectrum. The extensions made to the Debye theory given in refs. 9, 12, 20, and 21 may also be applied to NMR without great difficulty. We have mentioned that in the case of interactions between molecules, their relative translations as well as their characteristic rotations must be considered. If the rotations are neglected, the Smoluchowski equation for three-dimensional translational diffusion is used to calculate the translational relaxation time, except the displacement in the Smoluchowski equation for this problem is

---

*For a detailed review see H. Watanabe, A. Morita, *Adv. in Chem. Phys.* LVI 1984.

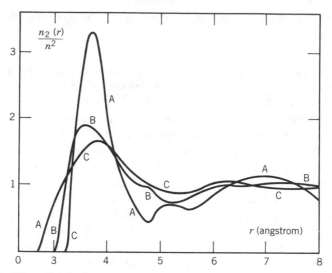

Figure 7.   The radial distribution function $g(r)$ in liquid argon. Curve A: 84.4 K, 0.8 atm. Curve B: 126.7 K, 18.3 atm. Curve C: 149.3 K, 43.8 atm. All are at the saturated vapor pressure, and normalized to unity at great distances. From H. S. Green, *The Molecular Theory of Fluids*, Dover, New York, 1969.

replaced by the distance between two identical molecules that diffuse relative to each other.

The application of the theory to incoherent scattering of slow neutrons is based on the discovery by Van Hove[32] that neutron scattering by an arbitrary system of atoms can be related to a time-displaced distribution function for *pairs* of atoms. The ordinary radial distribution function is a special case of this displacement. Vineyard[32] has suggested a basic approximation by which the time-displaced function can be calculated from the ordinary radial distribution function and a self-diffusion function that describes the wandering away of an atom from an arbitrary initial position. This approximation relates the time-displaced function to the convolution of the self-diffusion function with the radial distribution function. It is called the *convolution approximation*. Using this approximation, the cross section for neutron scattering in an arbitrary direction and with an arbitrary energy is found[32] to be a product of two functions, one expressing the effect of interference and giving the differential cross section* (as determined, for example, by X-rays), the other describing the spread in energy and arising directly from the self-diffusion function. In the present state of the theory of

* With neglect of energy change.

liquids, the self-diffusion function cannot be calculated rigorously. The function may be calculated only for simple models, which is the reason for the application of the theory of Brownian movement to the problem. Using this theory, it is possible to discover simple formulae for the scattering cross sections in the classical limit.

We follow below the discussion of Vineyard.[32] The following distribution functions in a liquid are of interest: $g(\mathbf{r})$, $G(\mathbf{r}, t)$, $G_s(\mathbf{r}, t)$, and $G_d(\mathbf{r}, t)$. The radial distribution function $g(\mathbf{r})$ is the probability that if the atom is at the origin, *another* distinct atom will be found simultaneously within unit volume at $\mathbf{r}$. The time-displaced pair distribution function $G(\mathbf{r}, t)$, introduced by Van Hove, is the probability that if an atom is at the origin at time 0, an atom will also be found within unit volume at $\mathbf{r}$ at time t. In the definition of $g(\mathbf{r})$, the atom at $\mathbf{r}$ *must* be distinct from that at the origin, but this need not be so in $G(\mathbf{r}, t)$. We thus divide $G(\mathbf{r}, t)$ into a *self* and a *distinct* part:

$$G(\mathbf{r}, t) = G_s(\mathbf{r}, t) + G_d(\mathbf{r}, t)$$

$G_s(\mathbf{r}, t)$ is the probability of finding at $\mathbf{r}$, at time t the atom that was at the origin at time 0. $G_d(\mathbf{r}, t)$ is the probability of finding at $\mathbf{r}$ at time t an atom *distinct* from the one that was at the origin at time 0. The convolution approximation[32] is

$$G(\mathbf{r}, t) \approx G_s(\mathbf{r}, t) + \int g(\mathbf{r}') G_s(\mathbf{r} - \mathbf{r}', t) \, d\mathbf{r}'$$

Thus G may be calculated from a knowledge of G_s and g.

The formulae derived by Van Hove for the differential cross sections per atom in the Born approximation for the scattering of neutrons from a system of atoms into a unit solid angle Ω and a unit range of energy ε are

$$\frac{d^2\sigma_{coh}}{d\Omega \, d\varepsilon} = \frac{a_{coh}^2 k}{hk_0} \int \int G(\mathbf{r}, t) e^{i(\mathbf{\kappa} \cdot \mathbf{r} - \omega t)} \, d\mathbf{r} \, dt$$

$$\frac{d^2\sigma_{inc}}{d\Omega \, d\varepsilon} = \frac{a_{inc}^2 k}{hk_0} \int \int G_s(\mathbf{r}, t) e^{i(\mathbf{\kappa} \cdot \mathbf{r} - \omega t)} \, d\mathbf{r} \, dt = \frac{a_{inc}^2 k}{hk_0} \Gamma_s(\mathbf{\kappa}, \omega)$$

The subscript "coh" refers to coherent scattering, the subscript "inc" to incoherent (which arises due to spin or isotope effects[32]); and a_{coh} and a_{inc} are the bound coherent and incoherent scattering lengths per atom. The incident neutrons have wavenumber $\mathbf{k}_0$ and energy ε_0. The outgoing neutrons have wavenumber $\mathbf{k}$ and energy ε. $\mathbf{\kappa}$ is the scattering vector and equals $\mathbf{k}_0 - \mathbf{k}$, $\hbar\omega (= \varepsilon_0 - \varepsilon)$ is the energy loss of the neutron on scattering, and $\Gamma_s(\mathbf{\kappa}, \omega)$ is

W. COFFEY

called the scattering function.* The functions G and G_s are properly defined[32] as averages of certain operators related to the scatterer. The classical definitions of G and G_s as probability density functions follow directly from the operator definitions when classical behavior is attributed to the scatterer. The function

$$F_s(\kappa, t) = \int G_s(\mathbf{r}, t) e^{i(\kappa \cdot \mathbf{r} - \omega t)} \, d\mathbf{r}$$

is also of interest and is referred to as the *intermediate* scattering function. Consider again the functions G_s and G. By definition,

$$G_d(\mathbf{r}, 0) = g(\mathbf{r})$$
$$G_s(\mathbf{r}, 0) = \delta(\mathbf{r})$$

Thus, if the theory of Brownian movement is used to model the scattering process, G_s corresponds to the *fundamental* solution of the Kramers equation when integrated over the velocities. Thus, any results obtained for the fundamental solution of the Kramers equation may be used to model G_s. (For some examples, see refs. 12 and 32.) The velocity autocorrelation function

$$C_v = \frac{\langle \mathbf{v}(0) \cdot \mathbf{v}(t) \rangle}{\langle v^2(0) \rangle}$$

is also of interest in the present case, since computation of it allows the complete determination of G_s provided the underlying Langevin equation is linear with constant coefficients (as in the Ornstein–Uhlenbeck process or the harmonic oscillator).

A superionic conductor[22] is a nearly fixed ion lattice in which some other ions are highly mobile. An example of such a conductor[22] is silver iodide (AgI). The lattice consists of the iodide ions, whereas the silver atoms are highly mobile. If an external field is applied, then the equation of motion of an ion in the periodic potential $V(x)$ provided by the lattice is (in one dimension)

$$m\ddot{x} + \zeta\dot{x} + \partial_x V = F + \lambda(t)$$

* $$\varepsilon_0 = \frac{\hbar^2 k_0^2}{2m}, \quad \varepsilon = \frac{\hbar^2 k^2}{2m}, \quad \kappa = k_0 \left[1 + \left(\frac{\varepsilon}{\varepsilon_0} \right) - 2 \left(\frac{\varepsilon}{\varepsilon_0} \right)^{1/2} \cos\phi \right]^{1/2}$$

m is the mass of the neutron, ϕ is the scattering angle, $k_0 = 2\pi/\lambda_0$, etc.

where λ is a white-noise driving force. The effect of the small lattice vibration terms on the motion of the mobile Ag^+ ions is modeled by the Brownian forces.[22] The frequency- (time-) dependent current is proportional to the velocity correlation function calculated from this equation. (More details are given in ref. 22.)

An ideal phase-locked loop is a circuit in which the phase of an oscillator is made to follow exactly the phase of a reference signal.[22] If noise is present in the circuit, then, the phase of the oscillator is no longer the same. A cycle slip is said to occur if the phase difference is 2π. The cycle-slip rate per second is a measure of the quality of the circuit. One may show[22] that the phase difference between the oscillator and the reference signal satisfies a Langevin equation of the form

$$\tau\ddot{\phi} + \dot{\phi} + \alpha AK \sin\phi = (\omega - \omega_0) - \alpha F(t)$$

where τ, α, A, K are constants; $\omega - \omega_0$ is the detuning (departure from the desired frequency); and $F(t)$ is the noise signal present in the circuit. The quantity we need to calculate is $\langle \dot{\phi} \rangle$ as a function of the detuning.[22]

The use of the theory for the description of quantum noise in ring laser gyros arises from the fact that the *phase difference* between the clockwise and counterclockwise running waves in such a gyro is.[25]

$$\dot{\phi}(t) = a + b\sin\phi(t)$$

where

$$a = -\frac{4}{\lambda L}A \cdot \Omega$$

$|A|$ is the area covered by the optical path of length L along the ring cavity; Ω is the rotation rate of the gyro; $\lambda = \lambda/2\pi$, where λ is the wavelength of the laser; and b is the back-scattering coefficient. The above equation is for the phase difference in the absence of noise. When noise is present, assuming the classical limit where all the quantum noise operators commute, one can easily show that the equation of motion for the phase difference is modified to[25]

$$\dot{\phi} = a + b\sin\phi + F(t)$$

where the noise source $F(t)$ satisfies

$$\langle F(t) \rangle = 0$$
$$\langle F(t)F(t') \rangle = 2D\delta(t - t')$$

The constant D is now

$$\frac{\bar{n}\nu}{2Q}$$

where $\bar{n}$ is the average number of photons in the field at steady state, ν is its frequency, and Q is the quality factor of the laser cavity. The quantity of interest in this application is the beat signal $\cos\phi(t)$, in particular, its spectrum, which may be calculated from the Wiener–Khinchin theorem:

$$\Phi(\omega) = \int_{-\infty}^{\infty} d\tau\, e^{i\omega\tau} \langle \cos\phi(\tau)\cos\phi(0)\rangle$$

Due to the nonlinear nature of the Langevin equation underlying the process, the Fokker–Planck (strictly the Smoluchowski) equation must again be applied in the form

$$\frac{\partial f}{\partial t} = -\frac{\partial}{\partial\phi}[(a + b\sin\phi)f] + D\frac{\partial^2 f}{\partial\phi^2}$$

with the periodic boundary condition

$$f(\phi, t_0|\phi_0, t_0) = \sum_{n=-\infty}^{n=\infty} \delta(\phi - \phi_0 + 2n\pi)$$

Cresser et al.[25] obtained approximate analytical formulae for the spectrum in the cases:

1. $|a| < b$, including $a = 0$.
2. $|a| \gtrsim b$ where distortion of the spectrum due to the sine term continues to be important.
3. $|a| \gg b$, the asymptotic limit in which a is so much larger than b that the $b\sin\phi$ term can be neglected.

In all cases only the weak noise (or strong back-scattering) limit (i.e., $D \ll b$) was considered.

Risken, Haken, and others[22] have also extensively investigated noise in the simple one-mode laser model, in the threshold region (see below) with adiabatic elimination of all variables save the laser field. Their starting point was the Van der Pol equation[22] for a self-sustained oscillator. In such an oscillator, for a small initial value the amplitude first increases exponentially and then oscillates with a finite fixed value. This equation is

$$\ddot{y} - 2\beta(d - y^2)\dot{y} + \omega^2 y = 0$$

(d in this case is called the pumping or control parameter; $d = 0$ is called the threshold region). They then supposed that the amplification is so small that the increase in amplification is small in one period $(1/\omega)$. This allowed them to make the rotating-wave approximation[22]

$$y(t) = b(t)e^{-i\omega t} + b*(t)e^{i\omega t}$$

where the asterisk denotes the complex conjugate. Neglecting terms in $\ddot{b}$ and $\dot{b}$ only, but not in $\omega \dot{b}$, leads to the first-order equation

$$\dot{b}(t) - \beta(d - b*b)b(t) = 0$$

This equation shows that for $d > 0$ a self-sustained oscillation will build up. The noise that arises from spontaneous emission due to the quantum nature of light is not included in this equation. Now, because of the large number (10^3) of photons in the laser cavity, one may neglect the operator character of the electric field[22] provided that an appropriate classical noise source is added. The strength of this source is chosen so as to lead to the correct spontaneous emission rate. One may then show in the classical limit[22] that the amplitude $b(t)$ of the laser field in the presence of noise satisfies the nonlinear Langevin equation

$$\dot{b} - \beta(d - b*b)b = \sqrt{q}\,\Gamma(t)$$

where $\Gamma(t)$ is a white-noise driving force satisfying

$$\langle \Gamma(t) \rangle = 0$$
$$\langle \Gamma(t)\Gamma*(t') \rangle = 4\delta(t - t')$$
$$\langle \Gamma(t)\Gamma(t') \rangle = 0$$

The constant q is chosen so as to give the correct quantum-mechanical spontaneous emission rate (details in ref. 22).

Since b and Γ are complex, we write the laser Langevin equation in the form[22]

$$b = b_1 + ib_2, \qquad \Gamma = \Gamma_1 + i\Gamma_2$$

so that it becomes*

$$\dot{b}_i - \beta(d - b_1^2 - b_2^2)b_i = \sqrt{q}\,\Gamma_i, \qquad i = 1,2, \quad j = 1,2$$
$$\langle \Gamma_i(t)\Gamma_j(t') \rangle = 2\delta_{ij}\delta(t - t')$$

*δ_{ij} denotes the kronecker delta.

The intensity and phase of b are then

$$I = b^*b = b_1^2 + b_2^2$$

$$\phi = \tan^{-1}\left(\frac{b_2}{b_1}\right)$$

The Fokker–Planck equation, usually called the laser Fokker–Planck equation, corresponding to our Langevin equation, is

$$\frac{\partial f}{\partial t} = \left[-\beta \sum_{i=1}^{2} \frac{\partial}{\partial b_i} (d - b_1^2 - b_2^2) b_i + q \sum_{i=1}^{2} \frac{\partial^2}{\partial b_i \, \partial b_i} \right] f$$

This is effectively the Smoluchowski equation for overdamped Brownian motion in the potential (e.g. ref 12 page 78)

$$V(b_1, b_2) = -\tfrac{1}{2}\beta d\left(b_1^2 + b_2^2\right) + \tfrac{1}{4}\beta\left(b_1^2 + b_2^2\right)$$

The parameters β and q are constants for a given laser. The parameter d, on the other hand, is variable and depends on the strength of the pumping ($d < 0$ below, $d = 0$ at, and $d > 0$ above threshold). An extensive account of the solution of the laser Fokker–Planck equation is given in Chapter 12 of ref. 22; our purpose here is merely to outline the applications.

The application of the theory to the thermalization of neutrons has its origin in that, for a heavy-gas moderator, the transport equation for the approach of fast neutrons to thermal equilibrium is[26] (in the absence of absorption and without a source term)

$$\frac{1}{v} \frac{\partial \Phi(E, t)}{\partial t} = \xi \sum_s \left(kTE \frac{\partial^2 \Phi}{\partial E^2} + E \frac{\partial \Phi}{\partial E} + \Phi \right)$$

where Φ is the neutron flux distribution, $E = mv^2/2$ is the energy of other neutrons with mass m and velocity v, t is the time, $\xi\sum_s$ is the slowing-down power, T is the temperature of a moderator, and k is the Boltzmann constant. This equation is based on the assumption[26] that the neutrons change energy only very slowly at a collision. The effect of a unit pulse of neutrons (i.e., a source term) of energy E_0 injected into the heavy-gas moderator at $t = 0$ may be incorporated into the equation by solving it subject to the initial condition[26]

$$\Phi(E, 0) = v\delta(E - E_0)$$

The additional $1/v$ absorption often used in nuclear physics may be taken account of by adding a term

$$-\frac{\lambda}{v}\Phi(E,t)$$

to the right-hand side of our starting equation. Risken and Voigtlaender[26] show, by means of a simple transformation, how the Green's function of the equation including this absorption term may be expressed in terms of the Green's function of the equation without the source term.

An equation similar to that for Φ above arises in the theory of photoelectromotive force in semiconductors.[26] Landau and Lifschitz show that the distribution function f, of particles created by monochromatic light satisfies[26]

$$A\hbar^{-1}8mW^2\sqrt{E}\left[2f+\frac{df}{dE}(2kT+E)+kTE\frac{d^2f}{dE^2}\right]=\lambda(f-f_0)$$

where f_0 is the Maxwell–Boltzmann distribution $\exp(-E/kT)$, λ describes the absorption, and A and W are constants. At energy $E = E_0$ particles are created by illumination with monochromatic light. The ensuing jump in the current is described by a delta-function source term. If we write

$$\xi\sum_s = A\hbar^{-1}8mW^2\sqrt{\frac{m}{kT}}$$

then the photoelectromotive-force equation corresponds to that for the neutron distribution. Risken and Voigtlaender then show that our initial equation for $\Phi(E, t)$ may be transformed into a Fokker–Planck equation. To do this, they introduce the normalized variables

$$\bar{E}=\frac{\bar{v}^2}{2}=\frac{E}{kT}$$

$$\bar{v}=v\sqrt{\frac{m}{kT}}$$

$$\bar{t}=t\xi\sum_s\sqrt{\frac{kT}{m}}$$

and normalize the neutron flux via

$$\int_0^\infty \frac{1}{v}\Phi(E,t)\,dE = \int_0^\infty \Phi(E,t)\,dv = 1$$

(From here on, we drop the bars over E, etc., for simplicity of notation.) The above transformation leads to the following Fokker–Planck equation for the probability density:[26]

$$\frac{\partial n(E,t)}{\partial t} = \left[-\frac{\partial}{\partial E}(2-E)\sqrt{2E}\,E + \frac{\partial^2}{\partial E^2}\sqrt{2E}\,E \right] n(E,t)$$

If we now write $v = \sqrt{2E}$ and note that

$$\frac{dE}{dv} = v$$

we have the following Fokker–Planck equation for the density $W(v,t)$ in $\sqrt{2E} = v$ space:

$$\frac{\partial W}{\partial t} = -\frac{\partial}{\partial v}\left(3 - \frac{v^2}{2}\right)W + \frac{\partial^2}{\partial v^2}\left(\frac{v}{2}W\right)$$

Since $dE/dv = v$, this probability density $W(v,t)$ is identical to the neutron flux density. The Langevin equation corresponding to this Fokker–Planck equation is

$$\dot{v}(t) + \tfrac{1}{2}(3 - v^2) - \tfrac{1}{4} + \sqrt{\frac{v}{2}}\,\Gamma(t) = 0$$

where

$$\langle \Gamma(t) \rangle = 0, \qquad \langle \Gamma(t)\Gamma(t') \rangle = 2\delta(t - t')$$

These equations are the starting point of the investigation of Risken and Voigtlaender.[26] They represent an interesting example of a stochastic differential equation with *variable* coefficients arising from a physical problem. Such equations have been extensively discussed in the mathematical literature (see ref. 33).

Our final physical example is the application of the theory to the escape of particles over potential barriers. This is the famous barrier-crossing problem of Kramers already alluded to. We shall therefore not describe the problem any further here, but refer the reader to Chapter V of ref. 22 and to a recent paper by Landauer et al.[24] in which the theory is applied to the Josephson junction.

XII.   INERTIAL EFFECTS IN THE DEBYE THEORY

The reader has already seen how the theory of Brownian motion was applied by Debye to the study of relaxation of an assembly of dipolar molecules. A brief account of the role of inertia in modifying the Debye theory is instructive here.

The absorption coefficient $A(\omega)$ (nepers cm^{-1}) used in dielectric measurements is[12]

$$A(\omega) = \frac{\omega \varepsilon''(\omega)}{n(\omega)c} \qquad (12.1)$$

where ε'' is the imaginary part of the complex permittivity, c is the velocity of light, and $n(\omega)$ is the refractive index. For the frequencies under consideration (10^{10}–10^{12} Hz), $n(\omega) \simeq 1$, and Eq. (12.1) becomes

$$A(\omega) = \frac{\omega \varepsilon''(\omega)}{c} \qquad (12.2)$$

According to the Debye equations, ε'' varies with frequency essentially as

$$\frac{\omega \tau_0}{1 + \omega^2 \tau_0^2}$$

and thus the absorption coefficient varies as

$$\frac{\omega^2 \tau_0^2}{1 + \omega^2 \tau_0^2} \qquad (12.3)$$

(see Fig. 8). Thus for $\omega \gg 1/\tau_0$ the absorption will be *constant*; that is, it reaches a plateau value. This is completely at variance with experiment, where a large peak is generally observed in the absorption in polar liquids at frequencies in the range 10^{11}–10^{12} Hz, which peak dies away to zero as the frequency is increased. Rocard[12] (1933) correctly surmised that the Debye equations must be incorrect in this high-frequency region because they did not include inertial effects. Including them essentially by adding an extra term to the Smoluchowski equation, he found that the polarizability $\alpha(\omega)$ becomes[20]

$$\frac{\alpha(\omega)}{\alpha'(0)} = \frac{1}{1 + i\omega\tau_0 - \dfrac{\omega^2 I \tau_0}{\zeta}} \qquad (12.4)$$

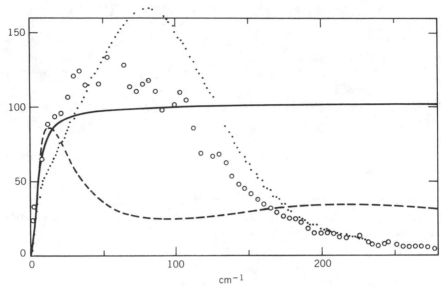

Figure 8.   O: Far-infrared power absorption of liquid CH_2Cl_2, measured with different pieces of equipment based on Michelson interferometry. ●: Computer-simulation result consisting of the numerical inverse Fourier transform of $\langle \dot{\mathbf{e}}_A(t)\cdot\dot{\mathbf{e}}_A(0)\rangle$, using the experimentally measured static permitivity ($\varepsilon_s = 9.08$) to scale the absolute intensity of $A(\bar{\nu})$ (neper cm^{-1}). Solid curve: Debye's theory of rotational diffusion, which produces a plateau in the power absorption. Dashed curve: Morita's theory of asymmetric-top diffusion, which is a generalized treatment of inertial effects with the Rocard equation as a special case[9]. $\mathbf{e}_A$ is a unit vector fixed in the molecule[12] (figure taken from ref. 12).

instead of

$$\frac{\alpha(\omega)}{\alpha'(0)} = \frac{1}{1 + i\omega\tau_0} \tag{12.5}$$

as predicted by the Debye theory when uncorrected for inertia. Equation (12.4) clearly makes the absorption coefficient $A(\omega)$ decrease to zero as $\omega \rightarrow \infty$, in accordance with experiment. (It does not, however, account for the observed resonant peak that may arise from dipole–dipole coupling. Accounting for this peak requires the inclusion of a potential in the Kramers equation.)

Gross[9] (1955) and Sack[9] (1957) undertook a rigorous investigation of inertial effects in the Debye theory based on the Kramers equation. They found that for the disk model of Debye, $\alpha(\omega)$ could be expressed as the infinite continued fraction[9] (Fig. 9)

$$\frac{\alpha(\omega)}{\alpha'(0)} = 1 - \frac{i\omega}{\beta}\left[\frac{1\ |}{|i\omega/\beta} + \frac{\gamma\ |}{|1+i\omega/\beta} + \frac{2\gamma\ |}{|2+i\omega/\beta} + \frac{3\gamma\ |}{|3+i\omega/\beta} + \cdots\right]$$

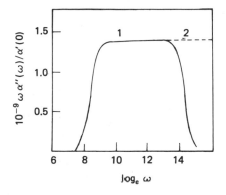

Figure 9. 1. Dashed line, Debye theory (uncorrected); solid line, corrected for inertia. The solid line curve also includes the effect of dipole–dipole coupling when the friction coefficient is large, see ref. 12, page 166.

where

$$\gamma = \frac{kT}{I\beta^2}$$

The first convergent of this yields the Debye relaxation formula

$$\frac{\alpha(\omega)}{\alpha'(0)} = \frac{1}{1 + i\omega\tau_0}$$

where

$$\tau_0 = \frac{\zeta}{kT} = \frac{I\beta}{kT}, \quad \beta = \frac{\zeta}{I}$$

while the second convergent yields Rocard's equation

$$\frac{\alpha(\omega)}{\alpha'(0)} = \frac{1}{1 + i\omega\tau_0 - \omega^2\tau_0/\beta}$$

Similar conclusions may be drawn for the 3D models. Thus the neglect of inertial effects renders the Debye theory of dielectric relaxation incorrect in the far-infrared band of frequencies. If inertial effects only are included, the opacity predicted by the Debye equations is removed and a return to transparency is achieved.

Recall that the complex polarizability of a macroscopic dielectric sphere is[12]

$$\left(\alpha(\omega) = \alpha'(\omega) - i\alpha'(\omega) = \frac{3v\varepsilon_0[\varepsilon(\omega) - 1]}{\varepsilon(\omega) + 2}, \quad \varepsilon(\omega) + 2 \approx 3\right)$$

However, the observed peak in the far infrared (termed the Poley absorption) can be explained only by including a potential-energy term in the

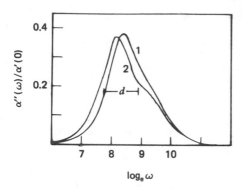

Figure 10. Broadening of microwave absorption caused by inclusion of dipole–dipole coupling as in Budó's calculation, d = Debye half-width. Normalized friction coefficient $\zeta = 0.5$ curve 1, 0.125 curve 2. (Potential strength)$/kT = 2.0$ in both cases.

Kramers equation to take account of molecular interactions. We next sketch briefly the results of including such a potential.

XIII.   INERTIAL EFFECTS AND BROWNIAN MOVEMENT IN A POTENTIAL

Acting on some earlier work of Hill[9,12] (1963) and Wyllie[9,12] (1971),* Calderwood and Coffey[12] proposed a simple model to explain the observed microwave and far-infrared dielectric absorptions of polar fluids. The real origin of this model lay in the suggestion of Poley[12] (1955) that the two dispersion regions may be explained by considering how the free rotation of a molecule is modified by an external mechanical potential. For example, if the strength of the potential is small in comparison with kT, the behavior of the molecule will be like that for free rotation, whereas if the potential is large compared with kT, the molecule will execute only small oscillations about an equilibrium position determined by the potential. The model proposed by Calderwood and Coffey is the itinerant oscillator[12] (see Fig. 14).

The parameters in the itinerant-oscillator model were originally somewhat arbitrarily defined; a more natural way of including the effect of an external potential on the relaxation process is simply to incorporate the electric dipole–dipole coupling directly in the equations of motion of the dipoles. This was first done by Budó[12] (1949), but he did not include the inertia of the rotating dipoles, so that his results hold only at low frequencies, that is, in the microwave region. He was able to show that the effect of including the coupling is to give rise to a *discrete set* of Debye-type relaxation mechanisms in the microwave region, rather than just the *single* mechanism predicted by the simple Debye theory. He obtained his result by writing down

*and in accordance with a suggestion of Scaife.

the Smoluchowski equation for the probability density of the reorienting dipoles under the influence of a time-varying electric field. He was then able to reduce the task of calculating the dipole moment to a Sturm–Liouville problem. The eigenvalues of the Sturm–Liouville equation yielded the relaxation times, while the eigenfunctions gave the magnitude of each component relaxation mechanism. The overall effect of the inclusion of the dipole coupling is to broaden the microwave absorption peak significantly from its zero coupling value. For two reorienting dipoles, the dipole-interaction potential will always have a cosine form. This implies that the Sturm–Liouville problem reduces to considering an equation of Hill's type.[12] In many cases, however, the potential height will be large compared with kT. The potential is then well approximated by a parabola and the Sturm–Liouville equation becomes Hermite's equation. The eigenvalues may be found in closed form. Indeed, closed-form expressions may be written down for the relaxation functions.[12] Here we shall first give the results of the *inclusion of inertial effects* in the *harmonic* approximation, and then briefly indicate the effect of including the *full* cosine potential. In order to clarify our argument we must say something about Onsager's 1936 method[9] of calculating the static permittivity of a polar fluid, which incorporates the effects of the long-range dipole–dipole coupling between the molecules.

Onsager's model[9] of a dipolar molecule in a dielectric is a point dipole situated at the center of an empty spherical cavity in a continuous dielectric with permittivity equal to the bulk permittivity ε_s (the permittivity at zero frequency) of the dielectric (Fig. 11). The volume of the cavity is the volume available to each molecule. The field of the dipole μ in the cavity polarizes its surroundings. The resulting polarization of the surroundings then induces a homogeneous field in the cavity, which is called the reaction field $\mathbf{R}$. Further, if a macroscopic negative potential gradient $\mathbf{E}$ is imposed on the dielectric by external sources, then a calculation in electrostatics shows that the field $\mathbf{G}$ in an empty cavity in the dielectric will not be equal to $\mathbf{E}$. This field $\mathbf{G}$ is called the cavity field. As in Evans et al.,[9] it is convenient to imagine

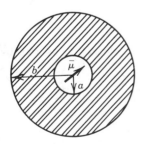

Figure 11. Onsager's model.

the dipole with its cavity of radius a placed at the center of a very large dielectric sphere of outer radius b and relative permittivity ε_s. (In this preliminary discussion we shall suppose that the dipole μ is rigid, that is, that displacement polarization is ignored.) The field within the cavity of such a sphere, when subjected to both a uniform external field $\mathscr{E}_0$ parallel to the z-axis and to the field of μ situated at the center of the cavity and making an angle ϑ with the z-axis, is*

$$\mathbf{F} = \mathbf{G} + \mathbf{R} = \frac{9\varepsilon_s \mathscr{E}_0 c}{(2\varepsilon_s + 1)(\varepsilon_s + 2)} + \frac{2(\varepsilon_s - 1)\mu c}{(2\varepsilon_s + 1)4\pi\varepsilon_0 a^3}$$

The field of μ itself is denoted by $\mathbf{D}(\mu)$. Note that

$$\mathbf{E} = \frac{3\mathscr{E}_0}{\varepsilon_s + 2}$$

These formulas will still hold in the dynamic case where ε_s is replaced by the frequency-dependent permittivity $\varepsilon(\omega)$, provided all wavelengths under consideration are much longer than the size of the sphere. The energy of the interaction between the dipole μ and $\mathscr{E}_0$ is

$$V = \frac{-9\varepsilon_s \mu \cdot \mathscr{E}_0}{(\varepsilon_s + 2)(2\varepsilon_s + 1)} + \text{terms independent of orientation}$$

One may then show that ε_s satisfies the equation

$$(\varepsilon_s - 1) = \frac{3\varepsilon_s}{2\varepsilon_s + 1} \frac{N}{v\varepsilon_0} \frac{\mu^2}{3kT}$$

where v is defined by

$$\frac{4}{3}\pi a^3 n = \frac{N}{v}$$

and n is the molecular number density.

Consider now the frequency-dependent case. The field acting on the dipole at any time t is

$$\mathbf{F}(t) = \mathbf{G}(t) + \mathbf{R}(t)$$

*c is a factor involving the ratio a^3/b^3, see ref. 9.

where $G(t)$ and $R(t)$ are the time-dependent cavity and reaction fields, respectively. The Langevin equation for the motion of μ considered as a sphere is

$$I\frac{d\omega(t)}{dt} + \zeta\omega(t) - \mu \times F(t) = \lambda(t)$$

where, as usual, $\lambda(t)$ is the white-noise driving torque arising from Brownian movement, and I is the moment of inertia of μ. In view of the formidable calculations associated with the spherical model[9,20] even in the absence of dipole–dipole coupling, we specialize the three-dimensional equation above to the case of the two-dimensional rotator, so that it reduces to

$$I\ddot{\theta} + \zeta\dot{\theta} - \mu F(t)\sin(\psi - \theta) = \lambda(t)$$

We will further suppose that the field $\mathscr{E}_0$ is switched off at time $t = 0$, so that

$$I\ddot{\theta} + \zeta\dot{\theta} - \mu R(t)\sin(\psi - \theta) = \lambda(t) \tag{13.1}$$

where θ is the angle between μ and a particular direction (which we will take as the direction $\mathscr{E}_0$ had before it was switched off), while ψ is the angle between the reaction field and this direction. The direction of R differs from that of μ because the surrounding medium is not loss-free. Solution of Eq. (13.1) presents formidable mathematical difficulties, mainly arising from the fluctuating field $R(t)$ and the lack of any knowledge of the time variation of $\psi(t)$. The result is that it is extremely difficult, if not impossible, to generalize Onsager's model to the dynamic case.

It may be possible to regard the two-rotating-dipole model of Budó as a very rough approximation to the behavior of the dipole in the cavity. The Budó model supposes that the dynamics of ψ are simply those of one other dipole that rotates about the same axis as the dipole under consideration. We denote the two dipoles by μ_1 and μ_2. We suppose that a steady field E has been applied for a long time in the same plane as that of the rotation of the dipoles. The dipoles are compelled to rotate about an axis through their common center and normal to the plane containing E. The equations of motion before removal of E are

$$I\ddot{\phi}_1 + \zeta\dot{\phi}_1 + V'(\phi_1 - \phi_2) + \mu_1 E \sin\phi_1 = \lambda_1(t)$$
$$I\ddot{\phi}_2 + \zeta\dot{\phi}_2 - V'(\phi_1 - \phi_2) + \mu_2 E \sin\phi_2 = \lambda_2(t)$$

where ϕ_1 and ϕ_2 are the angles μ_1 and μ_2 make with e, the direction of E at any time, and I is the moment of inertia of a dipole (the dipoles are assumed to be of equal size), λ_1 and λ_2 are the white-noise driving torques acting on both dipoles, and V is the potential energy of interaction (the primes denote differentiation of V with respect to its argument). In general

$$V(\phi_1 - \phi_2) = -V_0 \cos(\phi_1 - \phi_2)$$

It is convenient to rewrite these equations in the form (with $\mathbf{E}$ having been switched off at a time $t = 0$, so that we can get the decay of the polarization or mean dipole moment per unit volume)

$$\ddot{\phi}_1(t) + \frac{\zeta}{I}\dot{\phi}_1(t) + \frac{V'(\phi_1 - \phi_2)}{I} = \frac{\lambda_1(t)}{I} \tag{13.2}$$

$$\ddot{\phi}_2(t) + \frac{\zeta}{I}\dot{\phi}_2(t) - \frac{V'(\phi_1 - \phi_2)}{I} = \frac{\lambda_2(t)}{I} \tag{13.3}$$

The Kramers equation[12] corresponding to these Langevin equations is*

$$\frac{\partial W}{\partial t} + \dot{\phi}_1 \frac{\partial W}{\partial \phi_1} + \dot{\phi}_2 \frac{\partial W}{\partial \phi_2} - \frac{1}{I}\left(\frac{\partial W}{\partial \dot{\phi}_1}\frac{\partial V}{\partial \phi_1} + \frac{\partial W}{\partial \dot{\phi}_2}\frac{\partial V}{\partial \phi_2}\right)$$

$$= \frac{\zeta}{I}\left[\frac{\partial}{\partial \dot{\phi}_1}(\dot{\phi}_1 W) + \frac{\partial}{\partial \dot{\phi}_2}(\dot{\phi}_2 W) + \frac{kT}{I}\left(\frac{\partial^2 W}{\partial \dot{\phi}_1^2} + \frac{\partial^2 W}{\partial \dot{\phi}_2^2}\right)\right] \tag{13.4}$$

where $W(\phi_1, \phi_2, \dot{\phi}_1, \dot{\phi}_2, t)$ is the distribution function in configuration–angular-velocity space. Let us first suppose that the cosine potential above is replaced by its harmonic approximation (corresponding to the case of very strong dipole–dipole coupling). Then we may show, by solving either Eq. (13.4) or the Langevin equations (13.2) and (13.3), that the mean dipole moment in the direction the field had is[34]

$$\langle \mathbf{m} \cdot \mathbf{e} \rangle = \langle \mu_1 \cos \phi_1 + \mu_2 \cos \phi_2 \rangle = \mu \langle \cos \phi_1 + \cos \phi_2 \rangle$$

$$= \frac{2\mu^2 E}{kT} \exp\left[-\frac{\gamma}{2}y(t)\right] \exp(-\gamma_1)\cosh[\gamma_1 x(t)], \qquad t > 0 \tag{13.5}$$

where

$$y(t) = (\beta t - 1 + e^{-\beta t}), \qquad \beta = \frac{\zeta}{I}$$

$$\gamma = \frac{kT}{I\beta^2}$$

$$x(t) = \exp\left(-\frac{\beta t}{2}\right)\left(\cos \omega_1 t + \frac{\beta}{2\omega_1}\sin \omega_1 t\right)$$

$$\gamma_1 = \frac{kT}{8V_0}$$

*If the stochastic torques are weak compared to the deterministic ones these equations may be directly solved in terms of the Jacobian elliptic functions. The differential equation for half the sum of ϕ_1 and ϕ_2 is $I\ddot{\chi} = 0$, that for half the difference $I\ddot{\eta} + V_0 \sin \eta = 0$.

and

$$\omega_1 = \left(\frac{4V_0}{I} - \frac{\zeta^2}{4I^2} \right)^{1/2} = \left(\omega_0^2 - \frac{\beta^2}{4} \right)^{1/2}, \qquad \omega_0^2 > \frac{\beta^2}{4}$$

For the Debye disk model corrected for inertia, the corresponding expression for $\langle \mathbf{m} \cdot \mathbf{e} \rangle$, from which Eq. (12.4) is directly derived, is

$$\langle \mathbf{m} \cdot \mathbf{e} \rangle = \langle \mu \cos \theta \rangle = \frac{\mu^2 E}{2kT} \exp[-\gamma y(t)]$$

where θ is the angle μ makes with the field direction. We have written Eq. (13.5) for the underdamped motion where $4\omega_0^2 > \beta^2$. The corresponding expressions for the aperiodic and overdamped cases may also be written down, but these are not of as great physical interest as the underdamped case.*

We now note that Eq. (13.5), because it arises from a *two*-particle model, contains contributions to the polarization *both* from each dipole separately and from the *two acting together*. These are the *auto-* and *cross-correlation* function contributions to the polarization. The contribution of the autocorrelation-function portion to Eq. (13.5) is found by picking out the terms in that equation that reduce to μ^2/kT at time $t = 0$. Thus the autocorrelation function is (remember that $x(0) = 1$)

$$\frac{\mu^2 E}{kT} \exp\left[-\frac{\gamma}{2} y(t) \right] \exp\{ -\gamma_1 [1 - x(t)] \}$$

or

$$\frac{\mu^2 E}{2kT} \exp\left[-\frac{\gamma}{2} y(t) \right] \exp\{ -\gamma_1 [1 - x(t)] \} \qquad (13.6)$$

per dipole, while the cross-correlation function is

$$\frac{\mu^2 E}{kT} \exp\left[-\frac{\gamma}{2} y(t) \right] \exp\{ -\gamma_1 [1 + x(t)] \}$$

or

$$\frac{\mu^2 E}{2kT} \exp\left[-\frac{\gamma}{2} y(t) \right] \exp\{ -\gamma_1 [1 + x(t)] \} \qquad (13.7)$$

per dipole.

The behaviors of these two functions as a function of their arguments are interesting. The function $\exp[-(\gamma/2)y(t)]$ is essentially a *monotonically* decreasing function of its argument $(\gamma/2)y(t)$; in contrast, the argument of the exponential involving γ_1 can pass through maxima and minima. To see this,

---

*Note the potential energy in our model is $4V_0\eta^2$ following the value adopted by Budó for the elastic bond case. For small oscillations in the cosine potential case the potential energy is $V_0(1-\eta^2)$.

note that the argument $\gamma_1[1 - x(t)]$ of the second exponential in Eq. (13.6) is

$$\gamma_1\left[1 - \exp\left(-\frac{\beta t}{2}\right)\left(\cos \omega_1 t + \frac{\beta}{2\omega_1}\sin \omega_1 t\right)\right] \quad (13.8)$$

This is the usual expression for the response of a linear second-order differential equation with constant coefficients to a step input. This function has a maximum value at a time*

$$t_m = \frac{\pi}{\omega_1}$$

and the value of this maximum is

$$\gamma_1\left[1 + \exp\left(-\frac{\beta\pi}{2\omega_1}\right)\right] \quad (13.9)$$

When this maximum in the argument of the exponential occurs, the exponential itself passes through a minimum. This may explain the dips (glitches) in the correlation functions of orientation that are often observed in numerical computations.[35] These undulations will be observable only in the extremely underdamped case.

The above discussion applies to the autocorrelation function. For the cross-correlation function, on the other hand, the argument of the exponential is $(1 + x)$ instead of $(1 - x)$. Unlike the autocorrelation function, this function does not exhibit a maximum or overshoot.

Equation (13.5) has several well-known results as limiting cases. Let us consider the effects of varying the moment of inertia I and the potential strength V_0. Suppose V_0 becomes very strong. Then we find that Eq. (13.5) reduces to

$$\frac{\mu^2 E}{2kT}\exp\left[-\frac{\gamma}{2}y(t)\right] = \frac{\mu^2 E}{2kT}\exp\left[-\frac{\gamma}{2}(\beta t - 1 + e^{-\beta t})\right] \quad (13.10)$$

This is the same as the correlation function of a freely rotating dipole but with an extra factor of $\frac{1}{2}$ in the argument of the exponential. Let us now consider the opposite situation, where $V_0 \to 0$. We then find that Eq. (13.5) reduces to

$$\frac{\mu^2 E}{2kT}\exp\left[-\gamma(\beta t - 1 + e^{-\beta t})\right] \quad (13.11)$$

---

*The dependence of frequency on amplitude due to the cosine potential should cause this time to lengthen (see footnote to page 163 below). This dependence also dramatically alters the decay patterns of the various correlation functions from those predicted by the harmonic approxiamation, v. ref 12 chapter 9.

Thus when the dipoles rotate without any interaction, the relaxation is *twice* as fast as that when the dipoles are locked together. Suppose we let $I \to 0$ in both these equations. Then Eq. (13.10) becomes

$$\langle \mathbf{m} \cdot \mathbf{e} \rangle = \frac{\mu^2 E}{2kT} \exp\left(- \frac{kTt}{2I\beta} \right) \tag{13.12}$$

and Eq. (13.11) becomes

$$\langle \mathbf{m} \cdot \mathbf{e} \rangle = \frac{\mu^2 E}{2kT} \exp\left(- \frac{kTt}{I\beta} \right) \tag{13.13}$$

These equations represent pure Debye-type relaxation mechanisms. Equation (13.12) has a relaxation time

$$\tau_h = \frac{2I\beta}{kT} = \frac{2\zeta}{kT}$$

while Eq. (13.13) has a relaxation time

$$\tau_f = \frac{\zeta}{kT}$$

τ_h is the relaxation time for *pure hindered rotation*, that is, for when the dipoles lock together (very strong interaction); τ_f is the relaxation time for *free rotation*, that is, for when there is no interaction between the dipoles. The two relaxation times *differ* by a factor of 2. This behavior was first noted by Budó[12] in his analysis of dielectric relaxation of an assembly of molecules containing rotating polar groups.*

Another limiting case is that where V_0 is held fixed and the inertia is set equal to zero. We then find that

$$\langle \mathbf{m} \cdot \mathbf{e} \rangle = \frac{\mu^2 E}{kT} \left\{ \exp\left(- \left\{ \frac{kTt}{2\zeta} + \frac{kT}{8V_0}[1 - e^{-(4V_0/\zeta)t}] \right\} \right) \right.$$
$$\left. + \exp\left(- \left\{ \frac{kTt}{2\zeta} + \frac{kT}{8V_0}[1 + e^{-(4V_0/\zeta)t}] \right\} \right) \right\}$$

which is the result obtained by Coffey[12]

---

*Budó shows for a cosine potential that as the interaction energy is increased from zero, the polarizability at first has a discrete set of relaxation mechanisms which become a single one as V_0 gets large.

In analyzing the behavior of these correlation functions it is often useful to normalize them by their initial value. The normalized value of Eq. (13.5) is

$$R(t) = \frac{\langle \mathbf{m}(t) \cdot \mathbf{e} \rangle}{\langle \mathbf{m}(0) \cdot \mathbf{e} \rangle} = \frac{\exp\left[-\frac{\gamma}{2}y(t)\right]\cosh[\gamma_1 x(t)]}{\cosh\gamma_1} \tag{13.14}$$

The normalized autocorrelation function is, from Eq. (13.6),

$$\exp\left[-\frac{\gamma}{2}y(t)\right]\exp\{-\gamma_1[1-x(t)]\} = \rho_{11}(t) \tag{13.15}$$

The normalized cross-correlation function is

$$\frac{\exp\left[-\frac{\gamma}{2}y(t)\right]\exp\{-\gamma_1[1+x(t)]\}}{\exp(-2\gamma_1)} = \exp\left[-\frac{\gamma}{2}y(t)\right]\exp\{\gamma_1[1-x(t)]\}$$

$$= \rho_{12}(t) \tag{13.16}$$

Note that from the behavior of the function $[1 - x(t)]$ discussed earlier, one may immediately deduce that ρ_{12} goes through a maximum value when ρ_{11} goes through a minimum. Note also that

$$\rho_{11}(t)\exp\frac{\gamma}{2}y(t) = \exp\{-\gamma_1[1-x(t)]\}$$

$$\rho_{12}(t)\exp\frac{\gamma}{2}y(t) = \exp\{\gamma_1[1-x(t)]\}$$

More insight into the behavior of all these equations may be gained by expanding them in powers of the parameters γ_1 and γ. In general (because we are assuming that the potential may be approximated by a parabola) the condition $V_0 \gg kT$ will be satisfied. Thus by Taylor's theorem

$$R(t) \approx \exp\left[-\frac{\gamma}{2}y(t)\right]\left\{1 - \frac{\gamma_1^2}{2}[1-x^2(t)]\right\}$$

This will provide a reasonable approximation of $R(t)$.

$$x^2(t) = \frac{e^{-\beta t}}{2}\left\{\left(1 + \frac{\beta^2}{4\omega_1^2}\right) + \left(1 - \frac{\beta^2}{4\omega_1^2}\right)\left[\cos 2\omega_1 t + \frac{\beta}{\omega_1}\frac{\sin 2\omega_1 t}{\left(1 - \frac{\beta^2}{4\omega_1^2}\right)}\right]\right\} \tag{13.17}$$

The function $e^{-(\gamma/2)y(t)}$ has the following Taylor series expansion in powers of the parameter $\gamma/2 = kT/2I\beta^2$

$$\exp\left(-\frac{\gamma}{2}y(t)\right) = \exp\frac{\gamma\beta}{2} \sum_{p=0}^{\infty} \frac{\left(-\frac{\gamma}{2}\right)^p}{p!} \exp\left[-\left(\frac{\gamma}{2}+p\right)\beta t\right]$$

For $\gamma \lesssim 0.05$ this function is closely approximated by the function[21] (note that the Fourier transform of the time derivative of this function leads to an equation similar in form to the Rocard equation)

$$\left(1-\frac{\gamma}{2}\right)^{-1}\left(\exp\left(-\frac{\gamma}{2}\beta t\right) - \frac{\gamma}{2}\exp(-\beta t)\right),$$

$$\frac{\alpha(s)}{\alpha'(0)} \simeq -\int_0^{\infty} \frac{d}{dt}\left(1-\frac{\gamma}{2}\right)^{-1}\left[e^{-\gamma\beta t/2} - \frac{\gamma}{2}e^{-\beta t}\right]e^{-st}\,dt$$

$$= \frac{\frac{\gamma}{2}\beta^2}{\left(s+\frac{\gamma}{2}\beta\right)(s+\beta)} \simeq \frac{\frac{\gamma}{2}\beta^2}{\left(s^2+\beta s+\frac{\gamma}{2}\beta^2\right)}$$

$$s = i\omega, \qquad \beta s\left(1+\frac{\gamma}{2}\right) \simeq \beta s$$

$$\gamma \simeq 0.05$$

$$R(t) \simeq \left(1-\frac{\gamma}{2}\right)^{-1}\left(e^{-\gamma\beta t/2} - \frac{\gamma}{2}e^{-\beta t}\right)\left(1-\frac{\gamma_1^2}{2}\right)$$

$$\times\left\{1-\frac{e^{-\beta t}}{2}\left[\left(1+\frac{\beta^2}{4\omega_1^2}\right)+\left(1-\frac{\beta^2}{4\omega_1^2}\right)\left(\cos 2\omega_1 t\right.\right.\right.$$

$$\left.\left.\left.+\frac{\beta}{\omega_1}\frac{1}{(1-\beta^2/4\omega_1^2)}\sin 2\omega_1 t\right)\right]\right\}+O(\gamma_1^4)\right) \tag{13.18}$$

Just as we did for $R(t)$, we can write down approximate expressions for the normalized auto- and cross-correlation functions for sufficiently small γ_1. It is important to note that both the normalized autocorrelation and normalized cross-correlation functions will contain terms of the order of γ_1

rather than *just* γ_1^2 as in the complete correlation function. We find that

$$\left(\exp\frac{\gamma}{2}y(t)\right)\rho_{11}(t) \simeq 1 - \gamma_1[1-x(t)] + \frac{\gamma_1^2}{2}[1-x(t)]^2\ldots$$

$$= 1 - \gamma_1[1-(1-\gamma_1)x(t)] + \frac{\gamma_1^2}{2}[1+x^2(t)]\ldots$$

or

$$\rho_{11}(t) \simeq \left(1-\frac{\gamma}{2}\right)^{-1}\left(e^{-(\gamma/2)\beta t} - \frac{\gamma}{2}e^{-\beta t}\right)$$

$$\times\left\{1-\gamma_1\left[1-(1-\gamma_1)e^{-\beta t/2}\left(\cos\omega_1 t + \frac{\beta}{2\omega_1}\sin\omega_1 t\right)\right]\right.$$

$$+ \frac{\gamma_1^2}{2}\left[1+\frac{e^{-\beta t}}{2}\left\{\left(1+\frac{\beta^2}{4\omega_1^2}\right)+\left(1-\frac{\beta^2}{4\omega_1^2}\right)\right.\right.$$

$$\left.\left.\left.\times\left[\cos 2\omega_1 t + \frac{\beta}{\omega_1}\frac{\sin 2\omega_1 t}{1-\frac{\beta^2}{4\omega_1^2}}\right]\right\}\right]\right\}\ldots\right\} \qquad (13.19)$$

and

$$\rho_{12}(t) \simeq \left(1-\frac{\gamma}{2}\right)^{-1}\left(e^{-(\gamma/2)\beta t} - \frac{\gamma}{2}e^{-\beta t}\right)$$

$$\times\left\{1+\gamma_1\left[1-(1+\gamma_1)e^{-\beta t/2}\left(\cos\omega_1 t + \frac{\beta}{2\omega_1}\sin\omega_1 t\right)\right]\right.$$

$$+ \frac{\gamma_1^2}{2}\left[1+\frac{e^{-\beta t}}{2}\left\{\left(1+\frac{\beta^2}{4\omega_1^2}\right)+\left(1-\frac{\beta^2}{4\omega_1^2}\right)\right.\right.$$

$$\left.\left.\left.\times\left[\cos 2\omega_1 t + \frac{\beta}{\omega_1}\frac{\sin 2\omega_1 t}{1-\beta^2/4\omega_1^2}\right]\right\}\right]\right\}\ldots\right\} \qquad (13.20)$$

The angular-velocity correlation functions corresponding to these orientational correlation functions are the autocorrelation function (Figs. 12 and 13)

$$\langle \dot{\phi}_1(0)\dot{\phi}_1(t)\rangle = \frac{kT}{2I}\left[\exp\left(-\frac{\zeta}{I}t\right)+\exp\left(-\frac{\zeta t}{2I}\right)\left(\cos\omega_1 t + \frac{\beta}{2\omega_1}\sin\omega_1 t\right)\right]$$

(13.21)

which is an *oscillatory* function of t, and the cross-correlation function

$$\langle \dot{\phi}_1(0)\dot{\phi}_2(t)\rangle = \langle \dot{\phi}_2(0)\dot{\phi}_1(t)\rangle$$

$$= \frac{kT}{2I}\left[\exp\left(-\frac{\zeta t}{I}\right)-\exp\left(-\frac{\zeta t}{2I}\right)\left(\cos\omega_1 t + \frac{\beta}{2\omega_1}\sin\omega_1 t\right)\right]$$

(13.22)

This passes through a *maximum* before decaying to zero.

The parameters in our present model are the moment of inertia I of a dipole, the viscous drag coefficient ζ, and the potential strength V_0. This last parameter is directly related to the dipole moment, since for dipole–dipole coupling the strength of the coupling is proportional to μ^2, where μ is the dipole moment. Thus the parameters of the model may be readily estimated. This represents a considerable improvement over the itinerant-oscillator model, in which one always has to use the schematic picture of a disk of moment of inertia I_1 surrounded by an annulus of moment of inertia I_2. In all cases, to achieve reasonable agreement with experiment it was necessary to assume that the moment of inertia of the disk was greater than the moment of inertia of the annulus.* The present model obviates the need for the rather unsatisfactory disk–annulus picture, because its parameters arise naturally from the physics of the problem. Furthermore, the frequency of oscillation of the system may be *directly* related to the dipole moment and the moment of inertia.

The parameters γ and γ_1 are also of direct physical significance, since they may be related to the Q *factor* of the harmonic oscillator part of the response. The Q factor of a damped harmonic oscillator is defined by the relation

$$Q = \frac{2\pi \times \text{peak energy stored at resonance}}{\text{energy dissipated per cycle}}$$

For the system governed by the differential equation

$$\ddot{x} + \beta\dot{x} + \omega_0^2 x = F(t)$$

---

*In the original version of the i. o. model the contributions of the correlation functions $\langle\cos\psi(0)\cos\psi(t)\rangle_0$, $\langle\cos\theta(0)\cos\psi(t)\rangle_0$ to the polarization are not included, since the annulus is not thought of as a dipole. This will considerably affect the results of the model. ψ,θ are defined on page 164.

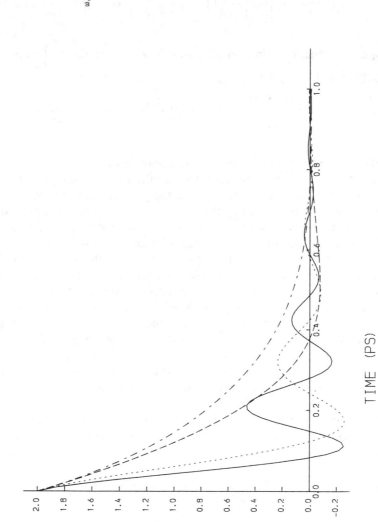

ANGULAR VEL. CF (UNDERDAMPED CASE) FOR β=10

TIME (PS)

$\omega_o =$

30
20
8
6

Figure 12. Angular velocity autocorrelation functions for two similar dipoles. Note that the function is not normalized, so that $\langle \dot{\phi}_1^2 \rangle = kT/I = 2$ for the values of parameters chosen.

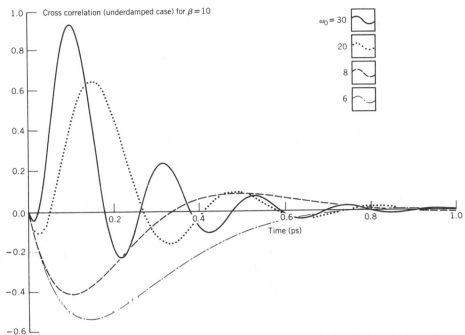

Figure 13. Angular velocity cross correlation functions for two similar dipoles. Dr. J. A. Abas is thanked for this figure and Fig. 12.

the Q factor is*

$$\frac{\omega_0}{\beta} = \sqrt{\frac{4V_0}{I\beta^2}} = \sqrt{\frac{\gamma}{2\gamma_1}}$$

The case in which $Q \to \infty$ (i.e., $\omega_0 \gg \beta$ or $\gamma \gg \gamma_1$) corresponds to a very sharply tuned oscillator; $Q \to \frac{1}{2}$ corresponds to the overdamped and aperiodic solutions. In the latter two cases there will be *no* resonant peak.

Consider now the Fourier transform $X(\omega)$ of the impulse response of such an oscillator. This is

$$X(\omega) = \frac{1}{\left(\omega_0^2 - \omega^2\right) + i\omega\beta}$$

It is usual to plot the function

$$\frac{|X(\omega)|}{|X(\omega_0)|}$$

---

*The inclusion of the full cosine potential makes Q decrease with temperature in accordance with the footnote to page 163 below. It will also reduce the oscillatory character of the motion for the same values of the parameters v. ref. 12 Figs 5.5.5 and 5.5.6, and C. J. Reid, ref. 12 loc. cit.

as a function of ω. At frequencies of, say, ω_1 and ω_2, the amplitude $|X(\omega)|$ will have fallen to 0.707 of its value at $\omega = \omega_0$. One may then show that the bandwidth B of such a system is*

$$B = \omega_2 - \omega_1 = \frac{\omega_0}{2\pi Q} = \frac{\beta}{2\pi}$$

which is a measure of the friction coefficient.

It is now appropriate to draw some conclusions from this analysis of dipole–dipole coupling and inertial effects. The results we have so far presented have been concerned with the response in the time domain. We may readily calculate the complex polarizability from our various expressions for $R(t)$, using the formulae of linear-response theory given in Chapter 6 of ref. 12. In all cases the leading term in the response will be the Rocard equation. The next term, which is of order γ_1^2, represents a *resonance absorption*. This will be the dominant term in the response in the far-infrared band of frequencies, and is a *direct consequence* of the *inclusion* of the *dipole–dipole interaction*. It has its origin in the oscillatory terms in $R(t)$, which arise from the dipole–dipole torques. Similar results should be obtained when a lattice of dipoles is considered.[12] To quantify this discussion we note that correct to terms $O(\gamma_1^2)$ the Laplace transform of $R(t)$ is

$$\left(1 - \frac{\gamma}{2}\right)^{-1}\left(1 - \frac{\gamma_1^2}{2}\right)\left(\frac{1}{s + \frac{\gamma}{2}\beta} - \frac{\gamma}{2}\frac{1}{s + \beta}\right)$$

$$+ \left(1 - \frac{\gamma}{2}\right)^{-1}\frac{\gamma_1^2}{4}\left(1 + \frac{\beta^2}{4\omega_1^2}\right)\left(\frac{1}{s + \frac{\gamma}{2} + 1\beta} - \frac{\frac{\gamma}{2}}{s + 2\beta}\right) + \frac{\gamma_1^2}{8}\left(1 - \frac{\gamma}{2}\right)^{-1}$$

$$\times \left[\left(1 - \frac{\beta^2}{4\omega_1^2} - \frac{\beta i}{\omega_1}\right)\left(\frac{1}{s - 2i\omega_1 + \frac{\gamma}{2} + 1\beta} - \frac{\gamma}{2}\frac{1}{s - 2i\omega_1 + 2\beta}\right)\right.$$

$$\left. + \left(1 - \frac{\beta^2}{4\omega_1^2} + \frac{\beta i}{\omega_1}\right)\left(\frac{1}{s + 2i\omega_1 + \frac{\gamma}{2} + 1\beta} - \frac{\gamma}{2}\frac{1}{s + 2i\omega_1 + 2\beta}\right)\right]$$

$$\tag{13.22a}$$

The first two terms represent inertia corrected Debye type absorption, (the Rocard equation) the last term represents high frequency resonance absorption.

---

*The bandwidth, of the f.i.r. peak being the ratio of frequency to Q factor, is not as directly affected by nonlinear affects due to the cosine potential as either ω_0 or Q. v. footnote to page 163. Compare Figs. 6.1.2.7 and 6.3.2.6 of ref. 9. This may also be proved using the concept of the equivalent linear system v. page 178.

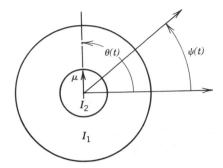

Figure 14.   Itinerant oscillator model.

We have assumed in obtaining these results that the cosine potential of dipole–dipole interaction may be replaced by a harmonic approximation.* This means that we are in effect leaving out one mode of relaxation in the system, namely, that due to dipoles jumping over the hills of the cosine potential. The effect of including the full cosine potential may be summarized by saying that for light damping (the case of interest), such a potential will *broaden* both the Debye and the far-infrared absorption peaks. This has been discussed in detail in the Smoluchowski approximation in Chapter 3 of ref. 12.

Calculations involving the cosine potential cannot be carried out directly using the Langevin equation; instead, appeal must be made to the Kramers equation. This may then be solved for the relaxation behavior using the matrix continued-fraction method described earlier. The process is detailed in ref. 12, Chapter 5. The main result is that the problem reduces to considering the Brownian movements of a free rotator and of a rotator in a $\cos 2\eta$ potential separately.

The analysis we have just given for two similar dipoles provides impetus for a reappraisal of the itinerant-oscillator model (Fig. 14). That model is fully described in ref. 12. We merely recall that in the model, a molecule and its cage of neighbors are envisaged as a disk of moment of inertia I_2 coupled via an interaction potential to an annulus of moment of inertia I_1. The inner disk carries a dipole moment μ. There is no dipole moment on the annulus, the Brownian torques are supposed to act only on the rim of the annulus, and the whole system is compelled to rotate about an axis through the center

---

*The cosine potential causes ω_0 to depend on the energy E as the formula for the lengthening of the periodic time τ of a pendulum viz (Coffey, Marchesoni, Vij, private communication.)

$$\tau = 2\pi\sqrt{\frac{I}{4V_0}}\left[1 + \left(\frac{1}{2}\right)^2\kappa^2 + \left(\frac{1.3}{2.4}\right)\kappa^4 + \dots\right], \quad \kappa = \frac{1}{2}\left(1 + \frac{E}{4V_0}\right)$$

so that $\omega_0 \to \omega_0[1 - \frac{1}{8}(1 + \frac{kT}{4V_0}) + 0(\kappa^4\dots)]$, if $E \simeq kT$, in qualitative agreement with experiment where frequency of maximum f.i.r. absorption decreases with T, v. ref. 9 page 424.

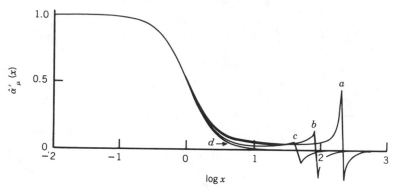

Figure 15.   Real part $\alpha'_\mu(x)/\alpha'_\mu(0)$, as calculated from the itinerant oscillator model.

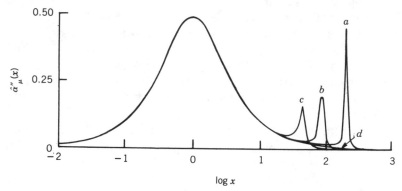

Figure 16.   Imaginary part, $\alpha''_\mu(x)/\alpha'_\mu(0)$, of normalized complex polarizability as calculated from the itinerant oscillator model as a function of $\log_{10} x$ where $x = \omega \tau$ for values of I_2/I_1 given by (a) 0.1, (b) 0.25, and (c) 0.5. The Debye equation is given by (d).

of the disk and normal to it. Thus the inner dipole is regarded as *shielded* from the surroundings by the annulus.

Let us now conceive of the system in another way. As far as the dynamics is concerned, the annulus may be replaced by a rigid rotator of moment of inertia I_1. Let us suppose that this rotator now has a dipole moment μ_1. Thus the equations of motion of the system in the presence of a steady electric field applied along the initial line are*

$$I_1\ddot{\psi} + \zeta\dot{\psi} - \mu_1 F_{12}(t)\sin[\theta - \psi] + \mu_1 E \sin\psi = \lambda(t) \qquad (13.23)$$

and

$$I_2\ddot{\theta} + \mu_2 E \sin\theta + \mu_2 F_{21}(t)\sin[\theta - \psi] = 0 \qquad (13.24)$$

*The conclusions drawn in the footnote to page 163 will apply to any other relaxation phenomena e.g. neutron scattering to which the i.o. model is applicable. The cosine potential also effectively destroys the harmonic character of the oscillations shown in Figs. 12 and 13. v. also discussion in pages 542–543. *Adv. in Chem. Phys.* LVI 1984.

where $\theta(t)$ is the angle μ_2 makes with $\mathbf{e}$, the direction of $\mathbf{E}(t)$; $\psi(t)$ is the angle between μ_1 and $\mathbf{e}$, and $\zeta\dot{\psi}(t)$ and $\lambda(t)$ are the Brownian torques on the annulus. $F(t)$ is the (generally) time-dependent field arising from the dipole–dipole coupling between μ_1 and μ_2.* It may be thought of as a crude representation of the reaction field. Equations (13.23) and (13.24) are similar to Eqs. (5.8.2) and (5.8.3) of ref. 12 with the exception of an extra term $\mu_1 E \sin \psi$ added to Eq. (5.8.2) to account for the fact that the annulus is now regarded as a dipole. Note that there is an error in Eq. (5.8.2): A superfluous term, $\mu E \sin \theta$, is present. This does not invalidate the subsequent analysis, however, since we supposed that $\mathbf{E}$ was switched off at $t = 0$ after having been applied at $t = -\infty$, so that steady conditions had been attained. For $t > 0$, Eqs. (13.23) and (13.24) become

$$I_1\ddot{\psi}(t)+\zeta\dot{\psi}(t)-\mu F(t)\sin[\theta(t)-\psi(t)] = \lambda(t) \qquad (13.25)$$

and

$$I_2\ddot{\theta}(t)+\mu F(t)\sin[\theta(t)-\psi(t)] = 0 \qquad (13.26)$$

These equations are the same as Eqs. (5.8.4) and (5.8.5) of ref. 12 and all the results of ref. 12 apply to them. Henceforth we assume $F(t) \simeq$ constant $= F$; otherwise the calculations become intractable.

Consider now the instantaneous dipole moment of the system. For $t > 0$, we have

$$\mathbf{m}(t)\cdot\mathbf{e} = \mu_1\cos\psi(t)+\mu_2\cos\theta(t) \qquad (13.27)$$

whence

$$\begin{aligned}
\langle(\mathbf{m}(0)\cdot\mathbf{e})(\mathbf{m}(t)\cdot\mathbf{e})\rangle_0 = &\ \mu_1^2\langle\cos\psi(0)\cos\psi(t)\rangle_0 \\
&+ \mu_1\mu_2\langle\cos\psi(0)\cos\theta(t)\rangle_0 \\
&+ \mu_2\mu_1\langle\cos\theta(0)\cos\psi(t)\rangle_0 \\
&+ \mu_2^2\langle\cos\theta(0)\cos\theta(t)\rangle_0 \qquad (13.28)
\end{aligned}$$

Equation (13.28) contains contributions to the polarization from both the *auto-* and *cross*-correlation functions of μ_1 and μ_2. In the original treatment by Calderwood and Coffey[12] of the small oscillation approximation, and in subsequent attempts to treat the problem for large values of $\theta - \psi$, the contributions $\mu_1\mu_2\langle\cdot\rangle$ and $\mu_1^2\langle\cdot\rangle$ to the polarization were not considered, since the outer disk (annulus) was not conceived of as a dipole. It is important to assess the effect of these terms. In view of our earlier experience

*Note $\mu_1 F_{12}(t) = \mu_2 F_{21}(t) = \mu F(t)$ say as required by Newton's third Law.

of the dynamics of two rotating dipoles (see ref. 12, Chapter 3), it is not un-reasonable to suppose (although this must be verified by direct calculation) that in Eq. (13.28)

$$\langle \cos\psi(0)\cos\theta(t)\rangle_0 = \langle \cos\theta(0)\cos\psi(t)\rangle_0 \tag{13.29}$$

so that

$$\langle (\mathbf{m}(0)\cdot\mathbf{e})(\mathbf{m}(t)\cdot\mathbf{e})\rangle_0 = \mu_1^2\langle \cos\psi(0)\cos\psi(t)\rangle_0$$
$$+ 2\mu_1\mu_2\langle \cos\theta(0)\cos\psi(t)\rangle_0$$
$$+ \mu_2^2\langle \cos\theta(0)\cos\theta(t)\rangle_0$$

The correlation functions may now be calculated from the differential-difference equations arising from the Kramers equation corresponding to Eqs. (13.25) and (13.26). These are (details in ref. 12)*

$$\frac{da_{q,p}^{m,n}(t)}{dt} + m\frac{\zeta}{I_1}a_{q,p}^{m,n}(t) + i\sqrt{\frac{kT}{I_1}}\left\{ q\left[a_{q,p}^{m-1,n}(t) + (m+1)a_{q,p}^{m+1,n}(t)\right]\right.$$

$$+ p\sqrt{\frac{I_1}{I_2}}\left[a_{q,p}^{m,n-1}(t) + (n+1)a_{q,p}^{m,n+1}(t)\right]\right\}$$

$$+ \frac{\mu Fi}{2\sqrt{I_1 kT}}\left\{ \left[a_{q+1,p-1}^{m-1,n}(t) - a_{q-1,p+1}^{m-1,n}(t)\right]\right.$$

$$\left. - \sqrt{\frac{I_1}{I_2}}\left[a_{q+1,p-1}^{m,n-1}(t) - a_{q-1,p+1}^{m,n-1}(t)\right]\right\} = 0,$$

$$m, n \geq 0, \quad m, n = 0,1,2,3,\dots$$
$$p = \dots, -2, -1, 0, 1, 2, \dots$$
$$q = \dots, -2, -1, 0, 1, 2, \dots \tag{13.30}$$

There is no small oscillation approximation made in deriving these equations so they automatically include relaxation due to crossing over potential hills. The autocorrelation function of μ_1 is gotten by picking off $a_{1,0}^{0,0}(t)$, that of μ_2 from $a_{0,1}^{0,0}(t)$, and the cross-correlation function from $a_{1,1}^{0,0}(t)$. The initial conditions for the computation of the a^s are gotten from the initial distribution in phase space,[12]

$$W(\theta, \dot{\theta}, \psi, \dot{\psi}, 0) = A\exp\left\{ -\frac{1}{kT}\left[\frac{1}{2}I_1\dot{\psi}^2 + \frac{1}{2}I_2\dot{\theta}^2 - \mu F\cos(\theta - \psi)\right]\right\}$$

$$\times\left[1 + \frac{\mu_1 E}{kT}\cos\psi + \frac{\mu_2 E}{kT}\cos\theta\right] \tag{13.31}$$

*If stochastic torques act on μ_2 in 13.26 these may be accounted for by adding a term $n\dfrac{\zeta_2}{I_2}a_{q,p}^{m,n}(t)$ to 13.30. This is needed when extending the Budó treatment to unequal dipoles. The version where $\mu_1 = \mu_2$ has recently been compared with experiment by Marchesoni, Coffey and Vij.

where the linear approximation for the applied field $\mathbf{E}$ has been used ($\mu E/kT \ll 1$). Having determined the a^s, we find the complex polarizability in the usual way from

$$kT\alpha(\omega) = - \int_0^\infty \frac{d}{dt} \left\{ \langle (\mathbf{m}(0)\cdot\mathbf{e})(\mathbf{m}(t)\cdot\mathbf{e}) \rangle_0 \right\} e^{-i\omega t} \, dt \qquad (13.32)$$

where the real and imaginary parts α' and α'' may be plotted as in Chapter 4 of ref. 12. This computation is currently being done; details may be had at a later stage. There is a considerable increase in difficulty over the two-similar-dipoles case, due to the fact that the a^s may no longer be factored. The parameters of the present model are I_1, I_2, μ_1, μ_2, and ζ.

A problem similar to those we have just discussed is the study of the line shape of single-mode semiconductor lasers.[47] The phenomenon to be studied is *phase noise*, in other words noise associated with fluctuations of the instantaneous frequency of the emitted radiation about its mean value.

Piazzolla and Spano[47] take into account the amplitude and frequency fluctuations of the laser field by writing the electric part of the radiation from a single mode semiconductor laser as

$$\mathscr{E}(t) = E(t)\exp[i\omega(t)t] \qquad (13.33)$$

$$= [E_0 + e(t)]\exp i[\omega_0 t + \phi(t)] \qquad (13.34)$$

where $e(t)$ ($\ll E_0$) is the deviation from the mean amplitude E_0 and $\phi(t)$ is the random phase of the field so that $\dot{\phi}(t) = \omega(t) - \omega_0$ represents the deviation of the instantaneous frequency $\omega(t)$ from its mean value ω_0. The spectrum (line shape) may now be found using the Wiener–Khinchin theorem. The autocorrelation function of $\mathscr{E}(t)$ is

$$\rho_{\mathscr{E}}(\tau) = \langle \mathscr{E}(t+\tau)\mathscr{E}^*(t) \rangle$$

$$\simeq E_0^2 \exp i\omega_0\tau \exp\left[-\tfrac{1}{2}\langle [\Delta\phi(\tau)]^2 \rangle \right]$$

For typical parameters encountered in the laser problem

$$\exp\left[-\tfrac{1}{2}\langle (\Delta\phi)^2 \rangle \right] \simeq \exp\left[-\frac{R_s}{2I_0}(1+\alpha^2)\tau \right]$$

$$\times \left\{ 1 - \frac{R_s\alpha^2}{2I_0\gamma_e}\left[1 - \exp\left(-\frac{1}{2}\gamma_e\tau \right)\left(\cos\Omega_0\tau + \frac{3\gamma_e}{2\Omega_0}\sin\Omega_0\tau \right) \right] \right\}$$

R_S is the average rate of spontaneous emission in the mode, $I_0 = VE_0^2$ the average number of photons belonging to the mode (V being the mode volume), α the enhancement line width factor, G_0 the average gain, $\gamma_e = \gamma + G_n I_0 / V$ (γ is the inverse spontaneous lifetime of the excited carriers and G_n is the derivative of gain with respect to carrier density), and $\Omega_R = G_n G_0 I_0 V^{-1}$) is the resonant peak frequency.

They find that the normalized line shape function $G(\omega)$ is

$$G(\omega) = \left(1 - \frac{R_s \alpha^2}{2 I_0 \gamma_e}\right) \frac{2 \Delta \Omega_1}{(\omega - \omega_0)^2 + (\Delta \Omega_1)^2} + \frac{R_s \alpha^2}{4 I_0}$$

$$\times \left\{ \frac{(2/\gamma_e)(\Delta \Omega_1 + \frac{1}{2}\gamma_e) + (3/\Omega_0)(\omega - \omega_0 + \Omega_0)}{(\omega - \omega_0 + \Omega_0)^2 + (\Delta \Omega_1 + \frac{1}{2}\gamma_e)^2} \right.$$

$$+ \left. \frac{(2/\gamma_e)(\Delta \Omega_1 + \frac{1}{2}\gamma_e) - (3/\Omega_0)(\omega - \omega_0 - \Omega_0)}{(\omega - \omega_0 - \Omega_0)^2 + (\Delta \Omega_1 + \frac{1}{2}\gamma_e)^2} \right\}$$

$$\Delta \Omega_1 = \frac{R_s}{2 I_0}(1 + \alpha^2)$$

This has a Lorentzian central lobe and sideband peaks shifted by an amount Ω_0, in accordance with experiment.

XIV. NONLINEAR EFFECTS IN DIELECTRIC RELAXATION— AN INTRODUCTION

The analysis we have presented for both the Debye disk model and the two-dipole model assumes that terms of the order of the square of the external field and higher may be neglected in the response. As an introduction to the analytical study of nonlinear response in the context of dielectric relaxation, we consider the Debye sphere model when a strong electric field is applied.

On the basis of the Debye sphere model without inertial effects, the variation in configuration space of the probability density function, $f(\theta, t)$, for a rigid dipole of moment μ undergoing rotational Brownian motion at absolute temperature T and under the influence of an external, time-varying electric field $\mathbf{E}(t)$ is governed by the equation[19]

$$2\tau \frac{\partial f(\theta, t)}{\partial t} = \frac{1}{\sin \theta} \frac{\partial}{\partial \theta} \left[\sin \theta \left(\frac{\partial f(\theta, t)}{\partial \theta} - \frac{f(\theta, t) M(\theta, t)}{kT} \right) \right] \quad (14.1)$$

where $f(\theta, t) \sin \theta \, d\theta$ is the probability that at time t the dipole has an orientation between θ and $\theta + d\theta$ relative to the direction of $\mathbf{E}(t)$, $M(\theta, t)$ is the magnitude of the torque acting on the dipole due to $\mathbf{E}(t)$, k is the Boltzmann constant, and τ is the Debye relaxation time.

We shall consider two particular types of applied field: (1) a strong alternating field

$$\mathbf{E}(t) = \mathbf{E}_0 \cos \omega t \qquad (14.2)$$

so that

$$M(\theta, t) = -\mu E_0 \cos \omega t \sin \theta \qquad (14.3)$$

and (2) a strong unidirectional field $\mathbf{E}_1$ together with a *weak* alternating field $\mathbf{E}_2 \cos \omega t$, so that

$$M(\theta, t) = -\mu(E_1 + E_2 \cos \omega t)\sin \theta = -\mu E_1(1 + \lambda \cos \omega t)\sin \theta \quad (14.4)$$

where

$$\lambda = \frac{E_2}{E_1} \ll 1 \qquad (14.5)$$

The general solution of Eq. (14.1) is of the form

$$f(\theta, t) = \sum_{n=0}^{\infty} a_n(t) P_n(\cos \theta) \qquad (14.6)$$

where the $a_n(t)$ are functions to be determined. On substituting Eq. (14.6) into Eq. (14.1) and making use of the recurrence relations for the Legendre polynomials $P_n(x)$, namely

$$(1 - x^2)\frac{dP_n(x)}{dx} = n\left(P_{n-1}(x) - xP_n(x)\right) \qquad (14.7)$$

$$nP_n(x) - (2n - 1)xP_{n-1}(x) + (n - 1)P_{n-2}(x) = 0 \qquad (14.8)$$

and the orthogonality property

$$\int_{-1}^{+1} P_n(x) P_m(x)\, dx = \frac{2}{2n + 1}\delta_{mn} \qquad (14.9)$$

we find that the $a_n(t)$ must satisfy the differential-difference equation

$$\dot{a}_n(t) = -\frac{n(n+1)}{2\tau}\left\{a_n(t) - \frac{\mu E(t)}{kT}\left[\frac{a_{n-1}(t)}{2n-1} - \frac{a_{n+1}(t)}{2n+3}\right]\right\} \quad (14.10)$$

We shall confine our attention to the behavior of $f(\theta, t)$ a long time after the field has been switched on, so that we may consider the stationary solution only. This solution is

$$a_n(t) = \frac{n(n+1)\gamma}{2\tau} \int_{-\infty}^{t} \exp\left[\frac{-n(n+1)}{2\tau}(t-u)\right]$$

$$\times e(u)\left[\frac{a_{n-1}(u)}{2n-1} - \frac{a_{n+1}(u)}{2n+3}\right] du \qquad (14.11)$$

in which for case 1,

$$\gamma = \gamma_1 = \frac{\mu E_0}{kT}; \qquad e(u) = \cos \omega u \qquad (14.12)$$

(Equation (14.10) has the same mathematical structure as the Brinkman equations when $V_0 = 0$), and for case 2,

$$\gamma = \gamma_2 = \frac{\mu E_1}{kT}; \qquad e(u) = (1 + \lambda \cos \omega u) \qquad (14.13)$$

One may now determine the ensemble averages

$$\langle \cos \theta \rangle = \langle P_1(\cos \theta) \rangle \quad \text{and} \quad \langle \tfrac{1}{2}(3\cos^2\theta - 1) \rangle = \langle P_2(\cos \theta) \rangle$$

appropriate to dielectric and Kerr-effect relaxation, respectively, since

$$\langle P_n(\cos \theta) \rangle = \frac{\int_{-1}^{+1} f(\theta, t) P_n(\cos \theta)\, d(\cos \theta)}{\int_{-1}^{+1} f(\theta, t)\, d(\cos \theta)} = \frac{1}{2n+1} \frac{a_n(t)}{a_0}$$

$$(14.14)$$

Equation (14.11) may be solved for $a_1(t)$, $a_2(t)$, and all the $a_n(t)$ as power series in γ by means of successive approximations. To demonstrate this, we write out Eq. (14.11) explicitly for the first few values of n:

$$a_0 = \text{constant} \qquad (14.15)$$

$$a_1(t) = \frac{\gamma}{\tau} \int_{-\infty}^{t} \exp\left[-\frac{(t-u)}{\tau}\right] e(u)\left(a_0 - \frac{a_2(u)}{5}\right) du \qquad (14.16)$$

$$a_2(t) = \frac{3\gamma}{\tau} \int_{-\infty}^{t} \exp\left[-\frac{3(t-u)}{\tau}\right] e(u)\left(\frac{a_1(u)}{3} - \frac{a_3(u)}{7}\right) du \qquad (14.17)$$

$$a_3(t) = \frac{6\gamma}{\tau} \int_{-\infty}^{t} \exp\left[-\frac{6(t-u)}{\tau}\right] e(u)\left(\frac{a_2(u)}{5} - \frac{a_4(u)}{9}\right) du \qquad (14.18)$$

To a first approximation in γ, the solution is (linear response)

$$a_1(t) = \frac{a_0 \gamma}{\tau} \int_{-\infty}^{t} \exp\left[-\frac{(t-u)}{\tau}\right] e(u)\, du \qquad (14.19)$$

To obtain a second approximation, we write down, from Eqs. (14.17) and (14.19),

$$a_2(t) = a_0 \frac{\gamma^2}{\tau^2} \int_{-\infty}^{t} du \int_{-\infty}^{u} \exp\left[-3\frac{(t-u)}{\tau}\right]$$

$$\times \exp\left[-\frac{(u-u_1)}{\tau}\right] e(u_1) e(u) du_1 \qquad (14.20)$$

and substitute this expression into Eq. (14.16). This leads to

$$a_1(t) = a_0 \left\{ \frac{\gamma}{\tau} \int_{-\infty}^{t} \exp\left[-\frac{(t-u)}{\tau}\right] e(u) du \right.$$

$$- \frac{\gamma^3}{5\tau^3} \iiint_{-\infty < u_2 \leq u_1 \leq u \leq t} \exp\left[-\frac{(t-u)}{\tau}\right] \exp\left[-\frac{3(u-u_1)}{\tau}\right]$$

$$\left. \times \exp\left[-\frac{(u_1-u_2)}{\tau}\right] e(u_2) e(u_1) e(u) du_2 du_1 du + \cdots \right\}$$

$$(14.21)$$

Continuing in this way we deduce, with the aid of Eq. (14.14), that

$$\langle P_1(\cos\theta) \rangle = \frac{1}{3} \left\{ \frac{\gamma}{\tau} \int_{-\infty}^{t} \exp\left[-\frac{(t-u)}{\tau}\right] e(u) du \right.$$

$$- \frac{\gamma^3}{5\tau^3} \iiint_{-\infty < u_2 \leq u_1 \leq u \leq t} \exp\left[-\frac{(t-u)}{\tau}\right] \exp\left[-3\frac{(u-u_1)}{\tau}\right]$$

$$\times \exp\left[-\frac{(u_1-u_2)}{\tau}\right] e(u_2) e(u_1) e(u) du_2 du_1 du$$

$$+ \frac{18}{175} \frac{\gamma^5}{\tau^5} \iiiint_{-\infty < u_4 \cdots \leq t} \exp\left[-\frac{(t-u)}{\tau}\right]$$

$$\times \exp\left[-3\frac{(u-u_1)}{\tau}\right] \exp\left[-6\frac{(u_1-u_2)}{\tau}\right]$$

$$\times \exp\left[-3\frac{(u_2-u_3)}{\tau}\right] \exp\left[-\frac{(u_3-u_4)}{\tau}\right]$$

$$\left. \times e(u_4) e(u_3) e(u_2) e(u_1) e(u)] du_4 \cdots du - \cdots \right\} \qquad (14.22)$$

and

$$\langle P_2(\cos\theta)\rangle = \frac{1}{5}\left\{\frac{\gamma^2}{\tau^2}\iint_{-\infty < u_1 \le u \le t} \exp\left[-3\frac{(t-u)}{\tau}\right]\exp\left[-\frac{(u-u_1)}{\tau}\right]\right.$$

$$\times e(u_1)e(u)\,du_1\,du$$

$$-\frac{18}{35}\frac{\gamma^4}{\tau^4}\iiiint_{-\infty < u_3 \cdots \le t}\exp\left[-3\frac{(t-u)}{\tau}\right]\exp\left[-6\frac{(u-u_1)}{\tau}\right]$$

$$\times\exp\left[-3\frac{(u_1-u_2)}{\tau}\right]\exp\left[-\frac{(u_2-u_3)}{\tau}\right]$$

$$\left.\times e(u_3)e(u_2)e(u_1)e(u)\,du_3\cdots du + \cdots\right\} \tag{14.23}$$

These equations allow us to calculate $\langle P_1(\cos\theta)\rangle$ and $\langle P_2(\cos\theta)\rangle$ correctly to any order in γ. In practice, one need go only as far as terms γ and γ^3 in Eq. (14.22) and as far as γ^2 in Eq. (14.23), since the higher-order terms in γ make negligible contributions to $\langle P_1\rangle$ and $\langle P_2\rangle$.

On writing $\gamma = \gamma_1$ and $e(u) = \cos\omega u$ in Eq. (14.22) and evaluating the terms in γ and γ^3 in that equation, we find, after considerable manipulation, the following expression for the mean moment:

$$\mu\langle P_1(\cos\theta)\rangle = \frac{\mu^2 E_0}{3kT}\left\{\left[1 - \frac{(27-13\omega^2\tau^2)\gamma_1^2}{540(1+\omega^2\tau^2)(1+4\omega^2\tau^2/9)}\right]\frac{\cos\omega t}{1+\omega^2\tau^2}\right.$$

$$+\left[1 - \frac{(21+\omega^2\tau^2)\gamma_1^2}{270(1+\omega^2\tau^2)(1+4\omega^2\tau^2/9)}\right]\frac{\omega\tau\sin\omega t}{1+\omega^2\tau^2}$$

$$\left.-\frac{\gamma_1^2}{90}\frac{\left[\frac{1}{2}(3-17\omega^2\tau^2)\cos 3\omega t + \omega\tau(7-3\omega^2\tau^2)\sin 3\omega t\right]}{(1+\omega^2\tau^2)(1+4\omega^2\tau^2/9)(1+9\omega^2\tau^2)}\right\}$$

$$\tag{14.24}$$

Setting ω equal to 0 in Eq. (14.24) restores the first two terms in the usual expansion of the Langevin function for the mean dipole moment in powers of γ_1 under the influence of a static field.[9] The terms independent of γ_1 in Eq. (14.24) are those obtained by Debye[19] in his treatment of dielectric relaxation under the influence of a weak periodic field, whereas the terms in

γ_1^2 in that equation represent relaxation due to the nonlinear behavior of the dielectric. We may immediately write down from Eq. (14.24) an expression for the polarization $\mathscr{P}(t, \gamma_1^2)$, namely,

$$\mathscr{P}(t, \gamma_1^2) = N\mu \langle P_1(\cos\theta) \rangle \tag{14.25}$$

where N is the number of dipoles per unit volume. Clearly the polarization will possess two distinct components, one having angular frequency ω and the other angular frequency 3ω. The component $\mathscr{P}_1(t, \gamma_1^2)$ having angular frequency ω may be written in the form

$$\mathscr{P}_1(t, \gamma_1^2) = \mathrm{Re}\left(\chi_1(\omega, \gamma_1^2) E_0 \exp i\omega t\right) \tag{14.26}$$

where "Re" means "real part of" and $\chi_1(\omega, \gamma_1^2)$ is a complex susceptibility defined by the equation

$$\chi_1(\omega, \gamma_1^2) = \chi_1'(\omega, \gamma_1^2) - i\chi_1''(\omega, \gamma_1^2) \tag{14.27}$$

in which

$$\chi_1'(\omega_1, \gamma_1^2) = \frac{N\mu^2}{3kT} \frac{1}{1+\omega^2\tau^2}\left[1 - \frac{(27-13\omega^2\tau^2)\gamma_1^2}{540(1+\omega^2\tau^2)(1+4\omega^2\tau^2/9)}\right] \tag{14.28}$$

$$\chi_1''(\omega_1, \gamma_1^2) = \frac{N\mu^2}{3kT} \frac{\omega\tau}{1+\omega^2\tau^2}\left[1 - \frac{(21+\omega^2\tau^2)\gamma_1^2}{270(1+\omega^2\tau^2)(1+4\omega^2\tau^2/9)}\right] \tag{14.29}$$

Because the ω and 3ω components of the polarization may always be separated experimentally, we shall here confine ourselves to a discussion of the implications of Eqs. (14.26)–(14.29). In virtue of these equations we may write down an expression for the loss tangent,[36] defined as

$$\tan\delta = \frac{\chi_1''(\omega_1, \gamma_1^2)}{\varepsilon_0 + \chi_1'(\omega_1, \gamma_1^2)} \tag{14.30}$$

On substituting Eqs. (14.28) and (14.29) into Eq. (14.30) we find, after considerable manipulation, the following equation for $\tan \delta$:

$$\tan \delta = (\tan \delta_0)\left[1 - \frac{\gamma_1^2(27 + 2\omega^2\tau^2 + 15\varepsilon_s)}{540(\varepsilon_s + \omega^2\tau^2)(1 + 4\omega^2\tau^2/9)}\right] \qquad (14.31)$$

where

$$\tan \delta_0 = \frac{\chi_1''(\omega,0)}{\varepsilon_0 + \chi_1'(\omega,0)} = \frac{(\varepsilon_s - 1)\omega\tau}{\varepsilon_s + \omega^2\tau^2} \qquad (14.32)$$

is the loss tangent for vanishingly small values of E_0, and ε_s, the relative permittivity at vanishingly small frequencies and low fields, is given (assuming that the effect of dipole–dipole coupling may be ignored) by the equation

$$(\varepsilon_s - 1) = \frac{N\mu^2}{\varepsilon_0 3kT} \qquad (14.33)$$

For $\langle P_2(\cos\theta)\rangle$, we find from Eq. (14.23) that

$$\langle P_2(\cos\theta)\rangle = \frac{\gamma_1^2}{30}\left[\frac{1}{1 + \omega^2\tau^2} + \frac{\cos(2\omega t - \phi)}{(1 + \omega^2\tau^2)^{1/2}(1 + 4\omega^2\tau^2/9)^{1/2}}\right] \qquad (14.34)$$

where

$$\phi = \tan^{-1}\left(\frac{5\omega\tau}{3 - 2\omega^2\tau^2}\right) \qquad (14.35)$$

Equations (14.34) and (14.35) are in full agreement with the work of Benoit.[37] Note that our τ corresponds to the $3\tau/2$ of his equations.

On writing $\gamma = \gamma_2$ and $e(u) = (1 + \lambda\cos\omega u)$ in Eq. (14.22) and proceeding as in the previous section, we find the following expression for the polarization:

$$\mathcal{P}(t,\gamma_2^2) = \frac{N\mu^2 E_1}{3kT}\left\{1 - \frac{\gamma_2^2}{15} + \frac{\lambda}{1 + \omega^2\tau^2}\left[1 + \frac{\gamma_2^2}{15}\frac{(2\omega^4\tau^4 - \omega^2\tau^2 - 27)}{(1 + \omega^2\tau^2)(9 + \omega^2\tau^2)}\right]\cos\omega t \right.$$
$$\left. + \frac{\lambda\omega\tau}{1 + \omega^2\tau^2}\left[1 - \frac{\gamma_2^2}{15}\frac{(\omega^4\tau^4 + 19\omega^2\tau^2 + 42)}{(1 + \omega^2\tau^2)(9 + \omega^2\tau^2)}\right]\sin\omega t + O(\lambda^2)\right\}$$
$$(14.36)$$

In this instance, provided terms $O(\lambda^2)$ may be neglected, only the fundamental frequency is present. The first two terms in Eq. (14.36) represent the expansion of the Langevin function for the polarization in the strong constant field E_1. The components of the last two terms that are independent of γ_2 are those due to dielectric relaxation in the field $E_2 \cos \omega t$ alone, whereas the two components prefixed by γ_2^2 represent dielectric relaxation induced by the nonlinear coupling between the constant field and the alternating field. Equation (14.36) may be rearranged to read as follows:

$$\mathscr{P}(t, \gamma_2^2) = \mathscr{P}_0 + \left[\chi'(\omega, \gamma_2^2) - i\chi''(\omega, \gamma_2^2) \right] E_2 \exp i\omega t \qquad (14.37)$$

where $\mathscr{P}_0$ denotes the constant portion of Eq. (14.36) and where

$$\chi'(\omega, \gamma_2^2) = \frac{N\mu^2}{3kT} \frac{\lambda}{1 + \omega^2\tau^2} \left[1 + \frac{\gamma_2^2}{15} \frac{(2\omega^4\tau^4 - \omega^2\tau^2 - 27)}{(1 + \omega^2\tau^2)(9 + \omega^2\tau^2)} \right] \qquad (14.38)$$

$$\chi''(\omega, \gamma_2^2) = \frac{N\mu^2}{3kT} \frac{\lambda\omega\tau}{1 + \omega^2\tau^2} \left[1 - \frac{\gamma_2^2(\omega^4\tau^4 + 19\omega^2\tau^2 + 12)}{15(1 + \omega^2\tau^2)(9 + \omega^2\tau^2)} \right] \qquad (14.39)$$

Equation (14.37) may be further rearranged to read

$$\mathscr{P}(t, \gamma_2^2) - \mathscr{P}_0 = \Delta\mathscr{P}(t, \gamma_2^2) = \left[\chi'(\omega, \gamma_2^2) - i\chi''(\omega, \gamma_2^2) \right] E_2 \exp i\omega t$$

$$(14.40)$$

where the quantity $\Delta\mathscr{P}(t, \gamma_2^2)$ may be regarded as the polarization arising from the alternating field and from the interaction of the alternating field with the constant field. As in the previous section, we may associate with Eq. (14.40) a loss tangent, $\tan\delta$, defined by the equation

$$\tan\delta = \frac{\chi''(\omega, \gamma_2^2)}{\varepsilon_0 + \chi'(\omega, \gamma_2^2)} \qquad (14.41)$$

which, with Eq. (14.38) and (14.39), may be written as

$$\tan\delta = (\tan\delta_0) \left\{ 1 - \frac{\gamma_2^2 \left[\omega^4\tau^4 + 16\omega^2\tau^2 + 27 + 3\varepsilon_0(\omega^2\tau^2 + 5) \right]}{15(9 + \omega^2\tau^2)(\varepsilon_s + \omega^2\tau^2)} \right\}$$

$$(14.42)$$

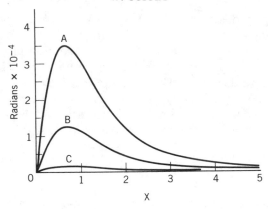

Figure 17. Δ_1 versus x for various values of E_0: curve A is for $E_0 = 5 \times 10^7$ V/m, curve B is for $E_0 = 3 \times 10^7$ V/m, curve C is for $E_0 = 10^7$ V/m. $\mu = 1.4$ D, $\varepsilon_s = 4.4, \varepsilon_\infty = 1$, $x = \omega\tau$. From Eq. 14.31, $\Delta_1 = \delta_0 - \delta$

$$\Delta_1 = \tan^{-1}\left[\frac{(\varepsilon_s - 1)x}{\varepsilon_s + x^2}\right] - \tan^{-1}\left\{\frac{(\varepsilon_s - 1)x}{\varepsilon_s + x^2}\left[1 - \frac{\gamma_1^2(27 + 2x^2 + 15\varepsilon_s)}{540(\varepsilon_s + x^2)(1 + 4x^2/9)}\right]\right\}.$$

in which

$$\tan\delta_0 = \frac{(\varepsilon_s - 1)\omega\tau}{\varepsilon_s + \omega^2\tau^2} \tag{14.43}$$

where

$$(\varepsilon_s - 1) = \frac{N}{\varepsilon_0}\frac{\mu^2}{3kT} \tag{14.44}$$

The average, $\langle P_2(\cos\theta)\rangle$, may be deduced in the same manner as before; the result is

$$\langle P_2(\cos\theta)\rangle = \frac{\gamma_2^2}{5}\left\{\frac{1}{3} + \lambda\frac{[2(3 + \omega^2\tau^2)\cos\omega t + \omega\tau(5 + \omega^2\tau^2)\sin\omega t]}{(1 + \omega^2\tau^2)(9 + \omega^2\tau^2)}\right\} \tag{14.45}$$

This brings to an end our short discussion of nonlinear effects in dielectric relaxation. A very complete analysis may be made by expressing the Laplace transform of the solution of Eq. (14.10) as a continued fraction.[9] The

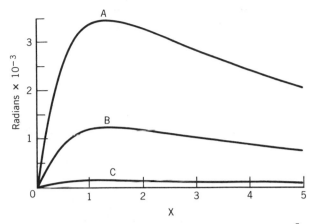

Figure 18. Δ_2 versus x for various values of E_1: curve A is for $E_1 = 5 \times 10^7$ V/m, curve B is for $E_1 = 3 \times 10^7$ V/m. Curve C is for $E_1 = 10^7$ V/m same values of μ, ε_s, ε_∞, $x = \omega\tau$

$$\Delta_2 = \delta_0 - \delta = \tan^{-1}\left[\frac{(\varepsilon_s - 1)x}{\varepsilon_s + x^2}\right]$$
$$- \tan^{-1}\left\{\frac{(\varepsilon_s - 1)x}{\varepsilon_s + x^2}\left[1 - \frac{\gamma_2^2\left[x^4 + 16x^2 + 27 + 3\varepsilon_s(x^2 + 5)\right]}{15(9 + x^2)(\varepsilon_s + x^2)}\right]\right\}.$$

Reproduced by permission from W. T. Coffey and B. V. Paranjape, *Proc. R. Ir. Acad.* **78A**, 17 (1978).

following conclusions may immediately be drawn:

1.   There is no longer the connection between the aftereffect and alternating-field solutions seen in the linear approximations.[9] The ratio of response to stimulus (transfer function or impulse response) now depends on the *form* of the applied field, and is *not independent* of it as in the linear case. A separate analysis must be carried out for every type of driving field. This consideration is particularly important when studying the response of the system to a transitory disturbance, such as a rectangular pulse[9,12] that is switched on and suddenly switched off a short time later.

2.   Application of strong alternating fields of frequency ω effectively allows the production of all harmonics of this frequency by the system.

3.   The analysis may also be carried out when inertial effects are taken into consideration, that is, when the underlying diffusion equation becomes the Kramers equation. However, the analysis does become more difficult. A striking conclusion is that inertial effects, when combined with a strong driving field, may in certain cases give rise to oscillations and resonance effects that are not present in the linear approximation to the solution. An example of such a case is the response of a Debye dielectric to a rectangular pulse. This is briefly treated in ref. 12. (Figures 19 and 20).

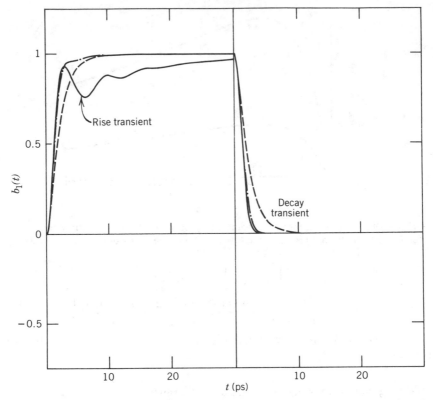

Figure 19.   Nonlinear effects in step response when treated using the Kramers equation: rise
transient of the dipole moment for varying field strengths, as field strength is increased.

4.   Analysis of the problem when the underlying equation is the Kramers
equation is best carried out by adopting the matrix continued-fraction
method already outlined. The method of successive approximations is entirely equivalent to the continued-fraction method, but is not simple to use
with the Kramers equation.

5.   For small nonlinearities, classical techniques for the analysis of nonlinear systems, such as use of the equivalent linear system,* may prove useful in gaining insight into the present problem.

A problem closely related to the previous one arises in the analysis of
Josephson junctions under the influence of an alternating current. (The
analysis that follows will in large part hold for quantum noise in ring laser
gyros and the first-order phase-locked loop.)

---

*v. N. W. Mclachlan, *Ordinary Nonlinear Differential Equations*, Oxford University Press
London. 1956, also T. K. Caughey J. Acoust. Soc. Amer. **35**, 1706, 1963, where the technique is
extended to stochastic differential equations.

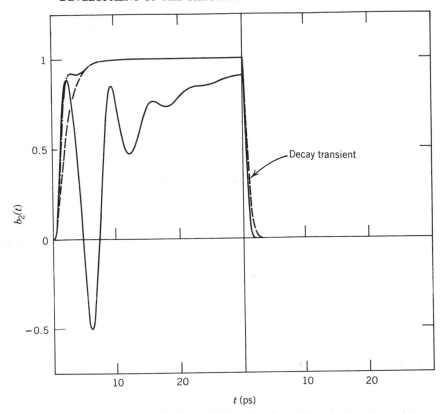

Figure 20.   Rise transient of the dynamic Kerr effect. As field strength is increased response becomes more and more oscillatory. (Decay transients are exponential or sums of ex-
ponentials.) Further details in ref. 12, page 151.

Returning to Eq. (11.15), we suppose that the noise current $i(t)$ in that
equation has superimposed on it an additional ac term $I_m\cos \omega t$. Thus Eq.
(11.17) for the phase becomes

$$\left(\frac{\hbar}{2e}\right)^2 C\phi(t) + \left(\frac{\hbar}{2e}\right)^2 \frac{1}{R}\dot{\phi}(t) + \frac{\hbar}{2e}I_1\sin\phi(t) = \frac{\hbar}{2e}\left(I_{dc} + I_m\cos \omega t + i(t)\right)$$

$$(14.46)$$

We now cast Eq. (14.46) into the form of a damped pendulum equation
driven by a constant torque on which is superimposed a weak alternating
torque and a white-noise torque:

$$J\ddot{\phi}(t) + \zeta\dot{\phi}(t) + V_0\sin\phi(t) = F_0 + F_1\cos \omega t + \lambda(t) \qquad (14.47)$$

where $J = (\hbar/2e)^2 C$, $\zeta = (\hbar/2e)^2(1/R)$, $F_0 = (\hbar/2e)I_{dc}$, $F_1 = (\hbar/2e)I_m$,
and $V_0 = (\hbar/2e)I_1$.

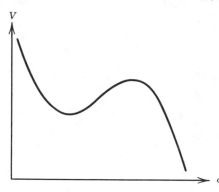

V

ϕ

Figure 21. Inclined cosine potential. This potential is discussed by Schleich et al.[49] for quantum noise in a dithered-ring laser gyro.

Eq. 14.47 may not be solved analytically and again recourse must be mode to the underlying probability density diffusion equation. To circumvent the difficulties posed by the Kramers equation, we shall for the present ignore "inertial effects"; thus

$$J\ddot{\phi} \approx 0$$

This corresponds to ignoring the effect of the capacitance of the junction. We may thus discuss the circuit using the Smoluchowski equation. Equation (14.47) is now written

$$\zeta\dot{\phi}(t) + V_0\sin\phi(t) - (F_0 + F_1\cos\omega t) = \lambda(t)$$

By definition

$$V_0\sin\phi(t) - (F_0 + F_1\cos\omega t) = \frac{\partial V}{\partial\phi}$$

and thus the potential energy of the system is (called the tilted cosine potential)

$$V(\phi, t) = -V_0\cos\phi(t) - (F_0 + F_1\cos\omega t)\phi(t)$$

The Smoluchowski equation for a particle rotating in two dimensions in a potential $V(\phi, t)$ is

$$\frac{\partial f(\phi, t)}{\partial t} = \frac{\partial}{\partial\phi}\left(\frac{f}{\zeta}\frac{\partial V}{\partial\phi}\right) + \frac{kT}{\zeta}\frac{\partial^2 f}{\partial\phi^2} \qquad (14.48)$$

We are interested in the physical situation where the bias current has been switched on in the infinite past, so that all transient effects due to the imposition of that current have died away. We therefore again seek the stationary solution of Eq. (14.48). We first assume that the probability density function $f(\phi, t)$ may be expanded in the Fourier series (we seek the periodic solutions of Eq. 14.48)

$$f = \sum_{p=-\infty}^{p=\infty} a_p(t) e^{ip\phi} \tag{14.49}$$

Now f must be entirely real, so the a_p must satisfy

$$\left.\begin{array}{c} a_p(t) = a^*_{-p}(t) \\ a_{-p}(t) = a^*_p(t) \end{array}\right\} \tag{14.50}$$

Ensemble averages such as the characteristic function $\langle e^{-ip\phi} \rangle$ may now be written down from Eq. (14.49), since

$$\langle e^{ip\psi} \rangle = \frac{\int_0^{2\pi} f e^{-ip\phi} \, d\psi}{\int_0^{2\pi} f d\phi} = \frac{a_p(t)}{a_0} \tag{14.51}$$

where a_0 is constant in time, as will become apparent later.

We note that

$$\langle \cos p\phi \rangle = \frac{a_{-p} + a_p}{2a_0} = \frac{a_p + a^*_p}{2a_0}$$

$$\langle \sin p\phi \rangle = \frac{a_{-p} - a_p}{2ia_0} = \frac{a^*_p - a_p}{2ia_0}$$

Writing Eq. (14.48) as

$$\frac{\partial f}{\partial t} = \frac{f}{\zeta} V_0 \cos\phi + \frac{1}{\zeta} \frac{\partial f}{\partial \phi} \left[V_0 \sin\phi - (F_0 + F_1 \cos\omega t) \right] + \frac{kT}{\zeta} \frac{\partial^2 f}{\partial \phi^2} \tag{14.52}$$

we substitute for f from Eq. (14.49), and on using the properties of the circular functions, find that the a_p satisfy

$$\dot{a}_p(t) + \frac{kT}{\zeta} \left[p^2 + \frac{(F_0 + F_1 \cos\omega t)}{kT} ip \right] a_p(t) = \frac{V_0 p}{2\zeta} \left[a_{p-1}(t) - a_{p+1}(t) \right] \tag{14.53}$$

This set of differential-difference equations may be rewritten as an integral equation by regarding the right-hand side as a forcing function just as we did in the dielectric case. To do this we recall that the first-order linear differential equation

$$\frac{dy(t)}{dt} + \alpha(t)y(t) = g(t) \tag{14.54}$$

has the stationary solution

$$y(t) = e^{-\int \alpha(t)\,dt} \int_{-\infty}^{t} e^{\int \alpha(u)\,du} g(u)\,du \tag{14.55}$$

where $e^{\int \alpha(t)\,dt}$ is termed the integrating factor. Hence this is

$$e^{(ip/\zeta)(F_0 t + F_1 \sin \omega t/\omega)} e^{p^2(kT/\zeta)t} \tag{14.56}$$

We recall that

$$\frac{F_0}{\zeta} = \frac{2e}{\hbar} I_{dc} R = \frac{2eV_{dc}}{\hbar} = \dot{\phi}_{ss} \tag{14.57}$$

which we may regard as the steady-state rate of change of phase, that is, the frequency of operation of the device in the absence of the alternating current. We write

$$\dot{\phi}_{ss} = \omega_0$$

so that expression (14.56) becomes

$$e^{i[p\omega_0 t + \gamma p(\omega_0/\omega)\sin \omega t]} e^{p^2(kT/\zeta)t}, \qquad \gamma = \frac{F_1}{F_0} \tag{14.58}$$

Thus the integrating factor has been expressed in terms of the ratio between the amplitudes of the bias current and of the ac current and the ratio of the steady-state rate of change of phase in the absence of an alternating current to the frequency of the alternating current. Expression (14.58), apart from the $\exp[p^2(kT/\zeta)t]$ term, represents a frequency-modulated signal with carrier frequency $p\omega_0$ and signal frequency ω. It is useful to write (14.58) in terms of a modulation index m_ω defined by

$$pm_\omega = \gamma \frac{\omega_0}{\omega} p, \qquad \gamma = \frac{I_m}{I_{dc}} \tag{14.59}$$

so that (14.58) becomes

$$e^{i(p\omega_0 t + pm_\omega \sin \omega t)} e^{p^2(kT/\zeta)t} \tag{14.60}$$

The other parameters in Eq. (14.53) are V_0/ζ and kT/ζ. We note that

$$\frac{V_0}{\zeta} = \frac{\dfrac{\hbar}{2e} I_1}{(\hbar/2e)^2/R} = \frac{2e I_1 R}{\hbar} \tag{14.61}$$

which again has the dimensions of frequency. We shall denote it by ω_1. The quantity

$$\tau_0 = \frac{\zeta}{kT} = \left(\frac{\hbar}{2e}\right)^2 \frac{1}{RkT} \tag{14.62}$$

plays the role of the relaxation time in the dielectric problem. Using all these parameters, we may now write the complete solution as

$$a_p(t) = \frac{\omega_1 p}{2} \int_{-\infty}^{t} e^{-(p^2/\tau_0)(t-u)} e^{-ip(\omega_0 t + m_\omega \sin \omega t)} e^{ip(\omega_0 u + m_\omega \sin \omega u)}$$
$$\times \left[a_{p-1}(u) - a_{p+1}(u) \right] du \tag{14.63}$$

This is an integral equation for the unknown functions $a_p(t)$ and is exact. We note that Eq. (14.53) implies that

$$\dot{a}_0(t) = 0$$

so that

$$a_0 = \text{constant}$$

This means that Eq. (14.63) may be written in terms of the characteristic function as

$$\left\langle e^{-ip\phi(t)} \right\rangle = \frac{\omega_1 p}{2} \int_{-\infty}^{t} e^{-(p^2/\tau_0)(t-u)} e^{-ip(\omega_0 t + m_\omega \sin \omega t)}$$
$$\times e^{ip(\omega_0 u + m_\omega \sin \omega u)} \left[\left\langle e^{-i(p-1)\phi(u)} \right\rangle - \left\langle e^{-i(p+1)\phi(u)} \right\rangle \right] du \tag{14.64}$$

It is interesting to compare this with the dielectric problem, where

$$\langle P_n(\cos\theta)\rangle = \frac{n(n+1)\gamma}{2\tau} \int_{-\infty}^{t} e^{-n(n+1)(t-u)/2\tau} e(u)$$

$$\times \left[\langle P_{n-1}(\cos\theta(u))\rangle - \langle P_{n+1}(\cos\theta(u))\rangle\right] du \quad (14.65)$$

The solution of Eq. (14.64) need only be determined for positive values of p on account of the relation

$$a_{-p} = a_p^* \qquad (14.66)$$

To proceed further we consider the definition of the Bessel functions $J_n(z)$:

$$\exp\frac{1}{2}z\left(\xi - \frac{1}{\xi}\right) = \sum_{n=-\infty}^{n=\infty} \xi^n J_n(z)$$

If we let

$$\exp\tfrac{1}{2}z(e^{i\theta} - e^{-i\theta}) = \exp iz\sin\theta = \sum_{n=-\infty}^{n=\infty} e^{in\theta} J_n(z)$$

and if we let $\theta = -\theta$

$$\exp(-iz\sin\theta) = \sum_{n=-\infty}^{n=\infty} e^{-in\theta} J_n(z)$$

The function

$$e^{-p^2 t/\tau_0} e^{-i(p\omega_0 t + pm_\omega \sin\omega t)}$$

may thus be written

$$e^{-p^2 t/\tau_0} e^{-ip\omega_0 t} \sum_{q=-\infty}^{q=\infty} e^{-iq\omega t} J_q(pm_\omega)$$

This is a sum of exponentially decaying cissoidal functions, with the amplitude of each function being determined by the Bessel function $J_q(pm_\omega)$, whose value depends on the modulation index pm_ω. Note that as p increases the depth of modulation increases. Since

$$e^{-ipm_\omega\sin\omega t} e^{ipm_\omega\sin\omega u} = \sum_{q,r} e^{-i\omega(qt-ru)} J_q(pm_\omega) J_r(pm_\omega) \quad (14.67)$$

Eq. (14.63) may be rewritten

$$a_p(t) = \frac{\omega_1 p}{2} \int_{-\infty}^{t} e^{-(p^2/\tau_0)(t-u)} \sum_{q,r} J_q(pm_\omega) J_r(pm_\omega) e^{-i[p\omega_0(t-u)+\omega(qt-ru)]}$$

$$\times \left[a_{p-1}(u) - a_{p+1}(u) \right] du \tag{14.68}$$

Due to our expansion in Bessel functions, we can perform any integration that may be required in the iterative solution of Eq. (14.68). To get a first approximation, we use Picard's method exactly as for the dielectric case. Thus we write down the scheme

$$a_0 = \text{constant} \tag{14.69}$$

$$a_1(t) = \frac{\omega_1}{2} \int_{-\infty}^{t} e^{-(t-u)/\tau_0} \sum_{q,r} J_q(m_\omega) J_r(m_\omega)$$

$$\times e^{-i[\omega_0(t-u)+\omega(qt-ru)]} \left[a_0 - a_2(u) \right] du, \tag{14.70}$$

$$a_2(t) = \frac{\omega_1}{2} \int_{-\infty}^{t} e^{-4(t-u)/\tau_0} \sum_{q,r} J_q(2m_\omega) J_r(2m_\omega)$$

$$\times e^{-i[2\omega_0(t-u)+\omega(qt-ru)]} \left[a_1(u) - a_3(u) \right] du, \tag{14.71}$$

and so on.

To a first approximation the solution of Eq. (14.68) is (we set $a_2 = 0$)

$$a_1(t) = \frac{\omega_1}{2} \int_{-\infty}^{t} e^{-(t-u)/\tau_0} \sum_{q,r} J_q(m_\omega) J_r(m_\omega) e^{-i[\omega_0(t-u)+\omega(qt-ru)]} a_0 \, du$$

$$\tag{14.72}$$

To obtain a second approximation we take Eq. (14.71) and substitute it into Eq. (14.70), ignoring the $a_3(u)$ term, and so on. This is the second order response. The process rapidly becomes very tedious due to the double sums of Bessel functions, so we content ourselves with giving the first approximation here. The approximation scheme is in fact more easily treated using two-sided Fourier transforms, as we show below.

Equation (14.72) yields on integration

$$\langle e^{-i\phi} \rangle = \frac{a_1(t)}{a_0} = \frac{1}{2} \sum_{q,r} \frac{\omega_1 \tau_0 J_r(m_\omega) J_q(m_\omega) e^{-i\omega(q-r)t}}{1 + i(\omega_0 + r\omega)\tau_0} \tag{14.73}$$

This should be compared with the first approximation in the absence of the alternating current, where

$$\langle e^{-i\phi}\rangle = \frac{a_1}{a_0} = \frac{1}{2}\frac{\omega_1\tau_0}{1+i\omega_0\tau_0} \tag{14.74}$$

which is, as it must be, *independent* of the time. It is helpful to recall that

$$\omega_1 = \frac{2eI_1R}{\hbar}, \qquad \omega_0 = \frac{2eI_{dc}R}{\hbar},$$

$$\tau_0 = \left(\frac{\hbar}{2e}\right)^2\frac{1}{R}kT, \qquad m_\omega = \gamma\frac{\omega_0}{\omega} \tag{14.75}$$

Note that from Eq. (14.73)

$$\langle\cos\phi\rangle = \frac{1}{2}\sum_{q,r}\frac{\omega_1\tau_0 J_r(m_\omega)J_q(m_\omega)}{1+(\omega_0+r\omega)^2\tau_0^2}$$

$$\times\left[\cos\omega(q-r)t-(\omega_0+r\omega)\tau_0\sin\omega(q-r)t\right] \tag{14.76}$$

$$\langle\sin\phi\rangle = \frac{1}{2}\sum_{q,r}\frac{\omega_1\tau_0 J_r(m_\omega)J_q(m_\omega)}{1+(\omega_0+r\omega)^2\tau_0^2}$$

$$\times\left[\sin\omega(q-r)t+(\omega_0+r\omega)\tau_0\cos\omega(q-r)t\right] \tag{14.77}$$

and that from Eq. (14.74)

$$\langle\cos\phi\rangle = \frac{1}{2}\frac{\omega_1\tau_0}{1+\omega_0^2\tau_0^2} \tag{14.78}$$

$$\langle\sin\phi\rangle = \frac{1}{2}\frac{\omega_0\omega_1\tau_0^2}{1+\omega_0^2\tau_0^2} \tag{14.79}$$

Equation (14.73), which is only the first approximation to $\langle e^{-i\phi}\rangle$, shows that the imposition of the alternating current has a very complicated effect on the steady-state solution. Essentially every harmonic of the fundamental is represented in the solution, and the amplitude of each frequency component is determined by the Bessel functions. Equation (14.73) will contain *time-independent* terms; these occur when $q=r$. For sufficient modulation depth these can be made to vanish, e.g. for $q=r=0$

$$J_0(2\cdot405) = 0$$

and so on.*

---

* These time (but not frequency) independent terms are the origin of the steps in the d. c. voltage/current characteristic of Josephson junctions when driven by alternating currents. They are generated by resonances of harmonics of the input signal arising from the nonlinear Eq. 14.47. If inertial effects are included subharmonic steps will also be generated.

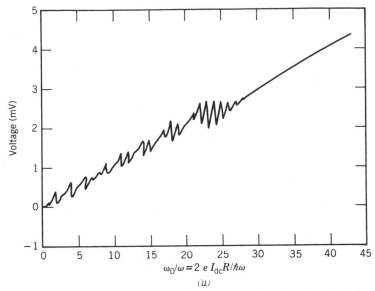

(a)

Figure 22. *Voltage–current characteristic for high-resistance Josephson junction in first approximation.*

(*a*) and (*b*) Depiction of the time-independent portion of the response for $T = 1.5$ and 4 K at 50 GHz. (*c*) and (*d*) The same response at 100 GHz. Note how the increase of T from 1.5 K to 4 K has considerably reduced the "sharpness" of the harmonic voltage spikes. (*e*), (*f*), and (*g*) Show the voltage–current characteristic at 200, 400, and 800 GHz, respectively. Note how the increase in frequency reduces the number of voltage steps (spikes). These characteristics were calculated by picking off the time-independent portion of Eq. (14.77), i.e.,

$$\frac{1}{2} \sum_q \frac{\omega_1^2 \tau_0^2 J_q^2 (m_\omega)[\omega_0 + r\omega]}{1 + (\omega_0 + r\omega)^2 \tau_0^2}$$

and using this expression for $\langle \sin \phi \rangle$ to calculate $\langle \dot{\phi} \rangle$ and thus $\langle v \rangle$ from the Josephson equation. A similar expression was obtained by Schleich et al.[49] for the Ring Laser Gyro. The x-axis units are dc-equivalent frequency normalized by the input ac frequency. These are the natural units to use because they allow the harmonic steps to be seen clearly.[49] These approximate analytic results are in keeping qualitatively with the experimental results of Danchi et al. [*Phys. Rev. B* **30**, 2503–2515, (1984)]. They find that the number of steps which may be obtained decreases at increased frequency and the size of the steps decreases proportionally with the frequency for high-resistance, small-area junctions. We have obtained harmonic spikes rather than steps because the expression cited above is only the first order approximation. It is hoped that iteration of further terms would improve the results shown. All figures assumed 176 ohms, high-resistance junction with a supercurrent of 2.5 μA and an alternating current amplitude of 15 μA. All figures, except (*a*) and (*c*), assumed $T = 4$ K and the summation of the Bessel functions was carried out over as large a number as was necessary. (The number needed depends on the modulation index m_ω. This varied from 15 to 120 Bessel functions). (Credit is due for this figure and Fig. 23 to P. M. Corcoran.) The above results are got from a power series expansion in V_0, cut off at the term **linear** in V_0. If a perturbation expansion of Eq. 14.53 in the ratio F_1 / V_0 is made the harmonic steps are well reproduced. (*v.* W. T. Coffey and P. M. Corcoran Proc. NASECODE IV Conference, Trinity College, Dublin 1985, Ed. J. J. H. Miller, Boole Press Dublin.)

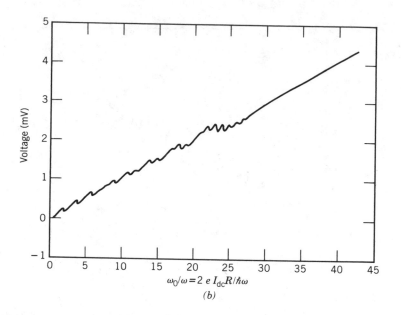

$\omega_0/\omega = 2\,e\,I_{dc}R/\hbar\omega$

(b)

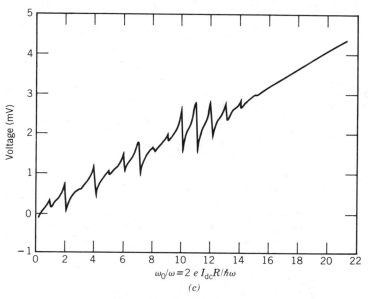

$\omega_0/\omega = 2\,e\,I_{dc}R/\hbar\omega$

(c)

Figure 22. Continued

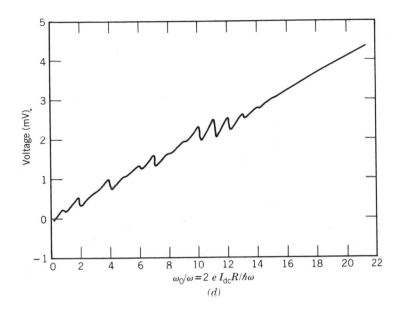

(d)

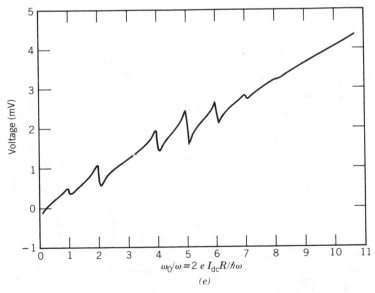

(e)

Figure 22. Continued

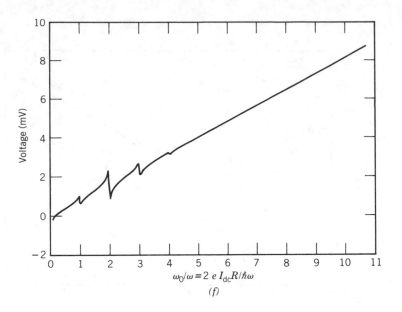

(f)

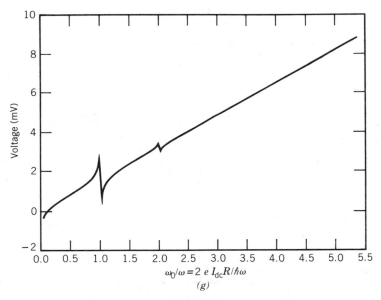

(g)

Figure 22. Continued

We have mentioned that successive iteration of Eq. (14.63) or (14.64) becomes extremely tedious. Perhaps a better way to approach the problem is by means of the two-sided Fourier transform. We rewrite Eq. (14.68) as

$$a_p(t) = \frac{\omega_1 p}{2} \sum_{q,r} J_q(pm_\omega) J_r(pm_\omega) e^{-(\alpha p^2 + ip\omega_0)t} e^{-iq\omega t}$$
$$\times \int_{-\infty}^{t} e^{(\alpha p^2 + ip\omega_0)u} e^{ir\omega u} \left[a_{p-1}(u) - a_{p+1}(u) \right] du \quad (14.80)$$

where for convenience we have written

$$\alpha = \frac{1}{\tau_0}$$

Let us now define a function

$$g(t) = e^{-(\alpha p^2 + ip\omega_0)t} \int_{-\infty}^{t} e^{(\alpha p^2 + ip\omega_0)u} e^{ir\omega u} \left[a_{p-1}(u) - a_{p+1}(u) \right] du$$
$$(14.81)$$

The function whose transform we wish to obtain is simply $e^{-iq\omega t} g(t)$.

Consider first the problem of finding the Fourier transform of $g(t)$. We note the theorem

$$\mathcal{F} \left\{ \exp(-at) \int_{-\infty}^{t} [\exp(au)] \chi(u) \, du \right\} = \frac{\mathcal{F}\{\chi(t)\}}{i\Omega + a} \quad (14.82)$$

where

$$\mathcal{F}\{\chi(t)\} = \int_{-\infty}^{\infty} \chi(t) e^{-i\Omega t} \, dt \quad (14.83)$$

and a is a constant*. This theorem should hold for complex a if $\mathrm{Re}(a) > 0$. We take $\lambda(u)$ as

$$e^{ir\omega u} \left[a_{p-1}(u) - a_{p+1}(u) \right]$$

Thus with theorem (14.82)

$$\mathcal{F}\{g(t)\} = \frac{1}{i\Omega + a} \left[\mathcal{F}\left\{ e^{ir\omega t} a_{p-1}(t) \right\} - \mathcal{F}\left\{ e^{ir\omega t} a_{p+1}(t) \right\} \right] \quad (14.84)$$

The two terms in square braces are convolution integrals, but they may be

*a is not to be confused with $a_p(t)$.

simplified using the shifting theorem of Fourier transformation. Since

$$\mathscr{F}\{h(t)e^{-i\xi t}\} = H(\xi + \Omega)$$

we then have (where A denotes transformed quantities)

$$\mathscr{F}\{g(t)\} = \frac{1}{i\Omega + a}\left[A_{p-1}(\Omega - r\omega) - A_{p+1}(\Omega - r\omega)\right] \qquad (14.85)$$

Applying the shifting theorem again, we obtain

$$\mathscr{F}\{e^{-iq\omega t}g(t)\} = \frac{1}{i(\Omega + q\omega) + a}\left[A_{p-1}(\Omega + \overline{q - r}\omega) - A_{p+1}(\Omega + \overline{q - r}\omega)\right]$$

$$(14.86)^*$$

where

$$\mathscr{F}\{a_p(t)\} = A_p(\Omega), \quad a = ip\omega_0 + \alpha p^2$$

and thus

$$A_p(\Omega) = \frac{\omega_1 p}{2}\sum_{q,r}\frac{J_q(pm_\omega)J_r(pm_\omega)}{i(\Omega + q\omega + p\omega_0) + \alpha p^2}$$

$$\times\left[A_{p-1}(\Omega + \overline{q - r}\omega) - A_{p+1}(\Omega + \overline{q - r}\omega)\right] \qquad (14.87)$$

which is the Fourier transform of Eq. (14.80). This is superior to the time-domain formulation of the problem, because it is purely algebraic and therefore much easier to work with.

It is helpful to check our calculations with those previously given. For $p = 1$ we have

$$A_1(\Omega) = \frac{\omega_1}{2}\sum_{q,r}\frac{J_q(m_\omega)J_r(m_\omega)}{i(\Omega + q\omega + \omega_0) + \alpha}A_0(\Omega + \overline{q - r}\omega) + \text{higher-order terms}$$

$$(14.88)$$

Some care must be exercised in finding the Fourier transform of a_0, namely A_0. We have to make use of the relation

$$\frac{1}{2\pi}\int_{-\infty}^{\infty}e^{\pm ixy}\,dx = \delta(y)$$

---

*The symbol — in these equations denotes a brace i.e. $\overline{q - r} \equiv q - r$.

where $\delta(y)$ is the Dirac delta function. Thus

$$\mathscr{F}\left\{e^{ir\omega t}a_0\right\} = 2\pi\delta(\Omega - r\omega)a_0$$

and so

$$\mathscr{F}\left\{e^{-iq\omega t}e^{ir\omega t}a_0\right\} = 2\pi\delta(\Omega + \overline{q - r\omega})a_0$$

whence

$$A_1(\Omega) = \frac{\omega_1}{2}\sum_{q,r}\frac{2\pi J_q(m_\omega)J_r(m_\omega)\delta(\Omega + \overline{q - r\omega})a_0}{i(\Omega + q\omega + \omega_0) + a} \qquad (14.89)$$

Equation (14.89) may be compared with our previous result for $a_1(t)$ obtained by direct integration [i.e., equation (14.73)]. We have

$$a_1(t) = \mathscr{F}^{-1}\{A_1(\Omega)\}$$

$$= \frac{\omega_1}{2}\sum_{q,r}\frac{2\pi J_q(m_\omega)J_r(m_\omega)}{2\pi}\int_{-\infty}^{\infty}\frac{\delta(\Omega + \overline{q - r\omega})e^{i\Omega t}\,d\Omega}{i(\Omega + q\omega + \omega_0) + \alpha}a_0$$

$$(14.90)$$

which with the delta-function formula

$$\int_{-\infty}^{\infty} f(x - a)\delta(x)\,dx = f(-a)$$

becomes

$$a_1(t) = \frac{\omega_1}{2}\sum_{q,r}\frac{J_q(m_\omega)J_r(m_\omega)e^{-i\omega(q-r)t}}{\alpha + i(\omega_0 + r\omega)}a_0 \qquad (14.91)$$

which is precisely Eq. (14.73). Note that in this expression for $a_1(t)$, and indeed in those for all the higher-order $a_p(t)$, the only time variation in the system is that arising from the imposed alternating current. Unlike in an FM signal, there is no time variation involving $e^{i\omega_0 t}$ present in Eq. (14.91). All the higher $A_p(\Omega)$, and hence $a_p(t)$, may be found by iteration of Eq. (14.87)* and inversion of the result back into the time domain using the properties of the delta function. The time variation of the phase and the voltage may now be readily calculated, since

$$\langle v(t)\rangle = \frac{\hbar}{2e}\langle\dot{\phi}\rangle$$

---

*In practice numerical calculation using this iterative method is hard to carry out due to the difficulty of multiple summation over large products of Bessel functions.

and from the original equation the ensemble averages are

$$\langle\dot\phi\rangle = -\frac{V_0}{\zeta}\langle\sin\phi\rangle + \frac{1}{\zeta}(F_0 + F_1\cos\omega t)$$

$$\langle\dot\phi\rangle = -\omega_1\langle\sin\phi\rangle + \omega_0(1+\gamma\cos\omega t),$$

from which we get

$$\langle\dot\phi\rangle = -\frac{\omega_1}{2}\sum_{q,r}\frac{\omega_1\tau_0 J_r J_q\left[\sin\omega(q-r)t + (\omega_0 + r\omega)\tau_0\cos\omega(q-r)t\right]}{1+(\omega_0 + r\omega)^2\tau_0^2}$$

$$+ \omega_0(1+\gamma\cos\omega t)$$

plus further frequency-dependent terms arising from the second- and higher-order iterations.

This completes our preliminary analysis* of the Josephson junction under the influence of an ac voltage and noise. It is evident that if an alternating current is injected into the junction, then all the harmonics of that alternating current will appear in the voltage across the junction.

It is interesting to consider how the problem of calculating the a_p may be simplified when the alternating current is absent. In this particular case, we may make use of continued-fraction methods to solve for all the a_p. If we return to Eq. (14.53) and set $\dot a_p = 0$ to obtain the steady-state solution. We find that

$$\frac{kT}{\zeta}\left(p^2 + F_0\frac{ip}{kT}\right)a_p = \frac{V_0 p}{2\zeta}(a_{p-1} - a_{p+1}) \qquad (14.92)$$

or in the usual notation

$$\left(\alpha p^2 + \omega_0 ip\right)a_p = \frac{\omega_1 p}{2}(a_{p-1} - a_{p+1})$$

We divide by a_{p-1} to get

$$\left(\alpha p^2 + ip\omega_0\right)\frac{a_p}{a_{p-1}} = \frac{\omega_1 p}{2}\left(1 - \frac{a_{p+1}}{a_{p-1}}\right) \qquad (14.93)$$

Let us write

$$S_p = \frac{a_p}{a_{p-1}}$$

*See also fig. 23.

Then

$$\left(\alpha p^2 + ip\omega_0\right)S_p = \frac{\omega_1 p}{2}\left(1 - S_p S_{p+1}\right)$$

or

$$S_p = \frac{\frac{1}{2}\omega_1 p}{\alpha p^2 + ip\omega_0 + \frac{\omega_1 p}{2}S_{p+1}} \tag{14.94}$$

that is

$$S_p = \frac{\frac{1}{2}\omega_1 p}{\alpha p^2 + ip\omega_0 + \dfrac{\dfrac{\omega_1 p}{2}\dfrac{\omega_1(p+1)}{2}}{\alpha(p+1)^2 + i\omega_0(p+1) + \dfrac{\omega_1(p+1)}{2}S_{p+2}}} \tag{14.95}$$

For example,

$$\frac{u_1}{a_0} = \frac{\frac{1}{2}\omega_1}{\alpha + i\omega_0 + \dfrac{\omega_1}{2}\dfrac{\omega_1}{2^2\alpha + 2i\omega_0 + \omega_1 S_3}} \tag{14.96}$$

If we truncate* Eq. (14.96) at S_3 we find that

$$S_1 = \frac{\frac{1}{2}\omega_1(2\alpha + i\omega_0)}{(\alpha + i\omega_0)(2\alpha + i\omega_0) + 0.25\omega_1^2} \tag{14.97}$$

or

$$S_1 = \frac{\dfrac{1}{2}\left(\dfrac{2kT}{V_0} + \dfrac{iF_0}{V_0}\right)}{\left[\left(\dfrac{kT}{V_0} + \dfrac{iF_0}{V_0}\right)\left(\dfrac{2kT}{V_0} + \dfrac{iF_0}{V_0}\right) + 0.25\right]} \tag{14.98}$$

Risken finds [Eq. (11.51) of his book]

$$\left\langle e^{-ix}\right\rangle = \frac{0.25}{\dfrac{\Theta}{d} + \dfrac{iF}{d}} + \frac{0.25}{\dfrac{2\Theta}{d} + \dfrac{iF}{d}} + \frac{0.25}{\dfrac{3\Theta}{d} + \dfrac{iF}{d}} \cdots \tag{14.99}$$

in his notation.[22] Equations (14.96) and (14.99) are in exact agreement.

*i.e. we set $S_3 = 0$

We thus see that the problem of calculating the steady-state averages may be treated very readily using continued fractions. This is much easier than solving the Smoluchowski equation directly using the modified Bessel functions, as was done by Stratonovitch[38] and Viterbi.[27]

The two-sided Fourier transform of Eq. (14.95), namely Eq. (14.87) with $\omega = 0$, may also be written as a continued fraction as follows. If $\omega = 0$, Eq. (14.87) becomes

$$A_p(\Omega) = \frac{\omega_1 p}{2} \frac{\left[A_{p-1}(\Omega) - A_{p+1}(\Omega)\right]}{i(\Omega + p\omega_0) + \alpha p^2} \qquad (14.100)$$

or

$$\left[i(\Omega + p\omega_0) + \alpha\right] \frac{A_p(\Omega)}{A_{p-1}(\Omega)} = \frac{\omega_1 p}{2} \left[1 - \frac{A_{p+1}(\Omega)}{A_{p-1}(\Omega)}\right] \qquad (14.101)$$

Writing

$$\bar{S}_p(\Omega) = \frac{A_p(\Omega)}{A_{p-1}(\Omega)}$$

we have

$$\left[i(\Omega + p\omega_0) + \alpha p^2\right] \bar{S}_p(\Omega) = \frac{\omega_1 p}{2} \left[1 - \bar{S}_p \bar{S}_{p+1}\right]$$

or

$$\bar{S}_p(\Omega) \left[i(\Omega + p\omega_0) + \alpha p^2 + \frac{\omega_1 p}{2} \bar{S}_{p+1}\right] = \frac{\omega_1 p}{2}$$

whence

$$\frac{A_p(\Omega)}{A_{p-1}(\Omega)} = \bar{S}_p(\Omega) = \frac{\dfrac{\omega_1 p}{2}}{i(\Omega + p\omega_0) + \alpha p^2 + \dfrac{\omega_1 p}{2} \bar{S}_{p+1}} \qquad (14.102)$$

This is the continued-fraction representation of the Fourier transform of a_p when $\omega = 0$. It would be very useful to extend this to the case where $\omega \neq 0$, that is, Eq. (14.87). It is not yet apparent how this might be achieved,* however-

---

*The solution has recently been discussed using matrix continued fractions see ref. 49. In the linear response case, the solution may be expressed as a scalar continued fraction see caption to fig. 23.

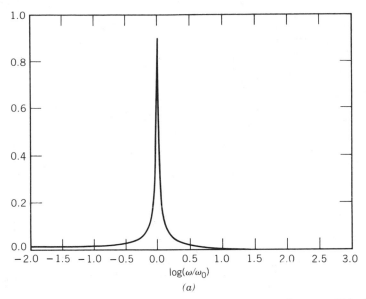

$$\log(\omega/\omega_0)$$

(a)

Figure 23. The *linear response* to F_1 may be written in terms of the $S_n(\omega)$. Write in (14.53)

$$a_p(t) = u_p^{(0)}(t) + a_p^{(F_1)} + O(F_1^2), \qquad a_p^{(0)} = (a_p)_{F_1 = 0}$$

Thus with (14.93) using stationarity

$$
\frac{a_1(t)}{a_0} = S_1(0)\left\{ 1 - \frac{iF_1}{\zeta\omega_1} \cdot S_1(\omega)[1 - S_2(0)S_2(\omega) \cdots] e^{i\omega t} \right.
$$

$$
\left. - \frac{iF_1}{\zeta\omega_1} S_1(-\omega)[1 - S_2(0)S_2(-\omega) \cdots] e^{-i\omega t} \right\} \qquad (14.102a)
$$

Figures 23a, 23b, and 23c show the magnitude of the time-dependent portion of the response term $\langle \sin\phi \rangle$ plotted as a function of the input ac frequency. This exhibits resonance behavior. These calculations have been made by truncation of the continued fraction expansion Eq. (14.102a) after (a) a single term, (b) three terms, and (c) seven terms. Frequency has been normalized about ω_0 and $\omega_1/\omega_0 = 1.2$. The ratio of alternating current to direct current is $1.5:10$ and 500 increments on the log frequency scale have been used. Note the secondary resonance effects that occur in (b) due to truncation of the CFE. In (c) they have been eliminated due to damping and the CFE may be assumed to have converged. The truncation after a single term may be written

$$
\langle \sin\psi \rangle = \frac{\omega_1}{2} \frac{\Omega(t)}{\alpha^2 + \omega_0^2}, \qquad \Omega(t) = \omega_0 \left[1 + \gamma \hat{A}_1 \cos(\omega t - \chi(\omega)) \right]
$$

$$
\hat{A}_1 = \left[\frac{(\alpha^2 + \omega_0^2)(\alpha^2 + \omega^2 - \omega_0^2) + \omega^2\alpha^2(\alpha^2 + \omega^2 - 3\omega_0^2)^2}{(\alpha^2 + \omega^2 - \omega_0^2)^2 + 4\alpha^2\omega_0^2} \right]^{1/2}
$$

$$
\tan\chi = \frac{\omega\alpha\left[1 - 3\omega_0^2/(\alpha^2 + \omega_0^2)\right]}{(\alpha^2 + \omega_0^2)\left[1 - \omega_0^2/(\alpha^2 + \omega_0^2)\right]}
$$

This clearly exhibits resonant behavior.

197

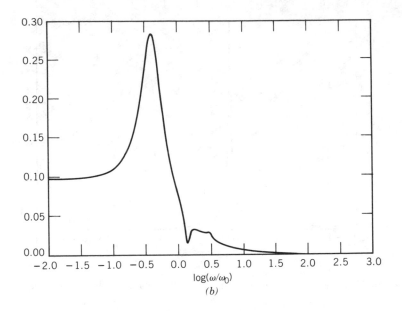

(b)

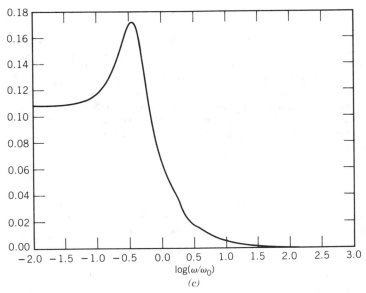

(c)

Figure 23.   Continued

ever, the main difficulty being that Eq. (14.87), in addition to being an equation in the differences p, $p-1$, and $p+1$, is *also* one in the differences Ω and $\overline{q - r\omega}$. (This implies a matrix cf).

There is one particular case of the dynamical problem that can be solved by continued fractions, however, namely the fundamental solution where we are given a sharp value for the phase at a time $t = 0$. We proceed as follows the equations of motion of the system at any time $t > 0$ are

$$\dot{a}_p(t) + \left(\alpha p^2 + ip\omega_0\right)a_p(t) = \frac{\omega_1 p}{2}\left[a_{p-1}(t) - a_{p+1}(t)\right], \qquad t > 0$$

$$(14.103)$$

Since we are now dealing with the time interval $(0, \infty)$ rather than $(-\infty, \infty)$ (as is appropriate to the steady state), we utilize the one-sided Fourier or the Laplace transform, rather than the two-sided Fourier transform of the previous examples. The Laplace transform of Eq. (14.103) is

$$sA_p(s) - a_p(0) + \left(\alpha p^2 + ip\omega_0\right)A_p(s) = \frac{\omega_1 p}{2}\left[A_{p-1}(s) - A_{p+1}(s)\right]$$

$$(14.104)$$

or

$$\left(s + \alpha p^2 + ip\omega_0\right)A_p(s) - a_p(0) = \frac{\omega_1 p}{2}\left[A_{p-1}(s) - A_{p+1}(s)\right]$$

$$(14.105)$$

Let us now divide by $A_{p-1}(s)$ and write

$$R_P(s) = \frac{A_p(s)}{A_{p-1}(s)} \qquad (14.106)$$

Equation (14.105) then becomes

$$\left(s + \alpha p^2 + ip\omega_0\right)R_p(s) - \frac{a_p(0)}{A_{p-1}(s)} = \frac{\omega_1 p}{2}\left[1 - R_p(s)R_{p+1}(s)\right]$$

$$(14.107)$$

or

$$\left[s + \alpha p^2 + ip\omega_0 + \frac{\omega_1 p}{2}R_{p+1}(s)\right]R_p(s) = \frac{\omega_1 p}{2} + \frac{a_p(0)}{A_{p-1}(s)}$$

$$(14.108)$$

By analogy with the theory of linear differential equations, we can regard Eq. (14.108) as having a "particular" solution and a "complementary" solution. Let us denote the complementary solution or solution to the homogeneous equation by $S_p(s)$. Then $S_p(s)$ satisfies

$$\left[s + \alpha p^2 + i p \omega_0 + \frac{\omega_1 p}{2} S_{p+1}(s) \right] S_p(s) = \frac{\omega_1 p}{2} \tag{14.109}$$

from which we obtain, just as in the steady-state solution [Eq. (14.102)],

$$S_p(s) = \frac{\dfrac{\omega_1 p}{2}}{s + \alpha p^2 + i p \omega_0 + \dfrac{\omega_1 p}{2} S_{p+1}(s)} \tag{14.110}$$

or

$$S_p(s) = \cfrac{\dfrac{\omega_1 p}{2}}{s + \alpha p^2 + i p \omega_0 + \cfrac{\dfrac{\omega_1 p}{2} \dfrac{\omega_1 (p+1)}{2}}{s + \alpha(p+1)^2 + i(p+1)\omega_0 + \omega_1 \dfrac{(p+1)}{2} S_{p+2}}} \tag{14.111}$$

and so on. To complete the solution we need the solution of the inhomogeneous recurrence relation

$$\left[s + \alpha p^2 + i p \omega_0 + \frac{\omega_1 p}{2} R_{p+1}(s) \right] R_p(s) = \frac{\omega_1 p}{2} + \frac{a_p(0)}{A_{p-1}(s)} \tag{14.112}$$

Let us write, following the procedure outlined in ref. 25,

$$R_p(s) = S_p(s) + Q_p(s) \tag{14.113}$$

Thus Eq. (14.112) becomes

$$\left\{ s + \alpha p^2 + i p \omega_0 + \frac{\omega_1 p}{2} \left[S_{p+1}(s) + Q_{p+1}(s) \right] \right\} \left[S_p(s) + Q_p(s) \right]$$

$$= \frac{\omega_1 p}{2} + \frac{a_p(0)}{A_{p-1}(s)} \tag{14.114}$$

We may now use the homogeneous solution, Eq. (14.110), to simplify Eq. (14.114). We find that

$$Q_p(s)(s + \alpha p^2 + ip\omega_0) + \frac{\omega_1 p}{2}\left[Q_{p+1}S_p + Q_p(S_{p+1} + Q_{p+1})\right] = \frac{a_p(0)}{A_{p-1}(s)}$$

(14.115)

We now multiply by $A_{p-1}(s)$ to get

$$(s + \alpha p^2 + ip\omega_0)A_{p-1}Q_p + \frac{\omega_1 p}{2}\left[Q_{p+1}S_p A_{p-1} + Q_p A_{p-1}(S_{p+1} + Q_{p+1})\right]$$
$$= a_p(0)$$

(14.116)

Let us write

$$A_{p-1}Q_p = q_p$$

(14.117)

Then

$$(s + \alpha p^2 + ip\omega_0)q_p + \frac{\omega_1 p}{2}\left[Q_{p+1}A_{p-1}(S_p + Q_p) + q_p S_{p+1}\right] = a_p(0)$$

(14.118)

Now

$$R_p = S_p + Q_p = \frac{A_p(s)}{A_{p-1}(s)}$$

(14.119)

and thus

$$R_p A_{p-1} = (S_p + Q_p)A_{p-1} = A_p(s)$$

(14.120)

Equation (14.118) then becomes

$$(s + \alpha p^2 + ip\omega_0)q_p + \frac{\omega_1 p}{2}(Q_{p+1}A_p + q_p S_{p+1}) = a_p(0) \quad (14.121)$$

Thus with Eq. (14.117),

$$(s + \alpha p^2 + ip\omega_0)q_p + \frac{\omega_1 p}{2}(q_{p+1} + q_p S_{p+1}) = a_p(0) \quad (14.122)$$

and solving for q_p we find that

$$q_p\left(s + \alpha p^2 + ip\omega_0 + \frac{\omega_1 p}{2} S_{p+1}\right) = a_p(0) - \frac{\omega_1 p}{2} q_{p+1}$$

or

$$q_p = \frac{a_p(0) - \dfrac{\omega_1 p}{2} q_{p+1}}{s + \alpha p^2 + ip\omega_0 + \dfrac{\omega_1 p}{2} S_{p+1}} \tag{14.123}$$

The complete solution is

$$R_p = S_p + Q_p, \qquad R_p = \frac{A_p}{A_{p-1}}$$

that is,

$$A_p = S_p A_{p-1} + Q_p A_{p-1}$$

or

$$A_p = S_p A_{p-1} + q_p \tag{14.124}$$

from which we get, with Eqs. (14.110) and (14.122),

$$A_p(s) = \frac{\dfrac{\omega_1 p}{2} A_{p-1}(s) + a_p(0) - \dfrac{\omega_1 p}{2} q_{p+1}}{s + \alpha p^2 + ip\omega_0 + \dfrac{\omega_1 p}{2} S_{p+1}(s)} \tag{14.125}$$

The Laplace transform of the density function is then

$$\mathscr{L}\{f(\phi, t)\} = \sum_{p=-\infty}^{p=\infty} \frac{\left\{\dfrac{\omega_1 p}{2}\left[A_{p-1}(s) - q_{p+1}(s)\right] + a_p(0)\right\}}{s + \alpha p^2 + i\omega_0 p + \dfrac{\omega_1 p}{2} S_{p+1}(s)} e^{ip\phi} \tag{14.126}$$

The frequency response of the system is obtained by letting $s = i\omega$ in Eq. (14.125):

$$A_p(\omega) = \frac{\dfrac{\omega_1 p}{2} A_{p-1}(\omega) + a_p(0) - \dfrac{\omega_1 p}{2} q_{p+1}}{i(\omega + p\omega_0) + \alpha p^2 + \dfrac{\omega_1 p}{2} S_{p+1}(\omega)} \tag{14.127}$$

We must next consider the initial conditions. We will in general have (on account of the periodicity of the solution)

$$f(\phi + 2\pi, t|\phi_0, t_0) = f(\phi, t|\phi_0, t_0)$$

Thus at an initial time $t = t_0$,

$$f(\phi, t_0|\phi_0, t_0) = \sum_{p=-\infty}^{p=\infty} \delta(\phi - \phi_0 + 2p\pi)$$

which is periodic in ϕ, as required.

For our present problem we shall take $t_0 = 0$. Now from our set of differential-difference equations for the a_p we know that $\dot{a}_0(t) = 0$ and we require that $f(\phi, t)$ be normalized to unity over $(0, 2\pi)$; thus

$$a_0(t) = \frac{1}{2\pi}$$

With the orthogonality property of the circular functions,

$$a_p(\phi_0, 0) = a_p(0) = \frac{1}{2\pi} e^{-ip\phi_0}$$

A further condition to be imposed on f (just as in the ac case) is that it should be entirely real. This requires that

$$a_p(t) = a^*_{-p}(t)$$

so that one need calculate the a_p only for positive values of p.

This analysis of the Green's function of the Josephson junction when "inertial effects" are ignored (i.e., at zero capacitance) is similar to that given by Cresser et al.[25] in their analysis of the effect of quantum noise on the operation of ring laser gyros. The analysis also applies without significant change to the corresponding problem for a first-order phase-locked loop.

We now resume our discussion of the solution of Eq. (14.103). To gain more insight into the nature of the recurrence relation given by Eq. 14.125, it is helpful to write that equation as

$$A_p(s) = S_p(s)A_{p-1}(s) + q_p(s) \tag{14.128}$$

or

$$\frac{A_p(s)}{A_{p-1}(s)} = S_p(s) + \frac{q_p(s)}{A_{p-1}(s)} \tag{14.129}$$

which for $p = 1$ becomes

$$\frac{A_1(s)}{A_0(s)} = S_1(s) + \frac{q_1(s)}{A_0(s)} \tag{14.130}$$

Note that $A_0(s) = a_0 s^{-1}$, so that Eq. (14.130) becomes

$$\frac{A_1(s)}{A_0(s)} = S_1(s) + \frac{sq_1(s)}{a_0} \tag{14.131}$$

It is helpful to write out the first few iterations of Eq. (14.131). We note that

$$q_1(s) = \frac{a_1(0) - \frac{\omega_1}{2} q_2(s)}{s + \alpha + i\omega_0 + \frac{\omega_1}{2} S_2(s)} \tag{14.132}$$

and writing out S_2 explicitly, we have

$$q_1(s) = \frac{a_1(0) - \frac{\omega_1}{2}\left(\dfrac{a_2(0) - \omega_1 q_3(s)}{s + 4\alpha + 2i\omega_0 + \omega_1 S_3(s)}\right)}{s + \alpha + i\omega_0 + \dfrac{\omega_1}{2} \dfrac{\omega_1}{s + 4\alpha + 2i\omega_0 + \omega_1 S_3(s)}} \tag{14.133}$$

The denominator of this expression is common to both terms in Eq. (14.131), so we may easily write down the complete solution as*

$$\frac{A_1(s)}{A_0(s)} = \frac{\dfrac{\omega_1}{2} + s\left[\hat{a}_1(0) - \dfrac{\omega_1}{2}\left(\dfrac{\hat{a}_2(0) - \omega_1 q_3(s) a_0^{-1}}{s + 4\alpha + 2i\omega_0 + \omega_1 S_3(s)}\right)\right]}{s + \alpha + i\omega_0 + \dfrac{\omega_1}{2} \dfrac{\omega_1}{s + 4\alpha + 2i\omega_0 + \omega_1 S_3(s)}} \tag{14.134}$$

*Eq. 14.134 may also be written in terms of the $S_p(s)$ only. We have with equation 14.134

$$2\pi A_1(s) = \frac{S_1(s)}{s} + \frac{2}{\omega_1} S_1(s) e^{-i\phi_0} - \frac{S_1(s) S_2(s) e^{-2i\phi_0}}{\omega_1} \cdots .$$

This provides a systematic way of truncating $A_1(s)$. For example the first truncation is

$$2\pi A_1(s) = \frac{S_1(s)}{s} + \frac{2}{\omega_1} S_1(s) e^{-i\phi_0}$$

$$= \frac{\omega_1 + s e^{-i\phi_0}}{2s(s + \alpha + i\omega_0)}$$

and so on.

where for notational convenience

$$\hat{a}_p(0) = \frac{a_p(0)}{a_0} \qquad (14.135)$$

The law of formation of successive terms in Eq. (14.134) is exceedingly simple; thus Eq. (14.134) may be evaluated to any desired degree of accuracy using a simple computer program. The frequency response may then be obtained by setting $s = i\omega$. It should be noted that $A_1(s)$ depends on every term $\hat{a}_p(0)$ of the initial conditions, which is a general characteristic of nonlinear systems. In the "linear response" approximation, one supposes that $\hat{a}_p(0) = 0$ for $p > 1$ and truncates the continued fraction at S_2. The time-domain behavior of the system may be found to any desired degree of accuracy from Eq. (14.134) simply by inverting the Laplace transform using Heaviside's method. This completes our analysis of the behavior of the junction given a delta function initial distribution for the phase. The behavior after a step rise in I_{dc} may likewise be computed.

Another situation of interest is that of the behavior of the system following the switching off of the bias current, that is, the decay transient. We will assume that the bias current was applied a long time before the switching-off time, so that steady-conditions will have been reached by that time. Thus we suppose for the present problem that the bias is switched on at $t = -\infty$ and switched off at $t = 0$. For $t > 0$, then, the equations of motion are

$$\dot{a}_p(t) + \alpha p^2 a_p(t) = \frac{\omega_1 p}{2} \left[a_{p-1}(t) - a_{p+1}(t) \right] \qquad (14.136)$$

The initial values of the a are found by setting $\dot{a}_p(t) = 0$ in Eq. (14.103). Any $a_p(0)$ may then be found in terms of continued fractions as a_1 in Eq. (14.96). Alternatively, if one does not wish to have the initial values of the a expressed as a continued fraction, one can obtain them by solving Eq. (14.48) with $\partial_t f = 0$, that is,

$$\frac{kT}{\varsigma} \frac{\partial^2 f}{\partial \phi^2} + \frac{\partial}{\partial \phi} \left(\frac{f}{\varsigma} \frac{\partial V}{\partial \phi} \right) = 0 \qquad (14.137)$$

Returning now to Eq. (14.136), we may again use the Laplace transform to write it as

$$sA_p(s) - a_p(0) + \alpha p^2 A_p(s) = \frac{\omega_1 p}{2} \left[A_{p-1}(s) - A_{p+1}(s) \right]$$

or

$$(s + \alpha p^2) A_p(s) = \frac{\omega_1 p}{2} \left[A_{p-1}(s) - A_{p+1}(s) \right] + a_p(0)$$

Again dividing by $A_{p-1}(s)$, we find that

$$(s + \alpha p^2) \frac{A_p(s)}{A_{p-1}(s)} = \frac{\omega_1 p}{2} \left[1 - \frac{A_{p+1}(s)}{A_{p-1}(s)} \right] + \frac{a_p(0)}{A_{p-1}(s)} \qquad (14.138)$$

Let us again write

$$R_p = \frac{A_p(s)}{A_{p-1}(s)}$$

Then

$$(s + \alpha p^2) R_p(s) = \frac{\omega_1 p}{2} (1 - R_p R_{p+1}) + \frac{a_p(0)}{A_{p-1}(s)} \qquad (14.139)$$

As before, this recurrence relation will have a particular and a complementary solution. Let us call the complementary solution $S_p(s)$. Then

$$(s + \alpha p^2) S_p(s) = \frac{\omega_1 p}{2} \left[1 - S_p S_{p+1} \right]$$

whence

$$\left[s + \alpha p^2 + \frac{\omega_1 p}{2} S_{p+1}(s) \right] S_p(s) = \frac{\omega_1 p}{2}$$

or

$$S_p(s) = \frac{\dfrac{\omega_1 p}{2}}{s + \alpha p^2 + \dfrac{\omega_1 p}{2} S_{p+1}(s)} \qquad (14.140)$$

from which it is evident that the analysis proceeds exactly as for the previous case (the fundamental solution where we are given a delta function initial distribution of phases). Indeed, the Fourier transform of the decay transient can be obtained directly by setting $\omega_0 = 0$ in Eq. (14.134), the only difference between the fundamental and decay transient solutions other than that $\omega_0 = 0$ being in the initial values $a_p(0)$. In the fundamental solution these are given by

$$a_p(0) = \frac{1}{2\pi} e^{-ip\phi_0}$$

whereas in the decay transient these are given by continued fractions of the form of Eq. (14.96).

The methods we have described here may be used when any type of input signal giving rise to a potential

$$V(\phi, t) = -V_0 \cos \phi(t) - [F_0 + h(t)] \phi(t)$$

is applied to the junction. Such a potential will automatically give rise in the Smoluchowski approximation to differential-difference equations of the form

$$\dot{a}_p(t) + \frac{kT}{\zeta} \left[p^2 + \frac{(F_0 + h(t))}{kT} ip \right] a_p(t) = \frac{V_0 p}{2\zeta} [a_{p-1}(t) - a_{p+1}(t)]$$

$$(14.141)$$

The ease or difficulty of solution depends entirely on the form of $h(t)$.

We have considered only the behavior of the junction in the overdamped case—the Smoluchowski limit, where inertial effects are ignored. If this approximation is discarded, we must solve the full Kramers equation corresponding to Eq. (14.47), which is

$$\frac{\partial W}{\partial t} + \dot{\phi} \frac{\partial W}{\partial \phi} - \frac{1}{J} \frac{\partial V}{\partial \phi} \frac{\partial W}{\partial \dot{\phi}} = \frac{\zeta}{J} \frac{\partial}{\partial \dot{\phi}} \left(\dot{\phi} W + \frac{kT}{J} \frac{\partial W}{\partial \dot{\phi}} \right) \quad (14.142)$$

where

$$W = W(\phi, \dot{\phi}, t), \qquad f = \int_{-\infty}^{\infty} W(\phi, \dot{\phi}, t) \, d\dot{\phi} \qquad (14.143)$$

This equation may be treated using the matrix continued-fraction methods described in refs. 12 and 22, and also in section XI of this article although the process of solution is very much more difficult than in the Smoluchowski approximation. It should be possible, however, to solve the decay-transient problem without any great difficulty by borrowing the results of the disk model for dielectric relaxation, with inertia included (see Chapter 2 of ref. 9 and Chapter 5 of ref. 12).

We remark that in the Smoluchowski approximation the problem of calculating the a_p values is clearly linked with that of finding the eigenvalues and eigenfunctions of a Sturm–Liouville equation of Hill's type. An example of the application of Sturm–Liouville theory to the Brownian motion in periodic potentials is given in Chapter 3 of ref. 12. The approach based on the Sturm–Liouville equation is particularly useful when one wishes to calculate the escape rate from one potential well into a neighboring one, since the escape rate depends on the lowest eigenvalue of the Sturm–Liouville equation. The essence of the Sturm–Liouville method is to take the Fourier or Laplace transform of the Smoluchowski equation directly over the time variables in that equation. This automatically converts the Smoluchowski equation into a second-order ordinary differential equation, or in certain cases, particularly those involving time-dependent potentials, into an integro-differential equation. In all cases where the Smoluchowski equation may be reduced to an ordinary differential equation, it may be further reduced to a Sturm–Liouville equation.

XV. FUNDAMENTAL PRINCIPLES OF STATISTICAL MECHANICS

A. The Concept of Phase Space

The concept of phase space evolved[1] mainly from the classical dynamics of particles and rigid bodies. The classical dynamics is concerned basically with determination of the motion of a system of particles when these are subject to certain constraints, such as rigid interconnections, sliding on perfectly smooth surfaces, and rolling on perfectly rough surfaces. The theory is based on Newton's laws of motion*. The instantaneous position of the system is given by r independent generalized coordinates $q_1, \ldots, q_r$. These are determined (in terms of the Cartesian coordinates of the pàrticles) by the geometry of the constraints. If the system is holonomic, that is, if there are no additional constraint relations between the time derivatives of the coordinates, the generalized coordinates are termed *degrees of freedom*.

Lagrange[1] (1778) showed that the equations of motion of a conservative holonomic system could be simplified to

$$\frac{d}{dt}\left(\frac{\partial L}{\partial \dot{q}_i}\right) - \frac{\partial L}{\partial q_i} = 0, \qquad L = T - V \tag{15.1}$$

where L is the Lagrangian of the system, T is the kinetic energy, V is the potential energy of the system, and

$$\dot{q}_i = \frac{dq_i}{dt}$$

Now $\partial L/\partial q_i$, $\partial L/\partial \dot{q}_i$ are known functions of the variables $(q_1, \ldots, q_r,$

*1687

$\dot{q}_1, \ldots, \dot{q}_r$). Thus Lagrange's equations will have the general form

$$\frac{d}{dt} A_i(q_1, \ldots, q_r, \dot{q}_1, \ldots, \dot{q}_r) - B_i(q_1, \ldots, q_r, \dot{q}_1, \ldots, \dot{q}_r) = 0$$

Differentiation of A_i with respect to time yields terms involving $(\ddot{q}_1, \ldots, \ddot{q}_r)$. Thus Lagrange's equations are a set of simultaneous second-order ordinary differential equations. The advantage these give over Newton's equations is that they no longer involve the reaction forces, which maintain the given constraints. There are r equations of the second order, so their solution requires $2r$ initial conditions. The complete state of the system is given by the instantaneous values of $(q_1, \ldots, q_r, \dot{q}_1, \ldots, \dot{q}_r)$. Hamilton showed* that the equations of motion can be further simplified by taking as coordinates

$$p_i = \frac{\partial L}{\partial \dot{q}_i}$$

so that Lagrange's equations are replaced by the following $2r$ equations, each of the first order:

$$\dot{q}_i = \frac{\partial H}{\partial p_i}, \qquad \dot{p}_i = -\frac{\partial H}{\partial q_i}$$

where the Hamiltonian (for a conservative holonomic system) is

$$H = T + V$$

the total energy.

$$H = H(\mathbf{q}, \mathbf{p}) = H(\mathbf{q}, \dot{\mathbf{q}})$$

where

$$\mathbf{q} = (q_1, \ldots, q_r), \qquad \mathbf{p} = (p_1, \ldots, p_r)$$

H is a known function of $(\mathbf{q}, \mathbf{p})$; thus Hamilton's equations may always be written

$$\frac{dq_i}{dt} = C_i(q_1, \ldots, q_r, p_1, \ldots, p_r)$$

$$\frac{dp_i}{dt} = D_i(q_1, \ldots, q_r, p_1, \ldots, p_r)$$

*1835

These $2r$ equations of the first order show that $2r$ variables are needed to specify the instantaneous state $(\mathbf{q}, \mathbf{p})$ of a system. H for a conservative holonomic system does not in general depend on the time. A simple generalization of the above is

$$\frac{dy_i}{dt} = f_i(y_1, \ldots, y_r, t) \tag{15.2}$$

These equations are more general then Hamilton's equations, since f_i is not plus or minus the partial derivative of some function H, and because t appears explicitly on the right-hand side. Subject to certain assumptions it may be proved* that the equations (15.2) have a unique solution in terms of the initial conditions, that is, in terms of the given values $y_1, \ldots, y_r$ at any time t_0, thus establishing that $y_1(t), \ldots, y_r(t)$ characterize completely the "state" at time t of the system represented by the set of equations (15.2).

Poincaré[1] (1881) stated that the variables $y_1, \ldots, y_r$ could be regarded as the coordinates of a point in-r-dimensional space and applied topological methods to the study of the trajectory that the point describes as time goes on. Gibbs[4] (1902), in applying the Hamiltonian method to statistical mechanics, thought of the instantaneous positions and velocities of N molecules as representing the coordinates of a point in space. He called the position of the point the *phase* of the system and the $6N$-dimensional space the *phase space*. This terminology has persisted, although it is now applied more generally to the $(y_1, \ldots, y_r)$-space of Poincaré. (A good account of the concept of phase space and its application in fields such as control engineering is given by Fuller.[1])

We must now consider the Liouville equation (Figs. 24 and 25). In the $6N$-dimensional phase space, the position of a point completely specifies the *microscopic* dynamical state of our N-particle system. Such a point is called a representative point or a phase point. It evolves in time according to Hamilton's equations. It is never possible, nor indeed necessary, to know the exact trajectory of such a point. All one can know is a few macroscopic properties of the system such as its energy and volume. Thus, quoting Tolman[3] (1938);

> The general nature of the statistical mechanical procedure for the treatment of complicated systems consists in abandoning the attempt to follow the precise changes in state that would take place in a particular system and in studying instead the behaviour of a collection or *ensemble* of systems of similar structure to the one of actual interest distributed over a range of different precise states. From a knowledge of the average behaviour of the system in a

*Cauchy[1] (1842)

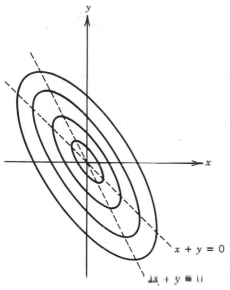

$x + y = 0$

$2x + y = 0$

Figure 24. Phase plane trajectories for the system $\dot{x} = x + y$, $x =$ displacement, $\dot{v} = 2x - y$.

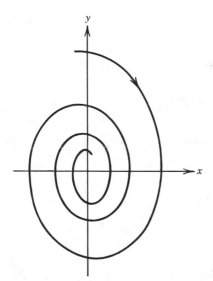

Figure 25. Phase plane trajectories of a damped harmonic oscillator.

211

representative ensemble appropriately chosen so as to correspond to the partial knowledge that we do have as to the initial state of the system of interest, we can then make predictions as to what may be expected on the average for the particular system which interests us.

We shall from now on regard a set of phase points as constituting an *ensemble* of systems (in the sense described in last paragraph). Clearly the set of phase points that could possibly describe the macroscopic behavior of our dynamical system becomes very dense. We may then speak of a *density* of phase points, or *distribution function*, as the *fraction* of phase points contained in a volume element. We shall denote this density by f. The volume element is

$$dv = dq_1 \ldots dq_{3N} \, dp_1 \ldots dp_{3N} = d\mathbf{p} \, d\mathbf{q},$$

we normalize f so that

$$\int f(\mathbf{p},\mathbf{q}, t) \, d\mathbf{p} \, d\mathbf{q} = 1$$

The phase points may be thought of as forming a sort of fine dust in the phase space; it is therefore reasonable to try to find a continuity equation for f. To do this, consider an arbitrary volume v. The number of phase points in v is

$$n = N \int_v f(\mathbf{q},\mathbf{p}, t) \, d\mathbf{q} \, d\mathbf{p} \tag{15.3}$$

The current (flux of phase points) through a surface S bounding v is

$$\frac{dn}{dt} = - N \int_S f\mathbf{u} \cdot d\mathbf{S}$$

where $\mathbf{u}$ denotes the $6N$-dimensional vector $(\dot{q}_1, \ldots, \dot{q}_{3N}, \dot{p}_1, \ldots, \dot{p}_{3N})$. Now by Gauss's divergence theorem

$$\int_S f\mathbf{u} \cdot d\mathbf{S} = \int_v \mathrm{div} f\mathbf{u} \, dv$$

and

$$\frac{dn}{dt} = - N \int_v \mathrm{div} f\mathbf{u} \, dv$$

or with Eq. (15.3),

$$\frac{\partial f}{\partial t} + \mathrm{div}\, f\mathbf{u} = 0 \tag{15.4}$$

or

$$\frac{Df}{Dt} = 0$$

where

$$\frac{D}{Dt} = \frac{\partial}{\partial t} + \mathbf{u}\cdot\nabla$$

denotes a hydrodynamical derivative. Equation (15.4) is the equation of *continuity* or *conservation* of phase points (called the Liouville equation).*

Consider now the term

$$\mathrm{div}\, f\mathbf{u}$$

This is, since $\mathbf{u} = (\dot{q}_1,\ldots,\dot{q}_{3N}, \dot{p}_1,\ldots,\dot{p}_{3N})$,

$$\mathrm{div}\, f\mathbf{u} = \sum_{i=1}^{3N} \frac{\partial}{\partial q_i}(f\dot{q}_i) + \sum_{i=1}^{3N} \frac{\partial}{\partial p_i}(f\dot{p}_i)$$

$$= \sum_{i=1}^{3N}\left\{\frac{\partial f}{\partial q_i}\dot{q}_i + \frac{\partial f}{\partial p_i}\dot{p}_i\right\} + \sum_{i=1}^{3N}\left\{\frac{\partial \dot{q}_i}{\partial q_i} + \frac{\partial \dot{p}_i}{\partial p_i}\right\}f$$

Now if H is the Hamiltonian of the system

$$\dot{q}_i = \frac{\partial H}{\partial p_i}, \qquad \dot{p}_i = -\frac{\partial H}{\partial q_i}$$

and thus

$$\frac{\partial \dot{q}_i}{\partial q_i} = -\frac{\partial \dot{p}_i}{\partial p_i}$$

from which we get

$$\sum_i\left(\frac{\partial \dot{q}_i}{\partial q_i} + \frac{\partial \dot{p}_i}{\partial p_i}\right) = 0$$

*Given by Liouville in 1838

Thus the Liouville equation is

$$\frac{\partial f}{\partial t} + \sum_{i=1}^{3N} \left(\frac{\partial H}{\partial p_i} \frac{\partial f}{\partial q_i} - \frac{\partial H}{\partial q_i} \frac{\partial f}{\partial p_i} \right) = 0 \tag{15.5}$$

or

$$\frac{Df}{Dt} = \frac{\partial f}{\partial t} + \{ f, H \} = 0$$

The summation is called a Poisson bracket.

Let us now return to our equation

$$\frac{Df}{Dt} = 0$$

This equation says that "the density in the neighbourhood of any selected moving representative point is constant along the *trajectory* of that point" or

$$f(\mathbf{p}, \mathbf{q}, t) = f(\mathbf{p}_0, \mathbf{q}_0, t_0)$$

if $\mathbf{p}_0, \mathbf{q}_0$ is an initial point. $t - t_0$ denotes the time elapsed in moving between positions $\mathbf{p}, \mathbf{q}$ and $\mathbf{p}_0, \mathbf{q}_0$. This is the *principle of conservation of density in phase*. Another way of saying this is that the cloud of phase points behaves as an incompressible fluid.

Yet another way of stating Liouville's theorem is to suppose that at an initial time a phase point lies in an infinitesimal volume v_0 of phase space and at a later time t it is in a volume v of phase space. Then $v = v_0$. The theorem in this form is known as the *principle of conservation of extension in phase space*. We express it mathematically by writing

$$\delta p \, \delta q = \delta p_0 \, \delta q_0$$

for all t; that is, the Jacobian of the set $(\mathbf{p}, \mathbf{q})$ to the set $(\mathbf{p}_0, \mathbf{q}_0)$ is unity. To prove the theorem in this fashion, consider a particle having coordinates q_i, p_i, with, as usual,

$$\dot{q}_i = \frac{\partial H}{\partial p_i}, \qquad \dot{p}_i = -\frac{\partial H}{\partial q_i}$$

After an infinitesimal time dt, q_i, p_i will become $q_i + \dot{q}_i \, dt$, $p_i + \dot{p}_i \, dt$, and v will change to $v + (dv/dt) \, dt$, where, by the rule for transforming small elements of volume,

$$\frac{1}{v}\left(v+\frac{dv}{dt}\,dt\right) = \frac{\partial(q_1+\dot{q}_1\,dt,\ldots,p_1+\dot{p}_1\,dt,\ldots)}{\partial(q_1,\ldots,p_1,\ldots)}$$

$$= \frac{\partial(\mathbf{q}+\dot{\mathbf{q}}\,dt,\mathbf{p}+\dot{\mathbf{p}}\,dt)}{\partial(\mathbf{q},\mathbf{p})}$$

This Jacobian is a determinant whose nondiagonal elements are all proportional to dt, and whose diagonal elements are

$$1+\frac{\partial\dot{q}_1}{\partial q_1}\,dt,\ldots,1+\frac{\partial\dot{p}_1}{\partial p_1}\,dt,\ldots$$

Thus, neglecting squares and higher powers of dt,

$$1+\frac{1}{v}\frac{dv}{dt}\,dt = 1+dt\left(\frac{\partial\dot{q}_1}{\partial q_1}+\frac{\partial\dot{q}_2}{\partial q_2}+\cdots\frac{\partial\dot{p}_1}{\partial p_1}+\cdots\right)$$

$$= 1+dt\left(\frac{\partial^2 H}{\partial q_1\,\partial p_1}+\frac{\partial^2 H}{\partial q_2\,\partial p_2}+\cdots-\frac{\partial^2 H}{\partial p_1\,\partial q_1}-\cdots\right)$$

$$= 1$$

so that

$$\frac{dv}{dt}=0$$

Hence

$$v = v_0 \tag{15.6}$$

The meaning of this is that even though the *shape* of the surface bounding that volume element in phase space may alter with time, the *volume* enclosed by that surface does not. Using classical mechanics, one may prove from this that if we are given two sets of coordinates and their conjugate momenta, say

$$q_1,\ldots,q_{3N},\,p_1,\ldots,\,p_{3N}$$
$$Q_1,\ldots,Q_{3N},\,P_1,\ldots,\,P_{3N}$$

which are equally good for the description of a system in phase space, then

$$dq_1\ldots dq_{3N}\,dp_1\ldots dp_{3N}=dQ_1\ldots dQ_{3N}\,dP_1\ldots dP_{3N}$$

The Liouville equation is often written in the operational form

$$\frac{\partial f}{\partial t} = - iLf$$

where

$$L = \sum_{i=1}^{3N} \left(\frac{\partial H}{\partial p_i} \frac{\partial}{\partial q_i} - \frac{\partial H}{\partial q_i} \frac{\partial}{\partial p_i} \right) \tag{15.7}$$

is called the Liouville operator. The formal solution is then

$$f(\mathbf{q},\mathbf{p},t) = \exp(- iLt)f(\mathbf{q},\mathbf{p},0)$$

It is sometimes useful to regard $(\mathbf{q},\mathbf{p})$ as Cartesian coordinates in the phase space. Then using Hamilton's equations, the Liouville equation becomes

$$\frac{\partial f}{\partial t} + \sum_{i=1}^{N} \frac{\mathbf{p}_i}{m_i} \cdot \mathrm{grad}_{\mathbf{r}_i} f + \sum_{i=1}^{N} \mathbf{F}_i \cdot \mathrm{grad}_{\mathbf{p}_i} f = 0$$

In this equation $\mathrm{grad}_{\mathbf{r}_i}$ denotes the gradient with respect to the spatial variables in f, and $\mathrm{grad}_{\mathbf{p}_i}$ denotes the gradient with respect to the momentum variables in f. $\mathbf{F}_i$ is the total force on the ith particle. This form of the equation will be useful when we discuss the Kramers equation for Brownian movement in a potential.

B. Calculation of Averages from the Liouville Equation

If we can determine f, we can compute the ensemble average of any dynamical variable $A(\mathbf{p},\mathbf{q},t)$ from the definition

$$\langle A(t) \rangle = \int A(\mathbf{q},\mathbf{p},t)f(\mathbf{q},\mathbf{p},t)\,d\mathbf{p}\,d\mathbf{q} \tag{15.8}$$

This is an integral over the *whole* volume of phase space. In most cases the variables of interest either are functions of the coordinates and momenta of just a *few* particles or may be written as a *sum* over such functions.[39] Thus the total intermolecular potential of a system may be written to a good approximation as a sum over *pair potentials*:

$$\langle U \rangle = \sum_{i,j} \int \cdots \int u(\mathbf{r}_i,\mathbf{r}_j)f(\mathbf{r}_1,\ldots,\mathbf{r}_N,\mathbf{p}_1,\ldots,\mathbf{p}_N,t)\,d\mathbf{r}_1 \cdots d\mathbf{p}_N$$

In the equilibrium theory of liquids one may integrate over all coordinates of all particles except i and j. The resulting function of $\mathbf{r}_i$ and $\mathbf{r}_j$ is called the *radial distribution function*.

This procedure may also be applied in the dynamical case.[39] Consider our expression for f. The probability that molecule 1 is in $d\mathbf{r}_1\,d\mathbf{p}_1$ at $\mathbf{r}_1,\mathbf{p}_1$, that molecule 2 is in $d\mathbf{r}_2\,d\mathbf{p}_2$ at $\mathbf{r}_2,\mathbf{p}_2$, and that molecule n in $d\mathbf{r}_n\,d\mathbf{p}_n$ at $\mathbf{r}_n,\mathbf{p}_n$, *irrespective* of the configuration of the remaining $N-n$ molecules, is

$$F^{(n)}(\mathbf{r}_1,\ldots,\mathbf{r}_n,\mathbf{p}_1,\ldots,\mathbf{p}_n,t)$$

where

$$F^{(n)}(\mathbf{r}_1,\ldots,\mathbf{r}_n,\mathbf{p}_1,\ldots,\mathbf{p}_n,t)$$

$$=\int\cdots\int f(\mathbf{r}_1,\ldots,\mathbf{p}_N,t)\,d\mathbf{r}_{n+1}\cdots d\mathbf{r}_N\,d\mathbf{p}_{n+1}\cdots d\mathbf{p}_N$$

But the probability that *any* molecule is in $d\mathbf{r}_1$ at $\mathbf{r}_1$, $d\mathbf{p}_1$ at $\mathbf{p}_1$, and $d\mathbf{r}_n\,d\mathbf{p}_n$ at $\mathbf{r}_n,\mathbf{p}_n$ *irrespective* of the configuration of the rest of the molecules is

$$f^{(n)}(\mathbf{r}_1,\ldots,\mathbf{r}_n,\mathbf{p}_1,\ldots,\mathbf{p}_n,t)$$

$$=\frac{N!}{(N-n)!}\int\cdots\int f(\mathbf{r}_1,\ldots,\mathbf{p}_N,t)\,d\mathbf{r}_{n+1}\cdots d\mathbf{r}_N\,d\mathbf{p}_{n+1}\cdots d\mathbf{p}_N$$

The function $f^{(n)}$ is called a *reduced distribution function*. The factor

$$\frac{N!}{(N-n)!}$$

arises because we have N choices for the first molecule, $N-1$ for the second, and so on. We now use the reduced distribution function combined with the Liouville equation to derive what is known as the Bogoliubov–Born–Green–Kirkwood–Yvon (BBGKY) hierarchy. Our method follows closely that of McQuarrie.[39] We first write the force $\mathbf{F}_j$ appearing in the Liouville equation* as the sum of the forces due to the other molecules in

---

*For convenience the Liouville equation in this part of the discussion is written

$$\frac{\partial f}{\partial t}+\sum_{j=1}^{N}\frac{\mathbf{p}_j}{m_j}\cdot\mathrm{grad}_{\mathbf{r}_j}f+\sum_{j=1}^{N}\mathbf{F}_j\cdot\mathrm{grad}_{\mathbf{p}_j}f=0$$

the system $\sum_j \mathbf{F}_{ij}$ and an external force $\mathbf{X}_j$. If we now multiply through by $N!/(N-n)!$ and integrate over $d\mathbf{r}_{n+1}\cdots d\mathbf{r}_N\, d\mathbf{p}_{n+1}\cdots d\mathbf{p}_N$ then

$$
\frac{\partial f^{(n)}}{\partial t} + \sum_{j=1}^{n} \frac{\mathbf{p}_j}{m_j}\cdot\nabla_{\mathbf{r}_j} f^{(n)} + \sum_{j=1}^{n} \mathbf{X}_j\cdot\nabla_{\mathbf{p}_j} f^{(n)} + \frac{N!}{(N-n)!}
$$

$$
\times \sum_{i,j=1}^{N} \int\cdots\int \mathbf{F}_{ij}\cdot\nabla_{\mathbf{p}_j} f\, d\mathbf{r}_{n+1}\cdots d\mathbf{r}_N\, d\mathbf{p}_{n+1}\cdots d\mathbf{p}_N = 0 \quad (15.9)
$$

In writing down equation (15.9) we have used the fact that f must vanish outside the vessel containing the molecules and when $\mathbf{p}_i = \pm\infty$. The last term may be broken up into two parts

$$
\sum_{i,j=1}^{n} \mathbf{F}_{ij}\cdot\nabla_{\mathbf{p}_j} f^{(n)} + \frac{N!}{(N-n)!}
$$

$$
\times \sum_{j=1}^{n} \sum_{i=n+1}^{N} \int\cdots\int \mathbf{F}_{ij}\cdot\nabla_{\mathbf{p}_j} f\, d\mathbf{r}_{n+1}\cdots d\mathbf{r}_N\, d\mathbf{p}_{n+1}\cdots d\mathbf{p}_N
$$

$$
(15.10)
$$

The second term may be written (on summation over i)

$$
\sum_{j=1}^{n} \int\int \mathbf{F}_{j,n+1}\cdot\nabla_{\mathbf{p}_j} f^{(n+1)}\, d\mathbf{r}_{n+1}\, d\mathbf{p}_{n+1}
$$

Combining the three preceding equations now gives

$$
\frac{\partial f^{(n)}}{\partial t} + \sum_{j=1}^{n} \frac{\mathbf{p}_j}{m_j}\cdot\nabla_{\mathbf{r}_j} f^{(n)} + \sum_{j=1}^{n} \mathbf{X}_j\cdot\nabla_{\mathbf{p}_j} f^{(n)} + \sum_{i,j=1}^{n} \mathbf{F}_{ij}\cdot\nabla_{\mathbf{p}_j} f^{(n)}
$$

$$
+ \sum_{j=1}^{n} \int\int \mathbf{F}_{j,n+1}\cdot\nabla_{\mathbf{p}_j} f^{(n+1)}\, d\mathbf{r}_{n+1}\, d\mathbf{p}_{n+1} = 0 \quad (15.11)
$$

This is the BBGKY hierarchy. This equation appears simpler than the Liouville equation. It would therefore seem possible to truncate it in some way to obtain equations for $f^{(1)}$ and $f^{(2)}$. However, no one has yet succeeded in uncoupling this hierarchy. We must therefore turn to other methods to obtain approximate equations for $f^{(1)}$ and $f^{(2)}$.

C. The Boltzmann Equation

In this section, we will follow closely the derivation of McQuarrie (1976).[39]

In a dilute gas it is reasonable to assume that only two-body interactions between molecules are ever important. Consider a phase point with coordinates $(\mathbf{r}, \mathbf{v}_j)$. The number of molecules of type j in $d\mathbf{r}\,d\mathbf{v}_j$ is $f_j\,d\mathbf{r}\,d\mathbf{v}_j$. Suppose there were no collisions in the gas and that an external force $\mathbf{X}_j$ is acting. The molecules then by Hamilton's equations arrive at the point $(\mathbf{r}+\mathbf{v}_j\,dt, \mathbf{v}_j +(1/m_j)\mathbf{X}_j\,dt)$ at time $(t + dt)$. Now by Liouville's theorem (principle of conservation of density in phase), all the points that start out end up at this point (in the absence of collisions), and thus

$$f_j(\mathbf{r}, \mathbf{v}_j, t) = f_j\left(\mathbf{r}+\mathbf{v}_j\,dt, \mathbf{v}_j + \frac{\mathbf{X}_j}{m_j}\,dt, t + dt\right) \qquad \text{(no collisions)}$$

Now collisions will occur in the gas. Thus all molecules that start out at $(\mathbf{r}, \mathbf{v}_j)$ at time t will *not* arrive at $(\mathbf{r}+\mathbf{v}_j\,dt, \mathbf{v}_j +(1/m_j)\mathbf{X}_j\,dt)$ at time $(t + dt)$. Some molecules will be *taken out* of the j stream because of collisions, while others will *enter* that stream, again because of collisions. We suppose that the number of molecules of type j *lost* from the velocity range $(\mathbf{v}_j, \mathbf{v}_j + d\mathbf{v}_j)$ and the position range $(\mathbf{r}, \mathbf{r}+ d\mathbf{r})$ because of collisions (with molecule of a type i, say) is $\Gamma_{ji}^{(-)}\,d\mathbf{r}\,d\mathbf{v}_j\,dt$ in $(t, t + dt)$. Similarly, we suppose that the number of molecules of type j that are *gained* by the group of molecules that starts at $(\mathbf{r}, \mathbf{v}_j)$ because of collisions with molecules of type i is $\Gamma_{ji}^{(+)}d\mathbf{r}\,d\mathbf{v}_j\,dt$, from which we obtain

$$f_j\left(\mathbf{r}+\mathbf{v}_j\,dt, \mathbf{v}_j + m_j^{-1}\mathbf{X}_j\,dt, t + dt\right)d\mathbf{r}\,d\mathbf{v}_j$$
$$= f_j(\mathbf{r}, \mathbf{v}_j, t)\,d\mathbf{r}\,d\mathbf{v}_j + \sum_i\left(\Gamma_{ji}^{(+)} - \Gamma_{ji}^{(-)}\right)d\mathbf{r}\,d\mathbf{v}_j\,dt$$

$$(15.12)$$

We now use Taylor's theorem to expand the left-hand side of Eq. (15.12). We find that

$$\frac{\partial f_j}{\partial t} + \mathbf{v}_j\cdot\nabla_r f_j + \frac{\mathbf{X}_j}{m_j}\cdot\nabla_{\mathbf{v}_j} f_j = \sum_i\left(\Gamma_{ji}^{(+)} - \Gamma_{ji}^{(-)}\right) \qquad (15.13)$$

or

$$\frac{Df_j}{Dt} = \sum_i\left(\Gamma_{ji}^{(+)} - \Gamma_{ji}^{(-)}\right)$$

The left-hand side is the change in f_j due to the *collisionless* motion of the molecules called *streaming*, and the right-hand side represents the change in f_j due to *collisions*. Note that $D_t f_j$ is *no longer zero* as in the Liouville equation *because of the collisions*.

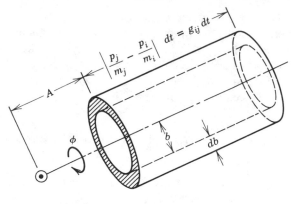

$$2\pi b\, db = d(\pi b^2)$$

Figure 26.   Collisions of molecules of type i with one molecule of type j. The distance A is essentially the intermolecular distance at which the potential begins to "take hold." (From J. O. Hirschfelder, C. F. Curtiss, and R. B. Bird, *Molecular Theory of Gases and Liquids*. New York: Wiley, 1954.)

We now have to try to find an explicit expression for the collision term in the above equation. Consider first the loss term

$$\Gamma_{ji}^{(-)}\, d\mathbf{r}\, d\mathbf{v}_j$$

Let us suppose we have a molecule of type j located at $\mathbf{r}$ with velocity $\mathbf{v}_j$. The *probability* that this molecule will collide with an i molecule in the time interval dt and with an impact parameter in a range db about b is obtained as follows: Suppose molecule j is fixed; then i approaches it with velocity $(\mathbf{v}_i - \mathbf{v}_j) = \mathbf{g}_{ij}$. If A in Fig. 26 is the range of the intermolecular potential, any i molecule in the cylindrical cell shown will collide with the fixed j molecule in a time interval dt. The probable number of i molecules in this cylindrical shell is

$$2\pi f_i(\mathbf{r}, \mathbf{v}_i, t)\, g_{ij} b\, db\, dt$$

where

$$g_{ij} = |\mathbf{g}_{ij}|$$

The total number of collisions that would occur with this one fixed j molecule is

$$2\pi\, dt \int \int f_i(\mathbf{r}, \mathbf{v}_i, t)\, g_{ij} b\, db\, d\mathbf{v}_i$$

The probable number of molecules of type j located in the volume element $d\mathbf{r}$ about $\mathbf{r}$ with velocity in the range $(\mathbf{v}_j, \mathbf{v}_j + d\mathbf{v}_j)$ is $f_j(\mathbf{r}, \mathbf{v}_j, t) \, d\mathbf{r} \, d\mathbf{v}_j$.

We now suppose that *the mean number of i molecules about the fixed j molecule is the product of f_i and f_j*. Thus we are assuming that *the positions and velocities are uncorrelated.* This is the *Stosszahlansatz* (collision-number hypothesis) or molecular-chaos assumption of Boltzmann. This is most important, and we shall refer to it later when we discuss Boltzmann's H theorem. Thus

$$\Gamma_{ji}^{(-)} \, d\mathbf{r} \, d\mathbf{v}_j \, dt = 2\pi \, d\mathbf{r} \, d\mathbf{v}_j \, dt \int \int f_j(\mathbf{r}, \mathbf{v}_j, t) f_i(\mathbf{r}, \mathbf{v}_i, t) g_{ij} b \, db \, d\mathbf{v}_i$$

and

$$\Gamma_{ji}^{(-)} = 2\pi \int \int f_j f_i g_{ij} b \, db \, d\mathbf{v}_i$$

$\Gamma_{ji}^{(-)}$ measures the number of molecules of type j lost by collisions with molecules of type i. We can use the foregoing argument to calculate $\Gamma_{ji}^{(+)}$, which measures the number of molecules of type j *gained* by collisions with molecules of type i. Thus $\Gamma_{ji}^{(+)}$ measures the number of particles scattered into the point $(\mathbf{r} + \mathbf{v}_j \, dt, \mathbf{v}_j + (1/m_j)\mathbf{X}_j \, dt)$. We may immediately write

$$\Gamma_{ji}^{(+)} \, d\mathbf{r} \, d\mathbf{v}_j \, dt = 2\pi \, d\mathbf{r} \, d\mathbf{v}_j^* \, dt \int \int f_i(\mathbf{r}, \mathbf{v}_i^*, t) f_j(\mathbf{r}, \mathbf{v}_j^*, t) g_{ij}^* b^* \, db^* \, d\mathbf{v}_i^*$$

The stars denote those quantities *before* the collision that become b, $\mathbf{v}_i$, and $\mathbf{v}_j$ afterward. The starred quantities are calculated from $b, \mathbf{v}_i, \mathbf{v}_j$ by solving the collisional equations of motion (in general, these are for motion under a central force). If we now apply the Liouville theorem to the element of phase space of the conservative system formed by the two molecules, we must have

$$d\mathbf{r} \, d\mathbf{v}_i \, d\mathbf{v}_j \, g_{ij} \, dt \, b \, db = d\mathbf{r} \, d\mathbf{v}_i^* \, d\mathbf{v}_j^* \, g_{ij}^* \, dt \, b^* \, db^*$$

and thus

$$\Gamma_{ji}^{(+)} = 2\pi \int \int f_i(\mathbf{r}, \mathbf{v}_i^*, t) f_j(\mathbf{r}, \mathbf{v}_j^*, t) g_{ij} b \, db \, d\mathbf{v}_i$$

$$= 2\pi \int \int f_i^* f_j^* g_{ij} b \, db \, d\mathbf{v}_i$$

where

$$f_i^* = f_i(\mathbf{r}, \mathbf{v}_i^*, t)$$

Thus

$$\frac{\partial f_j}{\partial t} + \mathbf{v}_j \cdot \nabla_\mathbf{r} f_j + \frac{1}{m_j} \mathbf{X}_j \cdot \nabla_{\mathbf{v}_j} f_j = 2\pi \sum_i \int \int \{ f_i^* f_j^* - f_i f_j \} g_{ij} b \, db \, d\mathbf{v}_i$$

$$(15.14)$$

This is the Boltzmann equation*. It is a nonlinear integro-differential equation for $f_j^{(1)}$. The equations of motion and hence the intermolecular potential enter this equation implicitly on the right-hand side because the functions f_i^* and f_j^* are related to f_i and f_j via the equations of motion governing the collision.‡ The process may be summarized by saying that collisions turn sets of molecules having velocities $\mathbf{v}_i$ and $\mathbf{v}_j$ into sets having velocities $\mathbf{v}_i^*$ and $\mathbf{v}_j^*$. The number of these collisions is governed by $f_i f_j$. Conversely, there is a *reverse* process whereby sets of molecules having velocities $\mathbf{v}_i^*$ and $\mathbf{v}_j^*$ are transformed into sets having velocities $\mathbf{v}_i$ and $\mathbf{v}_j$ by collision, the collision number being governed by $f_i^* f_j^*$. Since the right-hand side of the Boltzmann equation is integrated over b and $\mathbf{v}_i$, the only variables left are $\mathbf{r}$, $\mathbf{v}_j$, and t.

Let us consider

$$\frac{Df}{Dt} = 2\pi \sum_i \int \int \{ f_i^* f_j^* - f_i f_j \} g_{ij} b \, db \, d\mathbf{v}_i \qquad (15.15)$$

Df/Dt will vanish if on the right-hand side

$$f_i^* f_j^* = f_i f_j \qquad (15.16)$$

This is tantamount to saying that collisions that turn sets of molecules having velocities $\mathbf{v}_i$ and $\mathbf{v}_j$ into sets having velocities $\mathbf{v}_i^*$ and $\mathbf{v}_j^*$ are *balanced* by *reverse* collisions that create molecules having velocities $\mathbf{v}_i$ and $\mathbf{v}_j$ from those having velocities $\mathbf{v}_i^*$ and $\mathbf{v}_j^*$.

Equation (15.16) has wide implications, since it states that there is *detailed balancing* of the collisions as specified by the velocity exchanges. This

---

*Such an equation may be written for each component of the gas.

‡The equation of motion governing a molecular encounter or collision is the central-orbit equation, which is exhaustively treated in Chapter 3 of ref. 40. The application of that equation to the present problem, including the calculation of the impact parameter b, is discussed by Résibois and de Leener.[41] The special case of "Maxwellian" molecules,[2,41,42] where the gas is supposed to consist of centers of force repelling as r^{-5}, is particularly simple. In this case one may calculate transport coefficients *without* determining the velocity distribution function. Here the element $gb\,db$ does not depend on g. The central-force model for two interacting molecules was first suggested by Boscovich in 1758.[2]

condition means that just the same number of collisions of any one type and of the *reverse* type must occur per unit volume per second. This condition is naturally sufficient for the preservation of equilibrium. It does not follow that it is also *necessary* and therefore equivalent to the demand for preservation of equilibrium.

This leads us to Boltzmann's H theorem. His object was to show that molecular encounters would bring about a Maxwellian distribution of velocities in a gas left to itself, no matter what the initial distribution. By doing this he greatly strengthened the arguments for the Maxwellian distribution, which as Maxwell left them were still weak.[2] The idea of Boltzmann's proof was to attempt to construct a function H whose constancy *requires detailed balancing*. (H is not to be confused with the Hamiltonian.) The function $H(t)$ introduced by Boltzmann is defined for a *one-component* system as*

$$H(t) = \int \int f(\mathbf{r},\mathbf{v},t)\ln f(\mathbf{r},\mathbf{v},t)\,d\mathbf{r}\,d\mathbf{v} \qquad (15.17)$$

If we differentiate H with respect to t, then

$$\frac{dH}{dt} = \int\int \frac{\partial f}{\partial t}\ln f\,d\mathbf{r}\,d\mathbf{v} + \int\int \frac{\partial f}{\partial t}\,d\mathbf{r}\,d\mathbf{v} \qquad (15.18)$$

In Eq. (15.18)

$$\int\int \frac{\partial f}{\partial t}\,d\mathbf{r}\,d\mathbf{v} = \frac{d}{dt}\int\int f\,d\mathbf{r}\,d\mathbf{v} = \frac{dN}{dt} = 0$$

if the number of particles in the system is conserved. Thus

$$\frac{dH}{dt} = \int\int \frac{\partial f}{\partial t}\ln f\,d\mathbf{r}\,d\mathbf{v} \qquad (15.19)$$

We now multiply the Boltzmann equation by $\ln f$ and integrate over $d\mathbf{r}$ and $d\mathbf{v}$, obtaining

$$\int\int \frac{\partial f}{\partial t}\ln f\,d\mathbf{r}\,d\mathbf{v} = -\int\int(\ln f)\mathbf{v}\cdot\nabla_r f\,d\mathbf{r}\,d\mathbf{v} - \int\int(\ln f)\frac{\mathbf{X}}{m}\cdot\nabla_v f\,d\mathbf{r}\,d\mathbf{v}$$

$$+ 2\pi\int\int\int \ln f\{f^*f_1^* - ff_1\}\,gb\,db\,d\mathbf{v}\,d\mathbf{v}_1 \qquad (15.20)$$

*$H(t) = \langle \ln f \rangle$

The subscript "1" is used to distinguish between the two colliding molecules. Now f vanishes at the walls of the vessel containing the molecules and as $\mathbf{v} \to \pm \infty$; thus

$$\frac{dH}{dt} = 2\pi \int \int \int \ln f \{ f^* f_1^* - f f_1 \} \, gb \, db \, d\mathbf{v} \, d\mathbf{v}_1 \qquad (15.21)$$

But since f_1 and f refer to the *same* colliding molecules, we must also have

$$\frac{dH}{dt} = 2\pi \int \int \int \ln f_1 \{ f^* f_1^* - f f_1 \} \, gb \, db \, d\mathbf{v} \, d\mathbf{v}_1$$

Thus (by addition)

$$\frac{dH}{dt} = \pi \int \int \int \ln f f_1 \{ f^* f_1^* - f_1 f \} \, gb \, db \, d\mathbf{v} \, d\mathbf{v}_1$$

In the same way, for the *reverse* collisions

$$\frac{dH}{dt} = \pi \int \int \int \ln f^* f_1^* \{ f^* f_1^* - f_1 f \} \, g^* b^* \, db^* \, d\mathbf{v}^* \, d\mathbf{v}_1^*$$

Now by Liouville's theorem

$$g^* b^* \, db^* \, d\mathbf{v}^* \, d\mathbf{v}_1^* = gb \, db \, d\mathbf{v} \, d\mathbf{v}_1$$

and combining, we obtain

$$\frac{dH}{dt} = \frac{2\pi}{4} \int \int \int \ln \left\{ \frac{f f_1}{f^* f_1^*} \right\} \{ f^* f_1^* - f f_1 \} \, gb \, db \, d\mathbf{v} \, d\mathbf{v}_1 \qquad (15.22)$$

The integrand is of the form

$$-(x - y) \ln \frac{x}{y}$$

If $x > y$ this function is negative if $x < y$ it is also negative, and if $x = y$ it is zero. Therefore we have

$$\frac{dH}{dt} \leq 0$$

Now $H(t)$ is bounded and hence $H(t)$ must approach a limit as $t \to \infty$. Then

$$\frac{dH}{dt} = 0$$

So we have an equilibrium or steady-state solution when

$$f^*f_1^* = ff_1 \tag{15.23}$$

or

$$\ln f^* + \ln f_1^* = \ln f + \ln f_1 \tag{15.24}$$

Equation (15.23) is the *detailed balance condition*; that is, the effect of every type of encounter is *exactly balanced* by the effect of the *inverse* process. Detailed balancing is now adopted as a general principle in statistical mechanics.

To solve Eq. (15.23), let x be a function of the velocities (v_1, v_2, v_3) such that when two molecules collide the sum of the x's appropriate to the two molecules before impact is equal to the sum of the two x's after impact. Now $x = x(v_1, v_2, v_3)$; thus x will remain the same for every molecule unless altered by collision. Thus in the words of Jeans (1925),[42] "x is defined as being capable of exchange between molecules at a collision but is indestructible." That is, Σx remains the same throughout the motion where the summation extends over *all* the molecules of the gas. x is now called a *summational invariant*. Thus a particular solution is

$$\ln f = x$$

Suppose that the general solution may be written as a linear combination of x's

$$\ln f = a_1 x_1 + a_2 x_2 + a_3 x_3$$

When two molecules collide, five quantities are summationally invariant, namely the energy, the three components of the angular momentum, and the mass. These give five forms for x. There are no others, because the four constants involving the velocities give four independent relations between the six starred velocity components and the unstarred ones. Two relations must be left unfixed in order to depend essentially on the direction cosines of the line of centers at impact. Thus $\ln f$ is given by

$$\ln f = a_1 m + a_2 m v_1 + a_3 m v_2 + a_4 m v_3 + a_5 \frac{m}{2} \left(v_1^2 + v_2^2 + v_3^2 \right)$$

An easy way to evaluate the constants is to write $\ln f$ in vector form as

$$\ln f = a_1 m + \mathbf{a} \cdot m\mathbf{v} - \frac{\gamma m v^2}{2} = a_1 m + \frac{m}{2} \frac{\mathbf{a} \cdot \mathbf{a}}{\gamma} - \frac{m\gamma}{2} \left(\mathbf{v} - \frac{\mathbf{a}}{\gamma} \right)^2$$

Let

$$a_1 m + \frac{m}{2} \frac{\mathbf{a} \cdot \mathbf{a}}{\gamma} = \ln C$$

where C is constant; then

$$\ln f = \ln C - \frac{m\gamma}{2} \left(\mathbf{v} - \frac{\mathbf{a}}{\gamma} \right)^2$$

or

$$f(\mathbf{r}, \mathbf{v}) = C \exp \left[\frac{-m\gamma}{2} \left(\mathbf{v} - \frac{\mathbf{a}}{\gamma} \right)^2 \right]$$

Now the density $\rho(\mathbf{r}, t)$ is given by

$$\rho(\mathbf{r}, t) = \int f d\mathbf{v} = C \int \exp \left[-\frac{m\gamma}{2} \left(\mathbf{v} - \frac{\mathbf{a}}{\gamma} \right)^2 \right] d\mathbf{v}$$

Thus (using the properties of Gaussian distributions)

$$\rho(\mathbf{r}, t) = C \left(\frac{2\pi}{m\gamma} \right)^{1/2} \tag{15.25}$$

whence we have a value for C. Similarly

$$\mathbf{v}_0 = \langle \mathbf{v} \rangle = \frac{\mathbf{a}}{\gamma}$$

If we define temperature by

$$\tfrac{3}{2} kT = \frac{m}{2} \left\langle [\mathbf{v} - \mathbf{v}_0]^2 \right\rangle = \frac{m}{2} V^2$$

we have

$$\tfrac{3}{2} kT = \tfrac{3}{2} \gamma^{-1} \tag{15.26}$$

Hence

$$f(r, \mathbf{v}) = \rho \left(\frac{m}{2\pi kT} \right)^{3/2} \exp \left(-\frac{mV^2}{2kT} \right) \tag{15.27}$$

This is the classical expression for the Maxwellian distribution of velocities. Strictly speaking, this might be a local Maxwellian distribution, because ρ need not necessarily be in equilibrium (that is, the configurational part of f may not be in equilibrium). We must now discuss the physical meaning of the H theorem in more detail.

D. Physical Interpretation of the H Theorem

We first note that the H theorem predicts *a direction* for H, because dH/dt is either negative or zero. Since the theorem is derived by use of Boltzmann's equation, it predicts a *preferred* direction in *time* for that equation. We must now discuss the implications of this. To do so we must introduce the concepts of dynamical reversibility and the Poincaré recursion theorem. Consider now an isolated conservative dynamical system of n degrees of freedom corresponding to coordinates $(q_1, \ldots, q_n)$. Its behavior is governed by Lagrange's equation

$$\frac{d}{dt}\frac{\partial L}{\partial \dot{q}_i} - \frac{\partial L}{\partial q_i} = 0, \qquad i = 1, \ldots, n \qquad (15.28)$$

Because the system is conservative, L will not contain the time explicitly. Also, L will in general be a quadratic or even function of the velocities $\dot{q}_i$. Thus Eq. (15.28) could also be written

$$\frac{d}{d(-t)}\frac{\partial L}{\partial(-\dot{q}_i)} - \frac{\partial L}{\partial q_i} = 0, \qquad i = 1, \ldots, n \qquad (15.29)$$

Therefore either equation, (15.28) or (15.29), may be taken as giving a valid description of the motion. Corresponding to any particular solution of Eq. (15.28) that gives the coordinates q_i as a function of the time, say $\phi_i(t)$, there will be a solution of Eq. (15.29) that may be written

$$q_i' = \phi_i(-t)$$

We shall, following Tolman (1938),[3] describe $\phi_i(t)$ as giving the forward motion of the system, while $\phi_i(-t)$ gives the reverse motion. We shall suppose that the systems are labeled S and S', with the first carrying out the forward motion and the second the reverse motion. At $t = 0$ we have

$$q_i'(0) = q_i(0)$$

whereas their rates of change are opposite, that is

$$\dot{q}_i'(0) = -\dot{q}_i(0)$$

At any later time t the configuration of system S' would agree with that of S at an earlier time $-t$; thus

$$q_i'(t) = q_i(-t)$$

and the motion would be in the reverse direction to that which prevailed for S' at the earlier time, that is,

$$\dot{q}_i'(t) = -\dot{q}_i(-t)$$

Thus corresponding to any possible motion of a system of the kind mentioned would be a possible *reverse* motion in which the *same* values of the coordinates would be reached in the *reverse order*, with *reversed values* for the *velocities*. This is the principle of dynamical reversibility. This principle (a direct consequence of classical mechanics) shows that any motion of an isolated mechanical system and its reverse are *equally possible*.

We now give the *Poincaré recursion theorem*, often called the *Wiederkhersatz* [a proof can be found in Pars (1979)[43] or Chandrasekhar[11]]. This states that in a system of material particles under the influence of forces that depend only on the spatial coordinates, a given initial state defined by a representative point must in general recur *not exactly*, but to any *desired degree of accuracy*, infinitely often, provided the system always remains in the finite part of the phase space. Thus the trajectory described by the representative point in the phase space has a quasiperiodic character in the sense that after a finite interval (which can be specified), the system will return to the initial state to any desired degree of accuracy.

These two theorems (i.e., dynamical reversibility and the Poincaré recursion theorem) were used to point out apparent paradoxes in the H theorem. Loschmidt (1876, 1877)[3] argued that the H theorem is *incompatible* with the principle of dynamical reversibility, since according to that theorem, if all the molecules move in such a way as to make H decrease, there is a possible mechanical motion in which H must increase. This was the first objection to the H theorem. Zermelo (1896)[3] argued that the H theorem violates the Poincaré recursion theorem because the H theorem apparently states that H must evolve toward an equilibrium state and *stay there*. On the other hand, by Poincaré's theorem, every mechanical system (representative point) must retrace itself. Thus there would have to be increases as well as decreases in H.

The resolution of these paradoxes [Tolman (1938)[3]] is contained in the *statistical character* of the H theorem. First, it is especially important to recognize Boltzmann's result as a theorem of *statistical mechanics* that makes a reasonable prediction as to the behavior that may be expected in an *incom-*

pletely specified state, rather than as a theorem of exact mechanics that could furnish an *exact prediction* as to the *precise* behavior of a system starting from a *precisely* specified state. Thus the theorem is only a statement about the *probable* value of H. It does not rule out the possibility of *increases* as well as decreases of H. It is a statement of the fact that H decreases with a high probability. Occasional positive and negative fluctuations in H may occur, but there is an overwhelming probability that H will decrease from any high value to which it may fluctuate. The system will *on average* try to pass from the ordered to the less ordered state.

The following examples are of help in illustrating the points made above. Suppose we put red and blue ink in a vessel and stir them. The result is a uniform violet ink (a condition of the system). Let us suppose we start by stirring a uniform violet ink composed of red and blue inks. Then it is *possible* although *not probable* that the effect of the stirring will be to separate the inks of different colors so that one half of the vessel will be occupied solely by red ink and the other half solely by blue ink. From the point of view of the dynamics of the molecules composing the inks it is no less probable that this should occur than that we should be able to get a violet ink by stirring inks that were initially separated as regards color. The reason why it is not likely that a separation of violet ink into its constituent colors will occur is that there are *relatively few ways* for this event to occur. On the other hand, there are an *enormous number of ways* in which the red and blue inks may be combined to produce violet ink. The passage from separate red and blue inks to violet ink is an example of the transition from a *more* easily distinguishable to a *less* easily distinguishable arrangement of colored inks (i.e., from a higher to a lower value of H on the average).*

The following example (due to Jeans[42]) is also helpful. In dealing cards, it is just as likely that the dealer will have the 13 trumps as that he will have any other 13 cards that we might like to specify. The occurrence of a hand composed of 13 trumps might, however, be justly regarded as a "coincidence," whereas the occurrence of any specified hand in which the cards were more thoroughly mixed could not reasonably be so regarded. The explanation again is that there are comparatively few ways in which a hand that is all trumps can be dealt, but a great number of ways in which a *mixed* hand can be dealt.

As a further example,* let us consider a vessel full of water placed over a fire. It is only probable, and not certain, that the water will boil instead of freezing. If we attempt to boil the water a *sufficient* number of times, it is infinitely probable that the water will on some occasions freeze instead of

* This example is due to Willard Gibbs.
* Due to Lord Kelvin.

boil. The freezing of the water does not in any way imply a contravention of the laws of nature: The occurrence is merely a "coincidence" exactly the same as that which has taken place when the dealer in our game of whist finds that he has all the trumps in his hand.

A picturesque example of Maxwell's is of interest.[42,44] He wrote:

> One of the best established facts in thermodynamics is that it is impossible in a system enclosed in an envelope which permits neither change of volume nor passage of heat, and in which both the temperature and the pressure are everywhere the same, to produce any inequality of temperature or of pressure without the expenditure of work. This is the second law of thermodynamics, and it is undoubtedly true so long as we can deal with bodies only in mass and have no power of perceiving or handling the separate molecules of which they are made up. But if we conceive a being whose faculties are so sharpened that he can follow every molecule in its course, such a being, whose attributes are still as essentially finite as our own, would be able to do what is at present impossible to us. For we have seen that the molecules in a vessel full of air at uniform temperature are moving with velocities by no means uniform though the mean velocity of any great number of them, arbitrarily selected, is almost exactly uniform. Now let us suppose that such a vessel is divided into two portions A and B, by a division in which there is a small hole, and that a being, who can see the individual molecules, opens and closes this hole, so as to allow only the swifter molecules to pass from A to B, and only the slower ones to pass from B to A. He will thus, without expenditure of work, raise the temperature of B and lower that of A, in contradiction to the second law of thermodynamics.

Thus we may say with Jeans,[42] Maxwell's sorting demon could effect in a very short time what would probably take a very long time to come about if left to the play of chance. There would, however, be nothing contrary to natural laws in the one case any more than in the other.

Boltzmann[3,11] made some interesting estimates of the Poincaré recurrence time for some particular highly ordered states of a system. He concluded, for example, that times enormously great compared with $10^{10^{10}}$ years would be needed before an appreciable separation would occur spontaneously in a 100-cm^3 sample of two mixed gases. He also estimated that a system comprised of 10^{18} atoms cm^{-3} with an average velocity of 5×10^4 cm sec^{-1} would reproduce all of its coordinates to within 10^{-7} cm and all of its velocities to within 100 cm sec^{-1} in a time of $10^{10^{10^{19}}}$ years. For this system the length of the Poincaré recurrence time is such that to an observer it would appear irreversible, simply because the average time of recurrence is so very long compared with the time during which the system is under observation. Smoluchowski [see Chandrasekhar (1943)[11]] has stated this idea rather succinctly: "A process appears irreversible (or reversible) according to whether the initial state is characterized by a long (or short) average time of recurrence compared to the times during which the system is under observation."

A very illuminating discussion of reversibility and recurrence has been given by P. and T. Ehrenfest (1907,1959).[11,39] They imagine $2R$ balls numbered consecutively from 1 to $2R$ distributed between two urns (I and II) so that at the beginning there are $R + n$, $-R \leq n \leq R$, balls in urn I, select at random an integer between 1 and $2R$ (all these integers are assumed to be equally probable), and move the ball whose number has been drawn from the urn it is in to the other urn. This process is then repeated s times. We ask for the probability $Q(R + m, s|R + n,0)$ that after s drawings there are $R + m$ balls in urn I. This was used by the Ehrenfests as a simple and convenient model of heat exchange between two isolated bodies of unequal temperature [Kac (1947)[11]]. The temperatures were symbolized by the number of balls in the urns, and the heat exchange was not an orderly process, but a random one, as in the kinetic theory of matter.

To analyze the model we make use of the discrete form of the Smoluchowski equation. The continuous equation is often written in the form*

$$f(x_3, t_3) = f(x_3, t + \delta t) = \int_{-\infty}^{\infty} f(x_2, t) f(x_3, t + \delta t | x_2, t) \, dx_2$$

$$(15.30)$$

If we now introduce the variable $z = x_3 - x_2$ and fix x_2, then x_3 will determine z, so that

$$f(x_3, t + \delta t | x_2, t) = Q(z, \delta t | x_2, t)$$

or, if we drop the subscripts,

$$f(x, t + \delta t) = \int_{-\infty}^{\infty} f(x - z, t) Q(z, \delta t | x - z) \, dz \qquad (15.31)$$

The function Q is called the *transition probability density*. Equation (15.31) is perhaps more intuitively obvious than Eq. (15.30), it states that the probability that the system will be in a differential element at x at time $t + \delta t$ is equal to the probability that it changes from a differential element at $x - z$ at time t multiplied by the transition probability of the change z, summed with respect to all change values z. We will now give the discrete form of Eq. (15.31) for the discussion of the Ehrenfest model. The problem is always to find the probability that the system will be in a state m at some time given that it was in a state n at some earlier time. In the language of Eq. (5.32),

---

*Recall Eq. 5.32a earlier, x_1 and t_1 are dropped.

x_1 and x_3 can only have *discrete* values n and m, and the time t can only have discrete values $s\tau$ with $s = 1, 2, 3 \ldots$. The discrete form of the Smoluchowski equation is thus:

$$P(m, s\tau | n, \tau) = \sum_k P(k, (s-1)\tau | n) Q(m, k) \qquad (15.32)$$

where

$$Q(m, n) = P(m, \tau | n)$$

It is usual to drop the τ and just write the equation as

$$P(m, s | n) = \sum_k P(k, s-1 | n) Q(m, k) \qquad (15.33)$$

Now

$$\sum_m Q(k, m) = 1$$

thus

$$Q(k, k) + \sum_m{}' Q(m, k) = 1$$

where the prime means that the value $m = k$ must be omitted from the summation. If we use this in Eq. (15.33) and drop the initial value n, we can write Eq. (15.33) in the form

$$
\begin{aligned}
P(m, s) &- P(m, s-1) \\
&= -P(m, s-1) \sum_k{}' Q(m, k) + \sum_k{}' P(k, s-1) Q(k, m)
\end{aligned}
$$

$$(15.34)$$

According to Wang and Uhlenbeck,[11] one may interpret this by saying that the rate of change of $P(m, s)$ with time (where time is given by s) arises from the "gains" of P due to transitions from k to m minus the "losses" of P due to transitions from m to all possible k. This provides a complete analogue of the Boltzmann equation for the case where the molecules of the gas can collide only against fixed centers or against other molecules that have a *given* velocity distribution. Equation (15.34) must be solved for P given an initial distribution for P. Also, a "mechanism" or "physical cause" for the random process must be given; that is, Q must be specified. The initial condition for Eq. (15.34) is

$$P(m,0|n) = P(m,0) = \delta_{m,n} \qquad (15.35)$$

where δ is Kronecker's delta. This is just the mathematical statement of the fact that the particle was *certainly* in state n at the start of the process.

Before proceeding to discussing the Ehrenfest model, it will be helpful to consider two cases of random walks and the transition probabilities associated with these. We start with the random-walk problem in one dimension discussed in the introduction, but frame it in a slightly different manner here. We imagine a particle that moves along the x-axis in such a way that in each step it can move either Δ to the right or Δ to the left, the duration of each step being τ. We wish to evaluate

$$P(m\Delta; s\tau|n\Delta) = P(m; s|n) \qquad (15.36)$$

which is the probability that the particle is at $m\Delta$ at time $s\tau$ if at the beginning it was at $n\Delta$. The fact that the particle is free is now introduced by writing the transition probability Q as

$$Q(k,m) = \tfrac{1}{2}\delta_{m,k-1} + \tfrac{1}{2}\delta_{m,k+1} \qquad (15.37)$$

where δ is Kronecker's delta. This is the "mechanism" (analogous to the *Stosszahlansatz*). If we substitute Eq. (15.37) into Eq. (15.34) we find that $P(m, s)$ satisfies the difference equation (dropping the initial state n)*

$$P(m,s) = \tfrac{1}{2}P(m+1, s-1) + \tfrac{1}{2}P(m-1, s-1) \qquad (15.38)$$

This must be solved subject to the initial condition (15.35). The solution is

$$P(m, s|n) = \frac{s!\left(\tfrac{1}{2}\right)^s}{\left(\dfrac{s+|m-n|}{2}\right)! \left(\dfrac{s-|m-n|}{2}\right)!} \qquad (15.39)$$

where $|m - n| \le s$ and $|m - n| + s$ is even. Equation (15.39) is readily verified by direct substitution into Eq. (15.34).

Let us now consider an elastically bound particle. Again the particle can move either Δ to the right or Δ to the left, and the duration of each step is τ. However, the probability of moving in either direction is now made to depend on the *position* of the particle. More accurately, we say that if the par-

---

*This is the master equation for the problem

ticle is at $k\Delta$, the probabilities of moving right or left are

$$\tfrac{1}{2}\left(1-\frac{k}{R}\right), \qquad \tfrac{1}{2}\left(1+\frac{k}{R}\right)$$

respectively. R is a certain integer and possible positions of the particle are limited by the condition $-R \le k \le R$. The transition probability (the mechanism of the process) is now

$$Q(k,m) = \tfrac{1}{2}\left(1+\frac{k}{R}\right)\delta_{m,k-1} + \tfrac{1}{2}\left(1-\frac{k}{R}\right)\delta_{m,k+1} \qquad (15.40)$$

Thus there is a bias built into the process. The probabilities of making a step to the right or to the left are no longer equal. This is the *Stosszahlansatz*. Thus the particle will have a tendency to seek the position $k = 0$. If we substitute our expression for Q into Eq. (15.34), we find that the master equation is

$$P(m,s|n) = \frac{R+m+1}{2R}P(m+1,s-1|n) + \frac{R-m+1}{2R}P(m-1,s-1|n)$$

$$(15.41)$$

which again must be solved subject to the initial condition

$$P(m,0|n) = \delta_{m,n} \qquad (15.42)$$

It is not simple to solve Eq. (15.41), and for a detailed discussion the reader is referred to the article of Kac.[11] It is not, however, so difficult to calculate the moments of the distribution subject to the initial condition (15.42). We have

$$\langle m(s)\rangle = \sum_m mP(m,s) = \left(1-\frac{1}{R}\right)\langle m(s-1)\rangle$$

$$= n\left(1-\frac{1}{R}\right)^s$$

by virtue of Eq. (15.42). This shows how the average position of the point *goes to zero*. Note that in the limit $R \to \infty, (R\tau)^{-1} = \gamma, s\tau = t$

$$\langle m(t)\rangle = ne^{-\gamma t}$$

One may also show (by methods exactly similar to those used in the description of the free-particle random walk) that the stationary distribution of n is

$$f(n) = \frac{(2R)!}{N_+!N_-!}\left(\frac{1}{2}\right)^{2R}$$

where $N_+ = R + n$ is the number of moves to the right, and $N_- = R - n$ is the number of moves to the left. Thus

$$f(n) = \frac{(2R)!}{(R+n)!(R-n)!}\left(\frac{1}{2}\right)^{2R}$$

Note that

$$N_+ + N_- = 2R$$
$$N_+ - N_- = 2n$$

Furthermore, for very large R (i.e., for a very large number of moves) and small n (i.e., for little difference between the numbers of moves to the left and to the right)

$$f(n) \approx (\pi R)^{-1/2} e^{-n^2/R}$$

We shall now see how these results may be used to discuss the Ehrenfest model.

The random-walk problem for the elastically bound particle that we have just described is exactly equivalent to the Ehrenfest urn model if one interprets the *excess* over R of balls in urn I as the *displacement* of the particle $(\Delta = 1)$. The duration of each step τ is now the duration of each drawing; thus

$$Q(R + m, s | R + n) = P(m, s | n)$$

as in Eq. (15.36). The average excess over R of balls in urn I, namely

$$\langle m(t) \rangle = \overline{m(t)} = \sum_{m=-R}^{m=R} m P(m, s | n)$$

is

$$n\left(1 - \frac{1}{R}\right)^s$$

which in the limit where there is a very large number of balls $(R \to \infty)$ and the duration of a draw tends to zero, so that

$$(R\tau)^{-1} = \gamma, \qquad s\tau = t$$

goes over into

$$\langle m(t) \rangle = ne^{-\gamma t}$$

Thus the excess of balls in urn I *on the average* decreases exponentially to its equilibrium value of zero. Thus *on the average* the system appears to decay *irreversibly* toward its equilibrium state. The equilibrium mean value of the excess of balls is zero, since at equilibrium the probability distribution (in the limit of large numbers of balls) is

$$f(n) \approx (\pi R)^{-1/2} \exp\left(-\frac{n^2}{R}\right)$$

Now let $P'(m, s|n)$ denote the probability that after s drawings (the duration of a drawing being τ) $R + m$ balls will be observed for the *first* time in urn I, given that there were $R + n$ balls in that urn at the beginning. In particular, $P'(n, s|n)$ is the probability that the recurrence time of the state n is defined by the presence of $R + n$ balls in urn I. One may then show that (the proof is rather involved and is given by Kac[11])

$$\sum_{s=1}^{\infty} P'(n, s|n) = 1$$

which means that *each state of the system is bound to recur with probability 1.* This is the statistical analogue of the Poincaré recurrence theorem or *Wiederkehrsatz*. One may also show[11] that the mean recurrence time, namely

$$T_n = \sum_{s=1}^{\infty} s\tau P'(n, s|n)$$

is

$$T_n = \tau \frac{(R+n)!(R-n)!}{(2R)!} 2^{2R}$$

This is the statistical analogue of a Poincaré cycle. It tells us, roughly speaking, how long on the average one will have to wait for the state 'n' to recur.

Suppose $R + n$ and $R - n$ differ considerably; then T_n is enormous. If $R = 10,000$, $n = 10,000$, and $\tau = 1$s (this corresponds to a recurrence of that initial state where all the balls are in one urn), then

$$T_n = 2^{20,000} \text{ s}$$
$$= 10^{6000} \text{ years}$$

Thus the recurrence time for a "highly abnormal state" is enormously long.

If, on the other hand, $R + n$ and $R - n$ are nearly equal, T_n is quite short. If we set $n = 0$ in the above example, we get (using Stirling's formula)

$$T_n = 100\sqrt{\pi} \text{ s}$$
$$= 175 \text{ s}$$

This urn model provides an admirable example of the conclusion of Smoluchowski that *a process appears irreversible or reversible according to whether the initial state is characterized by a long or short average time of recurrence compared with the times during which the system is under observation.* Thus if in the urn model one urn contains 20,000 balls and the other zero, then *because of the length of the Poincaré recurrence time* one would observe what would appear to be an *irreversible* flow of balls from one urn to another. Put in another way, this process, although *reversible*, will have all the *appearances* of an *irreversible* process, simply because the *average time* of recurrence is so very long compared with the times during which the system is under observation. This is completely in accord with Boltzmann's view that the period of one of Poincaré's cycles is so enormously long, even for a cubic centimeter of gas, that the recurrence of an initially improbable state (i.e., the reversal to a state of lower entropy or greater H) is still so highly improbable that during the times normally available for observation the chance of witnessing such a reversal is extremely small. On the other hand, for states close to the equilibrium state in the urn model, the Poincaré recurrence time is quite short. Fluctuations in the number of balls in each urn occur to such an extent and with such frequency that there can no longer be any question of irreversibility.

It is appropriate to summarize this section with Tolman's[3] appreciation of the work of Boltzmann:

His fruitful discovery of a suitable function for measuring the displacement of a whole system of molecules from equilibrium and his elegant mastery of the complicated effects of collisions in making such displacements decrease with time, alike compel attention. His reconciliation of phenomenological irreversibility with the reversible character of the laws of exact mechanics and his understanding of the compatibility of continued fluctuations with a tendency towards equilibrium, are among the great achievements of theoretical physics. And his penetrating remarks on the great role that fluctuations might play in the long time behaviour of the Universe as a whole show Boltzmann's preoccupation with the deepest problems of Physics.

We now proceed to discuss the probability density diffusion equations associated with Brownian movement. We shall again start with the Chapman–Kolmogorov equation, but first we shall consider how probability density diffusion equations may arise from limiting cases of discrete distributions.

E. Examples of the Derivation of Probability Density Diffusion Equations from Limiting Cases of Discrete Distributions

We shall begin this section by again referring to the discrete random-walk models discussed earlier. For the random walk of the free particle we found that

$$P(m, s|n) = \tfrac{1}{2}P(m+1, s-1|n) + \tfrac{1}{2}P(m-1, s-1|n)$$

or, if Δ is the step length and τ the duration of the step,

$$P(m\Delta, s\tau|n\Delta) = \tfrac{1}{2}P((m+1)\Delta, (s-1)\tau|n\Delta) + \tfrac{1}{2}P((m-1)\Delta, (s-1)\tau|n\Delta)$$

Let us now subtract

$$P(m\Delta, (s-1)\tau|n\Delta)$$

from both sides of this equation, so it becomes

$$
\begin{aligned}
P(m\Delta, s\tau|n\Delta) - P(m\Delta, (s-1)\tau|n\Delta) = \\
\tfrac{1}{2}[P((m+1)\Delta, (s-1)\tau|n\Delta) \\
- 2P(m\Delta, (s-1)\tau|n\Delta) + P((m-1)\Delta, (s-1)\tau|n\Delta)] \quad (15.43)
\end{aligned}
$$

Equation (15.43) may be written in the equivalent form

$$
\frac{P(m\Delta, s\tau|n\Delta) - P(m\Delta, \overline{s-1}\tau|n\Delta)}{\tau} =
$$

$$
\frac{\Delta^2}{2\tau}\{ P(\overline{m+1}\Delta, \overline{s-1}\tau|n\Delta) - 2P(m\Delta, \overline{s-1}\tau|n\Delta)
$$

$$
+ P(\overline{m-1}\Delta, \overline{s-1}\tau|n\Delta) \} \frac{1}{\Delta^2} \qquad (15.44)^*
$$

_____

*The overbars denote round braces

Let us now consider a *large* number of *small* steps of *short* duration. More precisely, we suppose that Δ and τ approach zero in such a way that

$$\frac{\Delta^2}{2\tau} = D, n\Delta \to x_0, m\Delta \to x, s\tau = t$$

Then Eq. (15.44) (by the definition of the derivative) goes over formally into the partial differential equation

$$\frac{\partial P}{\partial t} = D\frac{\partial^2 P}{\partial x^2} \tag{15.45}$$

where P is now written

$$P(x, t | x_0, t_0)$$

This equation is the basis of Einstein's theory of Brownian movement. It shows how in a certain limit the solution of the random-walk problem may be reduced to solving an equation like Eq. (15.45). The conditions imposed on P, the probability density, are

$$P \geq 0 \tag{15.46}$$

$$\int_{-\infty}^{\infty} P(x, t | x_0)\, dx = 1 \tag{15.47}$$

and

$$\lim_{t \to 0} P(x, t | x_0) = \delta(x - x_0) \tag{15.48}$$

For convenience we have taken $t_0 = 0$. Conditions (15.46) and (15.47) are the usual ones that a probability density function must satisfy. Condition (15.48) expresses the *certainty* that at $t = 0$ the particle was at x_0. These conditions imply that

$$P(x, t, | x_0) = \frac{1}{2\sqrt{\pi Dt}} e^{-(x - x_0)^2 / 4Dt} \tag{15.49}$$

Thus the position $x(t)$ of the Brownian particle is a Gaussian random variable with mean value

$$\langle x(t) \rangle = x_0 \tag{15.50}$$

and variance

$$\sigma^2 = \langle [x(t) - x_0]^2 \rangle = 2Dt \tag{15.51}$$

We should again emphasize that these results are obtained only when in a short time τ the space coordinate can change only by small amounts. Compare the following remark of Einstein:[15] "We will introduce a time interval τ in our discussions which is *very small* compared with the *observed interval* of time (i.e. the interval of time between observations of the Brownian particle), but nevertheless of such a magnitude that the movements executed by a particle in two consecutive intervals of time τ are to be considered as mutually independent phenomena" (compare our earlier discussion). Einstein then goes on to consider the limiting case $\tau \to 0$. This is indirectly responsible for the erroneous conclusion that the velocity of a Brownian particle is infinite because if

$$\frac{\Delta^2}{2\tau} = D$$

then for finite D, as $\Delta \to 0$, the quantity (Δ / τ), which plays the role of the instantaneous velocity of the Brownian particle, must approach infinity. It must be strongly emphasized that a theory based on Eq. (15.45) is only approximate. Such a theory is valid only for relatively large t ($t \gg 1/2D$). We saw that such a limitation was well recognized by Einstein*, but has often been disregarded by writers, who *stress* that in Brownian motion the velocity of the particle is infinite. This paradoxical conclusion is a result of stretching the theory beyond the bounds of its applicability.

Before concluding, we note that the basic probability

$$P(m, s|n) = \frac{R+m+1}{2R} P(m+1, s-1|n) + \frac{R-m+1}{2R} P(m-1, s-1|n)$$

for the elastically bound particle in the limit

$$\Delta \to 0, \ \tau \to 0, \ R \to \infty, \ \frac{\Delta^2}{2\tau} = D, \ \frac{1}{R\tau} \to \gamma, \ s\tau = t, \ n\Delta \to x_0, \ m\Delta \to x$$

goes over formally into the equation

$$\frac{\partial P}{\partial t} = \gamma \frac{\partial}{\partial x}(xP) + D \frac{\partial^2 P}{\partial x^2}$$

This has fundamental solution

$$P(x, t|x_0) = \frac{\sqrt{\gamma} \exp\left[-\frac{\gamma(x - x_0 e^{-\gamma t})^2}{2D(1 - e^{-2\gamma t})} \right]}{\sqrt{2\pi D(1 - e^{-2\gamma t})}}$$

*v. discussion on p. 96.

Thus the mean value of $x(t)$ is

$$\bar{x}(t) = \langle x(t) \rangle = x_0 e^{-\gamma t}$$

in agreement with our earlier discussion of the Ehrenfest model.

F. Derivation of Probability Density Diffusion Equations from the Chapman–Kolmogorov Equation—the Kramers Equation

We have derived the Boltzmann equation

$$\frac{\partial f_j}{\partial t} + \mathbf{v}_j \cdot \nabla_r f_j + \frac{1}{m_j} \mathbf{X}_j \cdot \nabla_{\mathbf{v}_j} f_j = 2\pi \sum_i \int \int \{ f_i^* f_j^* - f_i f_j \} g_{ij} b \, db \, d\mathbf{v}_i$$

for the rate of change of the probability distribution due to collisions when only two-body interactions are ever important, as in gases. A related equation for f may be derived if one specifies the collision mechanism to be that of Brownian movement, that is, small collisions. To restate what we mean by such a mechanism, it is a mechanism in which the *positions* of the particles are *unchanged* and their velocities are altered by such small amounts that they can be treated as infinitesimal; correlation between successive impulses is negligible. Under the simplest assumptions the effect of such collisions appears as a *viscous* retardation

$$- m\beta \mathbf{v}(t)$$

superimposed on which are *random fluctuations* governed by a Wiener process

$$\mathbf{F}(t) = m\dot{\mathbf{B}}(t)$$

We start with the Langevin equation for a particle P of mass m moving in three dimensions:

$$m\frac{d\mathbf{v}}{dt} + m\beta \mathbf{v}(t) + \mathrm{grad}_r V = \mathbf{F}(t)$$

Then we have, according to Kramers[9] (although the equation was first derived by Klein[12]), the following generalization of the Liouville equation (the quantities (v_1, v_2, v_3) rather than (p_1, p_2, p_3) are chosen here as variables; This makes no difference in the form of the results):

$$\frac{\partial f}{\partial t} + \mathbf{v} \cdot \mathrm{grad}_r f - \frac{1}{m} \mathrm{grad}_v f \cdot \mathrm{grad}_r V = \beta \left(\mathrm{div}_v (\mathbf{v} f) + \frac{kT}{m} \nabla_v^2 f \right) \quad (15.52)$$

where $V(\mathbf{r})$ is the potential energy of the particle, T is the absolute temperature, and k is the Boltzmann constant. On the right-hand side, we have the terms arising from Brownian movement, while on the left-hand side we have the usual Stokes operator $D/Dt = \partial/\partial t + \mathbf{v}\cdot\text{grad}$ acting on f. We now show briefly how Eq. (15.52) may be heuristically derived from the Langevin equation.

The Langevin equation is written

$$d\mathbf{v}(t) = -\left[\frac{1}{m}(\nabla_{\mathbf{r}}V) + \beta\mathbf{v}(t)\right]dt + d\mathbf{B}(t) \tag{15.53}$$

If $[q_1(t), q_2(t), q_3(t)]$ are the Cartesian coordinates of the particle P of mass m at time t, then Eq. (15.53) may be written

$$d\begin{bmatrix} q_1 \\ q_2 \\ q_3 \\ v_1 \\ v_2 \\ v_3 \end{bmatrix} = \begin{bmatrix} v_1 \\ v_2 \\ v_3 \\ -(\beta v_1 + m^{-1}\partial V/\partial q_1) \\ -(\beta v_2 + m^{-1}\partial V/\partial q_2) \\ -(\beta v_3 + m^{-1}\partial V/\partial q_3) \end{bmatrix} dt + \begin{bmatrix} 0 \\ 0 \\ 0 \\ dB_1 \\ dB_2 \\ dB_3 \end{bmatrix} \tag{15.54}$$

which in turn has the form (interchanging 1, 2, 3, with 4, 5, 6 in the $d\mathbf{B}$ matrix)

$$d\begin{bmatrix} x_1 \\ \cdot \\ \cdot \\ \cdot \\ \cdot \\ x_6 \end{bmatrix} = \begin{bmatrix} g_1 \\ \cdot \\ \cdot \\ \cdot \\ \cdot \\ g_6 \end{bmatrix} dt + \begin{bmatrix} 0 & 0 & 0 & 0 & 0 & 0 \\ 0 & 0 & 0 & 0 & 0 & 0 \\ 0 & 0 & 0 & 0 & 0 & 0 \\ 0 & 0 & 0 & 1 & 0 & 0 \\ 0 & 0 & 0 & 0 & 1 & 0 \\ 0 & 0 & 0 & 0 & 0 & 1 \end{bmatrix} \begin{bmatrix} dB_1 \\ \cdot \\ \cdot \\ \cdot \\ \cdot \\ dB_6 \end{bmatrix} \tag{15.55}$$

Clearly

$$x_1 = q_1; x_2 = q_2;\ldots; x_6 = v_3; g_1 = v_1;\ldots; g_6 = -(\beta v_3 + m^{-1}\partial V/\partial q_3).$$

Thus our Langevin equation has the general form

$$d\mathbf{x}(t) = \mathbf{g}(t,\mathbf{x})\,dt + \mathbf{A}\,d\mathbf{B}(t) \tag{15.56}$$

where $\mathbf{x}$ is a column vector in $\mathbb{R}^6$ and

$$\mathbf{A} = \begin{bmatrix} \cdot & \cdot & 0 & \cdot & \cdot & \cdot \\ \cdot & \cdot & 0 & \cdot & \cdot & \cdot \\ 0 & 0 & 0 & 1 & 0 & 0 \\ \cdot & \cdot & \cdot & \cdot & \cdot & \cdot \end{bmatrix}$$

The Chapman–Kolmogorov (or Master) equation for the stochastic process defined by Eq. (15.56) is (under the *assumption* that that process is *Markovian*)

$$f(\mathbf{x}, t + \delta t) = \int_{-\infty}^{\infty} f(\mathbf{x}, t + \delta t | \mathbf{x}', t) f(\mathbf{x}', t) \, d\mathbf{x}', \qquad \int_{-\infty}^{\infty} \equiv \int_{-\infty}^{\infty} \cdots \int_{-\infty}^{\infty} = \int$$

$$(15.57)^*$$

We now write $\mathbf{x} - \mathbf{x}' = \mathbf{z}$, so that

$$f(\mathbf{x}, t + \delta t) = \int_{-\infty}^{\infty} Q(\mathbf{z}, \delta t | \mathbf{x} - \mathbf{z}, t) f(\mathbf{x} - \mathbf{z}, t) \, d\mathbf{z}$$

where the function Q is defined as follows:

$$Q(\mathbf{z}, \delta t | \mathbf{x} - \mathbf{z}, t) = f(\mathbf{x}, t + \delta t | \mathbf{x}', t)$$

We now assume that we may approximate the integrand by the first few terms of a Taylor series about the point $\mathbf{z} = \mathbf{x}$; to this end we define

$$h(\mathbf{x} - \mathbf{z}, t, \mathbf{z}, \delta t) = Q(\mathbf{z}, \delta t | \mathbf{x} - \mathbf{z}, t) f(\mathbf{x} - \mathbf{z}, t)$$
$$= h(\mathbf{x} - \mathbf{z})$$

to avoid unnecessary writing.

Clearly one may write

$$h(\mathbf{x} - \mathbf{z}) = \left\{ \left[\exp(-\mathbf{z} \cdot \nabla_{\mathbf{x}}) \right] h(\mathbf{x}) \right\}$$

where $\mathbf{z} \cdot \nabla_{\mathbf{x}}$ stands for the operator

$$z_1 \partial_{x_1} + \cdots + z_6 \partial_{x_6}$$

Thus

$$\exp(\mathbf{z} \cdot \nabla_{\mathbf{x}}) = 1 - \sum_{\ell=1}^{6} z_{\ell} \partial_{x_{\ell}}(\cdot) + \frac{1}{2!} \sum_{\ell=1}^{6} \sum_{m=1}^{6} z_{\ell} z_m \partial_{x_{\ell}} \partial_{x_m}(\cdot) + \cdots$$

---

*Eq. (15.57) is the same as Eq. 5.32a with the exception that primes are interchanged. The same assumption as that which led to Eq. 5.1 is made about the transition probability Q.

On utilizing these results we find that

$$f(\mathbf{x}, t+\delta t) = \left\{ \int h \, d\mathbf{z} - \sum_{\ell=1}^{6} \int z_\ell \partial_{x_\ell} h \, d\mathbf{z} \right.$$
$$\left. + \frac{1}{2!} \sum_{\ell=1}^{6} \sum_{m=1}^{6} \int z_\ell z_m \partial_{x_\ell} \partial_{x_m} h \, d\mathbf{z} + \cdots \right\}$$

Assuming that the *orders* of differentiation and integration may be interchanged in this equation, we have, on substituting for h,

$$f(\mathbf{x}, t+\delta t) = \left\{ \int Q(\mathbf{z}, \delta t | \mathbf{x}, t) f(\mathbf{x}, t) \, d\mathbf{z} \right.$$
$$- \sum_{\ell=1}^{6} \partial_{x_\ell} \int [z_\ell Q(\mathbf{z}, \delta t | \mathbf{x}, t) f(\mathbf{x}, t) \, d\mathbf{z}]$$
$$\left. + \frac{1}{2!} \sum_{\ell=1}^{6} \sum_{m=1}^{6} \partial_{x_\ell} \partial_{x_m} \int [z_\ell z_m Q f d\mathbf{z}] \cdots \right\}$$

Now

$$f(\mathbf{x}, t+\delta t) = f(\mathbf{x}, t) + \delta t \frac{\partial f}{\partial t} + O\big((\delta t)^2\big)$$

thus noting that

$$\int Q \, d\mathbf{z} = 1$$

$$\langle z_\ell \rangle = \int z_\ell Q(\mathbf{z}, \delta t | \mathbf{x}, t) \, d\mathbf{z},$$

we have

$$\delta t \frac{\partial f}{\partial t} = - \sum_{\ell=1}^{6} \partial_{x_\ell} [\langle z_\ell(\mathbf{x}, t, \delta t) \rangle f(\mathbf{x}, t)]$$
$$+ \frac{1}{2!} \sum_{\ell=1}^{6} \sum_{m=1}^{6} \partial_{x_\ell} \partial_{x_m} [\langle z_\ell z_m(\mathbf{x}, t, \delta t) \rangle f(\mathbf{x}, t)]$$

The task that now remains is to determine the averages $\langle z_\ell \rangle, \langle z_\ell z_m \rangle$, or more precisely,

$$\lim_{\delta t \to 0} \left[\frac{\langle z_\ell(\mathbf{x}, t, \delta t) \rangle}{\delta t} \right]$$

and

$$\underset{\delta t \to 0}{\text{Lim}} \left[\frac{\langle z_\ell z_m(\mathbf{x}, t, \delta t) \rangle}{\delta t} \right]$$

To do this we return to our initial stochastic differential equation (15.56), which may be written, on integrating between t and $t + \delta t$, as

$$\mathbf{x}(t + \delta t) - \mathbf{x}(t) = \int_t^{t + \delta t} \mathbf{g}(t', \mathbf{x}(t')) \, dt' + \int_t^{t + \delta t} \mathbf{A} \, d\mathbf{B}(t')$$

If we now approximate the first integral on the right-hand side of this equation by $\mathbf{g}(t, \mathbf{x}(t)) \, \delta t$, we find that the equation becomes

$$\mathbf{x}(t + \delta t) - \mathbf{x}(t) = g(t, \mathbf{x}(t)) \, \delta t + \mathbf{A}\xi(\delta t)$$

However,

$$\mathbf{x}(t + \delta t) - \mathbf{x}(t) = \mathbf{z}$$

by our earlier definition of $\mathbf{z}$, so that

$$\mathbf{z} = \mathbf{g}(t, \mathbf{x}(t)) \, \delta t + \mathbf{A}\xi(\delta t)$$

On utilizing the properties of the Wiener process ξ, we obtain

$$\langle \mathbf{z} \rangle = g(t, \mathbf{x}(t)) \, \delta t$$

and

$$\underset{\delta t \to 0}{\text{Lim}} \frac{\langle z_\ell(x, t, \delta t) \rangle}{\delta t} = g_\ell(x, t)$$

Now, for $\ell = 1, 2, 3$, we have

$$z_\ell = g_\ell(x(t), t) \, \delta t$$

and for $\ell = 4, 5, 6$,

$$z_\ell = g_\ell(x(t), t) \, \delta t + \xi(\delta t)$$

Let us now take two distinct components ℓ and m, $\ell, m > 3$. We then have

$$z_\ell z_m = g_\ell g_m(\delta t)^2 + g_\ell \xi(\delta t) \, \delta t + g_m \xi_\ell(\delta t) \, \delta t + \xi_\ell \xi_m \, \delta t$$

Ignoring terms $O(\delta t)^2$ and remembering that

$$\langle \xi_\ell(\delta t) \rangle = \langle \xi_m(\delta t) \rangle = 0$$

we have

$$\lim_{\delta t \to 0} \frac{\langle z_\ell z_m \rangle}{\delta t} = \lim_{\delta t \to 0} \frac{\langle \xi_m(\delta t) \xi_\ell(\delta t) \rangle}{\delta t} = c^2_{\ell m} \delta_{m\ell}$$

where $\delta_{m\ell}$ denotes Kronecker's delta. Thus the Kramers equation reduces to

$$\frac{\partial f(\mathbf{x}, t)}{\partial t} = - \sum_{\ell=1}^{6} \partial_{x_\ell}[g_\ell(\mathbf{x}, t) f(\mathbf{x}, t)] + \frac{1}{2!} \sum_{\ell=4}^{6} \sum_{m=4}^{6} c^2_{\ell m} \delta_{m\ell} \partial_{x_\ell} \partial_{x_m} f(\mathbf{x}, t)$$

which may be written

$$\frac{\partial f}{\partial t} + \mathbf{v} \cdot \nabla_r f - \frac{1}{m}(\nabla_v f) \cdot \nabla_r V = \beta \left[\nabla_v \cdot (\mathbf{v}f) + \frac{kT}{m} \nabla_v^2 f \right] = \frac{Df}{Dt} \quad (15.58)$$

where the constants $c^2_{\ell m}$ have been found by appealing to the fact that in the limit of *long* times the *Maxwell–Boltzmann distribution* must prevail, so that

$$\lim_{t \to \infty} f = A \exp\left\{ -\frac{1}{kT} \left[\frac{mv^2}{2} + V(\mathbf{r}) \right] \right\}$$

where A is a given constant. All physically significant questions concerning the translational Brownian movement of a particle under the influence of a potential $V(\mathbf{r})$ can be answered by finding the solution of Eq. (15.58).

We now go on to show how a differential equation (the Fokker–Planck equation) for the distribution of velocities in velocity space for a free Brownian particle may be derived.

G.   The Fokker–Planck Equation

The Langevin equation for a Brownian particle free to move in three dimensions is

$$m\frac{d\mathbf{v}}{dt} + m\beta\mathbf{v}(t) = m\frac{d\mathbf{B}(t)}{dt} \quad (15.59)$$

where the symbols have their usual meanings. Evidently Eq. (15.59) may be written in the form of Eq. (15.55), where this time $\mathbf{x}$ is a vector in velocity space $\mathbb{R}^3$. On proceeding exactly in the manner outlined in the previous section, we find the following differential equation for the velocity distribution $f = f(\mathbf{v}, t)$ in *velocity* space:

$$\frac{\partial f}{\partial t} = \beta\left(\text{div}_v(\mathbf{v}f) + \frac{kT}{m}\nabla_v^2 f \right) \tag{15.60}$$

This equation is known as the Fokker–Planck equation and it holds *only* for the velocity distribution of a *free* Brownian particle. When an *external* potential $V(\mathbf{r})$ is involved, the Kramers equation *must* be employed to determine the velocity distribution.*

H. The Smoluchowski Equation

As we have seen, the Kramers equation contains all possible information concerning the motion of a Brownian particle under the influence of a given potential. In many practical situations, however, we are interested only in the solution of the problems for time intervals that are *very large* compared with the "time of relaxation" β^{-1} (i.e., for times $t \gg \beta^{-1}$). This is tantamount to saying that the inertia of the Brownian particles has no *influence* on the time evolution of the density $f(\mathbf{r}, \mathbf{v}, t)$. Another way of saying this is that we apply the method originating in a Chapman–Kolmogorov equation to the configuration space *independently* of the velocity space. Thus, starting from the Langevin equation (where the term $m\ddot{\mathbf{r}}$ has been ignored)

$$m\beta\dot{\mathbf{r}} + \text{grad}_r V = m\frac{d\mathbf{B}(t)}{dt} \tag{15.61}$$

we find that the density $f = f(\mathbf{r}, t)$ in configuration space satisfies the following partial differential equation:

$$\frac{\partial f}{\partial t} = \text{div}_r\left(\frac{kT}{m\beta}\text{grad}_r f + \frac{f}{m\beta}\text{grad}_r V \right) \tag{15.62}$$

This equation is now generally called the Smoluchowski equation. It is simpler in form than the original Kramers equation; in many cases a solution can be easily effected where a solution of the corresponding Kramers equation would be extremely difficult. We should bear in mind that the basic equation underlying all these diffusion equations is Boltzmann's equation. The Kramers equation is a particular case of the Boltzmann equation for the case in which the velocity of the particle can change very little in each collision. Then the Boltzmann equation may be approximated by the Kramers equation. We now explore the connection between the two equations in more detail.

---

*The left hand side of the Kramers equation may be written as Df/Dt, hence it too is often known as the Fokker-Planck Equation.

The Boltzmann equation is a *closed* nonlinear integro-differential equation for the one-particle distribution function. The closed nature of the Boltzmann equation follows from the *Stosszahlansatz*, where the infinite hierarchy for the distribution functions is reduced to a *single* equation for the *one*-particle distribution function. This is greatly different from the Liouville equation reformulated in terms of the BBGKY hierarchy, where the one-particle distribution function depends on the pair-distribution function, which depends on the triple-distribution function, and so on.

The Kramers equation, like the Boltzmann equation, is a closed equation for the one-particle distribution function, but has the great advantage of being *linear*. The circumstances in which the Boltzmann equation may be approximated by the Kramers equation have been discussed by Chapman and Cowling,[2] by Balescu,[45] and by Résibois and de Leener.[41] The treatment in its essentials arises as a limiting case of a kinetic equation originally derived by Landau. We briefly sketch the derivation here. The fundamental assumption underlying our treatment is that in all encounters between two molecules the deflections are *small*. As Balescu[45] puts it, *the molecular interactions are very weak compared to the mean kinetic energy of the molecules*, meaning that the deviation in velocity resulting from a collision is small on the average as compared with the initial velocity of the particles. Following Chapman and Cowling,[2] we consider the encounter of two molecules m_1 and m_2 and suppose that the velocities v_1 and v_2 alter only by *small* quantities

$$\Delta v_1, \Delta v_2, \Delta g_{21} = \Delta v_1 - \Delta v_2$$

in an encounter (collision). The Boltzmann equation is, as usual,

$$\frac{Df_1}{Dt} = 2\pi \int \int (f_1^* f_2^* - f_1 f_2) gb\, db\, dv_2 \qquad (15.63)$$

We now make use of the assumption of small deflections to expand the functions f_1^* and f_2^* in Taylor series in the velocities, obtaining

$$f_1^* = f_1(v_1 + \Delta v_1) = f_1(v_1) + \Delta v_1 \cdot \frac{\partial f_1}{\partial v_1} + \tfrac{1}{2} \Delta v_1 \Delta v_1 : \frac{\partial^2 f_1}{\partial v_1 \, \partial v_1} + \cdots$$

$$(15.64)$$

with a similar expression for f_2^*. Retaining now only the first- and second-order terms in the velocities gives, with the Boltzmann equation,

$$\frac{Df_1}{Dt} = -\frac{\partial \mathbf{Q}_{12}}{\partial \mathbf{v}_1} = -\frac{1}{2}\frac{\partial}{\partial \mathbf{v}_1}\left\{\int\int\left[\left(\frac{m_1}{m_2}\frac{\partial}{\partial \mathbf{v}_2} - \frac{\partial}{\partial \mathbf{v}_1}\right)f_1 f_2\right]\Delta \mathbf{v}_1 \Delta \mathbf{v}_1 \, gb \, db \, d\mathbf{v}_2\right\}$$

(15.65)

Equation (15.65) is due to Landau. It shows that changes in f_1 due to encounters with molecules m_2 are the *same* as if there were a flow $\mathbf{Q}_{12}$ of molecules m_1 in velocity space. Equation (15.65) may be further transformed using the divergence theorem into

$$\frac{Df_1}{Dt} = \frac{\partial}{\partial \mathbf{v}_1}\left\{\frac{1}{2}\frac{\partial}{\partial \mathbf{v}_1}\left[f_1\int\int f_2 \,\Delta \mathbf{v}_1 \Delta \mathbf{v}_1 \, gb \, db \, d\mathbf{v}_2\right] - \int\int f_1 f_2 \,\Delta \mathbf{v}_1 \, gb \, db \, d\mathbf{v}_2\right\}$$

(15.66)

or* (interchanging the two terms on the right)

$$\frac{Df_1}{Dt} = -\frac{\partial}{\partial \mathbf{v}_1}(f_1\Sigma_2 \,\Delta \mathbf{v}_1) + \frac{1}{2}\frac{\partial^2}{\partial \mathbf{v}_1 \,\partial \mathbf{v}_2}:(f_1\Sigma_2 \,\Delta \mathbf{v}_1 \Delta \mathbf{v}_2)$$ (15.67)

where, if ϕ denotes a property

$$\Sigma_2\phi = \int\int f_2\phi \, gb \, db \, d\mathbf{v}_2$$ (15.68)

$\Sigma_2\phi$ denotes[2] the sum of the values of ϕ for all encounters per unit time of a molecule m_1 of velocity $\mathbf{v}_1$ with molecules m_2. Equation (15.67) has the form of the Kramers equation, but is still *nonlinear* and still an integro-differential equation; the method of writing it conceals integrations involving f_2. Equation (15.67) may be conveniently regarded as a Kramers equation with friction and diffusion coefficients that *both* depend on the distribution function and are not constant, as is usual.

Balescu[45] has given the conditions under which Eq. (15.67) reduces to the Kramers equation. Consider a large body of *weakly* coupled particles in thermal equilibrium at temperature T. Let us suppose that as a result of a fluctuation, there are a few particles (homogeneously distributed) that are not in equilibrium at time zero. Because of the rarity of these particles, we assume that they *never* collide *among themselves*, but *only* with particles of the background (heat bath or reservoir). We further assume that the global ther-

---

*The notation used is that of Reference 2 chapter 1.

mal equilibrium of the bath is not affected by these few particles. Under these assumptions, we may write for the distribution function of these particles a Landau equation in which we replace the distribution function of their collision partners by the time-independent Maxwellian distribution. This immediately reduces the Landau equation to a linear partial differential equation that is precisely the Kramers equation.

Finally, in connection with the Boltzmann equation, we mention the BGK model (after Bhatnagar, Gross, and Krook[22]). In this model, one assumes that after each encounter the velocity distribution becomes the Maxwell distribution. The Boltzmann equation then becomes (β^{-1} is the relaxation time)

$$\frac{Df_1}{Dt} = \beta \left[\sqrt[3]{\frac{m}{2\pi kT}} \, e^{-mv^2/2kT} \int f_1 \, d\mathbf{v} - f_1 \right]$$

This equation may also be treated using the continued-fraction methods developed for the Kramers equation.[9,20-22]

Acknowledgments

I would like to thank Professor George Papadopoulos and the University of Athens for their hospitality during the period in which this work was carried out. I thank the Royal Irish Academy for the award of a Senior Visiting Fellowship. I thank Trinity College Dublin Trust for a grant for this research and Ms. Anne Kenny and Ms. Olivia Musgrave for their preparation of the manuscript. I also thank Professor B. Scaife, Dr. R. Landauer and Dr. M. Büttiker for helpful conversations. Lastly I thank Mr. P. M. Corcoran for his checking of the manuscript.

References

1. A. T. Fuller, *Nonlinear Stochastic Control Systems*, Taylor and Francis, London, 1970.

2. S. Chapman and T. G. Cowling, *The Mathematical Theory of Non-Uniform Gases*, 3rd ed., Cambridge University Press, London, 1970.

3. R. C. Tolman, *The Principles of Statistical Mechanics*, Oxford University Press, London, 1938.

4. J. W. Gibbs, *Elementary Principles in Statistical Mechanics*, Dover, New York, 1960.

5. A. I. Khinchin, *Mathematical Foundations of Statistical Mechanics*, Dover, New York, 1949.

6. R. Barnes and S. Silverman, *Rev. Mod. Phys.*, **6**, 162 (1934).

7. M. Born, *Natural Philosophy of Cause and Chance*, Oxford University Press, London, 1948.

8. I. Todhunter, *A History of the Mathematical Theory of Probability from the Time of Pascal to That of Laplace*, Macmillan, London, 1865.

9. M. W. Evans, G. J. Evans, W. T. Coffey, and P. Grigolini, *Molecular Dynamics and the Theory of Broad Band Spectroscopy*, Wiley Interscience, New York, 1982.

10. H. Haken, *Synergetics*, 2nd ed., Springer-Verlag, Berlin, 1978.

11. N. Wax, ed., *Selected Papers on Noise and Stochastic Processes*, Dover, New York, 1954.

12. W. T. Coffey, M. W. Evans, and P. Grigolini, *Molecular Diffusion and Spectra*, Wiley Interscience, New York, 1984.

13. E. Nelson, *Dynamical Theories of Brownian Motion*, Princeton University Press, Princeton, 1967.

14. Sir Ralph Fowler, *Statistical Mechanics*, Cambridge University Press, London, 1936.

15. A. Einstein, *Investigations on the Theory of the Brownian Movement*, R. H. Fürth, ed., Dover, New York, 1956.

16. Lord Rayleigh, *The Theory of Sound*, Vol. 1, Macmillan, London, 1894.

17. P. I. Cootner, *The Random Character of Stock Market Prices*, The M.I.T. Press, Cambridge, Massachusetts, 1964.

18. G. Joos, *Theoretical Physics*, Blackie, Edinburgh, 1934.

19. P. Debye, *Polar Molecules*, Chemical Catalog, New York, 1929.

20. J. R. McConnell, *Rotational Brownian Motion and Dielectric Theory*, Academic, London, 1980.

21. B. K. P. Scaife, *Complex Permittivity*, The English Universities Press, London, 1971.

22. H. Risken, *The Fokker–Planck Equation*, Springer-Verlag, Berlin, 1984.

23. G. Barone and A. Paterno, *Physics and Applications of the Josephson Effect*, Wiley Interscience, New York, 1982.

24. M. Büttiker, E. P. Harris, R. Landauer, *Phys. Rev. B*, **28**, 1268 (1983).

25. J. D. Cresser, D. Hammonds, W. H. Louisell, P. Meystre, and H. Risken, *Phys. Rev. A*, **25**, 2226 (1982).

26. H. Risken and P. Voigtlaender, in press.

27. A. J. Viterbi, *Principles of Coherent Communication*, McGraw-Hill, New York, 1966.

28. R. Landauer, *Helv. Phys. Acta*, **56**, 847 (1982).

29. P. Hänggi and H. Thomas, *Phys. Rep.*, **88**, 207 (1982).

30. S. P. Davis, in *Encyclopaedia of Physics*, R. M. Besançon, ed., Rheinhold, New York, 1966.

31. A. Abragam, *The Principles of Nuclear Magnetism*, Oxford University Press, London, 1961.

32. G. H. Vineyard, *Phys. Rev.*, **110**, 999 (1958).

33. L. Arnold, *Stochastiche Differentialgleichungen*, Oldenbourg, Munich, 1973.

34. W. T. Coffey, J. A. Abas, and M. W. Evans, to be published.

35. R. W. Impey, P. A. Madden, and I. R. MacDonald, *Mol. Phys.*, **46**, 513 (1982).

36. B. K. P. Scaife, *Dielectric and Related Molecular Processes*, Vol. 1, The Chemical Society, 1972, p. 1.

37. H. Benoit, *Annes Phys.*, **6**, 561 (1951); *ibid.*, *J. Chim. Phys.*, **49**, 517 (1952).

38. R. L. Stratonovitch, *Topics in the Theory of Random Noise*, Vols. I and II, Gordon & Breach, New York, 1963, Vol. I, 1967, Vol. II.

39. D. A. McQuarrie, *Statistical Mechanics*, Harper & Row, New York, 1976.

40. H. Goldstein, *Classical Mechanics*, Addison-Wesley, Reading, Massachusetts, 1950.

41. P. Résibois and M. de Leener, *Classical Kinetic Theory of Fluids*, Wiley Interscience, New York, 1976.

42. Sir James Jeans, *Dynamical Theory of Gases*, Cambridge University Press, Cambridge, 1925.

43. L. A. Pars, *A Treatise on Analytical Dynamics*, Oxbow Press, Connecticut, 1979.

44. L. Brillouin, *Science and Information Theory*, Academic, New York, 1956.

45. R. Balescu, *Equilibrium and Nonequilibrium Statistical Mechanics*, Wiley Interscience, New York, 1975.

46. M. Büttiker and R. Landauer, *Phys. Rev. A*, **23**, 1397 (1980).
47. S. Piazzolla and P. Spano, *Optics. Comm.* **51**, 279, 1984.
48. M. Y. Azbel and P. Bak, *Phys. Rev. B* **30**, 3722, 1984.
49. W. Schleich, C. S. Cha and J. D. Cresser, *Phys. Rev. A*, **29**, 230, 1984.
50. W. W. Chow, J. Gea-Banacloche, L. M. Pedrotti, V. E. Sanders, W. Schleich and M. O. Scully, *The Ring Laser Gyro*, *Rev. Mod. Phys.* **57**, 61, 1985.

THE FADING OF MEMORY DURING THE REGRESSION OF STRUCTURAL FLUCTUATIONS

L. A. DISSADO

Department of Physics, Chelsea College‡, University of London, London, United Kingdom

R. R. NIGMATULLIN

The Department of Physics, Kazan State University, Kazan, U.S.S.R.
and

R. M. HILL

Department of Physics, Chelsea College‡, University of London, London, United Kingdom

CONTENTS

‡Since this manuscript was submitted Chelsea College has amalgamated with King's College London.

I. INTRODUCTION

The theoretical description of the way in which memory is lost as a disturbed system returns to equilibrium is a fundamental problem in response theory. It is discussed here in the context of structural fluctuations in condensed-phase materials, which we shall define as a set of displacements of the constituent elements away from their equilibrium positions. Since in condensed-phase systems all such elements vibrate/librate about their respective centers/axes of motion at all temperatures, two classes of displacements can be distinguished. In one class the centers/axes of motion are unchanged and the problem reduces to one concerning the generation of phonons, which can be solved exactly[1] if the system possesses translational symmetry. We are concerned here exclusively with the second class, in which the centers/axes of motion that define a structural configuration of the elements are themselves displaced.

The manner in which specified displacements regress to equilibrium values is quite clear in general physical terms. Immediately after the creation of such a fluctuation the elements start to vibrate/librate about the centers of motion of the displaced configuration, with force constants appropriate to that configuration. The dynamical behavior of the fluctuation must therefore possess a memory of the displaced configuration, at least initially. During the return to equilibrium, however, it is a necessary consequence of irreversible thermodynamics[2] that this memory be lost. The fundamental theoretical problem is to develop an adequate analytical description of the motion of the displacements during the period in which the memory is lost (the "fading of memory" mentioned in the title). Most attempts to obtain theoretical relaxation functions[3] fall into two categories that do not, in fact, address this problem directly. In one of these approaches the evolution of a system under the influence of a model Hamiltonian is determined. Expressions are then obtained that are valid over the short time[4-8] in which memory is important, and after long times, when memory is completely lost.[5,8] The alternative approach has been to assume from the outset a loss of memory in the components of the displacements of the structural elements, which are then combined in either an independent or a coupled manner to give an average description of the regression.[8,9]

The search for a theoretical description of regression with general applicability has received a powerful stimulus from the observation[10] that the regression of structural fluctuations exhibits a general form, as demonstrated by the similarity of the frequency dependences found in dielectric[11] or mechanical[12] response for a variety of materials. This leads one to the conclusion that the features governing regression, in general, possess a common physical interpretation. Taking this viewpoint, a number of authors have made approaches to the problem that can be seen to possess certain elements in common, namely the existence of correlated[13-20] or "locked-in"[17] motions and the presence of some local structural disorder. A more detailed physical basis for one of these models[14] has recently been described[20] and it is our intention here to outline the manner in which this particular model accounts for the memory-fading necessary to thermodynamic irreversibility. In the course of this development we will obtain the relationship between the model and thermodynamic quantities appropriate to relaxation. Using this relationship, the features common to the regression of any structural fluctuation can be given a single general formulation of wide applicability.

II. THE RESPONSE FUNCTION AND SUSCEPTIBILITY

A. Analytical Response Function

For any system in which a material property R can be defined only as a statistical average rather than possessing a unique value, a field E conjugate to R can couple to its fluctuations,[2,3,21] causing a change in its observed value. This is termed the response of the system to E, and the term linear in E is given formally[3] by

$$\Delta R(t) = \int_{-\infty}^{t} \phi(t - t') E(t') \, dt' \tag{1}$$

where $\phi(t)$ is the response function.

Structural fluctuations, which occur in all systems that do not possess an ideal lattice symmetry, will therefore respond to an applied stress through a generalized displacement ΔR, termed the strain. If, in addition, the displacements produce a change in dipole moment, the system will respond to an electric field E. Thus structural fluctuation may be observed through dielectric-response measurements, in addition to the more widely applicable mechanical[22] and acoustic[23,24] responses.

The response function $\phi(t)$ has a simple interpretation in terms of the observed response in two specific situations: a step-function change of field, or a delta-function application of field. Denoting the displacement $\Delta R(t)$ by $Q(t)$ for simplicity, the former case, which is defined by

$$E(t) = E_0 \theta(t) \tag{2}$$

with $\theta(t)$ the step-up function

$$\theta(t) = \begin{cases} 0 & t < 0 \\ 1 & t \geq 0 \end{cases} \tag{3}$$

gives a response whose rate of change is given by[3] $\phi(t)$:

$$\frac{dQ(t)}{dt} = \begin{cases} \phi(t)E_0 & t \geq 0 \\ 0 & t < 0 \end{cases} \tag{4}$$

In the second case the applied field has the form

$$E(t) = E_0\delta(t) \tag{5}$$

and $\phi(t)$ determines the regression of the displacement itself following its creation at zero time:

$$Q(t) = \begin{cases} \phi(t)E_0 & t \geq 0 \\ 0 & t < 0 \end{cases} \tag{6}$$

Basing their approach on this latter definition of $\phi(t)$, Dissado and Hill[14,20] have derived a general analytical expression for $\phi(t)$ in terms of the confluent hypergeometric function[25] $_1F_1(\ ;\ ;\ )$:

$$\frac{\phi(t)}{\phi(0)} = \Phi(t) = (\omega_P t)^{-n} e^{-\omega_P t} {}_1F_1(1 - m; 2 - n; \omega_P t) \tag{7}$$

where ω_P is the characteristic relaxation rate constant, and n and m are numerical constants in the range between 0 and 1. In the range of validity of this expression, which was found to be for times t in excess[15,20] of about $10\zeta^{-1}$ (where ζ is an oscillation frequency generally of the order of a long-wavelength lattice mode), use of Eqs. (6) and (7) shows that the unit displacement $\hat{Q}(t)$ [$= Q(t)/Q(0)$] obeys the equation of motion

$$\frac{d^2[\hat{Q}(t)]}{dt^2} + \frac{1}{t}(2 + n + \omega_P t)\frac{d\hat{Q}(t)}{dt} + \frac{1}{t^2}(n + \omega_P t\{1 + m\})\hat{Q}(t) = 0,$$

$$t\zeta \geq 10 \tag{8}$$

Since $\phi(t)$ is invalid for $t\zeta < 10$, the source of the displacement, $\delta(t)$, is missing and expression (8) is the homogeneous portion of the equation of motion. Use of Green's Function theory[26] then leads directly to Eq. (1) as the solution of the driven equation of motion for $t\zeta \geq 10$.

Expression (8) shows that the $\hat{Q}(t)$ are the "natural" or "spontaneous" fluctuations of the system and that in the region $\omega_P t < 1$ their equation of motion is second order and asymptotically approaches

$$\frac{d^2[\hat{Q}(t)]}{dt^2} + \frac{(2 + n)}{t}\frac{d[\hat{Q}(t)]}{dt} + \frac{n}{t^2}\hat{Q}(t) = 0 \tag{9}$$

for $\zeta^{-1} > t > \omega_p^{-1}$. The formal similarity of Eq. (9) to the equation of motion of a damped oscillator is obvious, with the essential difference that the "damping coefficient" and "oscillator frequency" are time dependent, which is the inevitable consequence of the self-energy approach adopted[14,15,20] in the derivation of $\hat{Q}(t)$.

For times in excess of ω_p^{-1} the coefficient of the $d[\hat{Q}(t)]/dt$ term approaches the constant value ω_p; however, in addition, the order of the equation of motion is reduced, since it asymptotically approaches

$$\frac{d[\hat{Q}(t)]}{dt} + \frac{(1+m)}{t}\hat{Q}(t) = 0, \qquad \omega_p t > 1 \tag{10}$$

where the relaxation rate is again time dependent.

A complete picture of the regression of the displacement fluctuation in this model can be obtained when it is realized that the same functions that give expression (7) retain validity as time approaches zero and show[14,15,20] that $\hat{Q}(t)$ initially behaves as an oscillator with a Gaussian decay envelope. Thus we see that immediately following its creation the displacement fluctuation $\hat{Q}(t)$ begins to oscillate. In doing so it influences the factors in its environment that determine the oscillation frequency and damping coefficient, which thereby become time dependent, losing memory of their initial values. The equation of motion is still, however, essentially second order and therefore some memory of the initial state is retained within the system. The damping coefficient, which self-consistently approaches a constant value,[16] ω_p, at $t \approx \omega_p^{-1}$, does not, however, determine a subsequent exponential regression, since this value is reached contemporaneously with a reduction of the order of the equation of motion.[18] In this final phase of behavior the decay of $\hat{Q}(t)$ is truly irreversible, but with a rate constant that is nonetheless also time dependent, indicating that a further loss of memory is required in the approach to equilibrium.

Qualitatively similar conclusions have been reached by Nigmatullin[27,28] for the case where a time power law is assumed for the memory kernel in the diffintegral equation of motion for the decay of polarization following a step-function change of field. This has been expressed in the form

$$\frac{d^{1+\alpha}\hat{Q}(t)}{dt^{1+\alpha}} = O_P\hat{Q}(t) \tag{11}$$

which for $\omega_p t < 1$ has $1 > \alpha > 0$ and O_P equivalent to the diffusion operator, ∇^2, but in the range $\omega_p t > 1$ approaches a form with O_P given by $-\omega_p$ and $-1 < \alpha < 0$.

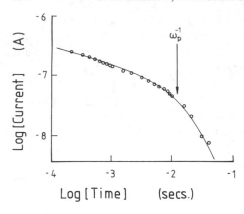

Figure 1. The relaxation current in a sample of amorphous gallium arsenide. Circles indicate the experimental data and the curve through the data is given by the response function of Eq. (7) with $m = 0.64$ and $n = 0.28$. The inverse of the characteristic relaxation rate is shown in the figure.

B.   The Experimental Validity of the Susceptibility Function

Before proceeding to explore the physical characteristics of this model and its relationship to the general thermodynamic features of regression, it is necessary to demonstrate the validity of its description of experimental reality.

The direct observation of response functions is common only in the pulse experiments of acoustic attenuation[29] and nuclear magnetic resonance,[30] the more widespread technique being the use of a step-function field change to obtain a response. In this case Eq. (4) shows that $\phi(t)$ is proportional to the relaxation current of the variable. A typical example from the field of dielectric relaxation is shown in Fig. 1 with expression (7), $\Phi(t)$, fitted to the data. The theoretical curve was obtained computationally, a procedure for which the hypergeometric functions are ideal; however, the asymptotic behavior at times long or short compared to ω_p^{-1} can be obtained analytically as

$$\Phi(t) = \begin{cases} (\omega_p t)^{-n} & \omega_p t < 1 & \text{(12a)} \\ (\omega_p t)^{-(1+m)} & \omega_p t > 1 & \text{(12b)} \end{cases}$$

The short-time behavior of $\Phi(t)$ is known as the Curie–von Schweidler Law[31] and is acknowledged to be typical of relaxation measurements.[32,33] The long-time behavior, however, although known in time-of-flight experiments,[34] has only recently been identified in the present form as applying to relaxation.[11,35]

The most common approach to experimental investigation of the response function is through the response to an oscillating applied field. Substitution in Eq. (1) then gives

$$Q(\omega) = \chi(\omega) E(\omega) \tag{13}$$

with the complex susceptibility $\chi(\omega)$ given by[3]

$$\chi(\omega) = \mathbf{L}_t \int_{\varepsilon \to 0}^{\infty} e^{-i\omega t} e^{-\varepsilon t} \phi(t) \, dt \qquad (14)$$

When expression (7) is used for $\phi(t)$ the analytical form for the susceptibility[14,15,20] is found to be

$$\frac{\chi(\omega)}{\chi(0)} = F\left(\frac{\omega}{\omega_P}\right) = \left(1 + \frac{i\omega}{\omega_P}\right)^{1-n} {}_2F_1\left(1-n, 1-m; 2-n; \frac{1}{1+i\omega/\omega_P}\right)$$

$$(15)$$

where ${}_2F_1(\cdot, \cdot; \cdot; \cdot)$ is the Gaussian hypergeometric function.[36]

At the limits of high and low frequency, asymptotic expressions for the susceptibility have been determined analytically[14,15,20] that are consistent with Eq. (12); these are given by

$$\chi'(\omega) \propto \chi''(\omega) \propto \left(\frac{\omega}{\omega_P}\right)^{n-1}, \qquad \omega > \omega_P \qquad (16a)$$

and

$$\chi''(\omega) \propto \chi'(0) - \chi'(\omega) \propto (\omega/\omega_P)^m, \qquad \omega < \omega_P \qquad (16b)$$

where

$$\chi(\omega) = \chi'(\omega) - i\chi''(\omega) \qquad (16c)$$

Thus in its asymptotic behavior at least the theoretical susceptibility function of Eq. (15) can be seen, if appropriate values for the indices m and n are chosen, to be consistent with the shape functions observed[11,12,37] in a variety of materials under a range of conditions. This is, however, insufficient to demonstrate completely the validity of the theoretical function, since the curvature of $\chi''(\omega)$ in the neighborhood of ω_P can be dependent on the values of m and n. The computation of the full function is extremely simple (a sample BASIC program is given in the Appendix), and the function has been shown to fit the dielectric or mechanical[12] response over the complete frequency range for a variety of materials. The published work demonstrating this fit is listed in Table I; however, it should be noted that ref. 37 contains a list of 100 materials whose responses were described by an empirical function[11] but 90% of which have subsequently been fitted by expression (15). No restrictions on the values of n and m within the range $(0,1)$ seem to be

TABLE I

Material	Response	n	m	Reference
Glycerol	Dielectric	0.48	0.68	
Impure quartz	—	0.15	0.69	
Er-doped CaF$_2$	—	0.05	0.9	20
PBLG (oriented film)	—	0.87	0.28	
PMLG (oriented film)	—	0.92	0.24	
Polyethylene	Mechanical	0.72	0.18	
Chlorocyclohexylacrylate	—	0.5	0.57	12
Carbon/iron	—	0.07	0.83	
Doped silicon	—	0.09	0.75	
Eugenol	Dielectric	0.57	0.9	15
Rochelle salt	Dielectric	0.04	1.0	
KD$_2$PO$_4$	—	0.1	0.9	38
AgNa(NO$_2$)$_2$	—	0.16	0.9	
Ceramic perovskite	—	0.88	0.41	
Chlorobenzene-pyridene (α)	Dielectric	0.36	0.84	39
Chlorobenzene-pyridene (β)	—	0.95	0.38	
Butyl rubber	Mechanical	0.41	0.39	
Crepe rubber	—	0.45	0.40	40
Polyvinyl acetate	—	0.14	0.01	
Poly-n-butyl methacrylate	—	0.3	0.3	
PBLG solution	Dielectric	0.5	1.0	41
		0.15	1.0	
YIG (doped samples)	Magnetic	0.20	1.0	42
	susceptibility	0.25	1.0	

observed. In Fig. 2 we show the agreement that is obtained for a number of different response shapes.

The theoretical function of Eq. (15) is intended to be totally general and therefore must include as limiting cases the empirical shape functions (e.g., the Cole–Davidson) in the ranges of their respective validities. A comparison between these functions and expression (15) has been carried out,[48] and in Fig. 3 we show comparative Cole–Cole plots for the theoretical function and the Havriliak–Negami[49] empirical function, which is also purported to have a general form. In Fig. 4 we compare the fit of the theoretical function with that suggested by Williams and Watts[52] for several shape functions typically observed in the dielectric response of polymers. The Williams and Watts function which has received some attention recently,[55,56] has the form

$$\Phi_{WW}(t) = (\omega_p t)^{\delta-1} \exp\left[-(\omega_p t)^\delta\right] \tag{17}$$

As revealed in the figure, this approaches expression (15) when $\omega_p t < 1$. It is

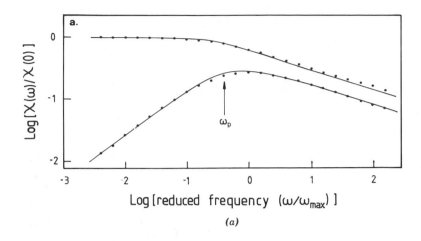

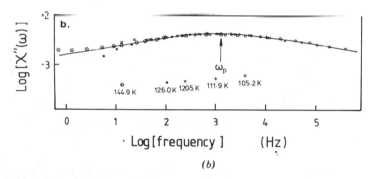

Figure 2. Examples of the fitting of the frequency-response function of Eq. (15) to a range of experimental susceptibility data. In order to obtain a wide experimental frequency range, each set of data has been normalized[11] with respect to the variable over which it was measured. (a) Relaxation in polypropene oxide from the normalized data of Williams[43] obtained in the temperature range 233–263 K and the pressure range 1–3000 atm. $m = 0.71$; $n = 0.67$. (b) Loss component of the susceptibility in the β-response region of nematic OHMBBA [N(o-hydroxy-p-methoxy)benzylidene-p-butylaniline], from the data of Johari.[44] The plot is scaled at 111.9 K. $m = 0.16$; $n = 0.14$. c) Loss component of the susceptibility (in units of tan δ) for 10% polypropene in polybutene, from the measurements of Boiteaux et al.[45] The plot is scaled for 263.8 K. $m = 0.35$; $n = 0.69$. (d) Susceptibility response of γ-irradiated triglycine sulfate at 327.5 K. The data were measured by Pawlaczyk et al.[46] $m = 0.32$; $n = 0.34$. (e) Permittivity response of the ferroelectric $CsH_{0.04}D_{1.96}PO_4$, normalized from the data of Deguchi et al.[47] The critical temperature lies between 267.0 and 267.9 K. The plot is scaled at 267.9 K. $m = 1.00$; $n = 0.036$.

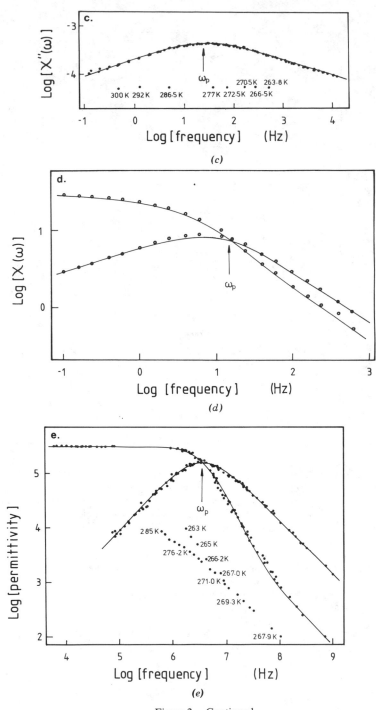

Figure 2. Continued

262

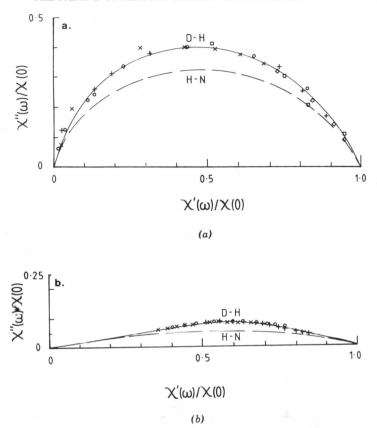

Figure 3. Cole–Cole plots of the normalized response of susceptibility. The data have been fitted at the low and high relative frequency asymptotes to both Eq. (15) and to the Havriliak–Negami spectral shape function.[49] The plots show that the former gives a better fit to the experimental data in the region of the characteristic relaxation frequency. (*a*) Data measured on a sample of impure natural quartz by Snow and Gibbs.[50] $m = 0.69$; $n = 0.15$. (*b*) Data measured in the β-response region of 43.4% chlorobenzene in pyridene by Johari and Goldstein.[51] $m = 0.15$; $n = 0.90$.

clear from the diagrams in Fig. 4, however, that this function cannot be correct in general for times greater than ω_P^{-1}, that is, for frequencies less than ω_P.

These comparisons show that expression (15) does indeed give an extremely good description of the different types of response shapes observed over a wide frequency/time range. Its claim to generality can therefore be held to be valid in detail, and it is thus acceptable to seek the basic principles of the framework on which the model is founded.

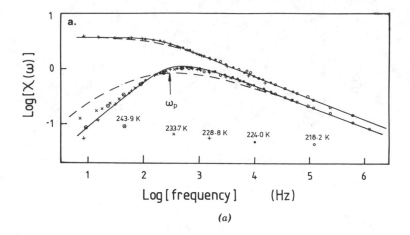

(a)

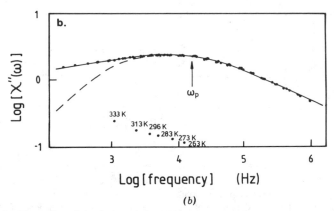

(b)

Figure 4. Examples of the fitting of the Williams and Watts susceptibility function[52] and of Eq. (15) to representative experimental data. As in Fig. 2, the data have been normalized to increase the effective frequency range. The Williams and Watts function is indicated by the dashed curves and Eq. (15) by the continuous curves. In all cases the former has been fitted to the high-frequency response and to either $\chi(0)$, where this is available, or to the curvature in the region of the characteristic relaxation rate ω_p. (a) The frequency dependence of the susceptibility in the α-response region of nematic OHMBBA as reported by Johari.[44] The plot is scaled for 224 K. $m = 0.85$; $n = 1 - \delta = 0.61$. (b) The loss component of the response for 15% poly-n-hexyl isocyanate in toluene in the isotropic phase, from the data of Moscicki et al.[53] The plot is scaled at 333 K. $m = 0.16$; $n = 1 - \delta = 0.54$. (c) The loss component of the susceptibility of polyallylbenzene, in units of $\tan \delta$, from the measurements of Boiteaux et al.[54] The plot is scaled at 353 K. $m = 0.12$; $n = 1 - \delta = 0.73$.

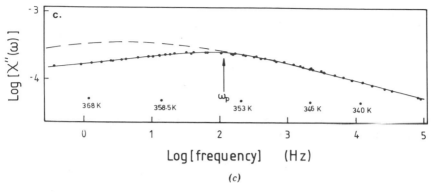

(c)

Figure 4.   Continued

III.   THEORETICAL DESCRIPTION OF REGRESSION

A.   Formal Features

We have seen that in the absence of an externally applied field, the response function, $\phi(t)$, is determined by the regression of spontaneous
fluctuations, $Q(t)$, under the influence of forces natural to the system.
Classically the time evolution is determined by the Liouville operator,
whereas quantally it is either the Hamiltonian, H (in the Heisenberg picture), or the wavefunction $\psi(t)$ (in the Schroedinger picture) that is time dependent, with $Q(t)$ being given by the matrix elements of the displacement
operator.

In the Schroedinger picture

$$\frac{d\psi(t)}{dt} = -iH\psi(t), \qquad \hbar = 1 \tag{18}$$

and the problem becomes one of determining an appropriate Hamiltonian.

A procedure common to all problems is to separate the Hamiltonian into
a component H_0, which is a constant of the motion, and a perturbation H_1,
which, in the present case, is responsible for the regression:

$$H = H_0 + H_1 \tag{19}$$

The wavefunction and Hamiltonian are now transformed into the interaction representation, giving

$$\frac{d\tilde{\psi}(t)}{dt} = -iH_1(t)\tilde{\psi}(t) \tag{20}$$

with

$$\tilde{\psi}(t) = e^{iH_0 t}\psi(t) \tag{21}$$

and[57]

$$H_1(t) = e^{iH_0 t}H_1 e^{-iH_0 t} \tag{22}$$

The formal solution to Eq. (20) can be obtained by iteration and expressed as

$$\tilde{\psi}(t) = \exp\left(-i\int_0^t H_1(t_1)\,dt_1\right)\psi(0) \tag{23}$$

and the diagonal matrix elements, in Dirac's bracket notation, become[57]

$$\langle\tilde{\psi}(t)\psi(0)\rangle = \left\langle\exp\left(-i\int_0^t H_1(t_1)\,dt_1\right)\right\rangle$$

$$= \exp(F(t)) \tag{24}$$

with

$$F(t) = \sum_{p=1}^{\infty} F_p(t)$$

$$= \sum_{p=1}^{\infty}(-i)^p\int_0^t dt_1\int_0^{t_1}dt_2\cdots\int_0^{t_{p-1}}dt_p\langle H_1(t_1)\cdots H_1(t_p)\rangle_c \tag{25}$$

where $\langle\;\rangle_c$ denotes a cumulant or connected diagram in a graphical representation of the infinite series of iterations.

Expression (25) shows that the evolution of a fluctuation is determined by the time-ordered correlations of the perturbation $H_1(t)$ involved in the series $F(t)$. Since the perturbation that gives rise to regression must necessarily couple different states, that is, is nondiagonal, the $F_1(t)$ is zero. It can be seen, therefore, that the existence of memory is an integral part of the quantal approach. Indeed, expression (25) can be obtained by iteration of the familiar rate equation with memory:

$$\frac{d}{dt}\langle\tilde{\psi}(t)\psi(0)\rangle = -\int_0^t \tilde{K}(t-t_1)\langle\tilde{\psi}(t_1)\psi(0)\rangle\,dt_1 \tag{26}$$

with

$$\langle\tilde{\psi}(t)\psi(0)\rangle = \begin{cases} 1 & t=0 \\ 0 & t<0 \end{cases} \tag{27}$$

if the memory kernel $\tilde{K}$ is identified as

$$-\tilde{K}(t_1 - t_2) = \sum_{p=2}^{\infty} (-i)^p \int_0^{t_2} dt_3 \cdots \int_0^{t_{p-1}} dt_p \langle H_1(t_1) H_1(t_2) \cdots H_1(t_p) \rangle_c.$$

$$F_1(t) = 0 \tag{28}$$

The problem involved in describing regression now becomes one of identifying the manner in which memory is lost in the correlations of the perturbation $H_1(t)$. At very short times memory will be conserved in any type of problem and

$$\langle H_1(t_1) H_1(t_2) \rangle \simeq D \tag{29}$$

which leads to the Gaussian result[58]

$$F_2(t) = -\tfrac{1}{2} D t^2 \tag{30}$$

Alternatively, we may take

$$\langle H_1(t_1) H_2(t_2) \rangle = \Gamma \delta(t_1 - t_2) \tag{31}$$

and find

$$F_2(t) = -\Gamma t,$$

an irreversible decay certainly, but one that arises from the assumed complete loss of memory in the perturbation. Sometimes the severity of expression (31) is modified by replacing the delta function by an exponential decay,[6,58] but this still begs the question by displacing it back one level of complexity. Indeed, the question of the manner in which memory is lost in quantal systems is not a trivial one. It is therefore instructive to take a look at the factors that lead to irreversibility in the damping approximation.

In this case H_1 is assumed to couple an isolated or prepared state to a continuum of states of a different type separated from the initial state by energies E. We thus find that

$$\langle H_1(t_1) H_1(t_2) \rangle = \int V^2 e^{-iE(t_1 - t_2)} \rho(E) \, dE \tag{32}$$

Performing the double time integration gives[59]

$$F_2(t) = -\int V^2 \rho(E) \left[t \frac{(1 - \cos Et)}{E^2 t} + it \left(1 - \frac{\sin Et}{Et} \right) \frac{1}{E} \right] dE \tag{33}$$

and making use of the relationships

$$\delta(x) = \pi^{-1} \, \mathbf{L}_{t \to \infty} \, \frac{1 - \cos xt}{x^2 t} \tag{34a}$$

$$\frac{P}{x} = \mathbf{L}_{t \to \infty} \left(1 - \frac{\sin xt}{xt}\right)\frac{1}{x} \tag{34b}$$

we find

$$F_2(t) = -\int \pi V^2 t \rho(E)\delta(E)\, dE + itP \int \frac{V^2}{E}\rho(E)\, dE \tag{35}$$

Thus, although conservation of energy between the prepared state and the connected state, expressed through the delta function $\delta(E)$, is necessary to loss of memory, it is not sufficient. In fact, a pair of coupled states in resonance modify the energy spectrum but retain perfect memory. The further requirement for irreversibility is the existence of a continuum of states arbitrarily close to resonance with, and coupled to, the isolated states. Under these conditions the total energy of the prepared state will be distributed over the continuum such that it will be impossible to recover the original state, barring the generation of a Poincaré cycle. This latter possibility is removed if the spectrum of excitations E is extended to $\pm\infty$, leading to a true irreversibility. These conditions represent the fundamental requirements for loss of memory in a quantal system, though the precise form[60] that this may take can differ strongly depending on the number of degrees of freedom and the energy range available to H_1.

B. Cluster Model

In constructing the cluster model[20] we made use of two features that are common to all systems in which structural fluctuations may occur: (1) the existence of regions the elements of which (atoms, ions, molecules, etc.) can adopt alternative configurations; and (2) that each region of the same type (e.g., interstitial impurity center, substitutional impurity, or topographical imperfection) can be characterized by properties appropriate to its own nature, such as its local structure and molar heat of formation in the host material. We have termed these centers "clusters," and because there will be many of any given type in a material sample, the cluster properties are to be thought of as bulk averages.

The details of this work have been given elsewhere;[20,61] we shall concern ourselves here only with the features pertinent to the present discussion. To begin with, we consider the structural fluctuation of an average cluster of the form that can couple to a spatially uniform field to be one in which the centers of motion of its elements are displaced to positions that exhibit an ideal lattice symmetry and possess its lattice vibrations. If this fluctuation were arbitrarily prevented from regressing, it would have an excess energy over the more disordered average center; this excess energy would take the form of an increase $\bar{E}$ in the frequency about the center of motion for each element, l. The fluctuation therefore can be regarded as defining a region of strain in which the local cluster structure has been "dissociated" into an ideal lattice. In the absence of any external constraint, the stress conjugate to this strain will act as the perturbation H_1 and drive the system back to its equilibrium structure. During the course of this regression the motions of the dissociated cluster elements will couple to form to the regionalized vibrations appropriate to the local cluster structure.

It is convenient here to adopt the language of second quantization, in which creation or annihilation operators create or destroy the wavefunction by operating on a state vector. We define a creation operator $A_f^+(l)$ that creates a fluctuation by displacing the center of motion of the element l and increasing its vibration frequency by $\bar{E}$. It therefore acts like an excitation operator in the sense that multiple creations at the same site are impossible, although identical states may be created at other sites. We note, however, that in the limit of the harmonic approximation in which displacements of the center of motion are not allowed, the excitation operator goes over to the site phonon operator. The corresponding annihilation operator is $A_f(l)$, and the Hamiltonian takes the form[20,61]

$$H = \sum_l \bar{E}A_f^+(l)A_f(l) + H_0 + \sum_c \sum_l \sum_{l'} V_{ll'}A_f^+(l)A_f(l)\left[A_c^+(l') + A_c(l')\right]$$

$$(36)$$

where H_0 contains the eigenvalues appropriate to the vibrations about the center of motion. The third term originates with the stress generated by the fluctuation and is thus zero in its absence, when $A_f^+(l)A_f(l)$ is zero. It is a two-center term, since it forms regionalized (cluster) vibrations in which centers are coupled and the index c runs over all the possible cluster eigenstates such that $\bar{E}$ is the average frequency change.

In the interaction representation, the equation of motion of $A_f^+(l, t)$ is given by

$$\frac{d\left[A_f^+(l, t)\right]}{dt} = \dot{A}_f^+(l, t)$$

$$= i\bar{E}A_f^+(l, t) + i\sum_c \sum_{l'} V_{ll'}\left[A_c^+(l', t) + A_c(l', t)\right]A_f^+(l, t)$$

$$(37)$$

with $\dot{A}_f(l, t)$ given by the Hermitean conjugate, and the operators by

$$A(t) = e^{iH_0t}Ae^{-iH_0t} \tag{38}$$

As a result we identify

$$\langle \tilde{\psi}(t)\psi(0)\rangle \equiv \langle A_f(l', t)A_f^+(l', 0)\rangle = e^{-i\bar{E}t}e^{F(t)}\langle A_f(l', 0)A_f^+(l', 0)\rangle \tag{39}$$

and

$$H_1(t) \equiv \sum_c \sum_{l'} V_{ll'}\left[A_c^+(l', t) + A_c(l', t)\right] \tag{40}$$

which allows us to determine $F(t)$.

Confining our attention to $F_2(t)$, we find

$$F_2(t) = -\int_0^t\int_0^{t_1}\sum_c\sum_{l'}|V_{ll'}|^2\exp\left[-iE_c(l')(t_1 - t_2)\right]dt_1\,dt_2 \tag{41}$$

where $E_c(l')$ is the frequency change suffered by the element l' when the cluster eigenstate c is annihilated on formation of the fluctuation. The sum over l' runs over only those elements whose motions are involved in the cluster eigenmode c. Thus the conjugate term in $F_2(t)$ that allows the fluctuation to be created at l by annihilation of a displacement at l' becomes zero.

The expression for $F_2(t)$ can be considerably simplified if we make use of some of the general features of the problem:

1.   All elements participating in the regionalized cluster vibration c must be in resonance and hence $E_c(l')$ is independent of l'. We can also take its value, E, for a particular cluster state to be the eigenfrequency of the displacement of the center of motion required to form this state from the ideal lattice.

2.   Energy conservation in the replacement of an element vibrating at an ideal lattice frequency ζ by a number of elements $N(E)$ whose centers of motion each vibrate with a frequency E requires that

$$EN(E) = \zeta \tag{42}$$

3.   The total stress involved in forming the cluster is a constant property of the type of center, and thus the perturbation strength per element arising

from this stress is

$$\sum_{l'} |V_{ll'}|^2 = \sum_l \sum_{l'} \frac{|V_{ll'}|^2}{N(E)} = \frac{U^2}{N(E)} \tag{43}$$

where the summation over l, l' runs over those elements involved in a particular center of motion displacement of frequency E.

4. The total energy carried by each mode of center of motion displacement is a constant, and hence each mode of frequency E is of equal *a priori* probability in the range $0-\zeta$, leading to a density of modes

$$\rho(E) = \zeta^{-1} \tag{44}$$

Applying these considerations to $F_2(t)$ gives

$$F_2(t) = -\int_0^t \int_0^{t_1} \int_0^{\zeta} e^{-iEt}\left(\frac{U^2}{\zeta^2}\right) E \, dE$$

$$= it\int_0^{\zeta} E\mathcal{N}(E) \, dE - \int_0^{\zeta}\mathcal{N}(E)(1 - e^{-iEt}) \, dE \tag{45}$$

where the sum over c has been converted into an integral, and

$$\mathcal{N}(E) = \frac{n}{E}, \qquad n = \frac{U^2}{\zeta^2} < 1 \tag{46}$$

is the number density of displacement oscillations.

The first term in Eq. (45) then gives the average excitation energy and allows us to identify $\bar{E}$ as $n\zeta$, and the second term defines the average number[62] of center of motion displacement modes that is required to form the fluctuation at l. Although we have extended the integrals to zero, it is clear from Eq. (42) that this implies that the fluctuation may generate displacement modes involving an infinite number of elements during its regression. For any finite-sized cluster, this integral must therefore be truncated at a lower bound; we shall see how this is done later. First, however, the integral will be completed as it stands to give

$$F_2(t) = i\bar{E}t + F_2'(t) \tag{47a}$$

$$F_2'(t) = -n\left[E_1(i\zeta t) + \ln(i\zeta t) + \gamma\right] \tag{47b}$$

where γ is Euler's constant and $E_1(\cdot)$ the exponential integral.[63] Using $F_2(t)$

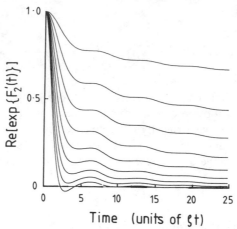

Figure 5.   The exact relaxation function, with $F_2'(t)$ given by Eq. (47b) plotted in linear response and time scales to show the initial Gaussian decay and the oscillations in the region $\zeta t > 1$. The plots are given for steps in n of 0.1 from $n = 0.1$, the uppermost curve, to $n = 0.9$, the lowest plot.

the displacement fluctuation

$$\tfrac{1}{2}\left[\left\langle A_f(l,t)A_f^+(l,0)\right\rangle + \left\langle A_f(l,t)A_f^+(l,0)\right\rangle^+\right] = \operatorname{Re}\left\{e^{F_2'(t)}\right\} \quad (48)$$

can be obtained.

For times $\zeta t < 1$, $F_2'(t)$ approaches zero as a Gaussian

$$F_2'(t) = -i\overline{E}t - \frac{n\zeta^2 t^2}{4}, \qquad \zeta t < 1 \tag{49a}$$

whereas for $\zeta t \gtrsim 5$ we find

$$F_2'(t) = -n\gamma - n\ln(i\zeta t), \qquad \zeta t \gtrsim 5 \tag{49b}$$

The exact displacement fluctuation is given in Fig. 5 for $\zeta t \leq 25$, and we can see that memory of the initial state is lost only slowly, with several near recurrences, which are particularly emphasized when the value of n approaches unity.[64] Analytically this behavior is the result of the progressive cancellation of the excitation energy $\overline{E}$ of Eq. (49a), which produces the oscillations, as the perturbing stress causes the cluster structure to approach equilibrium. The effective oscillation frequency therefore becomes a time-dependent property until finally its nature as an oscillation is lost entirely in the power law behavior of Eq. (49b). It is this latter result that determines the equation of motion of $\hat{Q}(t)$ in the form of approximation (9). Thus the origin of the behavior of $\hat{Q}(t)$ in this range lies in the time-dependent development of the self-energy[65] as a specified fluctuation evolves into a cluster.

There has, however, been no arbitrary assumption of irreversibility in this treatment, so we may ask: where does the loss of memory come from? The answer to this question may be found by comparing the quantal requirements for irreversibility with this cluster model. In doing so, we find that the unique initial fluctuation is represented as equivalent to a specific continuum of states each of which possesses the same total energy as the fluctuation, but which differ in the number of cluster elements that share it. It is therefore clear that the loss of memory originates in the distribution of the energy and structural information in the original fluctuation over the substantially larger, though not infinite, degrees of freedom belonging to the cluster.

C. Correlation Weakening

We have so far based our derivation on the behavior of the term $F_1(t)$ and ignored the effect of higher cumulants $[F_4(t)$, etc.]. Because of the apparent logarithmic divergence in $F_2(t)$ for $t > \zeta^{-1}$, it is necessary to demonstrate that the form of our solution remains unaffected by these higher terms. This is usually done by ensuring self-consistency using the renormalization group[65] technique. Nigmatullin and Hill,[19] however, have approached the problem in a different way and obtained an equivalent solution as a result of a correlation-weakening *Ansatz* that they have proposed. Defining $\langle A_f(t)A_f^+(0)\rangle$ to be $^0G(t)$, they use the Hermitean conjugate of Eq. (37) to write

$$\frac{d\,^0G(\tau)}{d\tau} = {}^1G(\tau) = X_1(\tau)\,^0G(\tau), \qquad \tau = it \qquad (50)$$

The equation of motion for $X_1(\tau)$ is then found to be

$$\frac{dX_1(\tau)}{d\tau} = X_1(\tau)[X_2(\tau) - X_1^{\cdot}(\tau)] \qquad (51)$$

and an endless chain of equations is obtained for $^PG(\tau)$ and $X_P(\tau)$ $(= {}^PG(\tau)/{}^{P-1}G(\tau))$. The *Ansatz* proposed to allow a solution to be obtained was to set

$$X_2(\tau) = \alpha_1 X_1(\tau) \qquad (52)$$

which in fact determines the dynamic behavior completely, since the motion of $X_2(\tau)$ given by Eq. (52) determines $X_3(\tau)$ through $dX_2(\tau)/d\tau$ and so on. The result is obtained in the general form

$$X_p(\tau) = \alpha_p X_{p-1}(\tau) \tag{53a}$$

with

$$\alpha_{p+1} = 2 - \frac{1}{\alpha_p} \tag{53b}$$

and self-consistently, $^0G(t)$ is found to be

$$^0G(t) = {}^0G(0)\exp\left(i\int_0^t \frac{X_1(0)\, dt_1}{[1 - i(\alpha_1 - 1)X_1(0)t_1]}\right) \tag{54}$$

We can compare this result with our previous derivation by inverting the order of integration to obtain

$$F_2(t) = -\int_0^t \int_0^{t_1} \tilde{K}(t_1 - t_2)\, dt_1\, dt_2 \tag{55}$$

with

$$\tilde{K}(x) = n\left(\frac{i\zeta e^{-i\zeta x}}{x} + \frac{e^{-i\zeta x} - 1}{x^2}\right) \tag{56}$$

which leads to

$$\ln\left[\frac{{}^0G(t)}{{}^0G(0)}\right] = F_2'(t) = -n\int_0^t \frac{1 - e^{-i\zeta t_1}}{t_1}\, dt_1$$

$$\simeq -n\int_0^t \frac{i\zeta\, dt_1}{1 + i\zeta t_1} \tag{57}$$

This latter approximation is valid in both the short-time ($\zeta t < 1$) and long-time ($\zeta t \geq 5$) limits and is identical to Eq. (54) if we identify

$$X_1(0) = -n\zeta \quad \text{and} \quad n = (\alpha_1 - 1)^{-1} \tag{58}$$

For $0 < n < 1$ Eq. (53a) shows that the driving force in the equation of motion for the correlations $X_p(\tau)$ progressively approaches zero as p increases, but that the series is never truncated. This correlation-weakening effect therefore gives a demonstrably self-consistent solution with correlations included to all orders.

The expression for $F_2(t)$ arrived at in the previous section should therefore be regarded as the self-consistent solution to $F(t)$ for the type of perturbation considered, with the self-consistency brought about through the features 1–4 used in the construction of the model.

D. The Trajectory of Regression

A somewhat different view of the regression can be obtained by regarding Eq. (48) as representing the overlap of an oscillator of frequency ζ with a continuum of oscillators with displaced centers of motion.[61] The identification of the overlap function with Eq. (48) can be made if we accept that the motion along the displacement coordinate is itself an oscillator of frequency E with a time-dependent correlation of displacement; hence

$$\exp\left[-\left(\frac{\Delta}{\beta}\right)^2\right] \rightarrow \exp\left[-\sum_E \left(\frac{\Delta(E)}{\beta}\right)^2\{1 - e^{-iEt}\}\right] \qquad (59)$$

where β is the root-mean-square amplitude of the oscillator of frequency ζ, and $\Delta(E)$, which is proportional to $E^{-1/2}$, is the root-mean-square amplitude of the displacement vibrations. The constant n is now given by the ratio between the number of mass equivalents per particle in the fluctuation and that in the regionalized displacement mode.

The regression can now be seen to be the result of a spontaneously created nonstationary state evolving into the packet of regionalized displacement motions of which it is composed, rather in the form of a Franck–Condon factor in optical spectroscopy. In this way the excess energy of the fluctuation, which at the time of its creation was stored as local lattice strain, is converted into the kinetic energy of local displacement motions.

As a result of this reconstruction it becomes possible to follow the trajectory of a cluster element as it progresses toward equilibrium from a stationary position in an initially strained lattice, if the forms of the local displacement coordinates are known. Although a detailed description will depend on the particular type of cluster, it is possible to elucidate some general features if we make recourse to some specific problems for which fractional power law regression is predicted.[65–67] The factor that these treatments have in common is the application of a scaling hypothesis. This means that the local structure of the fluctuation is related to dilation symmetry[67] rather than translational symmetry. A system exhibits an ideal dilation symmetry when the spatial relationship between a group of elements on one scale is reproduced identically when the scale is increased and the original group becomes merely a single element of the rescaled group. In an ideal system the process can be repeated indefinitely, an example being the Sierpinski Gasket described in ref. 67. Of course such a system is rigid and cannot entertain structural fluctuations. However, regionalized centers will possess imperfect dilation symmetry that merges with the symmetry of the host at large scales. This fading of the local symmetry can be described by defining a local property such as a position vector that is precisely defined on the smallest scale. On rescaling to the spatial limits of the cluster, the property is defined by a probability distribution, a description that has been used in phase-transition theory[68] and, in a somewhat empirical fashion, in dielectric relaxation.[69]

It is the vibration mode structure of the dilating system that is of interest to us here. Since each group of elements on one scale forms a new element on the expanded scale, they vibrate as a rigid body on that scale with a mass equivalent related to the number of subelements that make up the rescaled element. The specific feature of the cluster model, Eq. (42), is therefore a natural property of systems whose vibrations are described in terms of a local dilation symmetry. We therefore propose that the displaced element l relaxes by generating displacement modes of dilation symmetry each of which possesses the same total energy as the original element. The fraction of each mode generated is identical and equal to n, which therefore defines the average correlation of the displaced element with the cluster. The regression follows a sequence in which vibrations of progressively lower frequency and larger numbers of elements moving as a rigid body are mixed into the motion of the element l. As a result the displaced element moves along a classical trajectory away from its initial ideal lattice site, with the stored energy per site ($\equiv \bar{E}$) being converted into kinetic energy of motion rather than into oscillations about either the displaced center of motion or the average cluster site. The initial stages of this process are illustrated schematically in Fig. 6 for two types of system, namely, a cubic lattice distorted by an interstitial ion and a bent polymer chain. The latter example is conceptually simple because the dilation modes refer to progressively larger segments of the polymer chain moving as a rigid body. In this case the net result of the displacement trajectories of the polymer segments is to cause a twisting of the chain, which suggests that polymer reptation[70] can be brought into this framework. Other types of distortion can be constructed in a similar way, and it has been suggested that this application has a general structural validity.

E. Introduction of Thermodynamic Irreversibility

The path of regression that we have followed so far has seen the creation energy of the fluctuation, which is stored as local mechanical strain, being

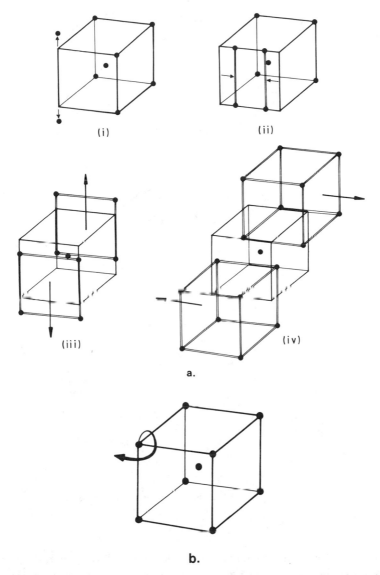

(i) (ii)

(iii) (iv)

a.

b.

Figure 6. (a) Sequence of dilation-related displacement modes of a distorting cube. (b) Example of a trajectory of a particle regressing by generating the modes indicated in (a). (c) Sequence of dilation-related motions for a polymer chain. The individual diagrams (i)–(iv) indicate how the motions expand along the chain. (d) trajectories of selected sites on a chain following the sequence indicated in (c).

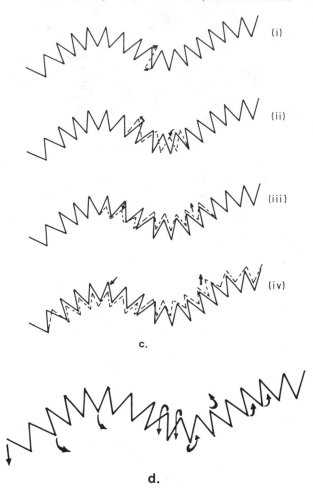

Figure 6.   Continued

converted to the kinetic energy of an element following a displacement trajectory under the influence of the conjugate stress. As yet, no thermodynamic irreversibility has occurred, since the energy is still located within the spatial extent of the cluster, although it is now present as cluster heat content rather than in the latent heat form of a modified structure. Irreversibility requires the removal of this excess local heat content and its dissipation in a "bath" with a number of degrees of freedom greatly in excess of that of the cluster.

The formal construction of the model allows the introduction of irreversibility in a natural manner. Consider the displacement modes with E approaching zero. As we have pointed out, these modes leave the cluster with almost the ideal structure appropriate to the fluctuation. In this state the vibrations about the center of motion approach those of the ideal lattice (i.e., phonons), and will become strongly involved in phonon-scattering processes, whether of single or multiple type.[71] Since the phonon manifold still represents the bulk of the degrees of freedom of the material and is directly coupled to the heat bath, any such processes will lead to an irreversible decay of these displacement modes ($E \to 0$), in the form described in Section III.A. They participate, therefore, in the regression of the structural fluctuation through their formation as virtual displacement (dilation) modes extending into ideal lattice regions outside the cluster, which are unstable with respect to the generation of phonons. In this manner a route is established for the irreversible removal of local kinetic energy from the cluster.

We can represent this behavior in the context of our model by describing the displacement modes as the solutions to an oscillator equation damped through participation in the phonon processes of the residual ideal-lattice portion of the cluster fluctuation; hence

$$E = -i\Gamma \pm \Omega \left(1 - \frac{\Gamma^2}{\Omega^2}\right)^{1/2} \qquad (60)$$

Modes for which $\Omega < \Gamma$ are phononlike and overdamped, generating the self-consistent damping constant, Γ, for the fluctuation itself,[58,72] leading to the replacement

$$\bar{E} \to \bar{E} - i\Gamma \qquad (61)$$

The adoption of a phenomenological form for Γ introduces no invalidity, since any such form must be describable quantum mechanically[71] through the relevant irreversible phonon processes. With these replacements the integrals in Eq. (45) now run over Ω and cover the range $\bar{\Gamma} - \bar{\zeta}$ ($0 \le \text{Re}\{E\} \le \zeta$). On completion of the integral the analytical form for $\tilde{K}(x)$ is found to be

$$\tilde{K}(x) = ne^{-\Gamma x}\left(\frac{i\bar{\zeta}e^{-i\bar{\zeta}x}}{x} + \frac{e^{-i\bar{\zeta}x} - 1}{x^2}\right) \qquad (62)$$

with

$$\bar{\zeta} = \zeta\left(1 - \frac{\Gamma^2}{\zeta^2}\right)^{1/2} \qquad (63)$$

The decay function can now be obtained through the double time integral of $F_2(t)$ [Eq. (55)], which gives

$$
\begin{aligned}
F_2(t) - i\bar{E}t - \Gamma t = {} & -\Gamma t - n\big[\{\,E_1[(\Gamma + i\bar{\zeta})t] + \ln[(\Gamma + i\bar{\zeta})t] + \gamma\,\} \\
& - \{\,E_1(\Gamma t) + \ln(\Gamma t) + \gamma\,\}\big] \\
& - n\Gamma\bigg[t\ln\bigg(\frac{\Gamma + i\bar{\zeta}}{\Gamma}\bigg) + \frac{1 - e^{-(\Gamma + i\bar{\zeta})t}}{\Gamma + i\bar{\zeta}} \\
& \quad - \frac{1 - e^{-\Gamma t}}{\Gamma} + tE_1[(\Gamma + i\bar{\zeta})t] - tE_1(\Gamma t)\bigg] \quad (64)
\end{aligned}
$$

Although it appears imposing, this expression is closely related to Eq. (47b), and if we take the usual experimental case of $\bar{\zeta} \gg \Gamma$, we obtain simple expressions in a number of limiting situations as follows:

1. For $\bar{\zeta}t \ll 1$, we recover Eq. (49a).
2. $\bar{\zeta}t > 1$, $\Gamma t \ll 1$ gives

$$
F_2(t) \simeq in\bar{\zeta}t - n\ln(\bar{\zeta}t) = in\bar{\zeta}t - n\ln(\Gamma t) - n\ln\bigg(\frac{\bar{\zeta}}{\Gamma}\bigg) \quad (65)
$$

3. $\bar{\zeta}t > 1$, $\Gamma t > 1$ gives

$$
F_2(t) \simeq in\bar{\zeta}t - n\Gamma t\ln\bigg(\frac{\bar{\zeta}}{\Gamma}\bigg) - n\ln\bigg(\frac{\bar{\zeta}}{\Gamma}\bigg) \quad (66)
$$

By writing the decay of the fluctuation as

$$
\tfrac{1}{2}\big[\langle A_f(l,t)A_f^+(l,0)\rangle + \langle A_f(l,t)A_f^+(l,0)\rangle^+\big] = \exp\big[-i\bar{E}t - \Gamma t + F_2(t)\big] \quad (67)
$$

we find a number of interesting features. The most important of these is a renormalization of the fluctuation amplitude, which is multiplied by

$$
\bigg(\frac{\bar{\zeta}}{\Gamma}\bigg)^{-n} = N(\Gamma)^{-n} \quad (68)
$$

where the equivalence is established by Eq. (46). This renormalization obtains even in the previous case and expresses the fact that as the fluctuation regresses, an element initially dissociated from, and therefore independent of, the cluster combines its motions into cluster-regionalized displacement modes. $N(\Gamma)$ defines the maximum number of elements that can establish a displacement mode oscillation in the average cluster prior to a transport of energy to the phonon bath, and can be thought of as the extent of the influence of the center. We note that the elements remain independent (dissociated) when n is zero, and form part of a fully correlated displacement (dilation mode) when n is unity.[20,61]

A second feature of interest is the renormalization of the damping constant[65] as t approaches Γ^{-1}, leading to a cluster-corrected[20] value $\tilde{\Gamma}$, given by

$$\tilde{\Gamma} = \Gamma\left(1 + n\ln\left(\frac{\bar{\xi}}{\Gamma}\right)\right) = \Gamma(1 + \ln[N(\Gamma)]) \tag{69}$$

and the replacement of the power law decay by an exponential damping when t exceeds Γ^{-1}. This is the advent of thermodynamic irreversibility in the regression, but we note that the amplitude renormalization is retained, indicating that the fluctuation loses energy through its participation in the displacement modes that cover the whole cluster.

We can now give a very general interpretation[61] of the path of regression followed so far. Returning to Eq. (45), we have already seen that the first term defines the average energy required to create a fluctuation by dissociation of an element from the average cluster, and that it is a fraction n of the energy of dissociation per element (U_d) from a cluster with ideal local symmetry. Since this is converted to kinetic energy during regression, we define the heat content per element of the fluctuation, ΔH_c, as

$$\Delta H_c = nU_d \propto \bar{E} \tag{70}$$

The second term describes the total number of ways that the energy available from a single element can be distributed among the cluster elements such that each displacement (dilation) mode contains the same total energy. It is therefore proportional to the configuration entropy[20,61] per element, ΔS_c, gained by the fluctuation during regression, and equal to that lost by the average cluster during reassociation. If we remember to truncate the integral at Γ as Ω approaches zero, a comparison with Eqs. (65) and (66) yields

$$\Delta S_c(t)/k = n\ln[\bar{\xi}t] = n\ln[\Gamma t N(\Gamma)], \qquad \Gamma t < 1 < \bar{\xi}t \tag{70a}$$

which saturates as

$$\Delta S/k = n\ln[N(\Gamma)], \qquad \Gamma t > 1 \tag{70b}$$

at time Γ^{-1}.

We see, therefore, that the configuration-entropy change increases from zero (at $t = 0$) as the fluctuation displacement travels its regression path. It is this behavior of the configuration entropy that causes the time dependence in both coefficients of Eq. (9), and modifies the damping constant to $\bar{\Gamma}$ ($= \omega_p$) as it approaches the average cluster value. It is also this change in ground-state configuration entropy that underlies the earlier pictorial representations of the theory,[14,35,73] in which the ideal lattice fluctuation constructed an intervalley system. In this case each level in the vibrational progression will relax identically as its zero-point motion is partially converted to the kinetic energy of displacement modes, and consequently activated rate processes[69,74] will include a contribution to the activation entropy from this ground-state change.

The configuration-entropy change for the average cluster [Eq. (70b)] is the same fraction of the value for an ideal cluster structure as is the enthalpy ΔH_c. Therefore we consider[61] that regression of an element, prior to time Γ^{-1}, follows a path of constant Gibbs free energy G such that

$$\Delta G = 0 = \Delta H_c(t) - T\Delta S_c(t), \qquad t < \Gamma^{-1} \tag{71}$$

In the course of this path, energy stored in the structure of the fluctuation is converted to kinetic energy, with the time development of $\bar{E}$ being contained in the exponential function of Eq. (47b) or its equivalent in Eq. (64). Simultaneously there is an equivalent change in configurational entropy, and it is the "evolution" of entropy that allows the process to be called memory-fading.[75] Following the removal of energy from the cluster, ΔH_c becomes zero, and the regression becomes thermodynamically irreversible with

$$\Delta G = - T\Delta S_c < 0 \tag{72}$$

F.   Remnant Memory

Up to this point we have ignored the term in $F_2(t)$, which allowed for the creation of a displacement fluctuation of our chosen element l at the expense of displacements at different elements. Since we have also considered our starting point for regression to be the spatially uniform fluctuation of an array of identical average, clusters,[20] it is clear that such terms have no influence on relaxation prior to time Γ^{-1}. Subsequently the introduction of these terms causes the transfer of a displacement from one cluster to another, a process that we have previously termed intercluster (IC) exchange. In general physical terms, this process feeds back displacements into the clusters that are partially relaxed and is responsible for the retardation of the approach to equilibrium in Eq. (10). A further consequence is that

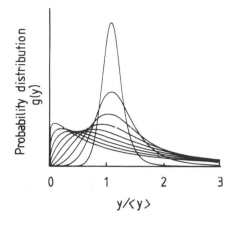

Figure 7. The probability distribution function $g(y)\,dy$ of eq. (74) for a range of values of m increasing from $m = 0.1$ (the sharply peaked curve) to $m = 0.9$ (the curve peaking closest to zero) in steps of 0.1. As m approaches zero, the distribution approaches a delta function, and as m approaches unity the distribution becomes an exponential-like decay. The parameter y has been normalized in terms of the average value $\langle y \rangle$.

the array is no longer composed of identical clusters and the equilibrium description must be in terms of a steady-state distribution.[20,21]

Our approach[15,20] to this problem has been to assume that IC exchanges are equivalent to a displacement on the cluster size scale. We have then redefined our coordinate system in terms of motions involving whole clusters as single elements and introduced a perturbation equivalent to V_{II}, that causes the "strained" ideal array of identical clusters to "displace" into a distribution of clusters. In this way we have followed the evolution of the cluster distribution from the ideal array with a technique similar to that of Section III.B. The important difference here is the inclusion in the equivalent F_2 term of both decay and restoration contributions of opposite sign, which must be convoluted together. Both contributions are of a power law type, with index $0 < m < 1$ determined by the strength of the force perturbing the ideal array, defined as a superlattice of clusters. The distribution[14,15,20,76] of clusters derived in this way is shown in Fig. 7 as a function of a reduced cluster variable y, which may be, for example, the number of elements. This form of distribution has been shown[76] to be independent of the choice of the original array, and persistent under multiple events. It therefore satisfies the requirements for a steady-state distribution. We note that a continuous range of shapes is obtained, ranging from a delta function when the original system is unperturbed ($m = 0$) to an exponentiallike form when the probability of each cluster structure is determined by a number of independent identical random events, that is, totally disordered ($m = 1$).

Our description of regression is now completed by averaging the decay of the displacement fluctuation of an ideal cluster, given by Eq. (67), over the randomly initiated decay and restoration events produced by the IC exchanges. In statistical-mechanical terms,[2,21] this gives us the motion of regression of a cluster averaged over the fluctuations in the configuration of the cluster array whose probability density is given in Fig. 7.

Since our initial state is defined[20] to be a unique displacement fluctuation of the whole system, the fluctuations in the configuration of the array must evolve contemporaneously with the local regression of the fluctuation toward the average cluster, and for $\bar{\xi}t > 1$ we find[14,15,20]

$$\Phi(t) = t^{-1} \int_0^t e^{-\tilde{\Gamma}(t-t_1)} (t-t_1)^{m-n} t_1^{-m} \tilde{\Gamma}^{-n} dt_1 \tag{73}$$

which is identical to Eq. (7) with ω_p equal to $\tilde{\Gamma}$. Because the array distribution is a steady state, expression (73) can be converted to the more revealing form of an average[21] of the cluster regression over the instantaneous cluster distribution $g(y)$ (Fig. 7), where

$$y \equiv (t-t_1)^m t_1^{-m}; \qquad g(y) = \frac{y^{(1-m)/m}}{m(1+y^{1/m})^2} \tag{74}$$

and for $\bar{\xi}t > 1$

$$\Phi(t) = \int_0^\infty y[\tilde{\Gamma}(y)t]^{-n} \exp[-\tilde{\Gamma}(y)t] g(y) \, dy \tag{75}$$

with

$$\tilde{\Gamma}(y) = \tilde{\Gamma} \frac{y^{1/m}}{1+y^{1/m}} \tag{76}$$

We therefore see that each cluster in the array contributes to the average in expression (75) an amplitude $y\phi(0)$ that regresses in the manner of Eq. (67) with a damping constant $\tilde{\Gamma}(y)$. Taking the amplitude to be that appropriate to a cluster in which a number of elements, N_e, different from the average number participates in the displacement mode at the instant it is involved in a thermal dissipation process, we obtain

$$y = \exp\{-n\ln(N_e) + n\ln(N(\tilde{\Gamma}))\}$$
$$= \exp\{-n\ln(N_e) + n\ln\langle N_e \rangle\} \tag{77}$$

where $\langle N_e \rangle$ is the array average value. Thus y describes, through the number of participating elements N_e, a specific cluster configuration-entropy fluctuation in the steady state, whose statistical weight is $g(y)$.

Equation (76) extends the relationship established between the damping constant and cluster entropy or number of elements to include array fluctuations, and we find

$$\tilde{\Gamma}(y) \to \tilde{\Gamma}, \qquad y \to \infty, \qquad N_e \to 0 \tag{78a}$$

and

$$\tilde{\Gamma}(y) \to 0, \qquad y \to 0, \qquad N_e \to \infty \tag{78b}$$

The origin of this behavior lies in the inclusion of terms that effectively couple different clusters at times $t > \Gamma^{-1}$ when nonvirtual IC exchanges can be realized. Such nondiagonal damping terms lead, in the case of the damped harmonic oscillator, to Eq. (60), which is the solution to the secular determinant

$$\begin{vmatrix} E - \Omega - i\Gamma & -i\Gamma \\ -i\Gamma & E + \Omega - i\Gamma \end{vmatrix} = 0 \tag{79}$$

When $\Omega < \Gamma$ the two-level system of Eq. (79) is coupled by damping[58] and one branch of the completely overdamped solutions follows a path similar to that of Eq. (78) as Ω approaches zero. The renormalization of Γ to $\tilde{\Gamma}$ has already allowed for processes equivalent to the alternative branch.

Therefore we find that when $t > \tilde{\Gamma}^{-1}$ the effect of array fluctuations is to couple displacement modes in different clusters through damping. In this time range a displacement mode of any number of elements has only a virtual existence, because it is unstable with respect to dissipation of energy. The coupling through damping,[58] however, renders mode displacements involving larger numbers of elements progressively more stable, which is to be expected, as it is these modes that approach most closely those of the ideal lattice. Although the *average* cluster value of the damping constant remains unchanged it applies as a local value only to those clusters with $N_e \approx 0$, where the elements are essentially dissociated.

Our theoretical description of regression is now completed by recognizing that for $t > \tilde{\Gamma}^{-1}$ virtual displacement (dilation) modes are generated involving numbers of elements in excess of the cluster average, which couple through damping to other clusters. This introduces a remnant memory of the fluctuation into clusters that have almost returned to their average equilibrium structures, retards the regression, and is responsible for the equation of motion, Eq. (10). During this phase of regression the average cluster entropy is redistributed over the cluster array as the steady-state description of the array evolves. This may be regarded as a secondary memory-fading process occurring as the memory of the unique superlattice array configuration is lost and the array configurational entropy expressed through the distribution $g(y) \, dy$ is evolved.

IV. CONCLUSIONS

We have described here an experimentally tested model for the regression of structural fluctuations in imperfectly structured materials in which a specified fluctuation relaxes by partially correlating the motions of an individual element with regionalized motions of a cluster of elements. During this process it follows a spatial trajectory whose path is defined by the increasingly larger number of elements involved in the regionalized motions. Finally the modes approach sufficiently close to lattice modes (phonons) to dissipate energy via them, or alternatively extend into other clusters, causing a retardation of regression through the resulting feedback.

These detailed physical concepts have also been given a thermodynamic interpretation, in which an initially stored (latent) energy is converted to local heat content and configuration entropy evolved as the system traverses a constant free energy path. Following the onset of energy dissipation, the local configuration entropy is redistributed on a nonlocal scale with the constraint that the average value remains constant (i.e., an isentropic path). We believe that this very basic formulation is the fundamental reason for the general applicability of our theoretical expressions and the ubiquity of form of these experimental features. As a result we expect our approach to have a wide applicability in chemical physics. Indeed, if stored energy is substituted for by information content in a system with mobile elements, then it is possible that these ideas may have relevance in general statistical theory to other than purely scientific matters.

APPENDIX A.   COMPUTER PROGRAM FOR THE DISSADO–HILL SPECTRAL FUNCTION IN NORTHSTAR FLOATING POINT BASIC

```
10   PRINT   "DISHIF—THE DISSADO–HILL FUNCTION"
20   PRINT   "NORMALIZED TO UNITY CHARACTERISTIC"
30   PRINT   "FREQUENCY AND UNITY MAGNITUDE"
40   PRINT   "A LOGARITHMIC FREQUENCY SCALE IS USED"
50   PRINT
60   PRINT   *1 "DISSADO–HILL FUNCTION"
70   PRINT   "PROGRAM OMITS FREQUENCIES BETWEEN 0.95 AND
     1.05"
80   PRINT   "PROGRAM ACCEPTS NEGATIVE VALUES OF M"
90   PRINT   "THE FREQUENCY REQUIRES TO BE INPUT AS A
     RATIO"
```

```
100 PRINT  "  FOR THE FOLLOWING STEPS PER DECADE"
110 PRINT  "  GIVE THE APPROPRIATE INCREMENT"
120 PRINT  " 3 − 2.1544; 5 − 1.5848; 10 − 1.2589; 20 − 1.12202"
130 PRINT
140 PRINT  "  INPUT THE CHARACTERISTIC PARAMETERS"
150 PRINT  " TYPE IN M, N FREQUENCY MIN., MAX., INCREMENT,
            AMPLITUDE"
160 INPUT  M0, N0, W1, W3, W2, A0
170 PRINT ∗1 " M N AMPLITUDE"
180 PRINT ∗1 "%Z.13E3, M0, N0, A0"
190 PRINT ∗1
200 PRINT ∗1 " FREQUENCY   CHI REAL    CHI IMAGINARY"
210 REM       CALCULATE THE AMPLITUDE NORMALIZER F0
220 S1 = 1
230 S2 = N0
240 S3 = 1 + M0
250 IF M0 < 0 THEN GOTO 260 ELSE GOTO 500
260 V = M0 − N0
270 IF V < 0 THEN GOTO 390
280 Z = 1 − N0
290 GOSUB 1270
300 H1 = G1
310 Z = 1 + M0
320 GOSUB 1270
330 H2 = G1/Z
340 Z = 1 + M0 − N0
350 GOSUB 1270
360 H3 = G1/Z
370 F0 = H1 ∗ H2/(H3 ∗ M0)
380 GOTO 650
390 Z = 1 − N0
400 GOSUB 1270
410 H1 = G1
420 Z = 1 + M0
430 GOSUB 1270
440 H2 = G1/Z
450 Z = 2 + M0 − N0
460 GOSUB 1270
470 H3 = G1/Z
480 F0 = H1 ∗ H2 ∗ (1 + M0 − N0)/(M0 ∗ H3)
490 GOTO 650
500 Z = M0
510 GOSUB 1270
```

```
520 H1 = G1/Z
530 Z = 1 − N0
540 GOSUB 1270
550 H2 = G1
560 Z = M0 − N0
570 If Z < 0 THEN GOTO 610
580 GOSUB 1270
590 H3 = G1
600 GOTO 640
610 Z = 1 + M0 − N0
620 GOSUB 1270
630 H3 = G1/Z
640 F0 = H2 * H1/H3
650 REM  F0 IS THE NORMALIZATION CONSTANT
660 PRINT %15E3, F0
670 REM  FINISHED CALCULATION OF F0
680 REM  CALCULATE HYPERGEOMETRIC SERIES
690 X = − W1
700 S4 = 1
710 S5 = 0
720 D6 = 1
730 D = 1
740 L = 1
750 A = S1
760 B = S2
770 C = S3
780 GOSUB 1330
790 X = − X
800 T1 = (X ↑ M0) * COS(M0 * 1.570796)
810 T2 = (X ↑ M0) * SIN(M0 * 1.570796)
820 T5 = ATN(X)
830 U1 = ((1 + X ↑ 2) ↑ ((N0 − M0)/2) * COS(T5 * (N0 − M0)))
840 U2 = ((1 + X ↑ 2) ↑ ((N0 − M0)/2) * SIN(T5 * (N0 − M0)))
850 R1 = F0 − (1 − N0) * (S4 * T1 * U1 − S5 * T2 * U1 − S5 * T1 * U2 −
S4 * T2 * U2)/M0
860 I1 = − (S5 * T1 * U1 + S4 * T2 * U1 + S4 * T1 * U2 − S5 * T2 * U2) *
    (1 − N0)/M0
870 R1 = R1 * A0/F0
880 I1 = I1 * A0/F0
890 PRINT %15E4, X, R1, I1
900 PRINT *1 %15E4, X, R1, I1
910 X = − X * W2
```

```
920 REM CHECK THAT THE NEW FREQUENCY DOES NOT LIE
BETWEEN
930 REM 0.95 AND 1.05; IF IT DOES THEN RE-INCREMENT
940 IF ABS(X) < 0.95 THEN GOTO 700
950 IF ABS(X) > 1.05 THEN GOTO 1000
960 PRINT *1 "CHI NOT CALCULABLE"
970 PRINT "CHI NOT CALCULABLE"
980 GOTO 910
990 REM CONTINUE THE CALCULATION FOR X GREATER THAN
    1.05
1000 X = ABS(1/X)
1010 S1 = 1 − N0
1020 S2 = 1 + M0 − N0
1030 S3 = 2 − N0
1040 S4 = 1
1050 S5 = 0
1060 D6 = 1
1070 D = 1
1080 L = I
1090 A = S1
1100 B = S2
1110 C = S3
1120 GOSUB 1330
1130 T1 = (X↑(1 − N0)) * COS((N0 − 1) * 1.570796)
1140 T2 = (X↑(1 − N0)) * SIN((N0 − 1) * 1.570796)
1150 R2 = S4 * T1 − S5 * T2
1160 I2 = S4 * T2 + S5 * T1
1170 Y = 1/X
1180 R2 = R2 * A0/F0
1190 I2 = I2 * A0/F0
1200 PRINT %15E4, Y, R2, I2
1210 PRINT *1 %15E4, Y, R2, I2
1220 X = 1/(W2 * Y)
1230 IF X > 1/W3 THEN GOTO 1040
1240 PRINT *1 " COMPLETED"
1250 GOTO 1490
1260 REM SUBPROGRAM TO CALCULATE GAMMA FUNCTION
1270 G2 = 1 − 0.5771916 * Z + 0.98820589 * (Z↑2) −
     0.897056937 * (Z↑3)
1280 G3 = 0.918206857 * (Z↑4) − 0.75670408 * (Z↑5) +
     0.48219939 * (Z↑6)
1290 G4 = − 0.193527818 * (Z↑7) + 0.035868343 * (Z↑8)
```

```
1300 G1 = G2 + G3 + G4
1310 RETURN
1320 REM SUBPROGRAM TO CALCULATE HYPERGEOMETRIC
     SERIES
1330 D6 = D6 * A * B * X/(C * D)
1340 IF L = 0 THEN GOTO 1380
1350 S5 = S5 + D6
1360 L = 0
1370 GOTO 1410
1380 D6 = - D6
1390 S4 = S4 + D6
1400 L = 1
1410 D7 = ABS(D6)
1420 IF D7 < 1E-7 THEN GOTO 1480
1430 A = A + 1
1440 B = B + 1
1450 C = C + 1
1460 D = D + 1
1470 GOTO 1330
1480 RETURN
1490 END.
```

The series used in the computation are poorly converging as the frequency approaches unity. It is for this reason that the frequency range 0.95–1.05 has been excluded. The range can be narrowed by altering the conditions in lines 940 and 950. The accuracy of the limited series summation is controlled by the instruction in line 1420 and can be increased at the expense of machine time.

References

1. A. A. Maradudin, E. W. Montroll, and G. H. Weiss, "Theory of Lattice Dynamics in the Harmonic Approximation," *Solid State Phys. Suppl.* **3** (1963).
2. W. Bernard and H. B. Callen, *Rev. Mod. Phys.*, **31**, 1017 (1959).
3. R. Kubo, *J. Phys. Soc. Jpn.*, **12**, 570 (1957).
4. W. T. Coffey, M. W. Evans, and G. J. Evans, *Mol. Phys.*, **38**, 477 (1979).
5. M. de Leener and P. Resibois, *Phys. Rev.*, **152**, 318 (1966).
6. M. de Leener and P. Resibois, *Phys. Rev.*, **178**, 819 (1969).
7. G. Williams, *Chem. Rev.*, **72**, 55 (1972).
8. C. J. F. Bottcher and P. Bordewijk, *Theory of Electric Polarisation*, Vol. 2, Elsevier, Oxford, 1978.
9. T. J. Lewis, J. McConnell, and B. K. P. Scaife, *Proc. R. Irish Acad.*, **76a**, 43 (1976).
10. A. K. Jonscher, *Nature*, **267**, 673 (1977).

11. R. M. Hill, *Nature*, **275**, 96 (1978).
12. R. M. Hill, *J. Mat. Sci.*, **17**, 3630 (1982).
13. K. L. Ngai, A. K. Jonscher, and C. T. White, *Nature*, **277**, 185 (1979).
14. L. A. Dissado and R. M. Hill, *Nature*, **279**, 685 (1979).
15. L. A. Dissado, *Physica Scripta*, **T1**, 110 (1982).
16. T. C. Guo and W. Guo, *J. Phys. C*, **16**, 1955 (1983).
17. J. L. Dote, D. Kivelson, and R. N. Schwartz, *J. Phys. Chem.*, **85**, 2169 (1981); P. J. Chappell and D. Kivelson, *J. Chem. Phys.*, **76**, 1742 (1982).
18. R. R. Nigmatullin and L. A. Dissado, *Chem. Phys.*, **79**, 455 (1983).
19. R. R. Nigmatullin and R. M. Hill, *Phys. Stat. Sol. b*, **118** 769 (1983).
20. L. A. Dissado and R. M. Hill, *Proc. R. Soc. Lond.*, **A390**, 131 (1983).
21. M. Lax, *Rev. Mod. Phys.*, **32**, 25 (1960).
22. J. D. Ferry, *Viscoelastic Properties of Polymers*, 2nd ed., Wiley, New York, 1970.
23. C. W. Garland, *Phys. Acoust.*, **VII**, 51 (1975).
24. I. L. Hopkins and C. R. Kurkjian, *Phys. Acoust.*, **IIB**, 91 (1965).
25. L. J. Slater, *Confluent Hypergeometric Functions*, Oxford University Press, Oxford, 1960.
26. G. F. Roach, *Greens Functions — Introductory Theory with Applications*, Van Nostrand Reinhold, London, 1970.
27. R. R. Nigmatullin, *Phys. Stat. Sol. b*, **123**, 739 (1984).
28. R. R. Nigmatullin, *Phys. Stat. Sol. b*, **124**, 389 (1984).
29. S. Hunklinger and W. Arnold, *Phys. Acoust.*, **XII**, 155 (1976).
30. A. Abragam, *The Principles of Nuclear Magnetism*, Oxford University Press, Oxford, 1961.
31. E. von Schweidler, *Annal. Phys.*, **40**, 817 (1907).
32. K. S. Cole and R. H. Cole, *J. Chem. Phys.*, **9**, 341 (1941).
33. A. K. Jonscher, *Nature*, **267**, 673 (1977).
34. H. Scher, in *Proceedings of the International Conference on Liquid and Amorphous Semiconductors*, W. E. Spear, ed., 1977, Centre for Industrial Consultancy and Liaison, Univ. of Edinborough, Edinborough, p. 209.
35. R. M. Hill and A. K. Jonscher, *Contemp. Phys.*, **24**, 75 (1983).
36. L. J. Slater, *Generalised Hypergeometric Functions*, Oxford University Press, Oxford, 1966.
37. R. M. Hill, *J. Mat. Sci.*, **16**, 118 (1981).
38. L. A. Dissado and R. M. Hill, *J. Phys. C*, **16**, 4023 (1983).
39. M. Shablakh, R. M. Hill, and L. A. Dissado, *J. Chem. Soc. Faraday Trans. 2*, **78**, 625 (1982).
40. R. M. Hill and L. A. Dissado, *J. Mat. Sci.*, **19**, 1576 (1983).
41. L. A. Dissado and R. M. Hill, *J. Chem. Soc. Faraday Trans. 2*, **78**, 81 (1982).
42. L. A. Dissado and R. M. Hill, *J. Phys. C*, **14**, L649 (1981).
43. G. Williams, *Trans. Faraday Soc.*, **61**, 1564 (1965).
44. G. P. Johari, *Philos. Mag. B*, **46**, 549 (1982).
45. G. Boiteaux, J.-C. Dalloz, A. Douillard, J. Guillet, and G. Seytre, *Eur. Polym. J.*, **16**, 489 (1980).
46. Cz. Pawlaczyk, G. Luther, and H. E. Müser, *Phys. Status Solidi b*, **91**, 627 (1979).
47. K. Deguchi, E. Nakamura, E. Okaue, and N. Aramaki, *J. Phys. Soc. Jpn.*, **11**, 3575 (1982).

48. R. M. Hill, *Phys. Status Solidi b*, **103**, 319 (1981).

49. S. Havriliak and S. Negami, *J. Polym. Sci. c*, **14**, 99 (1966).

50. E. H. Snow and P. Gibbs, *J. Appl. Phys.*, **35**, 2368 (1964).

51. G. P. Johari and M. Goldstein, *J. Chem. Phys.*, **53**, 2372 (1970).

52. G. Williams and D. C. Watts, *Trans. Faraday Soc.*, **66**, 80 (1970).

53. J. K. Moscicki, G. Williams, and S. M. Aharoni, *Polymer*, **22**, 571 (1981).

54. G. Boiteaux, G. Seytre, and P. Berticat, *Makromol. Chem.*, **180**, 761 (1979).

55. G. Williams and P. J. Haines, *Faraday Symp. Chem. Soc.*, **6**, 14 (1972).

56. J. Wong and C. A. Angell, *Glass Structure by Spectroscopy*, Marcel Dekker, New York, 1976. M. F. Schlesinger and E. W. Montvoll, *Proc. Natl. Acad. Sci.*, **81**, L80 (1984), R. G. Palmer, D. L. Stein, E. Abrahams, and P. W. Anderson, *Phys. Rev. Lett.*, **53**, 938 1984.

57. R. Kubo, *J. Phys. Soc. Jpn.*, **17**, 1100 (1962).

58. R. Kubo, *Fluctuation, Relaxation and Resonance in Magnetic Systems*, D. Ter Haar, ed., Oliver and Boyd, Edinburgh, 1961, p. 23.

59. See, for example, D. P. Craig and L. A. Dissado, *Chem. Phys.*, **14**, 89 (1976).

60. L. Fonda, G. C. Ghirardi, and A. Rimini, *Rep. Prog. Phys.*, **41**, 587 (1978).

61. L. A. Dissado, *Chem. Phys.*, **91**, 183, (1984).

62. J. J. Hopfield, *Comments. Solid State Phys.*, **11**, 40 (1969).

63. M. Abramowitz and I. A. Stegun, *Handbook of Mathematical Functions*, Dover, New York, 1965.

64. C. J. Reid and M. W. Evans. *Mol. Phys.*, **40**, 1357 (1980); G. W. Chantry, in *Infrared and Millimetre Waves*, Vol. 8, K. J. Button, ed., Academic, New York, 1983, p. 1.

65. P. C. Hohenberg and B. I. Halperin, *Rev. Mod. Phys.*, **49**, 435 (1977).

66. B. Shapiro and E. Abrahams, *Phys. Rev. B*, **24**, 4889 (1981).

67. R. Rammal and G. Toulouse, *J. Phys.— Lett.*, **44**, L-13 (1983).

68. K. G. Wilson and J. Kogut, *Phys. Rep.*, **12C**, 175 (1974).

69. M. Shablakh, L. A. Dissado, and R. M. Hill, *J. Chem. Soc. Faraday Trans. 2*, **79**, 369 (1983).

70. P. de-Gennes, *Scaling Concepts in Polymer Physics*, Cornell University, Ithaca, 1979.

71. J. H. Weiner, *Phys. Rev.*, **169**, 570 (1968); A. Nitzan, J. Jortner, and S. Mukamel, *J. Chem. Phys.*, **63**, 200 (1975).

72. M. Ferrario and M. W. Evans, *Adv. Mol. Rel. Int. Proc.*, **19**, 129 (1981).

73. L. A. Dissado and R. M. Hill, *Philos. Mag. B*, **41**, 625 (1980).

74. J.-P. Crine, *J. Macromol. Sci.— Phys.*, to be published.

75. W. Day, *The Thermodynamics of Simple Materials with Fading Memory*, Springer-Verlag, New York, 1972.

76. R. M. Hill, L. A. Dissado, and R. Jackson, *J. Phys. C*, **14**, 3925 (1981).

COOPERATIVE MOLECULAR BEHAVIOR AND FIELD EFFECTS ON LIQUIDS: EXPERIMENTAL CONSIDERATIONS

GARETH J. EVANS

Department of Chemistry, University College of Wales, Aberystwyth, SY23 1NE, Dyfed, Wales

CONTENTS

I. INTRODUCTION

We are taught at an early stage of our scientific training that there exist three states of matter—gas, liquid, and solid. Our understanding of the solid state of matter is advanced because the periodicity of molecules within the structured lattice simplifies solution of the equations involved. And, of course, it is possible to solve the Schrödinger equation itself for the dilute gas when the intermolecular interactions are insignificant. As we compress the gas, two- and three-body interactions become important and complicate such analyses. In the condensed liquid state the proximity of molecules is of the order of the internuclear separations within the molecules, the liquid is held together by attractive forces, and the whole ensemble behaves as an entity. A liquid flows freely, takes the shape of a containing vessel, and may even form droplets in free space held together by surface-tension forces in the peripheral regions, where the intermolecular forces are modified. Analyses are now enormously complicated. The intermolecular potentials are complex, time-varying quantities composed of attractive and repulsive forces. The phase is a cooperative molecular one in which a vast number of molecules move and interact cohesively.

The calculation of the intermolecular forces is considered to be at the root of the problem. But an intermolecular potential has been calculated for the simplest of liquid systems—argon and neon—only recently. This was achieved by measurement of the differential cross sections of argon and neon using modulated molecular beam techniques.[1] It is significant that the potentials calculated differed greatly from the Lennard-Jones potential, the potential that is even now being used extensively in the Monte Carlo and molecular-dynamics computer calculations of condensed-phase systems. In fact, the potentials for neon and argon themselves have different shapes and are not related by any simple scaling factors.

Thus, the elucidation of the liquid state of matter, the life-supporting state and the state in which most of our chemistry takes place, eludes us. As Berne and Harp[2] have said, "understanding the liquid state has proven to be both a challenge and an embarrassment to generations of outstanding chemists and physicists." Onsager's 50-year-old theory accounting for the dielectric constant of a polar fluid and the conductivity of a salt is still used extensively. Only recently has a full quantum calculation for the gas-phase exchange reaction

$$H + H_2 \rightarrow H_2 + H$$

been achieved. Obviously, we cannot yet contemplate extending such calculations to the solution state. This can be achieved only when we have detailed knowledge of the intermolecular potentials. In fact, the state of the art is such that we usually cannot calculate the solubility of one substrate in another *even where no reaction takes place*. Engineers and applied scientists, of course, would wish to be able to predict such solubilities, but it is only recently that theories of solution have reached the stage at which the right sign of volume change and a correct order of magnitude may be estimated.[3] As Hildebrand[4] has said, "of the various thermodynamic functions for the mixing process, the volume change on mixing at constant pressure has intrigued scientists for years. However, it is still little understood."

A.   Intermolecular Forces

The actual study of liquids has provided little information on intermolecular forces. The numerous polymorphs in the solid states of many substances testify to the complexity and diversity of these forces. One still often sees in the literature the use of hard sphere potentials that neglect attractive energy completely.

The ideas behind such potentials have been used for over a century.[5] The work of Van der Waals[6] in 1873 implied that the structure of a liquid is primarily determined by repulsive forces, with the attractive forces providing only a uniform background energy. The liquid could thus be regarded as a system of hard spheres. At that time the properties of such a system were not known. Such information has become available only with the recent rapid progress in computer methods for studying molecular dynamics. The properties of such a system are obviously not compatible with many of the pertinent facts. Though the Lennard-Jones interaction is a significant improvement, it does not represent satisfactorily the liquid-state interactions even between neon and argon. When we recall that the R-12 repulsive energy (unlike the R-6 attractive energy) has no physical justification, this is not so surprising.

The failure of liquid studies to provide information on potentials is exemplified by the fact that information for calculating the potentials is still commonly derived from the macroscopic properties of dilute gases: from the second virial coefficient (at low gas densities), from the transport properties of gases, from the scattering of molecular beams by gases, and from X-ray determinations (again at low densities). All of these techniques have limitations. The second-virial method is not discriminating in distinguishing between potentials. Even if B_2 is known for all temperatures, this determines only the positive part of the potential curve and the area beneath the "negative" part of the abscissa. The experiments should be carried out to the extremes of temperature.[7] At the lowest accessible temperatures below the boiling point, the 12-6 potential gives B_2 values that are less negative than the experimental ones, indicating that the minimum of the 12-6 potential is not deep enough and that the third virial coefficients, calculated using the Lennard-Jones potential derived from the second virial coefficients, are too small. As Prigogine[8] emphasizes, "we cannot determine the correct law of force using only the second virial coefficients."

Transport properties used to measure the potential suggest that model potentials cannot be used to fit data over a whole temperature range. Quoting Prigogine[8] again: "For high temperature, there is a significant deviation of the calculated transport properties from the Lennard Jones potential." There are severe problems with the scattering-of-beams technique, the most serious being the inapplicability of the Born approximation and the need to use complicated quantum-mechanical calculations.

Even assuming we have a pair potential between two molecules accurate enough for the calculation of measureable bulk quantities, we then have to assume that the potentials are additive, so that the potential energy of the bulk is the sum of interactions between all possible pairs of molecules in the ensemble. This is probably an oversimplification. Hildebrand[4] reminds us that "the molar volume V_t is defined as that volume at which soft molecules begin to be sufficiently separated between collisions to acquire fractions of their momentum in free space. At volumes smaller than V_t they are in fields of force not appropriately described by pair potentials. The role of temperature is only to determine volume. Values of V_t depend on the capacity of a molecular species to absorb collision momentum by bending, vibrating or rotating." Three-body forces are thought to be important even in the calculation of certain properties of rare-gas solids, such as the zero-point energy and the Debye temperature. In light of this it may seem surprising that two-body interactions can form the basis for a discussion of intermolecular forces in liquids. Buckingham[9] sums up the situation thus: "In spite of our increasing knowledge of interaction energies, we must confess that we do not know accurately the potentials for any organic systems, not even $(CH_4)_2$. So we are forced to use empirical or semi-empirical potentials." And these, of course, in spite of obvious deficiencies. For example, a long-standing objection to the Lennard-Jones potential is that an assembly of such molecules has a configuration of minimum potential energy in the hexagonal close-packed structure. However, all inert gases except helium crystallize in the cubic close-packed structure. The fault probably lies in the neglect of small nonadditive terms in the potential energy of the assembly. If we wish to consider polyatomic, polarizable molecular liquids, electrostatic terms have to be added and the shapes of molecules accounted for so that repulsive forces will change with the orientation[5] of the molecule. As discussed in the first article of this volume, the hydrogen atoms are important in a computer simulation of CH_2Cl_2 for describing its molecular dynamics. Narten et al.[10] reported a similar observation in their X-ray-diffraction study of liquid neopentane.

The influence of molecular shape on the intermolecular potentials may be observed straightforwardly by comparing different isomers of a given formula. For example, since the critical temperature (T_c) is roughly proportional to the pair potential, we can conclude that in the series of pentanes tabulated below cyclopentane has the largest energy of interaction and neopentane the smallest.

Account must also be made of the fact that the origins of the attractive and repulsive forces *need not be the geometrical centers of the molecules*. We

	n-Pentane	Isopentane	Neopentane	Cyclopentane
T_c	469.6	450.4	433.8	511.6
$\Delta S_M/R$	7	5.5	1.5	0.4

shall consider the consequences of this in discussing the coupling of a molecule's rotation with its own (or a neighbor's) translation in later sections. Buckingham[9] again sums up: "We are still ignorant of energies near to the minima in the potential wells, and we know little of the effects of many body interactions. It is commonly assumed that the medium diminishes the interaction between polar species, but in fact polar or polarizable molecules placed between opposite electric charges increase their attraction."

It may easily be demonstrated that an external field stabilizes a liquid drop, increasing the attraction between those molecules whose diameters are smaller than a critical diameter (characteristic of the liquid under study), if the dipole moment of the drop is larger than the corresponding dipole moment of the same number of noninteracting dipoles. If the solution is assumed ideal, it is found, analogously to the case of supersaturated vapors,[11] that[12]

$$\log\left(\frac{P(r)}{P(\infty)}\right) = \frac{1}{kT}\left[\frac{2\sigma}{r\rho} - \mathbf{E}\cdot\left(\mathbf{M}_1 - \mathbf{M}_2 + \frac{r}{3}\frac{d\mathbf{M}_1}{dr}\right)\right]$$

where $\mathbf{M}_1$ is the dipole moment in the absence of an electric field, $\mathbf{M}_2$ is the dipole moment in the presence of the field, $P(r) = -(\partial F/\partial V)_T$ is the "pressure" over the drop of radius r (F is the free energy and V the volume of the drop), $P(\infty)$ is the corresponding pressure over a plane interface of the same liquid whose density is $\rho = (3N_r/4\pi r^3)$, (N_r is the number of noninteracting dipoles), $\mathbf{E}$ is the external electric field, and T is the temperature.

This equation shows that the external electric field favors the growth of regions where the concentration of dipoles is increased. This can be seen in a simple laboratory experiment if a large field is applied across two parallel brass electrodes to a nondipolar liquid such as CCl_4. *Small droplets* of liquid are quickly expelled through the top of the two sealed electrodes[13]. The electric field also induces bulk translational motion in molecular liquids. We shall consider this phenomenon in a later section.

Computer simulation is presently being used to refine our "trial potentials" by comparison with experimental data. As we shall see, new experimental results may also provide direct access to details of the interactions between molecules in the liquid state.

B. The Liquid State and Its Range of Existence

Liquid argon is considered the model liquid because its dispersion energy is assumed to be pairwise additive and its atoms have spherical symmetry. In liquid argon the attractive energy is caused by mutual induction from the electrons fluctuating in the electron clouds. As we have said, in molecular liquids the pair potential depends on orientation and interactions between electrostatic moments may contribute to the attractive energy. If the polarizabilities are such that partial charges are located on the peripheries of the molecules, then temporary (chemical) bonds such as hydrogen bonds may form. The role of molecular shape in determining the properties of condensed-state systems is displayed even among the less complex molecular species. If molecules are flexible, coupling of intramolecular and intermolecular modes of motion may occur. Coupling of rotational and translational motions may be seen in fairly simple molecular liquids, so the two modes may not always be separated, as is customary in, for example, the calculation of equilibrium properties (ΔH, ΔS, ΔE, and density) from the relevant partition functions. The complexity of the interactions, and this coupling of modes of motion, has frustrated and delayed understanding of liquid-state behavior. As Kohler[14] says, "at present there exists no model which would reproduce satisfactorily the properties of liquids over the whole density range."

We will now ask some simple questions and see what answers may be provided. We must not confine ourselves to macroscopic, bulk considerations. Bulk properties have their origin at the molecular level and it is here that many of the subtleties of condensed-state matter may be observed. Bulk considerations alone obscure many of these details.

At the triple point of a substance gas, liquid, and solid phases may coexist—the states converge and are in equilibrium with each other. And the conventional belief that a sharp melting point indicates no continuous gradation of properties between liquid and crystal is not strictly true. It is possible for a liquid to be cooled below its normal melting point and to pass not into the crystalline solid but into a disordered solid in which there is *no long-range order*. The important variable in this process is the *rate of cooling*. If the rate is fast the liquid is changed into a glassy state with all its imperfections literally frozen in. The viscosity of the liquid increases steadily until it approaches the glass-phase transition. The short-range order in the glass is virtually perfect, but there is no definite long-range order. Glasses are intermediate between liquids and solids; their high viscosities prevent rearrangement of molecules into the completely regular crystalline form. It is now widely believed that it should be possible to cool any liquid into a glassy state providing the cooling rate is fast enough and the final temperature low enough to prevent the glass from recrystallizing. These requirements are easily met in computer simulation studies as Evans [15] has described for one liquid system. Stillinger[16] has done similarly for water in a computer simulation. They observe that the diffusions are markedly slower because the hydrogen bonds are frozen in and are stronger.

Glasses are interesting in their own right and offer challenging "test situations" for the molecular dynamicist. A few degrees' change in temperature near the glass-phase transition (T_g) may vary the viscosity of the system by several orders of magnitude. This has fascinating spectral consequences, as Evans et al.[17] have discussed. Recent interest has been directed toward quenching liquid metals into glassy states because the unique physical properties so acquired (e.g., the combination of high mechanical strength, ductility, and remarkable magnetic properties)[18] have technological importance. By changing alloy composition, one can make paramagnets, ferromagnets, ferrimagnets, and spin glasses. In ferromagnetic glasses consisting primarily of transition metals it is found that coercive forces are extremely low.[19] These materials make good soft ferromagnets. Another property they may exhibit is a high resistance to corrosion[20] compared with their crystalline equivalents.

As we have said, the numerous polymorphs of many solids testify to the significant range of structures, and consequently properties, that a solid may have. But what of a liquid? Does a similar gradation of molecular behavior exist in the liquid phase? Liquid helium has fascinated scientists more than any liquid apart from water and is the only known system in which two liquid phases almost certainly exist. Important quantum effects are found in this simple system (and some hydrogen-isotope systems). It remains a liquid down to 0 K and may be solidified only under considerable external pressure. Interestingly, Dyson[21] has suggested that all matter is "liquid" at 0 K because of quantum-mechanical barrier penetration. The lifetimes for such processes are so long that even the most rigid materials cannot preserve their shapes or their chemical structures for comparable times. Dyson says that on such a time scale "every piece of rock behaves like a liquid, flowing into a spherical shape under the influence of gravity. Its atoms and molecules will be ceaselessly diffusing around like the molecules in a drop of water."

Some liquids obviously have, at certain densities, properties similar to those of compressed and even dilute gases. Liquids are known in which *rotational spectral detail* may be resolved at submillimeter (far infrared) frequencies,[22] as in the gas phase. If HCl is dissolved in argon, spectral fine rotational detail is resolved for argon densities between 100 and 480 Amagat. The density dependence of the different linewidths is distinctly nonlinear and different for different lines. Such behavior is the same as in the compressed gas state, but in striking contrast to that observed in dilute gas systems. If HCl is dissolved in SF_6 and CCl_4, rotational fine structure is resolved in the former in the liquid state, but not in the latter, *even though* the Lennard-Jones parameters and masses of SF_6 and CCl_4 are quite similar.[23] Such observations provide vital clues to details of the dynamics. It is apparent that the ratios of mass and size of the probe to the mass and size of the solvent are significant in determining the dynamics, and that it is not just the translational motion but also the rotational motion of the solvent that modulates local density fluctuations. In fact it can be established straightforwardly that rotational and translational motions are strongly coupled even in such simple liquids. Expressed in classical terms, since translational frequencies in the liquid are usually comparable to rotational frequencies, these motions become easily coupled.

The simplest spectroscopic example that shows the effect of this coupling is the induced rotational absorption of HD in liquid argon.[24] The observed experimental rotational energy states of H_2 or D_2 in liquid argon are, within experimental error, equal to their gas-phase values. Because the electrical center and the center of interaction of the HD molecule are located halfway between the two nuclei *but the center of mass is somewhat displaced*, a HD molecule may rotate about its center of mass, in an argon liquid environment, only if the electrical center, which is displaced from the center of the cavity, simultaneously translates back into the center of the cavity and is restored to a position of minimum potential energy. As Ewing[25] says, "for HD or any molecule whose center of interaction does not coincide with its center of mass, rotational and translational motions are coupled by the solvent cavity." The spectral consequences are such that resolved far-infrared rotational details show large half-widths, erratic frequency shifts, and additional absorptions arising from the relaxation of the rotational selection rules. Transitions $+1$, $+2$, $+3$, and $+4$ are all observed for the HD–argon liquid system, a consequence of the rotation–translation coupling perturbation.

In such systems, when spectroscopic detail is resolved, molecular-dynamics analyses are greatly simplified. So Marteau et al.[26], examining the far-infrared collision-induced absorption of liquid Kr–Ar mixtures, were able to assign certain features of their spectra to *nonlocalized phonon* (*collective translational*) *modes in the liquid*. So too, Bulanin and Tonkov[27] have assigned a feature at 38 cm^{-1} in their spectra of D_2–argon solutions to phonon absorption by the liquid argon. And Chantry et al.[28] attributed features in the far-infrared spectra of CS_2 and benzene to lattice modes of the translational type. In recent years it has become customary to attribute the *broad absorptions* of these latter two liquids to rotational motions, that is, to the motions of dipoles induced by neighboring molecular quadrupoles or octopoles in the liquid. However, it is generally *not possible* to distinguish between rotational and translational molecular contributions, whether single-molecule or collective, to such broad and featureless profiles. By considering simple molecular liquid systems we may easily establish the importance of both rotational and translational modes of motion in determining the total dynamics of the system.

We may now envisage instances in which one of these two modes of motion (rotation and translation) is constrained in the liquid because of, for example, the shapes of the molecules. Consider long, rodlike molecules. Rotation about certain axes will be hindered because an orientational order is introduced into the system. The molecules may remain positionally disordered, with their centers of mass delocalized. What effect does such order impose on the liquid system, and do such systems exist in nature? Edgar Allan Poe wrote in 1837: [29]

> I am at a loss to give a distinct idea of the nature of this liquid, and cannot do so without many words. Although it flowed with rapidity in all declivities where common water would do so, yet never except when falling in a cascade, had it the customary appearance of limpidity. It was, nevertheless, in point of fact, as perfectly limpid as any limestone water in existence, the difference being only in appearance. At first sight, and especially in cases where little declivity was found, it bore resemblance, as regards consistency, to a thick infusion of gum arabic in common water. But this was only the least remarkable of its extraordinary qualities. It was not colourless, nor was it of any uniform colour—presenting to the eye, as it flowed, every possible shade of purple, like the hues of a changeable silk. The variation in shade was produced in a manner which excited as profound astonishment in the minds of our party as the mirror had done in the case of Too-Wit. Upon collecting a basinful, and allowing it to settle thoroughly, we perceived that the whole mass of the liquid was made up of a number of distinct veins, each of a distinct hue; that these veins did not commingle; and their cohesion was perfect in regards to their own particles among themselves, and imperfect in regard to neighbouring veins. Upon passing the blade of a knife athwart the veins, the "water" closed over it immediately, as with us, and also, in withdrawing it, all traces of the passage of the knife were instantly obliterated. If, however, the blade was passed down accurately between the two veins, a perfect separation was effected, which the power of cohesion did not immediately rectify.

This remarkable account of the character of *liquid crystals*, first brought to our attention by Gray,[30] introduces us to another state of matter—the liquid crystalline state. Liquid crystals in which the molecules are positionally disordered, but orientationally ordered, are called nematics. Because of the forces arising from their mutual interaction, the molecules adopt a common mean parallel orientation. Such systems have fascinating properties and technological applications, which are discussed at length in many texts.[31]

What if there is positional as well as rotational order, but only in one or two directions (and not three directions, as in the crystalline solid)? Again, such situations do exist in nature and are also classified as liquid crystals. With positional order in only one direction, they are called smectics. "In addition to possessing mean longitudinal parallelism (rotational order) each molecule conditions the mean position of its neighbour so that the mean distance between their centres (or ends) is minimized. This results in planes of structural discontinuity which are perpendicular to the direction of parallelism. The layers of molecules so constituted can slide with equal facility in all directions over the planes of discontinuity."[31(ii)]

There are tricks that we may play even on these liquid crystalline systems to produce additional phases of matter. Let us return to the nematic phase and recall some more elementary chemistry—the chemistry of chiral systems. Chiral molecules (containing, for example, an asymmetric carbon atom attached to four different elementary groups) display the fascinating property of handedness; that is, the separate atoms (or groups of atoms) may be arranged (if we envisage a planar representation of the molecule) in clockwise or anticlockwise directions. The two different molecules are then mirror images of each other; that is, they possess a plane of symmetry. If we dissolve such a molecule in a nematic liquid crystal we cause the structure to twist and to undergo a helical distortion. Some liquid crystals are twisted in this way in their natural form and may be considered distorted nematics; they are called cholesterics and are chiral liquid crystalline systems.

Optically active, isotropic molecular systems, even those consisting of small molecules, display more clearly than any others the importance of molecular shape and of the size of the atoms of a molecule in determining the packing and molecular dynamics of the liquid state, and consequently its physical properties.[32] A computer simulation of 1,1-fluorochloroethane reveals that spectra obtained with all of the spectroscopic techniques (dielectric, infrared, Raman, neutron scattering, Kerr-effect measurement, nuclear magnetic resonance, and light scattering) will be the same for the R and S enantiomers, but *may be different* for the racemic mixture. This is a consequence of a subtle detail of the molecular dynamics—the coupling of a molecule's rotation *to its own* translation. Certain cross elements of a mixed rotation–translation matrix are opposite in sign for the two enantiomers (because they are mirror images). However, this has no effect on the measured spectral properties of each individual enantiomer. If R and S species are mixed to form the corresponding racemic mixture, the elements of opposite sign must cancel and vanish, this being the only observable dynamical difference. This has remarkable effects on measured spectra and, indeed, on physical properties. In 1,1-fluorochloroethane the effect is sufficient to shift, for example, the far-infrared frequency of maximum absorption of the enantiomers (which occurs at 35 cm^{-1}) to ~ 65 cm^{-1} in the racemic mixture (the phenomenological Debye loss time is approximately halved). The differences in the dynamics are most pronounced. The role of the molecular atoms in determining condensed-state properties may now be easily examined, because if we replace the chlorine atom in this molecule with a larger iodine atom to obtain 1,1-fluoroiodoethane, the dynamics and physical properties are dominated by the mass and assumed sphericity of the iodine atom. That even physical properties may be modified in the racemate relative to the individual enantiomeric systems may be seen by considering lactic acid ($CHOHCH_3COOH$). At room temperature the racemic mixture is a liquid, whereas both enantiomers form solids (melting points 18°C and 53°C, respectively). A consideration of these optically active systems brings to the fore the subtlety of the roles of rotation and translation and the coupling between the two in determining condensed-state properties.

In a molecular crystal there may exist positional order (the centers of gravity of the molecules are located on a three-dimensional periodic lattice), but rotational disorder. Crystals so arranged are often called cubic mesophases or plastic crystals. The molecules are globular and may undergo thermal rotatory motions freely, in some cases more freely than in the liquid.[33] That these rotary motions are accompanied by translational displacements (the motions are coupled in the solid) is illustrated by the blurred X-ray-diffraction patterns that are always obtained in such systems. However, computer simulation shows[34] that in bromoform, for example, a solid rotor phase is formed below the freezing point at 281 K, rototranslation functions, which may be observed in a frame of reference with moving axes, are relatively small in these very liquids that form rotator-phase states. It is proposed, therefore, that a dynamical prerequisite for the existence of certain types of rotator phases is that rotation be largely decoupled from translation.[15(b)]

Submillimeter spectra (Fig. 1) firmly substantiate the existence of a solid rotator phase in bromoform. The rotator-phase spectrum is shifted slightly to higher frequencies relative to the liquid spectrum, but remains a broad and featureless band. No lattice modes, which are characteristic of the crystalline solid, are resolved until the sample is cooled below −2°C. In tertiary butyl chloride, in which two solid rotator phases exist between the isotropic liquid and nonrotator crystalline solid states, *a marked hysteresis may be observed as the sample is cooled and reheated through the phase-transition regions* (Fig. 2). Using a spot frequency at 84 cm^{-1} from a tunable far-infrared laser, spectroscopic changes were monitored as the sample was first slowly cooled through the transition region. The first transition, liquid to ro-

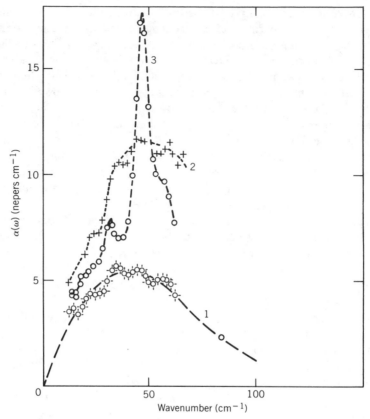

Figure 1.   Far-infrared absorption of bromoform in the liquid and rotator-phase states: (*1*)
⟡ , The liquid state at 295 K; O, laser point at 84 cm^{-1}. (*2*) +, The rotator phase at 273
K (as for the liquid, a broad, featureless band). (*3*) O, The crystalline solid (note that lattice
modes are resolved).

tator solid phase I, is barely observable at 247.53 K. The second transition,
rotator phase I to rotator phase II, is more pronounced at 219.25 K. The third
transition, rotator phase II to crystalline solid, involves the most significant
drop in throughput intensity, at 182.9 K. The most striking feature of these
curves is the marked hysteresis seen as the sample is reheated through the
phase-transitions region (slowly over about 1½ h). There is no indication
whatsoever that the crystalline solid passes into rotator phases II and I. Instead, the sample melts *at a higher temperature* (~ 268 K) than any of the
transition temperatures. Using such "molecular probes" (the various spectroscopies) to follow phase transitions provides us with valuable insight into
the freezing and melting processes.

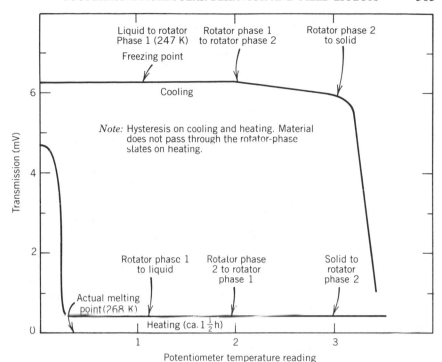

Figure 2.   Far-infrared microscopic probe of the freezing and melting processes in *t*-butyl chloride.

Sciesinska and Sciesinski reported similar observations for cyclohexanol.[35] As they said, "the phase situation in solid cyclohexanol still remains unclear. The number of ascertained solid phases varies from four to seven depending on the method of investigation and the authors." In Fig. 3 we reproduce a diagram of theirs that gives the temperature dependences of integral transmission (the transmitted power over a given frequency region) for cooling (b, c, e, g, and h) and heating slowly from liquid-nitrogen temperature (points f, d, and a). The various phase transitions are exposed distinctly because they most often cause a change of integral transmission due to increased scattering even when the absorptions are similar for both phases. Nor is this hysteresis confined to those substances that form rotator-phase solid states. Michel and Lippert[36] reported on "two ways of freezing acetonitrile" (CH_3CN). Their phase transitions for acetonitrile are reproduced in Figs. 4 and 5 which show the results of monitoring the ν_3 symmetric CH_3 deformation mode, the ν_4 C—C stretch, the ν_5 C—H antisymmetric stretch, and the combination modes $\nu_2 + \nu_4$ (C—N + C—C) and $\nu_3 + \nu_4$ (CH_3 def + C—C).

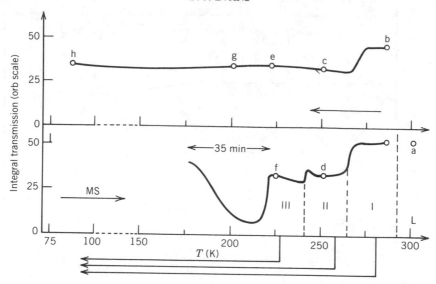

Figure 3. Integral transmission for cyclohexanol, obtained using an IRIS Spectrometer. Points b, c, e, g, and h represent the cooling process; points f, d, and a, the heating process. The phases are as follows: a, liquid; b, supercooled liquid; c and e, rotator phases; d, crystal II; f, crystal III; and h, crystal (metastable). *Note:* Again the material passes through rotator phases on cooling but not on reheating. (Reproduced by permission from ref. 35.)

Michel and Lippert stated that

> presolid states, metastable processes, and hysteresis can be observed as a continuous function of temperature and speed of cooling... hysteresis effects occur in both directions while acetonitrile is passing through presolid and metastable states. When cooling very slowly, one can observe a presolid state with a lifetime of less than half a second. It is characterized by a rapid decrease of the maximum intensities of the combinations modes. A definite melting point is not found when reheating, but rather a transition region with a shape similar to a titration curve. The first sign of the phase transition can be observed *eight degrees before melting.* The proper melting region is contained within two degrees. Directly after melting the liquid is different from the one before freezing at the same temperature. After some minutes or after reheating, the old situation returns.

We shall consider the nucleation process in molecular liquids in more detail in a later section, and shall see that *external electric fields can be used to induce nucleation.*

We have tended to label some condensed-phase properties as anomalous when they have failed to fit our restrictive preconceptions about condensed-

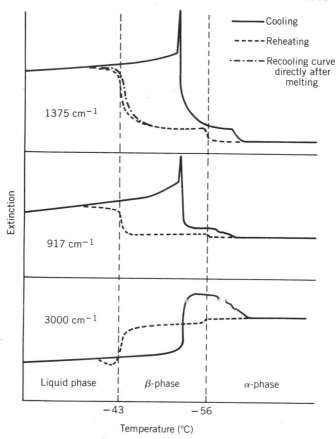

Figure 4. Acetonitrile phase transitions: ν_3, symmetric CH_3 deformation; ν_4, C—C stretch; and ν_5, C—H asymmetric stretch. Note the existence of hysteresis and presolid states. The diagrams show intensities at fixed wavenumbers on cooling and reheating. (Reproduced by permission from ref. 36.)

state matter. For example, the 10% contraction of ice on melting is not anomalous at all. As Eyring et al.[37] point out, "when a normal system, such as argon, melts it expands twelve per cent. Such an expansion is typical. The atypical ten percent contraction of ice upon melting can be made to disappear if one first applies some 2,500 atmospheres pressure at an appropriate temperature, to the hydrogen bonded tetrahedral ice I and so transform it into close packed ice III, which is a fifth more dense. Ice III then melts with a normal ten percent expansion." Problems have arisen because of our desire to label matter strictly as solid or liquid, making no allowance for intermediate states that may exist. The numerous ice polymorphs that exist attest

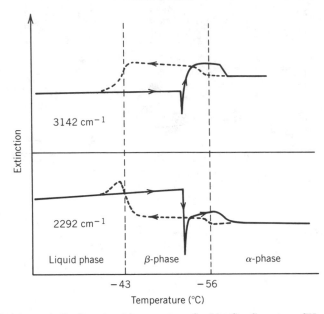

Figure 5.   Acetonitrile phase transitions: $\nu_2 + \nu_4$, C—N+C—C; $\nu_3 + \nu_4$, CH$_3$ def. + C—C. The combination modes show a rapid decrease in the presolid state. (Reproduced by permission from ref. 36.)

to the slight molecular readjustments that are necessary to produce different phases—a gradual gradation of properties through the liquid and into the solid is in fact apparent.

Water, of course, is the most extensively studied of all liquids because informed opinion holds that life may arise in, and be supported by, only a suitable aqueous medium.[38] Quoting Stillinger,[16] "it is only since 1960 that it has become technically feasible to produce a quantitative and deductive theory [see Vol. 1 of this issue] for liquid water without large elements of uncertainty. It is no coincidence that this period coincides with the general availability of rapid digital computers. These computers have provided essential numerical advances in both the quantum mechanical and statistical mechanical aspects of the field that underlie present understanding." Submillimeter spectra reveal the complexity of the water-molecule interactions, as well as subsequent dynamics that even now no theoretical molecular modeling successfully reproduces.[17] Rahman and Stillinger observed that the molecular dynamics depends strongly on the form of the hydrogen bonding in the liquid: "There was no evidence of a hopping process, but, rather, translational diffusion proceeds via individual molecules participating in the continual restructuring of the labile, random, hydrogen bond network." It is possible to break up the hydrogen-bonding network by dilution.[17] The effect of the hydrogen bonding is to broaden the zero-teraHertz spectral profile so that it spans eight or nine decades of frequency, compared with only two when the bonds are absent. This typifies the extremes of molecular behavior and properties that the isotropic liquid state may adopt. A full, consistent understanding in accord with the pertinent facts must start at the molecular level itself. It is not possible to explain many recent phenomena of the liquid state (some of which are discussed in later sections) with, for example, hydrodynamic theories that set out to explain liquid-state behavior in terms of macroscopic data. These do not consider the detailed molecular structure of the medium, let alone the structure of the molecules, as is so often necessary.

From the foregoing it is apparent that within the liquid state there may be many variations of molecular behavior. Near the freezing points considerable molecular rearrangement may occur, and pretransitional phases have been proposed. There may exist a considerable short-range structure that extends even to macroscopic distances in some liquid systems (liquid crystals). This structure disintegrates at the boiling point and some liquids even display the freedom of motion associated with the gas phase itself. Recently, another "new state of matter" has been proposed—a liquid crystalline state in which the sample is a *solid* film. Certain solid films of PBLG that show liquid crystalline properties represent "a new and very interesting phase of matter. It is quite possible that other rodlike molecules which exhibit a liquid crystal phase such as DNA can be induced to form the same state, in vitro or in vivo."[39]

II. COLLECTIVE MODES IN LIQUIDS

A. Liquid Crystalline Systems

We have seen that because of molecular shape a molecule's rotation may cause a simultaneous translation of neighboring molecules. Rotation and translation are mutually coupled and cooperative molecular behavior in liquids has its origin on the molecular level itself. If the center of mass is displaced from the center of interaction a correlation between translation and rotational degrees of freedom is introduced. This permits the explanation of some thermodynamic anomalies, such as the higher vapor pressure of HT compared with D_2 discussed by Babloyantz.[40] At 20.3 K the vapor pressure of HT is about 36% higher than that of D_2 and the molar volume and vapor pressure of HD are larger than the expected values obtained by interpolation from the values for H_2, D_2, and T_2.

Since rotation–translation coupling effects are so pronounced in these simple systems, they must also be present in other liquids. However, they are not easily distinguished using spectroscopic methods. Recent advances in computer technology are shedding new light on the problem. Various elements of the autocorrelation matrix $\langle \mathbf{p}(0)\mathbf{J}^T(t)\rangle$ show that the coupling of rotation with translation exists *for most molecules* and indirectly affects laboratory-frame autocorrelation functions measured with the spectroscopies. Here $\mathbf{p}$ is the *linear* center of mass momentum and $\mathbf{J}$ the molecular angular momentum. This mixed function vanishes for all times t in an isotropic molecular liquid in the laboratory frame because the parity of $\mathbf{P}$ to time reversal is opposite in sign to that of $\mathbf{J}$.[‡] The second-moment autocorrelation functions $\langle \mathbf{P}^2(0)\mathbf{J}^2(t)\rangle$ are invariant to frame transformation and may be observed in the laboratory frame. They provide us with a detailed description of molecular rototranslation, as may be seen in Fig. 8 for CH_3CN. Evans[15(a)] concludes that "the dynamics in the region of the boiling point may not be significantly affected by R–T interaction but the liquid–solid transition and molecular behavior in this region may be severely affected." These coupling effects are so important because, as Warner[41] points out in reference to liquid crystalline systems, "the ordering of rods under the influence of their packing density and shape is a purely entropic process.... ordering results from a coupling between the rotational and translational entropy of the rod.... such a coupling would be revealed down to wavelengths of the order of the rod dimensions." Our picture is therefore one of a local molecular order arising from steric constraints and from the tendency for molecules to order themselves over short distances as a consequence of their shapes and the requirement that "hard" molecules confined to a limited volume do not overlap. This short-range molecular ordering must arise to some extent in all condensed systems. And, as Warner points out, "this steric aspect of ordering is, given constancy of volume, independent of temperature. Experiments at constant volume confirm the constancy of the steric component of the ordering process."[42] The effect of molecular asymmetry in dominating correlations and leading to a residual short-range *orientational* order is discussed by Wulf.[43]

In the nematic phase of a liquid crystal, in addition to this short-range order (which must also be present in the isotropic phase), there exists a long-range order—molecules are oriented on the average with respect to a direction known as the "director." Long-range distortional fluctuations of this director exist, and new hydrodynamic modes (not present in normal liquids) associated with fluctuations of the director have been predicted.[44] Evans[45] has reported the resolution of new resonant modes in the nematic

[‡]Evans has recently computed laboratory-frame rotation—translation cross-correlation functions in a polyatomic, chiral molecular liquid (M. W. Evans, *Phys. Rev. Lett.*, accepted, 1985).

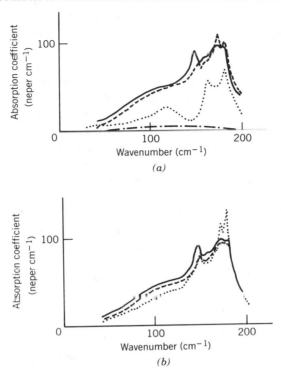

Figure 6. (*a*) Far-infrared spectrum of 4-cyano-4-*n*-heptyl biphenyl and related mole-cules: —— , 4-cyano-4-*n*-heptyl biphenyl at 296 K, nematic phase; ---, the same with a field of 7 kV cm^{-1} applied; $\cdots$, 2.52 mol dm^{-3} 4-cyanobiphenyl in dioxan at 296 K corrected for solvent absorption; $-\cdot-\cdot$, 0.9 mol dm^{-3} biphenyl in cyclohexane at 296 K corrected for solvent. (*b*) Far-infrared spectra of 4-cyano-4-*n*-heptyl biphenyl under external magnetic fields: —— , nematic phase with no field; ---, with a field of 400 G; $\cdots$, with a field of 800 G. Note the appearance of sharp resonant modes in the nematic phase of the liquid crystal in strong electric and magnetic fields. (Reproduced by permission from ref. 45.)

phase of a liquid crystal at submillimeter frequencies (Fig. 6).[‡] He wrote that "the intensity and frequency of these modes are controlled by externally ap-plied electric and magnetic fields. Their existence suggests that this spec-troscopy may serve as a convenient tool for the study of liquid crystals." However, continuum theory shows that external fields damp out fluctuations in the director and reorient it so that it is either parallel or perpendicular to the fields. In strong enough fields (magnetic fields of > 1000 G), the director motion is prevented. Evans observed no indication of any disappearance of the modes in his experiment, even though electric fields an order of magni-tude greater than the magnetic fields (a maximum of 1000 G) were used. In-terestingly, Bulkin[46] reported that some mid-infrared bands disappear at the crystal to nematic phase transition. Bulkin concluded that "these bands arise from sum and difference modes between lattice vibrations and internal vibrations...the disappearing bands are an indication of the coupling between molecules and hence are of importance in understanding ordering forces in the crystal as compared to those in the liquid crystal." Bulkin also reported pretransitional effects in the crystalline phase prior to melting, and that Schwartz and Wang[47] observed "striking changes in the relative intensity of several Raman bands as a function of applied external electric field strength in the nematic phase. Certain bands increased in integrated intensity, while others showed the opposite effect."

Evans has attributed his new modes to collective modes, that is, intermolecular or phonon vibrations. As Bulkin again says, "in the nematic phase we again have the possibility of observing 'pseudo-lattice' vibrations, i.e., intermolecular motions characteristic of the long range order in this phase...it may be possible to propagate phonons even in liquid crystals." There is not much evidence to support Evans's results in the nematic phase,[‡] but there is some in the case of more highly ordered smectic phases. Smectic phases, particularly such phases as smectics B and H, are, of course, much closer to crystalline phases than are nematics. So, Amer and Shen[48] observed that diethylazoxybenzoate and diethylazoxycinnamate, which have smectic A phases, each have a single low-frequency Raman mode in the crystalline phase, at 22 and 26 cm^{-1} respectively. In the smectic phase, this mode appears to shift, with the maximum at about 14 cm^{-1}. *It vanishes abruptly at the smectic A – liquid transition.* And Schnur and Fontana,[49] studying terpthal-bis-butyl aniline in the low-frequency, lattice-vibration spectrum region, observed that a number of bands in their spectra were almost the same in the smectic B phase as in the crystal. At the smectic B–smectic C transition most of this structure disappeared. Evans has reported the changes in the spectrum of 4-cyano-4-n-heptyl biphenyl as the sample was slowly cooled into the solid (Fig. 7). A constant magnetic field (600 G) was applied throughout the cooling process. It was anticipated that a uniaxial single crystal might be produced in this way. In this process the most intense absorption, at ca. 180 cm^{-1}, gradually shifts to the strongest lattice mode at ca. 165 cm^{-1} in the solid (Fig. 7 shows only two spectra), firmly establishing the common molecular origin of these two modes.

It is important to realize that the *local molecular order* may be as pronounced in the *isotropic* as in the nematic phase of a liquid crystal, and that the long-range order itself, which is characteristic of the nematic phase, is associated with and dependent on this well-established local order.

[‡]See also R. Chang, *Mol. Cryst. Liq. Cryst.* **12**, 105 (1971).

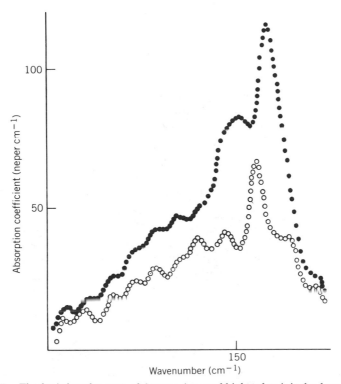

Figure 7.   The far-infrared spectra of 4-cyano-4-*n*-octyl biphenyl as it is slowly cooled into the solid phase, with a field of 600 G applied throughout the cooling process: ●, 8 CB near the freezing point; ○, a single cyrstal of 8 CB. (Reproduced by permission from ref. 45.)

B.   Isotropic Liquid Systems

What of isotropic, small-molecule molecular liquids? There is considerable evidence in the literature for some liquids that suggests that a significant local order exists extending to molecules that do not belong to the nearest-neighbor shell.[‡] For example, let us look at acetonitrile. It has a large dipole moment (3.9 D). Thermodynamic,[50] dielectric,[51] other spectroscopic,[52] and X-ray- and neutron-diffraction experiments[53] all reflect a strongly oriented structure in the pure liquid. For example, light scattering and nuclear magnetic[14] N relaxation probe the tumbling motion of the C_3 axis. In this

---

[‡]A. R. Ubbelhode (*The Molten State of Matter*, Wiley-Interscience, New York, 1978) presents a dearth of other precise experimental methods for elucidating molecular texture in quasi-crystalline assemblies such as the melt.

instance

$$\tau_{LS} = g_{00}^{(2)} \tau_S$$

where τ_S is the single-particle orientational time and $g_{00}^{(2)}$ is the second-rank structure factor. If τ_{LS} and τ_S are studied in a solution of acetonitrile in a solvent that shows weak depolarized scattering and does not form complexes with acetonitrile, τ_{LS} and τ_S should become equal at infinite dilution, when pair correlations are negligible. It is observed[54] that the factor $g_{00}^{(2)}$ (from the ratio τ_{LS}/τ_S) at lower concentrations is unity; that is, pair correlations vanish. In pure CH_3CN $g_{00}^{(2)}$ reaches a value of 1.4. This value of $g_{00}^{(2)}$ indicates that a more parallel or antiparallel configuration is favored (since $g_{00}^{(2)}$ is symmetrical with respect to $90°$, it is impossible to distinguish between the two cases). Further information comes from the ordinary $g_{00}^{(1)}$ factor, which is 0.78 for acetonitrile at room temperature. This negative deviation of $g_{00}^{(1)}$ from unity suggests that antiparallel alignment is the most likely structure. That $g_{00}^{(2)}$ deviates more from unity than does $g_{00}^{(1)}$ can be understood if pair ordering extends to neighbors *that do not belong* to the first shell of reference particles. That is, there is a suggestion of a significant local order in CH_3CN arising from molecular-shape considerations, as discussed by Warner and Wulf, and from the strong intermolecular forces due to the large dipole moments.

Information about pair correlations, of course, may be obtained from diffraction experiments, results of which have been reported for acetonitrile. To obtain this information the maximum number of diffraction experiments must be carried out. Neutron experiments on different isotopes of the substance must be performed together with X-ray-diffraction analysis. For acetonitrile X-ray data are available, as are neutron-diffraction data for CD_3CN, $CH_3C^{14}N$, and $CD_3C^{15}N$. These data allow three coefficients of the molecular pair-correlation function to be calculated: $g_{00}^{(000)}$, the center–center correlation term; and the orientational correlation terms $g_{00}^{(101)}$ and $g_{00}^{(202)}$. For acetonitrile $g_{00}^{(000)}$ is composed of two fluctuations: one at 4.7 Å, corresponding to a first-nearest-neighbor peak; and a second at 8.5 Å, corresponding to a second-neighbor peak. The orientational correlation terms give information concerning the orientation of a molecule relative to the center–center system irrespective of the orientation of the partner molecules. For $0° \le \gamma \le 90°$, $g_{00}^{(101)}$ is negative, and for $90° \le \gamma \le 180°$, $g_{00}^{(101)}$ is positive. Also, for $0° \le \gamma \le 54.7°$ and $125.3° \le \beta \le 180°$, $g_{00}^{(202)}$ is positive, but for $54.7° \le \gamma \le 125.3°$ it is negative. Thus, below 4.4 Å preferred orientations of the dipole axis relative to the center–center line are found in the range $90° \le \beta \le 125.3°$, whereas from ca. 5.2 Å preferred orientations of $0° \leq \beta \leq 54.7°$ dominate. A second reversal is observed at 6.8 Å that restores the initial situation.

Another X-ray investigation[53(ii)] revealed a short-range structure consisting of a bundle of five molecules in which a central molecule is surrounded by four antiparallel neighbors. A pronounced correlation is observed between these five molecules. The cluster diameters are 11 Å. Each cluster has eight molecules occupying sites in an orthorhombic unit cell with 95% site occupancy. This is in good agreement with the cluster diameter D of 13 Å estimated using a theory of Loshe[55] based on dipolar interactions in liquids.

P. Ignacz[56] has proposed a new theory for the liquid state based on this postulate of a significant local structure in liquids. The fundamental hypothesis is that the physical behavior of liquids is determined by their molecular structure. Both electrostatic and steric processes participate in the formation of the structure. Ignacz suggests that during melting the solid crystal decomposes into uniform octahedral groups of molecules. These groups or elementary crystals he calls "liquid grains," and he shows that the cubic lattice of molecules built up by octahedral formations is complete without any lack or surplus of molecules. The grains, and within them the molecules and atoms, are in thermal motion. The six peripheral molecules of a grain are in vibration relative to the central molecule. If the value of the vibrational energy exceeds the bond energy, then the grain disintegrates. The thermal motion of grains can be described in terms of the laws of thermodynamics. At the boiling point a second *crystalline* change is assumed to occur during which all the grains decompose totally. Below the boiling point, distances between the peripheral molecules of neighboring grains are not essentially larger than distances within the grains; the ratio is about 1:1.3. These distances exceed those between molecules in a solid crystal by less than 10%. This is in accord with diffraction studies on liquids in which local structure and the distances of neighbors appear to remain approximately constant relative to those in the corresponding solids. Since the grains get close to one another in the liquids, their peripheral molecules have a mutual influence. This influence can be shown statistically to be attractive, which explains the cohesion of liquids.

Let us return to our discussion of acetonitrile. The existence of this pronounced local structure must complicate significantly any attempts to obtain single-particle properties (e.g., single-particle correlation times) using the various spectroscopies. In fact several types of relaxation can and do contribute to the decay of the correlations observed in the various experiments. Furthermore, these relaxations contribute differently to correlation functions determined from different experiments. The experimental situation is a complex one, and considerable uncertainty is still associated with the experimental measurements, with the reduction of the data to a form suitable for comparison with theory, and with the theories themselves, which are often oversimplified. What we suggest now is a contribution of collective molecular motions to spectral profiles even in simple isotropic liquid systems.

Computer simulation (see the first article of this volume) shows that rotational motion is strongly coupled to translation in liquid CH_3CN. The moment of inertia of CH_3CN about the unique (C_{3V}) axis is about 18 times smaller than the other two (equal) principal moments of inertia. The anisotropy of rotational motion consequently is large and must involve simultaneous translation of the molecular center of mass. Mixed correlation functions with linear and angular momentum components exist in a moving frame of reference fixed in the molecule. For convenience this may be defined by the three principal moment of inertia axes. Vector quantities such v (the center of mass velocity), J (the angular momentum), F (force), and Tq (torque) may be defined in either the laboratory or the moving frame. We denote the components of v, for example, in the laboratory frame by v_x, v_y, and v_z; and in the moving frame by v_1, v_2, and v_3. If we define three unit vectors e_1, e_2, and e_3 with respect to the frame $(1,2,3)$, the velocity components are then related by

$$v_1 = v_x e_{1x} + v_y e_{1y} + v_z e_{1z}.$$
$$v_2 = v_x e_{2x} + v_y e_{2y} + v_z e_{2z}$$
$$v_3 = v_x e_{3x} + v_y e_{3y} + v_z e_{3z}$$

In symmetric top CH_3CN $|e_1| = |e_2|$, so that $v_1 = v_2$. In these relations e_{1x} is the x component of e_1 in the laboratory frame and so on.

Having made the transformation into the moving frame for each vector v_1, J, F, and Tq, we can simulate the various auto- and mixed correlation functions of interest, for example, the function $\langle v(t)J^T(0)\rangle / (\langle v_i^2(0)\rangle^{1/2} \langle J_i^2(0)\rangle^{1/2})$, where the angle brackets denote a running-time average over vector components defined with respect to the moving frame of reference $(1, 2, 3)$. The functions describe in detail the interaction of rotation with translation in acetonitrile molecules in the liquid state. By symmetry, all the elements of $\langle v(t)J(0)^T\rangle / (\langle v_i^2(0)\rangle^{1/2} \langle J_i^2(0)\rangle^{1/2})$ vanish except the $(1,2)$ and $(2,1)$ elements. Figure 8 shows that these elements are mirror images of each other peaking at ± 0.23 and vanishing by symmetry at $t = 0$.[57] The existence of these functions provides direct information on rototranslational and cooperative molecular motion. The molecular dynamics of a heteronuclear molecule such as CH_3CN in the condensed state *may not* be approximated by the customary, and still popular, theories based on rotational diffusion alone. Evans[58] and Ferrario[59] have recently developed the tradi-

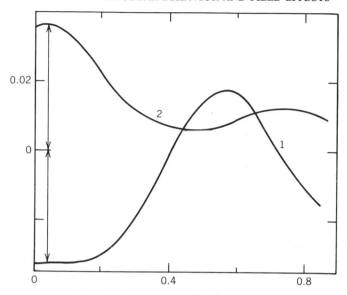

Figure 0. Pronounced rotation–translation coupling in CH_3CN as displayed by computer simulation, with $\langle \mathbf{v}(t)\mathbf{J}^T(o)\rangle$ in the moving frame. 1, (1,2) component; 2, (2,1) component. These functions provide directly information on rotation–translation coupling on the molecular scale. This coupling is the cause of cooperative molecular motions. (Reproduced by permission from ref. 57.)

tional Langevin and Fokker Planck equations so that they may be used with rototranslation. As Berne and Montgomery[60] say, "spectroscopic techniques rarely if ever determine the pure reorientational correlation functions

$$C_\ell(t) \equiv \langle P_\ell(\mathbf{u}(0)\cdot\mathbf{u}(t))\rangle,$$

where $P_\ell(x)$ is the Legendre polynomial of order ℓ and $\mathbf{u}$ is a unit vector embedded in the molecule. Instead the correlation functions

$$C_\ell(\mathbf{k}, t) \equiv \langle P_\ell(\mathbf{u}(t)\cdot\mathbf{u}(0))\exp[i\mathbf{k}\cdot\Delta\mathbf{r}(t)]\rangle$$

are determined where $\Delta\mathbf{r}(t) = \mathbf{r}(t) - \mathbf{r}(0)$ is the displacement of a molecule in a time t and $\mathbf{k}$ is the wave vector defining the scattering between the probing beam and the molecules of the fluid." Berne and Montgomery's calculations were based on a rough sphere model and did not take molecular shape implicitly into account. They add, therefore, that

real molecules, unlike rough spheres, sweep out a volume larger than the molecular volume when they rotate. Thus in dense fluids volume fluctuations must occur if these molecules are to rotate in certain directions. This should lead to a much stronger coupling between rotational and translational motions in real molecular liquids than in rough sphere fluids. And, by establishing the importance of coupling in rough sphere fluids [which they did], it becomes all the more obvious that these effects should be heeded in real molecular liquids.

So, there exists a pronounced interaction of rotation with translation and a significant local molecular structure in acetonitrile. This structure may persist in solutions of 10% and lower concentration, may actually be increased, and varies with the solvent used. For example, the relaxation time in benzene (25% volume/volume) is 6.7 ps at 297 K, decreasing to 3.1 ps at 333 K. In CCl_4 under the same conditions the relaxation times are 7.6 and 4.5 ps, respectively. Care must be exercised in using solvents to eliminate correlation effects, as is customary in, for example, nuclear magnetic resonance (NMR) experiments. Solvents may actually increase the apparent molecular aggregation.[15(b)] As Michel and Lippert[36] have said, "the environment of a single acetonitrile molecule is lattice-like and even solutions are influenced by this structure...in solutions in organic solvents a lattice-like orientation with mainly axial effects is dominant. Small molecules like LiBr, HBr, Br_2 and water take the lattice place of one acetonitrile molecule." Acetonitrile behaves as a waterlike organic solvent. Its local structure is extremely stable and survives even if the local lattice is disturbed by the interaction of ions in the system. The local lattice is stabilized by the strong electrostatic interactions arising from the large dipole moments of the molecules, just as the structure of water is stabilized by the three-dimensional hydrogen-bonded network. The structure of water is so stable that at ambient temperature and pressure a saturated solution of cesium fluoride in water contains ions and water molecules in the ratio 1 : 1.2. This extraordinary stability arises, of course, because the two particles are isomorphous and can be interchanged in the local lattice structure.

The stability of the CH_3CN local structure in CCl_4 is displayed in Fig. 9. The plot remains linear up to 20% (volume/volume) acetonitrile in CCl_4. Only thereafter is a striking nonlinearity observed as the local structure is disrupted. The frequency of maximum absorption of the far-infrared spectra is denoted as $\bar{\nu}_{max}$. If such spectra could be analyzed with purely rotational models for the molecular motions then, as Reid[61] has discussed, this feature of the spectrum could be related in a simple way to intermolecular mean-square torques. However, as we have established, such absorptions are in fact rototranslational in origin and, in the absence of a suitable theoretical model for such motion, $\bar{\nu}_{max}$ presently has an unknown dynamical significance. But

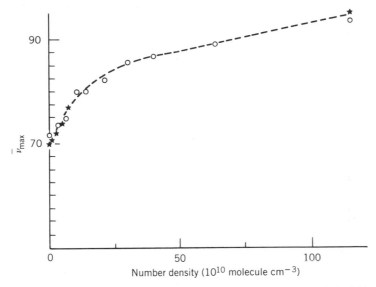

Figure 9   Variation of $\bar{\nu}_{max}$ with number density for CH_3CN in CCl_4: ●, data of Evans;[13] ○, data of Yarwood (personal communication). $\bar{\nu}_{max}$ is the frequency of maximum absorption of the far-infrared spectrum. The nonlinearity of the plot demonstrates the breakup of the liquid CH_3CN local structure in an approximately 20% solution. At this point CCl_4 molecules replace CH_3CN molecules at local lattice sites. (Reproduced by permission from ref. 13.)

we can be sure that it reflects in some way local changes in the intermolecular interactions, because the far-infrared frequency region is the energy range of intermolecular interactions and the time range of collisional and molecular relaxational processes.

Following the example of Lippert, it is revealing to compare the liquid and solid phases of CH_3CN. Two solid phases are known (α and β), though an additional γ-phase has been postulated. The α-phase is orthorhombic, with eight molecules per unit cell; the β-phase is probably a glassy phase. Comparisons of the α-phase and β-phase spectra indicate perturbations of the intermolecular symmetry. There is a loss of long-range order in the β-phase that reduces the degree of coupling of lattice vibrations. Lippert has calculated values for *inter*molecular vibrations for the α-, β-, and liquid phases. The intermolecular distances along two axes in the liquid are the same as those in the α-solid. In one perpendicular direction, however, the distance is significantly longer. Consequently, in this direction the frequency of the *inter*molecular vibration is reduced from 116 cm^{-1} in the α-solid phase to 87 cm^{-1} in the liquid. The frequencies corresponding to the two intermolecular distances that are approximately the same in the liquid and β-solid are the

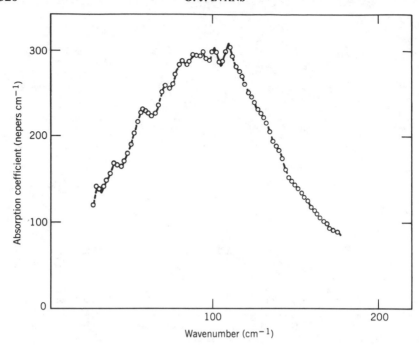

Figure 10.   Absorption spectrum of CH_3CN (isotropic liquid state) at far-infrared frequencies. Features are resolved on the low-frequency side of the spectrum. (Reproduced by permission from ref. 62.)

same (50 cm^{-1}). Are such *inter*molecular vibrations resolved in spectra of the liquid? If they were they would provide directly information on the intermolecular forces in the liquid.

Evans[13] has rerecorded the far-infrared spectrum of CH_3CN neat and in solution. He suggests that spectral details are resolved on the low-frequency side of the overall spectral contour for the neat solution and in the region where they have been estimated (Lippert, above) (Fig. 10). These details disappear at sufficiently low concentrations (< 10% volume/volume) of CH_3CN in CCl_4 (Fig. 11), as one would anticipate. It is encouraging that Evans's spectrum for the neat liquid is similar to that of Yarwood when the same window materials are used, even though the measurements were recorded on different makes of spectrometer.[62] It is also encouraging that the decay of the contour at higher frequencies is very smooth, which gives some indication of the low noise level of the instrument. However, having said this, we must emphasize that acetonitrile is a strongly absorbing liquid, so we must work at the absolute limits of present-day instrumentation. Only higher-precision apparatus will confirm or refute these results.

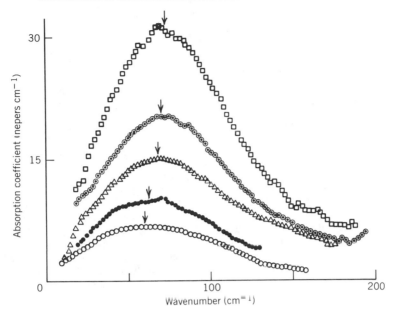

Figure 11. Far-infrared spectra of CH_3CN in CCl_4 at number densities between 7.2 and 0.6×10^{20} molecules cm^{-3}: $-\diamondsuit-$, 7.2×10^{20}; $\odot$, 4.9×10^{20}; $\triangle$, 3.3×10^{20}, $\bullet$, 1.2×10^{20}; $\bigcirc$, 0.6×10^{20}. *Note:* The resonant modes at lower frequency disappear in dilute solution as the pronounced local structure is disrupted. (Reproduced by permission from ref. 13.)

Gerschel[63] is presently reporting "the existence of an unexpected resonant absorption in CH_3Cl at millimetre frequencies" (Fig. 12). CH_3Cl is, of course, another strongly polar molecular liquid. Gershel's results, obtained using a carcinotron source available in Warsaw, are of high precision and high resolution. He explains his resonant millimeter absorption in terms of a dynamical coupling effect that is basically the same physical process as that assumed responsible for the appearance of long time tails in velocity autocorrelation functions (Fig. 13). The usual exponential behavior of the angular-velocity autocorrelation function at long times is replaced by one that varies roughly as $t^{-5/2}$. It is observed in the computer simulation of Tresser[64] that the more elongated the rotary molecule is, the more pronounced is the deviation from logarithmic decay. Because a molecule's rotation is coupled to its translation, the effect of a molecular rotation must be to cause neighboring molecules to translate in a coherent fashion; that is, a vortex is set up in the surrounding ensemble in which there is a marked degree of correlation between the linear velocities of the molecular centers of mass. Using generalized hydrodynamics, Ailawadi and Berne[65] have shown the normalized angular-velocity correlation function $\langle \omega(t) \cdot \omega(o) \rangle / \langle \omega^2 \rangle$ to decay

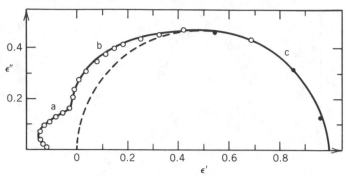

Figure 12. Normalized Cole–Cole plots of the complex permittivity for CH_3Cl at 273 K: —, experimental results; ---, predictions based on Debye's equation. Note the resolution of three absorptions, a, b, and c: a is the normal, far-infrared "Poley absorption"; b is a new resonant absorption in a polar liquid and is multimolecular in origin; and c is the Debye relaxational process. In terms of the normal absorption versus wavenumber spectra this resonant absorption peaks at < 20 cm^{-1}. It is important to realize that the frequency regions 1–10 cm^{-1} and 10–100 cm^{-1} are *two decades* of frequency. It is common practice, however, to characterize this whole frequency span with about ten data points, as above. High-precision, high-resolution results on such liquids are still required.

asymptotically as

$$C\omega(t) = \frac{d\pi I}{2mn}[4\pi(D+\nu)]^{-(d+2)/2}t^{-(d+2)/2}$$

where I, m, n, ν, D, and d are, respectively, the molecular moment of inertia, molecular mass, number density, kinematic shear viscosity ($\nu = \eta/mn$), self-diffusion coefficient, and dimensionality of the system. For planar motion, $d = 2$, and in space, $d = 3$. It follows that the orientational correlation functions

$$C_\ell(t) \equiv \langle P_\ell[\mathbf{u}(t)\cdot u(0)]\rangle$$

also behave asymptotically as $t^{-(d+2)/2}$, *if the effect of molecular translation on rotation is accounted for*. This asymptotic behavior of $C_\ell(t)$ was calculated from the "hydrodynamic equations" used to describe a feature of depolarized light scattering called the "shear doublet" or "Rytov dip," observable at low frequency (Fig. 14). It is proposed that this dip originates from local strains, set up by transverse shear waves, that are relieved by collective reorientations—a macroscopic or collective translation–rotation coupling. This shear dip does not appear to be related to a specific microscopic structure and has been observed in a number of molecular liquids, including

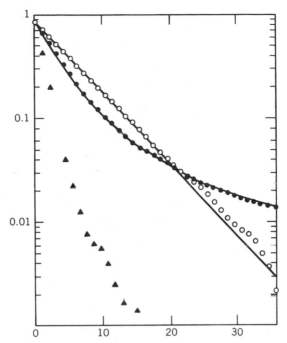

Figure 13. Semilogarithmic plots of the orientational autocorrelation function (▲), the velocity autocorrelation function (●), and the angular-velocity autocorrelation function (○) from a molecular-dynamics simulation of a two-dimensional system of 1600 diatomic molecules with periodic boundary conditions, atom–atom potential. (Reproduced by permission from ref. 64.)

polar and nondipolar liquids, and liquids containing planar and nonplanar molecules. It would be interesting to characterize the feature in terms of the symmetries of the molecules concerned and to discern whether in fact these observations have molecular origins. The shear waves are analogous to transverse phonons in a crystal.

Evans[62] reports that a *series* of absorptions appear on the low-frequency side of the far-infrared spectrum of CH_3CN. There also appear to be significant polarization effects. Such observations had been predicted by Ascarelli,[66] who pointed out that if a spectrum is composed of collective modes then there should be observable differences when different configurations of the polarized radiation are used. Also, a series of absorptions might be anticipated if the motions of aggregates of molecules are indeed contributing to the spectrum. A whole set of intermolecular vibrational transitions that may strongly but *incompletely* overlap is then possible. Michel and Lippert[36] propose the existence of collective modes in this liquid to explain the presolid state that exists just before freezing and does not depend on the

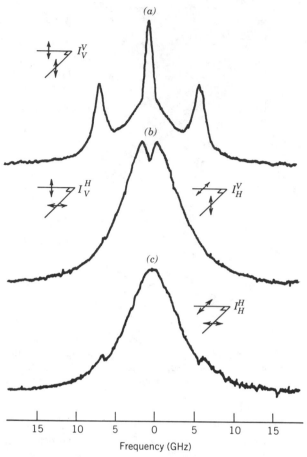

Figure 14.   High-resolution polarized and depolarized spectra of quinoline showing (c) deviation from a simple Lorentzian and the emergence of (b) a "shear wave" or "Rytov dip" in VH and HV spectra. [Reproduced by permission from G.I.A. Stegeman et al., *Phys. Rev.*, **A7**, 1160 (1973).]

temperature of freezing. As we saw in the last section, the freezing point varies and a significant hysteresis occurs on heating and cooling the material through the melting point. A sudden increase of the maximum intensities of the normal infrared modes coupled with a sudden decrease of the combination modes occurs near the freezing point. Michel and Lippert postulate the existence of strong cooperative phenomena originating from a coupling of the molecular oscillations and state that "collective modes might be the basis of the cooperative phenomena.... the discussion of collective modes leads us to a model of waves in liquid acetonitrile.... different waves must interfere and because of this the waves themselves will fluctuate. The dimension of the dynamical order in the liquid can be described through the size of the region of the expansion of the phonon-like waves." Fini and Mirone,[67] using isotropic and anisotropic Raman scattering, have also identified cluster vibrations on the shoulders of the ν_2 and ν_4 fundamentals.

Lobo et al.[68] have presented a theory of collective molecular behavior in the liquid state that predicts the existence of collective modes. Starting from the dynamical extension of the Onsager theory developed by Nee and Zwanzig,[69] but generalizing it so as to be applicable at high, far-infrared frequencies, they studied the dynamical dielectric response of a condensed system of molecules with permanent electric-dipole moments. The dielectric function so obtained, and consequently the far-infrared absorption, contains a diversity of both collective and single-particle behaviors. This supports the conclusions that Knozinger et al.,[70] Evans,[13] and Gerschel[63] arrived at from experimental considerations. The first two propose that *the submillimeter band is composite in nature*, with contributions arising from single-particle motions, the motions of dimers, and the motions of higher aggregates of molecules. Evans et al. discuss the first in detail in their text *Molecular Dynamics*.[11] The existence of dimers of CH_3CN has been postulated *even in the vapor phase*. Rowlinson's[71] second-virial-coefficient calculations for ΔU locate the maximum value of the dipole–dipole energy interaction at 4640 cal mol^{-1}, *which is larger than even that for water*, 4440 cal mol^{-1}. Interestingly, vibrations of the dimeric water band have been isolated, and at far-infrared frequencies. Harries et al.,[72] in a study of the Earth's atmosphere using the sun as a source, reported an absorption feature between 7 and 9 cm^{-1}, which they attributed to the dimeric water species. Dimers and higher clusters of water molecules are known to exist in isolation in the Earth's atmosphere. Jakobsen and Brasch[73] have also proposed a contribution to the far-infrared profile in acetonitrile from dimers of molecules. Bulkin[74] tested their hypothesis by measuring five aliphatic nitriles. He found that if the spectra of dilute solutions of CH_3CN were measured in nonpolar solvents and concentrations were such that the product obtained by multiplying concentration by pathlength was kept constant, then it was not possible to measure a decrease in the intensity of the absorption band. He explained this observation in terms of a simple monomer-dimer equilibrium.

A contribution to the spectral profile from higher aggregates of molecules is established if, following Knozinger et al., one studies the temperature dependence of the spectra at higher concentrations, when aggregates are most likely to exist. Both total band intensity and frequency of maximum absorption are seen to decrease when the temperature is raised between 263 and 313 K. In addition there is an increase in the band intensity at 263 K and a decrease in the total band intensity at 313 K when one increases the concentration; at the same time, an increase of concentration shifts the band maximum to higher wavenumbers independently of the temperature applied. All of this can be explained only if *at least two* different types of aggregates are present. A *whole set* of intermolecular vibrational transitions are then possible. Intermolecular vibrations associated with vibrations of hydrogen bonds, in structures stabilized by hydrogen bonds, are well known. But the hydrogen bond is only one of the known intermolecular links. The acetonitrile structure is stabilized by strong dipolar interactions. It is easy to show, for example, that the large dipole moment associated with the molecule is actually necessary for the existence of the liquid at room temperature. So, should we not then have anticipated the existence of absorptions arising from fluctuations of this significant local molecular order? This order is so well established that dielectric experiments suggest that the relaxation times of CH_3CN in CCl_4 and benzene may actually be increased initially upon dilution[79]—the solvent apparently *encourages the association of solute molecules*. In much the same way, in the classic example of 2,6-lutidine in water, the lutidine molecules *increase* the water structure. Frank and Evans in 1945 compared this increase in structure with "the formation of an iceberg around a hydrophobic molecule."

The contribution of collective modes to far-infrared spectra implies that the wavevector **k** remains finite at these frequencies, as Berne and Montgomery[60] have proposed. Following Lobo et al.,[68] if we deal with a Kirkwood correlation sphere containing a large number of molecules and do not restrict the cavity region to the volume around only one molecule, the relevant O-THz correlation function is

$$\frac{1}{k^2\mu^2}\left\langle \mathbf{k}\cdot\boldsymbol{\mu}_i(t)e^{i\mathbf{k}\cdot\mathbf{r}_i(t)}\sum_j \mathbf{k}\cdot\boldsymbol{\mu}_j(0)e^{-i\mathbf{k}\cdot\mathbf{r}_j(0)}\right\rangle$$

where the sums run over all the molecules within the Kirkwood sphere centered on $\mathbf{r}_i(0)$ and $\boldsymbol{\mu}_i(t)$ and $\mathbf{r}_i(t)$ are Heisenberg picture quantities. The $\mathbf{k}\cdot\boldsymbol{\mu}$ terms appear because it is the polarization charge density that now responds to our scalar probe *and not* the polarization, as tacitly assumed in more simplistic approaches. For $\mathbf{k} \to 0$ we are left with a *time-dependent Kirkwood correlation factor*

$$\left(\frac{1}{3\mu^2}\right)\left\langle \boldsymbol{\mu}_i(t)\cdot\sum_j \boldsymbol{\mu}_j(0)\right\rangle$$

This may be evaluated only by computer molecular dynamics. The familiar

Kirkwood static correlation factor

$$\left(\frac{1}{\mu^2}\right)\left\langle \mathbf{\mu}_i(0)\cdot\sum_j \mathbf{\mu}_j(0)\right\rangle$$

is recovered only at $\mathbf{k} = 0$. This gives us *no information concerning the dynamics*. As Lobo et al. say for H_2O, "symmetric orientation about the tetrahedral directions, by whatever mechanism and however slow, would suffice to give the Kirkwood value of the static correlation factor, which accordingly implies very little about the dynamics."

Consider a fluid made of ions. Here we may define a charge density (in c.g.s. units) via the Poisson equation

$$\nabla\cdot\mathbf{E} = 4\pi q\delta p$$

where q is the particle charge and $\mathbf{E}$ an external applied field. A term proportional to $\mathbf{E}(\mathbf{r}, t)$ must be applied to the Navier–Stokes equation. When solved, this leads directly to a plasma frequency

$$\omega_p^2 = \frac{4\pi pq^2}{M}$$

where M is the particle mass. The frequency of collective oscillations does not vanish as $|\mathbf{k}| \to 0$ in this case, but approaches ω_p. This collective oscillation in the ionic fluid is called the plasmon and may be simulated by computer molecular dynamics. In some molten salts it is possible to separate and identify the fluctuations in density and those in charge density. The former determine sound-wave propagation; the latter determine the plasmons and are the liquid-state analogues of the acoustic and optic longitudinal phonons in an ionic crystal. These show up in the far infrared for ionic crystals and have been identified, for example, by Wegdam and van der Elsken.[75]

There is a great similarity between ionic melts and molecular liquids. The dominant interaction in the ionic melt is the Coulomb interaction between the charged particles, which is attractive for unlike charges and repulsive for like charges. The induction effects must be larger than in an organic molecular liquid, because the dipole induced by a charge is larger than that induced by a dipole. Even in a dilute gas the ions are associated with dipoles, quadrupoles, or higher multipoles. It is because the interaction between the multipole associates dies off rapidly with distance that ionic melts and molecular liquids are so similar.

By similar analogy, therefore, if we consider a molecular liquid and substitute for the charge density the polarization charge density, as proposed above, then the plasmon frequency of polarization-charge-density fluctuations in the molecular liquid is the liquid-state analogue of the longitudinal optical phonon mode that may be observed in the far-infrared spectrum of a molecular crystal. Molecular crystal phonon modes are classifiable into those of translational and those of rotational origin.

The predicted frequencies of Lobo's longitudinal (ω_p) and transverse collective (ω_t) modes are

$$\omega_p^2 = \omega_t^2 + \omega_0^2$$

where

$$\omega_0^2 = 4\pi \left(\frac{N}{V}\right)\left(\frac{\mu_v^2}{I^*}\right)\frac{(\varepsilon_\infty + 2)^2}{9\varepsilon_\infty} \tag{I}$$

and

$$\omega_t^2 = \left(\frac{k_B T}{I^*}\right)\left(1 + \frac{2\tau_0}{\tau}\right) \tag{II}$$

where N/V is the number density of dipoles, μ_v is the value of the dipole moment in a vacuum, $(I^*)^{-1} = (I_1^{-1} + I_2^{-1})$ is the average moment of inertia for rotations of the dipolar molecule around an axis perpendicular to the dipole moment, and ε_∞ is the high-frequency dielectric constant of the liquid surrounding the dipole.

Ascarelli[66] studied nitromethane as a characteristic dipolar liquid with a large dipole moment (3.4 D) in which he anticipated collective modes might be isolated. The prerequisites laid down by Lobo et al. for the existence of collective modes were "physical systems composed of neutral molecules with large permanent dipole moments (~ 1D). They should also have small moments of inertia and large $\varepsilon_0/\varepsilon_\infty$." Ascarelli's results for the power reflection spectra of different solutions of nitromethane in CCl_4 are shown in Fig. 15. That the structure assigned to ω_p is collective was demonstrated by inserting a Mylar electret into the sample. The observed value of ω_p shifted from ca. 30 to 60 cm^{-1}, the Mylar electret producing a further aggregation of dipoles. Also, this feature completely disappeared when the light was polarized perpendicular to the plane of incidence, confirming its longitudinal oscillation assignment. The structure assigned to ω_t was present in all polarizations and was in excellent agreement with the absorption measured by Kroon and van der Elsken in a 0.7% solution of nitromethane in n-heptane by transmission methods. Insertion of the measured values of ω_p and ω_t for pure nitromethane into Eq. (1) gave a value for ω_0 that was in excel-

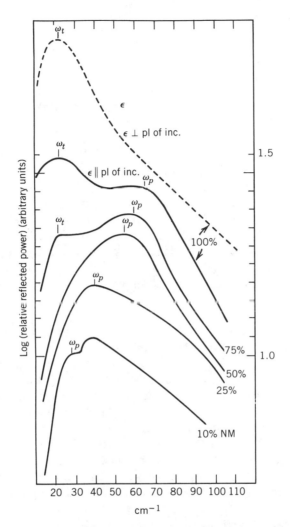

Figure 15. Power reflection spectra of different solutions of nitromethane in CCl_4. Polarization is either in the plane of incidence ($\parallel$) or perpendicular to the plane of incidence ($\perp$). The scale on the right side of the figure refers only to the spectrum of pure nitromethane with light polarized in the plane of incidence. Different spectra are displaced vertically for clarity. Note that ω_p shifts to lower frequency with dilution. ω_t remains approximately constant and in agreement with absorption measurements. (Reproduced by permission from ref. 66.)

lent agreement with the value calculated using independently measured parameters.

Two features of Ascarelli's measurements were not in accord with the theory of Lobo et al. First, the theory predicts sharp lines, whereas the observed spectra were much broader than this. Second, the spectra frequently showed extra structure that was not predicted by theory, in much the same way that Evans's spectra for CH_3CN showed a series of absorptions. It is important to recall, therefore, the assumptions used in Lobo's theory: "In any system, however strong the interactions, the 'elementary constituents' will evolve dynamically like free particles over some initial interval following an arbitrary zero of time. By elementary constituent we mean a physical entity whose internal structure is, or is treated as if it were, *fixed* and *constant*. In our present model, these elementary constituents are *rigid*, *nonpolarizable* molecules and this initial motion is free translation and free rotation." That is, flexibility of the molecules, the coupling of rotation and translation, and the polarizability of the molecules are not allowed for. Density fluctuations and the existence of intramolecular motions would certainly contribute to a broadening of the spectrum beyond the calculated value.

Before leaving this section, we must mention that several important new areas of research that may give information on these topics are now receiving considerable attention. The first is microcluster physics. Microclusters are defined as aggregates existing in appreciable proportions, where the aggregates may be atoms, ions, molecules, and so on that have lifetimes longer than the characteristic inverse frequencies for internal motions. This field is already causing a shift of interest in, for example, the very economically important field of catalysis from the study of surfaces as general adsorbing substrates with locally important electronic properties to the study of definite few atom features of cluster type known to be present on the surfaces of many active preparations. And insights are being gained into the nucleation process on an atomic level rather than through the more conventional and restricted continuum approach.

Microclusters composed of no more than a few molecules can now be prepared. As a consequence it is emerging that the thresholds for the onset of effectively "macroscopic" properties may, somewhat against our preconceptions, actually be in the microcluster size region. Melting transitions are surprisingly well-defined at the $N = 100$ level, and superconductivity and ferromagnetism have been shown to set in at cluster sizes that are surprisingly small to exhibit phenomena of long-range order. Wegener and co-workers[76] have carried out Rayleigh scattering measurements in controlled expansions of H_2O and observed that nucleation starts with a critical nucleus undoubtedly in the $N < 100$ range. It is interesting that helium clusters are the only species so far known unequivocally to be liquidlike under the conditions of nozzle expansion. (The clusters are made with high-velocity expanding jets. It is possible to produce selectively quite high concentrations of bound clusters from the dimer to the $N = 100$ range.) The result for helium is perhaps not so surprising if we recall from an earlier part of this review that it remains liquid down to 0 K and may be solidified only under considerable external pressure at this absolute temperature. Farges et al.[77] have produced argon clusters with size range below $N = 100$ and have determined a pentagonal symmetry giving way to a normal fcc structure at the level of some 1000 atoms.

The appearance of bulk properties on such microscopic scales is good news, of course, for the computer simulator, who, because of computer size and speed, is presently restricted to studying small ensembles of molecules. It is bad news for the hydrodynamicist, who does not consider the detailed structure of the medium, let alone the structure of molecules. Continuum theories *cannot* project short-range effects.

The second important new technique is the use of electrons as microscopic probes of the fluid.[78] The molecular structure in the fluid is reorganized only slightly to accommodate the electrons. It appears that even in condensed gases and supercritical vapors the electrons may be solvated and stabilized. On injection of the electrons, a cavity is created and the dipoles making up the inner walls of the cluster line themselves up in the field of the electron. Countereffects from repulsive dipole–dipole interaction also contribute to determining optimum cavity size, and the continuum, defined as the fluid outside the coordination shell, is polarized in response to this field of the electron. Thus, electrons residing in clusters can be viewed as microscopic probes of both the local structure and the molecular dynamics of the liquid. This enables us to study the evolution of the system from a single particle to a collective state. Trapped electrons in liquids exhibit far-infrared absorptions. Kenney-Wallace[78] says that "the evidence from solvated electron studies points toward a molecular model in which a delocalized electron is initially trapped within small molecular clusters, whose configuration relaxes under the influence of the electronic charge while the molecular structure grows. The excess charge may be distributed over a cluster of 4–12 molecules"—a conclusion in accord with many of the inferences we have made throughout this section. Kenney-Wallace also points out that "there is a symbiotic relationship between the excess electron and its supporting fluid, one which demands an understanding of the local structure prior to the addition of the electron. Not so surprisingly, such molecular details are seldom anticipated through a knowledge of only the bulk properties of the liquid."

In the presence of the solvated electron, inner molecules remain locked in their configuration for the lifetime of this electron, which at 295 K ranges from 10^{-9} to 10^{-4} s for most liquids—an indication of the lifetime of the local structure in a liquid. There is a characteristic vibration peaking from ca. 500 cm^{-1} upwards associated with the cluster as a whole. As an example of the results that may be obtained, the spectra and dynamics of a solvated electron in alcohol–alkene systems apparently remain unperturbed for dilutions $\chi_{ROH} < 0.15$, indicating that the structural integrity of the cluster is maintained. All absorptions undergo a shift to lower frequency with increasing temperature, consistent with the idea that the structure is loosened as the thermal energy is increased. When subjected to pressures on the order of kilobars, the solvated electron exhibits a shift to higher frequencies, implying a compression of the trap.

III. THE INTERACTION OF EXTERNAL ELECTRIC FIELDS WITH LIQUIDS

A. A New Crystal-Growing Technique and the Thermodielectric Effect

Nothing could be simpler, it would seem, than the description of an electric field between two large, identical, flat-ended, parallel metal electrodes separated by a highly insulating liquid and having different potentials. Conventional electrostatic theory leads us to expect a uniform field in such a situation. In reality, this ideal is but a fleeting transient condition, lasting perhaps a few nanoseconds. The field rapidly warps to a high degree of inhomogeneity in microseconds. This was shown by the workers at Grenoble[79] by using the Kerr effect on chlorobenzene. Pohl[80] emphasizes that at present, "experiment overshadows theory in the study of the behavior of real matter in the presence of real but non-uniform fields." We will discuss some instances of this in the following section, which is intended to catalyze interest in an exciting field still in its early stages of development. Quoting Pohl again:

> Non-uniform electrical fields produce unique, useful and frequently mystifying effects on matter—even on neutral matter. With non-uniform fields, for example, it is possible to classify and separate minerals, pump liquids or powders, produce images (xerography), provide an 'artificial gravity' useful in 'zero-g' conditions, clean up suspensions, classify microorganisms, and even separate live and dead cells. And this just starts the list. Applications in biophysics and cell physiology to studies of normal and abnormal cells are at an early but exciting phase. In colloid science the new technique is helping to resolve surface properties. At the molecular level, non-uniform field effects are seeing renewed use in determinations of molecular polarizabilities, in maser operations and in laser control.

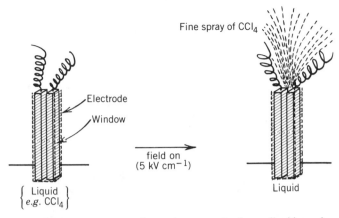

Figure 16. A simple laboratory experiment demonstrating how a liquid may be suspended against gravity or even pumped rapidly against gravity (as illustrated for CCl_4) in the presence of strong electric fields. Two brass electrodes are sealed with two insulating polymeric windows. (Reproduced by permission from ref. 13.)

Here we shall confine our discussions mostly to the author's own observations. We shall see that some of the phenomena we report may be explained in terms of ideas introduced in preceding sections of this review. We will confine ourselves to electric field effects but point out some similarities with magnetic field effects on magnetic fluids.

Figure 16 shows a simple electrode configuration that was used to suspend dipolar molecular liquids against gravity.[73] The two parallel brass electrodes were sealed with two insulating windows; the ends remained open. One end of this cell was immersed in a neat liquid and a static electric field (~ 5 kV cm^{-1}) was applied. Liquid was drawn up into the contained gap. For example, 11 cm of aniline was suspended in this way. For some nondipolar liquids (e.g., CCl_4) the effect is so strong that fine droplets of liquid are expelled very rapidly through the top of the two electrodes some 13 cm long. A liquid can be pumped against gravity in this way.

Insulating dielectric liquids are attracted from regions of lower to regions of higher electric field intensity *even in the absence of free electric charge.* This may be observed with a simple setup such as that shown in Fig. 17. In the presence of a field, liquid is drawn into the stronger-field region between the electrodes that are closest to each other. Jones et al.[81] have suggested that this effect occurs because of polarization charges. These authors demonstrated how the effect may be used to produce a dielectric syphon (Fig. 18). At large enough voltages, parametric surface instabilities set in. These are the electric analogue, in dielectric systems, of the surface instabilities of magnetic liquids in magnetic fields normal to the surface.[82] A destabilizing effect

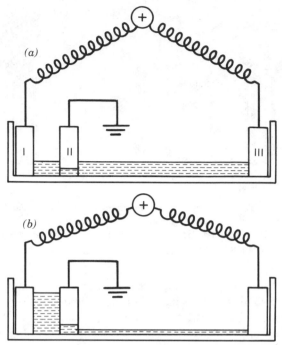

Figure 17.   An insulating liquid moves from a region of weaker to a region of stronger field strength. This may be demonstrated with a simple laboratory experiment. (*a*) Three brass electrodes (I, II, and III) are set in an insulating block of material. Electrodes I and III are fixed permanently in a rectangular hole machined in the block. Electrode II has a hole drilled in its base and may be moved between the two fixed electrodes. (*b*) When a field is applied, liquid fills the gap between the two electrodes in closest proximity to each other, where the field intensity is greatest.

is produced that results in the appearance of a periodic structure of liquid spikes.

Another well-known effect in magnetic liquids is fluid magnetic levitation. Consider the Bernoulli equation to which a magnetic energy term is added

$$p + \frac{pv^2}{2} + \rho gh - \mu_0 \int_0^H M\,dH = \text{constant}$$

where p is the pressure energy (work done by the liquid), $pv^2/2$ is the kinetic energy, ρgh is the potential energy, and the last term is the magnetic energy. Consider now the pressure term in conjunction with the magnetic term. The equation predicts that the greater the applied magnetic field is, the greater is

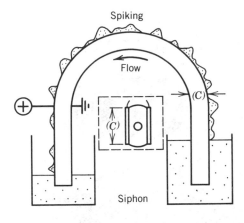

Figure 18. The dielectric siphon. The liquid flows in the direction shown. As it flows, the surface of the liquid exhibits parametric instability. The electrode arrangement is shown in inset (c). For photographs of the siphon in operation see ref. 81.

the pressure produced within the liquid. If a body is immersed in a magnetic liquid, it sinks or floats, depending on its density with respect to the liquid. Let us suppose it is more dense and sinks. If a magnetic field is now applied normal to the surface of the liquid and gradually increased, a gradually increasing field gradient is introduced, and at a strong enough field the body floats and eventually sits on the surface of the liquid, where the lowest pressure is found.

Evans[83] has used the electric analogue of these two effects, spiking and levitation, to induce the nucleation process. Consider the electrode arrangement schematically depicted in Fig. 19. For a liquid of nonzero conductivity, an electric field causes a current flow and, in our case, an accumulation of negative charge at the liquid surface. In addition, of course, polarized molecules travel to the surface of the liquid and reorient themselves in accordance with the direction of the field. As the electric field is increased, a force is exerted on the surface of the liquid. If the surface charge density is $\rho_s C/m^2$, this force is given simply by

$$F_s = \rho_s E$$

The force tries to pull the charges out of the surface; in a strong enough field, surface disruption occurs and, as illustrated in the figure, spikes appear.

If a saturated solution of a polarizable solid substance, for example, camphor, is prepared in nonpolar CCl_4 and an external electric field is applied using the electrode arrangement described above, the following occurs. The camphor molecules, or groups of molecules, are polarized and reoriented, and acquire, in effect, a negative charge on the side facing the upper positive electrode. The field is nonuniform, diverging across the individual (or groups of) particle and producing unequal forces on the two ends of the molecule(s).

(a)

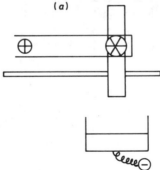

Figure 19. A new crystal-growing technique, as used for the growth of crystals of camphor. (a) The setup with no field. (b) A field is applied and a "spike" appears. Note that the PTFE rod used to induce this spiking acquires surface charge in the presence of the field. (c) The crystal growing. It may be drawn out of the solution or grown by steadily increasing the field. The maximum size of the crystal that may be grown depends on the size of and the distances between the electrodes.

The force on a small neutral body is given by[80]

$$\mathbf{F} = (\mathbf{P} \cdot \nabla) \varepsilon_e$$

where $\mathbf{F}$ is the net electric force on the body, $\mathbf{P}$ is the (constant) dipole vector, ∇ is the del operator, and ε_e is the external field. Assuming that the di-

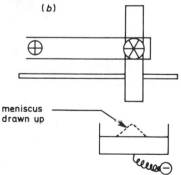

Figure 19.   (*Continued*)

electric body is linearly, homogeneously, and isotropically polarizable, that is, that

$$\mathbf{P} = \alpha \nu \varepsilon_e$$

where α is the polarizability (the dipole moment per unit volume in a unit

(c)

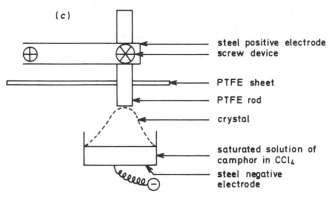

steel positive electrode
screw device

PTFE sheet

PTFE rod

crystal

saturated solution of
camphor in CCl₄

steel negative
electrode

Figure 19.   (*Continued*)

field) and v is the volume of the body, we can write

$$\mathbf{F} = \alpha v (\varepsilon_e \cdot \nabla) \varepsilon_e = \tfrac{1}{2} v \alpha \nabla |\varepsilon_e|^2$$

In a conservative (i.e., friction-free) field the force on the body is the negative of the gradient of its potential energy. Since $\mathbf{P}$ is a constant vector, and $\nabla \times \varkappa \varepsilon_e$ is zero in a static field, we have

$$\mathbf{F} = (\mathbf{P} \cdot \nabla) \varepsilon_e = \nabla (\mathbf{P} \cdot \varepsilon_e)$$

For an anisotropic body, of course, the polarizability α is a tensor and the calculation is more involved. However, for present purposes, this equation suffices to illustrate that a force exists impelling the particle into a region of stronger field.

Let us consider our saturated solution of solid camphor in the nonpolar solvent CCl_4. (The latter is chosen because of its strong response to an electric field, as already discussed.) The saturated solution is placed in a lower, cylindrical electrode made of stainless steel. The top electrode, also made of stainless steel, is isolated from the bottom one with an insulating sheet of PTFE (or similar material), (see photographs in Fig. 19). An insulating rod of the same material projects below this surface and is used as the source that induces the nucleation (Fig. 19); this rod acquires a surface charge. When the field is switched on (and gradually increased), particles of camphor may be seen floating on the surface of the solution—the electric levitation effect. Under strong enough fields a spike of the solution is drawn upward toward the PTFE rod. As the field is further increased, the spike grows, the CCl_4 is ejected out of the field region (just as it was in the first experiment discussed in this section), and a thin, uniform crystal of camphor remains. The samples are grown straightforwardly in a continuous manner using this technique. The crystals grow instantaneously when the field is increased, and instantaneously stop growing when it is removed. One therefore has great control over the growth of the crystal. There is a great potential for growing such samples *in situ* in many experiments.

If the drawn-up spike is allowed to touch the PTFE rod, the crystal may even be grown by keeping the field constant and raising the PTFE rod using a simple screw device. A crystal of required length may then be drawn out. The method resembles at this stage the Langmuir trough technique, developed by Irving Langmuir and Katharine Blodgett between the World Wars and now used to deposit monolayer films. This method has great potential in the field of electronics, where it is anticipated that new semiconducting materials can be developed to supplement the ubiquitous silicon. Some have speculated that our new crystal-growing technique may be used for such thin-film production.

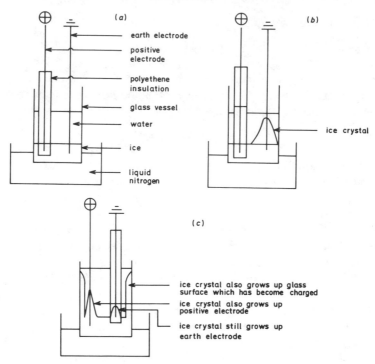

Figure 20.   Ice grows preferentially in the region of a strong field. (*a*) The setup with no field. A layer of ice is allowed to grow at the bottom of the glass vessel to seal the open end of the polyethylene insulation tube around the one electrode (thereby eliminating conduction processes) and to provide a liquid–solid interface before application of the field. (*b*) With a field applied and the positive electrode insulated, ice grows up the earth electrode. (*c*) With a field applied and the earth electrode insulated, ice grows up *both* electrodes and up the surface of the glass vessel, which must also become charged. [Reproduced by permission from G. J. Evans, *J. Chem. Soc., Faraday I*, **80**, 2343 (1984).

It was found necessary, at least for camphor, to insulate the top electrode from the bottom in the way described, because the crystals are easily broken and splinter if the fields are too strong. Evans[83] has even observed that ice crystals grow preferentially in a field region and around an immersed electrode (Fig. 20) if water is cooled too close to its freezing point in the presence of a field. And Rajeshwar[84] et al.[84] report that it is possible to "tune or even alter the thermal decomposition behavior of solid materials by imposition of electric fields across the heated samples." They found that the decomposition temperatures could be lowered by as much as ~100°C in the presence of the field. This is an important observation because of the significance of thermal decomposition of solid materials used in, for example, rocket propellants. Rajeshwar et al. report other observations:

Electric field effects have been implicated in various thermophysical and thermochemical phenomena in solids and in solid–gas or liquid interfaces. It has been reported that an electric field accelerates the rate of growth of $BaMoO_4$ crystals in silica gel media. A change in the sublimation pattern of KCl crystals has been observed in the presence of an electric field. The lower dielectric strengths observed for ionic metal azides relative to other more thermally stable materials have been attributed to 'electrocatalytic' effects induced by the presence of highly conducting metal nuclei. Selective phase formation at sample–electrode interfaces has been observed in a study of the effect of electric fields on the transformation of γ- to α-Al_2O_3. An electric field enhancement of dehydroxylation rates in magnesium and aluminum hydroxides, has also been reported.

Rajeshwar et al.'s differential scanning calorimetry (DSC) results for $KMnO_4$ in the presence and absence of dc fields are illustrated in Fig. 21. Note the drastic increase in the exothermicity of the autooxidation peak and the lowering (by $\sim 65°C$) of the peak temperature for the sample exposed to the electric field relative to that for the control sample. It is suggested that the efficacy of the external field in lowering the extrapolated onset temperature of the DSC peak is related to the extent to which charge transfer plays a rate-determining role in the very early stages of the decomposition, but this has not been proved.

Mechanisms for charging particles during the nucleation process have been proposed. For example, Workhom and Reynolds[86] found large potential differences of up to 230 V between the ice and water phases during the freezing of water. Costa Ribeiro[85] reports similarly that "in a condenser whose dielectric is partly in the solid and partly in the liquid state, one of the plates being in contact exclusively with the solid and the other with the liquid phase, if a phase change takes place at the boundary between the two phases, so that solidification or melting occurs at the interface, an electric current is produced as a consequence of the phase change." Costa Ribeiro asserts that electric charges always appear on the interface between the liquid and solid phases in a dielectric; he named this effect the thermodielectric effect. His results have similarities with those of Stephen Gray,[87] the discoverer of electrical conduction, who reported in 1732 that dielectrics acquire charge purely by melting and solidification.

Costa Ribeiro also claims that his thermodielectric effect may be observed not only in melting and solidification, but also in other changes of physical state in which one phase is a solid (e.g., similar potentials are produced in sublimation, as confirmed by the work of Rajeshwar et al. discussed above,

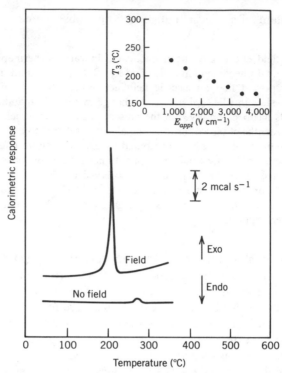

Figure 21. Electric field effects on the differential scanning calorimetry (DSC) results for KMnO₄. This is another example of the reciprocal effect of the thermodielectric effect of Costa Ribeiro.[85] The field applied was 2 kV cm⁻¹. The inset shows the dependence of the DSC peak extrapolated onset temperature, T_3, on the applied field, E_{appl}, for KMnO₄. (Reproduced by permission from ref. 84.)

and in the precipitation of substrates from saturated solutions, as in Evans's camphor–CCl_4 solution). Evans was unaware of Costa Ribeiro's experiments when reporting his induced nucleation effect.[83] However, Costa Ribeiro reported in his remarkable paper that "we have tried, but without success, to detect the reciprocal of the thermo-dielectric effect, that is: the production of melting or solidification, in a solid–liquid dielectric system, by application of a potential difference across the interface." This, the inverse of the thermodielectric effect, is precisely what Evans has now reported—the induction of solidification from a saturated solution by the application of strong external fields.

It is remarkable that Costa Ribeiro's work has not received the attention, scrutiny, and acclaim it deserves. His observations are of fundamental importance, yet little reference to his work is made in currently popular texts on liquids, solutions, and interfaces, or in the discussion of nucleation, where it is obviously of most importance. It is well known, however, that in crystal growth the structure at an interface has a major effect on the growth.[88] As Melcher et al.[89(a)] point out, an external field E "influences a liquid at an interface where the permittivity ε undergoes an abrupt change, and therefore $\Delta\varepsilon$ is not zero, as in the bulk. In some cases it is fruitful to think in terms of an analogy between dielectrophoretic forces (the force on a non-charged but polarizable molecule) and surface tension forces because, in a homogeneous liquid, the electric field produces only a surface force."

Costa Ribeiro proposed two laws concerning the thermodielectric effect and even postulates the existence of a new specific physical constant. His laws read thus:

1. *Law of Intensities*: "In isothermal melting or solidification, occurring at a constant rate, in a thermo-dielectric cell, the intensity of the current produced is proportional to the time rate of the phase change."

2. *Law of Charges*: "If a dielectric system passes from one state of equilibrium to another state of equilibrium, the total electric charge associated with the change of mass of one of the phases is proportional to this change of mass."

Costa Ribeiro proposes an internal, molecular mechanism for the phenomenon in terms of the Helmholtz double-layer hypothesis. When solid particles (e.g., camphor in Evans's experiment) are immersed in an insulating liquid (e.g., CCl_4), they may, because of the existence of a contact potential, become charged (if the conductivity of the liquid is nonzero) or polarized by selective adsorption of ions from the liquid or because of the adsorption of dipoles on the surfaces of the particles. Normally both processes are anticipated to contribute. When a particle has somehow acquired a charge or a dipole on its surface, ions or polarized molecules of the opposite sign will be attracted to the vicinity of the particle, but will be present in a diffuse layer around the particle. These two layers, a very thin, tightly bound primary layer and a loose secondary layer, together form the well-known Helmholtz double layer, which Helmholtz suggested is always formed at the surface between two phases of material. Ribeiro points out that a displacement of this double layer as a consequence of the phase change might account for the electric potentials observed during the phase change. As Ribeiro says, "there may be a difference in the electronic densities (cubic density of electrons loosely bound to the atoms) in the solid and liquid phases... this should permit a theoretical treatment of the subject similar to the method used in discussing cohesion of solids by means of the collective electron model of Bloch."

An excellent review on current approaches in applied quantum mechanics that adopt single-particle density as a basic variable, namely, density-functional theory, quantum fluid dynamics, and the study of the properties of a system through the study of "local" quantities in three-dimensional space, has recently appeared.[89(b)] The authors of this review write that

> in recent years electron density has attracted a great deal of attention in connection with its use as a basic variable in applied quantum mechanics. The appeal of this quantity as an attractive alternative to the quantum mechanical wave function is basically threefold:
>
> Firstly, it describes the 3D [three-dimensional] distribution of electrons in a system and hence is a function of only three coordinates irrespective of the number of electrons present. Thus density-based formulations offer a tremendous simplification over the usual wave function approach where the difficulty in solving the Schrodinger equation increases very rapidly with the number of electrons.
>
> Secondly, it is a fundamental physical observable and can be determined experimentally.[90] Thus the accuracy of quantum chemical calculations and approximations can be tested directly.
>
> Finally, being a function in 3D space it enables one to build up various interpretive models thus providing a 'classical' picture of quantum phenomena. The density plays a very important role[91] in many chemical and physical applications from both interpretational and computational points of view. Its vital significance has been emphasized by the statement:[92] 'A theory of chemistry and the chemical bond is primarily a theory of electron density.'

The interested reader is referred to this review for further details.

Because Costa Ribeiro observed electric charges in crystals formed by sublimation or by precipitation from a saturated solution, he was careful to emphasize that the thermodielectric effect occurs not only in processes of melting or solidification, but also *in other types of phase changes in which one of the phases is a solid*. He was unable to observe similar effects in changes of state not involving the solid state, such as vaporization or liquefaction, and so concluded that his effect must be related to the passage of the dielectric from a state of ordered structure, the crystalline state, to one of greater disorder, the liquid or gaseous state, or vice versa.

For some years after its appearance, no reference to Costa Ribeiro's work could be found in the literature. However, the effect reappears in the literature of the late 1960s. In brief, Dang Tran Quan[93] reported the existence of thermodielectric voltages and variations in the dielectric constants of some organic compounds near transition points when one of the states involved a solid. He observed that during cooling or heating at temperatures corresponding to transition points, radiofrequency measurements revealed abnormal variations of the dielectric constant. For 1-butanol and 1-bromobutane peaks corresponding to crystallization were observed at 134 and 127 K and peaks corresponding to melting at 183 and 162 K, respectively. For cyclopentanol and cyclohexanone, peaks were observed at 205 and 222 K, respectively, corresponding to solid-state molecular vibration–rotation transitions, and at 224 and 223 K, respectively, corresponding to solid–liquid transitions.

Cassettari and Salvetti[94] reported the design of an experimental setup that improves the study of the thermodielectric effect, and Mascarenhas[95] studied charge and polarization storage in solids over a wide range of phenomena and materials. Included was a study of the Costa Ribeiro effect, which he described as "charge storage during a phase transition."

Dias Tavares[96] provided experimental proof of production of the double space charge in dielectric and organic semiconductors due to the Costa Ribeiro effect and tested a theory of the distribution of such charges with two zero-field planes. He accomplished this by the liberation of charges inside the crystal using a focused beam of light that crossed thin layers of the crystal. The current produced changed sign twice when the beam swept over the thickness of the crystal, showing the existence of the two zero-field planes. Rozental and Cholin[97] showed that a reorientation of polar molecules determines the interfacial potentials and redistribution of ions in a two-phase system. The potentials detected by molecular polarization reached 10^5 V cm^{-1} for polar substances. Kapustin et al.[98] observed thermodielectric effects when nematic p-azoxyanisole crystallized. The space charge increased with the cooling rate, in accordance with Costa Ribeiro's conclusions, to a limit of 4×10^{-8} coulombs g^{-1}. And Garcio Francisco[99] reported use of the effect for solar-energy conversion. It is a fascinating thought that water, the most abundant of liquids, may yet contribute significantly to solving our world's energy problems. An efficient method for the sunlight-assisted electrolysis of water using a p-InP photocathode is already available. It uses 10.2 mW cm^{-2} of the incident 84.7 mW cm^{-2} sunlight for the production of hydrogen. The resulting engineering efficiency of 12% is the largest ever obtained for any scheme for conversion of sunlight to fuel. Green plants are able to convert 1–3% of the incident sunlight to combustible fuels. There is great interest in this area; the race is already underway to provide a scientific basis for a new large-scale energy resource option. The thermodielectric effect may make significant contributions to this race.

Eyerer[100] has reviewed the Costa Ribeiro and Workhom–Reynolds effects. He discusses experimental results, theoretical considerations, and potential applications of the thermodielectric effect, and reviews the Workhom–Reynolds effect in the water–ice system.

We will leave the last words on this particular effect to Costa Ribeiro himself: "From the standpoint of chemical-physics, the new possibility of associating a specific electric charge with changes of state of aggregation is certainly interesting... the existence of an electric double-layer in solid–liquid interfaces suggests also the possibility of investigating the correlation between the thermo-dielectric effect and other electrochemical phenomena of interfaces."

Colloidal solutions represent another interesting and important class of systems in which large electric field effects can be expected and from which crystals or films may be grown. Hauer and LeBeau[101] investigated the properties of films formed by the gradual extraction of water from benzonite solutions containing very small and uniform particles. They observed that a concentration was reached at which Brownian motion ceased and the particles tended to align themselves. On further removal of the water, the particles appeared to snap into position and form "crystallites" of highly anisometric shape. It would be interesting to carry out field experiments at various points of such an extraction procedure. Of course, colloidal particles are believed to be prevented from coagulating by potential barriers between them originating from their electrical charges—each particle is surrounded by an electric double layer that keeps the system as a whole electrically neutral. As Grisdale[102] says, "small amounts of ionic material, many times no more than are adventitiously present, suffice to create the double layers essential to long time stability of colloidal systems (*and also have profound influences on crystal growth*)."

Many biomolecular processes are consequences of high electric field effects. This follows from the role of the surface (the interface) in all cell structures, membranes, capillaries, and so on. Davies[103] writes that "such surfaces almost invariably carry, adsorbed, an excess of ions of one charge type: this could produce a field in excess of 10^6 V cm^{-1} at a distance of 5 Å from the surface." The molecular consequences can be considerable. Gregson and Krupkowski[104] studied lecithin, an important constituent of mammalian brain, of egg yolk, and of many other tissues. Its zwitterionic structure and long chains lead to its forming small inverted micelles in non-hydroxylic solvents. They found an interesting sequence of changes when the lecithin solution was subjected to a pattern of rectangular voltage pulses of width ca. 3×10^{-3} s and of amplitude varying up to 10^5 V cm^{-1}. Two differing perturbations of permittivity were observed. A small low-field perturbation was interpreted as a field-induced distortion of the quasispherical micelle whose polarity was so increased. The higher fields induced partial disruption, that is, fragmentation of the almost nonpolar micelle. These observations suggest a value for the effective field within the micelles (ca. 10^4 V cm^{-1}), offer the possibility of monitoring the kinetics of micelle formation, and illustrate the type of molecular-kinetic features that can be profitably explored by high-field methods.

Hirano[105] investigated crystal formation in a supersaturated electrolytic solution and analyzed the electrical behavior of the crystal–solution interface. Ubbelohde[106] has already postulated that charged microcrystals act as crystallization nuclei in such solutions and Hirano reported observations that appeared to support this view. Hirano reported the growth of crystals from a saturated solution of sodium nitrate (in water) that was electrolyzed between two platinum electrodes for 10 h in a field of 0.4–0.5 V cm^{-1}. He then charged a glass surface negatively and observed the "creeping phenomena."[107] He emphasized that silica glass, which has no ionic structure, failed to show the creeping. Note the analogies with the water-crystal growth in the presence of a field reported earlier (fig. 20). Crystals of ice grow up the sides of a glass vessel whose surface becomes charged by the high fields. Hirano also observed that crystallization occurred more rapidly in fields of high frequency, which seems to support the idea that charged nuclei exist in the solution and determine the crystal-growth process. The collision rate between such nuclei should be increased by an alternating field.

Ubbelohde[106] pointed out how important this mechanism for migration of nuclei may be in the deposition of crystalline masses at controlled places in animal systems. For example, it is possible that the deposition of crystals associated with certain diseases and the formation of teeth and bones are controlled by this phenomenon. That such a phenomenon is not confined to electrolytic solutions but also applies to organic molecular systems may have wide-ranging implications for plant and animal biology. It is interesting, for example, that citrus orchards have been protected[108] from low temperatures, which destroy unprotected control plants, by placement of high-frequency electric fields at the ends of the leaves so that high-frequency currents flow on the sensitive surface layers of the plants. A pronounced effect of electric fields on the growth of barley has been reported,[109] and it is well established that the common method of inhaling uncharged aerosols for the treatment of bronchial asthma, nasal and sinus conditions, bronchitis, emphysema, and other ailments of the respiratory tract is markedly enhanced by first charging the aerosols.[110]

External fields influence themselves most strongly at those interfacial regions in a liquid where the permittivity undergoes an abrupt change. However, electric field effects may also be observed in ordinary isotropic liquid systems, where interfacial effects would not be expected to be so pronounced. We have already considered some spectroscopic consequences of the interaction of external electric (and magnetic) fields on nematic liquids. Intensity changes occur, and at far-infrared frequencies, collective modes are resolved. But what of spectroscopic effects of electric fields in isotropic, small-molecule molecular liquids? In 1980 Evans and Evans[111] reported electric-field-induced intensity changes in the far-infrared spectrum of an isotropic liquid, aniline (Fig. 22). The effects are surprisingly large. And Michel and Lippert[36] reported that the π-electrons of the nitrile bond (in CH_3CN) may be polarized by electric fields in axial and perpendicular di-

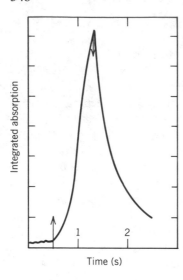

Figure 22.   The spectroscopic (far infrared) effect of an electric field on the transmission properties of liquid aniline. The transmission was monitored as the field was applied (first arrow) and removed (second arrow). Note that the effects were instantaneous and large. (Reproduced by permission from ref. 111.)

rections. The polarization is very sensitive and affects the intensity of the CN mode. Although the axial polarizability is higher than the perpendicular one, the CN intensity is more sensitive to perpendicular fields. Whereas axial fields increase the CN intensity, perpendicular fields decrease it. Michel and Lippert studied the polarization in static fields of up to 5×10^5 V cm^{-1}. With an increase in the electric field, a rapid decrease of the CN intensity was observed (Fig. 23). They suggested that the field effect points to a perpendicular excitation of the π-electrons.

Under the influence of applied external electric fields, matter may become electrified internally throughout its volume. This electrification or polarization may occur at the atomic, molecular, or bulk level, and particularly at an interface, which may be solid–liquid or solid–gas, as we have seen, or solid–solid and so on. Such interfaces are sites of natural transfer of electric charge. Solid–solid interfaces are well studied, of course, because of their importance in semiconductor devices. It is surprising that interfacial polarization effects between other surfaces have not been studied in such detail, and that the development of our understanding of polarization phenomena in such systems has not parallelled the rapid developments in solid state physics.

We can discern four basic types of polarization at the atomic or molecular level, all of which may occur to some extent in our molecular systems. The coexistence of all four types is particularly likely if charged carriers or impurities are also present in appreciable quantity. These types are as follows:

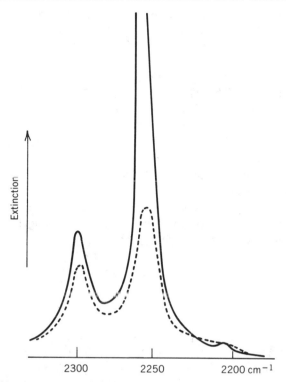

Figure 23.   This spectrum depicts similar effects on CH_3CN to those on aniline shown in Fig. 22 but at infrared frequencies. ---, No field; —— , with a field of 5×10^5 V cm^{-1} applied. (Reproduced by permission from ref. 36.)

1.   Electronic polarization: When an atom is placed in an external electric field, the electron cloud and positive nucleus shift in opposite directions and an electric dipole (which may be additional to an existing dipole) or "induced polarization" appears. Note that the effect involves the shift of the center of electrical interaction with respect to the center of mass, so we may predict a change in magnitude of rotation–translation coupling. This electronic shift (ca. 10^{-8} Å) is usually not large with respect to the nucleus and results in dielectric constants of 2–4 in organic solids and up to 20 in some inorganic solids.

2.   Ionic or atomic polarization: Ionic polarization occurs because if ions are present, the ions of different signs will be pulled apart by external electric fields. This effect, though most prominent in ionic dielectrics, may also occur in organic molecular systems if appreciable amounts of charged carriers or impurities are present. Even so, it is only expected to be 15–20% as large as the effect due to electronic polarization in such systems. It can be

very large in some solids; for example, it results in barium titanate having a dielectric constant of ca. 4000.

3. Orientation polarization: This occurs if the molecules have a permanent dipole. Under the influence of the electric field, all dipoles that are not lined up with the field experience a torque that tends to orient them in the direction of the field. This effect can be large, as, for example, in water, which has a dielectric constant of 80 at low frequencies and room temperature. It must be significant to some extent in all liquids, because even liquids composed of molecules with symmetrical charge distributions become distorted in a liquid environment and have temporary dipoles associated with them. Thus even "nonpolar" CCl_4, for example, has a rototranslation absorption at submillimeter frequencies.

4. Hyperelectronic polarization: This type of polarization occurs in large molecular domains, particularly domains that are elongated, as, for example, along polymeric chains, where long-range molecular orbital delocalization can occur. The electronic shift can then be 100–1000 Å, which is enormous in comparison with electronic polarization effects on atoms alone. Some polymers have dielectric constants of up to 300,000 because of this effect.

But polarization may also occur at multimolecular or bulk levels. One type of bulk polarization, flexoelectric polarization, results from distortions in the bulk of insulating nematic materials induced by external electric fields. Conversely, a splay or bending distortion can create a polarization. Schmidt et al.[113] observed that the long-range alignment does indeed become distorted in an electric field in a study on nematic MBBA.

The most well-known bulk polarization is interfacial polarization. This occurs more often than is generally realized. The crystal-growing procedure reported above and a phenomenon discussed in the following section, bubbling (the precursor to complete dielectric breakdown), both display elements of this polarization effect.

Interfacial polarization arises from the migration of charge carriers through the dielectric over some distance in the presence of a field and the appearance of "space charges" in the medium. Charge carriers may become trapped at impurity centers, at interfaces, or at electrodes if they are not freely discharged or replaced at the electrode. Pohl[80] reports the phenomenon of insulator-induced conduction, in which the current flow increases many-fold through a layer of highly purified dielectric liquid if one of the electrodes is covered with a thick layer of insulator particles, thus promoting the production of space charges that distort the macroscopic field. Interfacial polarization can become large at higher field strengths. Living matter and soils or earths consisting of conductive regions interleaved with barrier layers can appear to have enormous dielectric constants (10,000 or so). The technological applications and implications of different combinations of solid, liquid, and gas interfaces are widespread and varied.

It is not easy to separate all of these polarization mechanisms in a real system, nor to assign an observation to a particular polarization mechanism. The total induced polarization almost certainly arises from a number of contributory and competing factors. The spectroscopic effects on isotropic liquids alluded to above are interesting and need to be carefully analyzed. Ascarelli[66] also observed the pronounced effect of an electric field on the collective mode he resolved in nitromethane. The collective mode shifted considerably to higher frequencies when the sample was in contact with an electret (a permanently electrified body).

Interfacial polarization is certainly of vital importance in the nucleation process. Approaches to nucleation have emerged as a result of studies of interfacial problems.[116] In classical models interfacial energetics have been treated in terms of atomic bond-breaking and -making at the interface. This consideration conveniently leads us to another phenomenon of the liquid state—"bubbling" and dielectric breakdown. In this phenomenon bond-breaking is induced at the interface between solidlike and gaslike molecules in the liquid environment by strong electric fields. Bubbling always occurs before complete dielectric breakdown and demonstrates rather well the importance of interfacial phenomena even in isotropic "simple" molecular liquids.

B.   Bubbling in Liquids in Strong External Electric Fields and Liquid Structure

Ignacz[56] discusses bubbling in detail in his new text. His central hypothesis is that the physical behavior and dielectric breakdown properties of liquids are determined by their structure, and that microbubbles and bubbles are produced and stabilized by external electric fields. Ignacz proposes the existence of a local "crystalline structure" in liquids. "Crystal defects," he says, "lead to the generation of unattached gas-like molecules prior to complete dielectric breakdown." The thermal motions of aggregates of molecules lead to the formation of cavities in the liquids and the appearance of solid–gas interfaces. These cavities grow into bubbles in the presence of the external electric fields. Gassing, of course, is even known to occur (to a lesser extent) in solids. The author has observed bubbling in liquid acetonitrile under fields as small as 1 kV cm^{-1}.

Let us consider this phenomenon in some detail. According to Ignacz, bubbling occurs under both uniform and nonuniform electric fields in strongly stressed regions (interfaces) of nonpolar and polar liquids and at the interfaces with the metal electrodes. The bubbles become ionized by the external fields, and atomic bond-breaking occurs at the interface. Quoting Ignacz:

> The mean free path of electrons is 10^3 times longer in gases than in liquids, consequently, they can gain 1000 times more energy from a field in the gas. Becoming free, the electrons accelerate inside the bubble and they can break out into the parts of the liquid between the bubbles. Here, thin channels are formed by the kinetic energy of electrons and by several additional processes, e.g. electron multiplication, photo ionization etc. These channels connect the neighboring bubbles, thus preparing the total breakdown spark, which bridges the two electrodes. This spark does not form a continuous channel, it remains a chain of separate ionized bubbles connected by thinner channel parts.

Ignacz discusses the role of impurities and points out that their presence distorts the externally applied field: Locally the field becomes strongly non-uniform. If there is a great difference between the permittivities of the impurity and of the medium, then the local field intensity in the medium at the interface can be 20 to 50 times that of the original field. Bubbling occurs most easily in liquids containing impurities. Compare this situation with that in a solid. The microphysical influence of an impurity on the crystalline order in a solid is of very great importance, since the total physical equilibrium responsible for the strong bond between molecules can be ensured only by a perfect order. Every impurity, or boundary, disturbs this order and causes a decrease in the original polarization of the boundary molecules. So the possibility of bubbling (gassing) exists also in the solid. If there is a defect in the structure of a crystal, and some of the neighbors of a given molecule are missing, then the polarization of that molecule will decrease to a few percent of its original value or the bond may no longer exist. If the number of missing neighbors is as high as four the bond almost certainly will break. An externally applied electric field can easily break the bond for even fewer absent neighbors. When this happens the boundary molecule breaks out of the lattice and becomes free, entering into a gaslike state. The role of impurities and the defects they cause is as crucial in determining liquid-state properties as it is in determining solid-state properties. Recall the situation in colloidal systems, which may be stable over long periods only because of the presence of ionic material. As we have said, the amount of such material need be no more than is adventitiously present to create the double layers necessary for this stability.

Note how the phenomenon of bubbling requires a short range structure and demonstrates the role of interfacial phenomena in determining properties of the medium, which are made more pronounced by the presence of impurities. Let us consider the first aspect for nonpolar molecules, again following Ignacz's description. The nonpolar crystalline solid is stabilized by van der Waals forces. The field at a lattice point j generated by the molecules at other points of the lattice is

$$E_j = \sum_k{}' C_{jk} \frac{\alpha}{r_{jk}^3} E_k \qquad (2)$$

where r_{jk} denotes the distances between the points of the lattice, the factors C_{jk} depend on the angles between the directions of the radius vectors and that of Z, α is the polarizability, and E_k is the resultant field in the molecule at lattice point k. Calculations show that the moments and internal fields of the molecules along a given Z-line all have the same direction P or A, where in type P E_k and M_k (the dipole moment of the molecule) are parallel to Z and in type A both are antiparallel to Z. Along a line perpendicular to Z there are alternating P- and A-type moments. The moments pointing in the same direction in a Z-line attract each other, as do two moments of opposite directions in a line perpendicular to Z. "In fact, *only the above arrangement can give a cohesive structure.*"[56] The intermolecular bond energy of a chosen molecule is given by

$$V_j - - \alpha E_j^2$$

In the presence of an external field this becomes

$$V_j = - \alpha E_j (E_j - E)$$

A cubic or parallelepiped-based (if the molecules are elongated) lattice is identical to a lattice consisting of regular or irregular octahedrons. There is a molecule, c, in the center of each octahedron and its "nearest neighbors" are the molecules at the six vertices—two of P type and four of A type.

Thus, Ignacz proposes that during melting the solid crystal decomposes into uniform octahedral groups (liquid grains, the building blocks of condensed-state matter). The cubic lattice of molecules built up by octahedral groups is complete, with no lack or surplus of molecules. The system of equations (2), considered exclusively for the molecules of an octahedron, proves to be a stable formation in itself. As we have seen, the distances between molecules in the liquid, in some directions at least, are smaller than in the solid state. These can be calculated as for the solid crystalline state, with the important difference that we cannot assume that all of the local fields at a lattice point have the same absolute value. Formally the intermolecular bond energies are then the same in the liquid as in the solid state.

At the boiling point a second crystalline change occurs during which all of the grains totally decompose. The grains, and within them the molecules and atoms, are in thermal motion. The six peripheral molecules of a grain are in vibration relative to the central molecule. If the vibrational energy exceeds the bond energy, the grain disintegrates. This must occur at various places from time to time, producing gaslike molecules (it is an accepted fact that a liquid always contains a certain percentage of gas molecules). Gassing is produced within the bulk liquid itself by natural thermal motions and the subtleties of the intermolecular interactions. Ignacz calculates that "at room temperature about 10% of the molecules are unattached, and at a temperature somewhat below the boiling point this becomes about 25%." Hence, "there will always exist cavities in a liquid... a cavity which, for some reason, has survived for some time and exceeds a given size, shows a tendency to grow into a bubble."

An electric field contributes to the formation of cavities and their subsequent growth into bubbles. Recombination of unattached gaslike molecules to liquid grains is rendered difficult because they become polarized in the direction of the P-type molecules by the external field. Their dipole moments have the same direction as those of the grain fragments, contrary to the antiparallel direction that would be required for their recombination. We have already discussed the role of the external fields in breaking intermolecular bonds in cavity regions. The field may be able to detach A-type molecules of a complete grain. The accelerating effect of the field on ions and electrons causes the microbubbles to become charged and grow to macroscopic size. The fields must again have their most pronounced effects at interfacial regions where the change in permittivity is most significant, that is, across a solidlike–gaslike surface of a grain fragment. The electron distribution and its distortion at an interface is of paramount importance.

C.   The Induced Translation of Liquids in External Electric Fields

From the foregoing it is apparent that the intermolecular forces in the liquid state, like those of chemical bonding itself, are expressions of the electrical nature of matter and the distribution of electrons in the system. As Faraday taught us, magnetic aspects are also directly involved. These electronic distributions may be modified by external electric fields, particularly at interfacial regions in the medium. Quoting Davies:[103] "All our understanding of the behaviour of molecular systems, including biological systems, rests on our ability to represent interactions in an electric field, as there is, in these systems at the molecular level, nothing known to science other than atoms composed of localized positive and distributed negative charges."

Evans[13] has reported the observation of induced translation in *insulating* molecular liquids. Liquids may be suspended and, in some instances, even pumped rapidly against gravity. Induced translation occurs in both polar and nondipolar molecular liquids; the phenomenon is a general one. We have explained it in terms of nonuniform electric field effects on matter and dielectrophoretic forces. In one of our experiments the electric field between two parallel brass electrodes warps and rapidly becomes nonuniform in the region of the liquid–air interface (Fig. 16) and also within the bulk liquid itself.[79] The molecules become polarized, acquiring a negative charge on the side nearest the positive electrode and a positive charge on the side nearest the negative electrode. Because the field is nonuniform, it diverges across the molecules and produces unequal forces on the two effective ends of the molecules. The net effect, even on nonpolar molecules, is an overall force that results in the molecule being impelled into the region of stronger field. Ours is a paraelectric effect resulting from a two-step process, namely, polarization followed by the action of the nonuniform field on the polarized molecule producing a dielectrophoretic force toward the region of higher field intensity. Note that the effect of the first stage of this process at the molecular level may, therefore, increase the influence of rotation on translation, and vice versa.

As we have seen, we can study this influence with a molecular-dynamics computer simulation. In a simulation of 108 CHBrClF molecules, the application of a strong external electric field was seen to produce a net translation of the sample even though the direct effect of such a field on an isolated molecule is purely rotational. Computer simulation shows that on the molecular level, a redistribution of rotational to translational energy occurs in the presence of an external electric field. The statistical correlation between the rotational motion of a chiral molecule and its center-of-mass translation may therefore be used to separate, for example, the enantiomers of a racemic mixture by irradiation with an external field of force—a conclusion first arrived at by Baranova and Zeldovich[115] using hydrodynamic arguments based on the propeller effect in molecules that are left or right handed. The induced translation of one enantiomer, by symmetry-mirror images, must be opposite in direction to that of the other.

In the computer simulation[116] an electric field effective in the z-direction of the laboratory frame was applied[66] to the sample. The effect of the field was programmed into the algorithm in a variety of ways. The simplest was to take the net molecular dipole as the arm of the torque imposed on each molecule by the field of force $E_z \equiv |E|$. A second method was to decompose the electrostatic characteristics of each molecule into point charges located at each atomic site and to simulate the torque via the net force $\Sigma_i e_i E_z$, where e_i is the fractional charge on each atom of each of the (S) CHBrClF molecules. To emphasize the translation effect, a field strength E_z was chosen sufficient to saturate the system, producing an orientation rise transient as illustrated in Fig. 24. The mean molecular center-of-mass velocities $\langle v_x \rangle$, $\langle v_y \rangle$, and $\langle v_z \rangle$ were monitored in the presence of the field over some thousands of time steps using the methods of conventional constant-volume computer simulation.[117]

The effect of saturating the molecular ensemble with the strong field E_z is to force each and every one of the 108 molecules to rotate in the same direction against their natural thermal motions. An appropriate orientational

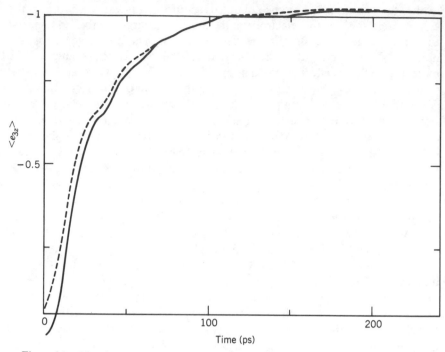

Figure 24. The rise transient $\langle e_{3z} \rangle$, where e_3 is a unit vector in the 3-axis of the principal moment-of-inertia frame of (3)-CHBrClF. The angle brackets denote averaging over the number of molecules in the ensemble at each time step. (Reproduced by permission from ref. 118.)

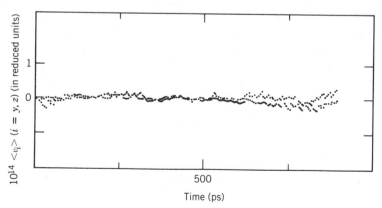

Figure 25. The behavior of $\langle v_z \rangle$ (straight line) and $\langle v_y \rangle$ (dotted line) in the absence of an applied field. (Reproduced by permission from ref. 118.)

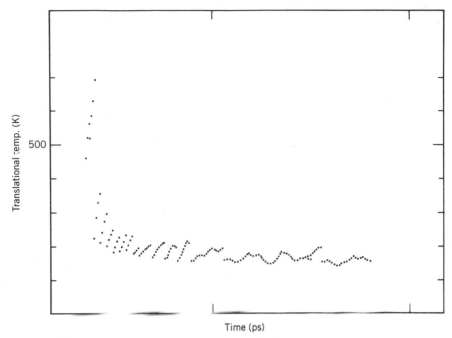

Figure 26.   The effect of the thermostat on the translational temperature. (Reproduced by permission from ref. 118.)

average such as $\langle e_{3z} \rangle$ then evolves as in Fig. 25; (that is, it rises from near 0 to near 1 in a finite interval of time—the rise transient. In our case this is about 0.6 ps from the instant the field is applied. Before application of the field, the system is in equilibrium with $\langle v_x \rangle = \langle v_y \rangle = \langle v_z \rangle \doteq 0$ (Fig. 25) and with a translational temperature of about 158 ± 25K. The sample is thermostatted by the conventional method of temperature rescaling. Figure 26 illustrates the work the thermostat does in restoring the translational temperature to 158 ± 25K after the sudden input of energy from the field E_z at $t = 0$. Immediately after $t = 0$, during the initial lifetime of the rise transient (about 0.07 ps), the translational temperature is very high, despite the fact that the effect of E_z on an isolated molecule would be purely rotational. After about 0.07 ps, the thermostat has effectively reduced the temperature once more to the required 158 ± 25K. The important point to note about Fig. 26 is that in the interval from about 0.07 ps to 5 ps the temperature is roughly constant at the same level as prior to the application of the field E_z.

In the interval from 0.07 ps after the field is applied the behavior of $\langle v_x \rangle$, $\langle v_y \rangle$, and $\langle v_z \rangle$ is illustrated in Fig. 27. It is clear that all three components of the mean molecular center-of-mass translational velocity of the 108 molecules gradually increase in magnitude over an interval of time much longer

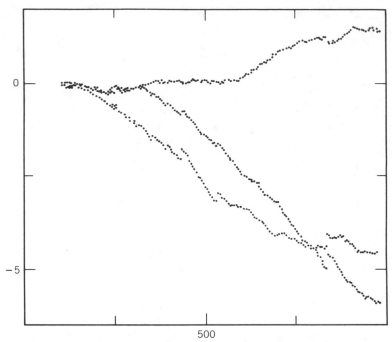

Figure 27. The behavior of (1) $\langle v_z \rangle$, (2) $\langle v_y \rangle$, and (3) $\langle v_x \rangle$ in the presence of a strong unidirectional field E_z. In contrast to the situation in Fig. 25, the averages are considerably different from zero. (Reproduced by permission from ref. 118.)

than that of the orientational rise transient. If the direction of E_z is reversed (i.e., $E_z \rightarrow -E_z$), the orientational rise transient is unaffected, but the drift $\langle \mathbf{v} \rangle$ is reversed.

These results can be explained only if rotational and translational motions are mutually coupled.[112] In simple terms, the field E_z forces the 108 molecules to rotate in the same way. Rotational freedom, at least in one direction, is restricted and there occurs a redistribution of the total thermal energy into other modes of motion. These may be translational modes—producing the observed translational motion. Because the induced translational motion for all the molecules is necessarily in the same direction (because of the unidirectional external field), the whole bulk sample must translate.

In their paper Baranova and Zeldovich[115] used the electric component of a circularly polarized electromagnetic field of force and showed that this type of field produces a translational effect similar to that illustrated above. Note that the orientational variable $\langle e_{3z} \rangle$ responds much more quickly to the field than does the translational variable $\mathbf{v}$. The extreme importance of rotation–translation interaction is established. So too vibration and rotation translation may be coupled. Van Woerkom et al.[118] were perhaps the first to question the separation of vibrational and reorientational functions in analysis of infrared and anisotropic Raman profiles for details of molecular dynamics. Lynden-Bell,[119] in reexamining the problem, showed theoretically that in many instances the line has a Lorentzian shape with a width equal to the sum of the vibrational and reorientational parts. Sometimes, however, these components become coupled to give complex, non-Lorentzian line shapes that are sometimes broader than expected, sometimes narrower, and on occasions showing a central dip. She also found that the linewidth sometimes varied with ℓ, that is, with the particular experiment. The coupling between *inter*molecular vibrations and other modes of motion must be significantly more pronounced.

IV. SOME CONCLUDING REMARKS—A CLASSICAL OR QUANTUM-MECHANICAL PROBLEM?

Some of the phenomena we have discussed are not predicted by and may not be explained using classical laws. Classical theories represent ensembles of molecules, as in a liquid, for example, as a system of particles or, at best, rigid bodies, and do not consider the detailed electrical nature of matter. When a body is considered as a particle, it is implicitly assumed that the mass is concentrated at a point and that all external forces *act through the same point*. As we have seen, in general, in heteronuclear molecules the electrical center (which is also the center of interaction) is displaced from the center of mass, which leads to the coupling of rotational and translational motions. Halfman[120] sums up the situation thus: "The particle approximation cannot serve even as a rough approximation in many situations. For example, it can give no information concerning the rotational motion of rigid or flexible bodies and by itself is useless in an attempt to study the dynamics of a fluid." Halfman discusses the analogies between fluid fields and electromagnetic fields and at the same time, by considering two charged particles in relative motion, provides some insight into the difficulties of applying classical Newtonian mechanics to the motion of charged particles. And Goldstein[121] warns that theorems derived for a system of particles should "be applied with due care to the electromagnetic forces between moving particles." Electromagnetic forces between moving particles do not generally obey Newtonian mechanics;[121] the third law is violated and the conservation of angular momentum, even in the absence of applied torques, is not valid. Internal forces of electromagnetic origin do not ordinarily have the convenient property of occurring in pairs that are collinear, equal, and opposite.

Molecules in a liquid are held together without being chemically bonded in the strictest sense. We say that residual forces hold them together—forces that must be electromagnetic in origin. The molecules and groups of molecules are polarizable; the repulsive forces change with the orientation of the molecule and are time-varying quantities; the centers of electrical interaction may not coincide with the center of mass and may be shifted with respect to this center by external electric fields; the distances between the atoms of the molecule are not constant because of intramolecular vibration; and the total potential energy of an ensemble of such molecules may not be approximated by the sum of interactions of all possible pairs.

Quoting Planck:[122]

Hitherto it has been believed that the only kind of causality with which any system of physics could operate was one in which all the events of the physical world might be explained as being composed of local events taking place in a number of individual and infinitely small parts of space. It was completely determined by a set of laws without respect to other events; and was determined exclusively by the local events in its immediate temporal and spatial vicinity. Let us take a concrete instance of sufficiently general application. We will assume that the physical system under consideration consists of a system of particles, moving in a conservative field of force of constant total energy. Then according to classical physics each individual particle at any time is in a definite state; that is, it has a definite position and a definite velocity, and its movement can be calculated with perfect exactness from its initial state and from the local properties of the field of force in those parts of the space through which the particle passes in the course of its movement. If these data are known, we need know nothing else about the remaining properties of the system of particles under consideration.

In modern mechanics matters are wholly different. According to modern mechanics, merely local relations are no more sufficient for the formulation of the laws of motion than would be the microscopic investigation of the different parts of a picture in order to make clear its meaning. On the contrary, it is impossible to obtain an adequate version of the laws for which we are looking, unless the physical system is regarded as a whole. According to modern mechanics, each individual particle of the system, in a certain sense, at any one time, exists simultaneously in every part of the space occupied by the system. This simultaneous existence applies not merely to the field of force with which it is surrounded, but also to its mass and charge.

Thus we see that nothing less is at stake here than the concept of a particle —the most elementary concept of classical mechanics. We are compelled to give up the earlier essential meaning of this idea; only in a number of borderline cases can we retain it.

We have tried to retain the concept of a particle in our treatments of liquid-state matter, with limited success. Our molecular theories are still not predictive; we still cannot calculate with any degree of precision seemingly simpler phenomena such as the solubility of one substrate in another. In this review I have presented experimental evidence that even requires that the concepts of collective molecular motions and distributed electric charge be introduced into our theories of liquid systems. Such concepts could certainly be used to explain some of the spectroscopic and electrical properties of liquids that have been presented.

Fortunately, to this end, it has recently been shown how the span between classical physics and quantum physics may be bridged using a mechanism derived only from classical physics. Piekara[123] considers the phenomenon of self-trapping (of, for example, a powerful laser beam in a liquid such as water), *a classical phenomenon based on the nonlinearity of the classical Maxwell equations*. As Piekara reminds us, "Maxwell's equations are essentially macroscopic, and are generally used in the linear approximation. However, these equations, with a change of name, 'microscopic Maxwell equations,'[124] are applied in the same linear form [incorrectly, Piekara believes] to isolated atoms and charges in any small volume." The main conclusion of Piekara's paper is that the quantum structure of waves is due to the nonlinearity of the wave equation. The nonlinear terms result in the self-trapping of waves, as, for example, in the liquid environment, and the production of photons. He provides us for the first time with a classical understanding of the energy quantum.

We have already made reference to the way in which the use of electron density as a basic variable in applied quantum mechanics may make possible the simplification of the usual wavefunction approach, the experimental determination of the physical variables, and the construction of various interpretive models, and thus may ultimately provide a "classical" picture of what are essentially quantum phenomena. We reiterate that the theory of chemistry and the chemical bond, whether intra- or intermolecular, permanent or temporary, is primarily a theory of electron density. The crystal-growing procedure and the thermodielectric effect certainly seem to be subtle consequences of electron redistribution across two phases of matter.

APPENDIX A: SOME POSSIBLE CONSEQUENCES OF THE THERMODIELECTRIC AND RECIPROCAL EFFECTS

As we have seen, the new crystal-growing technique reported in Section III.A is the reciprocal[125] of the little-known thermodielectric effect of Costa Ribeiro. Costa Ribeiro asserted that electric charges are always produced at the interface between the liquid and solid phases in a dielectric and may be observed not only in melting and solidification, but also in other changes of physical state in which one phase is a solid, such as sublimation and the precipitation of substrates from saturated solutions. These two effects demonstrate that mechanisms involving charge migration are involved in the nucleation process. Workhom and Reynolds[86] found potentials of greater than 200 V between ice and water phases *when a phase change was occurring*.

These two effects may explain many natural phenomena. For example, "before a thunderstorm, fair weather clouds suddenly grow dramatically in size and begin to exhibit *strong electric fields and precipitation*... almost without exception thunderclouds are characterized by convective instability with *strong up-drafts and down-drafts*... in a matter of less than 10^3 sec an innocuous cumulus cloud can suddenly change into a thundercloud producing both *heavy precipitation and lightning charges*.[126] These observations may now be explained because electric charges are always produced in changes of state involving a solid, including precipitation (the thermodielectric effect). Conversely, as the author has shown, electric fields may be used to induce phase changes at *an interface* between two states of matter when one of the states is a solid. Thus, in meteorological studies, "recent years have brought increasing evidence that electric fields can exert a pronounced effect on *the rate of growth of ice crystals in supercooled clouds*; indeed, it appears that under certain conditions *electric fields are responsible for ice nucleation phenomena* similar to those produced by cloud seeding techniques. Evidence is rapidly accumulating that electrostatic effects, far from being an incidental by-product of processes taking place in a thunderstorm, may be vitally important in determining the behaviour of the cloud."[126] The growth of ice crystals in field regions at an ice–water interface, reported by the present author, seems to substantiate this important hypothesis. Ice crystals may be grown from the melt in the laboratory in a simple manner using nonuniform electric fields.

It is known that in thunderstorms the intensity of electrical activity, as indicated by the number and repetition rate of lightning discharges, is closely related to the intensity of the convective activity. The existence of electrical forces in the atmosphere creates drafts of air, and it is estimated that in the strong electric fields in a thunderstorm the accelerations experienced by air might be equivalent to those resulting from temperature differences of as much as 5°C. Even so, movements of air in thunderstorms are for the most part the results of atmospheric temperature differences, which may be as large as 50°C.

Costa Ribeiro deduced that:

1. The current that flows in a thermodielectric cell on isothermal fusion or solidification occurring at a constant rate is proportional to the rate of migration of the phase boundary.
2. The total quantity of electric charge separated during the migration of the phase boundary in a two-phase dielectric system is proportional to the change of mass of one of the phases.

Considering these two laws in relation to thunderstorm activity, it is apparent that the convective activity resulting from temperature differences of as much as 50°C is so intense that the ice–water equilibrium in the atmosphere is continually and rapidly fluctuating. Law 1 tells us that large currents are consequently produced. In any natural phenomenon such as a thunderstorm, vast masses of ice and water are involved, and the changes of mass brought about by this convective activity are thus large. Very large electric charges are consequently produced (law 2), resulting in electrical breakdown of the atmosphere (lightning). Vonnegut[127] has estimated that "in the severe thunderstorms that produce tornadoes, lightning flashes at the rate of 10 or 20 per second may be capable of supplying power of the order of 10^8 kW" and has suggested that "if a column of air were heated by electrical discharges it might be capable of producing an updraft sufficient to cause a tornado."

There must also be biological consequences of these effects. Heinmets[128] pointed out the similarities between the Costa Ribeiro effect and electrical effects in biological systems. Many of the voltage curves recorded during the freezing of ice water, for example, resemble nervous impulses. We have already discussed other possible biological consequences in the text.

APPENDIX B: CRYSTAL GROWTH IN MAGNETIC FIELDS

Since writing this review, the author has observed the effects of magnetic fields[129] on the crystallization of 4n-octyl biphenyl from benzene solutions. 4n-Octyl biphenyl was specifically chosen for the experiment because of the existence in it of the magnetic moments associated with the π-electron clouds that are considered necessary for the observation of the effect. The author has not been able to observe similar effects of magnetic fields on the crystal growth of, for example, camphor from CCl_4 solution, as used in the electric field experiment described in Section III.A. It is important to realize that n-octyl biphenyl is not a liquid crystal in the melt (it is a solid at room temperature). However, 4-cyano-4n-octyl biphenyl forms both nematic and smectic phases. The effect of magnetic fields on the crystallization of the former is therefore all the more surprising.

Figures 28–30 show the effects of magnetic fields on the crystallization of 4-n-octyl biphenyl from a saturated solution in benzene. The magnetic field *induces a bulk translation of the sample — solution is expelled from the field region when a field is applied.* The author believes this is the first report of bulk translation in an isotropic liquid induced by an external magnetic field.

Figure 28 shows the crystals of n-octyl biphenyl grown from solution in benzene by evaporation in a darkened room. Note the random arrangement of crystalline masses. Figure 29 shows crystals of n-octyl biphenyl grown in

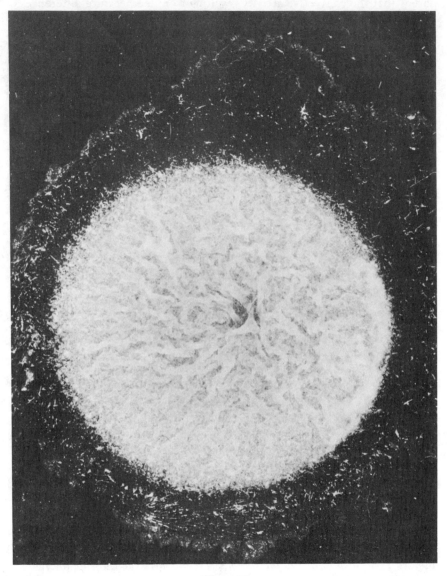

Figure 28.

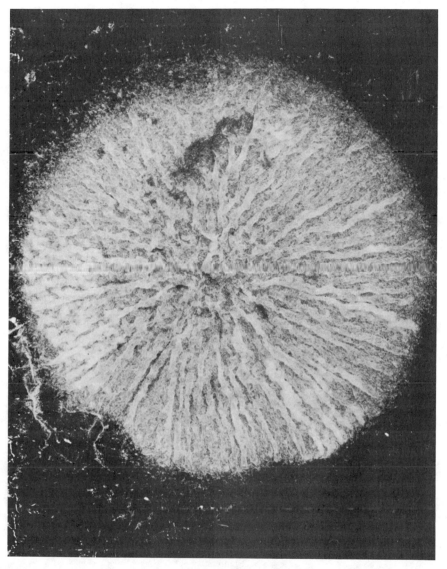

Figure 29.

365

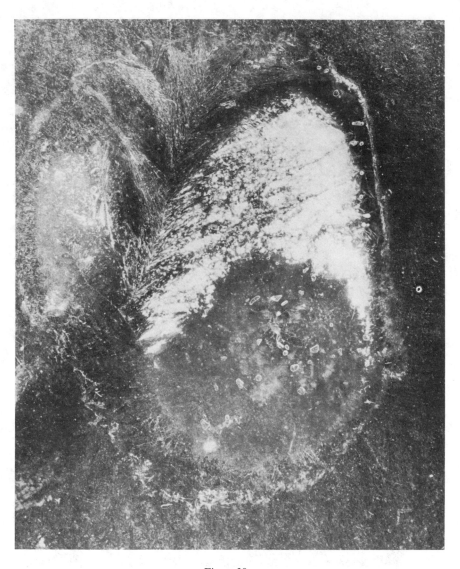

Figure 30.

a magnetic field of medium intensity (~ 5000 G) on a watch glass that retains the solution in the region of the field. Note the distinct veins following the magnetic lines of force running from the center to the perimeter. Single crystals, however, are not grown in this medium-intensity field. Figure 30 shows a single crystal of 4-*n*-octyl biphenyl grown in a strong magnetic field (> 1000 G). Arrows indicate the perimeter of the crystal (photographed on a darkened background), which is transparent in most regions.

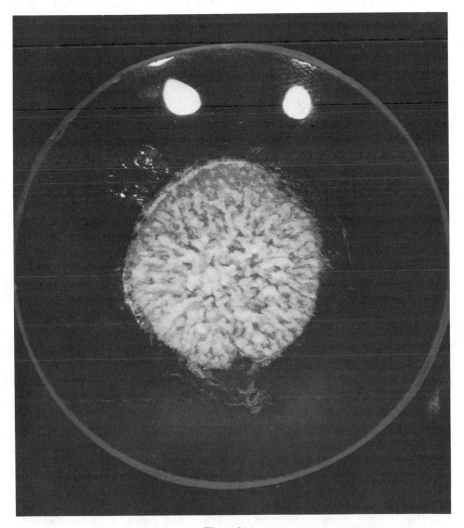

Figure 31.

It is interesting that in electric fields (Section III.A) crystals grow normal to the surface, whereas in magnetic fields the crystals grow parallel to the surface, at least at the field strengths currently used. However, crystals of azobenzene have recently been "pulled" out of solution in benzene by this author using similarly small magnetic fields. Note that the reciprocal, using magnetic fields, of the crystal-growing technique reported here may also exist, that is, the magnetic analogue of the thermodielectric effect of Costa Ribeiro.

It is important not to pretreat in any way the glass or plastic surfaces used to grow these crystals in magnetic fields. Even surfaces that are repeatedly used for crystallization may become charged. The author has successfully grown single crystals on charged glass and plastic surfaces. Also, it is essential to grow the crystals in darkened rooms, away from direct sunlight. The author believes that electromagnetic fields have similar influences on the crystallization of 4-n-octyl biphenyl. Figure 31 shows crystals of 4-n-octyl biphenyl grown in a darkened room under thermostatted temperature conditions. Note the random arrangement of crystalline masses, as in Fig. 28. Figure 32 shows crystals of n-octyl biphenyl grown in direct sunlight. The same solution (and same volume of solution) was used for Figs. 31 and 32. The crystals took significantly longer to grow in direct sunlight, even though the temperature was slightly higher and one would expect the evaporation rate to have been increased. In the crystals that grew in direct sunlight, distinct veins appear to follow the direction of incidence of the sunlight (indicated by an arrow). It would be desirable to repeat such experiments using electromagnetic fields under conditions controlled precisely using laser light sources and careful temperature control.

Living matter, of course, grows in electromagnetic fields. Plant life grows toward the light source itself, with growth originating at sensitive *surface regions*. As Calvin[130] says, in relation to the study of photosynthesis,

> we are now in the midst of trying to determine precisely what happens after chlorophyll has absorbed the quantum and has become an excited chlorophyll molecule, a problem that involves the physicist and physical chemist, as well as the organic and biochemists. The determination of the next stage in the energy-conversion process is one of our immediate concerns. Either it is *an electron transfer process*, and thus comes close in its further stages to the electron transfer processes which are being explored in mitochondria, or it is some independent non-redox method of energy conversion. This remains for the future to decide.

Present observations may be providing vital clues to the details of such processes. Studies should be extended gradually to more ordered liquid crystalline phases and then to biological materials themselves. Certainly it is

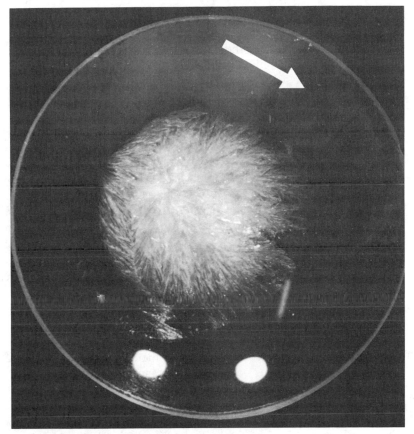

Figure 32.

now established that a specific electric charge is associated with a change of state or aggregation of matter, that is, with an order–disorder (or vice-versa) phase change (including a liquid crystalline-to-solid phase transition[98]). And it is established that external fields may influence and even cause such phase changes. A theory predicting such influences is abstracted in the Appendix C.

APPENDIX C: THE INFLUENCE OF ELECTRIC AND MAGNETIC FIELDS ON NUCLEATION KINETICS—THEORETICAL CONSIDERATIONS

Kashchiev[131] presents a theory explaining the effects of externally applied uniform electric fields on homogeneous nucleation as well as on nucleation on foreign completely wettable particles. He shows by means of a suitably defined effective supersaturation that if the dielectric permittivity of the new phase is smaller than that of the old phase, the electric field stimulates the nucleation, whereas in the reverse case it inhibits the process. He points out that his results apply equally to nucleation taking place in an external uniform magnetic field. Shablakh et al.[132] have observed in an investigation of the phase structures of the four cycloalcohols from cyclopentanol to cyclooctanol that time-dependent transformations occurred between some of these phases, and that at least one of these transformations was sensitive to the application of an electric field. They discuss their observations in terms of Kashchiev's theory.

Kashchiev derives equations for the nucleation rate and the time lag on the basis of thermodynamic considerations concerning the work required for the formation of a spherically shaped cluster. According to electrostatic theory, the free energies of the electric field before and after the formation of the cluster are given respectively by

$$W_1 = \frac{\varepsilon_m}{8\pi} \int_{\Omega_m} E^2 \, dv \tag{3}$$

$$W_2 = \frac{\varepsilon_c}{8\pi} \int_{\Omega_c} E_c^2 \, dv + \frac{\varepsilon_m}{8\pi} \int_{\Omega_m - \Omega_c} E_m^2 \, dv \tag{4}$$

where ε_c and ε_m are the dielectric constants of the new and old phases; Ω_c and Ω_m are the spatial regions occupied by the cluster and by the system; $E = |\mathbf{E}|$, $E_c = |\mathbf{E_c}|$, $E_m = |\mathbf{E_m}|$ and $\mathbf{E_c}$ and $\mathbf{E_m}$ are the electric fields inside and outside the cluster.

Assuming a spherically shaped cluster, and considering that the system too is a sphere with radius R obeying the inequality $R^3 \gg r_c^3$, where r_c is the radius of the cluster, then in spherical coordinates r, θ, ϕ,

$$E_c, r = \frac{3\varepsilon_m}{\varepsilon_c + 2\varepsilon_m} E \cos\theta \qquad 0 \le r < r_c$$

$$E_c, \theta = -\frac{3\varepsilon_m}{\varepsilon_c + 2\varepsilon_m} E \sin\theta \qquad 0 \le \theta < \pi \tag{5}$$

$$E_c, \phi = 0 \qquad 0 \le \phi < 2\pi$$

and

$$E_m, r = \left[2\frac{\varepsilon_c - \varepsilon_m}{\varepsilon_c + 2\varepsilon_m} \left(\frac{r_c}{r}\right)^3 + 1 \right] E \cos\theta \qquad r_c \le r < R$$

$$E_m, \theta = \left[\frac{\varepsilon_c - \varepsilon_m}{\varepsilon_c + 2\varepsilon_m} \left(\frac{r_c}{r}\right)^3 - 1 \right] E \sin\theta \qquad 0 \le \theta < \pi \tag{6}$$

$$E_m, \phi = 0 \qquad 0 \le \phi < 2\pi$$

where the center of the cluster is chosen as the origin of the coordinates and the polar axis has the direction of the applied electric field.

With a little mathematics, using the inequality $r_c^3 \leq R^3$ we find that the free-energy change due to the field is given by

$$\Delta W_E = W_2 - W_1 = -\frac{\varepsilon_m}{8\pi} f(\lambda) v_c E^2 x \tag{7}$$

where

$$f(\lambda) = \frac{1-\lambda}{2+\lambda}$$

$$\lambda = \frac{\varepsilon_c}{\varepsilon_m} \qquad 0 < \lambda < \infty \tag{8}$$

so that for $\varepsilon_c < \varepsilon_m$ the electric field stimulates cluster formation, whereas for $\varepsilon_c > \varepsilon_m$ the field inhibits it.

Using Eq. (7) and the expression of Volmer[133] for $\Delta G_0(x)$, the isothermal reversible work required for the formation of a cluster of x atoms (or molecules) in the absence of a field,

$$\Delta G_0(x) = -kTS_0 x + \frac{a\sigma x^{2}}{3} \tag{9}$$

we obtain

$$\Delta G(x) = -kTSx + \frac{a\sigma x^2}{3} \tag{10}$$

where $\Delta G(x)$ is the work expended in the presence of a field, $S_0 = \ln(P/P_e)$ and is the supersaturation, P_e is the equilibrium pressure, k is the Boltzmann constant, σ is the specific surface energy, $a = (4\pi)^{1/3}(3v_c)^{2/3}$ is a constant, and $\Delta G(x) = \Delta G_0(x) + \Delta W_E(x)$. Hence, the effective supersaturation at which cluster formation occurs is given by

$$S = S_0 + S_E \tag{11}$$

and the influence of the electric field on the process is taken into account through the additional term

$$S_E = cE^2$$

in which $c = \varepsilon_m f(\lambda) v_c / 8\pi kT$ is a constant.

The steady-state nucleation rate is given by

$$J_{ST} = \left(\frac{\Delta G_k}{3\pi kT x_k^2} \right)^{1/2} D_k N \exp\left(\frac{-\Delta G_k}{kT} \right) \tag{12}$$

where N is the total number of atoms in the system and $D_k = \alpha a x_k^{2/3}$ is the probability per unit time of an atom joining to the nucleus.

Assuming that the dependence of the frequency factor α on the electric field strength is weaker than an exponential one, then, following Volmer,[135] we can write

$$J_{ST} = A \exp\left[\frac{-B}{\left(S_0 + cE^2\right)^2}\right] \tag{13}$$

where $B = 4(a\sigma/3kT)^3$. A is also a constant. Thus, for $c > 0$ ($\varepsilon_c < \varepsilon_m$) the electric field stimulates nucleation, and for $c < 0$ ($\varepsilon_c > \varepsilon_m$) the field inhibits the process.

For $S_0 = 0$, $c > 0$, the system is saturated before application of the field and

$$J_{ST} = A \exp\left(\frac{-B}{c^2 E^4}\right)$$

that is, the field stimulates nucleation.

For $S_0 > 0$, $|S_E| < 0.1 S_0$, the system is supersaturated and the supersaturation is either enhanced or reduced by the field within the limits indicated.

From Eq. (13) it follows that E_{cr}^2 is a linear function of S_0.

Kashchiev also considers nucleation on a foreign particle. In obtaining these relations he assumes a spherically shaped cluster, but "in principle, the more realistic case of ellipsoidal or polyhedral shape may be treated in the same way, but then serious mathematical difficulties have to be overcome."[131] However, he concludes that "when a uniform electric field is externally applied to a nucleating system the initial supersaturation S_0 in it will be effectively either enhanced or reduced, depending on the sign of the constant multiplying $E^2 x$."

Kashchiev estimates that a field of $< 6.14 \times 10^3$ kV cm^{-1} is necessary to *inhibit* the formation of a water droplet. Such a field is orders of magnitude larger than the fields used by the present author for his camphor CCl$_4$ crystal-growth system, which are, in any case, nonuniform. The growth of camphor crystals from a saturated solution of camphor in CCl$_4$, for example, is observed in fields of only volts per centimeter. Crystals of camphor may also be grown away from a field region.[83] All of this seems to suggest that the phenomenon is an electrostatic one, involving charge migration as proposed by Costa Ribeiro.[85] Kashchiev points out that the results obtained in his paper "might be very useful in understanding some condensation phenomena occurring in the Earth's atmosphere in the presence of strong electric fields" (see Appendix A) and that "all results obtained hold good also when nucleation takes place in an externally applied uniform magnetic field" (see Appendix B).

Shablakh et al.[132] used their experiments on the cycloalcohols to examine the detailed predictions of Kashchiev's model, and in particular a square law dependence of some functions on the applied field. They observed that for such functions "the field power lies in the range 1.2 (cyclo-octanol) to 1.7 (cycloheptanol)." They propose that the discrepancies in the field power may have arisen because Kashchiev's calculation is based on the nucleation of a spherical droplet, a geometry which would be inappropriate to the solid–solid transition they examined. In cyclopentanol they observed a change in the field power exponent from 2.4 to 14 when the applied field was $> 5 \times 10^4$ kV cm^{-1}.

References

1. C. R. Mueller, B. Smith, P. McGuire, W. Williams, P. Chakraborti, and J. Penta, *Adv. Chem. Phys.*, **21**, 369 (1971).

2. B. J. Berne and G. D. Harp, *Adv. Chem. Phys.*, **17**, 63 (1970).

3. R. Batlino, *Chem. Rev.*, **71**, 5 (1971).

4. J. H. Hildebrand, *Faraday Disc.*, **66** (1978).

5. J. A. Barker and D. Henderson, *Acc. Chem. Res.*, 303 (1971).

6. J. D. van der Waals, Thesis, Leiden, The Netherlands, 1873.

7. E. A. Guggenheim and M. L. McGashen, *Proc. R. Soc. Lond.*, **A206**, 448 (1951).

8. I. Prigogine, *The Molecular Theory of Solutions*, North Holland, Amsterdam, 1957.

9. A. D. Buckingham, in *Organic Liquids*, A. D. Buckingham, ed., Wiley-Interscience, New York, 1978.

10. A. H. Narten, S. I. Sandler, and T. Rensi, *Faraday Disc.*, **66** (1978).

11. N. N. Das Gupta and S. K. Gosh, *Rev. Mod. Phys.*, **18**, 225 (1946).

12. G. Ascarelli, *Chem. Phys. Lett.*, **39**, 23 (1976).

13. G. J. Evans, *J. Chem. Soc., Faraday II*, **79**, 547 (1983).

14. F. Kohler, *The Liquid State*, Verlag Chemie, Wernheim, 1972.

15. a) M. W. Evans, *J. Chem. Soc., Faraday II*, **79**, 719 (1983). (b) See also the review by M. W. Evans and G. J. Evans in this volume.

16. F. H. Stillinger, *Adv. Chem. Phys.*, **31**, 1 (1975).

17. M. W. Evans, G. J. Evans, W. T. Coffey, and P. Grigolini, *Molecular Dynamics*, Wiley, New York, (1983), Chapter 7.

18. S. R. Nagel, *Adv. Chem. Phys.*, **51**, 227 (1982).

19. P. Chaudhari, J. J. Cromo, and R. J. Gambino, *IBM J. Res. Dev.*, **17**, 66 (1973).

20. T. Masumoto and K. Hashimoto, *J. Phys.*, **41**, C8-894 (1980).

21. F. J. Dyson, *Rev. Mod. Phys.*, **51**, 447 (1979).

22. G. J. Evans, Ph.D. Thesis, University of Wales, 1977, Chapter 3.

23. J. Van der Elsken and D. Frenkel, Faraday Symposium, no. 11, paper 9, December 1976, London.

24. G. Holleman and G. Ewing, *J. Chem. Phys.*, **47**, 571 (1967).

25. G. Ewing, *Acc. Chem. Res.*, **2**, 168 (1969).

26. P. Marteau, R. Grancer, H. Vu, and B. Voda, *C. R. Seances Acad. Sci.*, **265B**, 685 (1967).

27. M. Bulanin and M. Tonkov, *Phys. Lett.*, **26A**, 120 (1968).

28. G. Chantry, H. Gebbie, B. Lassier, and G. Wyllie, *Nature*, **214**, 163 (1967).

29. E. A. Poe, *Narrative of Arthur Gordon Pym*, Doubleday, New York, 1966, p. 706.

30. G. W. Gray, *Molecular Struct and Liquid Crystals*, Academic, New York, 1962.

31. For example, see (i) P. De Gennes, *The Physics of Liquid Crystals*, Clarendon Press, Oxford, 1974; and (ii) G. W. Gray and P. A. Winsor, *Liquid Crystals and Plastic Crystals*, Wiley, New York, 1974.

32. M. W. Evans, *Phys. Rev. Lett.*, **50**, 371 (1983); G. J. Evans and M. W. Evans, *J. Mol. Liq.*, **26**, 63 (1983).

33. J. G. Powles, D. E. Williams, and C. P. Smyth, *J. Chem. Phys.*, **21**, 136 (1951).

34. V. K. Agarwal, G. J. Evans, and M. W. Evans, *J. Chem. Soc., Faraday II*, **79(i)**, 137 (1983).

35. E. Sciesinska and J. Sciesinski, *Mol. Cryst. Liq. Cryst.*, **51**, 9 (1979).

36. H. Michel and E. Lippert, *Organic Liquids*, A. D. Buckingham, ed., Wiley, New York, 1978, Chapter 17.

37. H. Eyring, T. Ree, and N. Hirai, *Proc. Natl. Acad. Sci. USA*, **44**, 683 (1958).

38. G. Wald, *Proc. Natl. Acad. Sci., USA*, **52**, 1595 (1964).

39. A. V. Tobolsky and E. T. Samulski, *Adv. Chem. Phys.*, **21**, 529 (1971).

40. A. Babloyantz, *Mol. Phys.*, **2**, 39 (1959).

41. M. Warner, personal communication; also, paper submitted to *J. Chem. Phys.*, 1983.

42. J. R. McColl and C. S. Shih, *Phys. Rev. Lett.*, **29**, 85 (1972).

43. A. Wulf, *J. Chem. Phys.*, **67**, 2254 (1977).

44. M. J. Stephen and J. P. Straley, *Rev. Mod. Phys.*, **46**, 616 (1974).

45. G. J. Evans, *J. Chem. Soc., Faraday II*, **79**, 833 (1983).

46. B. J. Bulkin, in *Advances in Liquid Crystals*, Vol. 2, G. H. Brown, ed., Academic, New York, 1976, p. 199.

47. M. Schwartz and P. Wang, p. 206 in ref. 46.

48. N. M. Amer and Y. R. Shen, *Solid State Commun.*, **12**, 263 (1972).

49. J. M. Schnur and M. Fontana, *J. Phys. (Paris)*, **35**, L53 (1974).

50. W. E. Pulnam, D. E. McEachein, and J. E. Kilpatrik, *J. Chem. Phys.*, **42**, 749 (1965).

51. A. Krishnaji and A. Monsing, *J. Chem. Phys.*, **41**, 827 (1964).

52. E. Lippert, W. P. Shroer, H. Mahnke, and H. Michel, *Proceedings of an International Conference on Hydrogen Bonding*, H. J. Bernstein, ed., Ottawa, 1972.

53. (i) H. Bertagnoli, D. O. Leicht, and M. O. Zeidler, *Mol. Phys.*, **35**, 199 (1978). (ii) A. Kratochwill, J. V. Weidner, and H. Zimmerman, *Ber. Bunsenges Phys. Chem.*, **77**, 408 (1973)

54. T. Keyes and D. Kivelson, *J. Chem. Phys.*, **56**, 1057 (1972).

55. A. Loshe, *Z. Phys. Chem.*, **201**, 302 (1952).

56. P. Ignacz, a new book in press with the Hungarian Academic Press. See also P. Ignacz, *J. Electrostat.*, **7**, 309 (1979); A. R. Ubbelhode, *The Molten State of Matter*, Wiley-Interscience, New York, 1978; H. Eyring and M. S. Jhon, *Significant Liquid Structures*, Wiley, New York, 1969.

57. M. W. Evans, *J. Mol. Liq.*, **25**, 149 (1983).

58. M. W. Evans, *Phys. Rev. Lett.*, in press (1984).

59. M. Ferrario, CECAM Workshop, Paris, 1980 (and unpublished work).

60. B. J. Berne and J. A. Montgomery, Jr., *Mol. Phys.*, **32**, 363 (1976).

61. C. J. Reid, Ph.D. Thesis, University of Wales, 1979. See also ref. 17, Chapter 4.

62. G. J. Evans, *Chem. Phys. Lett.*, **99**, 173 (1983).

63. A. Gerschel, personal communications; also, see Z. Kioiel, K. Leibler, and A. Gerschel, *J. Phys. E: Sci. Instrum.*, **17**, 240 (1984).

64. G. S. Tresser, *J. Phys.*, **38**, 267 (1977).

65. N. K. Ailawadi and B. J. Berne, *Faraday Symp.*, **11** (1976).

66. G. Ascarelli, *Chem. Phys. Lett.*, **39**, 23 (1976).

67. G. Fini and P. Mirone, *Spectrochim. Acta*, **32A**, 439 (1976).

68. R. Lobo, J. E. Robinson, and S. Rodriguez, *J. Chem. Phys.*, **59**, 5992 (1973).

69. T. Nee and R. Zwanzig, *J. Chem. Phys.*, **52**, 6353 (1970).

70. E. Knozinger, D. Leutloff, and R. Wiltenbeck, *J. Mol. Struct.*, **60**, 115 (1980).

71. J. S. Rowlinson, *Trans. Faraday Soc.*, **45**, 974 (1949).

72. J. E. Harries, W. J. Burroughs, and H. A. Gebbie, *J. Quant. Spectrosc. Radiat. Transfer*, **9**, 799 (1969).

73. R. J. Jakobsen and J. W. Brasch, *J. Am. Chem. Soc.*, **86**, 3571 (1964).

74. B. J. Bulkin, *Helv. Chim. Acta*, **52**, 1348 (1969).

75. G. Wegdam and J. van der Elsken, *Phonons, Proceedings of an International Conference*, M. A. Nusimovici, ed., Flammarion Sci., Paris, 1971, p. 469.

76. P. Wegener and A. A. Pouring, *Phys. Fluids*, **7**, 352 (1964).

77. J. Farges, B. Raoult, and G. Toichect, *J. Chem. Phys.*, **59**, 3454 (1973)

78. G. A. Kenney-Wallace, *Acc. Chem. Res.*, **11**, 12, 433 (1978).

79. N. J. Felici, *Rev. Gen. Select.*, **76**, 786 (1967).

80. H. A. Pohl, "Non-Uniform Field Effects," in *Electrostatics and the Applications*, A. D. Moore, ed., Wiley, New York, 1973, Chapter 14.

81. T. B. Jones, M. P. Perry, and J. R. Melcher, *Science*, **136**, 1232 (1971).

82. S. W. Charles and J. Popplewell, *Endeavour*, **6**(4), 153 (1982).

83. G. J. Evans, *Materials Lett.*, **2**(SB), 420 (1984); *ibid., J. Chem. Soc., Faraday I*, **80**, 2043 (1984).

84. K. Rajeshwar, R. Rosenvold, and J. Du Bow, *Nature*, **301**, 48 (1983).

85. J. Costa Ribeiro, *Acad. Bras. Sci. An.*, **22**, 325 (1950).

86. E. J. Workhom and S. E. Reynolds, *Phys. Rev.*, **78**, 254 (1950); *ibid.*, **74** (1948).

87. S. Gray, *Philos. Trans.*, **37**, 285 (1732).

88. J. J. Burton and A. C. Zettlemoyer, eds., *Nucleation*, Elsevier, Amsterdam, 1977.

89. (a) J. R. Melcher, D. S. Guttman, and M. Hurintz, *J. Spacecraft*, **6**(1), 25 (1969). (b) S. K. Gosh and D. M. Deb, *Phys. Rep.*, **92**(1), 2 (1982).

90. P. Becker, ed., *Electron and Magnetization Densities in Molecules and Crystals* (*NATO Adv. Study Inst. Ser. B.*, **48**), Plenum, New York, 1980.

91. A. S. Bamzai and D. M. Deb, *Rev. Mod. Phys.*, **53**(95), 593 (1981).

92. V. H. Smith, Jr., *Physica Scripta*, **15**, 147 (1977).

93. Dang Tran Quan, *C. R. Seances Acad. Sci. Ser. B*, **271**(13), 604 (1970).

94. M. Cassettari and G. Salvetti, *Nuovo Cimento B*, **64**(1), 145 (1969).

95. Sergio Mascarenhas, *Radiat. Eff.*, **4**(3–4), 263 (1970).

96. A. Dias Tavares, *Phys. Stat. Solidi A*, **21**(2), 717 (1974).

97. O. Rozental and F. Chetin, *Electron Obsab. Mater.*, **6**, 51 (1971).

98. A. P. Kapustin, Z. Kh. Kuvalov, and A. N. Trofimov, *Kristallografiya*, **18**(3), 647 (1973).

99. J. Garcio Francisco, *Dissert. Abstr. Int. B*, **37**(7), 1366 (1976).

100. P. Eyerer, *Adv. Colloid Interface Sci.*, **3(3)**, 223 (1972).

101. E. A. Hauer and D. S. LeBeau, *J. Phys. Chem.*, **42**, 961 (1938).

102. R. O. Grisdale, in *The Art and Science of Growing Crystals*, Wiley, New York, 1963, Chapter 9.

103. M. Davies, *Nonlinear Behaviour of Molecules, Atoms and Ions in Electric, Magnetic or Electromagnetic Fields*, 1979, p. 301.

104. G. P. Jones, M. Gregson, and T. Krupkowski, *Chem. Phys. Lett.*, **13**, 266 (1972).

105. K. Hirano, *Nature*, **174**, 268 (1954).

106. A. R. Ubbelohde, *Trans. Faraday Soc.*, **36**, 863 (1940).

107. H. Dekameizer, *Helv. Chim. Acta*, **10**, 896 (1927).

108. G. E. Horonic and G. Yelonsky, *Proc. 1st Int. Atris Symp.*, **2**, 539 (1969).

109. M. Krole, J. Eichmeir, and R. W. Sciidn, in *Advances in Electronics and Electron Physics*, Academic, New York, 1964, p. 177.

110. A. P. Wiehner, *Am. J. Phys. Med.*, **48**, 3 (1961).

111. G. J. Evans and M. W. Evans, *J. Chem. Soc., Faraday II*, **76**, 667 (1980).

112. M. W. Evans, *Phys. Rev. Lett.*, in press, 1984.

113. D. Schmidt, M. Schadt, and W. Helfrich, *Nature*, **A27**, 277 (1972).

114. See A. G. Walton, in *Nucleation*, A. C. Zettlemayer, ed., Marcel Dekker, New York, 1969, Chapter 5.

115. N. B. Baranova and B. Y. Zeldovich, *Chem. Phys. Lett.*, **57**, 435 (1979).

116. M. W. Evans and G. J. Evans, *J. Mol. Liq.*, in press, 1984.

117. See, for example, SERC, *CCP5 Quarterly*, **Autumn** (1982).

118. P. C. M. van Woerkom, J. de Bleyser, M. de Zwart, and J. C. Leyte, *Chem. Phys.*, **4**, 236 (1974).

119. R. M. Lynden-Bell, *Mol. Phys.*, **33(4)**, 907 (1977).

120. R. L. Halfman, *Dynamics*, Addison-Wesley, Reading, Massachusetts, 1962.

121. H. Goldstein, *Classical Mechanics*, 1st ed., Addison-Wesley, Reading, Massachusetts, 1950.

122. M. Plank, *The Universe in the Light of Modern Physics*, George Allen & Unwin, 1931 (translated by W. H. Johnston from *Das Weltbeld der neuen Physik*, Joh. A. Borlk, Leipzig).

123. A. H. Piekara, *Optics and Laser Technology*, Butterworth and Co. Ltd., August, 1982, p. 207.

124. J. M. Stone, *Radiation and Optics*, New York, 1963.

125. G. J. Evans, *New Scientist*, submitted 1984.

126. B. Vonnegut, "Atmospheric Electrostatics," in *Electrostatics and its Applications*, Wiley, New York, 1977, Chapter 17.

127. B. Vonnegut, *J. Geophys. Res.*, **65(1)**, 203 (1960).

128. F. Heinmets, *Trans. Faraday Soc.*, **58(4)**, 788 (1962).

129. G. J. Evans, *Nature*, submitted (1984).

130. M. Calvin, *Les Prix Nobel*, 1961

131. D. Kashchiev, *Philos. Mag.*, **25**, 459 (1972).

132. M. Shablakh, L. A. Dissado, and R. M. Hill, *J. Chem. Soc., Faraday II*, **79**, 1443 (1983).

133. M. Volmer, *Kinetik der Phasenbildung*, Steinkopff, Dresden–Leipzig, 1939.

A REVIEW AND COMPUTER SIMULATION OF THE MOLECULAR DYNAMICS OF A SERIES OF SPECIFIC MOLECULAR LIQUIDS[‡]

M. W. EVANS

Adran Fathemateg Gymhwysol, Coleg Prifysgol Gogledd Cymru, Bangor, Cymru

and

GARETH J. EVANS

Department of Chemistry, University College of Wales, Aberystwyth, Dyfed, Wales

CONTENTS

[‡]Dedicated to the South Wales Area of the National Union of Mineworkers.

377

I. INTRODUCTION

In this review we will discuss the deficiencies and limitations of the traditional approach to the study of liquids by spectroscopy and stochastic analytical theory, and the extra depth of insight that may be obtained by means of computer simulation studies. We review the literature and compare results from different spectroscopies for a series of liquids ranging from liquids of C_{3v} symmetry, including two that form solid rotator phases, to molecules of lesser symmetry. (Optically active liquids are discussed in other articles of this volume.) We will see that results obtained for a particular liquid with different spectroscopies are sometimes contradictory. There are still obvious problems associated with the data-reduction processes. We will discuss the role of the various relaxation mechanisms that may contribute to a measured spectrum and show that experimental results are not always easily related to a purely orientational, single-particle mechanism. We will discuss the role of a distinct local structure, known to be present in some liquids, in distorting measured profiles, and we will see that in a liquid the various degrees of molecular freedom (rotation, translation, and vibration), rather than being decoupled, are more often coupled—and strongly so in some liquids.

In particular, computer simulation has shown the importance of rotation–translation interaction. The rotation and translation of the same molecule are normally coupled just as in a propeller, which, when it rotates, must simultaneously translate and move forward. We shall see that rotation–translation coupling, if strong, severely distorts spectral profiles. Consequently, our theories must be rototranslational in origin, and not purely rotational, as present theories are, if they are to be successful in reproducing experiments. In optically active liquids M. W. Evans has shown that rotation–translation interaction may be used to explain why the physical properties of enantiomers and of their racemic mixtures are different. (For example, at room temperature, a racemic mixture of lactic acid molecules is a liquid, whereas the individual enantiomers are solid.)

We have not had the foresight to envisage many of the subtleties of the liquid state. What follows will surprise some researchers, and, we hope, catalyze others into doing more precise and detailed experimentation using all of the dynamical probes in unison. Theoreticians should use the experimental and computer-simulated observations to develop more realistic analytical theories. Too often they have lost sight of the physical realities of liquid systems and, by increasing mathematical complexity, confused rather than clarified the situation. There is obviously still considerable progress to be made.

The set of operations that can be performed on a molecule constitutes a group in the mathematical sense. A set of operations is a group if the following are true:

1. The product of two or more operations is equivalent to a single operation that also belongs to the set.
2. The operations obey the associative law of multiplication.
3. The set includes the identity operator.
4. The inverse of every operation is also a member of the set.

In the first sections of this review we collate and compare the results from various spectroscopies for liquids composed of molecules of C_{3v} symmetry. We choose to classify our liquids in terms of the molecular symmetry because, as we shall see, the symmetry, and even the sizes of the atoms making up a molecule, determines the dynamics of a liquid system.

In later sections we extend our review and discussions to molecules of lower symmetry, and in other articles of the present volume we mention molecules with lower symmetry: the optically active liquids. The loss of symmetry affects the molecular dynamics. In particular, we shall see that it has a pronounced effect on at least one aspect of the dynamics, namely the way rotational and translational motions are mutually correlated.

The molecules of C_{3v} symmetry chosen for discussion have all been reported on extensively in the literature and are amenable to study by computer simulation using the new algorithms that have recently become available for polyatomic and "real" molecular liquids.

In reviewing the literature we look for insights into some of the following problems:

1. To what extent do collision-induced processes contribute to and distort measured spectral profiles? Depolarized-light-scattering and far-infrared absorption certainly exist for nondipolar spherical and tetrahedral molecules, in both the liquid and gaseous states. These spectra arise from a collision-induced mechanism: A fluctuating distortion of the symmetrical polarizability tensors of the isolated molecules is caused by the strong intermolecular interactions in the fluid. The interaction results in induced dipole moments (and consequently the far-infrared absorption) as well as asymmetry in the polarizability. It is probable that induced absorptions are present to some extent in all polar and nonpolar liquids and may give rise to measurable intensities in any of the spectroscopies. It has been postulated that for some liquids (e.g., the halogeno benzenes) the induced contributions make up as much as 50% of the total measured intensity.[1,2]

2. To what extent do rotational and translational motions contribute to the same profiles, and are they coupled? Evans and Evans[3] have shown through simulation and experiment how markedly the translation of a mole-

cule may affect *its own* rotation in the condensed state and have postulated that this effect explains the racemic modification in optically active liquids. The observation that a racemic mixture of lactic acid is a liquid at room temperature, yet its individual enantiomers are solid, may be explained in dynamical terms and, in particular, in terms of a rototranslational interaction that is modified in the racemic mixture relative to that in the individual enantiomers. These we term auto rotation–translation effects because they refer to the same molecule. But the rotation of a molecule may effect a simultaneous translation in neighboring molecules, so that rotation and translation may also be correlated in another sense; this we term cross rotation–translation. The significance of this effect has been clearly illustrated in a simple experiment by G. Ewing[4] on H_2 and HD in liquid argon. In H_2 in liquid argon, the rotational transitions observed in the far infrared do not differ from those calculated for the unperturbed gas. However, the spectrum of HD in liquid argon shows larger half-widths, erratic frequency shifts, and additional absorptions arising from the relaxation of the rotational selection rules, which is a consequence of rotation–translation coupling. The coupling perturbation is large in HD and small in H_2 because of the asymmetric mass distribution of the former—a small difference produces a pronounced effect. This cross rotation–translation coupling must also be significant in other liquids, but is not easily distinguished when the individual fine structure is not resolved, as is the situation in most liquids.

While discussing rotation–translation interaction we should also recall a hydrodynamic phenomenon in which local strains set up by transverse shear waves are assumed to be relieved by collective reorientations—a macroscopic translation–rotation interaction. This results in a shear-wave or "Rytov" dip observed in depolarized scattering. However, Berne and Pecora[5] point out the discrepancies that result from this hydrodynamical treatment. As molecular spectroscopists, we believe that all such phenomena have molecular origins, and indeed it is already more popular to treat this phenomenon with molecular models. Having said this, we emphasize that the phenomenon does not appear to be related to a specific microscopic structure and has been observed in a number of molecular liquids, including both polar and nonpolar liquids, both liquids composed of planar and those made of nonplanar molecules, and both liquids made of large and liquids made of small molecules. It would be interesting to characterize the phenomenon in terms of the molecular symmetry and of the atoms comprised by the molecule, following the procedure we shall adopt for the racemic modification in optically active liquids (discussed elsewhere in the present volume).

3.   Are other modes of motion coupled [e.g., rotation–vibration, vibration–vibration (of two neighboring molecules), etc.]? If modes of motion are coupled, then spectral profiles may be severely distorted and not amenable

to conventional data reduction and theoretical analysis, there being no theories for these coupled motions. Infrared and Raman profiles are certainly obtained by the convolution of a rotational and a vibrational profile. Lynden-Bell[6] has emphasized that is a problem of some magnitude, because these components may be coupled to give complex, non-Lorentzian line shapes that are sometimes broader, sometimes narrower, than expected and that on occasions show central dips. It is a challenge to experimentalists to observe this central dip, which may occur when the solute motion is significantly faster than that of the solvent it is dissolved in. We suggest that it may be observed when a small dipolar molecule is dissolved in a glass or liquid crystalline system. It may also be shown that the linewidths and line shapes depend on the type of spectrum observed; that is, the modifications of the Raman ($\ell = 2$) and infrared ($\ell = 1$) profiles may not be the same.

4. How significant are cooperative phenomena (cross correlations) and the internal field in modifying the observed spectrum? That they *are* significant is established by dilution studies, which may, *but do not always*, reduce the problem. The problem may actually be increased when a probe solute is dissolved in a nonpolar solvent. These phenomena are poorly understood, but have their origin on the molecular scale; the cross rotation–translation interaction discussed above is an obvious example. These phenomena are unquestionably strong in some liquids; we consider one (acetonitrile) in Section IV. There, the existence of dimers and larger aggregates of molecules is postulated.[7] If this postulate is true, contributions to spectral profiles may arise from the rotation, translation, and vibration of single molecules and of dimers and larger aggregates of molecules, all of which may be coupled. Evans has postulated the existence of collective modes in acetonitrile, as have other workers considering different liquids and using different experiments (see Section II). These studies will be recalled in relevant sections.

Molecular liquids obviously pose significant problems for the deciphering of dynamic motions. The experimental situation is a complex one; no one technique yields a complete picture of the dynamic process because each emphasizes particular aspects of the motion, which are not easily resolved.

Computer simulation, when used in conjunction with experiments, may clarify certain aspects of dynamic motions. In fact, as we shall see, computer simulation does indeed clarify the role of rotation–translation coupling, particularly in the study of optically active liquids. This role is pronounced, except in the few instances of liquids forming rotator phases. The coupling increases in magnitude near the melting point, but decreases near the boiling point. Its magnitude in the enantiomers of an optically active system may differ from that in a 50:50 racemic mixture. All of this will be discussed at length in the following sections.

II. CHLOROFORM (CHCl$_3$)

We will start these review sections with a study of liquid chloroform [see M. W. Evans, *J. Mol. Liq.* **25**, 211 (1983)]. As in subsequent sections, we will review the literature and discuss the conclusions reached from such a survey. We will compare and evaluate these literature results with our own computer molecular-dynamics calculations, and will see that the situation is, in general, a confused one. The simulations at least clarify certain aspects of the discrepancies. A full treatment of cooperative phenomena lies outside the scope of analytical theories and present-day simulations, but a pattern does emerge indicating in which liquids such phenomena are significant and to what degree the various experimental probes are affected. We will discuss the role of rotation–translation coupling in all of the liquids considered.

Chloroform must be the most extensively studied molecular liquid. Evans[8] analyzed over 100 papers on its structure and dynamics in compiling his review and making comparisons with new simulation data. The structure of the liquid has been investigated with atom–atom pair distribution functions, which are obtainable from neutron- and X-ray-scattering experiments and the dynamics has been researched with most of the spectroscopies. Results of the structure investigations suggest that on the local level, a significant oriented structure exists in the pure liquid.

In a neutron-scattering experiment, a double-differential scattering cross section, $d^2\sigma/d\Omega\, dE$, is measured. It represents the number of neutrons scattered per unit solid angle and unit energy interval. In a diffraction experiment, the scattered neutrons are collected for each scattering angle, but no energy analysis is performed. The diffraction cross section may be split into coherent and incoherent terms, and the former into a self and a distinct term. The distinct term is related to the atomic pair correlation function $g_{\alpha\beta}(r)$, which gives the probability of finding a nucleus β at a separation r from a nucleus α. To obtain information about the pair correlation function, the maximum possible number of diffraction studies must be carried out. For chloroform, results of four diffraction experiments are available: X-ray data for $CH^{35}Cl_3$ and neutron data for $CH^{35}Cl_3$, $CD^{35}Cl_3$, and $CD^{37}Cl_3$.

In molecular liquids it is more appropriate to interpret the coherent differential cross section in terms of molecular pair correlations. The intermolecular terms contributing to the coherent distinct differential cross section for a molecule of C_{3v} symmetry may be written as[9]

$$\left(\frac{d\sigma}{d\Omega}\right)_{\text{inter}} = a_0^{(0)}a_0^{(0)*}h_{00}^{(000)} + 0.385 a_0^{(1)}a_0^{(0)*}h_{00}^{(101)}$$
$$+ 0.179 a_0^{(2)*}h_{00}^{(202)} - 0.064 a_0^{(1)}a_0^{(1)}h_{00}^{(110)} \tag{1}$$

where the $a_0^{(\ell)}$ are molecular scattering factors that can be calculated from the atomic scattering lengths and the atomic positions within the molecule and $h_{m_i m_j}^{(\ell_i \ell_j \ell_{ij})}$ is the Fourier–Bessel transform of the coefficients of the molecular pair correlation with the termination $\ell_i = \ell_j = 2$ being made. The coefficients $h_{00}^{(000)}$, $h_{00}^{(101)}$, $h_{00}^{(202)}$, and $h_{00}^{(110)}$ were determined by carrying out the four experiments above and solving the corresponding four simultaneous equations. These allowed us to calculate four coefficients of the molecular pair correlation function: $g_{00}^{(000)}$, $g_{00}^{(101)}$, $g_{00}^{(202)}$, and $g_{00}^{(110)}$.

The center–center correlation term, $g_{00}^{(000)}$ shows a peak at 4.8 Å with an area corresponding to six nearest neighbors.

The coefficients $g_{00}^{(101)}$ and $g_{00}^{(202)}$ are orientational correlation terms giving information concerning the orientation of a molecule relative to the center–center system irrespective of the orientation of the partner molecules. Above 4 Å $g_{00}^{(101)}$ is negative and $g_{00}^{(202)}$ is positive. Consequently, stronger parallel alignment of the molecular C_3 axis along the center–center line must be present ($0° \leq \beta \leq 54.7°$).

If the z-axis of the system is chosen to coincide with the C_3 axis of one chloroform molecule and β to the the angle between this axis and either the line joining the centers of the two molecules or the C_3 axis of a second molecule, then it is shown that: (1) for $\beta < 90°$ (parallel alignment of the C_3 axis), a negative $g_{00}^{(110)}$ results; and (2) for $\beta > 90°$ (antiparallel alignment), a positive $g_{00}^{(110)}$ results. Thus parallel and antiparallel alignment changes from one coordinate sphere to another.

In dielectric and light-scattering experiments, the correlation parameters $g_{00}^{(\ell)}$ (where $\ell = 1$ and $\ell = 2$, respectively) occur. These are related to the dipole–dipole coefficient $g_{00}^{(\ell\ell 0)}$ of the molecular point correlation function by the relation

$$g_{00}^{(\ell)} - 1 = (-1)^{\ell} (2\ell + 1)^{-3/2} p \int_0^{\infty} g_{00}^{(\ell\ell 0)} 4\pi R^2 \, dR$$

$$= (-1)^{\ell} (2\ell + 1)^{-3/2} h_{00}^{(\ell\ell 0)} \qquad (\kappa = 0)$$

where for chloroform for $\ell = 1$, $g_{00}^{(1)} = 0.5 \pm 0.04$ from diffraction results. An estimation of $g_{00}^{(1)}$ from Kirkwood–Frohlich theory yields $g_{00}^{(1)} = 1.33 \pm 0.07$. It seems certain from neutron- and X-ray-scattering data that $g_{00}^{(1)}$ *must* be less than 1. The diffraction experimenters show that this discrepancy is a consequence of incorrect estimates of the dielectric constants or dipole moments in the fluid state. If we compare with the computer-simulated atom–atom pair distribution functions (pdfs) computed by M. W. Evans, we find that overall the pdfs are similar, although detailed agreement is not obtained (Figs. 1–3). The experimental results of Bertagnolli suggest that the

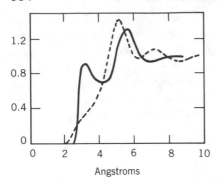

Figure 1. Atom–atom pair distribution functions: — , H—Cl; - - -, H—H. [Reproduced by permission from M. W. Evans, *J. Mol. Liq.*, **25**, 211 (1983).]

chloroform structure is more tightly packed than the simulation parameters allow for (Fig. 1). However, this picture is not corroborated by Figs. 2 and 3, because the positions of the first peaks in the experimental and these numerical data agree closely. Thereafter the two data sources are less consistent. The position of the first peak in each pdf is satisfactorily established, but the noise level of the experimental pdfs is too great for detailed comment beyond this.

We should remember that there are considerable uncertainties associated with the diffraction study. Diffraction data represent a one-dimensional quantity that is the average of a number of quantities that vary with the three dimensions of the system, with the relative orientations within the system, and with time. A one-dimensional experiment cannot map a multidimensional function. There are also numerous sources of systematic error in the intricate process of reducing from the experimentally observed function to the distribution function describing the liquid structure. The overall experimental precision of ~1% currently obtainable is barely adequate, and for

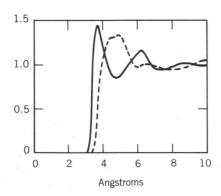

Figure 2. Atom–atom pair distribution functions: — , Cl—Cl; - - -, Cl—C. [Reproduced by permission from M. W. Evans, *J. Mol. Liq.*, **25**, 211 (1983).]

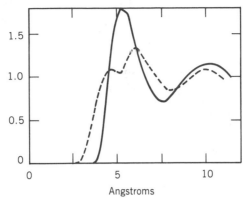

Figure 3.   Atom–atom pair distribution functions: — , C—C; ---, C—H. [Reproduced by permission from M. W. Evans, *J. Mol. Liq.*, **25**, 211 (1983).]

data prior to 1960 systematic errors of 10% and more were the rule rather than the exception.

As we shall see in the following sections, problems of data reduction and systematic errors also plague the experiments used to probe the molecular dynamics. Consequently, disagreement among various experimental results is often present. We propose that computer simulation used in conjunction with the various spectroscopies may help clarify at least certain aspects of the dynamics. But first we should consider the contradictions that result when spectroscopic experiments are used in isolation to study chloroform.

All of the techniques use various assumptions and approximations in the data-reduction processes, which are used for transforming raw spectral data into correlation functions, correlation times, or some other function amenable to comparison with theory.

The early papers in the fields of infrared absorption, Raman, and Rayleigh scattering have been summarized by Brodbeck et al.[10] All calculated correlation times are compared with the simulation results of M. W. Evans in Table I. [The correlation time is defined here as the area under the correlation function (though the existence of many other definitions testifies to the confused state of the art). Correlation times allow the convenient comparison of various techniques.] Brodbeck et al. discuss the spread in the first- and second-rank orientational correlation times from various sources and comment further on the need for a universal definition of correlation time. First-rank orientational correlation times vary from 2.3 to 4.0 ps depending on definition. According to Brodbeck et al., a comparison of relaxation times from the ν_1, ν_2, and ν_3 (A_1 symmetry) fundamentals of chloroform reveals rotation–vibration coupling to be insignificant.

TABLE I
Experimental and Simulated Correlation Times for Chloroform[a]

Technique	Correlation times (at ambient T unless otherwise indicated)	Computer simulated correlation times (293 K).
Dielectric relaxation	Pure liquid: 5.4–6.4 ps Dilute *n*-hexane solution: 2.9 ps Dilute CCl_4 solution: 6.1 ps Pure liquid: 4.7–6.2 ps Dilute cyclohexane soln: 3.2 ps 10% v/v decalin soln: 4.3 ± 0.3 ps Dilute CCl_4 soln: 5.0 ps Pure liquid: 6.0 ps (294 K) 17.0 ps (223 K)	Pure chloroform at 293 K, 1 bar. First rank *autocorrelation* time of the dipole (C_{3v} axis) unit vector $\mathbf{e}_3$ $\tau_1(\mathbf{e}_3) = 3.6$ ps
Nuclear magnetic resonance relaxation	^{35}Cl transverse nuclear quadrupole relaxation times 2.0 ps (pure liquid); 1.8 ps (1.3 *n*-hexane); 3.4 ps (1.3 CH_2I_2)	A weighted mean of three second-rank correlation times of unit vectors $\mathbf{e}_1$, $\mathbf{e}_2$ and $\mathbf{e}_3$ in the principal moment of inertia axes $\tau_2(\mathbf{e}_1) = \tau_2(\mathbf{e}_2) = 1.5$ ps $\tau_2(\mathbf{e}_3) = 1.3$ ps
	^{13}C spin–lattice relaxation time: 1.60 ps D nuclear quadrupole relaxation: 1.84 ps Spin-spin relaxation: 1.5 ps, 1.74 ps Proton spin–lattice relaxation: 1.39 ps (298 K); 3.0 ps (219 K); 1.07 ps (363 K)	$\tau_2(\mathbf{e}_3) = 1.3$ ps $\tau_2(\mathbf{e}_3) = 1.3$ ps $\tau_2(\mathbf{e}_3) = 1.3$ ps A mean of $\tau_2(\mathbf{e}_3) = 1.3$ ps $\tau_2(\mathbf{e}_1) = \tau_2(\mathbf{e}_2) = 1.5$ ps
	$\nu_3(A_1)$ (symm. CCl bend, $CDCl_3$): 5.3 ps, 6.8 ps, 6.6 ps. Variation of $\tau_2(\mathbf{e}_3)$ with def. 1.3–1.5 ps. Area defn. = 1.3 ps $\nu_1(A_1)$: 1.45 ± 0.15 ps (pure $CHCl_3$)$\nu_1(A_1)$: 1.20 ± 0.15 ps (20% mole fraction in CCl_4); 2.61 ± 0.15 ps (33% mole fraction in $(CD_3)_2CO$)	$\tau_2(\mathbf{e}_3) = 1.3$ ps (area defn.) $\tau_2(\mathbf{e}_3) = 1.3$ ps
Depolarized Rayleigh scattering	$\tau_r = 1.7$ ps $\tau_r(CDCl_3) = 2.9$ ps $\tau_r \pm 0.2$ ps (3.2) (1 bar) $\tau_r = 5.5(5) \pm 0.2$ ps (at 205 MPa $\tau_r = 2.95$ ps $\tau_r = 1.74$ ps (in 20% CCl_4 $\tau_g = 0.24$ ps (far wing Gaussian)	A weighted mean of $\tau_2(\mathbf{e}_1) = \tau_2$ $(\mathbf{e}_2) = 1.5$ ps $\tau_2(\mathbf{e}_3) = 1.3$ ps (N.B. these are *autocorrelation* times)

TABLE I (*Continued*)

Technique	Correlation times (at ambient T unless otherwise indicated)	Computer simulated correlation times (293 K).
Far infrared absorption	Experimental peak frequencies (cm^{-1}) $\bar{\nu}_{max} = 36$ cm^{-1} a range of $\quad\quad\quad = 45$ cm^{-1} 15 cm^{-1} $\quad\quad\quad = 38$ cm^{-1} in the $\bar{\nu}_{max} = 30$ cm^{-1} literature (in 10% decalin v/v) $\bar{\nu}_{max} = 39$ cm^{-1} (in 10% decalin v/v) $\bar{\nu}_{max} = 25$ cm^{-1} (in 10% n-pentane v/v) $\bar{\nu}_{max} = 37$ cm^{-1} (in 10% diphenylmethane v/v) $\bar{\nu}_{max} = 50$ cm^{-1} (in 10% v/v decalin, supercooled to 110 K)	Simulated peak frequency, $\bar{\nu}_{max}$(sim.) $= 31$ cm^{-1} (N.B. from the rotational velocity *autocorrelation* function) Simulated peak frequency in pure $CHCl_3$ at 293 K, 1 bar $= 31$ cm^{-1} (Fig. 4)
	Intermolecular (translational) correlation times τ_c: 80.5 ps (298 K); 263 ps (219 K); 52.2 ps (363 K)	Calculated in the extreme narrowing limit from the center-of-mass velocity correlation time (τ_v) using $\tau_c = ma^2/12kT\tau_v = 2.1$ ps m = mass of mol. a = eff. radius
	A combination of D and ^{35}Cl nuclear quadrupole relaxation $\tau_{\parallel}$ (spinning) $= 0.92$ ps $\tau_{\perp}$ (tumbling) $= 1.8$ ps ^{13}C, nuclear Overhauser enhancement ^{13}C—^{1}H: 1.26 ± 0.04 ps	$\tau_2(e_1) = \tau_2(e_2) = 1.5$ ps $\tau_2(e_3) = 1.3$ ps $\tau_2(e_3) = 1.3$ ps
Infrared bandshape analysis	N.B. These orientational times have been derived assuming that rotation–vibration coupling is negligible $\nu_1(A_1)$ (C—H stretch): 2.3–2.9 ps $\nu_1(A_1)$: variation with defn. of correlation time: 2.3–4.0 ps Area defn. $= 2.6$ ps	$\tau_1(e_3) = 3.6$ ps $\tau_1(e_3) = 3.6$ ps (area defn.)
Raman bandshape analysis	$\nu_1(A_1)$ (C—H stretch) 1.5 ps; 1.5 ps; literature 1.97 ps; 1.2 ps; variation 1.3 ps; 1.7 ps; 1.59 ps	$\tau_2(e_3) = 1.3$ ps

TABLE I (*Continued*)

Technique	Correlation times (at ambient T unless otherwise indicated)	Computer simulated correlation times (293 K).
	$\nu_1(A_1)$ (CDCl$_3$): 1.7 ps literature 1.96 ps; 1.8 ps; variation 5.1 ps; 2.0 ps 3.1 ps	$\tau_2(e_1) = \tau_2(e_2) = 1.5$ ps $\tau_2(e_3) = 1.3$ ps
	$\nu_2(A_1)$ (C—Cl stretch) literature 2.4 ps; 1.8 ps; variation 1.4 ps	
	$\nu_2(A_1)$ (C—Cl stretch literature CDCl$_3$): 2.4 ps [15]; variation 1.85 ps; 3.4 ps; 4.4 ps	
	Peak of Rayleigh second moment $= 45$ cm^{-1} Peak of $\nu_1(A_1)$ Raman second moment $= 50$ cm^{-1}	
Electric-field induced birefringence	Kirkwood factor $g_1 = 1.4$ Higher-order correlation factors: $g_2 = 1.7$, $g_3 = -0.4$ The factor g_1 compares with $g_1 = 1.32$ from far-infrared analysis and $g_1 = 1.25$ from dielectric permittivity	

[a]Compiled by M. W. Evans, *J. Mol. Liq.*, **25**, 211 (1983) and reproduced by permission. For references to original data sources, see this review in *J. Mol. Liq.*

The broadening of the ν_1 (C—H stretch) fundamental in liquid chloroform leads to an orientational autocorrelation function for the tumbling of this axis (the C_{3v}, or dipole, axis). Brodbeck et al. list first-rank P_1 (infrared) and second-rank P_2 (Raman) correlation times for this motion. For natural abundance CHCl$_3$ the former vary from 2.3 to 2.9 ps according to the depolarization ratio and the latter from 1.3 to 1.7 ps. The Raman orientational correlation times from the ν_2 (symmetric C—Cl stretch) range from 1.4 to 2.4 ps. These are affected by spinning and tumbling of the C_{3v} axis.

It is generally accepted that orientational correlation times from infrared and Raman bands are autocorrelation times, whereas the equivalent times from dielectric relaxation (first rank) and depolarized Rayleigh scattering (second rank) reflect the motions of many molecules, involving cross correlations. Soussen-Jacob et al.[11] have reported that the dielectric relaxation times varied from 5.4 to 6.4 ps in pure CHCl$_3$. They reported values of 2.9 and 6.1 ps for HCCl$_3$ diluted in *n*-hexane and CCl$_4$, respectively.

Systematic measurements have been reported in the dielectric/far-infrared region by Gerschel et al.[12] Gerschel and Brot[13] measured the static permittivity of chloroform along the gas–liquid coexistence curve to the critical point. The static Kirkwood g factor depends on the steric configuration, especially on the orientation of the dipole axis relative to the shape of the molecule. The value of g for the liquid phase of chloroform changes with temperature, and approaches unity at the critical point. Its value is the expected one of unity at all densities in the coexisting vapor, and about 1.25 in the liquid over a range of temperatures. The evolution of the dielectric relaxation time reported by Gerschel follows an Arrhenius law over a wide range of temperatures, but there is a definite temperature at which the liquid like rotational process evolves into a gaslike process in which the rotational motion is relatively freer. There is no particular critical phenomenon such as observable opalescence in the far infrared.

The study of a liquid in this way, over a range of state points, is more profitable than at one state point, as in the majority of the reported studies. Density has very rarely been used as a variable. An exception is the work of Jonas et al.,[14] who made a systematic study of liquid chloroform under hydrostatic pressure. They use second-moment analysis to reveal that "collision induced effects" play an important role in causing the second moment of the $\nu_1(A_1)$ band in liquid $CHCl_3$ and $CDCl_3$ to be density dependent. The collisional contribution to the second moment decreases with increasing density, and is revealed only by constructing the second-moment spectra using the far wings of the Raman line. These results, however, contradict those of Konynenberg et al.,[15] who suggest that there is no collisional contribution to the rotational second moment. Schroeder and Jonas[16] also measured the depolarized Rayleigh wing up to 4 kbar and out to about 200 cm^{-1}. Whereas the Raman bandshapes had dropped off fairly rapidly at 80 cm^{-1}, the Rayleigh bandshapes continue to about 180 cm^{-1} before decaying rapidly. They interpret this in terms of many-body collision effects at short times (in the range 0.03–0.5 ps).

The far infrared also produces the second moment, and Lund et al.[17] showed that Rayleigh and far-infrared second-moment spectra peak at the same frequency (45 cm^{-1}) and are similar in shape.

Claesson et al.[18] have also reported Rayleigh correlation times to 205 MPa. These vary linearly from 3.2 ± 0.3 ps at 0.1 MPa. The dependence of the rotational relaxation time on bulk viscosity at constant temperature is also linear, with a nonzero intercept. Claesson et al. compare their results with high-pressure nuclear magnetic resonance (NMR) and Raman data and discuss the effect of pressure on pair correlations. The NMR rotational relaxation times intersect the zero viscosity axis at different points from, and do not lie on the same straight line as, the Rayleigh correlation times. On the basis of these measurements pair correlations would seem to be significant in chloroform. Claesson et al. point out that the Raman and NMR correlation times do not agree except at atmospheric pressure. A comparison of the Rayleigh and Raman times shows that pair correlations decrease as hydrostatic pressure is increased, which conflicts with the view of Schroeder and co-workers. In the absence of collision-induced effects, τ_ω should increase faster than τ_2, according to Claesson, if only because of the increase in the number of scatterers (N) in the scattering volume. However, that this is not observed might be explained if the collision induction hypothesis of Schroeder is correct.

The NMR and Rayleigh scattering times of Alms et al.[18] are also shown in Claesson's plot. On the basis of the reported measurements, pair correlations seem to be significant in chloroform. At high pressures the Rayleigh light-scattering and Raman times approach each other, suggesting that the magnitudes of the static and dynamic contributions to the pair correlations change with pressure, that the approximations used to obtain single-particle relaxation times from these scattering techniques are not valid, or a combination of both. Alms et al. show that the depolarized Rayleigh times they calculated are strongly concentration dependent for chloroform, perhaps supporting their view that strong pair correlations affect the data at high concentrations. Or it might support the different interpretation, in terms of collision-induced absorption, of the same set of results by Jonas and co-workers. Alms et al. report a dynamic correlation factor of 1 for pure liquid chloroform, which may be compared with Gerschel's static Kirkwood factor of 1.25.

The confusion and contradiction in the literature concerning the dynamics from Raman and Rayleigh scattering alone is already apparent, and it continues. Kamagawa,[20] for example, has reported a depolarized Rayleigh study of $CHCl_3$ in CCl_4 in which the relaxation times are independent of the concentration of solution, in direct contradiction to the findings of Alms et al. Kamagawa analyzes his spectra in terms of several different relaxation times (overcomplicating the problem), calculated from the half-widths of the central and far wing positions. He comes to a conclusion opposite to that of Schroder et al. concerning the collision-induced effects, suggesting, on the basis of his observed relaxation times, that the effects cannot be important. He asserts that if the spectrum is caused by a binary collision process, A should be proportional to N^2, whereas he observes A to be linear with N (where A is the integrated intensity and N the number density).

Analysis by M. W. Evans of far-infrared collision-induced processes (ref. 21, Chapter 11) shows that collision-induced effects are not pairwise additive in molecular terms in the *liquid state* (in the gas phase, an N^2 dependence may exist). The far-infrared evidence does agree in one respect with

Kamagawa's analysis, in that A is linear in N. However, the problem is not easily resolved with spectroscopic evidence alone. There are examples of liquids in which the collision-induced contribution is displaced from the permanent dipole absorption (an example is N_2O). But there is not always such a displacement to ease the interpretation, and Lund et al.,[17] who construct the second moment of the Rayleigh wing spectrum and compare it with the far infrared, establish that there is no distortion of the bandshapes in chloroform (the bandshapes from both spectral sources are similar, and peak at 45 cm^{-1}). This is typical of the problems associated with analyzing a broad, featureless band. All we can really say based on these results is that in chloroform either collision-induced effects are small or they have the *same frequency dependence* as the underlying reorientational process at ambient temperature and pressure. Any far-infrared conclusion, of course, must directly contradict one of these two extreme viewpoints arrived at in the literature from different probe experiments on the same sample.

The induced contributions to the Raman spectra are likewise uncertain. Schroeder et al.[14] construct the second moment of the far wing of the Raman v_1 mode at pressures up to 4 kbar. At 30 bars this peaks at about 50 cm^{-1} (cf. the 15 cm^{-1} peak of Lund et al.[17] based on the Rayleigh wing), and the spectrum is noticeably similar in appearance to the zero-THz spectrum.[22] The area under this curve *decreases* rapidly with increasing number density as the hydrostatic pressure is increased to 5 kbar at 303 K. On this basis, Schroeder et al. conclude that multibody collision induction is responsible for the shape of the second-moment Raman spectrum of the chloroform v_1 mode. The Rayleigh wings reported by Schroeder et al. behave similarly to the Raman v_1 wing, and it is clear that the integrated intensity per molecule of the Rayleigh spectrum decreases in the wing portion as the pressure is increased to 4 kbar from 30 bars at 303 K. However, in neither the Raman nor the Rayleigh experiments does the behavior of the low-frequency part of the scattered intensity (close to the exciting line) as a function of pressure emerge. The *complete range* (0 THz dielectric, plus far infrared) should be the entity for analysis, as we have observed in our own spectroscopy. The low-frequency Raman and Rayleigh scattered intensities must dominate the overall integrated intensity. Conversely, if second moments are calculated (e.g., the far infrared), the low-frequency components are suppressed and the high-frequency components dominate. (We have discussed this at length elsewhere.[23]) Analyses should proceed through the zeroth, second, and higher moments if a satisfactory picture of the various contributory processes is to emerge.

It is possible, for example, to interpret the results of Schroeder in another way: in terms of dimer formation. If the population of dimers (which might even be weakly hydrogen bonded) were to increase with hydrostatic pressure, which is plausible, then the effective moment of inertia of C—H scattering units would increase. This would show up as a decrease in the second moment of the Raman ν_1 band. Suzuki et al.[24] report that the infrared ν_1 band is markedly asymmetric, with a tailing on the high-frequency side at ambient temperature and pressure. They point out that the integrated intensity of the chloroform infrared ν_1 band increases from 11 to 198 cm^2 mol^{-1} on passage from the gas to the liquid phase, and interpret this in terms of hydrogen bonding. Rothschild et al.[25] reject this hypothesis on the basis of their infrared and Raman data, and several subsequent studies, including those of Schroeder et al., have been based on the assumption that the ν_1 band is homogeneous (i.e., unaffected by hydrogen bonding). Suzuki et al. point out that the infrared ν_1 fundamental is strongly asymmetric in bulk CHCl$_3$ and CDCl$_3$ and that its width in CDCl$_3$ is significantly narrower. Moradi-Araghi et al.[26] have postulated that vibration–rotation coupling is stronger for the ν_1 mode in CHCl$_3$ because of anharmonic terms in the vibrational Hamiltonian, but Suzuki et al. reinterpret this as the formation of dimers through hydrogen bonding. They suggest that the ν_1 infrared profile is composite, with contributions arising from a dimer species, the monomer, and a combination mode or pseudolattice vibration. The intensity of the dimer band is reported to decrease relative to that of the monomer band on dilution in CCl$_4$, but their positions remain constant. The peak frequency and absolute intensity of the monomer band are closer to the values for gaseous chloroform. It is precisely this high-frequency ν_1 profile that is reported by Schroeder et al. in their Raman studies. Suzuki et al. claim that *44% of the molecules in pure liquid CHCl$_3$* at ambient temperature and pressure *are bound into dimers*. They also comment on the uncritical use of the intensity of vibrational modes for the study of intermolecular interactions, asserting that a vibrational spectrum in the condensed phase is composed, in principle, of *different* band maxima and intensities, corresponding to different species of monomer and dimer.

There is little corroborative evidence from other sources for the existence of hydrogen bonding in liquid chloroform. In fact, Gerschel's dielectric–far-infrared studies suggest that the Kirkwood g factor remains near unity along the gas–liquid coexistence curve, although higher values for this factor have been proposed, as discussed earlier. The value of ΔU, the maximum energy of dipole–dipole interaction, for chloroform in the gas phase is, at ca. 900 cal mol^{-1}, one of the smallest calculated. For H$_2$O, CH$_3$OH, and CH$_3$CN, ΔU is 5–6 times larger.[27] Also, Tanabe et al.[28] have investigated the Raman ν_1 stretch of CHCl$_3$ in different solvents. The bandwidths of solutions of CHCl$_3$ in water are about twice as large as those of neat liquid, or of CHCl$_3$ in CCl$_4$ or C$_6$D$_{12}$. This contrasts with the behavior of solutes such as CH$_3$OH, CH$_3$CN, CH$_3$NO$_2$, and acetone, and provides unambiguous evidence that the hydrogen bonding between CHCl$_3$ and H$_2$O solvent molecules is considerably weaker than in these other cases. On this evidence, the hydrogen bonding between $CHCl_3$ molecules in the pure liquid should be weak, amounting to no more than a slightly increased probability for certain configurations over others. There is no significant evidence for dimer formation in the X-ray- and neutron-diffraction atom–atom pdfs of Bertagnolli et al.[9]

It seems clear, however, that the infrared, and possibly the Raman, ν_1 fundamentals in $CHCl_3$ and $CDCl_3$ liquids are asymmetric, and consequently it is by no means unequivocal that the results of Schroeder et al. may be attributed to collision-induced absorption. Considering all of the evidence, it seems more plausible that the asymmetry of the ν_1 mode is caused by inhomogeneous broadening. This has recently been discussed by Laubereau et al.,[29] who find that rotational coupling with vibration, Fermi resonance, and resonance-energy transfer strongly affect the spontaneous Raman data. The second-moment behavior observed by Schroeder et al. may be explained in terms of decreased inhomogeneous broadening with increasing hydrostatic pressure. This assumes that the stimulated (Laubereau et al.) and spontaneous (Schroeder et al.) Raman processes behave similarly in this respect. In one sense, the inhomogeneous-broadening and hydrogen-bonding schools of thought can be reconciled as both being descriptions of the anisotropy of the local environment.

An interesting experiment is that of Moradi-Araghi et al.,[30] who have recently reported in detail a variety of reorientational correlation times obtained from the polarized and depolarized Raman spectra of the ν_1 band in neat $CHCl_3$, $CDCl_3$, and solutions of $CHCl_3$ in CCl_4, acetonitrile, and acetone. The second-rank orientational correlation functions decay more rapidly in CCl_4 than in the neat liquid, but more slowly in the two dipolar solvents. This is contrary to expectations based on classical rotational theory, according to Moradi-Araghi et al. The present authors feel that in saying this, Moradi-Araghi et al. are being overcritical of the diffusion models, which cannot be expected to follow the behavior of a polar solute in a more polar, strongly interacting solvent. Moradi-Araghi et al. proceed to interpret their results in terms of specific solute–solvent interactions causing a loss of reorientational correlation and vibrational dephasing in each solvent. They assume complete decoupling of rotation from vibration—a view shared by Brodbeck et al.[10] but not by van Woerkom et al.[31] If we accept this approximation, the second-rank correlation times vary from 1.45 ± 0.15 ps in pure $CHCl_3$ at room temperature and pressure to 1.20 ± 0.15 ps in 20% (mole fraction) $CHCl_3$ in CCl_4. These correlation times *are not* proportional to the bulk viscosity.

Van Woerkom et al. use isotope dilution methods to show that vibration is not statistically independent of rotation in the infrared spectra of liquid $CHCl_3$, $CDCl_3$, and $CHCl_3$–$CDCl_3$ mixtures. This point of view is shared by Laubereau et al. for stimulated Raman scattering. The conclusion reached by Brodbeck et al. that rotation–vibration coupling is negligible conflicts with the conclusions of van Woerkom et al.

Wertheimer[32] has recently questioned the interpretation of isotropic Raman band contours in terms of vibrational autocorrelation functions. He considers other processes that may affect the broadening of a Raman band, including vibrational decoupling from excited quantum states, resonance transfer (vibrational exciton hopping), and pure dephasing (transition-frequency fluctuations). If he is right, the rotovibrational correlation function from a Raman band is collective and *not* a pure autocorrelation function. The complexity does not end there, because the homogeneous bandwidths are then only interpretable in terms of a sum of these processes, *together with cross terms from the interference mixing* of pure dephasing and resonance-transfer processes and of resonance-transfer transitions involving *different* pairs of molecules. Resonance-transfer contributions are reduced on dilution (for experimental purposes, isotopic dilution is convenient). Recall that Suzuki et al. produced evidence suggesting that the v_1 mode of $CHCl_3$ is asymmetric and, in a sense, made up of weakly separated bands. For such circumstances Wertheimer provides a series of theoretical results tracing the origins of these weakly separated bands to the equivalent excitations in isotopic mixtures. The widths of the vibrational self-correlation functions are found to be essentially independent of concentration, because the nearly resonant transfer involving molecules of different types is almost as fast as the resonance transfer between molecules of the same species. The collective modes, involving two or more molecules, are dynamically coupled.

Wertheimer calls these processes "collision induced" processes. These differ from the collision-induced processes observed in the far-infrared or Rayleigh scattering experiments. The latter are multibody, non-pairwise-additive processes that involve the molecular polarizability anisotropy explicitly. Wertheimer assumes that the dynamic isotropic polarizability α of the liquid system is *a pairwise-additive* superposition of the polarizabilities of its individual molecules. This contradicts the Rayleigh–Schrödinger picture of basic quantum mechanics, in which polarizability is clearly not molecularly pair additive. Wertheimer's collision-induced processes are what we more normally call cross-correlations, that is, statistical influences of the motion of one molecule on that of another.

Wertheimer also summarizes the basic (and incorrect) assumptions about rotation–vibration coupling and cross-correlations made in the majority of papers in this field. Without these assumptions, the results of Schroeder et al. may be interpreted as indicative of the presence and significance of cross-correlations in the v_1 mode because an increasing number density with hydrostatic pressure would not, in principle, cause the second moment to decrease as observed. Döge et al.[33] have shown, in a careful study of chloroform dissolved in CH_3I solvent, how molecular-environment changes affect the ν_1 mode of $CHCl_3$: The band shape is "drastically" broadened.

Carlson et al [24] have concluded that half of the scattering for a system of independent, noninteracting molecules is affected by cross-correlations (in the sense of Wertheimer) and by transient, collision-induced perturbations. They also argue that the Kerr effect and Cotton–Mouton effect (which we shall consider later) are unaffected by collision-induced phenomena. They proceed to develop an experimental technique, based on interference filters, for estimating the collision-induced intensity of the depolarized Rayleigh spectrum. This is based on the theory of Bucaro et al.,[35] who assume the collision-induced processes to arise from isolated binary collisions. However, the Kerr and Colton–Mouton effects are probably *not* free of induced effects, because collision processes affect also μ, the permanent dipole moment. The effect may be observed directly in the far infrared, where nondipolar molecules absorb over a broad range of frequencies because of induced temporary dipoles.

All in all, there is considerable uncertainty about the role of collision-induced absorption in the Rayleigh and Raman spectra of $CHCl_3$. Some authors do not discuss it or claim it to be unimportant. Others claim it accounts for as much as one-half of the relevant spectral intensity. More work is required on intensity measurements by Rayleigh scattering, on compressed vapors such as those of $CHCl_3$, and along the gas–liquid coexistence line, following the excellent work of Gerschel. Ho and Tabisz[36] emphasize the need for a theory of "close encounters" in the collision-induced Rayleigh scattering, assumed to be present in liquid chloroform.

Lastly, there is the question of the internal field to be solved. Burnham et al.[37] have made a detailed study of the internal-field corrections available in the literature, and point out that for Raman and Rayleigh scattering both the choice of the model and the local-field corrections are in question. To explain both the observed isotropic intensity and the depolarization ratio, it is necessary to use an ellipsoidal model of the internal field. If we wish to be able to comment on collision-induced depolarized Rayleigh intensities, we must first be concerned with solving the problem for any molecule whose polarizability tensor has anisotropic components.

The list of far-infrared and dielectric relaxation measurements available for chloroform is almost as lengthy as that of infrared, Raman, and Rayleigh scattering experiments. The dielectric and far-infrared spectra, considered *as an entity*, we call the 0 THz spectrum. The far infrared is the high-frequency adjunct of the low-frequency loss first considered in detail by Debye. It is unsatisfactory to consider one part of the frequency range without the other, because both together define the total molecular-dynamical evolution of an ensemble of molecules in the condensed state. The two processes (the long- and short-time details) are separated by only a few decades of frequency in normal isotropic liquids, but may be separated by many decades of frequency (up to 14) in, for example, systems forming glassy or disordered states and liquid crystalline systems.

It is misleading to consider only a part of the total frequency span. Debye's model and extended rotational diffusion models, for example, work adequately at the lower frequencies, but may not even predict the existence of the high-frequency component. Debye's theory predicts that all liquids are more or less opaque at frequencies above 10^{12} Hz, including, of course, the visible-light frequencies. Yet it is still used frequently in the literature, either openly or in somewhat disguised form. Debye's model is a gross oversimplification the molecular dynamics that, particularly if used in data-reduction processes, confuses the interpretation of (and conclusions drawn from) experimental results. Experimentalists would not, in an ideal world, use these simplistic theories based on subjective ideas and incorporating so many *ad hoc* assumptions and often numerous adjustable parameters. As Gerschel has shown, it is far more profitable to expend one's energy in more detailed, precise, and extensive experimentation than to analyze a few room-temperature spectra with these models.

We have already reviewed Gerschel's studies of phenomena along the gas–liquid coexistence line. The rapid increase in interest in 0 THz spectrum is justified because of its ability to provide zeroth and second moments of the spectral system routinely and *without* the complication of mixing of intra- and intermolecular motions, which we have seen may severely distort Raman and infrared band profiles. Evans et al.[38] have indicated how one may obtain fourth and sixth moments from the high-frequency wing of the far-infrared band *providing* they are free of proper mode interference. Choosing the molecular liquid carefully is essential. An acceptable theory should be able to reproduce all of these spectral moments, not just the first one or two (the usual literature procedure).

In addition to the substantial shifts in the frequency of maximum absorption ($\bar{\nu}_{max}$) observed by Gerschel, $\bar{\nu}_{max}$ shifts occur on dilution in nonpolar solvents. According to Leroy et al.,[39] the pure-liquid spectrum peaks at 36 cm^{-1}, but the location of the peak varies from 25 to 33 cm^{-1} in various solvents (e.g., it is at 30 cm^{-1} in decalin). These shifts reflect directly the importance of the environment in determining the molecular dynamics, but there is *no straightforward and predictable dependence*. When chloroform is dissolved in decalin and *n*-pentane, $\bar{\nu}_{max}$ shifts from 36 cm^{-1} to lower frequencies, as anticipated and in accord with the predictions of molecular-dynamics theories. But in CCl_4, $\bar{\nu}_{max}$ actually shifts to *higher* frequencies. So too in diphenylmethane a smaller but discernible shift to higher frequen-

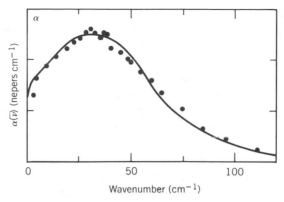

Figure 4.   The far-infrared spectrum of chloroform at 293 K, 1 bar: computed (●) and measured (—). [Reproduced by permission from M. W. Evans, *J. Mol. Liq.*, **25**, 211 (1983).]

cies occurs. The data show the extreme importance of the local molecular environment in determining the measured spectra. No current simple diffusional theory would predict these shifts to higher frequencies on dilution. The solvent is generally assumed to reduce dipolar cross-correlations, but apparently such cross-correlations may actually be increased in some solvents. (We will see further examples of this in the later sections on CH_3I and, in particular, CH_3CN dissolved in CCl_4.) In such solvent shifts, the reduction of cross-correlations and solvent rotational hindrance and interaction are contributory and competing factors that affect the overall dynamics of the solute probe. Nuclear magnetic resonance spectroscopists, who use dilution methods extensively to obtain correlation times, in particular should bear these factors in mind. Evans[22] compared far-infrared data with low-frequency results and the far-infrared results of other groups. His data are corrected for the contribution of an intermolecular mode at higher frequencies (Fig. 4).

The effect of hydrostatic pressure on this spectrum is to shift the peak to higher frequencies, as Gerschel showed by heating the liquid in a closed vessel. The effect of supercooling in decalin solvent is again to shift $\bar{\nu}_{max}$ to considerably higher frequencies and to sharpen the band profile.

The integrated intensity of the far-infrared band has been compared with the theoretical predictions based on the Gordon sum rule for the pure liquid by Hindle et al.[40] and for a 10% solution in decalin by Reid and Evans.[4] Both groups use a simple frequency-independent correction for the internal field—an oversimplification. However, Hindle et al. find that 88% of the observed intensity is produced by the sum rule, whereas Reid and Evans find that 73% of the intensity is produced by the sum rule in a 10% solution in decalin. This suggests that there is at least a 25% contribution to the measured intensities, even in a 10% solution, that *may not* be attributed to a purely rotational origin—assuming the validity of the Gordon sum rule.

Bossis[42] has discussed the application of this sum rule to highly dipolar molecules, in which in principle a mechanism of dipole-induced dipole induction should cause the theoretical intensity to be supplemented considerably. He elaborates on the original work of Gordon and concludes that induced effects are unlikely to be important in liquid chloroform and that, consistent with the results of Gerschel et al., they disappear at the critical point. We point out that the contour of any dipole-induced dipole absorption would be the same as that of the permanent dipole absorption—note the similarity of the bandshapes for polar liquids and nonpolar liquids in the far infrared.[21] Quadrupole-induced dipole absorption, on the other hand, would peak at a higher frequency.

Further evidence against a simple mechanism of collision-induced absorption contributing significantly to the chloroform spectrum is the apparent linearity of the integrated absorption intensity with dilution. However, such studies should be extended to concentrations well below 1%. In CH_3CN, for example, a distinct nonlinearity exists that is not observed until the concentration is approximately 10%. Dilution studies should cover the widest range possible.

However, to proceed it is necessary to assume that the far-infrared band of liquid chloroform is due to reorientation of the permanent dipole moment. Several groups have attempted to reproduce this unreduced band shape accurately, including Quentrec and Bezot,[43] Evans,[22] Reid and Evans,[41] and Evans et al.[38] A recent development in the field of theoretical modeling is the constrained librator model of Gaiduk and Kalmykov,[44] which was fitted to liquid-chloroform data over the 0 THz range using a minimum of adjustable variables. Hermans and Kestemont[45] have also considered theoretically the 0 THz chloroform absorption.

Nuclear magnetic resonance results are also extensive. We have reviewed some of these data already in the section on Rayleigh and Raman scattering. A disadvantage of NMR measurements is that they can produce only correlation times and not complete correlation functions (see ref. 21, Chapter 6). Forsen et al.[46] have discovered pronounced solvent effects on the chlorine magnetic-resonance relaxation of liquid chloroform (^{35}Cl and ^{37}Cl nuclei) at 303 K. They give ^{35}Cl transverse relaxation times for the neat liquid and for solutions. For the neat liquid this time is 2 ps, in a 1:3 solution in *n*-hexane it is 1.8 ps, and in a 1:3 solution in CH_2I_2 it is 3.4 ps. Shoup and Farrar[47] have measured the ^{13}C spin–lattice (R_1) and spin–spin (R_2) relaxation rates for 60% enriched chloroform. R_1 is dominated by the intramolecular dipole–dipole interaction with the proton, and R_2 by scalar coupling to the chlorine nuclei. The activation energies associated with the anisotropic molecular motion, and the rotation diffusion constant, were obtained from the ^{13}C relaxation data and found to agree well with those obtained from D and Cl studies by Huntress.[48] Spin-rotation contributions were assumed to be small, and intermolecular contributions to the total R_1 were not considered. R_1 is determined, therefore, by a correlation time τ_c characteristic of the tumbling motion of $CHCl_3$ (i.e., motion of the C_{3v} symmetry axis). The ^{13}C—H vector is on the axis, and spinning *around* C_{3v} does not contribute to R_1. R_1 is related to τ_c using the expression involving the dipole–dipole relaxation of two unlike spins, assuming the model of rotational diffusion for the molecular dynamics. This gives a correlation time of 1.6 ps, compared with a value of 1.74 ps from nuclear D quadrupole relaxation. These values are smaller than the transverse relaxation time (2 ps) reported by Forsen et al.[46] Two more values of τ_c are listed by Brodbeck et al.,[10] namely 1.5 and 1.73 ps at ambient temperature and pressure.

A rotating-frame spin–lattice relaxation time of 44 μs has recently been reported by Ohuchi et al.[49]; this compares with the ^{35}Cl nuclear quadrupole relaxation time of 21 μs in pure $CHCl_3$ reported by Forsen et al.[46] The spin-rotation mechanism is neglected by Shoup and Farrar.[47] The NMR relaxation time reported by Alms et al.[19] agrees with the Rayleigh-scattering correlation time at infinite dilution. Dinesh and Rogers[50] have also measured the proton spin–lattice relaxation in liquid chloroform from 219 to 363 K at 1 bar. According to these authors, the spin-rotational relaxation rate is *larger* than the intramolecular dipolar term over the entire temperature range, which finding conflicts with the view of Shoup and Farrar. Dinesh and Rogers point out the considerable disagreement in the NMR literature on chloroform, which extends to the actual value of the proton spin–lattice relaxation time and to the various contributory factors involved. They write the experimental spin–lattice relaxation rate as a sum of contributions from at least five different sources: intra- and intermolecular dipolar mechanisms, spin rotation, scalar coupling, and the anisotropy of the chemical shift tensor. The first three depend on temperature, the scalar-coupling term on temperature and frequency, and the chemical shift tensor on the strength of the magnetic field used. Dinesh and Rogers neglect the last two contributions and from the first three extract a rotational correlation time *using rotational diffusion theory*. They also derive a translational correlation time, but point out that they make many assumptions in data reduction before arriving at this parameter. The rotational correlation time varies from 3 ps at 219 K to 1.39 ps at 298 K to 1.07 ps at 363 K, and the translational correlation time from 263 ps at 219 K to 80.6 ps at 298 K to 52.2 ps at 363 K. The rotational correlation times are taken from Huntress,[48] and also quoted are diffusion coefficients for tumbling and spinning in $CDCl_3$. The anisotropy of the molecular diffusion is much greater at 219 K, and at 293 K and 1 bar the motion about the symmetry axis is faster than the motion about a perpendicular axis by a factor of about 2. In contrast, in the gas phase, $CDCl_3$ rotates about twice as fast about a perpendicular axis as it rotates about the C_{3v} axis. This conflicts with the data of Forsen et al. on ^{35}Cl transverse relaxation times. Huntress attributes the liquid-phase behavior to weak hydrogen bonding or self-association.

Farrar et al.[51] report the temperature dependence of ^{13}C relaxation studies in $CHCl_3$. Duplan et al.[52] report the Overhauser effect on the ^{13}C—^{1}H nuclei in chloroform, and give a relaxation time of 1.26 ± 0.04 ps for the motion of the axis perpendicular to the C—H axis. This compares with an equivalent value of 1.8 ps from Huntress and values of 1.6, 1.5, and 1.7 ps from other sources in the literature.

Nonlinear techniques have been used to study chloroform. Ho and Alfano[53] and Ratzch et al.[54] pioneered use of the "electrooptical Kerr effect," in which liquid-phase anisotropy is induced by electromagnetic radiation, enabling experimentation on a picosecond time scale using trains of laser pulses. Beevers and Khanarian[55] have used the electric-field-induced Kerr effect to study a number of liquids, including chloroform, over the temperature range 175–343 K. The Kerr effect is related to the polarization $\langle P_1(\cos\theta)\rangle_E$ and alignment $\langle P_2(\cos\theta)\rangle_E$, where P_1 and P_2 are the first and second Legendre polynomials, respectively, and θ is the angle between the dipole axis of the reference molecule and the applied field. Beevers and Khanarian combine Kerr-effect, dielectric, and light-scattering data to analyze the temperature dependence of $\langle P_2(\cos\theta)\rangle_E$. They compare their results with those obtained from studying the effect of an applied electric field on the NMR relaxation, and conclude that there is no hydrogen bonding of significance in chloroform and that orientational order is determined primarily by shape and electrodynamic effects. They estimate a value for the Kirkwood g factor of 1.4, which is larger than the estimate of Gerschel and deviates even further from the diffraction results, which demand that this should be less than 1. They estimate g_2 and g_3 to be 1.7 and -0.4, where

$$g_2 = 1 + \sum_{i \neq 1} \langle \tfrac{3}{2}\cos^2\theta_{1i} - \tfrac{1}{2} \rangle$$

$$g_3 = 1 + \sum_{i \neq 1} \sum_{i \neq 1 \neq j} \langle \tfrac{3}{2}\cos\theta_{1i}\cos\theta_{1j} - \tfrac{1}{2}\cos\theta_{ij} \rangle$$

The value of g_1 remains constant with temperature, but g_2 and g_3 show a weak temperature dependence. It is not clear why this temperature dependence should exist.

Beevers and Khanarian discuss some of the theoretical difficulties associated with the reduction of Kerr-effect data. We have already referred to the possible contribution from induced effects, but the main problem, as with so many of these experimental techniques, revolves around the contribution of the internal field. There is no easy or available solution to this problem. Proutière and Baudet[56] point out that in the Kerr-effect experiment the applied field is so strong that it is necessary to take into account nonlinear polarization phenomena and Buckingham et al.[57] have shown that hyperpolarizability effects are discernible in the Kerr effect when applied to gases. Proutière and Baudet compare their Kerr-effect results with those obtained from other sources, and show that the theoretical expression of Langevin and Born is not valid in the condensed state of matter. They derive an improved theoretical expression for the Kerr constant that still contains some of the uncertainties referred to above, and discuss the influence of electrostriction. Few seem to have discussed the bulk turbulence that is induced in the presence of these strong fields, as reported recently for nondipolar liquids by Evans.[58]

The nonlinear refractive index of liquid chloroform has been derived for the first time by Ho and Alfano[53] using the optical Kerr effect induced by picosecond laser pulses. This factor contains all the nonlinear contributions from electronic, librational, and reorientational motions, and the technique used to derive it is potentially of interest.

Brillouin scattering and sound dispersion can be used in combination to provide information on density fluctuations and vibration–translation coupling. Takagi et al.[59] have made ultrasonic measurements in liquid chloroform over the frequency range 3 MHz to 5 GHz using pulse–echo overlap, high-resolution Bragg reflection, and Brillouin scattering. The observed velocity dispersion revealed two relaxation processes, one at 650 MHz and the other at 5.1 GHz, both at 293 K. The lowest (261 cm^{-1}) and the second-lowest (366 cm^{-1}) fundamental vibrational modes have a common relaxation time of 50 ps. Every mode above the third has a relaxation time of 290 ps. These compare with the relaxation time of 104 ps measured by Samios et al.[60]

Yoshihara et al.[61] have measured the refractive indices and Brillouin frequencies for chloroform, which show significant deviations from values predicted based on the law of corresponding states. Altenberg[62] has discovered that the velocities of sound in a number of organic compounds, including chloroform, are smaller than in the deuterated analogues. Samios et al.[60] measured the temperature dependence of the relaxation strength, which is an indication of the nature of the observed relaxation process. They observed translation–vibration energy transfer involving the lowest and second-lowest energy level. The temperature dependence of the relaxation time provides information on the inelastic collision cross section and on the nature of the intermolecular potential. Vallauri and Zoppi[63] have analyzed the temperature variation of the relaxation time in liquid chloroform, which is described by these authors as "non-associated."

This then has been a short synopsis of a substantial literature search by M. W. Evans on one of the most studied of molecular liquids. It is difficult, as the reader may now appreciate, to explain this mass of experimental data consistently. We cannot even conclude that the liquid is nonassociated without having to explain some observations that strongly suggest otherwise. If there is association, then further analysis relating to single-molecule behavior is severely complicated, and perhaps even meaningless. Confusion exists concerning the role of collision-induced scattering, rotation–vibration coupling, vibration cross correlation, rotation–translation coupling, and even the anisotropy of diffusion in liquid chloroform. The NMR relaxation data, for example, support two irreconcilable conclusions about whether chloroform spins faster than it tumbles.

Let us see if the computer simulation by M. W. Evans helps reconcile any of these apparent discrepancies. In the simulation, the equations of translational motion are solved with a third-order predictor routine, and those of rotational motion are solved using as coordinates the angular momentum and the three unit vectors along the principle axes of the moment-of-inertia tensor. The Lennard-Jones parameters are as follows:

$$\sigma(\text{H—H}) = 2.75 \text{ Å}$$
$$\sigma(\text{Cl—Cl}) = 3.50 \text{ Å}$$
$$\sigma(\text{C—C}) = 3.20 \text{ Å}$$
$$\varepsilon/k(\text{H—H}) = 13.4 \text{ K}$$
$$\varepsilon/k(\text{Cl—Cl}) = 175.0 \text{ K}$$
$$\varepsilon/k(\text{C—C}) = 51.0 \text{ K}$$

Partial charges were added to each atomic site of

$$q_{\text{H}} = 0.131|e|$$
$$q_{\text{C}} = 0.056|e|$$
$$q_{\text{Cl}} = -0.063|e|$$

Correlation times (experimental and simulated) are compared in Table I. By comparison of the reorientational autocorrelation functions of $\mathbf{e}_3$ (a unit vector in the C_{3v} axis) and $\mathbf{e}_1$ (a unit vector in a perpendicular axis through the center of mass), the simulation yields the conclusion that the anisotropy of the rotational diffusion is smaller than that of Huntress and *opposite in sense*. That is, the spinning motion, characterized by $\langle \mathbf{e}_1(t)\cdot\mathbf{e}_1(0)\rangle$, has a longer correlation time than the tumbling, characterized by $\langle \mathbf{e}_3(t)\cdot\mathbf{e}_3(0)\rangle$. This is the "gas phase" result, and is the result suggested by Forsen et al.[46] based on ^{35}Cl measurements. It is also interesting to note that the Raman correlation times for the totally symmetric C—Cl band ($\nu_3(A_1)$) listed by Brodbeck et al.[10] are longer than those from the $\nu_1(A_1)$ and $\nu_2(A_2)$ modes, which again means that spinning is *slower* than tumbling, not faster as suggested by Huntress.

The first- (τ_1) and second- (τ_2) rank simulated autocorrelation times are as follows:

$$\tau_1(\mathbf{e}_1) = 3.9 \text{ ps}$$
$$\tau_2(\mathbf{e}_1) = 1.5 \text{ ps}$$
$$\tau_1(\mathbf{e}_3) = 3.6 \text{ ps}$$
$$\tau_2(\mathbf{e}_3) = 1.3 \text{ ps}$$

Dielectric results *do not* give single-particle correlation times, so the simulated times compare with those estimated from dielectric relaxation of 4.7–6.2 ps in neat solution and 3.2–5 ps (depending on the solvent) in solution. There is a large degree of uncertainty to the experimental correlation times, and they all tend to be longer than the simulated times.

As we have said, the far infrared is the high-frequency adjunct of the dielectric loss, and may be related via a Fourier transform to the rotational velocity autocorrelation function $\langle \dot{\mathbf{e}}_3(t)\cdot\dot{\mathbf{e}}_3(0)\rangle$. This is similar in shape to the orthogonal autocorrelation function $\langle \dot{\mathbf{e}}_1(t)\cdot\dot{\mathbf{e}}_1(0)\rangle$, indicating that at short times the computer predicts little diffusional anisotropy. The Fourier transform of $\langle \dot{\mathbf{e}}_3(t)\cdot\dot{\mathbf{e}}_3(0)\rangle$ peaks at 31 cm^{-1}, which compares with experimental values of 36 cm^{-1} in the pure liquid and 30 cm^{-1} in cyclohexane. The *shapes* of the simulated and experimental curves are similar (Fig. 4), so if the result is not simply fortuitous, collision-induced absorption, hydrogen bonding, and internal-field effects are all small. However, we must recall that the experimental spectrum is the result of the interaction of electromagnetic radiation with an ensemble of molecules, so that cross terms are certainly important. We would not have anticipated such good agreement between the experiment and simulation, because the two techniques are not strictly comparable—the simulation can produce *only* autocorrelation functions. A frequency-dependent correction has to be applied to the experimental spectral intensity to account for this dynamic internal field effect. After this correction has been applied, the resulting bandshape may be related to a correlation function, assuming the validity of the fluctuation–dissipation theorem in the adiabatic approximation. Chatzidimitriou-Dreismann and

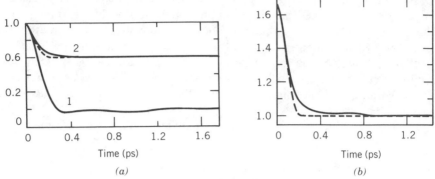

Figure 5. The non-Gaussian behavior of the pdfs governing the approach to equilibrium. (a) Linear velocity autocorrelation function: 1, $\langle \mathbf{v}(t)\cdot\mathbf{v}(0)\rangle/\langle \mathbf{v}(0)\cdot\mathbf{v}(0)\rangle$; 2, $\langle \mathbf{v}(t)\cdot\mathbf{v}(t)\mathbf{v}(0)\cdot\mathbf{v}(0)\rangle/\langle \mathbf{v}^4(0)\rangle$; ---, Eq. (4), the Gaussian result based on line 1. (b) Close-up of the non-Gaussian effect. — , $\langle \mathbf{v}(t)\cdot\mathbf{v}(t)\mathbf{v}(0)\cdot\mathbf{v}(0)\rangle/(\langle v^2(0)\rangle\langle v^2(0)\rangle)$; ---, the Gaussian result: $1 + \frac{2}{3}(\langle \mathbf{v}(t)\cdot\mathbf{v}(0)\rangle/\langle v^2(0)\rangle)^2$. [Reproduced by permission from M. W. Evans, J. Mol. Liq., **25**, 211 (1983).]

Lippert[64] have questioned the validity of this approximation. The use of a nonadiabatic fluctuation–dissipation theory opens up a range of possibilities, because several of the "symmetry theorems" on the fundamental properties of autocorrelation functions and spectral moments are valid only in the adiabatic approximation.

The computer simulation by M. W. Evans was not used to provide an accurate assessment of the intensity of the far infrared but rather to determine the shape and position of the spectrum. However the exercise of comparison needs to be repeated at state points under hydrostatic pressure and to the critical point.

The computer simulation suggests the following:

1. The pdfs governing the approach to equilibrium are not Gaussian (Fig. 5).
2. Rotation–translation coupling is significant (Fig. 6).

These observations cannot be accounted for with classical (purely rotational) theories for molecular diffusion based on the Langevin or Fokker–Planck equations. We shall see in the following sections concerned with other liquids that this conclusion is a valid and general one. Currently popular analytical theories are gross oversimplifications of the molecular dynamics.

III. IODOMETHANE (CH₃I)

Taking moments of inertia as a criterion, iodomethane [M. W. Evans and G. J. Evans, J. Mol. Liq., **25**, 177 (1983)] is almost a linear molecule and is

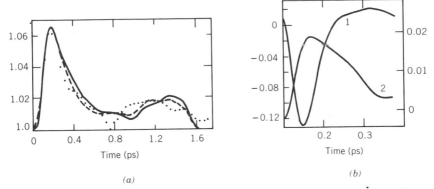

Figure 6. Rotation–translation coupling in chloroform. (a) ——— , $\langle v^2(0)J^2(t)\rangle/$ $(\langle v^2(0)\rangle\langle J^2(0)\rangle)$; ●, $\langle v^2(t)J^2(0)\rangle/(\langle v^2(0)\rangle\langle J^2(0)\rangle)$; ---, as for ———, but with 4560 time steps as opposed to 3600 time steps in the running time average $\langle\bullet\rangle$. (b) The rototranslational first-rank mixed autocorrelation functions $\langle v(0)J^T(t)\rangle$ in the molecule fixed frame: 1, $\langle v_1(0)J_2(t)\rangle/(\langle v_1^2\rangle^{1/2}\langle J_2^2\rangle^{1/2})$; 2, $\langle v_2(0)J_1(t)\rangle/(\langle v_2^2\rangle^{1/2}\langle J_1^2\rangle^{1/2})$. [Reproduced by permission from M. W. Evans, *J. Mol. Liq.* **25**, 211 (1983).]

therefore one of the simplest systems available for study. Good agreement between theory and experiment and among the various experiments is therefore expected, but, as we shall see, problems that are not easily resolved do exist in the reduction of the data. For example, Fig. 7 shows the infrared correlation function with various corrections applied. The decay and overall shape of the function may be changed considerably by such corrections. Also, the problem is not necessarily eased if we study the liquid in solution. For example, in CCl_4 the correlation function, as anticipated, decays somewhat faster, reflecting the more isotropic environment and the decrease in angular correlation between neighboring molecules. However, the viscosity of a mixture of CH_3I and CCl_4 may actually be greater than that of the neat liquid by a factor of as much as 1.7. The implications of this are significant, because the rotational diffusion models (of Debye and others) predict that the rotational diffusion constant is inversely proportional to the viscosity of the fluid. The models fail in this instance because they demand a larger value for the relaxation time of the mixture and thus a slower exponential decay of the correlation function.

Iodomethane has, in fact, nine vibrational modes, each of which may be used to obtain information on rotovibrational diffusion. There are three totally symmetric modes of symmetry species A_1 and three doubly degenerate E modes. All are both infrared and Raman active. A correct analysis of rotation–vibration coupling is necessary before accurate information can be obtained on rotational diffusion and its anisotropy. If we are to use classical dynamics these bands must be symmetrical. This must also be true of the

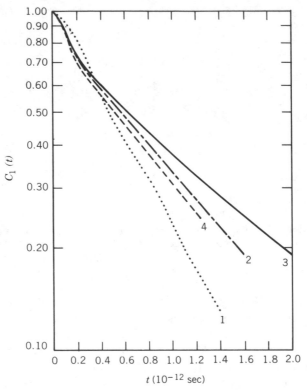

Figure 7.   Infrared correlation functions of CH_3I calculated from the ν_3 band. The curves show changes of the angular correlation function as various corrections are applied. Curve 1, Rothschild (1970) (see ref. 15); curve 2, Fulton (1970) (see ref. 15); curve 3, Fulton's results further corrected for hot bands and the presence of isotopes; curve 4, Favelukes et al. (1968) (see ref. 15). [Reproduced by permission from P. Van Konynenberg and W. A. Steele, *J. Chem. Phys.*, **56**, 5776 (1972).]

infrared E bands, which contain information about both spinning and tumbling (motions around and of the C_{3v} symmetry axis). The A_1 bands are symmetric, but the half-width of the ν_1 band is greater than those of the ν_2 and ν_3 bands, which are roughly the same. Rotovibrational effects are therefore present in the A_1 modes. Coriolis forces introduce strong vibration–rotation effects into the E-type modes because of the double degeneracy. The Coriolis force may also be transferred from an E mode to an A_1 type. The E bands are asymmetric, so that quantum effects are important because of spinning about the C_{3v} axis. The moment of inertia of the symmetry axis is approximately 20 times smaller than that about any perpendicular axis through the center of mass, and the free-rotational velocity is nearly 5 times that for motion of the C_{3v} axis (the dipole axis). The $\nu_4(E)$ band is the least affected by the force.

The results of several early investigations have been summarized by Goldberg and Persham[65] and are displayed in Table II. They conclude that not all of the bands are equally amenable to a simple analysis, and only by studying as many as possible can one expect to obtain a reasonable result from infrared and Raman spectroscopy. This viewpoint is extended by Hyde-Campbell et al.,[66] who investigated a combination of $\nu_3(A_1)$ and ^{2}D relaxation times. These authors report a large effect of density on the molecular motion of methyl iodide; their orientational (P_2) and angular-momentum relaxation times are listed in the table. Using ^{2}D and spin–lattice NMR relaxation times, they obtained diffusion coefficients for spinning and tumbling. As the density changes, however, the rotation of the methyl group about the C_{3v} axis remains largely unaffected, and it is pointed out that any "serious" attempt to characterize the molecular dynamics in a liquid must involve data taken under hydrostatic pressure. They suggest that the true effects of temperature show up only at constant density.

Döge and Schaeffer[67] calculate a Raman correlation time for spinning at 293 K using a relatively simple model for the bandshape. Arndt and Yarwood[68] extend the analysis to the overtones involved with $0 \rightarrow 1$ and $0 \rightarrow 2$ transitions. The effect of the intermolecular potential, and therefore the observed rate of vibrational relaxation, depends on the normal coordinates of the mode concerned. Several other groups have studied liquid methyl iodide by infrared, Raman, or Rayleigh spectroscopy. Their results are shown as correlation times in Table II. Cheung et al.[69] calculate the infinite-dilution intercept of the Rayleigh correlation time to be 1.8 ± 0.2 ps, which is roughly comparable to the Raman correlation times obtained from totally symmetric stretching. They compare their results with the Raman correlation times of 1.6 ps given by Wright et al.[70] and 2.2 ps given by Patterson and Griffiths.[71] Dill et al.[72] have derived angular-position and angular-velocity autocorrelation functions, and point out that the major axis of the rotational diffusion tensor coincides with that of the polarizability tensor, and that in this liquid depolarized Rayleigh scattering will measure an average reorientational time only.

An interesting result is that of Constant and Fauquembergue.[73] They list correlation times for solutions of CH_3I. The solvent used has a pronounced effect on the dynamics of the solute molecules. For example, the first-rank correlation time (infrared) for a 20% (mole fraction) solution of CH_3I in CCl_4 is 3.2 ± 0.1 ps, whereas in hexane it is 2 ± 0.1 ps. We saw for chloroform in the last section that the effect of a solvent is not always to reduce cross correlations. Our own far-infrared results for CH_3I (Fig. 8) show that the plot of $\bar{\nu}_{max}$ (the wavenumber of maximum absorption) versus concentration of

TABLE II
Experimental and Simulated Correlation Times for CH_3I^a

Some Relaxation Times for Liquid Methyl Iodide

Technique	Measured relaxation time (ps)	Computed autocorrelation time (ps) at 293 K
Dielectric	(i) 3.6_5 (pure liquid)	$\tau_1(e_3) = 1.5$
	(ii) 2.5 (infinite dilution in Cl_4)	$\tau_1(e_3) = 1.5$
NMR relaxation	(i) 0.27 ± 0.07 (weighted mean second-rank, pure)	$\tau_2(e_3) = 0.5$
	(ii) 0.50, as above	$\tau_2(e_1) = 0.05$
	(iii) 1.42 (of C_{3v} axis at 1 bar, 303 K)	$\tau_2(e_3) = 0.5$
	3.30 (of C_{3v} axis at 3 kbar, 303 K)	
	0.10 ($\perp$ to C_{3v} axis at 303 K, at both 1 and 2 kbar)	$\tau_2(e_1) = 0.05$
	(iv) Angular momentum correlation time, lab. frame	
	$\tau_J = 0.03_2$ at 1 bar, 303 K	$\tau_J = 0.01$
	$\tau_J = 0.01_3$ at 2 kbar, 303 K	
	(v) 1.4 (NMR dipole relaxation	$\tau_2(e_3) = 0.5$
	(vi) 1.4 (Debye–Stokes theory)	$\tau_2(e_3) = 0.5$
	(vii) 1.6 (of C_{3v} axis at 1 bar)	$\tau_2(e_3) = 0.5$
	0.07 to 0.08 ($\perp C_{3v}$ axis at 1 bar)	$\tau_2(e_1) = 0.05$
Infra-red Bandshape Analysis	(i) 3.6 (pure liquid)	$\tau_1(e_3) = 1.5$
	(ii) 3.1 (pure liquid)	$\tau_1(e_3) = 1.5$
	(iii) 3.2 ± 0.1 (20% mole fraction in CCl_4)	$\tau_1(e_3) = 1.5$
	2.0 ± 0.1 (20% mole fraction in hexane)	$\tau_1(e_3) = 1.5$
Raman bandshape analysis	(i) 0.8 (ν_1)	
	1.1 (ν_2)	$\tau_2(e_3) = 0.5$
	1.4 (ν_3)	
	1.4 (ν_2)	
	1.5 (ν_3)	$\tau_2(e_3) = 0.5$
	1.7 (ν_3)	
	(ii) 1.3 ± 0.1 (20% mole fraction in CCl_4)	
	0.8 ± 0.1 (20% mole fraction in hexane)	$\tau_2(e_3) = 0.5$
	(iii) 1.6 (ν_3 Raman and ^2D NMR spin-relaxation)	$\tau_2(e_3) = 0.5$
	0.07 to 0.08 ($\perp$ to C_{3v} axis, $\nu_3 + ^2D$ NMR)	$\tau_2(e_1) = 0.05$

408

TABLE II (*Continued*)

Some Relaxation Times for Liquid Methyl Iodide

Technique	Measured relaxation time (ps)	Computed autocorrelation time (ps) at 293 K
(iv)	0.05 (ν_4, $\perp$ to C_{3v} axis)	$\tau_2(e_1) = 0.05$
	1.6 ± 0.2 (mean value)	$\tau_2(e_3) = 0.5$
	In the range: 1.18 to 1.36	$\tau_2(e_3) = 0.5$
	(ν_3, depending on wing cut-off frequency)	
Depolarized Rayleigh scattering	0.94 (pure liquid)	$\tau_2(e_1) = 0.05$
		$\tau_2(e_3) = 0.5$
	3.1 (pure liquid)	$\tau_2(e_1) = 0.05$
	1.8 ± 0.2 (infinite dilution in 46:54 isopentane:CCl_4)	$\tau_2(e_3) = 0.5$
	2.2 (pure liquid)	A mean of three times as above
	1.0 (pure liquid)	A mean of three times as above
Far-infrared and second Rayleigh moment	$\bar{\nu}_{max} = 60 \pm 2$ cm^{-1} (100% v/v CH_3I in decalin) $T_q = 1980$ g $\bar{\nu}_{max} = 100$ cm^{-1} (pure liquid, this work)	
Rayleigh second moment analysis	$R(\bar{\nu}) = \bar{\nu}^2 I(\bar{\nu})$ = 64 cm^{-1} (pure CH_3I) = 58 cm^{-1} (pure CD_3I)	Compares with a far-infrared peak frequency of 100 cm^{-1} in the pure liquid (this work)
Incoherent neutron scattering	Studied in the gaseous, liquid, and solid states. Qualitative result: there is "freedom of translational motions" in the pure liquid.	Computer simulation (this work) provides quantitative evidence for rotation–translation coupling. Raman and infrared work provides evidence for rotation–vibration coupling
Coherent anti-Stokes Raman scattering (CARS)	Correlation time from all sources of 0.85 ± 0.2	
Vibrational relaxation	2.73 (ν_1, isotropic at 287 K) 2.81 (ν_2) 2.77 (ν_3)	

[a] Compiled by M. W. Evans *J. Mol. Liq.*, **25**, 177 (1983). Reproduced by permission. For references to original data sources, see this review in *J. Mol. Liq.*

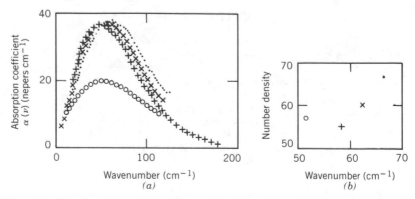

Figure 8. The plot of $\bar{\nu}_{max}$ (the frequency of maximum absorption) against concentration for solution of CH_3I in CCl_4. ●, Heat solution; ×, 75% solution; +, 50% solution; ○, 25% solution (all percentages volume/volume). [Reproduced by permission from M. W. Evans and G. J. Evans, *J. Mol. Liq.*, **25**, 177 (1983).]

CH_3I dissolved in CCl_4 is *not* linear, as would be anticipated if the effect of the dilution was only to remove cross correlations. The same nonlinearity is observed in other solvents.

It is already emerging that CH_3I is not such a simple and representative molecular liquid. There are large contributions to its measured properties from collision-induced absorption. Shermatov and Atakhodzhaev[74] have studied the ν_3 mode of methyl iodide to 150 cm^{-1}. At high values of $\Delta\nu$ a weak absorption maximum due to collision induction dominating the second-moment spectrum is observed. In addition, rotation–vibration coupling affects some, if not all, of the spectra to an unknown extent.

Laubereau and Kaiser[75] have postulated a new and useful way for studying this interaction directly. They have examined the vibrational modes of methyl iodide in the electronic ground state with picosecond laser-pulse spectroscopy. The vibrations are excited with an intense laser pulse via stimulated Raman scattering or by resonant infrared absorption. After the passage of the first pulse, the excitation process rapidly terminates and free precessional decay of the vibrational system occurs. The instantaneous state of the excited system is monitored by a second "interrogating" pulse after a variable delay. However, this method, which allows one to look directly at the vibration–rotation coupling of methyl iodide at short and intermediate times, awaits theoretical development for its proper interpretation.

Gburski and Szczepanski[76] have concluded that it is *impossible* to separate the different vibrational contributions to picosecond laser-pulse experiments using classical infrared and Raman spectroscopy. The isotropic part of a Raman line *cannot* be related to a single relaxation process. They provide a clear account of the problems involved in the interpretation of vibrational relaxation in picosecond laser-pulse experiments. A number of physical processes contribute to line shapes in picosecond dephasing. These include phase relaxation, rotational motion, isotope splitting, and inhomogeneous broadening due to a distribution of molecules with different resonance frequencies depending on the environment. Other problems include the coupling of vibrational and rotational motion due to transition-dipole to transition-dipole coupling and dephasing via vibrational anharmonicities not contained in the Laubereau–Fischer model (see ref. 77).

Gburski and Szczepanski also have considered the problems of interpreting the *vibrational* part of the overall motion and conclude that dephasing processes involving no energy transfer between the ensemble of molecular oscillation and the lattice usually dominate in the line-broadening process. This conflicts diametrically with a prediction of Gillen et al.[78] Their correlation times showed a large temperature dependence, larger than that observed experimentally, which implies that dipolar coupling is a minor effect and that resonant-energy exchange is the dominant factor. The true temperature dependence is not clear from the literature. The modulation times of Doege et al.[79] show a moderate temperature dependence, but this is not observed at all by Trisdale and Schwartz.[80] The difference between the two results is attributable to the methods of calculation. Trisdale and Schwartz used polarized and depolarized Raman spectra of the v_1 (2950 cm^{-1}), v_2 (1150 cm^{-1}), and v_3 (525 cm^{-1}) modes, calculating correlation functions only on the high-frequency side of the bands to get at the vibrational components. Kubo's line-shape theory was applied to describe the spectra in terms of relaxation through vibrational dephasing. This provides a bandshape of the form

$$G_{\text{iso}}(t) = \exp\left\{ - M_{2v}\left[\tau_m^2(e^{-t/\tau_m} - 1) + \tau_m\right]\right\}$$

where M_{2v} is the vibrational second moment and τ_m is a measure of the decay time of the stochastic perturbation. $M_{2v}^{1/2}\tau_m$ determines the nature of the vibrational relaxation (i.e., either motionally narrowed or inhomogeneous). Following evaluation of M_{2v} by numerical integration, Trisdale and Schwartz varied τ_m to obtain the best least-squares fit to the experimental and theoretical isotropic functions. M_{2v} is almost temperature independent for v_1 and v_3, but decreases markedly with increasing temperature for C—H bonding (v_2). The area beneath $G_{\text{iso}}(t)$ gives the isotropic relaxation times, which increase with temperature for all three vibrations, but much more sharply for v_2 than for the other two modes. The effect of dilution is to slow significantly the relaxation of the v_2 mode; the v_1 band is affected similarly by solution. This is attributed to intermolecular dipole or resonant-energy-transfer effects.

Döge[81] and van Woerkom et al.[82] have shown that for a long-range dipole–dipole potential, τ_m should be one-half of the dipole reorientational correlation time obtained from "either dielectric or infrared" spectroscopy. The predicted times of Gillen et al. are 5 to 10 times longer.

Trisdale and Schwartz[80] discuss the Laubereau–Fischer model.[83] This model assumes that vibrational dephasing results directly from binary hard-sphere collisions. The hard-sphere collision times (τ_c) are compared with those derived by Hyde-Campbell et al. from the Enskog model. Though qualitative agreement is obtained, recent studies of ν_1 in liquid CH_3I suggest that the principle decay mechanism of this mode is via population relaxation and not via resonance-energy transfer, and that intermolecular energy exchange dominates the relaxation of the ν_3 mode. Furthermore, if the primary mechanism of relaxation is by dephasing from short-range repulsive forces, it is expected that both τ_c and τ_m will represent the time between collisions in the liquid, yet the temperature trends for these two are not in agreement, particularly for ν_3.

Considerable confusion exists in the literature even if we consider vibrational relaxation alone for CH_3I. The situation is complicated further when rotovibrational experiments are analyzed. The experiments themselves produce composite results owing to a combination of relaxation processes that may be coupled and thereby distort each other considerably. Theoretical treatments are too simple; ideas are subjective and do not stand up to incisive analysis. Conclusions based on the "force fitting" of these models to Raman and infrared bandshapes are at best approximate and at worst misleading, causing greater confusion. Quoting Hildebrand:[84]

A model should be regarded as suspect if it yields inferences in serious conflict with any of the pertinent properties of a system, regardless of how closely it can be made to agree with some, *especially if there are adjustable parameters*. A model that is inconsistent with all properties, even if only approximately, can probably be made more precise, but if it is in irreconcilable conflict with any part of the evidence it is destined to be discarded, and in the meantime predictions and extrapolations based on it should be regarded as unreliable.

To test these models we need as much data over as wide a range of conditions as possible. Careful attention should be paid to the reduction of the data into forms suitable for comparison with theory and to the measurement of the data itself. Bansal et al.[85] emphasize the importance of looking well out into the wings of spectra. They find a marked disagreement between their own results on the $\nu_3(A_1)$ band profile (measured out to 175 cm^{-1}) of CH$_3$I and those of other workers. Their second-rank orientational relaxation times vary from 1.36 to 1.18 ps *according to where the band is truncated*. Their results are all markedly shorter than that of 1.6 ± 0.2 ps quoted by Steele.[86] Also, in infrared and Raman experiments the contribution from hot bands should be allowed for. Roland and Steele,[87] using coherent anti-Stokes Raman scattering to investigate polarized and depolarized bands for the symmetric stretch of liquid CH$_3$I at 526 cm^{-1}, found an appreciable hot-band intensity overlapping the fundamental. Their experimental correlation time is 0.85 ± 0.2 ps. They also quote the value of 1.6 ± 0.2 ps estimated by Steele from an analysis of the spontaneous Raman bands.

There have already been attempts to use computer simulation to unravel these complex experimental infrared and Raman results. Maple et al.[88] showed that the transient statistics of reorientation are non-Gaussian, that is, that nonequilibrium fluctuations approach equilibrium faster than the average correlation function does (this is referred to as "Gordon's equilibrium conjecture"). Riehl and Diestler[89] find that the vibrational correlation functions for their model of methyl iodide were highly oscillatory in the range up to 5 ps, whereas the normalized center-of-mass velocity autocorrelation functions were not exponential in their decay characteristics. The first-rank orientational autocorrelation functions obtained by Maple et al. were not oscillatory in the range up to 3 ps.

Once again there is contradiction and confusion. Spectroscopists should heed the advice of Goldberg and Persham[65] to study all nine vibrational modes available and the advice of Hyde-Campbell et al.[66] to look at the liquid under hydrostatic pressure over a temperature range. The mechanisms governing vibrational relaxation are not established with any degree of certainty, and no attempt has been made to produce the more revealing and appropriate higher spectral moments. Far-infrared spectra provide automatically the second spectral moment of the dielectric loss spectrum.

As an example of moment analysis, we recall the work of Nielsen et al.[90] on the depolarized Rayleigh wing spectrum of liquid methyl iodide. The higher moment $R(\bar{\nu}) = \bar{\nu}^2 I(\bar{\nu})$ was calculated from the experimental data. $R(\bar{\nu})$ peaks at about 64 cm^{-1} for CH$_3$I and 58 cm^{-1} for CD$_3$I. This shift in frequency was compared with a corresponding far-infrared frequency shift of ~ 6 cm^{-1}. Kubo's theory, as used by Trisdale, would leave these higher moments undefined, that is, produce a plateau in $R(\bar{\nu})$. The popular extended diffusion models of Gordon (M and J) are not able to shift the peak frequency of $R(\bar{\nu})$, as observed experimentally, when temperature is varied, when external pressure is applied, or when a solute is dissolved in a nonpolar and noninteracting solvent.

What of NMR relaxation for CH_3I? Discrepancies are again apparent when we consider NMR results. Nuclear quadrupolar relaxation times (of CD_3I), ^{13}C relaxation times, and the intramolecular part of the proton relaxation times have all been measured. Measurements of the nuclear Overhauser effect show that the last two are affected by a nondipolar relaxation mechanism that is spin rotational in origin. Schwartz[91] points out that the effective reorientational correlation time from ^{13}C NMR is a consequence of spin-rotation and dipolar coupling, the dominant mechanisms. At 301 K, this time is 0.27 ± 0.07 ps, a weighted mean of second-rank orientational correlation times (Table II) about all three principal moment-of-inertia axes. Schwartz obtained this value using Overhauser enhancement on natural-abundance $^{13}CH_3I$ with a $180° - \tau - 90°$ pulse sequence. He finds that extended diffusion models cannot characterize the reorientational dynamics. Steele,[86] summarizing the early work in this field, points out that temperature studies suggest that reorientation of the symmetry axis is governed by small-step diffusion. However, this result was arrived at despite the fact that the estimated ratio of P_2 to P_1 correlation times from Raman and infrared differs greatly from the value of $\frac{1}{3}$ necessary for rotational diffusion. The temperature dependence of the relaxation times for rotation about the symmetry axis, on the other hand, is quite minimal, so that Griffiths[92] suggests that motion about this axis is governed by inertial effects and almost free.

Heatley[93] produces a mean second-rank orientational correlation time τ_2 of 0.5 ps, which compares with the value of 0.27 ± 0.07 ps from Schwartz's work. Hyde-Campbell et al.[66] have produced ^2D relaxation times over a range of density and temperature, and derive resultant molecular angular-momentum correlation times of $\tau_J = 0.03$ ps at 1 bar, 303 K; and 0.01 ps at 2 kbar, 303 K. τ_J increases with temperature at constant density. Griffiths[92] combines ^2D NMR and diffusion coefficients parallel and perpendicular to the C_{2v} symmetry axis. The reorientational dynamics are highly anisotropic. These diffusion coefficients have been converted to relaxation times, which are tabulated in Table II. *Unequivocal NMR relaxation times cannot be calculated because of the uncertainties in the Raman times discussed at length above.*

Neutron-scattering studies on CH_3I have been reported for the gaseous, liquid, and crystalline solid states. Fischer[94] has measured the neutron-scattering cross section (in barn proton) for methyl iodide. This increases with increases in the neutron wavelength in the range 4–17 Å. The cross section for CH_3I at 17 Å is 254 barn proton and the slope (in barn Å) of σ versus neutron wavelength is 13.9. The latter is related to rotational "barriers" and decreases smoothly with increases in the NMR rotational relaxation time. The neutron method has some advantages, according to Fischer, for the separation of internal molecular rotations from reorientations of the whole molecule.

Janik et al.[95] have also studied CH_3I by neutron inelastic scattering. They studied the lattice molecular dynamics of solid and liquid CH_3I. The spectrum for the solid arose from lattice vibrations (below 120 cm^{-1}) and from intramolecular vibrations of the CH_3I molecules (above ~ 500 cm^{-1}). Peaks obtained in the intermediate region were higher harmonics of the torsional vibrations. The spectrum for the liquid was regarded as proof of the "freedom of translational motions" in the liquid state.

Apart from such semiqualitative results, it is not anticipated that significant insight into the molecular dynamics of CH_3I can be obtained by means of neutron scattering. The experimental information is a complex conglomerate, rich in information perhaps, but indecipherable with present techniques and knowledge.

Let us see what our computer-simulation results tells us about the dynamics of this molecular liquid. The interaction between two methyl iodide molecules was modeled with a 5×5 Lennard-Jones atom–atom "core" with point charges localized at each site. Multipole–multipole terms are therefore represented by means of charge–charge interactions.

The parameters used in the simulation were as follows:

$$\varepsilon/k(H—H) = 13.4 \text{ K}$$
$$\sigma(H—H) = 2.60 \text{ Å}$$
$$\varepsilon/k(C—C) = 51.0 \text{ K}$$
$$\sigma(C—C) = 3.20 \text{ Å}$$
$$\varepsilon/k(I—I) = 314.0 \text{ K}$$
$$\sigma(I—I) = 4.10 \text{ Å}$$

The cross terms were evaluated with Lorentz–Berthelot combining rules. The carbon and hydrogen parameters were the same as those used for $CHCl_3$. The I—I parameters were obtained using the molecular crystal data of Eliel et al.[96] The electrostatic interactions were represented by point charges, which were calculated using bond moments and bond distances. This provided the following values:

$$q_H = 0.055|e|$$
$$q_C = -0.043|e|$$
$$q_I = -0.122|e|$$

Charge–charge interactions are long ranged, but their relatively small magnitude in this instance allowed us to use periodic boundary conditions with 108 molecules.

The simulation run was initiated at 293 K, with a molar volume of 62.2 cm^3 at 1 bar. The 108 methyl iodide potentials were arranged initially on a face-centered cubic lattice in a cube with a half-side of 11.17 Å, the potential cut off distance. This is over twice the longest Lennard-Jones σ used. According to Bossis et al.,[97] medium-range correlations disappear at about 10 Å from a given molecule, even in intensely dipolar molecules.

All computed and experimental correlation times are collected in Table II. The computed times are generally shorter than their experimental counterparts. The computer-simulation method is, of course, pair additive. This may be an oversimplification in the case of CH_3I, because far-infrared results (Fig. 8), for example, show that the absorption cross sections of spectra for CH_3I dissolved in "noninteracting" solvents are clearly nonlinear. The CH_3I molecules are easily polarized, and induced effects, which are not pair additive, must be large. We have therefore also compared the simulated results with spectral results for CH_3I in dilute solutions, obtaining better experimental–simulated agreement, although the measured correlation times remain longer. For example, the dielectric relaxation time at infinite dilution is 2.5 ps, compared with a simulated time of 1.5 ps, and the infrared relaxation time of a 20% (mole fraction) solution in hexane is 2 ± 0.1 ps, compared with the computed time of 1.5 ps. Raman experimental times range from 0.8 to 1.7 ps, compared with a simulated time of 0.5 ps. Rayleigh correlation times range from 0.9 to 3.1 ps.

These are weighted averages of correlation times about each of the three principal axes in CH_3I. By using a combination of Raman and NMR relaxation, some authors have obtained estimates of second-rank correlation times for motion about the C_{3v} axis—the "spinning" of CH_3I, as opposed to the "tumbling" observed with dielectric spectroscopy. These times range from 0.05 to 0.08 ps and compare well with the simulated time of 0.05 ps. Seventy to ninety percent of the decay of these spinning autocorrelation functions is completed before they become exponential, so the CH_3I potential used in the computer simulation allows the molecule to rotate quite freely. We can conclude that the spinning motion of CH_3I is relatively unaffected by polarizability and induction effects, which are not accounted for in the simulation, and that the tumbling motion is severly affected by collision induction and other processes.

Dill et al.[72] have used Raman and Rayleigh scattering to obtain an angular-velocity correlation function, which is compared with the simulated function in Fig. 9. The latter is distinctly nonexponential. The influence of the spinning motion on the tumbling manifests itself through the oscillations superimposed on the basic structure of the autocorrelation function. This supports Steele's conclusion that rotational diffusion theory cannot describe the molecular motion in liquid CH_3I. The simulated autocorrelation func-

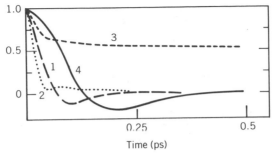

Figure 9.   Comparison of Raman and Rayleigh scattering angular-velocity correlation functions with simulated functions. 1, $\langle J(t) \cdot J(0) \rangle / \langle J^2(0) \rangle$; J = molecular angular momentum, laboratory frame. 2, $\langle \omega(t) \cdot \omega(0) \rangle / \langle \omega^2(0) \rangle$; ω = molecular angular velocity, laboratory frame. 3, $\langle J(t) \cdot J(t)J(0) \cdot J(0) \rangle / \langle J^4(0) \rangle$. 4, Normalized angular-velocity autocorrelation function from the work of Dill et al.[72] [Reproduced by permission from M. W. Evans and G. J. Evans, *J. Mol. Liq.*, **25**, 177 (1983).]

tion decays much more rapidly than the function derived by Dill. Kluk[98] shows that the angular-velocity autocorrelation function, obtained from the 525 cm^{-1} Raman (ν_2) band, has an even deeper negative overshoot than Dill's and intersects the time axis at about 0.2 ps.

Zero-terahertz spectroscopy unquestionably produces some of the most discriminating data for the evaluation of molecular models. In CH_3I it demonstrates quite clearly the deficiencies of the computer simulation. The rotational velocity autocorrelation function of e_3, where e_3 is a unit vector on the C_{3v} axis, may be related via a Fourier transform to the power absorption coefficient, $\alpha(\bar{\nu})$ (in neper cm^{-1}), of the far infrared. Straight Fourier transforms of our far-infrared data are compared with the rotational-velocity autocorrelation functions for spinning and tumbling in Fig. 10. These are, respectively, $\langle \dot{e}_1(t) \cdot \dot{e}_1(0) \rangle / \langle \dot{e}_1^2(0) \rangle$ and $\langle \dot{e}_3(t) \cdot \dot{e}_3(0) \rangle \cdot \langle \dot{e}_3^2(0) \rangle$. In CH_3I the C_{3v} axis is, of course, the dipole axis. There is a mismatch between the experimental and simulated functions. The experimental functions contain contributions from induced absorptions and multibody correlations that complicate their analyses. Some of these effects are lessened by dilution, but there is still a significant discrepancy between the computed and measured functions for 10% (volume/volume) CH_3I in decalin.

Figure 11 illustrates the center-of-mass linear-velocity autocorrelation functions for CH_3I computed in the laboratory and moving-axis frames of reference. The components of the velocity autocorrelation function are not isotropic in the frame of reference that moves with the molecule.[99] The laboratory-frame autocorrelation function has the characteristic long negative tail, and its second moment is transiently non-Gaussian, but attains the Gaussian equilibrium level of $\frac{3}{5}$. This is also the case with the computed sec-

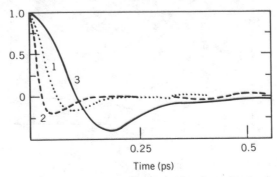

Figure 10. Comparison of far–infrared and simulated rotational velocity autocorrelation functions for CH_3I. 1, $\langle \dot{\mathbf{e}}_3(t) \cdot \dot{\mathbf{e}}_3(0) \rangle / \langle \dot{e}_3^2 \rangle$; 2, $\langle \dot{\mathbf{e}}_1(t) \cdot \dot{\mathbf{e}}_1(0) \rangle / \langle \dot{e}_1^2 \rangle$; 3, straight Fourier transform of the far-infrared power absorption of 10% CH_3I in decalin. [Reproduced by permission from M. W. Evans and G. J. Evans, *J. Mol. Liq.*, **25**, 177 (1983).]

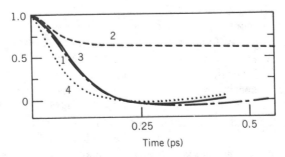

Figure 11. The simulated center-of-mass linear-velocity autocorrelation functions for CH_3I in laboratory and moving-axis frames of reference. 1, $\langle \mathbf{v}(t) \cdot \mathbf{v}(0) \rangle / \langle v^2(0) \rangle$, laboratory frame; 2, $\langle \mathbf{v}(t) \cdot \mathbf{v}(t)\mathbf{v}(0) \cdot \mathbf{v}(0) \rangle / \langle v^4(0) \rangle$, laboratory frame; 3, $\langle \mathbf{v}_C(t)\mathbf{v}_C(0) \rangle / \langle v_C^2(0) \rangle$, moving frame; 4, $\langle \mathbf{v}_A(t)\mathbf{v}_A(0) \rangle / \langle v_A^2(0) \rangle$, moving frame. [Reproduced by permission from M. W. Evans and G. J. Evans, *J. Mol. Liq.*, **25**, 177 (1983).]

ond-moment autocorrelation function $\langle \mathbf{J}(t) \cdot \mathbf{J}(t)\mathbf{J}(0) \cdot \mathbf{J}(0) \rangle / \langle J^4(0) \rangle$, which is initially non-Gaussian but finally reaches the equilibrium Gaussian level of about 0.5 for the C_{3v} symmetry of CH_3I. The angular-velocity second-moment autocorrelation function attains a final level of about 0.4, again the Gaussian result.

The (2,1) and (1,2) elements of the autocorrelation matrix $\langle \mathbf{v}(0)\mathbf{J}^T(t) \rangle$ show that the coupling of rotation with translation in this liquid is unexpectedly strong. This coupling will indirectly affect, to an unknown degree, the laboratory-frame autocorrelation functions measured in the spectroscopies. The calculated correlation times are disparate and must be regarded as subject to revision until methods enable us to estimate the non-rotational spectral intensities.

IV.   ACETONITRILE (CH$_3$CN)

Acetonitrile [M. W. Evans, *J. Mol. Liq.*, **25**, 149 (1983).] is an interesting molecule of C_{3v} symmetry because there is evidence from a number of spectroscopic sources that a strong local order exists in it that persists in solution down to concentrations of less than 10%. Acetonitrile has a large dipole moment (3.9D).[‡] When $g_{00}^{(2)}$, the second-rank structure factor, and $g_{00}^{(1)}$, the first-rank structure factor (or the Kirkwood g factor), are estimated using ^{14}N relaxation and dielectric spectroscopy, respectively, the value of $g_{00}^{(2)}$ is 1.4, indicative of parallel or antiparallel alignment on the local level, and $g_{00}^{(1)}$ is 0.78; the negative deviation from unity of the latter indicates that this pair ordering extends to neighbors *that do not belong* to the first shell of reference particles. The same is observed in a diffraction (combined X-ray and neutron) experiment. A first-nearest-neighbor peak is predicted at 4.7 Å and a second at 8.5 Å. Further, it is concluded that below 4.4 Å preferred orientations of the dipole axis relative to the center–center line are found in the range 90–125°, whereas from ca. 5.2 Å on up preferred orientations are in the range 0–54.7°. At 6.8 Å a second reversal occurs that restores the initial situation.[100]

The existence of this local structure should affect some if not all of the spectroscopies, and should certainly complicate attempts to obtain single-particle properties (e.g., single-particle correlation times) from the experiments.[101] Depolarized-light scattering and far-infrared spectroscopy results certainly reflect contributions from correlated mutual dipoles. With such a distinct local order, contributions to spectral profiles that may be rotational, translational, or vibrational in origin may arise from pairs or clusters of molecules, and, of course, these modes could be coupled, severely distorting band shapes.

It is not surprising, therefore, that M. W. Evans[102] finds, in reviewing the literature, that only the NMR results are even fairly consistent with his computer-simulated (6×6 site–site model) results. He attributes the discrepancies to (1) long-range orientational cross correlations due to association and (2) polarizability effects. The simulation clarifies certain aspects of the reported discrepancies that seem to be more numerous for CH$_3$CN than for CH$_3$I and CHCl$_3$, the two liquids already considered.

Bien et al.[103] used the techniques of NMR spin–lattice relaxation, Raman, and infrared spectroscopy to obtain orientational correlation functions from the degenerate (F) bands of acetonitrile. The shapes of the parallel bands [ν_1 (C—H) = 2944 cm^{-1}, ν_2 (C≡N) = 2253 cm^{-1}, ν_3 (C—H stretch) = 1376 cm^{-1}, and ν_4 (C—C) = 917.5 cm^{-1}] are distorted by hot bands. For perpendicular bands ν_5 (as C—H$_{stretch}$) = 3002 cm^{-1}, ν_6 (as C—H$_{bend}$) = 1444

[‡]1 D ≈ 3.3356 × 10^{-3} cm.

cm^{-1}, ν_7 (C—C$_{bend}$) = 1039 cm^{-1}, and ν_8 (C≡N$_{bend}$) = 378.5 cm^{-1}. The perpendicular bands provide a correlation time for motion about the C_{3v} symmetry axis. Bien et al. provide an angular-momentum correlation time for motion about the mutually perpendicular axes (of I_B or I_C) that differs from that obtained using NMR relaxation by Bull.[104] Their result is in better agreement with that of Schwartz[105] who gives an "effective" second-order orientational correlation time of 0.38 ps. Schwartz also obtained an effective laboratory-frame angular-momentum correlation time of 0.093 ps using NMR spin–lattice (^{13}C) relaxation.

Breulliard-Alliot and Soussen-Jacob[106] reported infrared correlation times for motion parallel and perpendicular to the C_{3v} symmetry axis. Fourier transforms of the raw data from all eight fundamentals produce eight different correlation functions. Vibrational relaxation causes the correlation function from ν_1 to decay much more rapidly than those from ν_2 or ν_4, which are similar. The ν_5, ν_7, and ν_8 correlation functions are different because of Coriolis coupling. They calculate P_1 infrared correlation times of 0.09 ps for motion around the C_{3v} symmetry axis and 5.7 ps for motion of the C_{3v} axis itself, and P_2 correlation times of 0.38 ps and 1.5 ps, respectively, for the equivalent motions.

Numerous papers have been devoted to the comparison of NMR, Rayleigh, and Raman correlation times. The attempts have been summarized by Tiffon et al.,[107] who plot single-particle ^{14}C NMR relaxation times versus η/T for CH$_3$CN in CCl$_4$ solutions. Here η is the bulk viscosity and T the absolute temperature. For neat CH$_3$CN, and for 0.0182 and 0.0037 mole fraction solutions in CCl$_4$, these plots are linear, having nearly the same intercept as $\eta/T \to 0$ of $\tau_0 = 0.29$ or 0.23 ps, depending on the value taken for the nuclear quadrupole coupling constant. This is in complete contradiction to the work of Whittenberg and Wang,[108] who find a zero intercept when assuming that $\tau_s = \tau_{Raman}$, the effective Raman-derived correlation time. The τ_0 value of Tiffon et al. is almost the same as the free rotor correlation time of the major axis of CH$_3$CN (0.33 ps at 295 K). The existence of the positive intercept has been discussed by Hynes et al.[109]

Tiffon et al. quote the single-particle Raman (ν_1) relaxation time of Patterson and Griffiths[110], $\tau_s = 0.9$ ps. These authors also report a Rayleigh correlation time τ_{LS} of 1.8 ps. Whittenberg and Wang also report $\tau_{LS} = 1.8$ ps, but, using the Raman ν_2 band, deduce that $\tau_s = 1.5$ ps. They find that both τ_{LS} and τ_s are independent of concentration in CCl$_4$, and that $\tau_s = \tau_{LS}$ at infinite dilution. They ascribe any discrepancy between τ_{LS} and τ_s to isotropic relaxation mechanisms affecting the Raman band. This conclusion is contested by Versmold,[111] who reports τ_{LS} for CH$_3$CN and CH$_3$CN in CCl$_4$ and τ_s from ^{13}C NMR. He relates a monotonic increase of τ_s with dilution directly to the shear viscosity. He concludes that the increase in τ_{LS}, however, is *not* related to the shear viscosity, and discusses the τ_{LS}/τ_s ratio in terms of a decrease in orientational correlations.

From ^{14}N NMR relaxation, Tiffon et al.[107] report $\tau_s = 1.29$ ps (solid-state NQC) and $\tau_s = 1.01$ ps (gas-state NQC) as limits between which the effective second-rank single-particle orientational correlation time may be defined at 295 K. They point out that the results of Whittenberg and Wang and of Versmold are contradictory. They show that the linear dependence of τ_s on viscosity is true only at high CH_3CN concentrations because association between CH_3CN molecules disappears at roughly a 0.2 mole fraction of CH_3CN, the concentration at which the single-particle correlation times dramatically decrease. This is consistent with the far-infrared results of Knozinger et al.[112] and G. J. Evans,[101] who find that CH_3CN clusters remain in CCl_4 down to concentrations of less than 0.2 of a mole fraction. Fini and Mirone[113] arrive at the same conclusion by studying weak, high-frequency anisotropic components in the totally symmetric Raman vibrational bands of CH_3CN. They identified cluster vibration using isotropic and anisotropic Raman scattering on the shoulders on the ν_2 and ν_4 fundamentals. G. J. Evans[101] has also postulated the existence of cluster vibrations (collective modes) in his far-infrared spectrum.

Amorim da Costa et al.[114] conclude that energy transfer from excited vibrational states to rotational motion affects the Raman bandwidths so severely in liquid acetonitrile as to make impossible any interpretation of the differences between Raman and Rayleigh bandwidths. Yarwood et al.[115] also emphasize the difficulty of effectively separating out vibrational effects in the ν_1 and ν_3 infrared and Raman bands of CH_3CN in dilute CCl_4 solution. In another paper,[116] they conclude from a study of the ν_1, ν_3, and $2\nu_3$ bands that the second-rank (Raman) reorientational correlation times (ν_3 band) do not agree with the literature values from other bands, even though the ν_2 and ν_3 times are self-consistent.

NMR relaxation shows that the rotational motion is anisotropic. A molecule diffuses about its C_{3v} axis about 10 times more easily than about the mutually perpendicular axes of the principal moment-of-inertia frame. Rayleigh and dielectric correlation times are necessarily weighted averages of the three diffusion coefficients corresponding to these three axes of the frame. The anisotropy may be established by using a combination, for example, of ^{14}N and ^2D relaxation times obtained by means of nuclear quadrupole relaxation. ^{13}C and ^1H NMR relaxation can also be employed for this purpose, as demonstrated by Heatley[117] and Lyerla et al.[118] Heatley produces an effective (averaged) second-rank correlation time of 0.32 ± 0.06 ps with a ratio of 9.6 between the diffusion constant for spinning of the C_{2v} axis and those governing tumbling around perpendicular axes. Leipert et al.,[117] using ^{13}C spin–lattice relaxation times and nuclear Overhauser enhancement in the range 239–314 K, conclude that dipole–dipole and spin–rotation effects are important.

It should be noted that almost all of the NMR papers use the theory of rotational diffusion of Debye or, at best, M and J diffusion theory in their data-reduction processes. These models are oversimplified[21] and can only produce approximately correct correlation times.

The parameters used in M. W. Evans's 6×6 atom–atom representation[102] are those available in the literature, namely

$$\sigma(H\!-\!H) = 2.75 \text{ Å}$$
$$\sigma(C\!-\!C) = 3.20 \text{ Å}$$
$$\sigma(N\!-\!N) = 3.31 \text{ Å}$$
$$\varepsilon/k(H\!-\!H) = 13.4 \text{ K}$$
$$\varepsilon/k(C\!-\!C) = 51.0 \text{ K}$$
$$\varepsilon/k(N\!-\!N) = 37.3 \text{ K}$$

The point charges were estimated by M. W. Evans in two ways. First, the dipole and quadrupole moments were calculated directly following Stucky et al.[120] using experimental results from X-ray and neutron diffraction on the electron-density distribution of acetonitrile. He thus obtained values of

$$q_{C_1} = 0.01|e|$$
$$q_{C_2} = 0.16|e|$$
$$q_N = -0.20|e|$$
$$q_H = 0.01|e|$$

where C_1 is the methyl and C_2 the nitrile carbon atom. Second, he used the CNDO/2 calculation of Pople and Beveridge,[121] which produces values of

$$q_{C_1} = -2.02|e|$$
$$q_{C_2} = 0.09|e|$$
$$q_N = -0.16|e|$$
$$q_H = 0.03|e|$$

The values used in the simulation were averages from both sources.

The moment of inertia of CH_3CN about the unique (C_{3v}) axis is about 18 times smaller than the other two (equal) principal moments of inertia. It is anticipated, therefore, that the anisotropy of the rotational motion must be large and must involve translation of the molecular center of mass—the coupling of rotation with translation should be pronounced. The interpretation with analytical theories of the spectroscopic results discussed above ignores this interaction.

Some of the simulated correlation times are compared with available experimentally derived correlation times in Table III. Our main conclusions are as follows:

1.   The NMR times are similar to those of Bien et al.[103] Yarwood et al.,[122] and Schwartz.[105] The correlation times from spin–lattice relaxation are shorter than the simulated counterparts (in both static and moving frames). Tiffon's[107] results are significantly in error. It is worth recalling, however, that Tiffon finds that his correlation times decrease "dramatically" at a concentration of less than a 20% mole fraction of CH_3CN in CCl_4, suggesting that strong cross correlations affected his results for the pure liquid.

2.   The simulated correlation times for the infrared and Raman spectra are nearly always shorter than the experimental times. The only exception, in fact, is the P_1 correlation time of Breulliard-Alliot and Soussen-Jacob.[106] The overall discrepancy is much greater than those for the liquids $CHCl_3$ and CH_3I, discussed earlier (Sections II and III). The discrepancies are so large and so inconsistent from experimentalist to experimentalist that we must conclude that severe errors arise in the data-reduction processes. For example, Amorim da Costa[114] believes that energy transfer from excited vibrational states to rotational motion affects the Raman bandwidths severely. The table shows a considerable spread in the experimental times themselves (3.2–5.7 ps) and that the experimental range straddles the computed range (0.09–5.7 ps, as compared with 0.3–0.7 ps).

3.   In both Rayleigh and dielectric cases the discrepancy between the experimental and simulated correlation times is very large. It might be more meaningful to extrapolate the available Rayleigh and dielectric relaxation times to infinite dilution but care must be taken. There is evidence that some "non-interacting" solvents actually encourage association. The dielectric relaxation times of acetonitrile actually increase in both CCl_4 and benzene.

We have carefully reexamined the far-infrared spectrum of CH_3CN with the aim of isolating the effects, if any, of a well-defined local structure on the spectra of the liquid.[101] Some dielectric work has been reported, so that the complete 0 THz profiles are available. The dielectric relaxation time of acetonitrile in CCl_4 (25% volume/volume) has been measured as 3.3 ps at 303 K by Eloranta and Kadaba,[123] who also measure the far-infrared part of the frequency-dependent loss. Their values compares with the relaxation time of 3.8 ps measured by Krishnaji and Mansingh.[124] These relaxation times happen to agree with the reorientational P_1 correlation time of 3.2 ps

TABLE III
Experimental and Simulated Correlation Times for CH_3CN^a

Technique	Correlation times	Computer simulation (293 K, 1 bar)
Infrared absorption	Reorientation of C_{3v} axis, $\tau_1 = 3.2$ ps $\tau_1 = 5.7$ ps	$\tau_1 (e_3) = 0.7$ ps
	Reorientation about C_{3v} axis, $\tau_1 = 0.09$ ps	$\tau_1 (e_1) = 0.3$ ps
Dielectric relaxation	Pure acetonitrile at 303 K, $\tau_1 = 3.3$ ps	$\tau_1 (e_3) = 0.7$ ps
	Pure acetonitrile, $\tau_1 = 3.8$ ps 25% v/v in benzene, $\tau_1 = 6.7$ ps 25% v/v in CCl_4, $\tau_1 = 7.6$ ps	
Raman scattering	Reorientation of C_{3v} axis, $\tau_2 = 1.5$ ps	$\tau_2 (e_3) = 0.4$ ps
	Reorientation about C_{3v} axis $\tau_2 = 0.38$ ps (symmetric A_1 vibration–rotation),	$\tau_2 (e_1) = 0.2$ ps
	Reorientation of C_{3v} axis $\tau_2 = 0.9$ ps (symmetric ν_1 stretch),	$\tau_2 (e_3) = 0.4$ ps
	Reorientation of C_{3v} axis $\tau_2 = 1.5$ ps (symmetric ν_2 stretch),	$\tau_2 (e_3) = 0.4$ ps
Rayleigh scattering	Reorientational correlation time, $\tau_2 = 1.8$ ps	A weighted mean of:
	Reorientational correlation time (second-rank many-particle orientational correlation times), $\tau_2 = 1.8$ ps	$\tau_2 (e_2) = 0.4$ ps; $\tau_2 (e_2) = \tau_2 (e_1) = 0.2$ ps
NMR relaxation	Spin lattice relaxation, angular momentum correlation times in the principal moment of inertia frame (297 K):	
	$\tau_w = 0.063$ ps (NQC = 172.5 kHz)	0.19 ps Both acfs
	$\tau_w = 0.044$ ps (NQC = 160.0 kHz)	0.12 ps nonexponential
	$\tau_w = 0.025$ ps	
	Spin-lattice relaxation (^{13}C), effective lab. frame angular momentum correlation time, $\tau_J = 0.093$ ps	0.15 ps (nonexponential acf)
	^{14}N nuclear-quadrupole relaxation, mean second rank orientational correlation time,	Mean of $\tau_2 (e_3) = 0.4$ ps;
	$\tau_2 = 1.29$ ps (NQC = solid state) $\tau_2 = 1.01$ ps (NQC = gas)	$\tau_2 (e_2) = \tau_2 (e_1) = 0.2$ ps
	^{13}C to ^{1}H spin–spin relaxation, mean (i.e., isotropic) second-rank orientational correlation time,	
	$\tau_2 = 0.43 \pm 0.06$ ps	Mean of $\tau_2 (e_3) = 0.4$ ps;
	$\tau_2 = 0.38$ ps	$\tau_2 (e_2) = \tau_2 (e_1) = 0.2$ ps
	The molecule spins about 10 times faster than it tumbles	

^{a}Compiled by M. W. Evans *J. Mol. Liq.*, **25**, 149 (1983). Reproduced by permission. For reference to original data sources, see this review in *J. Mol. Liq.*

measured by Rothschild[125] from the ν_4 band, but the infrared correlation time of Breulliard-Alliot and Soussen-Jacob[106] is 5.7 ps.

That not all solvents are "noninteracting" is established by the fact that the relaxation time in benzene (25% volume/volume) decreases from 6.7 ps at 297 K to 3.1 ps at 333 K. In carbon tetrachloride under the same conditions, the relaxation times are 7.6 ps and 4.5 ps, respectively.

Burnham and Gierke[126] have observed a strong local order using the Kerr-effect experiment, the Cotton–Mouton effect, and light scattering to obtain orientational pair correlation functions that suggest a strong antiparallel alignment. Beevers[127] has also used the optical Kerr effect, and points out that acetonitrile is strongly dipolar and optically anisotropic, so that the Kerr effect can be traced back to the orientation of the dipoles.

Some papers have been devoted to the equilibrium structure of liquid acetonitrile. Kratochwill[128] determined the ^{13}C relaxation rate of the nitrile carbon and the self-diffusion coefficient of acetonitrile over a wide temperature range. The ^{13}C—H intermolecular part of the total relaxation rate was determined by isotope substitution and dilution, and combined with the known intermolecular proton relaxation rate to yield the orientation-dependent molecular pair distribution function. In the first coordination sphere, the molecules are arranged antiparallel. Kratochwill provides atom–atom pair correlations derived from X-ray and NMR measurements for the methyl-to-nitrile and the methyl-to-methyl carbon pairs of an acetonitrile molecule described by the rod I—S—T, where I is the methyl group, S is carbon, and T is nitrogen.

Hsu and Chandler[129] have used the X-ray-scattering data of Kratochwill in a RISM calculation of molecular distribution functions. They conclude that although the molecule has a large electric dipole moment, dipole-dipole interactions are unimportant in determining the microscopic structure. In Fig. 12 we compare the simulated atom—atom pdfs for nitrogen–nitrogen, carbon–carbon, and hydrogen–hydrogen. Though the overall shapes are similar, it is clear that the RISM theory tends to overestimate the intensity of the first peak and underestimate its position. The best agreement is found for the carbon–carbon pdfs. The worst mismatch is for the hydrogen–hydrogen pdfs: The pronounced first peak of RISM theory is reduced to a flat shoulder in the computer simulation. This mismatch is *not* caused by electrostatic interactions, because the essential shape of the computed pdfs is retained when we remove the point charges and work with the Lennard–Jones atom–atom cores.

Hsu and Chandler point out that scattering experiments can probe linear combinations of the Fourier transforms of these pair distribution functions and can be used to determine the structure factor. Their simple RISM theory does describe the experimental results of Kratochwill[128] and Bertagnolli

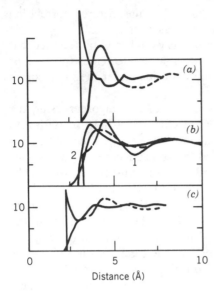

Figure 12. Comparison of simulated atom–atom pdfs for nitrogen–nitrogen, carbon–carbon, and hydrogen–hydrogen with those calculated using RISM theory: — , RISM theory; ---, computer simulation. (*a*) Nitrogen–nitrogen. (*b*) Carbon–carbon: (1, nitrile carbon; 2, methyl carbon. (*c*) Hydrogen–hydrogen. [Reproduced by permission from M. W. Evans, *J. Mol. Liq.*, **25**, 149 (1983).]

et al.[100] to within the uncertainty of the data. They point out the interesting fact that when two CH_3 groups come together, the hydrogen atoms must interlock, and there are two well-resolved intermolecular H—H lengths associated with the same pair of neighboring molecules. So even though they consider dipolar interactions to be of "little importance," there are strong pair correlations between neighboring molecules. The most probable location of a neighboring CH_3CN molecule is at right angles at a distance of 3.5–4.5 Å. There is, they conclude, a "mild tendency" for the angle between the principal (C_{3v}) axes to be between 60° and 120°. The most preferred orientation is near $\theta = 0°$ or $\theta = 180°$ (i.e., parallel alignment), but relatively few pairs of molecules are close enough to be so oriented. There are slightly more than 12 RISM neighbors in the first coordination shell. The coupling of orientational and translational coordinates is negligible outside the first coordination shell, but inside it is fairly strong. This, at least, is in accord with the simulation by M. W. Evans.[102] As we have said, we observe in CH_3CN some of the strongest $R-T$ coupling in the series of polyatomic molecular liquids we have so far simulated (Fig. 13), and certainly the strongest interaction for a molecule of C_{3v} symmetry.

We have seen postulates of parallel alignment and antiparallel alignment, postulates of insignificant pair correlations inside the first coordination shell, and postulates of the existence of clusters of molecules—cluster vibrations have even been assigned. Lippert et al.[130] have reported strong effects of

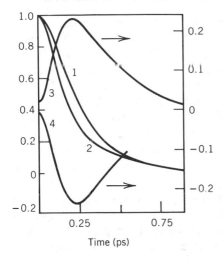

Figure 13. Rotation–translation coupling in CH_3CN. (1) Center-of-mass velocity auto-correlation function component $\langle v_3(t)v_3(0)\rangle/\langle v_3^2(0)\rangle$ in the moving frame of reference. (2) $\langle v_2(t)v_2(0)\rangle/\langle v_2^2(0)\rangle$, which is not isotropic with (1) in the moving frame. (3) $\langle v_2(t)J_1(0)\rangle/[\langle v_2^2\rangle^{1/2}\langle J_1^2\rangle^{1/2}]$, the $(2,1)$ component of the mixed linear velocity–angular momentum (J) correlation function in the moving frame of reference. (4) $\langle v_1(t)J_2(0)\rangle/[\langle v_1^2\rangle^{1/2}\langle J_2^2\rangle^{1/2}]$. [Reproduced by permission from M. W. Evans, *J. Mol. Liq.*, **25**, 149 (1983).]

external electric fields on the infrared spectra of CH_3CN, which seems to indicate that the *electrodynamic interactions can be transmitted and enhanced over a macroscopic distance*.

Let us look at the far-infrared spectrum of acetonitrile again. Figure 14 shows typical spectra of CH_3CN.[131] The neat liquid is extremely absorbing, so that only small path lengths (0.03 mm) of the liquid can be studied by

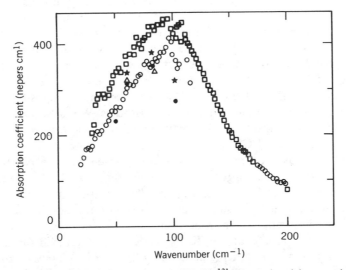

Figure 14. The far-infrared spectrum of CH_3CN.[131] [Reproduced by permission from M. W. Evans, *J. Mol. Liq.*, **25**, 149 (1983).]

normal transmission methods. We therefore estimate an uncertainty of at least 10% for measurement with the interferometric technique. Two runs, obtained using beam dividers that cover the spectral ranges 10–100 and 60–240 cm^{-1}, are shown. The spectra are deliberately not superimposed in order that the reader may judge the typical uncertainties involved. We note the appearance of structure on the low-frequency sides of the spectra. It is encouraging that the high-frequency sides decay smoothly and that spectra of solutions of CH_3CN in CCl_4 (10% and less) all produce smooth profiles. This detail cannot be attributed to channel spectra (internal reflections), which for acetonitrile would be separated by much larger frequencies and would anyway span the whole frequency range.

One feature in the spectrum, in particular, appears to be substantiated: a laser point at 103 cm^{-1}, chosen to coincide with the apparent minimum of the strongest line near the frequency of maximum absorption. Four laser points in all are shown. These are considered to define more precisely the absolute intensity. Three at lower frequencies agree well with the result obtained using a 100-gauge beam divider (interferometric spectroscopy). The fourth laser point seems to be at a lower intensity than that measured with the interferometric spectrometer, suggesting perhaps that at 4-cm^{-1} resolution the spectral detail is incompletely resolved.

One study in the literature may give some support to this result. Knozinger et al.[112] have reported spectra for the neat liquid and for the liquid in solution. Their spectrum for the former also shows signs of a "dip" at the frequency of peak intensity. Their frequency of maximum absorption is, at ca. 80 cm^{-1}, lower than our own (ca. 95 cm^{-1}), although ours agrees well with the measurement of ca. 93 cm^{-1} by Yarwood et al.[122] In their study Yarwood et al. do not appear to resolve the same detail. This could be a consequence of the averaging procedure used in obtaining the spectrum as well as of any subsequent data reduction.

The far-infrared spectrum for CH_3CN is certainly composite in nature. This is conclusively established in dilution studies. We have made our own measurements at concentrations of between 0.5 and 10% volume/volume (or, as number densities of CH_3CN in CCl_4, between 0.5 and 7.5×10^{20} molecules cm^{-3}). This concentration range was chosen because NMR results, discussed above, have suggested an association of CH_3CN molecules at higher concentrations (> 20%) that *gradually* disappears at roughly 20%, the concentration at which the single-particle correlation times dramatically decrease. This observation has now been shown to be consistent with our own[131] and those of Yarwood et al. and Knozinger et al. in the far infrared. We observe an apparent linear dependence of $\bar{\nu}_{max}$ (the frequency of maximum absorption) with number density up to a concentration of ca. 25%, at which distinct nonlinearity occurs; our and Yarwood's data agree well. This nonlinearity tells us that the absorptions are not solely single particle in nature, and that other contributions to the profiles exist.

We have postulated, as did Knozinger et al., that the band is composite in nature, with contributions arising from:

1. Single particle motions (the "Poley absorption," which is characteristic of all molecular liquids in the far infrared and has been measured in a computer-simulation experiment).

2. Absorption by dimers of molecules.

3. Absorption by larger aggregates of molecules.

Contributions (2) and (3) may be rotational, translational, or vibrational in origin.

The existence of the Poley absorption is established in the dilution studies. The frequency of maximum absorption shifts gradually to lower frequencies at higher dilution, so that a 0.5% solution has a maximum absorption at 70 cm^{-1} compared with (95 cm^{-1} in the neat liquid). There is also a lack of any temperature dependence of the spectra at the lowest concentration, although a distinct temperature dependence exists at higher concentrations. Also, this absorption at 70 cm^{-1} is much closer to that predicted in our 6×6 site–site model molecular-dynamics simulation.

The existence of a contribution from dimers of molecules (item 2 above) is more difficult to establish. It was first proposed by Jakobsen and Brasch.[132] To test the hypothesis, Bulkin[133] measured the spectra of five aliphatic nitriles. He measured the spectra of dilute solutions of CH_3CN in nonpolar solvents (CCl_4, benzene, and cyclohexane), and varied concentrations such that the product of multiplying concentration by pathlength was kept constant. It was not then possible to measure a decrease in the intensity of the absorption band. To explain this, he considered a monomer–dimer equilibrium, which revealed that if the association constant is ca. 10^3, the pure liquids are almost completely associated and there will be a decrease in dimer concentration of only ca. 5% on 500-fold dilution. This seems to be the case in CH_3CN. The equilibrium also reveals that if a system is only partially associated, dilution will eventually effect complete dissociation into monomers. Saum[134] had already proposed that butyronitrile (C_3H_7N) represented a case of only 75% association, so that if the dimer hypothesis held, any dimer absorption should disappear in a sufficiently dilute solution of this liquid. This was indeed found to be the case in a 0.25% solution of butyronitrile in cyclohexane: At this concentration the intermolecular dimer vibration band disappeared, leaving only an intramolecular vibration band that Saum assumed was a fundamental mode of the isolated molecule. But perhaps the strongest evidence of all is that dimers of molecules of CH_3CN

have long been postulated to exist even in the vapor phase.[27] Rowlinson's[27] second-virial-coefficient calculations for ΔU, the maximum value of the dipole–dipole energy of interaction, gave a value for CH_3CN of 4640 cal mol^{-1}, which is larger than that for water, 4440 cal mol^{-1}.

The contribution from larger aggregates of molecules (item 3 above) has been considered carefully by Knozinger et al.[112] They observed the temperature dependence of the spectra at higher concentrations, at which larger aggregates of molecules might be expected to exist. Both total band intensity and frequency of maximum absorption decrease when the temperature is raised between 263 and 313 K. In addition, there is an increase in the band intensity at 263 K and a decrease in the total band intensity at 313 K when the concentration is increased; an increase of concentration also shifts the band maximum to higher wavenumbers independently of the temperature applied. All of this is explained only if at least two different types of aggregates are present. A whole *set* of intermolecular vibrational transitions or rotations and translations of aggregates of molecules may then exist and strongly, though perhaps not completely, overlap. It may be this incomplete overlap of contributory absorptions that we now observe at low frequencies in the far infrared.

There is other evidence supporting the existence of these collective, intermolecular modes, in an isotropic liquid, and G. J. Evans[101] has postulated the existence of such in liquid crystalline systems. Lobo et al.[135] propose that collective modes arising from oscillations of the long-range interactions between electric dipoles could exist in certain liquids. Ascarelli[136] studied nitromethane as a characteristic dipolar liquid with a large dipole moment (3.4 D) and observed the existence of these collective modes with a power reflection technique. When a mylar electret was inserted into his sample, the frequency of maximum absorption of the collective mode shifted by some 30 cm^{-1}. The mylar electret, he supposed, produced a further aggregation of dipoles. A model calculation showed that an externally applied field stabilizes aggregates of molecules the radii of which are above a certain value— aggregates that would normally be unstable. The electric field favors the growth of regions where the concentration of dipoles is increased. This may be observed with a simple experiment in which the gap between a pair of square brass electrodes is sealed with two transparent (polymeric) windows. The complete cell is immersed into the pure (conducting impurities removed), neat liquid and a field is applied. The liquid is drawn into the enclosed gap. Aniline, for example, could be suspended to a height of 11 cm in this way. Some of the most striking effects are to be observed on nondipolar solvent liquids such as CCl_4. The effect is so strong in this instance that fine droplets of liquid are sprayed rapidly through the top of two parallel electrodes 13 cm long.[101] Lippert et al.[130] and Evans and Evans[137] have both reported spectroscopic effects of electric fields on isotropic molecular liquids. One of us (M.W.E.) reports on electric-field effects in computer simulation experiments elsewhere in this volume.

In concluding we emphasize that it is not possible to analyze these complex bandshapes with present dynamic theories for the liquid state. The best models for the rotary motions and currently available computer simulations can be expected to reproduce only the "Poley absorption"—that arising from single-particle orientation motions. Translational effects are normally ignored, and rototranslational theories are either intractable or involve unacceptably large numbers of adjustable parameters. Cooperative behavior cannot yet be simulated because of the number of molecules involved and because of the limitations imposed by the speeds of present-day computers.

This situation discussed in this section applies *to all* the spectroscopies. The association is a natural property of the liquid that persists to concentrations of less than 10% and may even be enhanced in some solvents at higher dilution. It is not a trivial problem, therefore, to obtain single-particle properties for CH_3CN, as the discrepancies and contradictions in the literature show. It is questionable if it is meaningful and desirable even to attempt to do so in such an unassociated system.

V. BROMOFORM (CHBr$_3$) [‡]

Bromoform is an interesting example of a molecule with C_{3v} symmetry because it possesses a solid rotator-phase state, the liquid freezing at 281 K. We define such a state as one in which rotational motion is comparatively free and translational freedom is restricted—the molecules are constrained to sites in the solid lattice. Consequently, in this state of matter we anticipate a small coupling of rotation with translation that may persist even in the liquid state.

Brodbeck et al.,[138] in reviewing the literature on bromoform, notice again the difficulty of comparing results obtained with different techniques by way of measured or calculated correlation times because of the variety of definitions of correlation time: "The literature results may double according to definition." Boldeskal et al.[139] define the correlation time as the time taken by the normalized autocorrelation function in question to fall to $1/e$. Other authors[138] use the bandwidth $\Delta\omega$ to calculate the relaxation time τ as proportional to $\Delta\omega^{-1}$ of the zeroth moment of the particular autocorrelation function under question. As we shall see, no overall viewpoint is attained from the comparison of these correlation times for bromoform.

[‡] This section is based on a computer simulation and literature search by M. W. Evans, reported as ref. 147 (editor's note).

We will use our proposed methodology to see if computer simulation of the various correlation functions can help bring some consistency to the results and, in particular, to prove our hypothesis that the interaction of rotation with translation must be small. We use again the atom—atom Lennard-Jones positive-charges framework and the available intermolecular-potential parameters to construct spectra of various kinds.

The interaction between $CHBr_3$ molecules is modeled by a 5×5 atom—atom Lennard-Jones "core" with point charges localized at each atomic site. The Lennard-Jones parameters are as follows:

$$\sigma(H—H) = 2.75 \text{ Å}$$
$$\sigma(Cl—Cl) = 3.50 \text{ Å}$$
$$\sigma(C—C) = 3.20 \text{ Å}$$
$$\varepsilon/k(H—H) = 13.4 \text{ K}$$
$$\varepsilon/k(Br—Br) = 263.0 \text{ K}$$
$$\sigma(Br—Br) = 3.7 \text{ Å}$$

The H and C parameters are as for chloroform (Section II). The parameters for Br were taken from the values of Eliel et al.,[140] which successfully reproduce crystal-phase properties such as heat of sublimation. The calculation of partial charges is based on the very simple lcao technique of del Re.[141] Taking into account the slightly smaller dipole moment of $CHBr_3$ in comparison with $CHCl_3$, the increase in bond length from C—Cl to C—Br, and the slightly smaller electronegativity of Br in comparison with Cl, M. W. Evans arrives at the following estimate of partial charges for $CHBr_3$:

$$q_H = 0.021|e|$$
$$q_C = 0.055|e|$$
$$q_{Br} = -0.059|e|$$

To compare results from the simulation with those available from experiments, we adopt once more the definition of a correlation time used throughout this article.

The moments of inertia of bromoform calculated by M. W. Evans with Hirschfelder's[142] dyadic minimization technique are

$$I_A = 6.87 \times 10^{-38} \text{g cm}^2$$
$$I_B = I_C = 1.352 \times 10^{-37} \text{g cm}^2$$

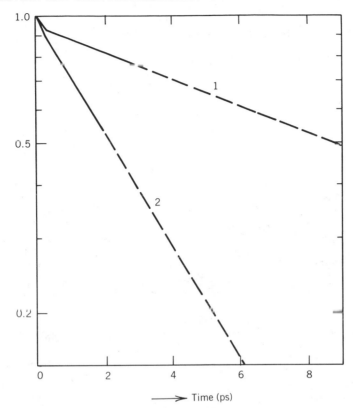

Figure 15.   Computer-simulated orientational autocorrelation functions for bromoform: 1, $P_1(e_A)$; 2, $P_2(e_A)$. (Reproduced by permission from ref. 147.)

The dipole unit vector is e_A. We define the first-rank, P_1, orientational correlation time as the area beneath the autocorrelation function $\langle e_A(t) \cdot e_A(0) \rangle$, and the second-rank, P_2, orientational correlation time as the area beneath the autocorrelation function $\frac{1}{2}\langle 3[e_A(t) \cdot e_A(0)]^2 - 1 \rangle$. The simulated autocorrelation functions are shown in Fig. 15. The latter correlation time can also be obtained from NMR dipole–dipole relaxation because the dipole–dipole ^{13}C NMR relaxation time refers to the ^{1}H-to-^{13}C vector of $CHBr_3$, which is directionally the same as e_A.

We will use only spectroscopic data that we consider to have been correctly reduced, and make allowances for the various factors that contribute to spectral profiles discussed at length in Section I. For example, Sandhu[143] finds that spin rotation and inter- and intra-molecular dipole–dipole relaxation contribute to the observable nuclear spin relaxation. He calculates a

TABLE IV
Experimental and Simulated Correlation Times for Bromoform[a]

Technique	Experimental correlation time (ps)[b]	Simulated autocorrelation time (ps)[c]
Dielectric relaxation	19.0	11.0 (e_A)
Infrared		
ν_1 stretch (C—H)	8.2 ± 3.3	11.0 (e_A)
Raman		
ν_1 stretch (C—H)	5.1 ± 1.0, 2.0 $(1/e)$, 3.1, 4.4,	2.8 (e_A)
ν_2 (CBr$_3$ symmetric stretch)	3.4, 4.4	Weighted mean of 2.8 (e_A), 4.5 (e_B, e_C)
ν_3 (CBr$_3$ symmetric bend)	5.3, 6.6, 6.8	4.5 (e_B, e_C)
Rayleigh scattering	10.1	Weighted mean of 2.8 (e_A), 4.5 (e_B, e_C)
NMR		
intramolecular $\mu - \mu$	4.1 (average)	Weighted mean of 2.8 (e_A), 4.5 (e_B, e_C)
Intermolecular $\mu - \mu$	$\tau_c = 10.4$	$\tau_v < 0.1$[d]
translational correlation time		$\tau_r \approx 11.3$

[a] Compiled by M. W. Evans, and reported in *J. Chem. Soc., Faraday II*, **79**, 137 (1983). Reproduced by permission. See this paper for original data sources.

[b] Multiple values represent literature variation.

[c] In the rotator phase at 273 K, $\tau_1(e_A)$ (first rank) is 18.0 ps and $\tau_2(e_A)$ (second rank) is 5.7 ps from the computer simulation.

[d] Center-of-mass velocity.

translational correlation time of 10.4 ps at 298 K, making the assumption that the diffusion tensor is related to the effective molecular radius by Stokes's law. He calculates a rotational (P_2) correlation time of 4.1 ps, which is much less than one-third of the dielectric (P_1) relaxation times of 19 ps measured by Soussen-Jacob et al.[144] and of 24.4 ps at 293 K decreasing to 13.3 ps at 323 K measured more recently by Sharma and Agarwal.[145] This is therefore inconsistent with the theory of rotational diffusion, which nevertheless was used by all three sources in deriving their correlation times.

In Table IV we compare the simulated correlation times with these experimental results. The agreement is satisfactory for the "single particle" infrared and Raman correlation times. The infrared (C—H stretch) rotational correlation time, for example, is 8.2 ± 3.3 ps, compared with a computer-simulated time of 11.0 ps. There is satisfactory agreement between the NMR rotational mean correlation time of Sandhu and the three computed times (one for each axis, i.e., $P_2(e_A)$, $P_2(e_B)$, and $P_2(e_C)$). However, there is at first sight a serious difference between the intermolecular NMR translational correlation time of 10.4 ps and the computer-simulated time of < 0.1 ps.

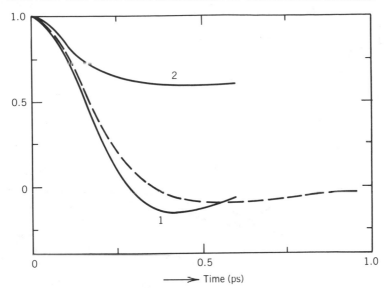

Figure 16. Computer-simulated center-of-mass velocity autocorrelation functions for CHBr$_3$, 293 K, 1 bar: 1, $\langle \mathbf{v}(t) \cdot \mathbf{v}(0) \rangle / \langle v^2(0) \rangle$; 2, $\langle \mathbf{v}(t) \cdot \mathbf{v}(t) \mathbf{v}(0) \cdot \mathbf{v}(0) \rangle / \langle v''(0) \rangle$; - - -, $\langle \mathbf{v}_3(t) \mathbf{v}_3(0) \rangle / \langle v_3^2(0) \rangle$, molecule fixed frame. (Reproduced by permission from ref. 147.)

However, Sandhu assumes that the motion is describable by the diffusion equation and an exponentially decaying $\langle \mathbf{v}(t) \cdot \mathbf{v}(0) \rangle$; thus τ_c as measured by Sandhu is given by

$$\tau_c = \frac{a^2}{12D} = \frac{Ma^2}{12kT\tau_v}$$

where a is the molecular radius, M the molecular mass, and τ_v the velocity correlation time. We must adjust τ_v of the molecular dynamics accordingly and relate it to τ_c as above. Assuming a molecular radius of 3.6 Å and using the molecular mass, we obtain a "simulated" τ_c of 11.3 ps. However, this point is academic, because the computer-simulated center-of-mass velocity autocorrelation function is not exponential, as assumed by Sandhu in his derivation, and has a characteristic long time tail (Fig. 16).

Agreement between simulation and experimental data is not satisfactory for dielectric relaxation, far-infrared absorption, and depolarized Rayleigh scattering techniques (Table IV). This is again anticipated, because the multimolecular counterparts of $\langle \mathbf{e}_A(t) \cdot \mathbf{e}_A(0) \rangle$ and $\frac{1}{2} \langle 3[\mathbf{e}_A(t) \cdot \mathbf{e}_A(0)]^2 - 1 \rangle$ are obtained from these techniques. These data give information on groups or

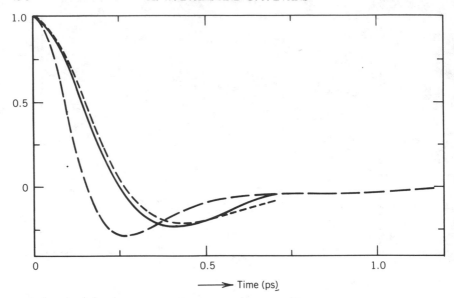

Figure 17. Rotational velocity correlation functions: – – –, experimental; ——— , simulation $\langle \dot{e}_A(t) \cdot \dot{e}_A(0) \rangle / \langle \dot{e}_A^2(0) \rangle$; ------, $\langle \dot{e}_C(t) \cdot \dot{e}_C(0) \rangle / \langle \dot{e}_C^2(0) \rangle$. e_A is the dipole vector. [Reproduced by permission from ref. 147.]

ensembles of molecules, and the cross correlation functions involved are much more difficult to simulate. The evidence provided by Brodbeck et al. seems to indicate that the correlation times involved in cross correlation are longer than those of the autocorrelation functions. For example, the dielectric relaxation time measured by Soussen-Jacob is 19 ps, and the infrared rotational correlation time measured by Brodbeck is 4.1 or 8.2 ps (depending on definition). Similarly, the ν_1 stretch Raman correlation times from various sources lie in the range 2–5.1 ps, and the depolarized Rayleigh correlation time of Patterson and Griffiths[146] is 10.1 ps.

The discrepancy is clearly displayed in Fig. 17, which shows the simulated rotational velocity autocorrelation function and the Fourier transform of the $\alpha(\omega)$ far-infrared power absorption of liquid bromoform.[147] We have therefore also simulated a "multimolecular rotational velocity correlation function" for bromoform using subspheres built up of three or four CHBr$_3$ molecules. However, this does not produce results significantly different from the computed correlation function of Fig. 16.

The far-infrared spectra (Fig. 18) firmly substantiate the existence of the solid rotator phase. The rotator-phase spectrum is shifted slightly to higher

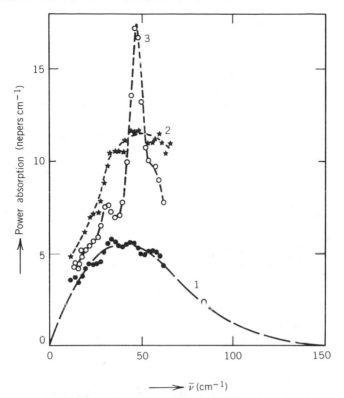

Figure 18.   Far-infrared absorption of bromoform in liquid and rotator-phase states: 1, liquid state at 295 K, ○ laser point (84 cm^{-1}); 2, rotator phase at 273 K (as for the liquid, a broad, featureless band); 3, crystalline solid (note that the lattice modes are resolved). (Reproduced by permission from ref. 147.)

frequencies, but remains a broad and featureless band. No lattice modes, characteristic of the crystalline solid, are resolved until the sample is cooled to below $-2°$C. The solid rotator phase exists between -2 and $+8°$C. Depending on the rates of heating and cooling, a marked hysteresis may be observed as the sample is cooled and reheated through this phase.

In a moving frame of reference, discussed at length in this volume, M. W. Evans observes directly the coupling of rotation with translation by constructing the matrix $\langle \mathbf{v}(t)\mathbf{J}^T(0)\rangle$. By symmetry, in this molecule the only nonvanishing elements of this matrix are $(1,2)$ and $(2,1)$. These are illustrated in Fig. 19 and are very small in the liquid state (the same is true, as anticipated, in the rotator-phase solid).

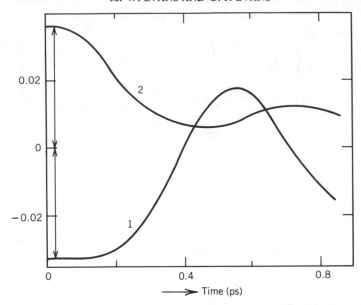

Figure 19.   Rotation–translation autocorrelation function $\langle \mathbf{v}(t)\mathbf{J}^{T}(0)\rangle$ in the moving frame: 1, the (1,2) component; 2, the (2,1) component. The noise level is indicated on the ordinate axis by arrows, showing that the extent of coupling is small and within the noise level of the simulation. (Reproduced by permission from ref. 147.)

VI.   TERTIARY BUTYL CHLORIDE

Tertiary butyl chloride [M. W. Evans et al., *J. Chem. Soc., Faraday II*, **79**, 767 (1983)] is an example of a molecule of C_{3v} symmetry for which two solid rotator phases are known to exist between the isotropic liquid and crystalline solid states. The transition temperatures are

$$T_c \rightarrow \text{II} = 182.9 \text{ K}$$

$$T_{\text{II}} \rightarrow \text{I} = 219.25 \text{ K}$$

$$T_{\text{I}} \rightarrow \text{liq} = 247.53 \text{ K}$$

where "c" signifies the crystalline state, liq. the isotropic liquid, and II and I are the two rotator-phase states.

Lassier and Brot[148] have studied *t*-butyl chloride in its crystalline and rotator-phase states using models of molecular rotational diffusion to interpret their dielectric and far-infrared results. They calculated lattice energies for phases II and I. The difference between the computed energies of phases I and II was found to be equal to the observed enthalpy of transition, which is configurational rather than purely electrostatic in origin. Lassier and Brot calculated the configurational energy and energy of transition using:

1. The energy of repulsion.
2. The London attraction energy.
3. The electrostatic energy between permanent charges.
4. The energy of formation of the induced polarization.
5. The charge-induced dipole energy.
6. The energy between induced dipoles.

The first two terms were found to be some 10 times as large as the electrostatic energy, which in turn was an order of magnitude greater than the polarization and induction energies.

Larson and Mansson[149] have studied the rotational motions that dominate the elastic and quasielastic scattering of the liquid. None of the models used satisfactorily reproduced the measured data, even though models ranging from free rotation to "undamped libration" were considered. Goyal et al.,[150] reporting on neutron-scattering studies in the rotator-phase state, attribute its origin to molecular orientation about the C—C axis. According to Goyal, the dipole axis is frozen and reorientation between individual sites occurs at 12-ps intervals. The different depths of the energy well cause Davydov splitting of the infrared spectrum.

Both these groups consider rotation in t-butyl chloride to be wholly decoupled from translation. The agreement of their results with those from other techniques is not good; for example, the P_1 correlation time from neutron scattering is 3 ps at 325 K, compared with the dielectric relaxation time at 233 K of 7.7 ps. The P_2 orientational correlation time from neutron scattering is 1.2 ps.

Heatley[117] provides a proton–^{13}C magnetic resonance relaxation time for (assumed) isotropic rotational diffusion of 1.05 ± 0.10 ps in the liquid state at 308 K. Boguslavskii et al.[151] have measured the temperature dependence of the solid-state NMR relaxation, and find an effect of large-amplitude reorientational and translational motions in the electric field gradient of the resonant nucleus that persists over the temperature range of the phase transition. Koeksal,[152] using ^{3}H spin–lattice relaxation at 60 MHz between 100 and 330 K, discerned three types of motion: center-of-mass translation, molecular tumbling, and methyl-group torsion.

Constant and Fauquembergue[153] have studied the Raman C—Cl stretch over a wide temperature range for liquid t-butyl chloride neat and in CCl_4 and n-hexane solution. They compare their results with Heatley's, assuming in the data-reduction process that there is (1) no rotation–vibration coupling, (2) no vibrational cross correlations between molecules, and (3) no collision-induced scattering. Their correlation time of 1.2 ± 0.1 ps is only in fair agreement with Heatley's value. They also made a direct comparison of the Raman and Rayleigh correlation functions in pure liquid t-butyl chloride at room temperature (298 K). The P_2 correlation time is, at 1.3 ± 0.05 ps, slightly longer than the C–Cl Raman correlation time. They attribute this to cross-correlation effects, concluding that collision-induced effects are too weak to be detected. Carlson and Flory[154] contest this conclusion, reporting that induced scattering is very significant over a range of frequencies. They assert that no less than 50% of the absolute intensity of the depolarized Rayleigh spectrum is collision induced. Czarniecka et al.[155] have also compared their Raman correlation times with those of Constant and Fauquembergue, and find better agreement.

As we compared P_2 correlation times above, so we may compare the P_1 correlation times from infrared bands and 0 THz absorption or with the Rayleigh "power spectrum" (obtained by multiplying the Rayleigh band by $\bar{\nu}^2$, where $\bar{\nu}^2$ is the wavenumber).

Constant and Fauquembergue report a C–Cl stretch correlation time of 4.2 ps at 288 K decreasing to 2.8 ps in 20% (mole fraction) n-hexane solution. The dielectric time of Czarniecka et al. is 4.9 ps. Reid and Evans[156] find that in a 10% solution in decalin the dielectric relaxation time is reduced to 3.6 ps. There is a better agreement between the "P_1 experiments" for this particular liquid. The agreement suggests that cross correlations must be small, a suggestion supported by the similarity of the Raman and Rayleigh correlation times. Reid and Evans analyze their data using a model of itinerant oscillation that predicts a correlation time of 3 ps for t-butyl chloride in decalin, in fair agreement with the experimental value. They also observe that in the glassy state this is increased to microsecond magnitude, and proceed to discuss the validity of the Gordon sum rule in this instance (the integrated intensity of the experimental band is 325 cm^{-2}, whereas the Gordon sum rule gives 266 cm^{-2}).

All in all, the conclusions to be reached from the large number of studies on this system, particularly those conclusions relating to the extent of collision-induced absorption, are again inconsistent. t-butyl chloride is therefore another liquid that we hope to gain further insight into using computer simulation. It is also ideal for simulation work because, as Lassier and Brot have shown, the electrostatic energy is so much smaller than the effective Lennard-Jones energy that periodic boundary conditions are not likely to be a problem. The experiments above also suggest that it is safe to ignore polarization and induction effects.

M. W. Evans[137] has reproduced a range of spectral data in a computer simulation using the atom–atom Lennard-Jones and partial-charge parameters of Lassier and Brot and ignoring polarizability and induction effects. The Lennard-Jones parameters are as follows:

$$\sigma(C-C) = 3.4 \text{ Å}$$
$$\sigma(CH_3-CH_3) = 4.0 \text{ Å}$$
$$\sigma(Cl-Cl) = 3.6 \text{ Å}$$
$$\varepsilon/k(C-C) = 35.8 \text{ K}$$
$$\varepsilon/k(CH_3-CH_3) = 158.6 \text{ K}$$
$$\varepsilon/k(Cl-Cl) = 127.9 \text{ K}$$

The Lorentz–Berthelot combining rules used by Lassier and Brot provide the cross terms. The partial-charge parameters of Lassier and Brot are:

$$q_{CH_3} = 0.054|e|$$
$$q_C = 0.038|e|$$
$$q_{Cl} = -0.201|e|$$

The simulations were carried out at 293 K, 1 bar, in the liquid and at 228 K, 1 bar, in rotator phase I. Haffmans' and Larkin[157] obtained a density of 0.96 g cm^{-3} for rotator phase I at 228 K (fcc, $a = 8.62$ Å, four molecules per unit cell) and this was used by M. W. Evans to calculate the input molar volume for the computer simulation.

Table V compares some of the correlation times from the simulation with experimental results at 293 K, 1 bar; and 228 K, 1 bar. Figure 20 shows the results of computer simulation[137] compared with the infrared, Raman, and Rayleigh orientational correlation functions of Constant and Fauquembergue. The agreement of the correlation functions is fair and within experimental error in some cases. It is probable that (1) reorientational effects have been factored out correctly from vibrational dephasing effects in the liquid, and (2) the same is true of the Raman second-rank orientational autocorrelation function.

Carlson and Flory[34] have suggested that a large amount of collision-induced scattering occurs in t-butyl chloride, which conflicts with the view of Constant and Fauquembergue. The results of the computer simulation indicate that the theoretical, $P_2(e_A)$ autocorrelation function has a similar time dependence similar to that of the multimolecule correlation function from the Rayleigh wing of Constant and Fauquembergue. So, we can at least say that the time dependence of the collision-induced effects, if they are significant, is the same as that of the scattering due to the permanent molecular polarizability anisotropy.

TABLE V

Experimental and Simulated Correlation Times for t-Butyl Chloride[a]

Technique	Experimental correlation times	Simulated autocorrelation times (293 and 228 K) ps
Infrared	C—Cl stretch, 288 K	$\tau_1 (e_c) = 4.0$
	Pure liquid TBC: 4.2 ps	A weighted mean of:
	20% mole fraction TBC in n-hexane: 2.0 ps	$\tau_1 (e_C) = 4.0; \ \tau_1 (e_A) = \tau_1 (e_B) = 3.2$
	Totally symmetric C—CH$_3$ stretch: 1.49 ps	
Raman	C—Cl stretch: 1.2 ps	$\tau_2 (e_C) = 1.5; \ \tau_2 (e_A) = \tau_2 (e_B) = 1.2$
	20% mole fraction in n-hexane: 0.8 ps	
Rayleigh	Pure liquid at 298 K: 1.3 ps	A weighted mean of:
		$\tau_2 (e_C) = 1.5; \ \tau_2 (e_A) = \tau_2 (e_B) = 1.2$
Dielectric	Pure liquid: 4.9 ps	$\tau_1 (e_C) = 4.0; \ \tau_1 (e_A) = \tau_1 (e_B) = 3.2$
Relaxation	10% decalin solution: 3.6 ps	
	Rotator phase I: 274 K: 5.6 ps	
	Rotator phase I: 238 K: 7.4 ps	
	Rotator phase I: 233 K: 7.7 ps	$\tau_1 (e_C) = 9.5; \ \tau_1 (e_A) = \tau_1 (e_B) = 5.2$
NMR relaxation	^{1}H—^{13}C spin–spin: 1.05 ± 0.1 ps	$\tau_2 (e_C) = 1.5; \ \tau_2 (e_A) = 1.2$
Neutron	Mean time between successive reorientations:	
scattering	10 ps at 208 K	
	(Rotator II): 14 ps at 193 K	
	P$_1$ (dipole vector): 235 K: 3.0 ps	$\tau_1 = 9.5$
	P$_2$ (dipole vector): 235 K: 1.2 ps	$\tau_2 = 5.2$ at 228 K

[a] Compiled by M. W. Evans, and reported in *J. Chem. Soc. Faraday Trans. II*, **79**, 767 (1983). Reproduced by permission. See this paper for original data sources.

We can investigate further the role of orientational cross correlations using far-infrared data. Larkin,[158] Reid,[159] and Evans and Evans[137] have reported far-infrared studies. In Fig. 21 we compare simulated and measured rotational velocity correlation functions. Initially, the decays of the two functions are the same, but then the experimental function decays more quickly and has a slightly deeper overshoot, indicating that cross-correlation functions are significant but not markedly so. The effect of dilution corroborates this. The effect is small and the integrated intensity is linear in the molecular number density [at least in the range studied, but again there is a need to extend these dilution studies to concentrations of 1% and less (see Section IV)]. We can attribute the discrepancy between the measured integrated intensity and that obtained using Gordon's sum rule, referred to earlier, only to some limitation of the sum rule. Reid and Evans[156] have discussed the validity of the Gordon sum rule in this context, and Bossis[161] has extended the analysis to include anisotropic cavities and the internal-field effect.

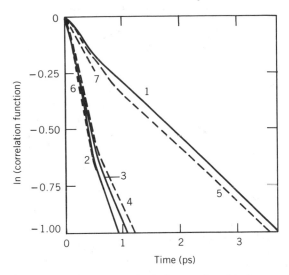

Figure 20. Comparison of computer-simulation results with the experimental data for liquid t-butyl chloride at 293 K. 1, P_1 orientational autocorrelation function from the infrared bandshape; 2, P_2 orientational autocorrelation function from Raman scattering; 3, P_2 orientational autocorrelation function from Rayleigh scattering; 4, Simulated orientational autocorrelation function of the dipole vector (C_{3v} symmetry axis), P_2 autocorrelation function; 5, as for 4, P_1 autocorrelation function; 6, as for 4, vector $\perp C_{3v}$ axis; 7, as for 5, vector $\perp C_{3v}$ axis. [Reproduced by permission from M. W. Evans et al., *J. Chem. Soc., Faraday II*, **79**, 767 (1983).]

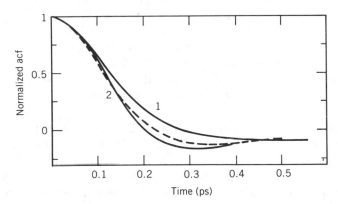

Figure 21. Comparison of simulated and measured rotational velocity correlation functions for t-butyl chloride: 1, computer-simulated $\langle \dot{\mathbf{e}}_C(t)\cdot\dot{\mathbf{e}}_C(0)\rangle/\langle\dot{\mathbf{e}}_C^2\rangle$; 2, Fourier transform of far-infrared spectrum; ---, $\langle \dot{\mathbf{e}}_A(t)\cdot\dot{\mathbf{e}}_A(0)\rangle/\langle\dot{\mathbf{e}}_A^2\rangle$, illustrating the anisotropy of the rotational motion. [Reproduced by permission from M. W. Evans et al., *J. Chem. Soc., Faraday II*, **79**, 767 (1983).]

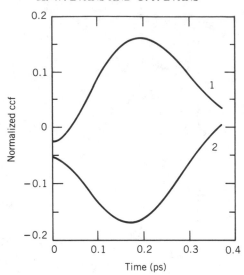

Figure 22.   Elements of $\langle \mathbf{v}(t)\mathbf{J}^T(0)\rangle$ in the moving frame of reference. 1, $\langle v_2(t)J_1(0)\rangle/$ $[\langle v_2^2\rangle^{1/2}\langle J_1^2\rangle^{1/2}]$, the (2,1) element; 2, the (1,2) element. The noise level is demonstrated by the $t = 0$ intercept, which is zero by symmetry in the absence of noise. [Reproduced by permission from M. W. Evans et al., *J. Chem. Soc., Faraday II*, **79**, 767 (1983).]

M. W. Evans[137] uses the numerical method to investigate dynamical properties that are not directly observable by experiment. First he considers rotation–translation interaction, which we recall, was insignificant in bromoform. Again, it must be investigated in a moving frame of reference. For *t*-butyl chloride the (1,2) and (2,1) elements of the rototranslation matrix $\langle \mathbf{v}(t)\mathbf{J}^T(0)\rangle$ do not vanish by symmetry for $t > 0$. The result is shown, suitably normalized, in Fig. 22. Despite the statistical noise, it is clear that there is a real, but small, statistical correlation between the two modes of motion.

It is again interesting to examine what differences exist between the solid rotator phases and the liquid phase. In particular, we may compare the liquid and rotator phases over a short (molecular) range (less than 10 Å, within which distance correlations persist) by constructing atom–atom pdfs. The functions are illustrated in Fig. 23. The functions measure the probability of finding an atom of molecule B at a distance *r* from an atom of molecule A given the position of the latter atom relative to the origin in the laboratory frame of reference. The correlations disappear at 10 Å, the range of the first coordinate shell, in both instances (rotator and liquid). The lack of long-range order in the rotator phase suggests why this phase is incapable of supporting phonon modes such as may be observed in the crystalline phase below 182.9 K.

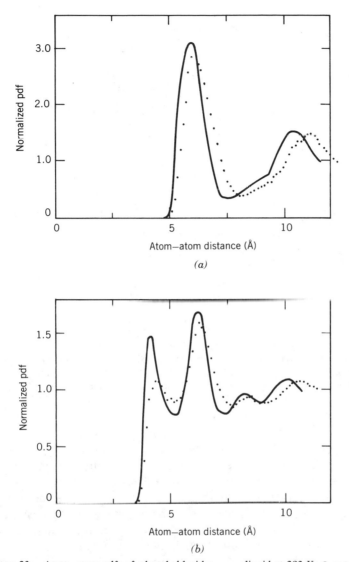

Figure 23. Atom–atom pdfs of *t*-butyl chloride: —— , liquid at 293 K; ●, rotator phase I at 228 K. (*a*) Carbon–carbon function. (*b*) Methyl–methyl function. [Reproduced by permission from M. W. Evans et al., *J. Chem. Soc., Faraday II*, **79**, 767 (1983).]

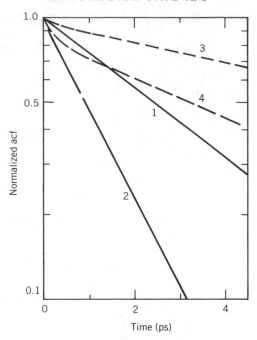

Figure 24.   Comparison of neutron-scattering and simulation results. 1, P_1 autocorrelation function from neutron inelastic scattering in rotator phase I at 235 K; 2, as for 1, but P_2 autocorrelation function; 3, $P_1(e_C)$, computer simulation; 4, $P_2(e_C)$, computer simulation. [Reproduced by permission from M. W. Evans et al., *J. Chem. Soc., Faraday II*, **79**, 767 (1983).]

The rotator phases I and II of *t*-butyl chloride have been studied in depth with thermal neutron scattering.[149] In Fig. 24 we compare their first- and second-rank orientational correlation functions with those from the computer simulation by M. W. Evans.[137] The functions derived by neutron scattering decay far more quickly than the computer-simulated autocorrelation functions. The dielectric relaxation time of 7 ps at 283 K measured by Lassier and Brot is also longer than time derived from the area beneath the P_1 function obtained with neutron scattering. The discrepancy may partly be attributed to the slight persistence of rotation–translation interaction even into the rotator phases. The area beneath the P_1 function at 228 K agrees better (Table V) with the measured dielectric relaxation time (8.5 ps).

Finally, in Fig. 25 we compare the rotational velocity autocorrelation function computed at 228 K with the straight Fourier transform of the far-infrared spectrum of the rotator phase measured by Reid.[159] The simulated autocorrelation function decays, as in the liquid, faster than the measured cross-correlation function, although the shift in the time at which these

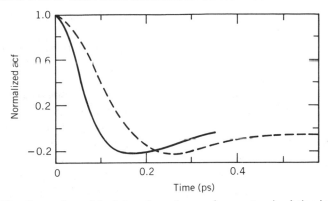

Figure 25. Comparison of far-infrared spectrum and computer simulation in the rotator phase. ----, $\langle \dot{e}_C(t) \cdot \dot{e}_C(0) \rangle \langle \dot{e}_C^2(0) \rangle$, rotator phase at 228 K, computer simulation; —, Fourier transform of the far-infrared spectrum. [Reproduced by permission from M. W. Evans et al., *J. Chem. Soc., Faraday II*, **79**, 767 (1983).]

functions intersect the abscissa is the same in both experimental and theoretical cases, about 50%.

We conclude this section by emphasizing that the experimental techniques themselves do not provide us with a consistent view of the orientational dynamics of molecules of C_{3v} symmetry. A series of liquids have been considered, and it is established that we do obtain a more consistent view with use of the simulation data now becoming available. This is particularly so if we choose our molecules carefully. For example, in choosing *t*-butyl chloride, we were aware that the atom–atom and charge–charge parameters that were to be used in the simulation produce the correct enthalpy of transition for more than one phase change in *t*-butyl chloride. The thermodynamicist obviously has a major role to play in any future progress in this field. Methods like the ones we are now using to gain insights into the interaction of rotation with translation may be developed to gain similar insights into the interaction of other modes of motion with each other.

In general we must conclude that the results of the present literature search, and of the many years of experimentation and theoretical analysis that have enabled such a search, leave many of the questions raised in Section I unanswered for molecules of C_{3v} symmetry. Experimental spectra are not easily reduced to functions that may be compared with theory. Oversimplistic theories are used in the data-reduction processes, and the same theories are then used to interpret the "reduced" profiles. A concerted experimental effect, using in conjunction the technique of computer simulation and better reduction processes, is required if significant inroads into the molecular dynamics of such "simple" liquids are to be made.

VII. DICHLOROMETHANE

We now proceed to molecules of lower symmetry, in particular CH_2Cl_2. Acetone (C_{2v} symmetry) and ethyl chloride (C_1 symmetry) are considered in subsequent sections.

Dichloromethane is mechanically a near-prolate symmetrical top. Hence, there are infrared active vibrations with transition moment vectors parallel to any one of the axes of inertia. Spectroscopic studies on this molecule have already been reviewed.[162-164] It is relevant here to recall some pertinent facts.

There are nine infrared active vibrations in dichloromethane with transition moment vectors parallel to any one of the axes of inertia. By evaluating the bandshapes of different vibrational modes we should, in principle, be able to discern whether the motion is anisotropic and reflects the asymmetry of the moments of inertia, as in the gas phase, or if dipole–dipole interactions in the condensed state distort this rotational motion.

About 70 papers have been published on infrared studies of CH_2Cl_2 alone.[162] It has been made clear[165] for CH_2Cl_2 that it is not possible to factor the relevant autocorrelation functions into a vibrational autocorrelation function and a rotational autocorrelation function. However, this factorization has been done *a priori* in many papers and therefore the conclusions in these papers *are not* meaningful.

Generally, in choosing a set of axes for a molecule of arbitrary geometry, the principal axes of the rotational diffusion tensor are not related to those of the inertia tensor. However, the C_{2v} symmetry of CH_2Cl_2 ensures that these directions coincide and the molecule is consequently a favorable molecule for study. Inertial or molecular-shape considerations would predict symmetrical top behavior, but the complex nature of the intermolecular interactions arising from the effects of the dipole moment along the b-axis may invalidate such simple arguments. However, infrared evidence suggests that the angular motion is indeed axially symmetric.

Since the transition moment vector studied in the infrared is always parallel to only one of the inertial axes of the molecules, the rotational motion of each axis is studied by choosing an appropriate vibrational band. Figure 26 shows the motions of inertial axes a, b, and c. The correlation functions decay more slowly for motion around the larger axis of inertia. The rotational motion in the liquid is anisotropically about the same as the motion of freely rotating molecules, so that it appears that the forces that tend to twist the axis containing the dipole moment are not markedly different from those acting on the other two axes.

Brier and Perry[164] emphasize that the infrared and NMR data provide almost diametrically conflicting pictures of the anisotropy of the motion. In comparing these experimental results, care must be taken. Correlation times

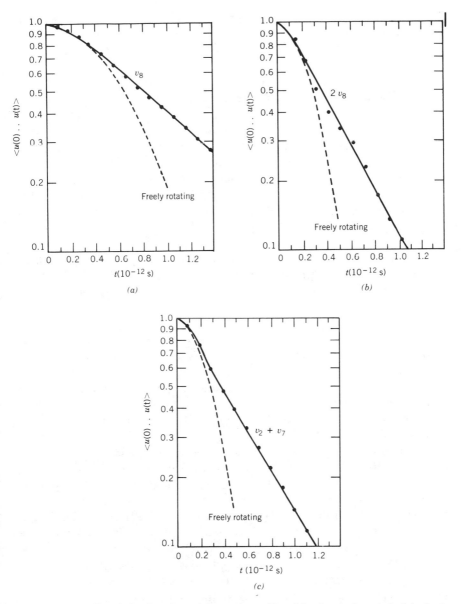

Figure 26. Correlation functions of the motions of inertial axies A, B, and C of the freely rotating and liquid-phase CH_2Cl_2 molecules. (a) Rotational motion of axis A (symmetry species B_2). (b) Rotational motion of axis B (symmetry species A). (c) Rotational motion of axis C (symmetry species B_1).

from the nuclear resonance experiment are calculated using a rotational diffusion model, so experimental spin–lattice relaxation times of the equivalent protons of CH_2Cl_2, extrapolated to zero proton concentration, can be equated with the rotational spin–lattice relaxation time of a *single* CH_2Cl_2 molecule in the liquid by means of *a simple expression*. For CH_2Cl_2 in CD_2Cl_2 solution, the relaxation time estimated in this way is 0.5 ps. The infrared experiment gives a correlation time of 1.1 ps. A discrepancy is evident, because if the rotational process is indeed Debye-like, the latter should be 3 times the former, which is clearly not the case.

For CH_2Cl_2 it is possible to look at the movements of the carbon, hydrogen, and chlorine atoms separately and to evaluate the anisotropy of the reorientational motion with NMR spectroscopy. The studies that are available typify the problems associated with intertechnique comparisons, which are more acute for CH_2Cl_2 than for the molecules of C_{3v} symmetry considered earlier.

O'Reilly et al.[166] measured the intramolecular ^1H, ^2H, and ^{35}Cl relaxation rates over a range of temperature. For orientation of the C—H vector, O'Reilly reports a correlation time of 0.64 ps, which compares well with Rothschild's[167] measurement of 0.66 ps for CH_2Cl_2 at various dilutions in CD_2Cl_2. Unfortunately, Rothschild's measurement was carried out at a temperature 11 K higher. Adjusting the two to the same temperature reveals a considerable discrepancy. In addition, we have to correct the two sets of correlation times to the same concentration, which increases the discrepancy further. It becomes clear that experimentalists should make more effort to collect their results at the same state points to aid such intercomparison.

In another investigation of the proton τ_1, Heatley[168] made measurements on ^{13}C satellites. His experiment gave $\tau_2(H) = 0.48 \pm 0.06$ ps at 308 K or 0.53 ps at 300 K. This value for $\tau_2(H)$ was a key result in Brier and Perry's discussion. If Heatley's measurements are analyzed using a zero contribution from spin–rotation relaxation $(1/\tau_1(S-R) = 0)$, the correlation time $\tau_2(H)$ for the H—H vector is found to be 0.66 ps at 310 K, 0.73 ps at 300 K, in perfect agreement with the adjusted values of both O'Reilly and Rothschild.

On analyzing his value for $\tau_2(H)$ in terms of the small-step rotational diffusion model for a *symmetric top* (CH_2Cl_2 is in fact an inertial asymmetric top), Heatley observed that stochastic models for large angular steps are generally experimentally *indistinguishable* from small-step theory unless the angular steps are larger than 30°. In contrast to this analysis, O'Reilly et al. tried to interpret their NMR data in terms of models involving the idea that significant rotational motion occurs during the period in which a molecule is excited to an "interstitial site" in the "liquid lattice" by "hard" collisions. They assumed that during this time the anisotropy of the reorientational motion was determined entirely by the inertial properties of the molecule.

The model predicts a $\tau_2(^{13}C)/\tau_2(D)$ ratio for the C—H/C—D vector in CH_2Cl_2 and CD_2Cl_2 of 0.87, which compares favorably with the value of 0.88 ± 0.14 estimated experimentally. However, within experimental uncertainty, the measured ratio also agrees with the predictions of stochastic models for which the correlation times are independent of inertial changes produced bydeuteration.

Brier and Perry considered the anisotropy implied by the NMR results. Heatley's value of $\tau_2(H) = 0.53$ ps does not support the infrared result of axial symmetry, because $\tau_2(H—H) \neq \tau_2(C—H)$. Using a formula given by Woessner,[169] they found that no single set of values for the three adjustable parameters in the model would produce the NMR results (viz. $\tau_2(H—H) = 0.53$, $\tau_2(C—H) = 0.75$, and $\tau(C—Cl) = 1.2$ ps) simultaneously. The value for $\tau_2(H—H)$ is incompatible with the value of $\tau_2(C—H)$ in this model. Also, if the three variables of the model were fitted to experimental data, the "best fit" values of the $\ell = 2$ NMR data predicted an anisotropy completely different from that implied by the infrared data and the $\ell = 1$ NMR data. If these best-fit data are compared with available neutron-scattering results, they turn out to have only the most approximate validity. Brier and Perry analyzed four different dynamical models with the neutron-scattering data. They assumed that rotational and translational motions are decoupled. For the translational motions they considered only the Egelstaff–Schofield (E–S) modification of the simple diffusion model. Though this, in contrast to the simple Fick's law, gives the correct short-time behavior ($\approx t^2$), it predicts a velocity autocorrelation function that monotonically decreases to zero. This is *not* in accord with molecular-dynamics results, which show that the correlation function has a negative overshoot.

For rotational motion Brier and Perry considered four models. The first two were empirical, allowing for axial symmetry about the a-axis. No detailed physical interpretation of the reorientational dynamics or even an estimate of the degree of anisotropy of the motion can be obtained from such an empirical approach. They also considered the M and J diffusion models. The agreement for both was poor. Changing the one adjustable parameter of the models did little to improve the situation: It altered the magnitude but not the position of maximum absorption. Brier and Perry increased the anisotropy of the motion many times in an attempt to improve the agreement with experiment. However, the agreement was only slightly improved. This does not mean that neutron-scattering data are insensitive to the anisotropy of the motion. Rather, it indicates how insensitive the models used are to the anisotropy of the motion. Indeed, we have found similar results using our own 0 THz spectroscopy.[170]

Thus, from a literature search we conclude that the lack of any coordination in the research has resulted in a situation in which we are unable to state, with any degree of certainty, the anisotropy of the molecular motion of CH_2Cl_2. There have been no spectroscopic studies on the liquid under applied external pressure. There appears to be a strong interaction of rotation with vibration that affects infrared and Raman studies to unknown degrees. Baranov[171] has established the predominant role of dipole–dipole interaction in his Raman study, and Nestor and Lippincott[172] have considered the effect of the internal field by comparing of gas- and liquid-phase spectra. They conclude that the cross section for each Raman band is greater for molecules in the liquid than in the gas because of strong internal-field effects. These effects should also distort Rayleigh and 0 THz band profiles.

We can see how changing this field affects the far-infrared spectrum dramatically by extending the spectroscopic work to supercooled and vitreous solutions. For example, if we study a glassy solution of CH_2Cl_2 in decalin, the low-frequency part of the loss curve exhibits a peak that shifts upward by about two decades with a 4 K increase in temperature and at the glass-to-liquid transition temperature moves very quickly out of the audiofrequency range toward the microwave. The far-infrared peak in the loss curve is displaced by 30 cm^{-1} to 90 cm^{-1} in the glass, as compared with 60 cm^{-1} in the liquid solution at room temperature, indicating that the molecular-dynamical evolution in CH_2Cl_2 in the glassy state starts on the picosecond time scale and evolves gradually into a process occurring on immensely longer time scales (seconds and much longer). The whole process should, in principle, be described by the orientational correlation function of the resultant dipole in the sample. This would require a molecular-dynamics simulation (on present-day computers) lasting approximately 10^9 years!

Brier and Perry conclude that "despite the numerous studies on CH_2Cl_2, no clear picture of the reorientational dynamics has yet emerged, even of a semi-quantitative nature." If we consider our own 0 THz studies,[162] a similar picture emerges. In particular, currently popular models of the liquid state give *poor* representations of the measured quantities. The far-infrared profile provides, even at ambient temperature and pressure, a highly discriminating test for all models, and has revealed the deficiencies of the currently popular molecular models, including Debye's, the extended diffusion models, models derived from the Kubo–Mori formalism, and even the models with more realistic physical interpretations—the itinerant oscillator and its various extensions. In addition to these theoretical problems (quoting Brier and Perry again) "there are limitations and assumptions involved in both the measurement and subsequent analysis of the data itself... and the available data may not be sufficiently varied and accurate to make a critical test of any model of the liquid dynamics."

Let us consider inferences from a molecular-dynamics simulation of this liquid. The molecular dynamics has been simulated with two model representations of the intermolecular potential. These consist of 3×3 and 5×5 atom–atom simulations with and without fractional charges at atomic sites. These empirical forms for the potential energy of two interacting CH_2Cl_2 molecules are used in the absence of a more acceptable quantitative expression. In the 3×3 representation the CH_2 group is taken as a moiety and is developed from an algorithm of Singer et al.[173] The core atom–atom interaction is Lennard-Jones in type, with the following parameters:

$$\sigma(Cl—Cl) = 3.35 \text{ Å}$$
$$\sigma(CH_2—CH_2) = 3.96 \text{ Å}$$
$$\varepsilon/k(Cl—Cl) = 173.5 \text{ K}$$
$$\varepsilon/k(CH_2—CH_2) = 70.5 \text{ K}$$

The $Cl—CH_2$ interaction is evaluated using the equations

$$\sigma(Cl—CH_2) = \tfrac{1}{2}(\sigma(Cl—Cl) + \sigma(CH_2—CH_2))$$
$$\frac{\varepsilon}{k}(Cl—CH_2) = \left(\frac{\varepsilon}{k}(Cl—Cl)\frac{\varepsilon}{k}(CH_2—CH_2)\right)^{1/2}$$

Partial charges are added to reproduce the total dipole of 1.6 D, so that the charge in the Cl unit is $-0.151|e|$ and that on the CH_2 units is $+0.302|e|$. The full potential (atom–atom + charges) was tested by McDonald[174] at 287 K (a molar volume of 62.92 cm^3 mol^{-1}), and gave a mean potential energy of -6.2 kcal mol^{-1}, which compares with a measured value[‡] of -6.2 to -6.3 kcal mol^{-1} estimated from experimental ΔH values of 6.69–6.83 kcal mol^{-1}. Approximately 0.5 kcal mol^{-1} has been allowed for the $\Delta(PV)$ term. From this indication it seems that the main features of the force field are correct. There are *no free parameters* in this 3×3 model.

The 5×5 simulation algorithm, originally written by Singer et al.,[173] has been modified by Ferrario and Evans[175] to include a charge–charge interaction and a force-cutoff criterion based on molecule center of mass-to-center of mass distance (cutoff radius $= 11.28$ Å). The Lennard-Jones parameters are:

$$\sigma(H—H) = 2.75 \text{ Å}$$
$$\sigma(Cl—Cl) = 3.35 \text{ Å}$$
$$\sigma(C—C) = 3.2 \text{ Å}$$
$$\varepsilon/k(H—H) = 13.4 \text{ K}$$
$$\varepsilon/k(Cl—Cl) = 175.0 \text{ K}$$
$$\varepsilon/k(C—C) = 51.0 \text{ K}$$

[‡] For energy of vaporization see "selected values of chemical thermodynamic properties," 1961, p. 588. At 760 mm Hg, 313 K $\Delta H = 6.69$ kcal mol^{-1}; $\Delta S = 21.4$ cal (mole K)$^{-1}$.

TABLE VI
Experimental and Simulated Correlation times for Dichloromethane[a]

Technique	Vector	Correlation Time (ps)	
[1]H (intra)	H—H ($\parallel$ to e_C)	0.53 ± 0.06	
[2]D (quadrupole)	C—D	0.80 ± 0.10	
[13]C—H (dipolar)	C—H	0.70 ± 0.07	P_2
[35]Cl (quadrupole relaxation)	C—Cl (approx. $\parallel e_B$)	1.20 ± 1.10	
Computer simulation	e_A	0.50	
Computer simulation	e_B	0.9	
Computer simulation	e_C	0.51	
Neutron scattering	Center of mass to H	0.56	
Dielectric	e_A	1.45	
relaxation	e_A	0.5	
Infrared (Rothschild)	e_B	1.1	P_1
Infrared	e_A	1.1	
(van Konynenberg and Steele)[15]	e_A	1.2	
Computer simulation	e_B	3.8	
Computer simulation	e_C	1.21	
Rayleigh scattering (van Konynenberg and Steele)	e_A	1.85	P_2

[a] Compiled by M. W. Evans, and reported in *J. Mol. Liq.*, **23**, 113 (1982). Reproduced by permission. See this paper for original data sources.

with fractional charges of $0.98|e|$ on H, $-0.109|e|$ on Cl, and $0.022|e|$ on C. The former were chosen to optimize the thermodynamic conditions and the latter from a molecular-orbital calculation by del Re (see ref. 176). Again there are no free parameters to be varied.

The 5×5 and 3×3 algorithms produce data that are directly comparable because the thermodynamic conditions are the same (293 K, 1 bar, $V_m = 64.0$ cm^3 mol^{-1}). Any difference in the resulting dynamical functions may therefore be attributed only to the difference in the pairwise-additive force fields used. Simulated and experimental correlation times are compared in Table VI.

We will consider briefly the results of the structural functions before considering the dynamics. Unlike chloroform (Section II), no diffraction results with which to compare the simulated distribution functions are available for CH_2Cl_2. However, it is of interest to compare the simulated distribution functions for the 5×5 algorithm with and without charges. Any differences will reflect the sensitivity of the pdf structure to electrodynamic parts of the pair potential. If the pair distribution is structured, this implies that the liquid has a fair degree of residual ordering—local order. Based on the pair distribution function describing the H II atom–atom positions (Fig. 21), it is by no means certain that the structure may be attributed solely to repulsive parts of the intermolecular potential (as in RISM theory) because the addition of charges clearly inhibits the first peak at about 2.8 Å while enhancing the second.

For the dynamics, experimental results are available with which to make comparisons. We are also now able to establish which of the two models (the 3×3 or 5×5) best approximates the true intermolecular potential. For example, far-infrared spectra are most discriminating. The simulated and experimental results are compared in Fig. 28. The 5×5 is obviously the most realistic potential, since the 3×3 produces a result that both is too sharp and peaks at too low a frequency. The profile in fact resembles very closely the result we anticipate for the extended diffusion models, which are unable to shift the far-infrared profile from the free rotor maximum. The 5×5, though better, produces a profile still displaced some 30–40 cm^{-1} from the measured spectrum for the neat liquid. Either the potential is still an oversimplification or other factors (induced absorption, cross correlations, etc.) contribute significantly to the far-infrared profile. Whichever of these is the case, Fig. 28 shows just how discriminating a test 0 THz spectra provide for any model of the liquid state.

We referred above to contributions from cross correlation (i.e., the mutual dependence of the motion of one molecule on that of its immediate neighbors). Such correlations are anticipated to be removed (gradually) on dilution—providing the solvent is noninteracting. Certainly, if CH_2Cl_2 is dissolved in CCl_4 or decalin, the far-infrared peak frequency shifts gradually to a lower frequency. It is interesting that the simulated spectrum for a 10% solution of CH_2Cl_2 in CCl_4 reproduces well the measured spectrum (Fig. 28b). It is debatable whether such a comparison is meaningful, because the simulation involves a neat solution of the liquid in which the "probe" molecule is surrounded by similar and not solvent molecules. Also, at a concentration of 10% CH_2Cl_2 in CCl_4 the far-infrared spectrum is still in the process of shifting to lower frequencies. The plot of $\bar{\nu}_{max}$ versus concentration (Fig. 29) shows a linear dependence in the concentration range from neat solution to 10%, the shift continuing below 10%. It may even emerge that the shift *does not* remain linear below 10%, as for CH_3CN (Section IV), if and when the necessary experimentation is done. So the simulated function, though agreeing well with the spectrum for a 10% solution, must in fact peak at higher frequencies than does the measured spectrum of, say, a 1% solution.

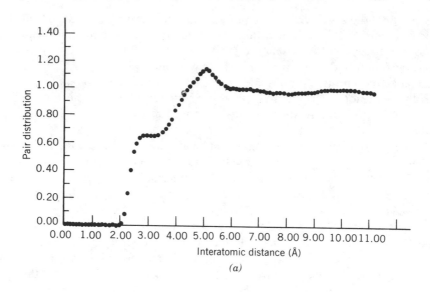

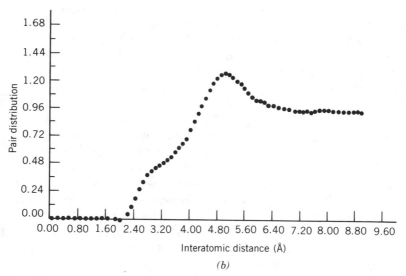

Figure 27. (*a*) Hydrogen-to-hydrogen atom–atom pair distribution function extracted as a mean over the equilibrium run. 5×5 potential, no charges, 293 K, 1 bar. (*b*) Same as *a*, but with charges included. [Reproduced by permission from M. W. Evans and M. Ferrario, *Adv. Mol. Rel. Int. Proc.*, **24**, 75 (1982).]

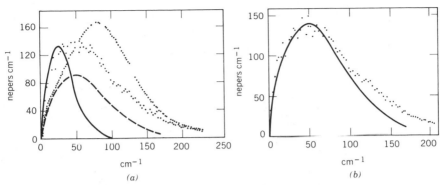

Figure 28. (a) Comparison of molecular-dynamics simulation and far-infrared spectra of CH_2Cl_2 and solution in CCl_4. O, —O—, □, measured data, G. J. Evans and M. W. Evans; ●, computer simulation, 5×5 potential, no charges; —, computer simulation, 3×3 potential, no charges; ---, 10% solution in CCl_4, experimental. (b) —, Scaled-up spectrum of CH_2Cl_2 in CCl_4; ·····, simulation of same. [Reproduced by permission from M. W. Evans and M. Ferrario, *Adv. Mol. Rel. Int. Proc.*, **24**, 75 (1982).]

The discrepancy is not easily explained. However, we may at least conclude that since the 5×5 is a better representation than the 3×3 simulation, the *hydrogen atoms play an important part in the dynamics of* CH_2Cl_2. This is a surprising result and emphasizes to the exponents of molecular and hydrodynamic theories the need to represent the shape of the molecules carefully in their modeling procedures and also to account for the smallest atoms

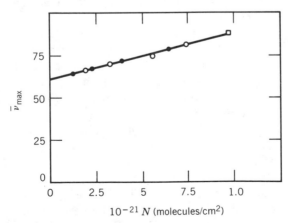

Figure 29. Variation of peak frequency $\bar{\nu}_{max}$ with concentration for CH_2Cl_2 in CCl_4 (O) and decalin (●). [Reproduced by permission from M. W. Evans and M. Ferrario, *Adv. Mol. Rel. Int. Proc.*, **24**, 75 (1982).]

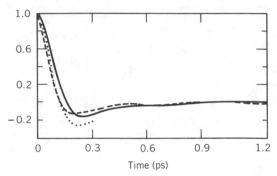

Figure 30. Rotational velocity autocorrelation functions (of $\dot{e}_A(t)$): —— , 3×3 potential, no charges; - - -, 5×5 potential, no charges; · · · · ·, for 10% (volume/volume) CH_2Cl_2 in CCl_4. [Reproduced by permission from M. W. Evans and M. Ferrario, *Adv. Mol. Rel. Int. Proc.*, **24**, 75 (1982).]

of the molecule. The extensive use of spheres, needles, and the like to represent molecules, and the complete neglect of molecules (as in the hydrodynamic theories), are obvious oversimplifications.

It is also revealing to compare actual correlation functions. Those of $\langle e_A(t) \cdot e_A(0) \rangle$, for example, are shown in Fig. 30. The experimental curve is in fact deeper and slightly more oscillatory than both the 3×3 and 5×5 simulated functions. We emphasize that it is difficult to Fourier transform from time to frequency domains because of the observable long-time tails (Fig. 31). The angular momentum autocorrelation function has a small but long positive tail, and the rotational velocity autocorrelation function a correspondingly long negative tail.

When charges are incorporated into the algorithms, the effect on the equilibrium time correlation functions is not pronounced but is nevertheless significant. This indicates that the dynamics of the liquid are only approximately describable in terms of Lennard-Jones parameters. A full description requires the inclusion of electrodynamic terms. The inclusion of charge–charge interaction modifies the P_1 and P_2 correlation functions. The far-infrared spectrum, for example, is shifted to higher frequency. Consequently, the P_1 and P_2 correlation times are increased; in particular, the microwave and NMR relaxation times are increased by the long-range terms of this nature, often to a significant degree. Figures 32 and 33 show the effect of including charges using the 3×3 potential—the algorithm is still not capable of reproducing the observed spectra. The effect of adding charges to the 5×5 simulation is illustrated in Figs. 34–36. The far-infrared spectrum is shifted slightly to higher frequencies (Fig. 37), but is still below the measured spectrum, which peaks at > 80 cm^{-1}.

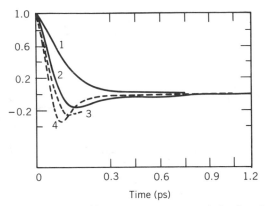

Figure 31.   Comparison of (1) angular momentum autocorrelation function with (2) the autocorrelation function of $\dot{e}_A(t)$ and (3) and (4) the rotational velocity autocorrelation functions from data for CH_2Cl_2 in solution and in the pure liquid state, respectively. Normalized to 1 at $t = 0.3 \times 3$ potential, no charges. [Reproduced by permission from M. W. Evans and M. Ferrario, *Adv. Mol. Rel. Int. Proc.*, **24**, 75 (1982).]

It is interesting to observe how the effect of rototranslational interaction changes as we move from molecules of C_{3v} symmetry to those of lower symmetry. Up to nine elements of the mixed autocorrelation function matrices (discussed elsewhere in this volume) are observable, depending on the molecular symmetry. As we have already emphasized, it is a truism that the outcome of every experiment on liquid-state molecular motion is the observation of rototranslation in a time-averaged form. Even though it is an orientational function that we may measure how these functions evolve in a system in which rotation and translation motions interact to varying degrees.

For CH_2Cl_2, of the nine cross elements already referred to, only two exist by symmetry in the molecule-frame autocorrelation function of the *linear*

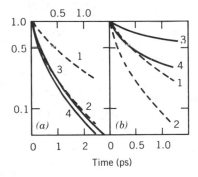

Figure 32.   (a) 1 and 2, P_1 and P_2 autocorrelation functions 3×3 potential including charges (upper abscissa); 3 and 4, P_1 and P_2 autocorrelation functions of e_A, 3×3 potential, no charges (lower abscissa). Both at 293 K, 1 bar. (b) 1 and 2, as for a; 3 and 4, 3×3 potential including charges at 177 K, 1 bar. [Reproduced by permission from M. W. Evans and M. Ferrario, *Adv. Mol. Rel. Int. Proc.*, **24**, 75 (1982).]

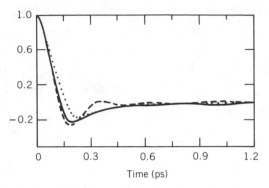

Figure 33. Rotational velocity autocorrelation functions of $\dot{\mathbf{e}}_A(t)$: ——, 3×3, including charges, at 293 K, 1 bar; ----, 3×3, including charges, at 177 K, 1 bar; ·····, 3×3, no charges, at 293 K, 1 bar. [Reproduced by permission from M. W. Evans and M. Ferrario, *Adv. Mol. Rel. Int. Proc.*, **24**, 75 (1982).]

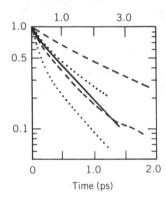

Figure 34. ——, P_1 autocorrelation function of $\mathbf{e}_A$, 5 ×5 no charges, 293 K, 1 bar (upper abscissa); ---, P_1 and P_2 autocorrelation functions of $\mathbf{e}_A$, 5×5, including charges, 293 K, 1 bar (lower abscissa); ·····, P_1 and P_2 autocorrelation functions of $\mathbf{e}_A$, 3×3, including charges, 293 K, 1 bar (lower abscissa). [Reproduced by permission from M. W. Evans and M. Ferrario, *Adv. Mol. Rel. Int. Proc.*, **24**, 75 (1982).]

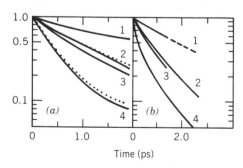

Figure 35. (*a*) Anisotropy of rotational diffusion: 5×5 potential, including charges, 293 K, 1 bar. 1, P_1 autocorrelation function of $\mathbf{e}_B$; 2, P_1 autocorrelation function of $\mathbf{e}_A$ and (●) of $\mathbf{e}_C$; 3, P_2 autocorrelation function of $\mathbf{e}_B$; 4, P_2 autocorrelation function of $\mathbf{e}_B$ and (●) of $\mathbf{e}_C$. (*b*) As for *a*, but no charges. Note that P_1 and P_2 of $\mathbf{e}_C$ are not shown for clarity, as they are similar to P_1 and P_2 of $\mathbf{e}_A$. [Reproduced by permission from M. W. Evans and M. Ferrario, *Adv. Mol. Rel. Int. Proc.*, **24**, 75 (1982).]

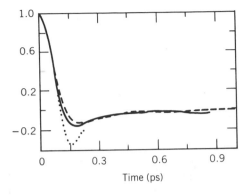

Figure 36. Rotational velocity autocorrelation functions (of $e_A(t)$): ---, 5×5, no charges, 293 K, 1 bar; —, 5×5, with charges, 293 K, 1 bar; ····, computed from far-infrared data on pure liquid CH_2Cl_2. [Reproduced by permission from M. W. Evans and M. Ferrario, *Adv. Mol. Rel. Int. Proc.*, **24**, 75 (1982).]

center of mass momentum **p** with the resultant molecular *angular* momentum **J** at the instant t in time. The mixed laboratory-frame autocorrelation function $\langle \mathbf{p}(0)\cdot\mathbf{J}(t)\rangle_{lab.}$ vanishes for all t in an *isotropic* molecular liquid because the parity of **p** to time reversal is opposite in sign to that of **J**. We wish to emphasize that these cross interactions and their extent of influence on the molecular dynamics are not in any sense critically dependent on the

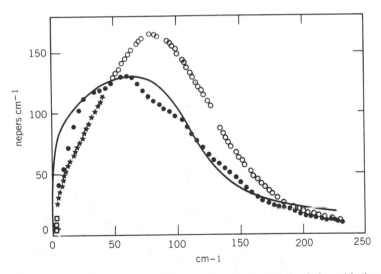

Figure 37. Far-infrared spectrum of dichloromethane from 5×5 simulation with charges.

●, simulation; □, —O—, O, experimental (neat solution); — , experimental (10% solution in CCl_4). [Reproduced by permission from M. W. Evans and M. Ferrario, *Adv. Mol. Rel. Int. Proc.*, **24**, 75 (1982).]

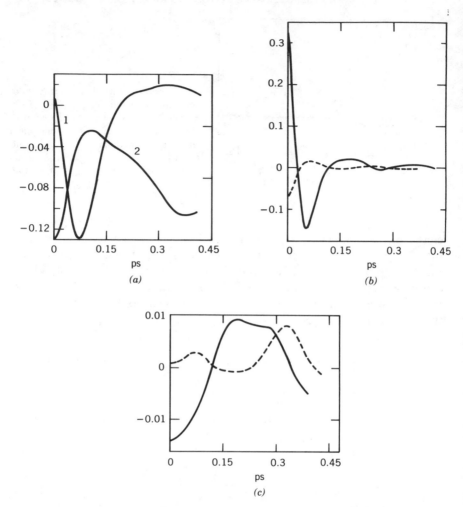

Figure 38. (*a*) The nonvanishing elements of the molecule-frame correlation matrix $\langle \mathbf{p}(0)\mathbf{J}^T(t)\rangle_{\text{mol}}$. 1, $[\langle p_1(0)J_2(t)\rangle + \langle p_1(t)J_2(0)\rangle]_{\text{mol}}/2\langle p_1^2\rangle^{1/2}\langle J_2^2\rangle^{1/2}$ (left-hand scale); 2, $[\langle p_2(0)J_1(t)\rangle + \langle p_2(t)J_1(0)\rangle]_{\text{mol}}/2\langle p_2^2\rangle^{1/2}\langle J_1^2\rangle^{1/2}$ (right-hand scale). (*b*) —— , (1,2) element of the molecule-frame force–torque mixed autocorrelation function, normalized as in *a*; ---, (2,1) element. (*c*) Illustration of the noise level in *a* and *b*. —— , (3,2) element of the linear-angular momentum correlation matrix; ---, (2,2) element. Both of these autocorrelation functions should vanish by symmetry. [Reproduced by permission from M. W. Evans and M. Ferrario, *Adv. Mol. Rel. Int. Proc.*, **24**, 75 (1982).]

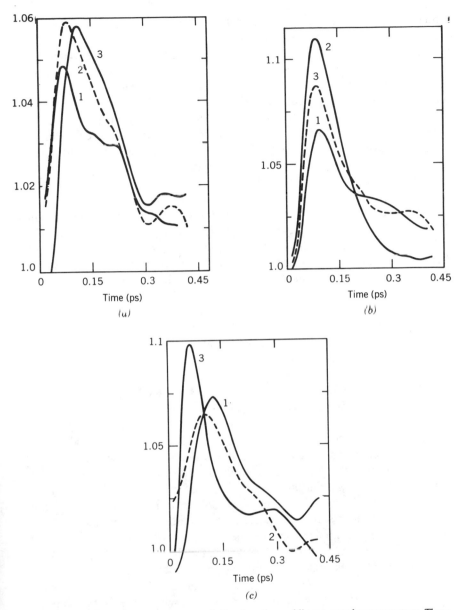

Figure 39.   Second-moment autocorrelation functions of linear–angular momentum. These are invariant to frame transformation and *do not* vanish in the laboratory frame. (*a*) 1, $[\langle p_3^2(0)J_2^2(t)\rangle + \langle p_3^2(t)J_2^2(0)\rangle]/[2\langle p_3^2\rangle\langle J_2^2\rangle]$; 2, (2,2) element; 3, (2,3) element. (*b*) 1, (2,1) element; 2, (1,3) element; 3, (3,1) element. (*c*) 1, (1,1) element; 2, (3,3) element; 3, (1,2) element. [Reproduced by permission from M. W. Evans and M. Ferrario, *Adv. Mol. Rel. Int. Proc.*, **24**, 75 (1982).]

simulation model, so long as that model has elements of realism (i.e., reflects the known C_{2v} symmetry in some sense).

Figure 38 shows that the (2,1) and (1,2) elements of the cross matrix *are not* symmetric for either the linear–angular momentum or force–torque mixed autocorrelation functions ("auto" because they refer to the rotation and translation of the *same* molecule).

The second-moment autocorrelation functions $\langle p^2(0)J^2(t)\rangle$ and $\langle F^2(0) T_q^2(t)\rangle$ are invariant to frame transformation and may be observed in the laboratory frame. All elements (i, j) exist. These functions are illustrated in Figs. 39 and 40 and provide a detailed description of molecular rototranslation. Obviously the rototranslation coupling perturbation in CH_2Cl_2, one of the most extensively studied molecules of C_{2v} symmetry, is pronounced.

Rototranslational equations have no known analytical solutions. The phenomenological theories, as developed by Debye and his contemporaries and extended by many others in ensuing years, consider rotational motion of (in the case of Debye) spherical entities. Rototranslation leads to a morass of supermatrices with too many adjustable coefficients to be useful.

The influence of rototranslation is not straightforwardly related to the molecular symmetry because the sizes of the atoms making up a molecule are also significant. It appears that a full understanding of liquid-state dynamics requires starting at the level of the atoms or even of the constituents of the atoms themselves.

VIII. ACETONE [M. W. EVANS AND G. J. EVANS, *J. CHEM. SOC., FARADAY II*, 79, 153 (1983).]

Many of the spectroscopic techniques available molecular motion by bandshape transformation have been applied to liquid acetone. Koga et al.[177] used infrared bandshape analysis to study the anisotropy of rotational diffusion through three correlation times, τ_A, τ_B, and τ_C, defined about the three inertial axes A, B, and C. In nondipolar solvents, the ratio $\tau_A : \tau_B : \tau_C$ does not vary much from solvent to solvent. Reorientation about the B-axis is more restricted. In dipolar solvents the correlation times are longer, especially τ_B. The Favro[178] model of rotational diffusion produces the result $\tau_A : \tau_B : \tau_C = 1.02 : 1.00 : 0.88$, in contrast to the observed (except in CS_2) $1.3 : 1 : 1.1$. Using van der Waals radii the excluded volumes about each axis are $V_A = 55.1 \text{ Å}^3$, $V_B = 44.4 \text{ Å}^3$, and $V_C = 48.5 \text{ Å}^3$, which does not explain the order of the three correlation times. Koga et al. observe that only motions about the A- and C-axes are restricted by dipole–dipole interactions.

The paper of Dill et al.[179] on the Rayleigh scattered bandshape of liquid acetone under hydrostatic pressure is one of the most interesting in the literature on any isotropic liquid. These authors have provided angular-position

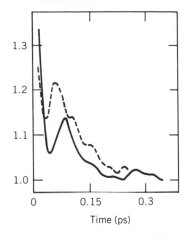

Figure 40.   Two elements of the mixed force–torque molecule-frame autocorrelation function matrix: ——— , (1,1) element; ----, (2,2) element. [Reproduced by permission from M. W. Evans and M. Ferrario, *Adv. Mol. Rel. Int. Proc.*, **24**, 75 (1982).]

correlation functions and angular-velocity correlation functions for liquid acetone at 1 bar, 293 K. Dill et al. compare their light-scattering data with the NMR relaxation results on liquid acetone of Jonas and Bull,[100] who estimated a translational correlation time $\langle \tau_t \rangle_{NMR}$ a rotational correlation time $\langle \tau_2 \rangle_{NMR}$, and their density dependence. The NMR data in acetone show that $\langle \tau_t \rangle_{NMR}$ has the same density dependence as the viscosity, η, but that $\langle \tau_2 \rangle_{NMR}$ and the Rayleigh-scattering correlation time are much less density dependent than $\langle \tau_t \rangle_{NMR}$. To a good approximation $\langle \tau_2 \rangle_{NMR} = \langle \tau_2 \rangle_{\text{light scattering}}$. This can be interpreted in two ways: Either two components of the diffusion tensor are equal, or the light-scattering correlation time is related to a mixture of the same two.

Schindler et al.[181] have reported a Raman study of pressure effects in liquid acetone. The Raman bandshape of the symmetric C=O stretch at 1710 cm^{-1} was measured up to 4 kbar over a wide range of temperature. They conclude that intermolecular dipole–dipole coupling is responsible for unusual effects as the hydrostatic pressure is increased. The frequencies of the polarized (VV) and depolarized (VH) bands in acetone differ by several wavenumbers. These authors point out that hydrodynamic theories usually consider repulsive forces as the main source of line broadening, but in reality the increase or decrease of half-width (and frequency) with pressure depends on the relative importance of attractive and repulsive intermolecular forces that influence a specific vibration. Schindler et al. expected that repulsive forces play a minor role in acetone compared with the attractive forces. However, their interpretation was based on the *simple* Kubo stochastic bandshape theory containing one parameter τ_m, defined as the modulation time. Oxtoby et al.[182] have equated this parameter to the duration of a collision; Döge[183] and Wang,[184] to the rotational correlation time; and Lynden-Bell,[185] to the correlation time governing translational diffusion. Schindler et al. defined it as the correlation time for frequency fluctuations in their calculations.

Perrot et al.[186] have studied liquid acetone using depolarized Rayleigh scattering. The function $\omega^2 I(\omega)$ (the second moment) peaks at 64.5 cm^{-1}, with a half-width of 92 cm^{-1}. This may be compared with the far-infrared spectrum for a 10% (volume/volume) in decalin solution of Reid and Evans,[187] in which the peak absorption is at 52 cm^{-1}. The integrated absorption intensity (A) of the $\alpha(\omega)$ in the far infrared is linearly dependent on the number density (N) of acetone solute molecules in nondipolar solvents such as decalin. This seems to imply the absence or insignificance of induced effects, but to confirm this the dilution studies should be extended to well below 10% (see the review by Vij and Hufnagel in this volume).

Electrooptic techniques have been used to study acetone. Burnham and Gierke[188] have used results from the optical Kerr effect, Cotton–Mouton effect, and light scattering to obtain orientational pair correlation functions from the theory of Laudanyi and Keyes.[189] They also estimated orientational pair correlation parameters for liquid acetone. The values range from 0.5 (Kerr effect) to 1.4 (anisotropic Rayleigh scattering), which values yield contradictory implications concerning the local structure.

A contradiction is observed in dielectric studies. At 293 K the relaxation time of pure liquid acetone is 3.1 ± 0.1 ps, which decreases to 2.5 ± 0.3 ps at 317 K. In 0.19 (mole fraction) CCl_4 solution, the relaxation time decreases slightly to 2.9 ± 0.3 ps at 293 K. The effect of dilution is small, which agrees with far-infrared observation. However, in dilute CCl_4 the three correlation times calculated by Koga et al.[177] are $\tau_A = 1.29$ ps, $\tau_B = 1.01$ ps, and $\tau_C = 1.11$ ps. These contrast with the dielectric relaxation time of acetone in CCl_4 of 2.9 ± 0.3 ps.

Jonas and Bull[180] have calculated a reorientational spin–spin NMR correlation for acetone of 0.75 ps at 290 K, 1 bar, which increases to 1 ps at 296 K, 2 kbar. The translational correlation times under the same conditions are 4.8 and 11.0 ps, respectively, showing that these times are more dependent on density. In deriving these times, Jonas and Bull again made use of rotational diffusion theory with the assumption of isotropic diffusion. This theory implies that the mean first-rank correlation time should be 3 times the mean second-rank correlation time. Comparing the dielectric relaxation time in the same liquid (3.1 ± 0.1 ps) with the NMR spin–spin time (0.75 ps) shows this to be approximately the case. However, such a comparison is not meaningful, because the measurements of Koga et al. and Dill et al. clearly show that the theory of rotational diffusion does not explain the molecular-dynamical properties in liquid acetone.

We may also compare depolarized Rayleigh correlation times with Raman and NMR correlation times. Dill et al.[179] report that the Rayleigh and NMR correlation times for pure liquid acetone are the same at 1 bar and 2 kbar. This result contrasts markedly with that of the equivalent "first rank" procedure, for which we have seen already that the dielectric relaxation time is 4 times the infrared correlation time at 293 K and 1 bar in dilute CCl_4 solution. The Raman correlation times of Schindler et al.[181] are 0.27 ps for pure liquid acetone at 1 bar and 298 K, and 0.55 ps at 2 kbar and 298 K. These are considerably shorter than those from spin–spin NMR and depolarized Rayleigh scattering and seem to have little significance in terms of rotational dynamics, except that they are considerably more dependent on density. We have already commented on the interpretation of the correlation times measured by Schindler et al. It seems certain that they cannot be translational in origin, because the NMR results yield translational correlation times of 4.8 ps at 296 K and 1 bar and 11 ps at 2 kbar. These are an order of magnitude longer than the corresponding Raman correlation times of Schindler et al., but interestingly, have a similar density dependence. They are too short to be purely rotational in origin.

There is inconsistency in the literature on acetone, as indeed there has been for all the liquids we have considered in this review. No overall viewpoint concerning the molecular dynamics of liquid acetone is obtained from the many studies reported. In such circumstances a computer simulation study must contribute to our understanding of the liquid-state dynamics; it will certainly again clarify, for example, the anisotropy of the rotational diffusion. In this respect, the picosecond laser-induced inhomogeneous broadening of Raman bands reported by George et al.[190] has clearly shown that rotovibrational diffusion in liquid acetone is anisotropic.

In the simulation the interaction between $(CH_3)_2CO$ molecules is modeled with a 4×4 Lennard-Jones atom–atom "core" with point charges localized at each site. The CH_3 group is taken as an entity (an oversimplification —see Section VII for CH_2Cl_2, which revealed the significance of the hydrogen atoms in determining the molecular dynamics) and the complete set of parameters is as follows:

$$\sigma(CH_3-CH_3) = 3.92 \text{ Å}$$
$$\sigma(C-C) = 3.00 \text{ Å}$$
$$\sigma(O-O) = 2.80 \text{ Å}$$
$$\varepsilon/k(CH_3-CH_3) = 72.0 \text{ K}$$
$$\varepsilon/k(C-C) = 50 \text{ K}$$
$$\varepsilon/k(O-O) = 58.4 \text{ K}$$

The cross terms were evaluated by Lorentz–Berthelot combining rules. The $-CH_2$ parameters were taken from a paper by Bellemans et al.[191] on the molecular-dynamics simulation of n-alkanes, the O—O parameters from molecular crystal data, and the C–C parameters from our previous work on CH_2Cl_2, $CHCl_3$, and $CHBr_3$. We adhere to our method of using literature values without adjustment so that future comparisons with more precise algorithms may be made.

We represent electrostatic interactions with point charges taken from a calculation by Wellington and Khouwaiter.[192] These are $q_{CH_3} = -0.032|e|$, $q_C = 0.566|e|$, and $q_O = -0.502|e|$. The three principal moments of inertia used in the computer-simulation program were calculated from structural data. These are $I_A = 71.2 \times 10^{-40}$ g cm^2, $I_B = 83.0 \times 10^{-40}$ g cm^2, and $I_C = 154.2 \times 10^{-40}$ g cm^2. The dipole axis in this notation is that of I_B and the dipole unit vector is e_B.

TABLE VII
Experimental and Simulated Correlation Times for Acetone[a]

Technique	Experimental correlation time (ps)	Simulated autocorrelation time (ps)[b] (293 K, 1 bar)
Dielectric relaxation		
(i) pure acetone	(i) 3.1 ± 0.1	$\tau_{1A} = 3.2$, $\tau_{1B} = 2.2$, $\tau_{1C} = 2.2$
(ii) 20% (mole fraction in CCl$_4$	(ii) 2.9 ± 0.3	
Infrared bandshapes in dilute solution	$\tau_{1A} = 1.29$, $\tau_{1B} = 1.01$, $\tau_{1C} = 1.11$	$\tau_{1A} = 3.2$, $\tau_{1B} = 3.3$, $\tau_{1C} = 2.2$
Spin–spin NMR, rotational correlation time	0.75	$\tau_{2A} = 0.75$, $\tau_{2B} = 0.6$, $\tau_{2C} = 0.6$
NMR translational	4.80	$\tau_t \approx 1.0$
Raman, C=O stretch	0.27	$\tau_{2B} = 0.6$
Rayleigh scattering	0.75	$\tau_{2A} = 0.75$, $\tau_{2B} = 0.6$, $\tau_{2C} = 0.6$
	Frequency maximum/cm^{-1}	
Far-infrared power absorption		
(i) pure acetone (this work)	(i) 57	
(ii) 10% acetone in decalin	(ii) 52	
Rayleigh scattering	64.5	

[a] Compiled by M. W. Evans and reported in *J. Chem. Soc., Faraday II*, **79**, 153 (1983); reproduced by permission. See this paper for original data sources.
[b] The dipole unit vector is e_B.

Computed and experimental correlation times are compared in Table VII. The simulated infrared rotational correlation time τ_B is 2.2 ps, which is compared with the infrared and dielectric relaxation measurements in the table. Note that the dielectric relaxation time is a weighted mean of the three simulated correlation times about the three principal moment-of-inertia axes. The experimentally measured dielectric relaxation time is longer than the simulated times. This conforms with the pattern that emerged in our discussions of other liquids in the preceding sections.

The anisotropy of the rotational diffusion in acetone from infrared band-shape analysis suggested a ratio $\tau_A : \tau_B : \tau_C$ of $1.3 : 1 : 1.1$, which compares

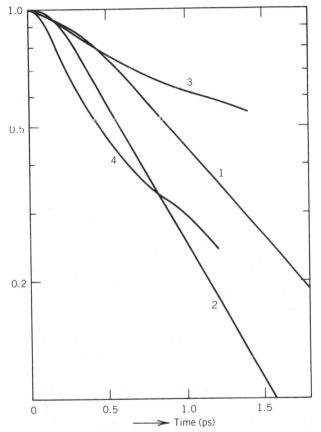

Figure 41. First-rank orientational autocorrelation functions $P_1(\mathbf{e}_B)$. 1, Acetone in acetonitrile, infrared bandshape analysis; 2, as for 1, but in n-hexane; 3, computer simulation; 4, as for 3, but second-rank orientational autocorrelation function $P_2(\mathbf{e}_B)$. [Reproduced by permission from M. W. Evans et al., *J. Chem. Soc., Faraday II*, **79**, 153 (1983).]

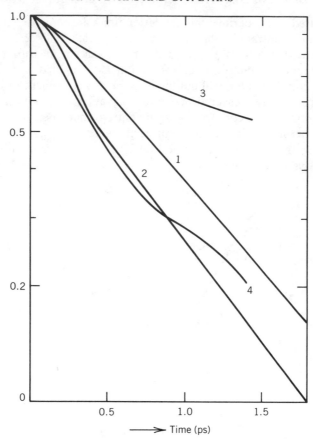

Figure 42.   As for Fig. 41, but for the unit vector $\mathbf{e}_C$. 1, infrared bandshape analysis, acetone in CCl_4; 2, as for 1, but in cyclohexane; 3, computer simulation; 4, second-rank orientational autocorrelation function $P_2(\mathbf{e}_C)$. [Reproduced by permission from M. W. Evans et al., *J. Chem. Soc., Faraday II,* **79,** 153 (1983).]

favorably with that of $1.5:1:1$ from the simulation. Figure 41 compares the simulated dipole autocorrelation function $\langle \mathbf{e}_B(t)\cdot\mathbf{e}_B(0)\rangle$ with the results of Koga et al.[177] in acetonitrile and *n*-hexane, which exhibited the extremes of molecular motion for the solvents they considered. The simulated result decays more slowly than either of the experimental functions. We also show the P_2 function $\frac{1}{2}\langle 3\mathbf{e}_B(t)\cdot\mathbf{e}_B(0)^2-1)\rangle$ for comparison. Figure 42 is the same as Fig. 41 except that it is for the $\mathbf{e}_C$ vector autocorrelation function. The experimental results refer to carbon tetrachloride and cyclohexane, and again results for other solvents fall between these two extremes. The simulated function also decays more slowly for this vector.

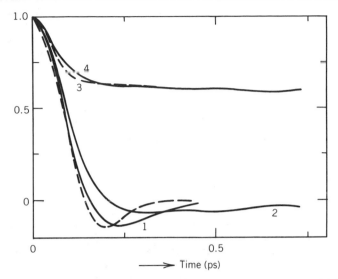

Figure 43.   ---, Angular-velocity correlation function from depolarized Rayleigh scattering; 1, simulated angular-velocity autocorrelation function; 2, simulated angular-momentum autocorrelation function; 3, simulated second moment angular-velocity autocorrelation function, $\langle \omega(t) \cdot \omega(t)\omega(0) \cdot \omega(0)\rangle / \langle \omega^4(0)\rangle$; 4, simulated second-moment angular-momentum autocorrelation function, $\langle J(t) \cdot J(t)J(0) \cdot J(0)\rangle / \langle J^4(0)\rangle$. [Reproduced by permission from M. W. Evans et al., *J. Chem. Soc., Faraday II*, **79**, 153 (1983).]

In Fig. 43 we compare the simulated function with the angular-velocity correlation function from the depolarized Rayleigh wing of acetone as measured by Dill et al.[179] Note that the angular-velocity and angular-momentum autocorrelation functions *do not* have the same time dependence in an asymmetric top. Up to 0.1 ps the experimental and simulated functions decay similarly; thereafter the experimental function decays faster than the simulated one. The simulated second-moment autocorrelation functions, also shown in the figure, are transiently non-Gaussian, but go to the correct Gaussian limit of ca. 0.6.

In Fig. 44 we compare simulated results with the Fourier transforms of the far-infrared spectra of pure liquid acetone and acetone in 10% (volume/ volume) decalin. The experimental functions decay more quickly. The shapes of the functions show the characteristic long negative tails. However, the far-infrared bandshapes and simulated bandshapes are similar when scaled, so that contributions such as induced absorption and internal-field effects, if they do exist, appear to have time dependences similar to that of the pure rotational functions.

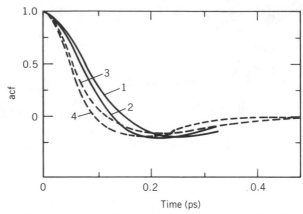

Figure 44. 1, Computer-simulated rotational-velocity autocorrelation function of the dipole unit vector, $\langle \dot{\mathbf{e}}_B(t)\cdot\dot{\mathbf{e}}_B(0)\rangle / \langle \dot{e}_B^2(0)\rangle$; 2, same as 1, but $\langle \dot{\mathbf{e}}_C(t)\cdot\dot{\mathbf{e}}_C(0)\rangle / \langle \dot{e}_C^2(0)\rangle$; 3, Fourier transform of the far-infrared power coefficient of a 10% solution of acetone in decalin; 4, Fourier transform of the far-infrared band of pure liquid acetone. [Reproduced by permission from M. W. Evans et al., *J. Chem. Soc., Faraday II*, **79**, 153 (1983).]

There are other inconsistencies in the table. The most serious is represented by the translational correlation time from NMR spectroscopy by Jonas and Bull,[180] 4.80 ps. The simulated laboratory-frame velocity autocorrelation function is not exponential, as this function is assumed to be by Jonas and Bull. We also observed this in $CHBr_3$ (Section V), for which the NMR correlation time measured by Sandhu was far longer than the computer-simulated center-of-mass velocity correlation time. This may be due to definition. In the extreme narrowing limit, the translational NMR correlation time is sometimes quoted in the literature as τ_t, defined as

$$\tau_t = \frac{ma}{12kT\tau_v}$$

where m is the molecular mass and a the radius. Using $a = 2.0$ Å and our simulated τ_v of ca. 0.08 ps, we obtain $\tau_t = 1.0$ ps, which is still significantly shorter than that measured. We should remember that the center-of-mass velocity autocorrelation function is oscillatory, with a long negative tail and positive longer time tail decaying as $t^{-3/2}$. The NMR data-reduction process, however, assumes a Stokes's law behavior, implying that the autocorrelation function is an exponential with, of course, a finite correlation time.

Another inconsistency is that the Raman C=O stretch band, after data reduction, produces a correlation time of 0.27 ps, which compares with a simulated τ_B value of 0.6 ps and a spin–spin NMR rotational correlation

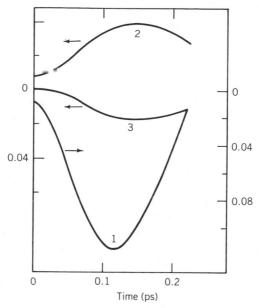

Figure 45. Elements of the rototranslational matrix: 1, $[\langle v_C(0)J_A(t)\rangle + \langle v_C(t)J_A(0)\rangle]/(\langle v_C^2\rangle^{1/2}\langle J_A^2\rangle^{1/2})$; 2, as for 1, but (A, C) element; 3, as for 1, but (B, A) element. All other elements vanish for all t. [Reproduced by permission from M. W. Evans et al., J. Chem. Soc., Faraday II, **79**, 153 (1983).]

time of 0.75 ps. The Raman correlation time is already too short because of the data-reduction process. It is more realistic to take the correlation time as the inverse half-width of the band, carefully decoupling rotational and vibrational effects (though this is by no means a trivial matter).

If we use our simulation to give details of the interaction of rotation with translation for this molecule of C_{2v} symmetry, the $(1, 2)$, $(2, 1)$, and $(3, 1)$ elements are nonvanishing (Fig. 45). Our moving frame is defined as that of the principal moments of inertia, so that $(1, 2) \equiv (A, C)$, $(2, 1) \equiv (C, A)$, and $(3, 1) \equiv (B, A)$. We remark that the elements are not symmetric with respect to each other (e.g., $(A, C) \neq (C, A)$, $(B, A) \neq (A, B)$). In the simulations discussed earlier for the molecules of C_{3v} symmetry, the (A, C) and (C, A) elements were symmetric.

The existence of a phase change in $CHBr_3$ (liquid to rotator-phase solid) was dependent on the fact that the (A, B) and (B, A) elements were very small (i.e., that rotation was almost wholly decoupled from translation). In acetonitrile, which is also of C_{3v} symmetry, the same normalized elements were an order of magnitude larger, peaking at ± 0.21 in amplitude. In acetone and CH_2Cl_2, the (C, A) elements peak at -0.11 and -0.12, respectively. CH_3CN, CH_2Cl_2, and acetone, of course, do not form rotator phases because of this pronounced coupling. In the next section, we consider a molecule of yet lower symmetry in which the rotation–translation interaction is again increased significantly.

IX. ETHYL CHLORIDE

Ethyl chloride [M. W. Evans, *J. Chem. Soc., Faraday II*, **79**, 719 (1983)] is an example of a low-symmetry (asymmetric) top with a dipole moment that does not lie on an axis of the principal-moment-of-inertia frame. At 1 bar the liquid boils at 285 K. For either or both of these reasons, the liquid has not been investigated in any depth for details of its molecular dynamics.

However, ethyl chloride is a favorable molecule for study because the second virial coefficient of gaseous ethyl chloride is known, suggesting that there is little or no association or dimerization. The existence of this coefficient can complicate spectral profiles considerably, as we have seen. In ethyl chloride the molecular interactions are probably dominated by Lennard–Jones-type repulsion and dispersion, and not by strong electrodynamic and polarizability effects. It is interesting that unlike molecules of C_{3v} symmetry, which *may form* rotator phases, the liquids of molecules of lower symmetry can often be vitrified, and the liquid supercooled. That is, translational motions may continue when rotational motion is considerably hindered. We recall that in the rotator phase the reverse is true—translational freedom is constrained (the molecules are often confined to solid lattice sites), yet rotational freedom remains. In molecules of C_{3v} symmetry the rotation–translation coupling is small. Our computer simulation will indicate the importance of this interaction as a prerequisite for the formation of a glass phase.

When glassy C_2H_5Cl is heated, an endothermic process begins at a temperature T_v at which the glass softens. The glass is transformed into a supercooled liquid by the appearance of free volume (holes). At a temperature T_r the supercooled liquid recrystallizes. It would be interesting to try to follow all of this in a computer simulation—to see just how this complex liquid- to solid-state phase transition is controlled by features of the molecular dynamics. We consider, in other sections of this volume, the significant role of $R-T$ coupling in determining details of the melting process in optically active liquids. This coupling may cause individual enantiomers to remain liquid at a temperature at which an equimolar racemic mixture is a solid, or vice versa. It appears to depend on the nature of the *intra* cross-rototranslation matrix and the signs of individual elements of this matrix. This phenomenon illustrates how mathematical laws may account for features of real physical systems in one of the most subtle ways the authors have encountered. Certainly it seems to be established that the melting process is controlled almost entirely by the molecular symmetry properties and sizes of the atoms of the molecule, which in turn determine the molecular dynamics of the condensed phase itself.

If for the present we return to ethyl chloride, X-ray powder spectra indicate that there is no crystallization below T_r. Infrared spectra prove that in the glassy and supercooled-liquid states the same rotational isomers coexist as may be observed in the liquid state above the melting point—only the *trans* isomer exists. NMR spectra show that between T_v and T_r a dynamic reorientation of the different segments of the hydrocarbon chain occurs that ccases in the crystal. The existence of an intense dielectric absorption between T_v and T_r confirms the reorientation of the polar group CH_2X, which reorientation is related to the appearance of "holes" in the lattice.

The total electric polarization of C_2H_5Cl has been measured in the gaseous state by Barnes et al.[193] at various temperatures, and the density and complex permittivity have been measured over the liquid range by McMullen et al.[194] The density ranges from 0.9214 g cm^{-3} at 273 K to 1.1281 g cm^{-3} in the supercooled liquid at 118.3 K. In the solid at 112.9 K the density increases to 1.139 g cm^{-3}. The dielectric loss above 203 K is small up to 8 MHz, but at lower temperatures the loss process begins to appear at the highest frequencies and increases with decreasing temperature until freezing.

Neutron-scattering spectroscopy has been used[195] to observe the C–C–Cl deformation and CH_3 torsional modes of ethyl chloride in the liquid and solid states, and the torsional frequency of the methyl group internal rotation, which is assigned at 278 cm^{-1}.

The potential between two ethyl chloride molecules is represented in the simulation by a 5×5 site–site model of atom–atom Lennard-Jones interactions with partial charges localized on the atom sites. The partial charges were obtained from a paper by Mark and Sutton,[196] who estimated a dipole of 1.86 D from values of $q_{CH_3} = 0.0465$, $q_C = -0.0502$, $q_H = 0.0808$, and $q_{Cl} = -0.157$ in units of $|e|$. These values compare with the previous set derived by del Re using a linear combination of atomic orbitals: $q_{CH_3} = 0.040$, $q_C = 0.001$, $q_H = 0.068$, and $q_{Cl} = -0.177$. The Lennard-Jones parameters are

$$\varepsilon/k(Cl-Cl) = 127.9 \text{ K}$$
$$\sigma(Cl-Cl) = 3.6 \text{ Å}$$
$$\varepsilon/k(C-C) = 35.8 \text{ K}$$
$$\sigma(C-C) = 3.4 \text{ Å}$$
$$\varepsilon/k(CH_3-CH_3) = 158.6 \text{ K}$$
$$\sigma(CH_3-CH_3) = 4.0 \text{ Å}$$
$$\varepsilon/k(H-H) = 10.0 \text{ K}$$
$$\sigma(H-H) = 2.8 \text{ Å}$$

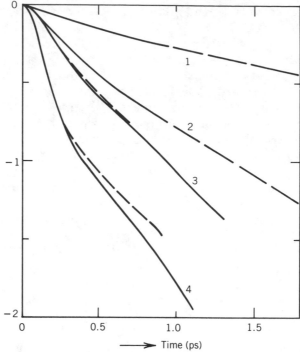

Figure 46. Orientational autocorrelation functions at 293 K; first- (P_1) and second- (P_2) rank Legendre polynomials. 1, $P_1(e_3)$; 2, $P_2(e_3)$; 3, $P_1(e_1)$ (- - -, $P_1(e_2)$); 4, $P_2(e_1)$ (- - -, $P_2(e_2)$). [Reproduced by permission from M. W. Evans, *J. Chem. Soc., Faraday II*, **79**, 719 (1983).]

Simulations were carried out at room temperature and, by simulating sudden drops in the temperature, also in the supercooled or vitreous states. The room-temperature liquid was simulated at 293 K, d (density) = 0.8978 g cm^{-3}; the supercooled liquid at 118 K, d = 1.1281 g cm^{-3}. The liquid boils at 285 K, 1 bar, so that the simulated liquid state at 293 K is under slightly more than an atmosphere of its vapor pressure.

The only available literature relaxation time for liquid ethyl chloride is the NMR spin–spin time of Miller and Gordon.[197] The derived second-rank orientational correlation time is 0.7 ± 0.1 ps. It is not clear to which vector this refers; consequently there are no acceptable data available with which to compare our simulated results.

The first- (P_1) and second- (P_2) rank orientational autocorrelation functions of the three unit vectors e_1, e_2, and e_3 are illustrated in Fig. 46. These unit vectors are in the three axes of the principal moment-of-inertia frame. There is a simple relationship between the unit vector $u = \mu/|\mu|$ (the dipole

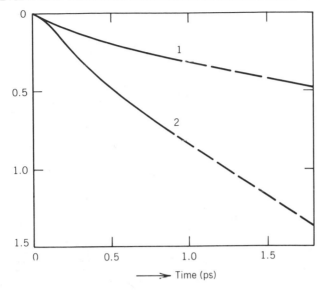

Figure 47. As for Fig. 46. 1, $P_1(\mathbf{u})$; 2, $P_2(\mathbf{u})$, where $\mathbf{u}$ is the dipole unit vector. [Reproduced by permission from M. W. Evans, *J. Chem. Soc., Faraday II*, **79**, 719 (1983).]

unit vector) and $\mathbf{e}_1$ and $\mathbf{e}_3$. This takes the form

$$\mathbf{u} = x\mathbf{e}_1 + y\mathbf{e}_3$$

where x and y are numbers, if we assume that $\mathbf{u}$ lies in the C—Cl bond. So it is straightforward to compute the first- and second-rank orientational autocorrelation functions of $\mathbf{u}$ (Fig. 47). Neglecting cross correlations and other contributions, the Fourier transform of $P_1(\mathbf{u})$ is a measure of the dielectric loss spectrum, and that of $\langle \dot{\mathbf{u}}(t) \cdot \dot{\mathbf{u}}(0) \rangle / \langle \dot{\mathbf{u}}(0) \cdot \dot{\mathbf{u}}(0) \rangle$ is the far-infrared spectrum. $P_2(\mathbf{u})$ can be obtained from the Raman spectrum or from the depolarized Rayleigh spectrum. All simulated functions are now available for comparison with future experimental results.

Simulation results show that the anisotropy of the rotational diffusion in the moment-of-inertia frame is very large—a result that NMR spectroscopy should be able to reproduce. The "tumbling" of the ethyl chloride molecule (reorientation of the $\mathbf{e}_3$ axis) is slower than its spinning (reorientation of either the $\mathbf{e}_1$ or $\mathbf{e}_2$ axes). The spinning motion is "inertia dominated"; that is, the P_1 and P_2 autocorrelation functions of $\mathbf{e}_1$ or $\mathbf{e}_2$ are not exponential initially, but become so after about 0.7 ps. The $P_1(\mathbf{e}_3)$ and $P_2(\mathbf{e}_3)$ functions, on the other hand, become exponential after the autocorrelation function has dropped only to about 80% of its initial value. The motion of $\mathbf{u}$, and conse-

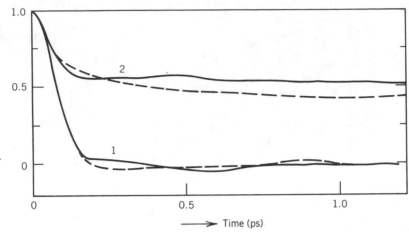

Figure 48.   —— , Angular-velocity autocorrelation function at 293 K, $\langle \omega(t)\cdot\omega(0)\rangle/\langle\omega^2\rangle$:
1, $(\omega(t)\cdot\omega(t)\omega(0)\cdot\omega(0))/\langle\omega^4\rangle$; 2, $\langle\omega(t)\cdot\omega(0)\rangle/\langle\omega^2\rangle$. ---, Angular-momentum autocorrelation function: 1 and 2 as above. [Reproduced by permission from M. W. Evans, *J. Chem. Soc.*, *Faraday II*, **79**, 719 (1983).]

quently the experimentally measured functions, reflects a combination of tumbling and spinning in ethyl chloride, as is the case for most asymmetric tops of C_{2v} symmetry.

The angular-velocity and angular-momentum autocorrelation functions (Fig. 48) do not decay in the same way, as is true for molecules of any symmetry lower than T_d. In Fig. 48 the second-moment autocorrelation functions are also shown. These decay to a constant level depending on the nature of the equilibrium statistics. They attain a Gaussian limit as $t \to \infty$, but are transiently non-Gaussian. The same is true for the linear center-of-mass velocity autocorrelation function and its second moment. Both linear- and angular-velocity autocorrelation functions have the characteristic long negative tail at intermediate time, as discussed in the last section for acetone.

It is interesting to observe that rotation–translation interaction is larger in this molecular liquid than in any other liquid we have simulated. There are four finite elements of the autocorrelation matrix $\langle v(t)J^T(0)\rangle$ for $t > 0$. They are illustrated in Fig. 49. The $(1,2)$ and $(2,1)$ elements are mirror images and greater in intensity than the $(3,2)$ and $(2,3)$ elements. The significance of the rotation–translation interaction may be attributed to the fact that the relatively heavy Cl atom is so close to the center of mass. It is not possible for the molecule to rotate without simultaneously translating a good deal, and vice versa.

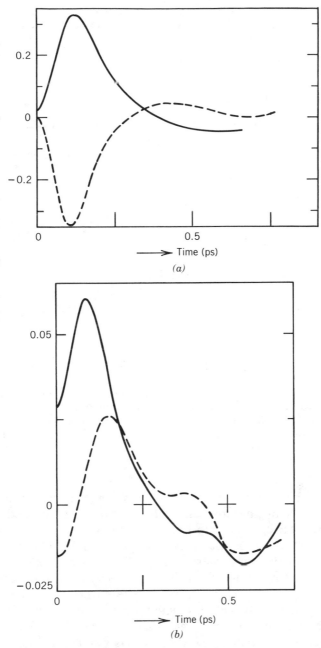

Figure 49. (a) ——, (1,2) element of the rotating-frame matrix $\langle \mathbf{v}(t)\mathbf{J}^T(0)\rangle$ at 293 K, normalized as $\langle v_1(t)J_2(t)\rangle/(\langle v_1^2\rangle^{1/2}\langle J_2^2\rangle^{1/2};$ ---, (2,1) element. (b) As for a. ——, (2,3) element; ---, (3,2) element. Note that the latter is very much smaller than the (1,2) or (2,1) elements. [Reproduced by permission from M. W. Evans, *J. Chem. Soc., Faraday II*, **79**, 719 (1983).]

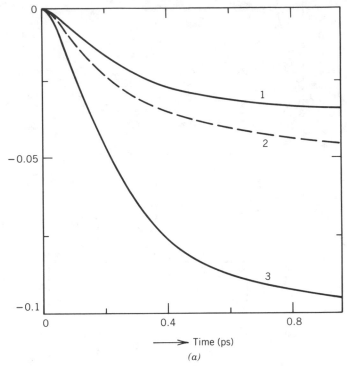

Figure 50. Orientational autocorrelation functions for supercooled ethyl chloride at 118 K, plotted as log (correlation function) versus time; $\mathbf{u}$ = dipole unit vector. (a) 1, $P_1(\mathbf{e}_3)$; 2, $P_1(\mathbf{u})$; 3, $P_2(\mathbf{e}_3)$. Note that the slope at long times (>1.0 ps) is almost parallel to the time axis. (b) 1, $P_1(\mathbf{e}_2) = P_1(\mathbf{e}_1)$ (on this scale); 2, $P_2(\mathbf{e}_2) = P_2(\mathbf{e}_1)$. [Reproduced by permission from M. W. Evans, *J. Chem. Soc., Faraday II*, **79**, 719 (1983).]

Results in the supercooled states are shown in Figs. 50 and 51. The P_1 and P_2 orientational autocorrelation functions of $\mathbf{e}_1$, $\mathbf{e}_2$, $\mathbf{e}_3$, and $\mathbf{u}$ are *not* exponential in the interval up to 1 ps, and for $t > 1$ ps they tail on this scale parallel to the time axis. This implies that the correlation times are effectively infinite (on the picosecond time scale of the abscissa), or in other words, that the dynamical process originating in the picosecond time scale evolves into one that spans a complete time scale covering many decades of frequency. Note how this contrasts with the rotator-phase systems, in which rotation–translation interaction was almost absent. Rotation–translation coupling becomes exceedingly large in the glassy state (Fig. 51). Such interactions are undoubtedly effective in setting up long-lived vortices in the supercooled-liquid state—the interactions become coherent on a macroscopic scale. The function $\langle \dot{\mathbf{u}}(t)\cdot\dot{\mathbf{u}}(0)\rangle / \langle \dot{u}^2\rangle$, when Fourier transformed, provides

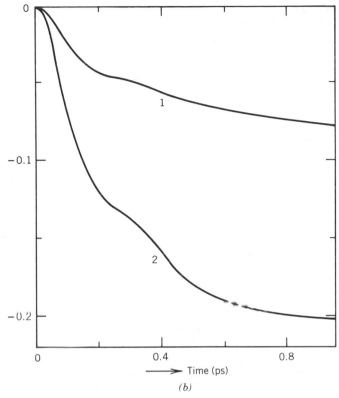

Figure 50.   (*Continued*)

the far-infrared spectrum. It intersects the time axis (Fig. 52*b*) earlier than in the liquid (Fig. 52*a*) at room temperature, which means that the far-infrared spectrum is shifted to higher frequencies and considerably sharpened in the supercooled liquid. This is precisely what is observed experimentally (see ref. 162, Chapter 7) for a variety of dipolar solutes in supercooled decalin, which seems to corroborate the (α, β, γ) hypothesis of Evans and Reid.[198] This states that far-infrared process is the high-frequency (γ) adjunct of a multidecade relaxation representing the evolution of the molecular dynamics from picosecond time scales to time scales effectively on the order of years (recall the slow flow of common window glass).

So it is established that these molecules of low symmetry cannot possibly rotate without simultaneously displacing their *own* centers of mass. It should also be noted that the molecule must simultaneously displace neighboring molecules. These are the rotation–translation coupling effects observed by Ewing et al. (1966) in a series of experiments. The existence of rotation–

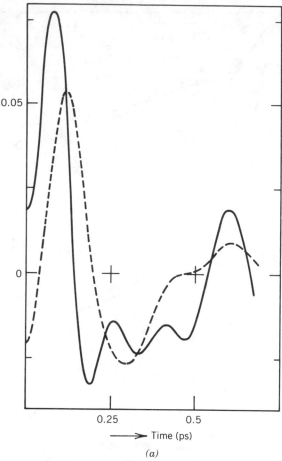

Figure 51.   Elements of the rotating-frame matrix $\langle \mathbf{v}(t)\mathbf{J}^T(0)\rangle$ at 118 K, supercooled liquid, showing very strong rotation–translation coupling. (a) ---, (1,2) element;   — , (2,1) element. (b) ---, (3,2) element;   — , (2,3) element. [Reproduced by permission from M. W. Evans, *J. Chem. Soc., Faraday II, 79*, 719 (1983).]

translation interaction seems to explain many properties of the liquid state, ranging from the existence of rotator-phase states to the formation of vitreous or glassy states. As we consider elsewhere in this volume, it also explains other basic phenomena, relating to optically active systems of yet lower symmetry. For example, it explains for the first time *in terms of the molecular dynamics* why the melting point of a racemic mixture of lactic acid is 18°C, yet that of the pure R and S enantiomers is 53°C. In addition, we may

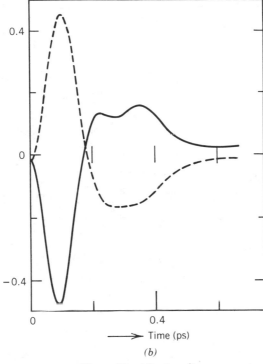

Figure 51. (*Continued*)

now try to explain why the solubilities of some optically active species in common solvents (e.g., *n*-hexane) are 9 times larger for the racemic mixture than the enantiomers.

We should note at this stage that the boiling point is not significantly affected by this rotation–translation interaction, but that the liquid–solid transition (i.e., the melting point) may be greatly affected.

X. CONCLUSION

As we have seen, computer simulation can be used to clarify and even to provide new insights into certain aspects of the molecular dynamics of the condensed state of matter. As computing power grows, so too the sizes of the molecules and the ensembles that may be studied will grow, intermolecular potentials will be improved (the Lennard-Jones interactions currently used in computer simulations are already orders of magnitude more realistic than those implied by the most advanced analytical theories), and other de-

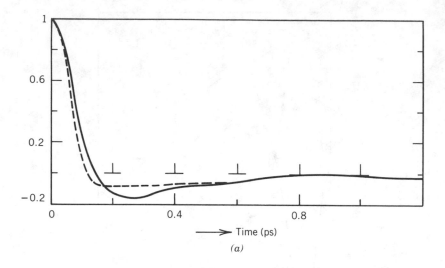

(a)

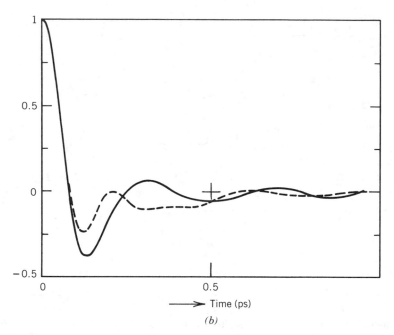

(b)

Figure 52. (*a*) Rotational velocity autocorrelation functions at 293 K: —— , $\langle \dot{\mathbf{e}}_1(t) \cdot \dot{\mathbf{e}}_1(0) \rangle / \langle \dot{e}_1^2 \rangle$; ---, $\langle \dot{\mathbf{u}}(t) \cdot \dot{\mathbf{u}}(0) \rangle / \langle \dot{u}^2 \rangle$. (*b*) Rotational velocity autocorrelation functions at 118 K; ---, $\langle \dot{\mathbf{e}}_3(t) \cdot \dot{\mathbf{e}}_3(0) \rangle / \langle \dot{e}_3^2 \rangle$; —— , $\langle \dot{\mathbf{e}}_2(t) \cdot \dot{\mathbf{e}}_2(0) \rangle / \langle \dot{e}_2^2 \rangle$. (*c*) Same as *b*, only for: ---, $\langle \dot{\mathbf{u}}(t) \cdot \dot{\mathbf{u}}(0) \rangle / \langle \dot{u}^2 \rangle$; —— , $\langle \dot{\mathbf{e}}_1(t) \cdot \dot{\mathbf{e}}_1(0) \rangle / \langle \dot{e}_1^2 \rangle$. [Reproduced by permission from M. W. Evans, *J. Chem. Soc., Faraday II*, **79**, 719 (1983).]

484

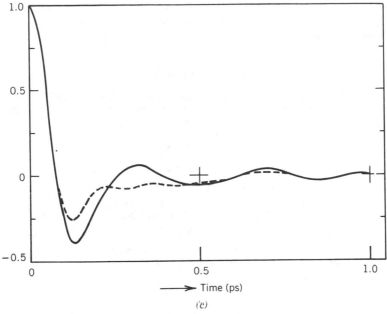

Figure 52. (*Continued*)

tails of the simulations that are presently questioned (e.g., the assumption of pair additivity) will be better understood. At this stage the foundations are being laid. Already it is apparent that we have not had the foresight to envisage many of the subtleties of the liquid state, and particularly the role of molecular shape and the constituent atoms in determining condensed-state properties. Molecular theories for liquid-state dynamics are oversimplified and restricted and have tended to lead us into great morasses of mathematical complexity without solving or shedding new light on the problems at hand. Hydrodynamic theories, by definition (they aim to explain liquid-state behavior in terms of macroscopic data), do not consider the detailed molecular structure of the medium, let alone the structure of the molecules.

Computer simulation of the molecular dynamics by-passes many of the problems that plague these analytical theories because the postulates are comparatively few. Gordon[199] once said that "it becomes impractical to follow the dynamics at long times because of the complexity of the molecular trajectories," and so analytical theories have evolved dependent on the methods of statistical mechanics, in which the number of dynamical variables was reduced from Avogadro's order of magnitude (the real number involved) to only a few. Modern-day simulations are doing just what was considered impossible then, and follow the complete time evolution of the molecular trajectories. In so doing they change the very basis of our subject. We believe we have entered an exciting new era of liquid-state molecular dynamics.

Acknowledgments

The SERC is thanked for two Advanced Fellowships and the University of Wales for University Fellowships at Bangor and Swansea to M. W. E. after contract termination at Aberystwyth.

References

1. G. W. F. Pardoe, Ph.D. Thesis, University of Wales, Aberystwyth, 1969.

2. G. J. Evans, Ph.D. Thesis, University of Wales, Aberystwyth, 1977.

3. G. J. Evans and M. W. Evans, *J. Mol. Liq.*, **26**, 63 (1983); M. W. Evans, *Phys. Rev. Lett.*, **50**, 371 (1983); G. J. Evans and M. W. Evans, *Chem. Phys. Lett.*, **96**, 416 (1983).

4. G. Ewing, *Acc. Chem. Res.*, **2(6)**, 168 (1966).

5. B. J. Berne and R. Pecora, *Dynamic Light Scattering*, Wiley, New York, 1975.

6. R. M. Lynden-Bell, *Mol. Phys.*, **33(4)**, 907 (1977).

7. G. J. Evans, *J. Chem. Soc., Faraday II*, **79**, 547 (1983).

8. M. W. Evans, *Adv. Mol. Rel. Int. Proc.*, **24**, 123 (1982).

9. H. Bertagnolli, D. O. Leicht, and M. D. Zeidler, *Mol. Phys.*, **35**, 199 (1978).

10. C. Brodbeck, I. Rossi, N. van Thanh, and A. Ruoff, *Mol. Phys.*, **32**, 71 (1976).

11. J. Soussen-Jacob, E. Devril, and J. Vincent-Geisse, *Mol. Phys.*, **28**, 935 (1974).

12. A. Gerschel, I. Darmon, and C. Brot, *Mol. Phys.*, **23**, 317 (1972).

13. A. Gerschel and C. Brot, *Mol. Phys.*, **20**, 279 (1971).

14. J. Schroeder, V. H. Schiemann, and J. Jonas, *Mol. Phys.*, **34**, 1501 (1977).

15. P. van Konynenberg and W. A. Steele, *J. Chem. Phys.*, **56**, 4776 (1972).

16. J. Schroeder and J. Jonas, *Chem. Phys.*, **34**, 11 (1978).

17. P. A. Lund, O. Faurskov-Nielsen, and E. Praestgaard, *Chem. Phys.*, **28**, 167 (1978).

18. S. Claesson and D. R. Jones, *Chem. Scripta*, **9**, 103 (1976).

19. G. R. Alms, D. R. Bauer, J. I. Brauman, and R. Pecora, *J. Chem. Phys.*, **59**, 5310 (1973).

20. K. Kamagawa, *Bull. Chem. Soc. Jpn.*, **51**, 3475 (1978).

21. M. W. Evans, G. J. Evans, W. T. Coffey, and P. Grigolini, *Molecular Dynamics*, Wiley, New York, 1982.

22. G. J. Evans, *Spectrochim. Acta*, **33A**, 699 (1977).

23. G. J. Evans and M. W. Evans, *Chem. Phys. Lett.*, **45**, 454 (1977).

24. T. Suzuki, Y. K. Tsutsui, and T. Fujiyama, *Bull. Chem. Soc. Jpn.*, **53**, 1931 (1980).

25. W. G. Rothschild, G. J. Rosasco, and R. C. Livingston, *J. Chem. Phys.*, **62**, 1253 (1975).

26. A. Moradi-Araghi and M. Schwartz, *J. Chem. Phys.*, **68**, 5548 (1978).

27. J. S. Rowlinson, *Trans. Faraday Soc.*, **45**, 974 (1949).

28. K. Tanabe and J. Higashi, *Chem. Phys. Lett.*, **71**, 46 (1980).

29. A. Laubereau, G. Wochner, and W. Kaiser, *Chem. Phys.*, **28**, 363 (1978); A. Laubereau, *C. R. Int. Conf. Sp. Raman 7th*, p. 450, (1980).

30. A. Moradi-Araghi and M. Schwartz, *J. Chem. Phys.*, **71**, 166 (1979).

31. P. C. M. van Woerkom, J. de Bleisser, M. de Zwart, P. M. J. Burgers, and J. C. Leyte, *Ber. Bunsenges. Phys. Chem.*, **78**, 1303 (1974).

32. R. K. Wertheimer, *Mol. Phys.*, **36**, 1631 (1978); *ibid.*, **35**, 257 (1978).

33. G. Döege, I. Bieh, and M. Possiel, *Ber. Bunsenges. Phys. Chem.*, **85**, 1074 (1981).

34. C. W. Carlson and P. J. Flory, *J. Chem. Soc.*, *Faraday Trans. II*, **73**, 1729 (1977).

35. J. A. Bucaro and T. A. Litovitz, *J. Chem. Phys.*, **54**, 3846 (1971); *ibid.*, **55**, 3585 (1971).

36. J. H. K. Ho and G. C. Tabisz, *Can. J. Phys.*, **51**, 2025 (1973).

37. A. K. Burnham, G. Alms, and W. H. Flygare, *J. Chem. Phys.*, **62**, 3289 (1975).

38. M. W. Evans, G. J. Evans, and B. Janik, *Spectrochim. Acta*, **38A**, 423 (1982).

39. Y. Leroy, E. Constant, C. Abbar, and P. Desplanques, *Adv. Mol. Rel. Proc.*, **1**, 273 (1967).

40. P. Hindle, S. Walker, and J. Warren, *J. Chem. Phys.*, **62**, 3230 (1975).

41. C. J. Reid and M. W. Evans, *J. Chem. Soc.*, *Faraday II*, **76**, 286 (1980).

42. G. Bossis, These d'Etat, Université de Nice, Nice, 1981.

43. B. Quentrec and P. Bezot, *Mol. Phys.*, **27**, 879 (1974).

44. V. I. Gaiduk and Y. P. Kalmykov, *J. Chem. Soc.*, *Faraday II*, **77**, 929 (1981).

45. F. Hermans and E. Kestemont, *Chem. Phys. Lett.*, **55**, 305 (1978).

46. S. Forsen, H. Gustavsson, B. Lindman, and N. O. Persson, *J. Mag. Res.*, **23**, 515 (1976).

47. R. R. Shoup and T. C. Farrar, *J. Mag. Res.*, **7**, 48 (1972).

48. W. T. Huntress, *J. Phys. Chem.*, **73**, 103 (1969).

49. M. Ohuchi, T. Fujito, and M. Imanari, *J. Mag. Res.*, **35**, 415 (1979).

50. Dinesh and M. T. Rogers, *Chem. Phys. Lett.*, **12**, 352 (1971).

51. T. C. Farrar, S. J. Druck, R. R. Shoup, and E. D. Becker, *J. Am. Chem. Soc.*, **94**, 699 (1972).

52. J. C. Duplan, A. Briguet, and J. Delmar, *J. Chem. Phys.*, **59**, 6269 (1973).

53. P. P. Ho and R. R. Alfano, *Phys. Rev.*, **20**, 2170 (1979).

54. M. T. Ratzch, E. Rickelt, and H. Rosner, *Z. Phys. Chem.*, **256**, 349 (1975).

55. M. S. Beevers and G. Khanarian, *Aust. J. Chem.*, **32**, 263 (1979); *ibid.*, **33**, 2585 (1980).

56. A. Proutière and J. G. Baudet, *Ann. Univ. Abidjan, Ser. C*, **11**, 13 (1975).

57. M. P. Bogaard, A. D. Buckingham, and G. L. D. Ritchie, *Mol. Phys.*, **18**, 575 (1970).

58. G. J. Evans, *J. Chem. Soc.*, *Faraday II*, **79**, 547 (1983).

59. K. Takagi, P. K. Choi, and K. Negishi, *J. Chem. Phys.*, **74**, 1424 (1981).

60. D. Samios, T. Dorfmuller, and A. Asembaum, *Chem. Phys.*, **65**, 305 (1982).

61. A. Yoshihara, A. Anderson, R. A. Aziz, and C. C. Lim, *Chem. Phys.*, **61**, 1 (1981).

62. K. Altenburg, *Z. Phys. Chem.*, **250**, 399 (1972).

63. R. Vallauri and M. Zoppi, *Lett. Nuovo Cimento So. Ital. Fis.*, **9**, 447 (1974).

64. C. A. Chatzidimitriou-Dreismann and E. Lippert, *Ber. Bunsenges. Phys. Chem.*, **84**, 775 (1980); *ibid.*, *Croatica Chimica Acta*, in press.

65. H. S. Goldberg and P. S. Persham, *J. Chem. Phys.*, **58**, 3816 (1973).

66. J. Hyde-Campbell, J. F. Fisher, and J. Jonas, *J. Chem. Phys.*, **61**, 346 (1974).

67. G. Döge and A. Schaeffer, *Ber. Bunsenges. Phys. Chem.*, **77(a)**, 682 (1973).

68. R. Arndt and J. Yarwood, *Chem. Phys. Lett.*, **45**, 155 (1977).

69. C. K. Cheung, D. R. Jones, and C. H. Wang, *J. Chem. Phys.*, **64**, 3567 (1976).

70. R. B. Wright, M. Schwartz, and C. H. Wang, *J. Chem. Phys.*, **58**, 5125 (1973).

71. G. D. Patterson and J. Griffiths, *J. Chem. Phys.*, **63**, 2406 (1975).

72. J. F. Dill, T. A. Litovitz, and J. A. Bucaro, *J. Chem. Phys.*, **62**, 3839 (1975).

73. M. Constant and M. Fauquembergue, *C. R. Seances Acad. Sci.*, **272**, 31293 (1971).

74. E. N. Shermatov and A. K. Atakhodzhaev, *Theor. Spektrosk.*, 29–30 (1977).

75. A. Laubereau and W. Kaiser, *NATO Adv. Study Inst.-Ser.*, **B37**, 329 (1977).

76. Z. Gburski and W. Szczepanski, *Mol. Phys.*, **40**, 649 (1980).

77. D. W. Oxtoby, D. Levesque, and J. J. Weiss, *J. Chem. Phys.*, **68**, 5528 (1978).

78. K. T. Gillen, M. Schwartz, and J. H. Noggle, *Mol. Phys.*, **20**, 199 (1971).

79. G. Döge, R. Arndt, and A. Khuen, *Chem. Phys.*, **21**, 53 (1977).

80. N. Trisdale and M. Schwartz, *Chem. Phys. Lett.*, **68**, 461 (1979).

81. G. Döge, *Z. Naturforsch.*, **28A**, 919 (1973).

82. P. C. M. van Woerkom, J. de Bleyser, M. de Zwart, and J. C. Leyte, *Chem. Phys.*, **4**, 236 (1974).

83. S. F. Fisher and A. Laubereau, *Chem. Phys. Lett.*, **35**, 6 (1975).

84. J. H. Hildebrand, *Faraday Disc.*, **66** (1978).

85. M. L. Bansal, S. K. Debard, and A. P. Roy, *Chem. Phys. Lett.*, **83**, 83 (1981).

86. W. A. Steele, *Adv. Chem. Phys.*, **34** (1976).

87. C. M. Roland and W. A. Steele, *J. Chem. Phys.*, **73**, 5924 (1980).

88. J. R. Maple, R. S. Wilson, and J. T. Knudtson, *J. Chem. Phys.*, **73**, 3346 (1980).

89. J. P. Riehl and D. J. Diestler, *J. Chem. Phys.*, **64**, 2593 (1976).

90. O. F. Nielsen, D. H. Christensen, P. A. Lund, and E. Praestgaard, *Proc. Int. Conf. Raman Spect.*, **2**, 208 (1978).

91. M. Schwartz, *Chem. Phys. Lett.*, **73**, 127 (1980).

92. J. E. Griffiths, *J. Chem. Phys.*, **59**, 751 (1973).

93. F. Heatley, *J. Chem. Soc., Faraday II*, **70**, 148 (1970).

94. C. O. Fischer, *Ber. Bunsenges. Phys. Chem.*, **75**, 361 (1971).

95. J. A. Janik, J. M. Janik, A. Bajorek, K. Parlinski, and M. Sudnik-Hoynkiewicz, *Physica*, **35**, 4 (1967).

96. E. K. Eliel, N. L. Allinger, S. J. Angyal, and G. A. Morrison, *Conformational Analysis*, Wiley, New York, 1965.

97. G. Bossis, B. Quentrec, and C. Brot, *Mol. Phys.*, **39**, 123 (1980).

98. E. Kluk, T. W. Zerda, and J. Zerda, *Acta Phys. Polonica*, **56A**, 121 (1979).

99. G. Ciccotti, J. P. Ryckaert, and A. Bellemans, *Mol. Phys.*, **44**, 979 (1981).

100. H. Bertagnolli, D. O. Leicht, and M. D. Zeidler, *Mol. Phys.*, **35**, 199 (1978).

101. G. J. Evans, *J. Chem. Soc. Faraday II*, **79**, 547 (1983).

102. M. W. Evans, *J. Mol. Liq.*, **25**, 149 (1983).

103. T. Bien, M. Possiel, G. Döge, J. Yarwood, and K. Arnold, *Chem. Phys.*, **56**, 203 (1981).

104. T. E. Bull, *Chem. Phys. Lett.*, **73**, 127 (1980).

105. M. Schwartz, *Chem. Phys. Lett.*, **73**, 127 (1980).

106. C. Breulliard-Alliot and J. Soussen-Jacob, *Mol. Phys.*, **28**, 903 (1974).

107. B. Tiffon, B. Ancian, and J. E. Dubois, *J. Chem. Phys.*, **74**, 6981 (1981).

108. S. L. Whittenberg and C. H. Wang, *J. Chem. Phys.*, **66**, 5138, (1977).

109. J. T. Hynes, R. Kapral, and M. Weinberg, *J. Chem. Phys.*, **69**, 2725 (1978).

110. G. Patterson and J. E. Griffiths, *J. Chem. Phys.*, **63**, 2406 (1975).

111. H. Versmold, *Ber. Bunsenges. Ges. Phys. Chem.*, **82**, 451 (1978).

112. E. Knozinger, D. Leutloff, and R. Wiltenbeck, *J. Mol. Struct.*, **60**, 115 (1980).

113. G. Fini and P. Mirone, *Spectrochim. Acta*, **32A**, 439 (1976).

114. A. M. Amorim da Costa, M. A. Norman, and J. H. R. Clarke, *Mol. Phys.*, **29**, 191 (1975).

115. J. Yarwood, P. L. James, G. Döge, and R. Arndt, *Disc. Faraday Soc.*, **66**, 252 (1972).

116. J. Yarwood, R. Arndt, and C. Döge, *Chem. Phys.*, **25**, 387 (1977).

117. F. Heatley, *J. Chem. Soc.*, *Faraday II*, **70**, 148 (1974).

118. J. R. Lyerla, D. M. Grant, and C. H. Wang, *J. Chem. Phys.*, **55**, 4670 (1971).

119. T. K. Leipert, J. H. Noggle, and K. T. Gillen, *J. Mag. Res.*, **13**, 158 (1974).

120. G. D. Stucky, D. A. Matthews, J. Hedman, M. Klasson, and C. Nordling, *J. Am. Chem. Soc.*, **94**, 8009 (1972).

121. J. A. Pople and K. Beveridge, *J. Am. Chem. Soc.*, **92**, 5298 (1970).

122. J. Yarwood, R. Ackroyd, K. E. Arnold, G. Döge, and R. Arndt, *Chem. Phys. Lett.*, **77**, 239 (1981).

123. J. K. Eloranta and P. K. Kadaba, *Mat. Sci. Eng.*, **8(4)**, 203 (1971).

124. E. Krishnaji and A. Mansigh, *J. Chem. Phys.*, **41**, 827 (1964).

125. W. G. Rothschild, *J. Chem. Phys.*, **57**, 991 (1972).

126. A. K. Burnham and T. D. Gierke, *J. Chem. Phys.*, **73**, 4822 (1980).

127. M. S. Beevers and D. A. Elliott, *Mol. Cryst. Liq. Cryst.*, **26**, 411 (1979).

128. A. Kratochwill, *Ber. Bunsenges. Ges. Phys. Chem.*, **82**, 783 (1978).

129. C. S. Hsu and D. Chandler, *Mol. Phys.*, **36**, 215 (1978).

130. E. Lippert, W. Schroer, H. Mahnke, and H. Michel, in *International Conference on Hydrogen Bonding*, H. J. Bernstein, ed., Ottowa, 1972, p. 22.

131. G. J. Evans, *Chem. Phys. Lett.*, in press (1983).

132. R. J. Jakobsen and J. W. Brasch, *J. Am. Chem. Soc.*, **86**, 3571 (1964).

133. B. J. Bulkin, *Helv. Chim. Acta*, **52**, 1348 (1969).

134. A. M. Saum, *J. Polym. Sci.*, **42**, 57 (1960).

135. R. Lobo, J. E. Robinson and S. Rodriguez, *J. Chem. Phys.*, **59**, 5992 (1973).

136. G. Ascarelli, *Chem. Phys. Lett.*, **39**, 23 (1976).

137. M. W. Evans and G. J. Evans, *J. Chem. Soc.*, Faraday II, **76**, 767 (1982).

138. C. Brodbeck, I. Rossi, N. van Thanh, and A. Ruoff, *Mol. Phys.*, **32**, 71 (1976).

139. A. E. Boldeskal, S. S. Esman, and V. E. Pogornelov, *Opt. Spectrosc.*, **37**, 521 (1974).

140. E. K. Eliel, N. L. Allinger, S. J. Angyal, and G. A. Morrison, *Conformational Analysis*, Wiley, New York, 1965.

141. R. Del Re, *J. Chem. Soc.*, **43**, (1958).

142. J. O. Hirschfelder, *J. Chem. Phys.*, **8**, 431 (1940).

143. H. S. Sandhu, *J. Mag. Reson.*, **34**, 141 (1979).

144. J. Soussen-Jacob, E. Devvil, and J. Vincent-Geisse, *Mol. Phys.*, **28**, 935 (1974).

145. V. K. Agarwal, A. K. Sharma, and A. Mansingh, *Chem. Phys. Lett.*, **68**, 151 (1979).

146. G. D. Patterson and J. E. Griffiths, *J. Chem. Phys.*, **63**, 2406 (1975).

147. V. K. Agarwal, G. J. Evans, and M. W. Evans, *J. Chem. Soc.*, *Faraday II*, **79**, 137 (1983).

148. B. Lassier and C. Brot, *J. Chim. Phys.*, **65**, 1723 (1968).

149. K. E. Larson and T. Mansson, *J. Chem. Phys.*, **67**, 4995 (1977).

150. P. S. Goyal, W. Navrocik, S. Urban, J. Dornosiawski, and I. Nathaniel, *Proc. Nucl. Phys.*, *Solid State Phys. Symp.*, **16C**, 193 (1973); *ibid.*, *Acta Phys. Polonica*, **A46 141**, 399 (1974).

151. A. A. Boguslavskii, R. S. Lotfullin, and G. K. Senun, *Phys. Stat. Solidi*, **66(2)**, K95 (1974).

152. F. Koeksal, *J. Chem. Soc.*, *Faraday II*, **76**, 55P (1980).

153. M. Constant and R. Fauquembergue, *J. Chem. Phys.*, **72**, 2459 (1980).

154. C. W. Carlson and P. J. Flory, *J. Chem. Soc.*, *Faraday II*, **73**, 1505 (1977).

155. K. Czarniecka, J. M. Janik, and J. A. Janik, *Acta Phys. Polonica*, **55A**, 421 (1979).

156. C. J. Reid and M. W. Evans, *Mol. Phys.*, **40**, 1357 (1980).

157. R. Haffmans and I. W. Larkin, *J. Chem. Soc.*, *Faraday II*, **68**, 1729 (1972).

158. I. W. Larkin, *J. Chem. Soc.*, *Faraday II*, **69**, 1379 (1973).

159. C. J. Reid, Thesis, University of Wales, 1979.

160. M. W. Evans, *J. Mol. Liq.*, **27**, 11, 19 (1983).

161. G. Bossis, These d'Etat, Université de Nice, Nice, 1981.

162. M. W. Evans, G. J. Evans, W. T. Coffey, and P. Grigolini, *Molecular Dynamics*, Wiley-Interscience, New York, 1982, Chaps. 6 and 12.

163. M. W. Evans and J. Yarwood, *Adv. Mol. Rel. Int. Proc.* **21**, 2 (1981).

164. P. N. Brier and A. Perry, *Adv. Mol. Rel. Int. Proc.*, **13**, 46 (1978).

165. P. C. M. van Woerkom, J. de Bleyser, M. de Zwart, P. M. Burgers, and J. C. Leyte, *Ber. Bunsenges. Phys. Chem.*, **78**, 1303 (1979).

166. D. E. O'Reilly, E. M. Peterson, and E. L. Yasaites, *J. Chem. Phys.*, **57**, 890 (1972).

167. W. G. Rothschild, *J. Chem. Phys.*, **53**, 990, 3265 (1970).

168. F. Heatley, *J. Chem. Soc.*, *Faraday II*, **70**, 148 (1974).

169. D. E. Woessner, *J. Chem. Phys.*, **37**, 647 (1962).

170. M. W. Evans and G. J. Evans, *J. Chim. Phys.* (*Paris*), **72**, 522 (1978).

171. V. F. Baranov, *Vopt. Mol. Specktrosk.*, **89**, (1974).

172. J. R. Nestor and E. R. Lippincott, *J. Raman Spectrosc.*, **1(3)**, 305 (1973).

173. K. Singer, J. V. L. Singer, and A. J. Taylor, *Mol. Phys.*, **37**, 1239 (1979).

174. I. R. McDonald, personal communication.

175. M. Ferrario and M. W. Evans, *Adv. Mol. Rel. Int. Proc.*, **24**, 139 (1982).

176. P. S. Y. Cheung, *Mol. Phys.*, **3**, 519 (1977).

177. K. Koga, Y. Kamazawa, and H. Shimizu, *J. Mol. Spectrosc.*, **47**, 107 (1973).

178. L. D. Favro, *Phys. Rev.*, **53**, 119 (1960).

179. J. F. Dill, T. A. Litovitz, and J. A. Bucaro, *J. Chem. Phys.*, **62**, 3839 (1975).

180. J. Jonas and T. E. Bull, *J. Chem. Phys.*, **52**, 4553 (1972).

181. W. Schindler, P. T. Sharko, and J. Jonas, *J. Chem. Phys.*, **76**, 3493 (1982).

182. D. W. Oxtoby, D. Levesque, and J. J. Weis, *J. Chem. Phys.*, **68**, 5528 (1978).

183. C. Döge, *Z. Naturforsch. A*, **28**, 919 (1973).

184. C. H. Wang, *Mol. Phys.*, **33**, 207 (1977).

185. R. M. Lynden-Bell, *Mol. Phys.*, **33**, 907 (1977).

186. M. Perrot, M. H. Brooker, and J. Lascombe, *J. Chem. Phys.*, **74**, 2787 (1981).

187. C. J. Reid and M. W. Evans, *Mol. Phys.*, **40**, 1357 (1980)

188. A. K. Burnham and T. D. Gierke, *J. Chem. Phys.*, **73**, 4822 (1980).

189. B. M. Laudanyi and T. Keyes, *Mol. Phys.*, **37**, 1413 (1979).

190. S. M. George, H. Auweter, and C. B. Harris, *J. Chem. Phys.*, **73**, 5573 (1976).

191. A. Bellemans, G. Ciccotti, and J-P. Ryckaert, *Mol. Phys.*, **44**, 979 (1981).

192. C. A. Wellington and S. H. Khouwaiter, *Tetrahedron*, **34**, 2183 (1978).

193. A. N. M. Barnes, D. J. Turner, and L. E. Sutton, *Trans. Faraday Soc.*, **67**, 2902 (1971).

194. T. McMullen, E. D. Crozier, and R. McIntosh, *Can. J. Chem.*, **46**, 2945 (1968).

195. K. A. Strong, R. M. Brugger, and R. J. Pugmire, *J. Chem. Phys.*, **52**, 2277 (1970).

196. J. E. Mark and C. Sutton, *J. Am. Chem. Soc.*, **94**, 1083 (1972).

197. C. M. Miller and S. L. Gordon, *J. Chem. Phys.*, **53**, 3531 (1970).

198. M. W. Evans and C. J. Reid, *J. Chem. Phys.*, **76**, 2576 (1982).

199. R. G. Gordon, *J. Chem. Phys.*, **45**, 1649 (1966).

RECENT ADVANCES IN MOLECULAR-DYNAMICS COMPUTER SIMULATION

D. FINCHAM

DAP Support Unit, Queen Mary College, University of London, Mile End Road, London E1 4NS, United Kingdom

and

D. M. HEYES

Department of Chemistry, Royal Holloway College, University of London, Egham, Surrey TW20 OEX, United Kingdom

CONTENTS

493

I. INTRODUCTION

Molecular dynamics (MD) is a computer-based technique for modeling fluids and solids at microscopic levels of distance and time, and is therefore the ideal technique for gaining a molecular appreciation of many physical processes.[1,2] It is concerned with discrete model molecules and should be distinguished from the many continuum models of fluids.[3] Because the movements of molecules in condensed phases are highly cooperative, analytic theories of molecular motion are unsatisfactory[4] and a computer is required to provide the exact trajectories through time and space (which ideally covers all phase space). Consequently, theories of condensed phases that are based on the same pair potential but take the cooperative ordering and motion into account in some approximate manner can be tested in this manner. This is an invaluable aid to the devising of models for the fluid state because it indicates the most significant parameters needed in the model.

Molecular-dynamics calculations have shown that in many cases very simple pairwise-additive interaction potentials (and hence forces) are responsible for many of the important effects observed experimentally.[5] In particular, the short-range repulsive part of the intermolecular potential is dominant in determining the short-range and even long-range order.[6] The choice of a reasonable intermolecular pair potential can enable one to obtain directly experimental solid and fluid properties. By gradually increasing the sophistication of the pair potential, one can gain insights into what features in this potential give rise to the observed effects of interest. Often simple spherical molecules can account for most of the gross features, and therefore invaluable information as to the "true" causes of phenomena can often be gained.[7] Exact agreement with experiment is often difficult to achieve, because real intermolecular interactions are many body and therefore three-molecule or even higher-order potential terms are needed to provide a full treatment. These are often too computationally expensive to include in MD simulations. Therefore the MD pair potentials are often derived from experimental solid-state properties for which analytic expressions have been cast in terms of two-body potentials.[8] The MD pair potentials are consequently only "effective"; that is, they include the average effects of higher-body forces. In fact, if one were to use a gas-phase or quantum-mechanically derived "true" pair potential, agreement with experimental condensed-phase properties would probably be poor in most cases.[9]

Another problem frequently encountered in MD is insufficient evolution time to follow completely the event of interest. The time window is too narrow to observe all the relevant processes. This is because the time step in a MD simulation is quite rigidly fixed, being linked to the thermal collisions between molecular fragments. More complicated inter- and intra-molecular reorganization can be many orders of magnitude slower. It is possible to freeze out certain less important degrees of freedom to lengthen the accessible time scale, but only with the inclusion of associated methodological errors.[10-13]

Many excellent reviews of MD have appeared in recent years[1,2] and so we confine our discussion to those aspects that are relatively new or have received little specific attention in the literature. Our emphasis is on providing technical examples in FORTRAN of operations that are becoming increasingly more valuable additions to the simulator's armory. Our discussion is therefore eclectic and unashamedly detailed in parts. We concentrate on those areas in which further elucidation would be timely. Subjects treated are integration algorithms for the equations of motion; methods for dealing with large numbers of molecules; and nonstandard dynamics as routes to different ensembles, transport properties, and second-order thermodynamic quantities.

As more complex systems requiring larger numbers of molecules and longer run times come to demand study, computational efficiency becomes the criterion for a project's feasibility. Consequently, we make frequent reference to computational speed and in particular any adaptations to the discussed techniques needed for performance on the new generation of vector and parallel computers (the so-called supercomputers). These new machines gain speed by use of radically different architectures, as increased performance by means of improvements in circuit technology is more and more subject to the law of diminishing returns. The computer-oriented scientist therefore needs to be aware, perhaps for the first time, of these changes and to reexamine traditional methodology and algorithms. It may be necessary to solve an MD model in a completely new way to take advantage of these new computer performance features.

II. EQUATIONS OF MOTION

A. Force and Torque Evaluation

The principles of MD are extremely simple. The evolution of a system of a few hundred to thousands of model molecules is followed in time by numerical integration of the classical equations of motion of the molecules. To make the system pseudoinfinite we suppose the molecules to be confined within a box that is surrounded in all three dimensions (in bulk calculations) by periodic replicas of itself. This produces only small errors in many cases.[14,15] The simulation proceeds in a series of time steps. At each step the force on each molecule (and the torque, if the molecule is nonspherical) is evaluated from its interactions with its neighbors, utilizing some form of intermolecular pair potential. The positions of the molecules can then be integrated over the time step Δt to get a new configuration of the system. This is repeated usually at least several thousand times, and sometimes hundreds of thousands of times, depending on how slow the relaxations in the modeled system are. Long runs need to be performed for interfacial work[16] and in the regions of phase changes.[17] Once the system has reached thermal equilibrium, we can determine equilibrium properties of the simulated ensemble, which are equivalent to averages over time, that is, over time steps of the simulation. As specified here the equations of motion will (or should) conserve total energy, E, so that we are working in the constant NEV or microcanonical ensemble (where N is the number of molecules and V is the volume of the MD cell).

Each time step begins with calculations of the force (and torque) on each molecule. In the case of potentials of reasonably short range, that is, for molecules with no regions of high net electronic charge density, this is straightforward. For spherical molecules with an interaction pair potential $\phi(r)$, the force on molecule i at position $\mathbf{r}_i$ is (apart from the effects of the periodic boundaries) given by

$$\mathbf{F}_i = - \sum_{j \neq i}^{N} \left(\frac{\mathbf{r}_{ij}}{r_{ij}} \right) \frac{d\phi}{dr} \bigg|_{r_{ij}} = \sum_{j \neq i}^{N} f_{ij} \tag{1}$$

where $\mathbf{r}_{ij} = \mathbf{r}_i - \mathbf{r}_j$. The pair forces obey Newton's third law, which fixes $\mathbf{f}_{ji} = -\mathbf{f}_{ij}$, so that each pair needs to be considered only once. In the case of nonspherical but rigid molecules, the forces are chosen to act between sites fixed within the molecules. Let $\phi_{\alpha\beta}(r)$ be the interaction potential between site α on one molecule and site β on another molecule. Let $\mathbf{d}_{i\alpha}$ represent the position of site α of molecule i relative to its center of mass $\mathbf{r}_i$. Then the force on this site is

$$\mathbf{f}_{i\alpha} = -\sum_{j\neq i}^{N}\sum_{\beta}\left(\frac{\mathbf{r}_{ij\alpha\beta}}{r_{ij\alpha\beta}}\right)\frac{d\phi_{\alpha\beta}}{dr}\bigg|_{r_{ij\alpha\beta}} \tag{2}$$

with

$$\mathbf{r}_{ij\alpha\beta} = (\mathbf{r}_i + \mathbf{d}_{i\alpha}) - (\mathbf{r}_j + \mathbf{d}_{j\beta})$$
$$= \mathbf{r}_{ij} + d_{i\alpha} - d_{j\beta} \tag{3}$$

The total force on molecule i is

$$\mathbf{F}_i = \sum_{\alpha}\mathbf{f}_{i\alpha} \tag{4}$$

and the torque is

$$\mathbf{T}_i = \sum_{\alpha}\mathbf{d}_{i\alpha}\times\mathbf{f}_{i\alpha} \tag{5}$$

To take account of the periodic images we adopt the nearest-image convention; that is, we assume that because of the limited range of the pair potential we need to consider only the interactions between a molecule i and another molecule j or its nearest image, whichever is closer. This is illustrated for the case of a cubic computational box in Fig. 1. If in the above equations the distance $\mathbf{r}_{ij}$ between the molecules in the box is greater than that between one of them and the other's image (as shown in the figure), then the latter vector is taken. The pair potential has a limited range, so we can ignore completely the interaction between molecules whose nearest-image separation is greater than some cutoff radius r_c. For the cubic box, the value of r_c should be no greater than half the side length of the box for consistency with the nearest-image convention, since the separation between all images except the nearest must then be greater than r_c.

Systems with long-range forces arising from regions of high charge density in the species simulated are more difficult to treat because the nearest-image and spherical cutoff techniques are no longer valid.[18-22] All images then need to be included. This necessitates use of lattice summation techniques or rethinking in terms of a new, effective, but short-ranged pair potential.[23] The former approach has been reviewed by one of us[24,25] for bulk and interfacial systems. In practice, these approaches need to be applied to molecules with large dipole moments.[26] However, molecules with only small dipole moments or with higher moments can be handled within the present

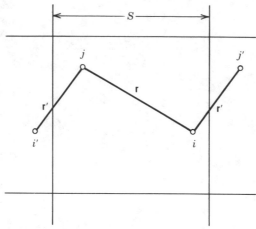

Figure 1. Sketch demonstrating the application of periodic boundary conditions. The nearest distance between molecules i and j in this example is $\mathbf{r}'$ *not* $\mathbf{r}$, as $\mathbf{r}' < \mathbf{r}$. Therefore i interacts with j' and not j. Also j interacts with i' and not i. S is the side length of the cubic MD cell.

framework, whether these moments are represented by fractional charges on sites within the molecule (which need not coincide with atom sites) or by point multipoles at its center.

B.   Center-of-Mass Motion

Despite the fundamental importance of the form of the algorithm used to integrate the equations of motion of the molecules, there has only been sporadic specific interest in this topic. This is perhaps due to the great success early in the history of MD of a leapfrog form of algorithm introduced by Verlet.[2,27] It is stable at large time steps, is easy to program, and has minimal storage requirements (e.g., it requires six vectors of dimension N for a monoatomic system). This is the form we recommend and use regularly; we will describe it in detail below.

For rigid molecules the translational motions of the center of mass separate from the rotational motions about the center of mass. We first consider the translational components. The equations of motion of a particular molecule are governed by the formula

$$\frac{d^2\mathbf{r}_i}{dt^2} = \frac{\mathbf{F}_i}{m_i} \qquad (6)$$

or

$$\dot{V}_i = \frac{F_i}{m_i} \tag{7}$$

where $\dot{r}_i = V_i$ is the velocity of molecule i and m_i is the mass of molecule i. Since the force on molecule i depends not only on its own coordinates but also on those of all other molecules, we will have a set of coupled equations. We have already implicitly proposed that these will be solved in terms of independent molecular motion over the time step Δt; this is a necessary approximation. We can then use a Taylor series expansion for each molecule as if it were moving in a fixed field of force.[2] The most useful algorithm is derived as follows.

Let

$$r(t + \Delta t) = r(t) + \dot{r}(t)\Delta t + \frac{1}{2}\frac{F(t)\Delta t^2}{m} + O(\Delta t^3) \tag{8}$$

and

$$r(t - \Delta t) = r(t) - \dot{r}(t)\Delta t + \frac{1}{2}\frac{F(t)\Delta t^2}{m} - O(\Delta t^3) \tag{9}$$

where we have dropped the subscript i and vector notation (which are nevertheless implicitly assumed). Adding Eqs. (8) and (9) gives

$$r(t + \Delta t) = r(t) + \delta r(t) \tag{10}$$

where

$$\delta r(t) = \delta r(t - \Delta t) + \frac{F(t)\Delta t^2}{m} \tag{11}$$

The terms $\delta r(t)$ and $\delta r(t - \Delta t)$ are known as position increments. If the velocity is defined to be

$$\dot{r}(t) = \frac{r(t + \Delta t) - r(t - \Delta t)}{2\Delta t} \tag{12}$$

then

$$\dot{r}(t) = \frac{\delta r(t - \Delta t)}{\Delta t} + \frac{1}{2}\frac{F(t)\Delta t^2}{m} \tag{13}$$

By making a forward and backward time expansion of $\frac{1}{2}\Delta t$ in a Taylor series expansion of the velocity we can identify the terms $\delta r(t)\Delta t^{-1}$ and $\delta r(t - \Delta t)\Delta t^{-1}$ with the half-time-step velocities $V^{n+1/2}$ and $V^{n-1/2}$ frequently encountered in the literature.[28] The force at time t ($= n\Delta t$) is calculated from the coordinates at time $n\Delta t$. It is then used to update the position increment from time $(n - 1)\Delta t$ to time $n\Delta t$; this new "velocity" is then used to update the coordinates to time $(n + 1)\Delta t$. It is obvious why this is known as the leapfrog algorithm. Verlet[29] introduced an alternative formulation to the field of MD simulation in which the coordinates from the previous step are used, rather than the midstep velocities (or position increments). However, this formulation is less convenient in practice.

After the coordinates are updated, the periodic boundary conditions are applied to ensure that all particle coordinates lie within the computational box. Parallels can be drawn between this procedure and the numerical integration of equidistantly spaced points. Verlet's algorithm can be thought of as equivalent to the trapezoid-rule integration.

More accurate higher-order expansions may be used, but have the disadvantage that they often require the particle coordinates to be stored for a larger number of previous configurations.

Comparisons between the leapfrog algorithm and a high-order predictor–corrector algorithm, the Gear four-level formula,[30] reveal that the latter gives a slightly better root-mean-square (RMS) total energy fluctuation at normal time steps (~ 10 fs for triple-point model argon). But its errors increase very rapidly with moderate increases in the time-step duration, whereas the errors in the leapfrog model increase comparatively slowly with increasing Δt (i.e., a ratio of $\sim 5:1$ at 30 fs). On theoretical grounds the Gear formulas are expected to be very accurate and stable, but the theoretical analysis applies strictly only to a particle moving in a fixed force field. This is not the case in a dynamic simulation, where each particle moves in the fluctuating force field produced by its neighbors. The high-order algorithms predict the time derivatives of the force from its value at previous time steps, but it would appear that such predictions are invalidated by the fluctuations in the force field in the liquid at larger time steps, and indeed have a destabilizing effect on the algorithm. This is important, because we want to use as large a time step as possible in order to sample phase space most efficiently. The leapfrog formulation of Verlet's algorithm is therefore the most suitable algorithm.

C. Rotational Motion

There is much evidence that in a molecular liquid the fluctuations in torque act on a very similar time scale to that of the force fluctuations, suggesting that a simple low-order algorithm might also be suitable for modeling rotational motion.[31-33]

The first question to consider in modeling rotational motion is the representation of the orientations of a rigid molecule. For nonlinear molecules there are three degrees of rotational freedom, so the orientations can be specified in terms of three parameters, such as Euler angles. Unfortunately, equations of motion written in terms of these parameters contain singularities. Orientations are better specified in terms of a four-parameter quaternion $\mathbf{q} = (\xi, \eta, \zeta, \chi)^T$ that is subject to the constraint

$$\xi^2 + \eta^2 + \zeta^2 + \chi^2 = 1 \tag{14}$$

In terms of the quaternion parameters, the relation between the components of a vector $\mathbf{a}$ in space-fixed coordinates and its components $\tilde{\mathbf{a}}$ in the principal axis system of the molecule is

$$\tilde{\mathbf{a}} = \mathbf{A}\mathbf{a} \tag{15}$$

where the rotation matrix $\mathbf{A}$ is given in terms of quaternions by

$$\mathbf{A}(q) = \begin{pmatrix} -\xi^2 + \eta^2 - \zeta^2 + \chi^2 & 2(\zeta\chi - \xi\eta) & 2(\eta\zeta + \xi\chi) \\ -2(\xi\eta + \zeta\chi) & \xi^2 - \eta^2 - \zeta^2 + \chi^2 & 2(\eta\chi - \xi\zeta) \\ 2(\eta\zeta - \xi\chi) & -2(\xi\zeta + \eta\chi) & -\xi^2 - \eta^2 + \zeta^2 + \chi^2 \end{pmatrix} \tag{16}$$

The rotational matrix is evaluated at each step of the calculation and used to determine the site positions $\mathbf{d}_\alpha$ as used in Eqs. (2) and (4) from the principal-axis site coordinates, which specify the molecule's geometry:

$$\mathbf{d}_\alpha = \mathbf{A}^{-1}(\mathbf{q})\tilde{\mathbf{d}}_\alpha \tag{17}$$

Since $\mathbf{A}$ is an orthogonal matrix, its inverse is equal to its transpose. The basic equations governing the rotational motions are

$$\frac{d\mathbf{J}}{dt} = \mathbf{T} \tag{18}$$

where $\mathbf{J}$ is the angular momentum and $\mathbf{T}$ is the torque, and

$$\frac{d\mathbf{q}}{dt} = \mathbf{Q}(\tilde{\omega}, 0)^T \tag{19}$$

an equation in a four-dimensional space, where $\tilde{\omega}$ is the angular momentum in the principal-axis system and the matrix $\mathbf{Q}$ is

$$\mathbf{Q}(\mathbf{q}) = \frac{1}{2} \begin{pmatrix} -\zeta & -\chi & \eta & \xi \\ \chi & -\zeta & -\xi & \eta \\ \xi & \eta & \chi & \zeta \\ -\eta & \xi & -\zeta & \chi \end{pmatrix} \tag{20}$$

We also need the relationship between angular momentum and angular velocity, which can be expressed most simply in the principal-axis system as

$$\tilde{J}_j = I_j \tilde{\omega}_j \tag{21}$$

where $j = x, y, z$ and the I_j are the principal moments of inertia. For example,

$$I_x = \sum_\alpha m_\alpha \left[\left(\tilde{d}_{\alpha y}^m \right)^2 + \left(\tilde{d}_{\alpha z}^m \right)^2 \right] \tag{22}$$

where the cycle α is taken over x, y, and z. Here $\tilde{\mathbf{d}}_\alpha^m$ is the position of mass center α in the molecule relative to the center of mass in the principal-axis system. Note that the mass centers can coincide with the force centers, but need not do so. To review these equations, we have eliminated any explicit inclusion of angles by expressing the orientations in terms of so-called quaternion parameters, which enable the equations of rotational motion to be expressed in a singularity-free form.

We now describe a new solution of Eq. (18) in a modified leapfrog expansion that is, like the analogous translational equations, simple to implement, reasonably accurate, and stable. We first note that the quaternion "velocity" depends not only on $\mathbf{J}$ but also on the quaternions themselves (i.e., on the orientation), both directly in Eq. (19) and through the rotation matrix. This is a situation in which a simple leapfrog scheme cannot be used. To see the reason for this, consider the case of a simple first-order differential equation

$$\dot{x} = V(x, t) \tag{23}$$

A leapfrog formulation for this equation would have the form

$$x^{n+1/2} = x^{n-1/2} + \Delta t V(x^n, t^n) \tag{24}$$

$$x^{n+1} = x^n + \Delta t V(x^{n+1/2}, t^{n+1/2}) \tag{25}$$

where both the "step" and "midstep" coordinates are required, since $x^{n+1/2}$ appears in the equation for x^n. The problem with this algorithm is that the step and midstep equations are only weakly coupled through the velocity term; numerical errors cause them to decouple and we get two solutions that oscillate unstably about the correct solution. It is possible to remedy this by using an auxiliary equation that propagates x from time $n\Delta t$ to time $(n + \frac{1}{2})\Delta t$ by means of a first-order Taylor expansion:

$$x^{n+1/2} = x^n + \tfrac{1}{2}\Delta t V(x^n, t^n) \tag{26}$$

The main equation "leapfrogs" from n to $n+1$, employing $x^{n+1/2}$ in the velocity terms:

$$x^{n+1} = x^n + \Delta t V(x^{n+1/2}, t^{n+1/2}). \tag{27}$$

The first-order midstep coordinate $x^{n+1/2}$ only has a temporary stabilizing role and is not saved. Overall, the algorithm is second-order accurate. The following modified leapfrog algorithm for rotational motion, which is based on this principle, can then be employed. The step n commences with values stored for $\mathbf{q}^n$ and $\mathbf{J}^{n-1/2}$. We use the auxiliary part of the algorithm to generate $\mathbf{J}^n$:

$$\mathbf{J}^n = \mathbf{J}^{n-1/2} + \tfrac{1}{2}\Delta t \mathbf{T}^n \tag{28}$$

From this we determine the principal components of angular velocity,

$$\tilde{\mathbf{J}}^n = \mathbf{A}(\mathbf{q}^n)\mathbf{J}^n \tag{29}$$

$$\tilde{\omega}_j^n = \frac{\tilde{J}_j^n}{I_j} \tag{30}$$

for $j = x, y, z$. We then propagate the quaternion over half a step,

$$\mathbf{q}^{n+1/2} = \mathbf{q}^n + \tfrac{1}{2}\Delta t \mathbf{Q}(\mathbf{q}^n)(\tilde{\omega}, 0)^T \tag{31}$$

and use $\mathbf{q}^{n+1/2}$ to calculate $\mathbf{A}^{n+1/2}$ and $\mathbf{Q}^{n+1/2}$. The main part of the algorithm updates the angular momentum in the usual leapfrog way,

$$\mathbf{J}^{n+1/2} = \mathbf{J}^{n-1/2} + \Delta t \mathbf{T}^n \tag{32}$$

and then uses this result, together with the midstep quaternions from the auxiliary step, to calculate the midstep angular velocity,

$$\tilde{\mathbf{J}}^{n+1/2} = \mathbf{A}(\mathbf{q}^{n+1/2})\mathbf{J}^{n+1/2} \tag{33}$$

$$\tilde{\omega}^{n+1/2} = \frac{\tilde{J}_j^{n+1/2}}{I_j} \tag{34}$$

for $j = x, y, z$. Hence to increment the quaternion, we use

$$\mathbf{q}^{n+1} = \mathbf{q}^n + \Delta t \mathbf{Q}(\mathbf{q}^{n+1/2})\tilde{\omega}^{n+1/2} \tag{35}$$

It is useful at this stage to renormalize the quaternion to ensure that it obeys the constraint on its norm. $\mathbf{J}^{n+1/2}$ and $\mathbf{q}^{n+1}$ are then stored for the next time step.

Calculations have been carried out that compare the degree of energy conservation of this algorithm with that of the fourth-order predictor–corrector, and also with that of the constraint method[34] for simulating rigid polyatomics. In the last of these the "free-flight" phase used a leapfrog algorithm. Calculations were performed taking an 8-fs time step on a three-center model for cyclopropane. The RMS fluctuations in total energy, expressed as percentages of the fluctuations in potential energy, are 4.7, 4.2, and 2.8%, whereas with a 12-fs time step these figures are 20, 85, and 37%, respectively. Therefore, although the fourth-order algorithm is slightly more accurate at small time steps, its errors increase much more rapidly as the time step increases than do those of this new algorithm, which is also better in this respect than the constraint algorithm. The new algorithm is simple to implement, needs minimal storage (three components of $\mathbf{J}$ and four of $\mathbf{q}$), and is self-starting.

We recommend an alternative approach for linear molecules. Because there are only two degrees of freedom, a second implicit constraint among the four quaternions exists that numerical errors could violate. We thus represent the orientation by a unit vector $\mathbf{e}$ along the axis, which has three components and is subject to the constraint

$$e_x^2 + e_y^2 + e_z^2 = 1 \tag{36}$$

The positions of the sites are given in terms of their distances from the center of mass along the axis,

$$\mathbf{d}_\alpha = \tilde{d}_\alpha \mathbf{e} \tag{37}$$

and the moment of inertia about any rotational axis through the center of mass and perpendicular to $\mathbf{e}$ is

$$I = \sum_\alpha m_\alpha \tilde{d}_\alpha^m \tag{38}$$

Again the mass centers need not coincide with the force centers. The torque on the molecule is

$$\mathbf{T} = \mathbf{e} \times \sum_\alpha d_\alpha \mathbf{f}_\alpha \tag{39}$$

We integrate the rotational motions by studying an equivalent diatomic "pseudomolecule." Let the pseudomolecule have unit length and masses m at each end on which forces $\mathbf{G}$ and $-\mathbf{G}$ act. Its rotational motion will be the same as that of the actual linear molecule, provided that it has the same moment of inertia and that the same torque acts on it. These conditions are satisfied if

$$m = 2I \tag{40}$$

and

$$\mathbf{G} = \sum_\alpha d_\alpha \mathbf{f}_\alpha - \beta \mathbf{e} \tag{41}$$

where β can have any value, since any component of $\mathbf{G}$ parallel to $\mathbf{e}$ cannot affect the rotational motion. A particular choice of β leads to a vector $\mathbf{G}$ that has no component in the direction of $\mathbf{e}$; we call this $\mathbf{G}_\perp$:

$$\mathbf{G}_\perp = \sum_\alpha d_\alpha \mathbf{f}_\alpha - \mathbf{e}\left(\mathbf{e} \cdot \sum_\alpha d_\alpha \mathbf{f}_\alpha\right) \tag{42}$$

Note that with this choice we have the relationship

$$\mathbf{G}_\perp^2 = \mathbf{T}^2 \tag{43}$$

so that if we use $\mathbf{G}_\perp$ we can obtain a useful quantity, the mean square torque.

By considering the motions of the two "atoms" of the pseudodiatomic, we easily obtain its equations of motion, written as two first-order equations,

$$\dot{\mathbf{e}} = \mathbf{u} \tag{44}$$

$$\dot{\mathbf{u}} = \frac{\mathbf{G}}{I} + \lambda \mathbf{e} \tag{45}$$

where λ is an undetermined multiplier for the constraint force along the bond axis. It is possible to solve these equations by using a simple leapfrog and then obtaining a quadratic equation for λ using the constraint that the length of $\mathbf{e}$ should remain unity.[35] However, there is an alternative, simpler algorithm that gives results identical to those of the more complicated constraint algorithm if we adopt $\mathbf{G}_\perp$ as our choice for $\mathbf{G}$.[35] Writing Eqs. (44) and (45) in a leapfrog form,

$$\mathbf{u}^{n+1/2} = \mathbf{u}^{n-1/2} + \Delta t \frac{\mathbf{G}_\perp^n}{I} + \lambda \mathbf{e}^n \tag{46}$$

$$\mathbf{e}^{n+1} = \mathbf{e}^n + \Delta t \mathbf{u}^{n+1/2} \tag{47}$$

Because $\mathbf{G}_\perp$ has no component along $\mathbf{e}$, the constraint force is only required to produce the centripetal acceleration, and in fact it can be shown that $\lambda = \mathbf{u}^2$. However, it is not necessary to use this. The constraint on the bond length requires that $\mathbf{u} \cdot \mathbf{e} = 0$, which can be used because we know $\mathbf{e}^n$, and $\mathbf{u}^n$ can be estimated. As explained above (see also ref. 36), for nonlinear molecules the correct procedure is to use a first-order expansion over $\frac{1}{2}\Delta t$ to obtain this estimate. Thus,

$$\mathbf{u}^n = \mathbf{u}^{n-1/2} + \tfrac{1}{2}\Delta t \left[\frac{\mathbf{G}_\perp^n}{I} + \lambda e^n \right] \tag{48}$$

Then using $\mathbf{G}_\perp^n \cdot \mathbf{e}^n = 0$ and $\mathbf{e}^n \cdot \mathbf{e}^n = 1$, and applying the constraint $\mathbf{u}^n \cdot \mathbf{e}^n = 0$, we have

$$0 = \mathbf{u}^{n-1/2} \cdot \mathbf{e}^n + \tfrac{1}{2}\Delta t \lambda \tag{49}$$

which on substituting back for λ in Eq. (46) yields

$$\mathbf{u}^{n+1/2} = \mathbf{u}^{n-1/2} - 2\mathbf{u}^{n-1/2} \cdot \mathbf{e}^n + \Delta t \frac{\mathbf{G}_\perp^n}{I} \tag{50}$$

Equations (47) and (50) constitute a leapfrog algorithm for the rotational motion in which we have expressed the centripetal acceleration in terms of the known quantities $\mathbf{u}^{n-1/2}$ and $\mathbf{e}^n$.

D. Toxvaerd Algorithm

There are potential advantages to going to a higher-order algorithm to solve the equations of motion (i.e., better energy conservation and larger time steps). Exploratory calculations in this direction have already been carried out by the authors, who concluded that to be successful any high-order algorithm must incorporate the many-body rearrangement of molecular positions close to each molecule.[28] Predictor–corrector algorithms such as the type due to Nordsieck[37] do not incorporate this feature and offer no significant advantages over the considerably simpler to implement leapfrog algorithm already described. A recent algorithm proposed by Toxvaerd[38,39] does have the desired feature of improving energy conservation (but not of increasing the length of the time step). The Toxvaerd algorithm involves two passages through an MD two-particle double loop at each time step. This usually less than doubles the CPU time per step, but can give more than two orders of magnitude of improvement in RMS fluctuation in total energy for a NEV program run near to the triple point.[40] Let

$$\dot{\mathbf{f}} = \dot{\mathbf{F}}_i \frac{\Delta t^3}{m_i} \tag{51}$$

$$\ddot{\mathbf{f}} = \ddot{\mathbf{F}}_i \frac{\Delta t^4}{m_i} \tag{52}$$

and

$$\ddot{\mathbf{f}} = \ddot{\mathbf{F}}_i \frac{\Delta t^5}{m_i} \tag{53}$$

Also let the subscripts $+$ and $-$ define quantities evaluated at times $(t + \Delta t)$ and $(t - \Delta t)$, respectively. Taking the Taylor expansions of Eqs. (8) and (9) to the next two orders gives

$$\mathbf{r}_+ = \mathbf{r} + \dot{\mathbf{r}}\Delta t + \frac{\mathbf{f}}{2} + \frac{\dot{\mathbf{f}}}{6} + \frac{\ddot{\mathbf{f}}}{24} + O(\Delta t^5) \tag{54}$$

Also, if we expand backwards in time,

$$\mathbf{r}_- = \mathbf{r} - \dot{\mathbf{r}}\Delta t + \frac{\mathbf{f}}{2} - \frac{\dot{\mathbf{f}}}{6} + \frac{\ddot{\mathbf{f}}}{24} - O(\Delta t^5) \tag{55}$$

These yield

$$\delta\mathbf{r} = \delta\mathbf{r}_- + \mathbf{f} + \frac{\ddot{\mathbf{f}}}{12} + O(\Delta t^6) \tag{56}$$

Also,

$$\dot{\mathbf{r}} = \frac{\delta\mathbf{r}_- + \mathbf{f}/2 - \dot{\mathbf{f}}/6 + \ddot{\mathbf{f}}/24}{\Delta t} + O(\Delta t^4) \tag{57}$$

The first MD loop gives $\mathbf{f}$, $\dot{\mathbf{f}}$, and part of $\ddot{\mathbf{f}}$. The second MD loop for the time step completes the calculation of $\ddot{\mathbf{f}}$.

If

$$G_{ij} = \frac{1}{r}\frac{d}{dr}\phi(r)\Big|_{r=r_{ij}} \tag{58}$$

$$G_{ij}^A = \frac{1}{r}\frac{d}{dr}G_{ij}\Big|_{r=r_{ij}} \tag{59}$$

and

$$G_{ij}^B = \frac{1}{r}\frac{d}{dr}G_{ij}^A\Big|_{r=r_{ij}} \tag{60}$$

then

$$\dot{G}_{ij} = G_{ij}^A(\mathbf{r}_{ij}\cdot\mathbf{r}_{ij}) \tag{61}$$

Also,

$$\dot{\mathbf{F}}_i = -\sum_{j\neq i}^N \left\{ G_{ij}\dot{\mathbf{r}}_{ij} + G_{ij}^A(\mathbf{r}_{ij}\cdot\dot{\mathbf{r}}_{ij})\mathbf{r}_{ij} \right\} \tag{62}$$

Similarly,

$$\ddot{\mathbf{F}}_i = \sum_{j \neq i}^{N} \left\{ G_{ij}\left(\frac{\mathbf{F}_i}{m_i} + \frac{\mathbf{F}_j}{m_j} \right) \right.$$

$$+ G_{ij}^A \mathbf{r}_{ij}\left[(\dot{\mathbf{r}}_{ij} \cdot \dot{\mathbf{r}}_{ij}) + \mathbf{r}_{ij} \cdot \left(\frac{\mathbf{F}_i}{m_i} - \frac{\mathbf{F}_j}{m_j} \right) \right]$$

$$\left. + 2G_{ij}^A (\mathbf{r}_{ij} \cdot \dot{\mathbf{r}}_{ij})\dot{\mathbf{r}}_{ij} + G_{ij}^B (\mathbf{r}_{ij} \cdot \dot{\mathbf{r}}_{ij})^2 \mathbf{r}_{ij} \right\} \tag{63}$$

To obtain $\dot{\mathbf{f}}$ and $\ddot{\mathbf{f}}$ it is necessary to have $\dot{\mathbf{r}}$ already. This cannot be obtained from Eq. (57) because $\dot{\mathbf{f}}$ is not yet known. Consequently, the following Taylor expansion is used as an adequate approximation:

$$\dot{\mathbf{r}} \approx \dot{\mathbf{r}}_- + \frac{\left\{ \mathbf{f}_- + \dot{\mathbf{f}}_-/2! + \ddot{\mathbf{f}}_-/3! + \dddot{\mathbf{f}}_-/4! \right\}}{\Delta t} \tag{64}$$

where $\dot{\mathbf{r}}_-$ is obtained using Eq. (57) at the previous time step. We estimate $\ddot{\mathbf{f}}_-$ as follows. Let

$$\mathbf{f}^P = \mathbf{f}_- + \dot{\mathbf{f}}_- + \frac{1}{2!}\ddot{\mathbf{f}}_- \tag{65}$$

But we already know the force at time t, $\mathbf{f}$, which is in reality produced by an implicit algorithm of the form

$$\mathbf{f} = \mathbf{f}_- + \dot{\mathbf{f}}_- + \frac{\ddot{\mathbf{f}}_-}{2!} + \frac{\dddot{\mathbf{f}}_-}{3!} + \cdots \tag{66}$$

Then

$$\dddot{\mathbf{f}}_- \approx 3!(\mathbf{f} - \mathbf{f}^P) \tag{67}$$

Hence $\dddot{\mathbf{f}}_-$ is obtained from the difference between the actual force at time t and the force predicted at time t based on Eq. (65).

Although the Toxvaerd algorithm does improve the degree of total energy conservation from time step to time step, it has still to be proved that it leads to results for other properties significantly different from those obtained with the leapfrog formula.[41] Although we would not recommend this method for general MD work until further tests have been performed, we do believe that it offers the most encouraging direction for improvements on the Verlet leapfrog algorithm; consequently we include it here.

This raises a very important practical question that we believe has not been studied with the thoroughness it perhaps deserves: How accurate should we make our MD calculations? In a MD experiment, errors in the trajectories of the particles arise because of the use of a noninfinitesimal time step and, to a lesser extent, through rounding. We lack a good measure of these errors. Measuring the fluctuation in total energy in a so-called constant energy calculation is one crude means of assessment. This is an important subject because doubling the length of a MD time step allows us to sample phase space twice as efficiently and halve our computer time. As we have seen, a simple algorithm such as the Verlet leapfrog gives rise to errors that increase only moderately and linearly with Δt, whereas higher-order algorithms can be more accurate at small time steps but become unstable when the time step is lengthened. Knowing which algorithm to choose is a dilemma. We need to know how the errors in the trajectories feed through into the measured properties of the simulated system, presumably affecting some more than others.

III. COMPUTATIONAL CONSIDERATIONS

A. Computer Hardware

We now turn to methods for implementing the above algorithms in a computer program. Although the basic equations are very simple, this is a nontrivial task, even before we raise the question of analysis of results. The reason is simply computer cost. A single simulation on a system of 256 molecules with four interaction sites could take approximately an hour on a multimillion-dollar supercomputer, a day on a fifty thousand-dollar super minicomputer and a week on a ten thousand-dollar super microcomputer. (Incidentally, these rough figures indicate that there is not much difference in cost effectiveness between the various classes of machine.) An active researcher can very easily use computer resources amounting to several times his or her annual salary. In these circumstances it is important to make the programs as efficient as possible, and this requires some understanding of recent developments in computer hardware.

Molecular-dynamics simulations do not make extravagant demands on memory or input–output facilities. Arithmetic processing is the main cause of concern and can be reduced by choosing internal physical units that simplify the equations. Such a choice may also be necessary to avoid over- or underflows, which could occur if we attempted to calculate microscopic quantities using macroscopic physical units. There are many ways in which to approach this problem. We can adopt internal units that bear an explicit relationship to a real solid or liquid. Distance and time would be in angstroms and picoseconds, for example. Alkali halides are suitable for this treatment.[42] Alternatively, we could take units that are related to the interaction potential.[43] The Lennard–Jones potential is an obvious example of this approach; in it the well depth ε is the unit of energy and the diameter σ is the unit of length. This latter method is perhaps more suited to theoretical studies that aim to determine equations of corresponding states, which can then be applied to a wide range of similar molecules.[2] With either method it is sometimes convenient to superimpose another tier of program units in which S (the side of the cubic computational box, say) is equal to either 1 or 2 as internal units.

There are two broad classes of computer hardware to be considered. Conventional computers have arithmetic processors capable of performing only one operation at a time, and are therefore known as serial computers. They can only do intensive computational tasks if the single processors work quickly. This involves "leading edge" technology, and hence such processors are expensive. Parallel computers have many processors and can therefore perform many operations simultaneously, providing that they are logically independent. By replicating cheap processors it is possible to have high performance at a moderate cost. At the moment there is only one true parallel computer available, the ICL Distributed Array Processor (or DAP). There is, however, a widely used type of computer, often called a vector processor, that is intermediate between these two types. It obtains high performance by overlapping the various stages of the arithmetic operations on successive operands. Since this is possible only if the operations on different operands are logically independent, the algorithms that work well on such a machine are very similar to those most suitable for the true parallel computer. The vector computer is more flexible, in that the "vectors" of operands can be of any length, whereas the parallel computer can perform operations simultaneously only on as many operands as it has processors. However, the vector computer is not as cost effective as the true parallel computer. The CRAY-1S and CDC CYBER-205 are examples of commercially available vector processors.

The most computationally expensive part of the calculations is the force/torque evaluation, and the choice of the most appropriate algorithm for this depends not only on the type of computer involved but also on the range of the intermolecular potential and the ranges of the correlations to be studied. For molecules without significant electric moments, the intermolecular potentials of which are usually modeled by Lennard–Jones interactions between internal sites, the effective range of the potential is no more than about three molecular diameters, and the range of structural correlations is similarly short. In such a case reliable results can be obtained, at least for thermodynamic and structural properties, on simulated systems containing surprisingly small numbers of molecules; 108 and 256 are popular choices.

(These are 4×3^3 and 4×4^3, respectively, which values are produced by expanding a basic cubic unit cell containing four atoms into an fcc lattice. The atoms are disposed at coordinates (x, y, z) of $(0.25, 0.25, 0.25)$, $(0.75, 0.75, 0.25)$, $(0.75, 0.25, 0.75)$, and $(0.25, 0.75, 0.75)$ within the unit cell of side length unity, and therefore form a tetrahedron.) The most frequently encountered algorithm is outlined in Table I. The force arrays FX, FY, and FZ are each dimensioned to the number of molecules. The double loop examines each pair of molecules once, and applies the nearest-image transformation to the vector separating them before finding the squared distance. The pair force is then evaluated, but only for those pairs separated by less than the distance beyond which the explicit interactions are chosen to be ignored. It is added to the force accumulator for particle i, and subtracted from the accumulator for particle j (making use of Newton's third law). This algorithm is written for a serial computer, so that the pair interactions are evaluated one by one. However, because these $N(N-1)/2$ calculations are logically independent, they could be performed simultaneously on a parallel computer with enough processors. The ICL DAP has 4096 processors configured as a 64×64 array. Each processor is capable of performing single-bit arithmetic operations. The processors operate simultaneously, executing the same instruction on their own data. The store associated with each processor consists of 4096 bits. Therefore the DAP could calculate all the pair interactions simultaneously for a system of 64 molecules. It should be noted, however, that this involves more work than is really necessary, since both the $i-j$ and $j-i$ interactions would be calculated. For larger numbers of molecules the pair interactions are calculated in a number of 64×64 blocks, and then it is possible to use the antisymmetry of the pair forces to reduce the number of blocks.

The serial algorithm needs to examine all pairs in order to apply the nearest-image boundary conditions and determine the pair separations, but thereafter it can avoid the evaluation of forces for pairs separated by more than the cutoff distance. This is not true for the parallel computer, which cannot save execution time by avoiding some of the calculations but must perform the same arithmetic operation on all processors. A similar remark applies to vector computers. The DO loop over pairs can be performed in a "pipeline" (vectorized) manner, but only if the IF statement is removed. In practice, it is possible to ensure that the out-of-range interactions are set to zero (using built-in functions) so that the corrections for the long-range interactions, which rely on a spherical cutoff, come out easily, but this has *no* effect on execution time. An example of a vectorized forces loop suitable for the CRAY-1S is given in Table II.

On the CRAY the performance is fairly independent of the vector length, whereas in the DAP blocks of data of size 4096 give maximum efficiency. In other respects the DAP is more flexible and easier to program. The indexing

TABLE I

Conventional (*NEV*) MD Using the Verlet Leapfrog Algorithm: FORTRAN Code
Appropriate to the Lennard-Jones System

```
C       NOTE THAT ONLY THE ESSENTIAL FORTRAN CODE IS GIVEN.
C       N IS THE NUMBER OF LJ ATOMS.
C       DT IS THE TIME STEP DURATION IN LJ UNITS.
C       RHO IS THE DENSITY IN LJ UNITS.
C       SG IS THE LJ 'SIGMA' IN MD BOX UNITS.
C       CU IS THE INTERACTION TRUNCATION DISTANCE IN UNITS OF SIGMA SQUARED.
C       RX(I),RY(I),RZ(I) ARE THE LJ COORDINATES IN PROGRAM UNITS WHICH RANGE
C       FROM -1 TO 1.
C       DX(I),DY(I),DZ(I) ARE THE POSITION INCREMENTS IN VERLET'S LEAPFROG
C       ALGORITHM.
C       NR IS THE NUMBER OF TIME STEPS TO BE PERFORMED THIS RUN.
C       NTOT IS THE TOTAL NUMBER OF TIME STEPS FROM ALL RUNS UNTIL
C       THE END OF THIS RUN.
C       NT IS THE NUMBER OF TIME STEPS READ IN FROM THE PREVIOUS RUN.
C
C
C
        TEMP=0.722
        RMSV=SQRT(TEMP)
        SG=(RHO/FLOAT(N))**(1.0/3.0)*2.0
        SG2=SG*SG
        SG2I=1.0/SG2
        SG3=SG2*SG
        CFV=1.0/(SG*DT)
        VOL=FLOAT(N)/RHO
C       THE INTERACTION TRUNCATION DISTANCE IS HALF THE MD CUBE SIDELENGTH
        CU=1.0/(2.0*SG)
        CU2=CU*CU
        PI=3.14159265359
C       CONFIGURATIONAL ENERGY LONG RANGE CORRECTION
        CPR=8.0*PI*RHO/(9.0*CU**9)
        CPA=-8.0*PI*RHO/(3.0*CU**3)
C       CONFIGURATIONAL PRESSURE LONG RANGE CORRECTION
        CPRR=32.0*PI*RHO**2/(9.0*CU**9)
        CPRA=-16.0*PI*RHO**2/(3.0*CU**3)
        READ(3)RX,RY,RZ,DX,......
C       IF IP EQUALS 0 ZERO  ACCUMULATORS
        IF (IP.NE.0) GOTO 99
        NT=0
        ...         .
        ...         .
        ...         .
99      CONTINUE
        NTOT=NT+NR
C
C       MAIN PROGRAM STARTS HERE
C
6       CONTINUE
C       ZERO FORCE COMPONENT ARRAYS
        DO 110 I=1,N
        FX(I)=0.0
        FY(I)=0.0
        FZ(I)=0.0
110     CONTINUE
        ..  .
C       EVALUATE FORCES, POTENTIAL ENERGY AND THE STRUCTURAL PART OF THE PRESSURE
        PS=0.0
        PES=0.0
```

512

TABLE I   (*Continued*)

```
        DO 120 I=1,N-1
        DO 130 J=I+1,N
        X=RX(I)-RX(J)
        Y=RY(I)-RY(J)
        Z=RZ(I)-RZ(J)
C       APPLY PERIODIC BOUNDARY CONDITIONS
        X=X-2*INT(X)
        Y=Y-2*INT(Y)
        Z=Z-2*INT(X)
C       EVALUATE THE SEPARATION SQUARED IN UNITS OF SIGMA SQUARED
        RR=(X*X+Y*Y+Z*Z)*SG2I
        IF (RR.GT.CU2) GOTO 140
        RRI=1.0/RR
        R6I=RRI*RRI*RRI
        R12I=R6I*R6I
        FF=(R12I+R12I-R6I)*RRI
C       ACCUMULATE POTENTIAL ENERGY COMPONENTS
        PES=R12I-R6I
        PS=PS+RR*FF
        FX(I)=FX(I)+X*FF
        FY(I)=FY(I)+Y*FF
        FZ(I)=FZ(I)+Z*FF
        FX(J)=FX(J)-X*FF
        FY(J)=FY(J)-Y*FF
        FZ(J)=FZ(J)-Z*FF
140     CONTINUE
130     CONTINUE
120     CONTINUE
C       EVALUATE THE KINETIC ENERGY
        EK=0.0
        DO 150 I=1,N
        FX(I)=FX(I)*24.0*DT*DT
        FY(I)=FY(I)*24.0*DT*DT
        FZ(I)=FZ(I)*24.0*DT*DT
        VELX=DX(I)+0.5*FX(I)
        VELY=DY(I)+0.5*FY(I)
        VELZ=DZ(I)+0.5*FZ(I)
        EK=EK+(VELX*VELX+VELY*VELY+VELZ*VELZ)*CFV*CFV
C       UPDATE POSITION INCREMENTS
        DX(I)=DX(I)+FX(I)
        DY(I)=DY(I)+FY(I)
        DZ(I)=DZ(I)+FZ(I)
C       UPDATE POSITIONS AND APPLY PERIODIC BOUNDARY CONDITIONS
        RX(I)=RX(I)-2*INT(RX(I))
        RY(I)=RY(I)-2*INT(RY(I))
        RZ(I)=RZ(I)-2*INT(RZ(I))
150     CONTINUE
C       CALCULATE THE TOTAL ENERGY IN LJ UNITS
        PE=PES+EK+CPR+CPA
C       CALCULATE THE PRESSURE IN LJ UNITS
        P=(PS*24.0+EK)/3.0/VOL+CPRR+CPRA
        ...      .
        IT=IT+1
        NT=NT+1
        IF (NT.LT.NTOT) GOTO 6
        ...      .
        ...      .
        STOP
        END
```

TABLE II

Cray Vectorized FORTRAN for the Forces Loop Applied to Lennard-Jones Molecules

```
C      CVMGP(X,0.0,Y) RETURNS X IF Y ⩾0 BUT 0.0 IF Y <0.
C      SSUM(M,A,1) SUMS THE FIRST M ELEMENTS OF ARRAY A.
C      FXI,FYI,FZI ARE SCRATCH ARRAYS EACH OF DIMENSION N.
C
       ...      .
       DO 110 I=1,N
       FX(I)=0.0
       FY(I)=0.0
       FZ(I)=0.0
110    CONTINUE
       ...      .
       DO 120 I=1,N-1
       RXI=RX(I)
       RYI=RY(I)
       RZI=RZ(I)
       M=0
       DO 130 J=I+1,N
       M=M+1
       X=RXI-RX(J)
       Y=RYI-RY(J)
       Z=RZI-RZ(J)
C      APPLY PERIODIC BOUNDARY CONDITIONS
       X=X-2*INT(X)
       Y=Y-2*INT(Y)
       Z=Z-2*INT(Z)
       RR=(X*X+Y*Y+Z*Z)*SG2I
       R2I=1.0/RR
       RRI=CVMGP(R2I,0.0,CU2-RR)
       R6I=RRI*RRI*RRI
       R12I=R6I*R6I
       FF=(R12I+R12I-R6I)*RRI
       FXI(M)=X*FF
       FYI(M)=Y*FF
       FZI(M)=Z*FF
       FX(J)=FX(J)-X*FF
       FY(J)=FY(J)-Y*FF
       FZ(J)=FZ(J)-Z*FF
130    CONTINUE
       MAX=M
       FX(I)=FX(I)+SSUM(MAX,FXI,1)
       FY(I)=FY(I)+SSUM(MAX,FYI,1)
       FZ(I)=FZ(I)+SSUM(MAX,FZI,1)
120    CONTINUE
       ...      .
       ...      .
```

techniques can be used to handle conditionals, whereas these are difficult on the CRAY, the only possibility being the choice of one of two numerical values depending on the sign of a third. Also, on the CRAY the common operation of summing the elements of a vector does not "vectorize," whereas the bit serial arithmetic on the DAP makes it a very rapid operation. The bit serial arithmetic has other surprising effects; for example, "SQRT" is faster than a multiply! Overall, for floating-point arithmetic the performance of the DAP lies between those of the CDC 7600 and the CRAY-1S.

B. Molecular-Dynamics Cell Shape

There are other factors that affect the efficiency of the program. The shape of the computational cell should be chosen so that for a given range of interaction and range of correlations to be studied, the number of particles will be a minimum. Also, the application of the nearest-image transformation should be as efficient as possible. As we shall see, these requirements can conflict, and the balance between them will be different for the different classes of computer.

On conventional computers most programs have employed cubic boundary conditions. The nearest-image transformation (see Fig. 1) can in this case be formally written in pseudo-FORTRAN as

$$\text{if} \left(r_{ij\alpha} > \frac{S}{2} \right) \quad \text{then} \ r_{ij\alpha} := r_{ij\alpha} - S$$

$$\text{if} \left(r_{ij\alpha} < \frac{-S}{2} \right) \quad \text{then} \ r_{ij\alpha} := r_{ij\alpha} + S$$

where $\mathbf{r}_{ij} = \mathbf{r}_i - \mathbf{r}_j$, $\alpha = x, y, z$, and S is the length of the side of the cube. However, the use of two IF statements per Cartesian component is unlikely to be the most efficient way of programming this transformation. A popular alternative is to arrange that, $S = 2$ in internal (or program) units. Then the two IF statements can be replaced by the single FORTRAN statement

```
RX = RX - 2 * INT(RX)
```

This works well on most computers. Yet more "tricks" are possible; the integer multiplication by 2 can be done by a simple shift along the word on some machines.[44] On CDC computers we convert the real coordinates into integer coordinates by means of the statement

```
IX(I) = INT(RX(I) * TWOTW)
```

where $TWOTW = 1048576.0 = 2^{20}$. This power of 2 is chosen so that the truncation introduces negligible errors, but integer overflows are avoided. The nearest-image transformation can then be programmed using shift operations:

```
IRX = IX(I) - IY(I)
IRX = IRX - SHIFT(SHIFT(IRX, -20),21)
```

Then $RX = IRX/TWOTW$ completes the transformation. It is quicker to

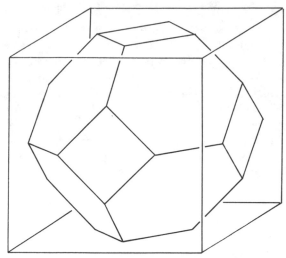

Figure 2. The 14-hedron or truncated octahedron MD cell. It has 36 edges, 24 vertices, and 14 faces, and packs as body-centered cubic. It is obtained by chopping off the corners of a cube, in such a way that the symmetry of the cube is maintained, until one half of the original cube's volume is left. The original cube is shown surrounding the truncated octahedron. The ratio of inscribed-sphere volume to total volume is 0.68 for this cell, whereas it is 0.52 for the original cube.

continue to use IRX until the cutoff has been applied, to take advantage of the greater speed of integer arithmetic. For 108 argon atoms this method reduces the CPU time per step of the simulation from 36 to 29 ms on a CDC 7600. On the DAP, integer operations are very fast and even faster, bit-level operations can also be used to achieve the same result.

When cubic boundary conditions are used, the cutoff radius and the distance to which we study structural correlations are limited to $S/2$, the radius of a sphere inscribed in the cube. The ratio of the total number of neighbors of a given particle to the "useful" number of neighbors, within the cutoff, is the ratio of the cube volume to the inscribed-sphere volume, or almost $2:1$. Adopting a more nearly spherical shape for the computational box enables the number of particles to be reduced for a given cutoff radius. There are a variety of space-filling shapes, which have been thoroughly reviewed by Adams[45] and Smith.[46] However, overall, the most suitable shape seems to be the truncated octahedron (TO). This is obtained by symmetrically cutting off the corners of a cube until its volume has been reduced by half, as shown in Fig. 2. The ratio of the octahedron volume to that of the inscribed sphere is 1.47. Thus there are considerably fewer out-of-range neighbors between the cutoff sphere and the edges of the computational box. For example, a simulation employing 256 molecules in cubic periodic boundaries requires only 197 molecules for the same cutoff with TO boundaries. On a serial computer the saving in computer time achieved by means of this reduction in cell volume is not very great, since out-of-range pairs are jumped over for the actual force evaluation, as shown in Table I. Furthermore, there is an overhead involved in performing the minimum image transformation with the more complicated boundary conditions. One of the reasons for choosing the TO rather than, say, the rhombic dodecahedron is that the TO transformation can be expressed reasonably simply. Again taking coordinates of the containing cube of the TO to be between -1 and $+1$, the following FORTRAN code can be used:

```
RX = RX − 2 ∗ INT(RX)
RY = RY − 2 ∗ INT(RY)
RZ = RZ − 2 ∗ INT(RZ)
IF(ABS(RX) + ABS(RY) + ABS(RZ).GE.1.5)THEN
RX = RX − SIGN(1.0,RX)
RY = RY − SIGN(1.0,RY)
RZ = RZ − SIGN(1.0,RZ)
ENDIF
```

where the FORTRAN function SIGN(1.0,RX) returns 1.0 if RX is positive and -1.0 if RX is negative.

On the parallel computer the situation is different. The pair force is evaluated for all pairs in the system. The saving involved in using TO boundary conditions, and hence a smaller number of particles, is much greater, because the number of pairs is proportional to the square of the number of particles. On the DAP the TO boundary conditions can be applied very efficiently, because the taking of absolute values and the picking up of signs are very rapid single-bit operations. For these reasons we expect the use of TO boundaries to become standard on vector and parallel computers.

C. Neighborhood Lists

In what we have said so far we have assumed that no more than a few hundred molecules are required in the simulation. If, for example, the range of the correlations in the solid or liquid is large, then it is necessary to have many more particles in the simulation. It then becomes essential to use a list technique so that only in-range pairs will be considered in the forces loop. For moderately large numbers of molecules (precise values will be given later), a list can be constructed that contains the neighbors of each particle found within some distance r_2 that is somewhat larger than the cutoff distance for the interactions, r_c.[47] Only pairs taken from this list need be considered in the forces loop. Whenever any particle has moved a distance greater than $\frac{1}{2}(r_2 - r_c)$ since the "neighbor" lists (NL) were last formed, the lists are calculated again. Consequently there is no chance that a particle will be missed, because a particle cannot cross the boundary between outer and inner truncation spheres between NL updates. The factor of $\frac{1}{2}$ allows for a similar motion of the origin molecule within this time (a necessarily pessimistic assumption). As r_2 increases, the frequency with which the lists need to be recreated obviously decreases, but this increases the time spent in the force evaluation of each step. More elements are also included in the list array, which can be prohibitively expensive in terms of computer memory. In practice, the neighborhood tables are typically remade about every 10 time steps.

An example of FORTRAN code for formation of the lists, evaluation of the forces, and updating the NL is given in Table III. The code is suitable for a serial computer. The NL method accesses pairs in the MD forces loop whose absolute indices are not sequentially related. This prevents vectorization. It can be made partially vectorizable by so-called GATHER AND SCATTER routines, as explained elsewhere.[47,48] However, this somewhat lessens the advantages when one implements this method on parallel machines. Calculations for a 500-atom Lennard–Jones (LJ) system take only 27% of the previously required CPU time when one implements the NL method in a conventional MD program on a CDC 7600 computer, whereas the drop in time is only 58% for a CRAY. The NL technique has been implemented on the DAP[49] using a logical mask rather than an actual list of indices; on the DAP logicals are single-bit quantities, so the storage problems that have made the technique unpopular on the CDC 7600 do not arise. The in-range interactions then need to be "packed" onto the DAP in 64×64 blocks, a process akin to the "GATHER" operation on the CRAY.

D. Link Cells

The so-called link cell (LC) method popularized by Hockney[50] and Woodcock[51] is particularly useful for very large numbers (i.e., many thousands) of molecules. When it is used, the computer time increases approximately in proportion to N instead of $\sim N^2$ as for the conventional method. This is achieved by eliminating the more distant, noncontributing interactions before the interparticle separations are calculated. The molecules are assigned to NLC smaller subcells, called link cells, that completely fill the original MD cell. It is arranged that the minimum side length of a LC is *less* than the truncation range of the intermolecular interaction. Particles within a LC therefore need to interact only with those in the adjacent link cells. Consequently systems interacting through long-range (e.g., Coulomb) forces

TABLE III
FORTRAN Code for Implementing the Neighborhood List Method

```
       ...         .
       ...         .
C      THE MAIN PROGRAM LOOP STARTS HERE
6      CONTINUE
       ...         .
       ...         .
       CALL NBLIST
C      PERFORM THE FORCES LOOP USING NEIGHBOR TABLES


C
       MF=1
       N1=N-1
       DO 50 I=1,N1
       ...         .
       ...         .
C      ADDRESS OF LAST ENTRY IN NEIGHBOR LIST
       ML=LAST(I)
       IF (MF.GT.ML) GOTO 60
C      LOOP OVER PARTICLES J
       DO 70 K=MF,ML
       J=LIST(K)
       ..          .
C      PERFORM PAIRWISE ADDITIVE INTERACTIONS
       ..          .
70     CONTINUE
       MF=ML+1
60     CONTINUE
50     CONTINUE
       ...         .
       ...         .
       GOTO 6
       ...         .
       ...         .
       STOP
       END
       SUBROUTINE NBLIST


C      USED TO CHECK EACH TIME STEP IF THE NEIGHBORHOOD LISTS
C      NEED UPDATING
C      RX(I),RY(I),RZ(I) LIE BETWEEN -1 AND 1
C      N IS THE NUMBER OF MOLECULES IN THE MD CELL
C      ALL THE RELEVANT I,J PAIRS ARE IN  ARRAY LIST. IT STARTS WITH THE NEAREST
C      NEIGHBORS OF I=1, THEN I=2...ETC.
C      LAST(I) GIVES THE LAST ENTRANT IN  ARRAY LIST FOR THE NEIGHBORS OF
C      MOLECULE I.
C      RCUT2D IS THE OUTER CUT-OFF RADIUS
C      RCUT1D IS THE INNER (INTERACTION) CUT-OFF RADIUS
C      SQUARE OF CRITICAL MOVEMENT EQUALS RSQCRT=(0.5*(RCUT2D-RCUT1D))**2
C
C      ACCUMULATE MOVEMENTS UPDATED EACH TIME STEP.
       DO 10 I=1,N
       QX(I)=QX(I)+DX(I)
       QY(I)=QY(I)+DY(I)
       QZ(I)=QZ(I)+DZ(I)
10     CONTINUE
C      TEST DISTANCE MOVED BY EACH MOLECULE
```

519

TABLE III (*Continued*)

```
      DO 11 I=1,N
      RSQ=QX(I)**2+QY(I)**2+QZ(I)**2
      IF (RSQ.GT.RSQCRT) GOTO 12
11    CONTINUE
      RETURN
C
C     SQUARE OF DIMENSIONLESS RADIUS OF OUTER SPHERE
12    RCT2SD=RCUT2D**2
C     NO INTERACTIONS ARE CONSIDERED BEYOND THIS SEPARATION.
      L=1
C     LOOP OVER MOLECULE I
      DO 20 I=1,N1
C     LOOP OVER MOLECULE J
      I1=I+1
      DO 30 J=I1,N
C     TEST SEPARATION OF PAIR
      X=RX(I)-RX(J)
      Y=RY(I)-RY(J)
      Z=RZ(I)-RZ(J)
      X=X-2*INT(X)
      Y=Y-2*INT(Y)
      Z=Z-2*INT(Z)
      RSQ=X*X+Y*Y+Z*Z
      IF (RSQ.GT.RCT2SD) GOTO 30
C     MAKE ENTRY IN LIST
      LIST(L)=J
      L=L+1
30    CONTINUE
C     SET ADDRESS OF LAST NEIGHBOR OF I
      LAST(I)=L-1
20    CONTINUE
C     SET MOVEMENT ACCUMULATORS TO ZERO
      DO 40 I=1,N
      QX(I)=0.0
      QY(I)=0.0
      QZ(I)=0.0
40    CONTINUE
      RETURN
      END
```

cannot easily be treated by this method. (Ironically, it was originally used in the P^3M method for Coulomb systems.[50])

The memory requirements of the LC method are rather modest; it demands an increase of approximately $3N$ over those of the conventional MD technique. The main difference in approach is that LC pairs rather than molecular pairs form the backbone of the force-evaluation loops. The periodic boundary conditions are applied to the coordinates of each LC and then automatically to their contents. There is a two-step decrease in the search for the relevant nearest images. First, interactions between nonadjacent LCs are automatically eliminated. Second, the *same* nearest-image transformation is applied to all the molecules in the two LCs considered at any one time.

A problem with this method may lie in the scheme for determining which molecules are in each LC. One possible solution would be to have a two-dimensional array in which the first index is for the LC number and the second for the molecule's index. However, for some computers this would be prohibitively expensive in terms of computer memory. This potential drawback is circumvented by the following ingenious scheme: As in conventional MD, each molecule retains an inviolable index I, with values of I ranging from 1 to N. This we will call the absolute index. These are sorted into another vector of size N, which "chains" them together. Each element in this additional array, which we call "LINK" here for obvious reasons, contains the absolute index of the next particle in the chain. For example, "$J = LINK(K)$" states that the molecule with the absolute index J is the next molecule in the chain after the particle with absolute index K. The size of LINK is N elements and it contains a consecutive series of closed chains of absolute indices; each chain is associated with a particular LC.

Another common feature of a LC program is a facility for expanding a read-in configuration by $2 \times 2 \times 2$ (i.e., to 8 times its original size). This explains the common use of $N = 108$, 864, and 6912, and $N = 256$ and 2048. In both of these sets, the higher numbers are multiples by 8 the next smaller N. This provides an efficient mechanism for achieving a large equilibrated system. Small equilibrated samples (at least as far as short-wavelength density fluctuations are concerned) are packed together so that effectively only the lower-frequency and longer-wavelength fluctuations need to develop during subsequent equilibration of this large system. It is essential to reassign the velocities (usually taken from a Maxwell–Boltzmann distribution) after this expansion, because otherwise the original system would in fact be simulated, even though a bigger cell would be nominally followed. The FORTRAN code for performing this operation is given in Table IV.

The molecules are linked together at the beginning of the simulation as presented in Table V. Also given in this table is the outer framework of the forces loop, which enables only pairs within the same or adjacent LCs to interact. Note that each relevant LC pair is treated only once. After the molecules have been moved in the integration algorithm, there is a possibility that some will have passed from one LC to another. All molecules could be relinked together every time step, but because reassignment is likely to affect only a few molecules each time step, the quicker scheme outlined at the end of Table V is employed by the authors instead. Because of the frequent but unavoidable use of IF statements and noncontiguous array element accessing in the LC scheme, the gains are again greatest for a serial machine when this method is used.

To conclude this section on computational considerations, we bring together a number of themes discussed. We give here a comparison between

TABLE IV

FORTRAN Code for Expanding the MD Cell by $2 \times 2 \times 2$ and Reassigning the Velocities
from a Maxwell–Boltzmann Velocity Distribution

```
C       MOLECULAR COORDINATES: RX,RY,RZ RANGING FROM -1 TO 1
C       VERLET LEAPFROG POSITION INCREMENTS: DX,DY,DZ
C       RMSV IS THE ROOT MEAN SQUARE VELOCITY IN ANY DIRECTION
C       NOLD IS THE NUMBER OF MOLECULES FROM THE LAST RUN
C       ALL MOLECULES HAVE THE SAME MASS
C       N IS THE NEW NUMBER OF MOLECULES = 8*NOLD
C
C
C       INITIALIZE THE RANDOM NUMBER GENERATOR-NAG LIBRARY ROUTINE
        CALL G05CBF(0)
        DO 10 I=1,NOLD
        RX(I)=(RX(I)-1.0)*0.5
        RY(I)=(RY(I)-1.0)*0.5
        RZ(I)=(RZ(I)-1.0)*0.5
10      CONTINUE
        DO 20 I=1,NOLD
C       FIRST EXPAND THE OLD CELL IN THE X AND Y DIRECTIONS
C       TO FORM THE BOTTOM LAYER OF THE NEW CELL
        DO 30 K=1,4
        J=I+(K-1)*NOLD
        RX(J)=RX(I)
        RY(J)=RY(I)
        RZ(J)=RZ(I)
        IF (K.EQ.2) RX(J)=RX(J)+1.0
        IF (K.EQ.3) RY(J)=RY(J)+1.0
        IF (K.EQ.4) RX(J)=RX(J)+1.0
        IF (K.EQ.4) RY(J)=RY(J)+1.0
C       SELECT POSITION INCREMENTS FROM A NORMAL DISTRIBUTION
        DX(J)=G05DDF(0.0,RMSV)/CFV
        DY(J)=G05DDF(0.0,RMSV)/CFV
        DZ(J)=G05DDF(0.0,RMSV)/CFV
30      CONTINUE
        DO 40 K=5,8
        J=I+(K-5)*NOLD
        L=I+(K-1)*NOLD
        RX(L)=RX(J)
        RY(L)=RY(J)
        RZ(L)=RZ(J)+1.0
        DX(L)=G05DDF(0.0,RMSV)/CFV
        DY(L)=G05DDF(0.0,RMSV)/CFV
        DZ(L)=G05DDF(0.0,RMSV)/CFV
40      CONTINUE
20      CONTINUE
C       FIND MOMENTA
        X=0.0
        Y=0.0
        Z=0.0
        DO 50 I=1,N
        X=X+DX(I)/FN
        Y=Y+DY(I)/FN
        Z=Z+DZ(I)/FN
50      CONTINUE
C       ZERO MOMENTA AND CALCULATE KINETIC ENERGIES
        EK=0.0
        DO 60 I=1,N
        DX(I)=DX(I)-X
        DY(I)=DY(I)-Y
        DZ(I)=DZ(I)-Z
```

TABLE IV (*Continued*)

```
        EK=EK+0.5*(DX(I)**2+DY(I)**2+DZ(I)**2)*CFV*CFV/FN
60      CONTINUE
        H=SQRT(1.5*TEMP/EK)
        DO 70 I=1,N
        DX(I)=DX(I)*H
        DY(I)=DY(I)*H
        DZ(I)=DZ(I)*H
70      CONTINUE
        ...        .
        ...        .
```

conventional, NL, and LC methods on a series of computers. We have considered the simple MD system consisting of N spherically symmetric LJ molecules enclosed in a cubic cell. The truncation radius for all interactions was 2.5 σ. No pair radial distribution function was evaluated, since it is not really necessary in many applications and can increase the computer time by about 25%. We have confined the forces loop calculations to the attractive and repulsive components of the potential energy and virial. Two densities and two temperatures were considered: the near-triple-point state of reduced number density $\rho^* = 0.8442$ and reduced temperature $T^* = 0.722$, and another state at $\rho^* = 0.5$ and $T^* = 1.5$. The CPU times per time step in seconds for the near-triple-point state using the conventional method are as follows for an ICL 2980 DAP, VAX 11/780, AMDAHL 470V/8, CDC 6600, CDC 7600, and CRAY-1S: 0.05, (0.95, 3.6*), 0.10*, 0.23, 0.04, and 0.007 for $N = 108$; 0.11, (4.4, 13*), 0.48*, 1.0, 0.19, and 0.03 for $N = 256$; and 1.09, (39, 126*), 4.7*, 11, 1.8, and 0.28 for $N = 864$, respectively. Here * denotes double precision. It is recommended that MD on the AMDAHL 470V/8 and possibly the VAX 11/780 be performed in double precision because of the 32-bit words used in single-precision mode on these computers. Although these are probably adequate for force evaluation and position updates (which are limited mainly by update algorithm errors), the accumulators, especially those for fluctuation properties, need to be more accurately maintained. The above timings are not significantly different for the low-density state. An ICL PERQ workstation running the POS operating system runs at roughly the speed of a VAX 11/780 in double-precision mode. The timings reveal that the CRAY-1S is approximately 100 times faster than the VAX 11/780 operating in single precision for N up to 864. Note that for a vectorized code on the CRAY-1S the time per time step is independent of density because all interactions in the MD cell (even between molecules in opposite corners of the cube) are evaluated.

The NL method has been implemented on the VAX 11/780 and CRAY-1S by the authors. Table VI summarizes the times, memory requirements, and optimum parameters for the implementation of this "bookkeeping"

TABLE V
Outline of a FORTRAN Program that Incorporates Link Cells

```
C       FIRST LINK THE MOLECULES TOGETHER


C       NLC IS THE NUMBER OF LINK CELLS IN THE CUBIC MD CELL.
C       L0(NLC) CONTAINS THE ABSOLUTE INDEX OF THE START OF THE
C       CLOSED CHAIN IN EACH LINK CELL
C       L1(NLC) IS USED AT THE BEGINNING OF THE PROGRAM FOR
C       FORMING THE CLOSED CHAINS AND CONTAINS THE END
C       MOLECULE IN EACH LINK CELL'S CHAIN
C       IC(N) CONTAINS THE LINK CELL INDEX, IP, OF EACH MOLECULE
C       LINK(N) CONTAINS THE ABSOLUTE INDEX I OF EACH MOLECULE
C       NEXT ALONG THE CHAIN
C
        NLC=NLCX*NLCX*NLCX
        FNLCX=FLOAT(NLCX)
        DO 100 I=1,NLC
        L0(I)=0
        L1(I)=0
100     CONTINUE
        DO 110 I=1,N
        IX=INT((RX(I)+1.0)*FNLCX*0.5+1.0)
        IY=INT((RY(I)+1.0)*FNLCX*0.5)
        IZ=INT((RZ(I)+1.0)*FNLCX*0.5)
        IP=IX+NLCX*IY+NLCX*NLCX*IZ
C       ASSIGN MOLECULE I TO LINK CELL IP
        IC(I)=IP
        ME=L1(IP)
        IF (ME.EQ.0) GOTO 120
C       MAKE I THE END OF THE CHAIN SEGMENT FOR LINK CELL IP
        LINK(ME)=I
        L1(IP)=I
        GOTO 130
120     CONTINUE
        L0(IP)=I
        L1(IP)=I
130     CONTINUE
110     CONTINUE
C       JOIN TOGETHER THE CHAIN ENDS IN EACH LINK CELL
        DO 200 IP=1,NLC
        ME=L1(IP)
        IF (ME.EQ.0) GOTO 210
C       CONNECT THE LAST ELEMENT IN THE CHAIN TO THE FIRST ELEMENT
        LINK(ME)=L0(IP)
210     CONTINUE
200     CONTINUE
        ...        .
        ...        .
C       THE MAIN PROGRAM STARTS HERE
6       CONTINUE
        ...        .
        ...        .
C       NOW FIND CLOSE PARTICLE PAIRS FOR FORCE EVALUATION
        IP=0
        DO 300 IZ=1,NLCX
        DO 300 IY=1,NLCX
        DO 300 IX=1,NLCX
        IP=IP+1
        I=L0(IP)
C       IF THERE ARE NO MOLECULES IN THE LC OF INDEX IP THEN GO TO THE NEXT LC
        IF (I.EQ.0) GOTO 310
C       THE FOLLOWING LINK CELLS ARE CONSIDERED: THE 9 LINK CELLS ABOVE IP
```

TABLE V   *(Continued)*

```
C       AND IN THE PLANE, (-1,1),(1,0),(1,1),(0,0) AND (0,1)
C       RELATIVE LC SHIFTS WITH RESPECT TO IP.
C       TRANSPOSE THE LINK CELLS FROM THE OPPOSITE FACES
C       IF NECESSARY TO PARTIALLY COMPLETE A FIRST SHELL OF LINK CELLS.
C
        DO 320 KZ=2,3
        JZ=IZ+KZ-2
        CZ=0.0
        IF ((JZ-NLCX-1).NE.0) GOTO 340
        JZ=1
        CZ=2.0
340     CONTINUE
        KK=2
        IF (KZ.EQ.3) KK=1
        JZN=(JZ-1)*NLCX*NLCX
        DO 320 KY=KK,3
        JY=IY+KY-2
        CY=0.0
        IF (JY.NE.0) GOTO 360
        JY=NLCX
        CY=-2.0
360     IF ((JY-NLCX-1).NE.0) GOTO 370
        JY=1
        CY=2.0
370     CONTINUE
        KL=1
        IF (KZ.EQ.2.AND.KY.EQ.2) KL=2
        JYZN=JZN+(JY-1)*NLCX
        DO 320 KX=KL,3
        CX=0.0
        JX=IX+KX-2
        IF (JX.NE.0) GOTO 390
        JX=NLCX
        CX=-2.0
390     IF ((JX-NLCX-1).NE.0) GOTO 400
        JX=1
        CX=2.0
400     CONTINUE
        JP=JX+JYZN
        IF (IP.EQ.JP) GOTO 410
420     J=LO(JP)
C       IF CELL JP IS EMPTY THEN TRY THE NEXT LINK CELL JP
        IF (J.EQ.0) GOTO 430
        GOTO 440
410     J=LINK(I)
C       IF THERE IS ONLY 1 MOLECULE IN CELL JP THEN TRY THE NEXT LINK CELL JP
        IF (J.EQ.LO(IP)) GOTO 450
        GOTO 440
C       INCREMENT I
450     I=LO(IP)
        GOTO 430
440     CONTINUE
460     CONTINUE
C       PERFORM (I,J) INTERACTION AS IN CONVENTIONAL MD
        X=RX(I)-RX(J)-CX
        Y=RY(I)-RY(J)-CY
        Z=RZ(I)-RZ(J)-CZ
        ...          .
C       NOW EVALUATE PAIRWISE ADDITIVE PROPERTIES
        ...          .
C       TRY NEXT J IN THE CHAIN
        J=LINK(J)
        IF (J.EQ.LO(JP)) GOTO 470
        GOTO 460
```

525

TABLE V   (*Continued*)

```
470     CONTINUE
C       GO THROUGH ALL I IN LINK CELL IP
        I=LINK(I)
        FXI=0.0
        FYI=0.0
        FZI=0.0
        ...     .
        ...     .
        IF (I.EQ.LO(IP)) GOTO 430
        IF (IP.EQ.JP) GOTO 410
        GOTO 420
430     CONTINUE
320     CONTINUE
310     CONTINUE
300     CONTINUE
        ...     .
        ...     .
C       UPDATE MOLECULAR POSITIONS
        ...     .
C       REASSIGN MOLECULES TO LINK CELLS

        DO 500 I=1,N
C       INCREMENT POSITIONS
        ...     .
        ...     .
C       SORT THE MOLECULES INTO THEIR NEW  LINK CELLS
510     IX=INT((RX(I)+1.0)*FNLCX*0.5+1.0)
        IY=INT((RY(I)+1.0)*FNLCX*0.5)
        IZ=INT((RZ(I)+1.0)*FNLCX*0.5)
        IP=IX+NLCX*IY+NLCX*NLCX*IZ
C       CHECK IF LINK CELL FOR PARTICLE I HAS CHANGED
        IF (IP.EQ.IC(I)) GOTO 520
C       TAKE MOLECULE I OUT OF THE OLD LINK CELL, ME
        ME=IC(I)
C       J IS THE STARTER MOLECULE IN LINK CELL ME
        J=LO(ME)
C       IS I THE NEXT MOLECULE AFTER J IN THE CHAIN ?
530     IF (LINK(J).EQ.I) GOTO 540
C       CONTINUE SEARCHING THROUGH THE CHAIN SEGMENT FOR CELL ME UNTIL THE J
C       BEFORE I IN THE CHAIN IS FOUND
        J=LINK(J)
        GOTO 530
C       J IS LINKED TO THE MOLECULE AFTER I IN THE CHAIN
540     LINK(J)=LINK(I)
C       THE NEXT ATOM I IN THE CHAIN NOW BECOMES THE STARTER (*N.B. NOT J)
        LO(ME)=LINK(J)
        IF (I.EQ.J) LO(ME)=0
C       START TO INSERT I INTO THE NEW LINK CELL
C       J IS THE STARTER OF CELL IP IN WHICH I IS NOW FOUND
        J=LO(IP)
C       I IS PUT BETWEEN J AND LINK(J)
        IF (LO(IP).EQ.0) J=I
        LINK(I)=LINK(J)
        LINK(J)=I
        IC(I)=IP
C       I IS THE NEW STARTER IN CELL IP
        LO(IP)=I
520     CONTINUE
        ...     .
        ...     .
500     CONTINUE
        ...     .
        ...     .
        GOTO 6
```

TABLE VI

Neighborhood Lists on a VAX 11/780 and a CRAY-1S[a]

	Computer	Method	Mode of Arithmetic	N	$\frac{\Delta r}{\sigma}$	N_e	N_f	$\frac{\delta t}{sec}$
(a)	VAX	C	SP	256				3.9
	VAX	N.T.	SP	256	0.323	6000	10	1.41
	VAX	N.T.	SP	256	0.587	7000	18	1.44
	VAX	C	DP	256				12
	VAX	N.T.	DP	256	0.323	6000	10	4.8
	VAX	N.T.	DP	256	0.587	8000	18	5.4
	CRAY	C	SP/S	256				0.12
	CRAY	N.T.	SP/S	256	0.323	7000	10	0.05
	CRAY	N.T.	SP/S	256	0.587	9000	20	0.05
	CRAY	C	SP/V	256				0.029
	CRAY	N.T.	SP/V	256	0.323	7000	10	0.021
	CRAY	N.T.	SP/V	256	0.587	10000	26	0.024
	VAX	C	SP	864				37
	VAX	N.T.	SP	864	0.323	20000	20	8
	CRAY	C	SP/S	864				1.3
	CRAY	N.T.	SP/S	864	0.323	20000	16	0.22
	CRAY	N.T.	SP/S	864	0.734	34000	28	0.23
	CRAY	C	SP/V	864				0.28
	CRAY	N.T.	SP/V	864	0.323	20000	16	0.09
	CRAY	N.T.	SP/V	864	0.734	34000	24	0.10
(b)	VAX	C	SP	256				4.4
	VAX	N.T.	SP	256	0.323	10000	12	2.5
	VAX	N.T.	SP	256	0.734	17000	30	3.2
	VAX	C	DP	256				13
	VAX	N.T.	DP	256	0.323	10000	12	6.9
	VAX	N.T.	DP	256	0.734	17000	30	9.5
	CRAY	C	SP/S	256				0.12
	CRAY	N.T.	SP/S	256	0.323	11000	18	0.06
	CRAY	N.T.	SP/S	256	0.734	17000	50	0.08
	CRAY	C	SP/V	256				0.029
	CRAY	N.T.	SP/V	256	0.323	11000	20	0.027
	CRAY	N.T.	SP/V	256	0.734	17000	50	0.038
	VAX	C	SP	864				39
	VAX	N.T.	SP	864	0.323	34000	14	10
	CRAY	C	SP/S	864				1.3
	CRAY	N.T.	SP/S	864	0.323	34000	20	0.25
	CRAY	N.T.	SP/S	864	0.734	60000	30	0.33
	CRAY	C	SP/V	864				0.28
	CRAY	N.T.	SP/V	864	0.323	34000	20	0.11
	CRAY	N.T.	SP/V	864	0.734	60000	30	0.15

[a] This table illustrates the relative advantages of using NLs on a serial and vector computer. C denotes conventional MD; Δr is the thickness of the boundary shell between the interaction sphere and NL outer sphere; SP and DP denote single-and double precision arithmetic, respectively; N_e is the average number of entrants in the NL; N_f is the average number of time steps in between the re-creation of the NLs; δt is the computer time consumed on average for each time step; V and S denote vector and scalar (serial) optimization modes on the CRAY-1S. LJ state points (a) $\rho^* = 0.5$, $T^* = 1.5$ and (b) $\rho^* = 0.8442$, $T^* = 0.722$ are considered. The time step is $0.005\sigma(m/\varepsilon)^{1/2}$. N is the number of LJ molecules in the MD cell.

scheme on these machines. The NL method improves matters most for large N, at which there are more interactions to be (justifiably) ignored, and for serial machines. The gains in speed on going from conventional to NL algorithms are greater for the VAX 11/780 than the CRAY-1S. Fascinatingly, the CPU times per time step are relatively insensitive to the thickness of the shell between interaction radius r_c and the outer sphere radius r_2, which indicates that there is compensation between the frequency of NL updates and number of NL entrants. As r_2 increases, the number of entrants in the NL increases. This lengthens the force loop. The NL need to be updated less frequently, however, because $\frac{1}{2}(r_2 - r_c)$ is larger and therefore the probability of a molecule pair crossing between these shells is smaller. The conventional method is too time consuming for values of N larger than approximately 1000. Although the NL method is much faster than the conventional method in this region, it too is not practicable for N greater than approximately 1000 because of its considerable memory requirements ($\approx 40N$). Even if this problem is overcome, the fact that all interactions must be considered approximately once every 10 time steps to create the NLs would be a prohibitive factor in its implementation. Timings for the conventional method indicate the long time required for this operation.

Test runs on the ULCC CDC 7600 using the LC method on a LJ state with $\rho^* = 0.8552$ and $T^* = 0.705$ illustrate that although the LC method provides rather unspectacular gains in speed for moderately large values of N ($\sim 1000-2000$), it becomes a progressively more effective method when very large N values (~ 7000) are considered. The conventional-method LC timings per time step are 1.57 and 0.834 (64) for $N = 864$; 8.46 and 2.32 (125) for $N = 2048$; and 94.2 and 6.67 (512) for $N = 6912$. The numbers in brackets are the number of LCs in the MD cell. Insufficient computer memory on the CDC 7600 prevented the NL method from being used for $N = 2048$ and greater; for $N = 864$, the NL timing per time step was 0.31. It is possible that a hybrid NL–LC method would be successful in the intermediate regime of only a few thousand particles. The infrequent update of the NL recreation would be performed with the aid of LC, so that this step would be much faster than if implemented via a conventional algorithm. There is a trend toward simulating systems with inherently large dependences on N. Liquid crystals,[52,53] solid–solid phase changes,[54] interfaces,[55,56] and adsorbate layers[57] are examples. It is therefore more important to use large MD cells for such systems.

Undoubtedly the last word has not been written on these "bookkeeping" schemes. The authors believe that none developed so far takes account of the fact that the configurations evolve only slowly. The collection of interaction pairs changes by only a few elements from time step to time step. Therefore the NL *should* need only a continual, but small, modification and not frequent total reconstruction as at present. The LC method has the disadvantage that by working with cubes about twice the number of molecule pairs are considered than is necessary. The spherical symmetry of the NL method is not a feature of the LC scheme. Possibly a LC method using TO cells could be devised to circumvent this problem.

IV. THERMODYNAMIC PROPERTIES

A. First-Order Properties

Thermodynamic quantities are primary output data from a simulation, and are usually printed out at selected time steps during a run. At the end of the calculation, or at selected intervals in a run, an average of a property X, denoted by $\langle X \rangle$, is often printed for use in subsequent analysis. The translational kinetic energy $K_{tr.}$, for example, is calculated according to

$$K_{tr.} = \frac{1}{2} \sum_{i}^{N} m_i \dot{\mathbf{r}}_i \cdot \dot{\mathbf{r}}_i \tag{68}$$

The temperature follows from this:

$$T = \frac{\frac{2}{3} \langle K_{tr.} \rangle}{N k_B} \tag{69}$$

The configurational part of the internal energy is calculated from

$$\langle U \rangle = \left\langle \sum \sum_{i<j} \phi_{ij} \right\rangle + U_c \tag{70}$$

The sum in Eq. (70) is taken over all those pairs for which $r_{ij} < r_c$. U_c, a correction term based on a uniform distribution of molecules beyond r_c, is added to compensate, to a certain extent, for this truncation. U_c is given by

$$U_c = 2\pi \frac{N^2}{V} \int_{r_c}^{\infty} r^2 \phi(r) g(r) \, dr \tag{71}$$

To illustrate the density dependence of a popular simple fluid we have performed MD simulations of the soft-sphere fluid, characterized by the potential $\phi(r) = 4\varepsilon(\sigma/r)^{12}$. The derived U values are presented in Fig. 3. A closely related function is the virial $\langle \psi \rangle$, which is used to evaluate the con-

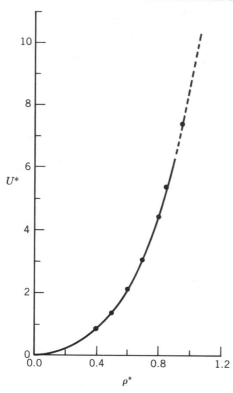

Figure 3. Comparison between simulated configurational energies per molecule obtained from Eq. (70) for the soft-sphere fluid using MD for the potential $\phi(r) = 4\varepsilon(\sigma/r)^{12}$. Reduced energy U^* is in ε and reduced number density ρ^* is in σ^{-3}. The solid line is the Cape and Woodcock equation of state for the soft-sphere fluid. $N = 256$, $\Delta t = 0.005\sigma$ $(m/\varepsilon)^{1/2}$.

figurational part of the internal pressure, P, and is given by

$$\langle\psi\rangle = \left\langle \sum_{i<j}\sum r_{ij}\frac{\partial}{\partial r_{ij}}\phi_{ij}(r)\right\rangle + \psi_c \qquad (72)$$

where

$$\psi_c = 2\pi\frac{N^2}{V}\int_{r_c}^{\infty}r^3\phi'(r)g(r)\,dr \qquad (73)$$

From this follows the internal pressure:

$$\langle P\rangle = \frac{N}{V}k_B\langle T\rangle - \frac{1}{3V}\langle\psi\rangle \qquad (74)$$

Note that the thermodynamic pressure p is given by

$$p = P - T\left(\frac{\partial P}{\partial T}\right)_{N/V} \qquad (75)$$

TABLE VII
Thermodynamic Characteristics of the Soft-Sphere Fluid[a]

$\frac{NT}{10^3}$	T^*	ρ^*	u^* (MD)	u^* (EQ)	P^* (MD)	P^* (EQ)	$\langle F_x^{*2}\rangle$
60	0.968	0.4	0.88	0.88	1.79	1.79	0.74
	1.0[b	0.403		0.91		1.86	
60	0.969	0.5	1.39	1.39	3.26	3.26	1.41
	1.0[b	0.504		1.43		3.39	
60	1.00	0.6	2.13	2.13	5.71	5.71	2.63
60	0.97	0.7	3.08	3.08	9.29	9.31	4.17
	1.0[b	0.705		3.18		9.66	
60	1.00	0.8	4.44	4.45	15.01	15.05	6.99
20	1.22	0.9	6.02	6.58	22.76	24.78	12.88
	1.0[b	0.8564		5.39		19.33	
60	1.21	1.0	8.30	8.94	34.41	36.97	19.30
	1.0[b	0.9535		7.39		29.13	

[a] This table gives the average configuration energies, $U^*(=U/\varepsilon)$, and pressures, $P^*(=P/\varepsilon\sigma^{-3})$, for the soft-sphere system, which is characterized by the pair potential $\phi(r) = 4\varepsilon(\sigma/r)^{12}$. The simulation was done on 256 molecules in a cubic MD cell for NT time steps, each of duration 0.005 $\sigma(m/\varepsilon)^{1/2}$, where m is the soft-sphere mass. EQ denotes the prediction of the Cape and Woodcock[51] equation of state, which is based on $N = 4000$ MD simulations. The mean-square x-component force on each molecule, $\langle F_x^{*2}\rangle$, is also presented.
[b] Adjusted state points for $T^* = 1$, obtained using data from the row above.

the second term of which is called the thermal pressure. In contrast to the situation for configurational energy, the tail correction to the virial can make a significant contribution to $\langle\psi\rangle$ at reasonably dense fluid states. Obviously the importance of U_c and ψ_c is minimized by taking a large cutoff distance. Some examples of U and P for the soft-sphere fluid are given in Table VII.

B. Second-Order Properties

Second-order thermodynamic quantities such as specific heats and thermal-pressure coefficients are discussed in detail by Cheung.[58] These quantities are useful in providing evidence for phase changes. They can be obtained from a differentiation of the equation of state with respect to the appropriate variable (e.g., temperature for specific heats and thermal-pressure coefficients). A formally equivalent but more direct method is to use the fluctuations in first-order thermodynamic properties, which can be related to the second-order quantities.[59] These fluctuations can be obtained from a single simulation, whereas the first method requires many calculations in the region of the state point of interest to establish an equation of state there. Unfortunately, these fluctuations often have a strong cutoff dependence[60] and N dependence.[28] Most fluctuations can be written as the sum of two-, three-, and four-particle contributions. The last should die away rapidly with distance, but because of periodic boundary conditions there is always a spurious correlation between pairs at opposite sides of the MD cell. The relative importance of these interactions should diminish as $\sim 1/N$, and therefore working with very large systems is advised when using the fluctuation approach.

C. Free Energy

It is often difficult to locate phase boundaries by MD. To do so involves determining the temperature, density, and pressure at which one phase changes into another. It is accepted that MD is unsuited to doing this using a direct approach. Superheating and glass formation are two unavoidable features of the crystal–liquid phase change that hinder these efforts. A large N dependence and slow relaxation times in these regions are other troublesome associated features. A more indirect approach may be employed that involves determining the free energies of the two phases as a function of state point. A phase change will occur when these two are equal. Because thermodynamic properties are readily evaluated by even small periodic systems, this offers a more convenient approach. Current methods for obtaining free energy have certain problems associated with them, especially at high densities close to the phase transitions themselves. Widom's[61-63] method for obtaining the absolute free energy is particularly susceptible to these drawbacks. In brief, it involves establishing an $m \times m \times m$ grid within the MD cell ($m \approx 10$). At each time step, say, the potential energy, u_t, of a test particle of the host fluid is evaluated on each of the grid points. A running average of the Boltzmann factor of this potential leads directly to the chemical potential,

$$\mu = -k_B \langle T \rangle \ln \left[\langle \exp(-\beta u_t) \rangle \right] \tag{76}$$

where $\beta = (k_B T)^{-1}$. Examples of $\mu(t)$ for a number of LJ state points are presented in Fig. 4. The effects of state point on the convergence character-

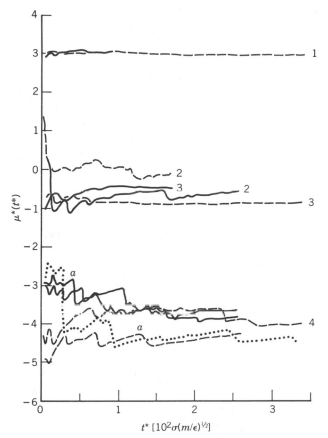

Figure 4.   The chemical potential function $\mu(t)$ obtained from Eq. (76) for various LJ states run under NEV ($\cdots$), NPH (—), and NPT (----) conditions. The approximate state points are (1) $\rho^* = 0.64$, $T^* = 2.5$, $P^* = 3.34$; (2) $\rho^* = 0.85$, $T^* = 1.1$, $P^* = 2.67$; (3) $\rho^* = 0.75$, $T^* = 1.3$, $P^* = 1.6$; and (4) $\rho^* = 0.84$, $T^* = 0.72$, $P^* = 0.0$. The Andersen constant pressure mass M_p^* is $(N/\rho^*)^{1/3}/45$, except where indicated by the letter a, where $M_p^* = (N/\rho^*)^{1/3}/15$. $N = 256$, $\Delta t = 0.005\sigma(m/\varepsilon)^{1/2}$.

istics of $\mu(t)$ are clear from this figure. Near the triple point, convergence is slow. At higher temperature and lower density, then, convergence is much more rapid, giving a well-defined free energy.

D.   Statistical Analysis

Because one of the main uses of simulation is to determine thermodynamic properties of model fluids, it is important to assess statistical accuracy. We measure the instantaneous values of u and P, say, at each time step and determine the mean $\overline{X}_N$ (where $X = u$ or P) and variance σ_N^2 over the run of N steps. Suppose that the "true" distribution of X has a mean $\overline{X}$ and variance σ^2. An important theorem[64] states that if we take random samples of length n from such a distribution, the means $\overline{X}_n$ of such samples themselves have a distribution with a mean of $\overline{X}$, but a variance of σ^2/n. Furthermore, if n is large enough, they are normally distributed, whatever the underlying distribution of X. If the measurements from the N timesteps were regarded as forming a single random sample of length N, then the best estimate for $\overline{X}$ would be $\overline{X}_N$, with an error of $\sigma/N^{1/2}$. Using σ_N as an estimate for σ, we would then obtain the so-called standard error (SE):

$$SE = \frac{\sigma_N}{N^{1/2}} \tag{77}$$

However, the instantaneous values from successive time steps are not independent but rather have some correlation. This means that the effective number of independent measurements is less than N, say N/t, where t is called the statistical inefficiency. It can also be regarded as a correlation length, since the inefficiency would be t if every t successive time steps were identical, but were independent of the next set of t steps. Then the standard error in the mean becomes

$$SE = \sigma_N \left(\frac{t}{N} \right)^{1/2} \tag{78}$$

We can find t by taking samples of varying lengths n from a single long run and finding their variances σ_n^2. As long as $n \gg t$, we expect

$$\sigma_n = \sigma \left(\frac{t}{n} \right)^{1/2} \tag{79}$$

whereas for $n \ll t$ we expect

$$\sigma_n = \frac{\sigma}{n^{1/2}} \tag{80}$$

since, by definition, $\sigma = \sigma_1$. Thus

$$t = \lim_{n \to \infty} \left(\frac{n\sigma_n^2}{\sigma_1^2} \right) \tag{81}$$

and we can find the limit by looking for a plateau in a graph of $n\sigma_n^2$ versus

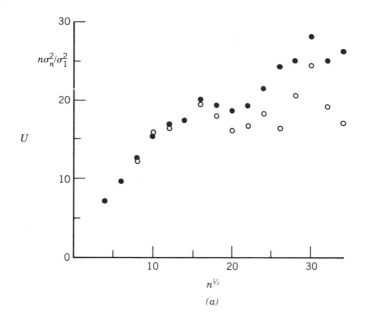

(a)

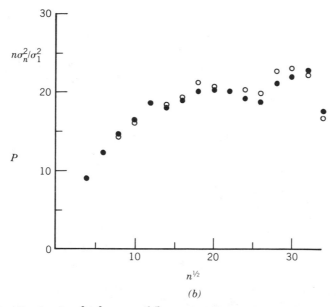

(b)

Figure 5. Graphs of $n\sigma_n^2/\sigma_1^2$ versus $n^{1/2}$ [see Eq. (81)] for (a) potential energy and (b) pressure. Solid circles are for a run of 70,000 time steps; open circles are for the first 40,000 steps. The simulation was done on a two-center LJ model of ethane. The time step was 10 fs. $T = 268$ K, molar volume = 72.31 cm^3 mol^{-1}, bond length = 2.344 Å, $\sigma = 3.56$ Å, and $\varepsilon = 137.5$ K.

535

n. Typical graphs are shown in Fig. 5.[65] Results are shown after runs of 40,000 steps and 70,000 steps as a check that a true plateau has been reached. The plateau is less clear in the case of the potential energy than in that of pressure because the fluctuations in energy are small and the energy drift due to algorithm errors becomes comparable to these fluctuations over this length of run. For both the energy and pressure, this run gives values of *t* of about 20. The values for this particular simulation, along with their errors [obtained using Eq. (78)], are

$$U = -10.266 \pm 0.002 \text{ kJ mol}^{-1}$$
$$P = 3.84 \pm 0.18 \text{ MPa}$$

which indicates the good statistical accuracy obtainable by simulation. Unfortunately, it tells us nothing about systematic errors due to size effects and algorithm errors. Research into these topics is urgently needed.

V. STRUCTURAL PROPERTIES

A useful quantity to calculate is the pair radial distribution, which forms the underlying structural basis for first-order thermodynamic quantities.[66] During the force-evaluation loop, configuration space around each molecule is divided into concentric spherical shells of thickness δr. The number of neighbors in each shell is counted, going usually as far out as half the side width of the box. If the average number of neighbors in a slice of inner radius r_- and outer radius r_+ (where $r_\pm = r \pm \delta r/2$) is $n(r)$, then the radial distribution function $g(r)$ can be computed via

$$n(r) = \frac{N}{V} V(r) g(r) \qquad (82)$$

where $V(r)$ is the volume of the shell bounded by $r = r_+$ and $r = r_-$:

$$V(r) = \frac{4\pi}{3} \left(r_+^3 - r_-^3 \right)$$
$$= 4\pi\delta r \left(r^2 + \frac{\delta^2 r}{12} \right) \approx 4\pi r^2 \delta r \qquad (83)$$

For each neighbor j of molecule i, the shell index is k_{ij} where $k_{ij} = \text{INT}(r_{ij}/\delta r + 0.5)$. This corresponds to a radius r of $k_{ij}\delta r$. Note that in determining k_{ij} a square root has to be evaluated, so this procedure can significantly slow down the MD program. This has recently prompted interest

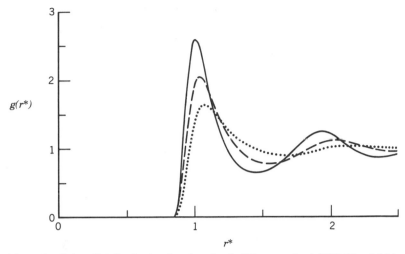

Figure 6.   Pair radial distribution functions for the LJ states $\rho^* = 1.0397$, $T^* = 3.35$ (——); $\rho^* = 0.8029$, $T^* = 3.54$ (----); and $\rho^* = 0.5$, $T^* = 3.47$ (· · · ·). The simulations were for 60,000 time steps and each of the 256 molecules was used as a center for $g(r)$ every five time steps. $\Delta t = 0.0025\sigma(m/\varepsilon)^{1/2}$.

in fast square root routines. The accuracy of the FORTRAN SQRT function is usually far greater than is required for the pair radial distribution function (i.e., where $r_{ij}^2 \rightarrow r_{ij}$). An iterative procedure can be constructed that has sufficient precision for this operation and is typically twice as fast.[67]

Some examples of $g(r)$ for supercritical fluids are presented in Fig. 6. Over a wide density range the position of the first peak in $g(r)$ hardly moves from $r^* \approx 1$. With decreasing density, there is a slow drift out, however, and the height diminishes. A second coordination shell becomes less evident as well.

One can also determine "second-order" structural properties such as the pair radial fluctuation function[28]

$$W(r) = \left[\left\langle m(r)^2 \right\rangle - \left\langle m(r) \right\rangle^2 \right] \Big/ \left(\tfrac{4}{3}\pi r^3 N V^{-1}\right) \tag{84}$$

where $m(r)$ is the average number of particles within a sphere of radius r around a reference particle. This function provides a close link with "second-order" thermodynamic quantities. Table VIII gives some of the main features of $g(r)$ and $W(r)$ for the soft-sphere fluid, obtained for this work using a CRAY-1S.

TABLE VIII

Principal Features of the Soft-Sphere Radial Pair Distribution and Pair Fluctuation Functions[a]

T^*	ρ^*	r_1	$g(r_1)$	r_2	$g(r_2)$	r_1	$W(r_1)$	r_2	$W(r_2)$	n
0.968	0.4	1.25	1.63	1.94	0.87	1.35	0.39	2.15	0.18	11.6
0.969	0.5	1.23	1.87	1.84	0.80	1.25	0.38	1.85	0.16	12.4
1.00	0.6	1.18	2.14	1.75	0.73	1.21	0.38	1.75	0.11	12.9
0.97	0.7	1.15	2.48	1.66	0.64	1.20	0.36	1.75	0.10	13.0
1.00	0.8	1.13	2.82	1.59	0.56	1.14	0.39	1.66	0.09	13.0
1.22	0.9	1.11	3.12	1.55	0.37	1.10	0.35	1.58	0.04	13.6
1.21	1.0	1.09	3.73	1.40	0.34	1.08	0.35	1.39	0.03	12.2

[a] The soft-sphere calculations of Table VII were analyzed in terms of $g(r)$ from Eq. (82) and $W(r)$ from Eq. (84). The first maximum of $g(r)$ or $W(r)$ occurs at a radius of r_1. The first minimum of $g(r)$ or $W(r)$ occurs at radius r_2. The number of molecules within the first coordination shell of defined to be $n = 4\pi\rho\int_0^{r_2} r^2 g(r)\,dr$.

VI.   TRANSPORT COEFFICIENTS BY EQUILIBRIUM MOLECULAR DYNAMICS

A.   Self Diffusion

We now consider the application of MD to the examination of transport processes. This is a somewhat neglected area compared with the vast literature that exists on thermodynamic behavior. Perhaps the most easily evaluated, and surely the most useful, transport parameter is the self-diffusion coefficient.[68] If r_{in} is the position vector of a molecule i at time step n, and r_{im} is the absolute position vector of the same molecule i at time step m, then the mean-square displacement after $(m - n)$ steps is

$$m(t) = \frac{1}{N} \sum_{i=1}^{N} (r_{im} - r_{in})^2 \qquad (85)$$

where $t = (m - n)\Delta t$. It is important not to apply the periodic boundary conditions when evolving r_{im}. This means that we follow the same particle even if it crosses the cell boundary and becomes an image molecule. This is simple to do, because the coordinates of this image molecule are directly related to those of its "real" counterpart. The quantity $m(t)$ is calculated and

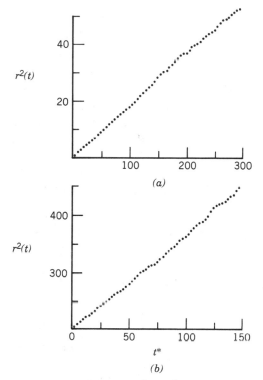

Figure 7.   The mean-square displacements, $r^2(t)/\sigma^2$, given by Eq. (85), of LJ molecules interacting via the potential $\phi(r) = 4\varepsilon[(\sigma/r)^{12} - (\sigma/r)^6]$. The reduced time t^* is in $\sigma(m/\varepsilon)^{1/2}$. (a) $\rho^* = 0.8442$, $T^* = 0.73$, $D^* = 0.030$; (b) $\rho^* = 0.6$, $T^* = 4.53$, $D = 0.56$. $N = 256$.

printed at selected intervals during an MD simulation. At sufficiently long times, the mean-square displacement becomes linear. The diffusion coefficient D is then given by a fit to the equation

$$m(t) = A(t) + B + 6Dt \qquad (86)$$

where

$$\lim_{t \to \infty} A(t) = 0 \qquad (87)$$

and B is a constant that depends on the thermodynamic state. Some examples of $m(t)$ for a LJ system at two state points were calculated for this work and are presented in Fig. 7. It is quite common for $m(t)$ to exhibit oscillations superimposed on the theoretical linear rise. These have a period of approximately 30 ps in the examples shown in the figure.

B.   Correlation Functions

Perhaps the most informative way of analyzing liquid-state dynamics is via time correlation functions,[69-71] which are used to represent the relationships between the value of a system property at one time and its value or that of another property at a later time. Because a liquid has no lasting permanent structure, the relative properties, whether in distance space (e.g., pair radial distribution function) or in the time domain, are the most sensible means of characterization. To illustrate the implementation of time correlation functions, we consider the specific example of the velocity autocorrelation function[71-73] (VACF), although the treatment is very similar for most other correlation functions.

The VACF is defined by

$$C_v(k) = \frac{1}{(N_T - k)} \sum_{n=0}^{N_T - k - 1} V^*(n)V(n+k) \tag{88}$$

where $k = 0, 1, 2, \ldots, N_T - 1$, and N_T is the number of time steps in the simulation. This is concisely expressed as

$$C_v(t) = \sum_{\alpha} \langle V_{\alpha i}(0) V_{\alpha i}(t) \rangle \tag{89}$$

where $\alpha = x$, y, z, and $t = (k-1)\Delta t$. If we treat only N_c molecules in the MD cell to generate the VACF, then

$$C_v(t) = \frac{1}{3} \left\{ \sum_{i=1}^{N_c} \left[\langle V_{xi}(0) V_{xi}(t) \rangle + \langle V_{yi}(0) V_{yi}(t) \rangle \right. \right.$$
$$\left. \left. + \langle V_{zi}(0) V_{zi}(t) \rangle \right] \right\} \tag{90}$$

$V_{\alpha i}(t)$, and consequently $C_v(t)$, are evaluated at times separated by the same time interval Δt. The VACF is evaluated at times $t = (k-1)\Delta t$, where k ranges from 1 to N_T. Here $\langle \ldots \rangle$ takes on the added significance of an average over time origins. Quite often N_c is small (~ 10) because of computer memory limitations. The velocities need to be stored for the previous N_A (≈ 200) time steps. Arrays VXO(N_c, N_A),... are needed as temporary storage in the algorithm for calculating the VACF. We advocate calculating the VACF during the MD run and *not* as a postsimulation calculation from catalogued velocities. This is mainly because any unnecessary data manipulation after a MD run often involves much time-consuming human effort!

The practical details of the recommended scheme are outlined in Table IX. In this application a time origin (i.e., $t = 0$) is started every time step. This allows the most efficient use of the velocities, which can contribute to a number of overlapping individual VACF determinations. The necessary programming falls into three distinct regions:

1.   Read in parameters associated with the VACF and also array CV containing the digitized discrete unnormalized VACF, if the calculation is to be continued from a previous subaverage simulation. Initialize the VACF accumulators if required.

2.   Enter the main body of the MD program, in which the positions are updated at time intervals. At each time step multiply the current velocity components of N_c molecules by their previous velocities going back a maximum of $(N_T - 1)$ time steps to the immediate past.

3.   At the end of the current subaverage simulation, print out the VACF derived so far. $CV(N_A)$ is divided by the number of time origins N_{TO} and by N_c.

In the presented formulation it is required that the number of time steps in the computer run, N_R, be an integer multiple of N_A. This is rarely a serious restriction, and it prevents problems associated with normalization (to achieve $C_v(0) = 1$) and the cataloguing of problematical large data sets that would need to contain the two-dimensional arrays VXO,.... It is achieved by treating certain time steps toward the end of the run in such a way that they cannot contribute to the VACF. They then are not assigned as time origins, because if they were a complete correlation function time span of N_A time steps would not be covered before the end of the run. This does not usually seriously diminish the statistical accuracy, provided the complete MD simulation is broken into as large subaverage runs as possible.

To aid understanding of the logic involved in this algorithm, consider $N_A = 5$, $N_R = 10$, and $N_T = 10$. The history of the integer accumulators generated in part 2 of the program is presented in Table X. The assignment of an arbitrarily large negative number (-3000 here) to NST(IOR) for $N_T > 6$ affects JOR so as to prevent certain time steps at the end of the run from being assigned as time origins.

As examples of the applications of this approach the VACFs and force autocorrelation functions,[28] (FACFs) for three LJ states at $T^* = 3.5$ are presented in Fig. 8. At low density the absence of frequent "collisions" produces a velocity development that changes only slowly. At high liquid densities, collisions evident in the FACFs show up also in the VACFs. A reversal

TABLE IX
FORTRAN Code for Evaluating a Velocity Autocorrelation Function

```
C       NA IS THE NUMBER OF TIME STEPS IN THE VACF
C       NC IS THE NUMBER OF  MOLECULES USED IN CALCULATING
C       THE VACF
C       NOR INDICATES HOW MANY TIME STEPS BACK THE VACF
C       ACCUMULATORS ARE ACCESSED
C       IOR IS THE ARGUMENT OF NST(..)
C       NST(IOR) HOLDS THE TIME STEP INDICATOR IT
C       FOR TIME ORIGIN IOR
C       NT IS THE NUMBER OF PREVIOUSLY ACCUMULATED TIME STEPS
C       NR IS THE NUMBER OF TIME STEPS THIS RUN
        N=108
        NR=10000
        NA=200
        NC=20
        ...      .
        ...      .
        DO 10 I=1,NA
        CV(I)=0.0
10      CONTINUE
        NTO=0
        IOR=0
        IT=1
        NOR=0
        NTOT=NT+NR
6       CONTINUE
C       MAIN PROGRAM STARTS HERE
        ...        .
        ...        .
C       CALCULATE THE VELOCITY AUTOCORRELATION FUNCTION
        NOR=NOR+1
        IF (NOR.GT.NA) NOR=NA
        IOR=IOR+1
        IF (IOR.GT.NA) IOR=1
        DO 20 I=1,NC
C       USE THE VERLET POSITION INCREMENTS AS GOOD APPROXIMATIONS TO THE VELOCITY
        DX0(I,IOR)=DX(I)
        DY0(I,IOR)=DY(I)
        DZ0(I,IOR)=DZ(I)
20      CONTINUE
        NST(IOR)=IT
        IF (IT.GT.(NR-NA+1)) NST(IOR)=-3000
        IF (IT.LE.(NR-NA+1)) NTO=NTO+1
        DO 30 IS=1,NOR
        JOR=IT+1-NST(IS)
        IF (JOR.GT.NA) GOTO 40
        DO 50 I=1,NC
        CV(JOR)=CV(JOR)+DX0(I,IS)*DX(I)+
1       DY0(I,IS)*DY(I)+DZ0(I,IS)*DZ(I)
50      CONTINUE
40      CONTINUE
30      CONTINUE
        ...      .
        ...      .
        IT=IT+1
        NT=NT+1
        IF (NT.LT.NTOT) GOTO 6
        ...      .
        ...      .
```

542

```
C       AT END OF THE RUN :
        CVS=CV(1)/(FLOAT(NTO)*FLOAT(NC)*3.0)
        DO 60 I=1,NA
        CV1=CV(I)/CV(1)
        T=FLOAT(I-1)*DT
        WRITE(6,70)I,T,CV1
70      FORMAT(I7,2F12.8)
60      CONTINUE
        STOP
        END
```

TABLE X

Typical Parameter Trends During the Evaluation of the Velocity Autocorrelation Function Outlined in Table IX[a]

NT	IT	NOR	IOR	NST(IOR)	IS	:	JOR
0	1	1	1	1	1	:	1
1	2	2	2	2	1,2	:	2,1
2	3	3	3	3	1,2,3	:	3,2,1
3	4	4	4	4	1,2,3,4	:	4,3,2,1
4	5	5	5	5	1,2,3,4,5	:	5,4,3,2,1
5	6	5	1	6	1,2,3,4,5	:	1,5,4,3,2
6	7	5	2	-3000	1,2,3,4,5	:	2,3008,5,4,3
7	8	5	3	-3000	1,2,3,4,5	:	3,3009,3009,5,4
8	9	5	4	-3000	1,2,3,4,5	:	4,3010,3010,3010,5
9	10	5	5	-3000	1,2,3,4,5	:	5,3011,3011,3011,3011

[a] The assignment of parameters is given in Table IX. NST(IOR) is set to -3000 for NT ≥ 6 in order to make JOR is too large to contribute to the correlation function.

of velocity within $t^* \approx 0.1$ (on average) reflects "hard" collisions with neighbors in the surrounding cage. The FACF (the second time derivative of the VACF) delineates these collisions more clearly.

An alternative route to correlation functions (CFs) is by fast Fourier transformation (FFT) of the individual components of the CF. Fourier transformation of integrals of the type discretized in Eq. (88) produces a simple product in frequency space. The two Fourier transforms (FTs) are those of the two components of Eq. (88), namely $V_{ai}^*(t)$ and $V_{ai}(t)$. The inverse FT of this product retrieves the CF of interest. Although this may seem

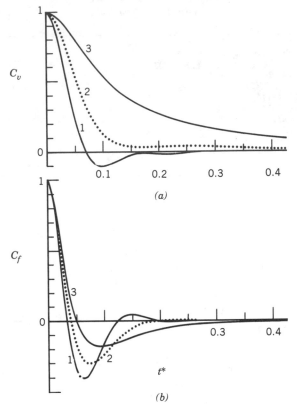

Figure 8. Single-particle time correlation functions for three LJ states, obtained from 256-molecule MD simulations. (1) $\rho^* = 1.0397$, $T^* = 3.35$; (2) $\rho^* = 0.8029$, $T^* = 3.54$; (3) $\rho^* = 0.5$, $T^* = 3.47$. All simulations were performed for 60,000 time steps each of duration $0.0025\sigma(m/\varepsilon)^{1/2}$, 59,801 time origins were established for 10 of the molecules. (*a*) Velocity autocorrelation function $C_v(t)$ [see Eq. (88)]; (*b*) force autocorrelation function $C_f(t)$.

a "roundabout" way of obtaining the CF, each step can be accomplished very efficiently, and this method turns the number of floating-point operations from one proportional to N_T^2 (in the direct method already described) into $N_T \log_2 N_T$. The theory behind this method has been reviewed by Smith.[74] The FFT procedure is summarized by the expression

$$C_v(k) = \frac{1}{(N_T - k)(2N_T)} \sum_{j=0}^{2N_T - 1} V^*(j)V(j)\exp\left(\frac{i2\pi jk}{2N_T}\right) \qquad (91)$$

where again $k = 0, 1, \ldots, N_T - 1$. Here j is the index that characterizes the frequency, which is a multiple of increments of magnitude $(N\Delta t)^{-1}$.

$$V*\left(\frac{j}{N_T\Delta t}\right) = \sum_{k=0}^{N_T-1} V*(k\Delta t)\exp\left(\frac{-i2\pi jk}{N_T}\right) \tag{92}$$

The FFT acts on the time-dependent velocity, DX(I),..., which is discrete in time (i.e., separated by time intervals of Δt). This distorts the "true" FT, which would be produced from a continuous function in time, by producing an infinite series of the "true" FT separated by frequency intervals Δt^{-1} apart. This periodic replication of the ideal time function can overlap in some applications of these formulas—an effect known as *aliasing*. However, this is not a serious problem for MD, because Δt is very small when compared with typical relaxation times. As a finite set of data points in time is Fourier transformed, ripples are introduced into the FT—a distortion called *leakage*. It is known that if the finite time window is of duration t_o, then the spurious ripples have a periodicity of frequency t_o^{-1}. The FT is also analyzed at discrete frequency intervals. It is most satisfactory to choose the sampling interval in the frequency domain such that the same number of data points are considered in the time and frequency domains.

Rarely is the VACF a major consumer of computer time when compared with that used to produce the configurations from which it is derived. Consequently, until recently there has been little motivation to adopt the FFT route to time CFs, despite the undoubted gains in execution speed over the direct method (usually by at least a factor of 20) thus achieved.[74] However, it is probable that in the near future the more adventurous study of aspects of slowly relaxing systems (e.g., spectral line shapes[75]) will provide workers with good cause to adopt the FFT method.

VII. MOLECULAR DYNAMICS OF DIFFERENT ENSEMBLES

Another fairly recent development is the generation of techniques for modeling liquids and solids under conditions more frequently encountered experimentally.[76-79] Constant temperature and pressure are usually fixed parameters in experiments, whereas until recently MD was performed under conditions of constant number N of molecules in the MD cell, constant cell volume V, and constant total energy E, where

$$E = \sum_{i=1}^{N} \frac{1}{2}m_i\dot{r}_i^2 + \frac{1}{2}\sum_{i=1}^{N}\sum_{j=1\neq i}^{N} \phi_{ij} \tag{93}$$

This is called a microcanonical or NEV ensemble.

A. Constant Temperature

A constant-temperature method was proposed by Woodcock[80] over 10 years ago and is recommended. Although other methods have been proposed since then that are more soundly linked (from a statistical-mechanical viewpoint) to a canonical ensemble, there is no evidence to suggest that they produce different results if one accounts for the noise inherent in any MD experiment.[81,82] In fact, they frequently disturb individual molecules' dynamics to a greater extent and therefore care must be exercised in implementing them when one is trying to obtain realistic dynamical information. Because the Woodcock method is the simplest approach to adopt (surely an important factor in choosing a technique), we describe it below. Briefly, the method involves calculating the temperature at each time step (something one does anyway) and using it to define a velocity, or more correctly, position-increment scaling factor. The position increments are then multiplied by this scaling factor at the next time step. Although there is a one-time-step lag between "cause and effect," in practice the average temperature obtained corresponds well to that prespecified, T_o. At very high perturbation rates some deviation from T_o is observed; for example, at shear rates of ≥ 2 in LJ reduced units, deviation values of 10% are typical,[83] in which case the Hoover method would more rigorously fix the temperature.[82] The modified Verlet leapfrog equations of motion are

$$\dot{\mathbf{r}}_i(t) = \delta\mathbf{r}_i \frac{(t - \Delta t)}{\Delta t} + \Delta t \frac{\mathbf{F}_i(t)}{2m_i} \tag{94}$$

$$T = \frac{1}{3} \sum_{i=1}^{N} \frac{m_i \dot{\mathbf{r}}_i^2(t)}{k_B N} \tag{95}$$

$$f = \left(\frac{T_o}{T} \right)^{1/2} \tag{96}$$

$$\delta\mathbf{r}_i(t + \Delta t) = \delta\mathbf{r}_i(t) {}^* f + \frac{\mathbf{F}_i(t)\Delta t^2}{m_i} \tag{97}$$

Note that f affects $\delta\mathbf{r}_i$ only—obviously an *ad hoc* procedure, but one that in practice frequently is adequate to produce the desired result of a constant prespecified average temperature. The success of this approximation hinges on the fact that on average, the force does not change the velocity too much between time steps. Table XI contains the essential FORTRAN code for performing this operation during a LJ MD simulation.

TABLE XI

Verlet *NVT* MD by Momentum Scaling: FORTRAN Code Appropriate to the
Lennard–Jones System

```
      F=1.0
      TEMP=0.722
      READ(3)RX,RY,RZ,DX,....F
C     IF IP EQUALS 0 ZERO  ACCUMULATORS
      NTOT=NT+NR
C
C     MAIN PROGRAM STARTS HERE
C
6     CONTINUE
C     ZERO FORCE COMPONENT ARRAYS
      DO 110 I=1,N
      FX(I)=0.0
      FY(I)=0.0
      FZ(I)=0.0
110   CONTINUE
      ...        .
C     EVALUATE FORCES, POTENTIAL ENERGY AND THE STRUCTURAL PART OF THE PRESSURE
      DO 120 I=1,N-1
      DO 130 J=I+1,N
      X=RX(I)-RX(J)
      Y=RY(I)-RY(J)
      Z=RZ(I)-RZ(J)
C     APPLY PERIODIC BOUNDARY CONDITIONS
      X=X-2*INT(X)
      Y=Y-2*INT(Y)
      Z=Z-2*INT(Z)
      ...        .
      ...        .
      ...        .
130   CONTINUE
120   CONTINUE
C     EVALUATE THE KINETIC ENERGY
      EK=0.0
      DO 150 I=1,N
      FX(I)=FX(I)*24.0*DT*DT
      FY(I)=FY(I)*24.0*DT*DT
      FZ(I)=FZ(I)*24.0*DT*DT
      VELX=DX(I)*F+0.5*FX(I)
      VELY=DY(I)*F+0.5*FY(I)
      VELZ=DZ(I)*F+0.5*FZ(I)
      EK=EK+0.5*(VELX*VELX+VELY*VELY+VELZ*VELZ)*CFV*CFV
C     UPDATE POSITION INCREMENTS
      DX(I)=DX(I)*F+FX(I)
      DY(I)=DY(I)*F+FY(I)
      DZ(I)=DZ(I)*F+FZ(I)
150   CONTINUE
C     CALCULATE THE TOTAL ENERGY
      F=SQRT(1.5*TEMP*FN/EK)
      DO 160 I=1,N
      RX(I)=RX(I)+DX(I)
      RY(I)=RY(I)+DY(I)
      RZ(I)=RZ(I)+DZ(I)
      RX(I)=RX(I)-2*INT(RX(I))
      RY(I)=RY(I)-2*INT(RY(I))
      RZ(I)=RZ(I)-2*INT(RZ(I))
150   CONTINUE
      NT=NT+1
      IF (NT.LT.NTOT) GOTO 6
      ...        .
      ...        .
      STOP
      END
```

B. Constant Pressure

More recently, a number of mechanical methods have been proposed for also achieving a prespecified average pressure.[81,84-86] Although in certain cases the thermodynamics can be claimed to be that of an isobaric ensemble, the dynamics are not so soundly based. Of the constant-pressure methods, only the recent method of Evans and Morriss[81] is claimed to constrain the pressure to the preset value at each time step by means of an iterative procedure performed on the volume. Usually there is a degree of slackness in both the MD cell volume and the pressure, so that both fluctuate. In fact, this pointedly reflects the problem of working stably at constant pressure. This problem occurs because, since the pressure depends on the first derivative of the potential with respect to distance for molecular pairs, it changes more rapidly with configurational adjustments than, for example, the average potential energy does. This makes pressure a particularly difficult quantity to constrain.

The constant-pressure method discussed in detail here is that proposed by Andersen.[84,87] Both volume and pressure are allowed to fluctuate, in the latter case about a fixed mean. Contrast this with the conventional constant-energy MD scheme, in which the pressure is not predetermined, but the volume is rigidly kept to a constant value. In all the "constant pressure" algorithms the cell volume is expanded if the "pressure" is greater than the desired value P_o, and contracted if the "pressure" is less than P_o. This reflects a positive bulk modulus for all liquids. The pressure at liquid densities is very sensitive to volume, and consequently the pressure is not an easy quantity with which to drive the system to preset conditions in a stable manner. Therefore an "effective" pressure is usually introduced that is constructed so as to vary less rapidly with time than does the instantaneous pressure of the system. Alternatively, the MD cell volume is forced to be less sensitive to the instantaneous pressure than it would be if the bulk modulus and instantaneous pressure determined entirely the volume change at each time step. Andersen uses the second approach to damp the MD box volume by introducing an equation of motion for V using the following governing formulas:

$$\dot{\mathbf{R}}_i(t) = \dot{\mathbf{r}}_i(t) + \frac{1}{3}\mathbf{R}_i\frac{d}{dt}\left(\ln V(t)\right) \tag{98}$$

$$\ddot{\mathbf{r}}_i(t) = -\sum_{j \neq i}\mathbf{r}_{ij}\cdot\frac{d\phi}{dr}(r_{ij}) - \frac{1}{3}\dot{\mathbf{r}}_i\frac{d}{dt}\left(\ln V(r)\right) \tag{99}$$

$$V^{-1}M_p\frac{d^2}{dt^2}V(t) = \left(P(t) - P_o\right)S(t) \tag{100}$$

where $V(t)$ is the volume of the MD box at time t and $S(t)$ is the side length, so

$$V(t) = S^3(t) \tag{101}$$

M_p is the so-called piston mass. It is useful to consider $\dot{\mathbf{R}}_i(t)$ to be the actual velocity of molecule i and $\dot{\mathbf{r}}_i(t)$ as the "thermodynamic" velocity, which is used to obtain the temperature and other thermodynamic properties of the system. According to a Verlet leapfrog algorithm the following equation of motion can be established for the box side length, as derived in detail elsewhere: [87]

$$S(t + \Delta t) = S(t) + \delta S(t) \tag{102}$$

where

$$\delta S(t) = \delta S(t - \Delta t) + \Delta t^2 S(t) \ddot{\varepsilon}(t) \tag{103}$$

with

$$\ddot{\varepsilon}(t) = \frac{S(t)}{3M_p} [P(t) - P_o] \tag{104}$$

as

$$\frac{d}{dt} [\ln V(t)] = 3\dot{\varepsilon} \tag{105}$$

where $\dot{\varepsilon} = \dot{L}/L$ and $\ddot{\varepsilon}$ is the compressional–dilatational strain acceleration along each side of the MD cell. The instantaneous pressure $P(t)$ is given at the microscopic level by

$$P(t) = \frac{1}{3V(t)} \left\{ \sum_{i=1}^{N} m_i \dot{\mathbf{r}}_i^2(t) - \frac{1}{2} \sum_{i=1}^{N} \sum_{j=1 \neq i}^{N} \frac{(\mathbf{R}_{ij} \cdot \mathbf{R}_{ij})}{R_{ij}} \phi'_{ij} \right\} + P_r + P_a \tag{106}$$

where P_r and P_a are the long-range "tail" corrections to the pressure from interactions beyond the truncation sphere.

Equations (102) and (103) can be reduced to a form suitable for computer coding as follows: Let

$$S(t + \Delta t) = S(t)(1 + \delta \varepsilon(t)) \tag{107}$$

where

$$\delta \varepsilon(t) = \frac{\delta S(t)}{S(t)} \tag{108}$$

Similarly, Eq. (98) becomes

$$\frac{\delta S(t)}{S(t)} = \frac{\delta S(t - \Delta t)}{S(t)} + \Delta t^2 \ddot{\varepsilon}(t) \tag{109}$$

so that

$$\delta \varepsilon(t) \approx \delta \varepsilon(t - \Delta t) + \Delta t^2 \ddot{\varepsilon}(t) \tag{110}$$

The difference between $\delta S(t - \Delta t)/S(t)$ and $\delta S(t - \Delta t)/S(t - \Delta t)$ is assumed to be negligible. These contractions and expansions are also performed uniformly on the contents of the MD cell. This motion is superimposed on that generated by the internal forces as follows:

$$\mathbf{R}_i(t + \Delta t) = \mathbf{R}_i(t) + \delta \mathbf{r}_i(t) + \delta \mathbf{R}_i(t) \tag{111}$$

where

$$\delta \mathbf{r}_i(t) = \delta \mathbf{r}_i(t - \Delta t) + \frac{\Delta t^2 \mathbf{F}_i(t)}{m_i} \tag{112}$$

$$\mathbf{F}_i(t) = -\sum_{j \neq i}^{N} \frac{\dfrac{\mathbf{R}_{ij}}{R_{ij}} \dfrac{d\phi}{dR_{ij}}}{m_i} \tag{113}$$

and

$$\delta \mathbf{R}_i(t) = \delta \mathbf{R}_i(t - \Delta t) + \Delta t^2 \mathbf{R}_i(t) \dot{\varepsilon}(t) \tag{114}$$

where $\mathbf{R}_i(t)$ is the so-called macroscopic coordinate of i, which changes as a result of both interparticle and externally applied forces. Equation (111) can be reduced, as for S before, to

$$\mathbf{R}_i(t + \Delta t) = \mathbf{R}_i(t)(1 + \delta \varepsilon(t)) \tag{115}$$

These equations of motion produce a system with constant enthalpy H as well as constant pressure.

Now we discuss the system of program units and some technical details of implementation. Distance is in internal units of the box side length, which equals 2 or $(N/\rho)^{1/3}\sigma$. Energy is in ε, the LJ potential-well depth. Temperature, pressure, and time are in ε/k_B, $\varepsilon\sigma^{-3}$, and $\sigma(m/\varepsilon)^{1/2}$, respectively. The Verlet scheme for implementing NEV dynamics is outlined in Table I. Two equivalent routes can be chosen to modify an NEV program to a NPH program. In the first, the molecular diameter in program units, SG, is made time dependent. This allows the box side length to remain 2 throughout the simulation. Alternatively, SG could be maintained at the value compatible with $S = 2$ and an arbitrary read-in density (in LJ reduced units). Then S would need to be time dependent. The latter approach is adopted here.

Two parts of the NEV code need to be modified. First, the variation in S needs to be built into the nearest-image convention. The search for nearest images in the forces loop and the movement of molecules from one side of the cell to the other through a real–image molecule interchange are two points of application of the periodic boundary conditions. Second, because the printed-out quantities are most usefully expressed in LJ reduced units (and not program units), these changes in S must continually alter the relevant formulas transferred from an NEV code (which assumes $S = 2$ and an SG compatible with a read-in density). The pressure has a volume term, and so do the long-range corrections to the configurational energy and virial:

$$U_r = \frac{8\pi\rho}{9r_c^9} \tag{116}$$

$$U_a = \frac{-8\pi\rho}{3r_c^3} \tag{117}$$

$$P_r = \frac{32\pi\rho^2}{9r_c^9} \tag{118}$$

$$P_a = \frac{-16\pi\rho^2}{3r_c^3} \tag{119}$$

where r_c is again the truncation separation for interactions. Because the MD cell volume fluctuates, corrections need to be made to these terms so that they respond to the changing conditions. Again, there are several ways of doing this: (1) One could keep the cutoff at the same number (e.g., 2.5) of σ. (2) Alternatively, one could let r_c change in proportion to the MD cell side length. The cutoff would remain the same fraction of $S(t)$, whatever its value. In previous studies one of us adopted method (1).[88] If ρ_o and r_{co} correspond to the read-in number density and truncation distance for interactions, then

$$U_{ro} = \frac{8\pi\rho_o}{9r_{co}^9} \tag{120}$$

$$U_{ao} = \frac{-8\pi\rho_o}{3r_{co}^3} \tag{121}$$

$$P_{ro} = \frac{32\pi\rho_o^2}{9r_{co}^9} \tag{122}$$

$$P_{ao} = \frac{-16\pi\rho_o^2}{3r_{co}^3} \tag{123}$$

and

$$U_r = \frac{U_{ro}}{S^3(t)} \tag{124}$$

$$U_a = \frac{U_{ao}}{S^3(t)} \tag{125}$$

$$P_r = \frac{P_{ro}}{S^6(t)} \tag{126}$$

$$P_a = \frac{P_{ao}}{S^6(t)} \tag{127}$$

for method (1). However, method (2) is now favored by the authors because it keeps the same number of molecules within the cutoff sphere around each particle. Consequently, spurious movements of molecules in and out of the specific interaction volume due to MD cell volume fluctuations are avoided. Hence, because,

$$r_c = r_{co} S(t) \tag{128}$$

and

$$\rho = \frac{\rho_o}{S^3(t)} \tag{129}$$

then

$$U_r = \frac{U_{ro}}{S^{12}(t)} \tag{130}$$

$$U_a = \frac{U_{ao}}{S^6(t)} \tag{131}$$

$$P_r = \frac{P_{ro}}{S^{15}(t)} \tag{132}$$

$$P_a = \frac{P_{ao}}{S^9(t)} \tag{133}$$

for method (2).

Table XII outlines the NPH scheme of method (2), which, it will be noted, is very similar to that of a NEV algorithm. A portion of a 256-LJ-atom simulation carried out under NPH conditions is shown in Fig. 9. The time

TABLE XII

Verlet *NPH* MD: FORTRAN Code Appropriate to the Lennard-Jones System

```
      S=2.0
      CONP=15.0
C     CONP=D/(3*Np)
      SG=(RHO/FLOAT(N))**(1.0/3.0)*2.0
      SG2=SG*SG
      SG2I=1.0/SG2
      SG3=SG2*SG
      CFV=1.0/(SG*DT)
      VOL=FLOAT(N)/RHO
C     THE INTERACTION TRUNCATION DISTANCE IS HALF THE MD CUBE SIDELENGTH
      CU=1.0/(2.0*SG)
      CU2=CU*CU
      PI=3.14159265359
      CPR=8.0*PI*RHO/(9.0*CU**9)
      CPA=-8.0*PI*RHO/(3.0*CU**3)
      CPRR=32.0*PI*RHO**2/(9.0*CU**9)
      CPRA=-16.0*PI*RHO**2/(3.0*CU**3)
      READ(3)RX,RY,RZ,DX,....EPSDD,EPS,S
C     IF IP EQUALS 0 ZERO  ACCUMULATORS
      IF (IP.NE.0) GOTO 99
      NT=0
      ...         .
      ...         .
      ...         .
99    CONTINUE
      NTOT=NT+NR
C
C     MAIN PROGRAM STARTS HERE
C
6     CONTINUE
C     ZERO FORCE COMPONENT ARRAYS
      DO 110 I=1,N
      FX(I)=0.0
      FY(I)=0.0
      FZ(I)=0.0
110   CONTINUE
      ...         .
      S2=0.5*S
      S22=S2*S2
      TSI=1.0/S2
C     EVALUATE FORCES, POTENTIAL ENERGY AND THE STRUCTURAL PART OF THE PRESSURE
      PS=0.0
      PES=0.0
      DO 120 I=1,N-1
      DO 130 J=I+1,N
      X=RX(I)-RX(J)
      Y=RY(I)-RY(J)
      Z=RZ(I)-RZ(J)
C     APPLY PERIODIC BOUNDARY CONDITIONS
      X=X-S*INT(TSI*X)
      Y=Y-S*INT(TSI*Y)
      Z=Z-S*INT(TSI*Z)
```

TABLE XII   (*Continued*)

```
C       EVALUATE THE SEPARATION SQUARED IN UNITS OF SIGMA SQUARED
        RR=(X*X+Y*Y+Z*Z)*SG2I
        IF (RR.GT.S22) GOTO 140
        RRI=1.0/RR
        R6I=RRI*RRI*RRI
        R12I=R6I*R6I
        FF=(R12I+R12I-R6I)*RRI
        PES=R12I-R6I
        PS=PS+RR*FF
        FX(I)=FX(I)+X*FF
        FY(I)=FY(I)+Y*FF
        FZ(I)=FZ(I)+Z*FF
        FX(J)=FX(J)-X*FF
        FY(J)=FY(J)-Y*FF
        FZ(J)=FZ(J)-Z*FF
140     CONTINUE
130     CONTINUE
120     CONTINUE
C       EVALUATE THE KINETIC ENERGY
        EK=0.0
        DO 150 I=1,N
        FX(I)=FX(I)*24.0*DT*DT
        FY(I)=FY(I)*24.0*DT*DT
        FZ(I)=FZ(I)*24.0*DT*DT
        VELX=DX(I)+0.5*FX(I)
        VELY=DY(I)+0.5*FY(I)
        VELZ=DZ(I)+0.5*FZ(I)
        EK=EK+(VELX*VELX+VELY*VELY+VELZ*VELZ)*CFV*CFV
C       UPDATE POSITION INCREMENTS
        DX(I)=DX(I)+FX(I)
        DY(I)=DY(I)+FY(I)
        DZ(I)=DZ(I)+FZ(I)
150     CONTINUE
C       CALCULATE THE TOTAL ENERGY
        TE=EK/FN+PE+CPR/S**12+CPA/S**6
        VOLI=S**3/VOL/8.0
C       CALCULATE THE PRESSURE
        P=(PS*24.0+EK)*VOLI/3.0+CPRR/S**15+CPRA/S**9
        EPSDD=-(PINP-P)*CONP*DT*DT
        EPS=EPS+EPSDD
        S=S*(1.0+EPS)
        S2=0.5*S
        TSI=1.0/S2
        DO 160 I=1,N
C       UPDATE POSITIONS AND APPLY PERIODIC BOUNDARY
C       CONDITIONS
        RX(I)=RX(I)*(1.0+EPS)+DX(I)
        RY(I)=RY(I)*(1.0+EPS)+DY(I)
        RZ(I)=RZ(I)*(1.0+EPS)+DZ(I)
        RX(I)=RX(I)-S*INT(RX(I)*TSI)
        RY(I)=RY(I)-S*INT(RY(I)*TSI)
        RZ(I)=RZ(I)-S*INT(RZ(I)*TSI)
150     CONTINUE
        NT=NT+1
        IF (NT.LT.NTOT) GOTO 6
        ...      .
        ...      .
C       AT THE END OF THE RUN:
        STOP
        END
```

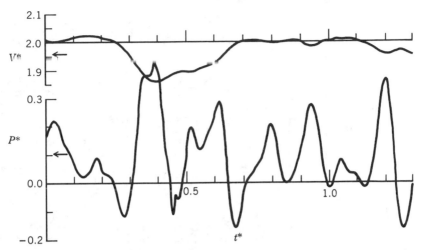

Figure 9.   Variation in reduced volume per molecule, V^*, and reduced pressure P^* for a 256-molecule LJ system run under NPH conditions using the Andersen constant-pressure algorithm as described by Eqs. (102)–(115). $\langle P^* \rangle = 0.103$, $\langle T^* \rangle = 1.28$, $\langle \rho^* \rangle = 0.510$. The reduced piston mass is 0.178. The reduced time step is $\Delta t^* = 0.005$. The averages are taken over a 600-time-step subaverage run, and the figure represents time developments over the first 260 time steps. The arrows mark the average volume in the run, $\langle V^* \rangle$ ($= 1.960$), and average pressure as above. The Daresbury Laboratory CRAY-1S computer was used.

dependence of the volume per molecule and the instantaneous pressure are shown. The value of the piston mass M_p in Eq. (100), which determines the constant CONP in the NPH program, governs the time scale of the volume fluctuations. These are less frequent (but not exceedingly more so) than the pressure fluctuations. If the piston mass is made too large, then the volume fluctuations will be on an even slower time scale, so that regions of microcanonical ensemble behavior will be observed if the simulation is short in comparison with this time scale. The greater the piston mass, then, the longer the simulation has to be conducted for to cover the possible configurations. There are advantages to this limit, however, because it means that the dynamics of the molecules (both single particle and collective particle) are likely to be more "realistic" than for smaller M_p values.

The disturbances to the system show up in the RMS fluctuations in the thermodynamic quantities. Those are shown in Table XIII for a wide range of LJ states in a simulation performed on 256 LJ atoms. The RMS total energy fluctuations per molecule differ among the ensembles NEV, NPH, and NPT by orders of magnitude. Of course, this value should be zero for the NEV ensemble, but due to algorithm and round-off errors, it is $\leq 4 \times 10^{-4}$. The NPH and NPT values are $\sim 10^{-2}$ and 10^{-1}, respectively. The dis-

TABLE XIII
Root-Mean-Square Fluctuations of Quantities During MD Simulations on Different Ensembles[a]

Number of time steps (10^3)	Ensemble	ρ^*	T^*	k'	u_r'	u_a'	u'	P_k'	P_r'	P_a'	p'	V'
18	NEV	0.64747	1.21	0.050	0.121	0.091	0.0003	0.021	0.0314	0.118	0.191	0
21.	NVT [b]	0.64747	1.20	0.009	0.127	0.094	0.054	0.004	0.328	0.121	0.221	0
32	NPH	0.6269	1.23	0.072	0.193	0.239	0.015	0.043	0.655	0.501	0.224	0.033
9	NPT	0.6421	1.21	0.085	0.234	0.351	0.177	0.042	0.895	0.784	0.227	0.051
12	NEV	0.5	1.26	0.051	0.132	0.133	0.0003	0.017	0.264	0.133	0.142	0.0
12	NPH	0.499	1.27	0.072	0.189	0.262	0.041	0.045	0.559	0.488	0.160	0.078
13.8	NPT	0.125	1.29	0.094	0.127	0.244	0.160	0.023	0.110	0.120	0.024	1.078
15	NEV	0.600	1.25	0.047	0.136	0.118	0.0003	0.0190	0.327	0.142	0.186	0
21	NPH	0.583	1.32	0.078	0.193	0.246	0.020	0.047	0.621	0.495	0.206	0.043
12	NPT	0.590	1.29	0.098	0.266	0.430	0.225	0.049	0.986	0.908	0.212	0.075
31.8	NEV	0.800	0.71	0.036	0.084	0.058	0.0004	0.019	0.268	0.093	0.167	0
27	NPH	0.8226	0.68	0.042	0.171	0.198	0.013	0.027	0.727	0.511	0.253	0.012
15.6	NPT	0.802	0.71	0.052	0.187	0.255	0.107	0.029	0.832	0.674	0.235	0.019
24	NEV	0.72	1.00	0.042	0.097	0.071	0.0003	0.020	0.279	0.103	0.172	0
18	NPH	0.72	0.98	0.056	0.177	0.210	0.012	0.034	0.672	0.488	0.236	0.019
17	NPT	0.71	1.01	0.078	0.239	0.359	0.178	0.040	1.011	0.888	0.249	0.041
66	NEV	0.8442	0.71	0.036	0.079	0.052	0.0004	0.020	0.268	0.088	0.170	0
51	NPH [c]	0.8442	0.71	0.043	0.167	0.186	0.006	0.028	0.726	0.491	0.270	0.011
51	NPH	0.8442	0.71	0.045	0.180	0.202	0.006	0.029	0.783	0.531	0.288	0.012
68	NPT	0.8387	0.72	0.055	0.201	0.265	0.114	0.031	0.926	0.726	0.286	0.017

556

TABLE XIII (Continued)

51	NPT [c]	0.8371	0.73	0.054	0.197	0.253	0.103	0.031	0.900	0.690	0.286	0.017
51	NPH	0.8466	1.14	0.067	0.217	0.217	0.021	0.045	0.941	0.572	0.414	0.012
34	NPT	0.8501	1.13	0.085	0.227	0.259	0.130	0.048	1.039	0.716	0.410	0.016
34	NPH	0.7621	1.26	0.074	0.216	0.235	0.012	0.047	0.855	0.559	0.353	0.017
68	NPΣ	0.7395	1.34	0.098	0.242	0.307	0.161	0.050	1.003	0.763	0.350	0.029
17	NPH	0.641	2.51	0.122	0.271	0.257	0.050	0.075	0.888	0.524	0.437	0.029
68	NPT	0.6372	2.55	0.189	0.294	0.327	0.236	0.087	1.026	0.701	0.448	0.042

[a] A 256-molecule LJ system was considered. $k + U_a + U_r = U$; $P_k + P_a + P_r = P$, where k, a, and r refer to kinetic, attractive, and repulsive components of the energy U and pressure P. Quantities are in LJ reduced units. Refer to ref. 85 for other details of these states. For conciseness, let $X' = \langle\langle X^2\rangle - \langle X\rangle^2\rangle^{1/2}$. $\Delta t = 0.005\,(m/\varepsilon)^{1/2}$. V is the volume per molecule, ρ^{*-1}.

[b] Isothermal method used was T_3 of ref. 87; everywhere else, T_4 was used.

[c] $M_p^* = (N/\rho^*)^{1/3}/15$; otherwise $M_p^* = (N/\rho^*)^{1/3}/45$.

557

turbances caused by the extra constraints are therefore reflected in fu
departures from NEV dynamics. Perhaps surprisingly, the RMS fluctua
in component energies per molecule are not too different among the en
bles, increasing only slightly from ~0.1 to 0.2 along this sequence.
reverse trend is observed for pressure. The total pressure RMS fluctua
increase by only ~50% along the ensemble sequence. The repulsive an
tractive components, in contrast, go up by factors of 2–4, typically.

As so often happens in the pursuit of a worthwhile aim, the study of
techniques for achieving isothermal and isobaric ensembles has becom
most an end in itself. There is an ever-growing literature on this subject.
worth noting that for small periodic systems the thermodynamics of
mally equivalent ensembles are not exactly the same.[59,87] This particu
shows up in second-order thermodynamic quantities, especially at mod
fluid densities.[87]

VIII.   TRANSPORT AND THERMODYNAMIC PROPERTIES BY NONSTANDARD MOLECULAR DYNAMICS

A.   Theory

Historically, liquid transport coefficients were first obtained by MD u
time CFs via Green–Kubo integrands.[89] However, in the last 10 years c
a formally equivalent, but in practice more efficient, method has come
prominence. This approach directly calculates the response of the molec
to the application of a perturbation.[90] The ratio of the response to the ex
nal field gives the transport coefficient of interest; for example, the s
stress divided by the applied shear rate is the shear viscosity. The princ
of this technique is that the response of some observable $A(\mathbf{k})$ at waveve
k is related to the equilibrium correlation function $\langle \ldots \rangle_o$ by[91]

$$\langle A(\mathbf{k}) \rangle_t = \frac{V}{k_B T} \int_{-\infty}^{t} dt' \langle A(\mathbf{k}, t) J(-k, t') \rangle_o \times ik\phi(\mathbf{k}, t') \qquad (1$$

where the very small perturbation $\phi(\mathbf{k}, t')$ couples to a dynamical vari
to produce a corresponding isothermal current J. The response of the
rent to a delta-function perturbation is proportional to the correlation fu
tion,

$$\langle J(\mathbf{k}) \rangle_t = \frac{V}{k_B T} \langle J(\mathbf{k}, t) J(-\mathbf{k}, 0) \rangle_o \qquad (1$$

whereas a step function gives the integral

$$\langle J(\mathbf{k}) \rangle_t = \frac{V}{k_B T} \int_0^t \langle J(\mathbf{k}, t') J(-\mathbf{k}, 0) \rangle_o dt' \qquad (1$$

In the limit of infinite ensemble averaging, integration of the delta-func response should give a result identical to the observed response to the step-function perturbation. Linear-response theory has been applied extensively to the calculation of shear viscosity. Here we concentrate on new applications for obtaining bulk and shear viscosity and thermal conductivity.

B. Bulk Viscosity

Consider the response of a system subjected to a small volume strain rate $3V\dot{\varepsilon} = \dot{V}$. The resulting pressure change is

$$\langle \delta P(t) \rangle = \frac{V}{k_B T} \int_{-\infty}^{t} \langle \delta P(t)\delta P(t') \rangle 3\dot{\varepsilon}\, dt' \tag{137}$$

This is modeled in MD by scaling all MD side lengths and the coordinates of the resident molecules by the same factor $(1 + \delta\varepsilon)$, just as in the constant-pressure scheme described in the last section. This is described in detail in Table XIV. The small pressure change resulting from the volume change is distinguished from the much larger "background" pressure fluctuations by performing two simulations of the molecules through time and space, once with and once without the perturbation. In both cases the system is allowed to evolve from the same initial configuration. The developments differ because in one an externally imposed constraint (a volume change here) is applied to the system, whereas in the other no such perturbation is present. Subtraction of properties of interest (the pressure here) of the unperturbed state from those of the perturbed state gives the desired response. The accuracy is improved by eliminating the noise produced by poor statistics and the algorithm's numerical errors. Significantly, these are present in both halves of the process to much the same extent.[92] An alternative approach is sometimes used in which these differential trajectories are produced by an expansion of virtual responses about the equilibrium system.[90] This method covers the same portion of phase space once only, unlike the two times of the difference method. The difference algorithm is favored (despite rounding errors), because it is readily adaptable to new situations and is easier to program. It need not be half as fast as the expansion method, because the latter involves more array and analytic manipulations.

We define the path through phase space, with and without the external change (usually for several hundred time steps), as a *segment*. Both parts of the segment last for the same number of time steps. Usually at least 40 segments are needed to give acceptable statistics for the near-triple-point LJ liquid, for example. The bulk viscosity is obtained from Eq. (137), taking the limit of $t \to \infty$ for the volume step-function response $\delta P(t)$ (which is the integral of the volume step function):

$$\eta_B = \lim_{t \to \infty} \frac{\langle \delta P(t) \rangle}{3\dot{\varepsilon}} \tag{138}$$

TABLE XIV
Small Step in Compression Under Isothermal Conditions Using Difference-in-Trajectories MD:
FORTRAM Code Appropriate to the Lennard-Jones System

```
C       EPS IS THE FRACTIONAL CHANGE IN MD CELL VOLUME.
C       ISEG IS THE NUMBER OF TIME STEPS IN THE SEGMENT.
C       NSEG IS THE NUMBER OF SEGMENTS THIS RUN.
C       NSEGT IS THE NUMBER OF SEGMENTS FOR ALL RUNS ACCUMULATED.
C       DPRESS CONTAINS THE PRESSURES AT TIMES IN THE SEGMENT, SEPARATED INTO
C       THOSE FROM THE TWO VOLUME CONSTRAINTS.
        TEMP=0.722
        EPS0=-0.00001
        ISEG=200
        NSEG=40
        NSEGT=0
        SG=(RHO/FLOAT(N))**(1.0/3.0)*2.0
        SG2=SG*SG
        SG2I=1.0/SG2
        SG3=SG2*SG
        CFV=1.0/(SG*DT)
        VOL=FLOAT(N)/RHO
C       THE INTERACTION TRUNCATION DISTANCE IS HALF THE MD CUBE SIDELENGTH
        CU=1.0/(2.0*SG)
        PI=3.14159265359
        CPR=8.0*PI*RHO/(9.0*CU**9)
        CPA=-8.0*PI*RHO/(3.0*CU**3)
        CPRR=32.0*PI*RHO**2/(9.0*CU**9)
        CPRA=-16.0*PI*RHO**2/(3.0*CU**3)
        READ(3)RX,RY,RZ,DX,....NSEGT,DPRESS,EPS,S,F
C       IF IP EQUALS THEN 0 ZERO ACCUMULATORS
        IF (IP.NE.0) GOTO 199
        NT=0
        DO 99 K=1,2
        DO 99 I=1,ISEG
        DPRESS(K,I)=0.0
99      CONTINUE
        NSEGT=0
199     CONTINUE
        NTOT=NT+NR
        NSEGT=NSEGT+NSEG
C
C       MAIN PROGRAM STARTS HERE
C
6       CONTINUE
        DO 10 IS1=1,NSEG
        DO 11 JSG=1,2
        DO 12 I=1,N
        IF (JSG.EQ.2) GOTO 13
C       PERFORM THE PERTURBED PART OF THE SEGMENT PAIR
C       STORE THE POSITIONS AND POSITION INCREMENTS FROM THE LAST
C       EQUILIBRIUM RUN
        RXP(I)=RX(I)
        RYP(I)=RY(I)
        RZP(I)=RZ(I)
        DXP(I)=DX(I)
        DYP(I)=DY(I)
```

TABLE XIV   (*Continued*)

```
           DZP(I)=DZ(I)
           GOTO 14
13         CONTINUE
C          PERFORM THE EQUILIBRIUM RUN PART OF THE SEGMENT
           RX(I)=RXP(I)
           RY(I)=RYP(I)
           RZ(I)=RZP(I)
           DX(I)=DXP(I)
           DY(I)=DYP(I)
           DZ(I)=DZP(I)
14         CONTINUE
12         CONTINUE
           EPS=0.0
           S=2.0
           SI=1.0/S
           TSI=2.0*SI
           IF (JSG.EQ.1) EPS=EPS0
           DO 15 IT=1,ISEG
C          ONLY THE FIRST TIME STEP IN THE PERTURBED HALF OF THE SEGMENT
C          IS COMPRESSED
           IF (IT.GT.1) EPS=0.0
C          ZERO FORCE COMPONENT ARRAYS
           DO 110 I=1,N
           FX(I)=0.0
           FY(I)=0.0
           FZ(I)=0.0
110        CONTINUE
           ...           .
           S2=0.5*S
           S22=S2*S2
           TSI=1.0/S2
C          EVALUATE FORCES, POTENTIAL ENERGY AND THE STRUCTURAL PART OF THE PRESSURE
           PS=0.0
           PES=0.0
           DO 120 I=1,N-1
           DO 130 J=I+1,N
           X=RX(I)-RX(J)
           Y=RY(I)-RY(J)
           Z=RZ(I)-RZ(J)
C          APPLY PERIODIC BOUNDARY CONDITIONS
           X=X-S*INT(TSI*X)
           Y=Y-S*INT(TSI*Y)
           Z=Z-S*INT(TSI*Z)
C          EVALUATE THE SEPARATION SQUARED IN UNITS OF SIGMA SQUARED
           RR=(X*X+Y*Y+Z*Z)*SG2I
           IF (RR.GT.S22) GOTO 140
           RRI=1.0/RR
           R6I=RRI*RRI*RRI
           R12I=R6I*R6I
           FF=(R12I+R12I-R6I)*RRI
           PES=R12I-R6I
           PS=PS+RR*FF
           FX(I)=FX(I)+X*FF
           FY(I)=FY(I)+Y*FF
           FZ(I)=FZ(I)+Z*FF
           FX(J)=FX(J)-X*FF
           FY(J)=FY(J)-Y*FF
           FZ(J)=FZ(J)-Z*FF
140        CONTINUE
130        CONTINUE
```

561

TABLE XIV (*Continued*)

```
120     CONTINUE
        S0=S
        S=S*(1.0+EPS)
        S2=S*0.5
        SI=1.0/S
C       EVALUATE THE KINETIC ENERGY
        EK=0.0
        DO 150 I=1,N
        FX(I)=FX(I)*24.0*DT*DT
        FY(I)=FY(I)*24.0*DT*DT
        FZ(I)=FZ(I)*24.0*DT*DT
        VELX=DX(I)*F+0.5*FX(I)
        VELY=DY(I)*F+0.5*FY(I)
        VELZ=DZ(I)*F+0.5*FZ(I)
        EK=EK+0.5*(VELX*VELX+VELY*VELY+VELZ*VELZ)*CFV*CFV
C       UPDATE POSITION INCREMENTS
        DX(I)=DX(I)*F+FX(I)
        DY(I)=DY(I)*F+FY(I)
        DZ(I)=DZ(I)*F+FZ(I)
150     CONTINUE
C       CALCULATE THE TOTAL ENERGY
        TE=EK/FN+PE+CPR/S0**12+CPA/S0**6
C       CALCULATE THE VELOCITY SCALING FACTOR
        F=SQRT(1.5*TEMP*FN/EK)
        VOLI=S0**3/VOL
C       CALCULATE THE PRESSURE
        P=(PS*24.0+EK)*VOLI/3.0+CPRR/S0**15+CPRA/S0**9
        DPRESS(JSG,IT)=DPRESS(JSG,IT)+P
        DO 160 I=1,N
C       UPDATE POSITIONS AND APPLY PERIODIC BOUNDARY
C       CONDITIONS
        RX(I)=RX(I)*(1.0+EPS)+DX(I)
        RY(I)=RY(I)*(1.0+EPS)+DY(I)
        RZ(I)=RZ(I)*(1.0+EPS)+DZ(I)
        RX(I)=RX(I)-S*INT(RX(I)*TSI)
        RY(I)=RY(I)-S*INT(RY(I)*TSI)
        RZ(I)=RZ(I)-S*INT(RZ(I)*TSI)
160     CONTINUE
11      CONTINUE
10      CONTINUE
        IF (NT.LT.NTOT) GOTO 6
        ...           .
        ...           .
C       AT THE END OF THE RUN:
        FNS=FLOAT(NSEGT)
        DPS=0.0
        DO 200 I=1,ISEG
        T=DT*FLOAT(I-1)
C       DP IS THE DIFFERENCE IN PRESSURE BETWEEN PERTURBED AND
C       UNPERTURBED RUNS
        DP=(DPRESS(1,I)-DPRESS(2,I))/FNS
        WRITE(6,210)I,T,DP
210     FORMAT(I6,1X,F12.5,2E14.6)
        STOP
        END
```

562

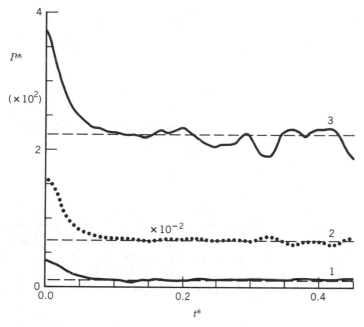

Figure 10.   Isothermal pressure responses to a uniform contraction of the MD cell and molecular coordinates by $(1 - \delta\varepsilon)$ applied to 256 LJ molecules at the states (1) $\rho^* = 0.5$, $T^* = 3.50$, $\delta\varepsilon = 0.00006$; (2) $\rho^* = 0.8029$, $T^* = 3.50$, $\delta\varepsilon = 0.006$; and (3) $\rho^* = 1.0397$, $T^* = 3.50$, $\delta\varepsilon = 0.00006$. All runs were for 180 segments. The following isothermal coefficients are derived from these curves (K_∞, K_0, η_B, respectively): (1) 22.86, 5.83, 0.48; (2) 85.91, 38.33, 1.34; and (3) 204.67, 116.67, 2.46.

Examples of the appropriate relaxation function obtained using a delta-function strain rate perturbation are given in Fig. 10 for a range of LJ densities between $\rho^* = 0.5$ and $\rho^* = 1.04$ at $T^* = 3.5$. The area between the pressure-decay curve and the dashed line in the figure is proportional to the bulk viscosity for each state. Sometimes this can present a problem, because the asymptotic value of the delta-function response is not easy to locate, yet has a large impact of η_B.

C.   Shear Viscosity

In contrast to the bulk viscosity, the shear viscosity can be obtained by means of an imposed shear-velocity profile[93-98] or by means of asymmetric distortions of the MD cell.[41] The logic behind this latter method is that simultaneous longitudinal strain rates in the x- and z-directions produced by imposing very small velocity profiles, dV_x/dx and dV_z/dz, respectively, are given by

$$- P_{xx} = - P + \left(\eta_B + \tfrac{4}{3}\eta \right) \frac{dV_x}{dx} + \left(\eta_B - \tfrac{2}{3}\eta \right) \frac{dV_z}{dz} \tag{139}$$

$$- P_{zz} = - P + \left(\eta_B + \tfrac{4}{3}\eta \right) \frac{dV_z}{dz} + \left(\eta_B - \tfrac{2}{3}\eta \right) \frac{dV_x}{dx} \tag{140}$$

where η_B is the bulk viscosity. Now if $dV_x/dx = - dV_z/dz$, then

$$\eta = \frac{P_{zz} - P_{xx}}{4(dV_x/dx)} \tag{141}$$

In practice this can be implemented as follows. All coordinates in the x-direction are scaled by the factor $(1 + \delta\varepsilon)$ and all coordinates in the z-direction by $(1 - \delta\varepsilon)$. An example of a shear-stress response measured by this route is given in Fig. 11. Here the above equations were modified to predict the response to a step in shear strain (not strain rate). A typical strain of $\delta\varepsilon \approx 10^{-6}$ was applied to a LJ state at $\rho^* = 0.4$ and $T^* = 2.47$. The time decay of the difference in normal pressure components,

$$C(t) = \frac{P_{zz}(t) - P_{xx}(t)}{4\delta\varepsilon} \tag{142}$$

leads to η via

$$\eta = \int_0^t C(t') \, dt' \tag{143}$$
$$\scriptstyle t \to \infty$$

The rapid decay of $C(t)$ within $t^* = 0.2$ (or 0.43 ps for argon) is disturbed at later times ($t^* > 0.6$) by the onset of statistical noise. The generating curves of $\delta P_{\alpha\alpha}(t)$, which are also shown in this figure, are useful because they provide extra information about the origins of potential long time tails in $C(t)$. The inherent fluctuations in this case for $t^* \gtrsim 0.6$ are probably the result of fortuitous additions of component noise in $\delta P_{\alpha\alpha}(t)$. The departures from zero in $\delta P_{yy}(t)$ reflect the viscoelastic nature of a fluid, which is not accounted for in Eqs. (139) and (140). Figure 12 gives a similar set of curves for a soft-sphere state at $\rho^* = 0.4$ and $T^* = 0.968$. Note the smoother profiles in this case, which result from the shorter interaction range of this potential. Although these response functions were evaluated for less than 1 reduced time unit, it would not be practical to continue for a longer time because of an unfavorable signal-to-noise ratio. This can pose an insurmountable problem at high densities, where these functions do not decay appreciably within this

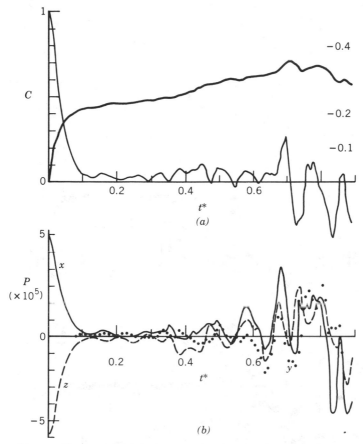

Figure 11.   (a) The normalized shear-stress relaxation function $C(t)$, from Eq. (142), and time-dependent effective viscosity η^* (right-hand ordinate), from Eq. (143), calculated by MD using 256 LJ molecules at the state $\rho^* = 0.4$ and $T^* = 2.47$. The derived quantities are $G_\infty^* = 6.41$ and $\eta = 0.31$. (b) The pressure relaxations used to calculate $C(t)$: ——, $\delta P_{xx}^*(t)$; ----, $\delta P_{zz}^*(t)$; and $\cdots$, $\delta P_{yy}(t)$. The dimensions of the MD side were contracted by a factor of 0.999994 in the x-direction and expanded by 1.000006 in the z-direction.

time. The soft-sphere state $\rho^* = 1.0$ and $T^* = 1.22$ shows this behavior forcefully (Fig. 13). The integrated $C(t)$ is a linear curve that reflects an elastic rather than a viscous overall response within the measured time scale. Just what density it is advisable not to exceed to obtain reasonable viscosity estimates is illustrated in Fig. 14, which shows several integrated $C(t)$ curves for soft-sphere fluids in the density range 0.5–0.9 at $T^* = 1.0$. It is obvious that the boundary line for this fluid is near $\rho^* = 0.7$.

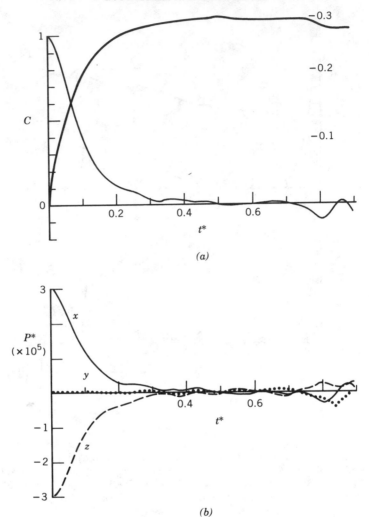

Figure 12. As for Fig. 11, except that a soft-sphere system at $\rho^* = 0.4$ and $T^* = 0.968$ was considered. $G_\infty^* = 2.94$; $\eta = 0.27$.

D. Second-Order Properties

Nonstandard MD techniques are not restricted to use for the evaluation of transport coefficients, but also can be used to obtain second-order thermodynamic quantities. Obtaining such quantities can be time consuming with other, more conventional methods (e.g., from the fluctuations in first-

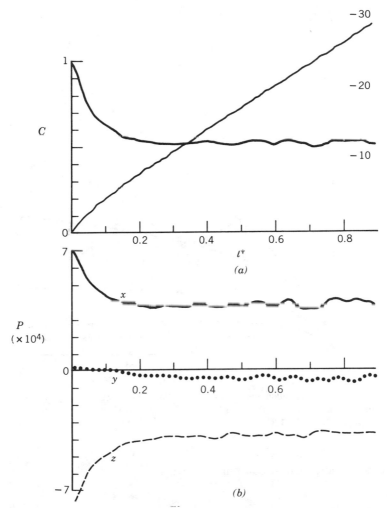

Figure 13.   As for Fig. 12, except that $\rho^* = 1.0$ and $T^* = 1.22$. $G_\infty^* = 62.68$; $\eta > 31$.

order quantities or from derivatives of MD-derived equations of state[87]). As an illustration of this capability we consider the response of a constant-volume MD cell to a temperature pulse. All the molecules in the MD cell have their velocities multiplied by a factor ε for one time step only. There is an initial rise of temperature of $\sim 2\varepsilon - 1$.[92] Some of this kinetic energy is used in changing the liquid structure, so that there is a short-lived descent in kinetic energy to a new stable value. The ratio of these two kinetic energies

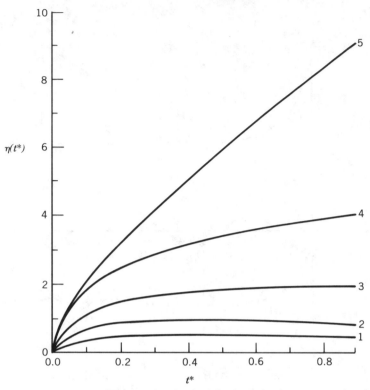

Figure 14.   Time-dependent viscosities for the following soft-sphere states (values for curves 1–5, respectively): $T = 0.969, 1.0, 0.97, 1.00, 1.22$; $\rho = 0.5, 0.6, 0.7, 0.8, 0.9$; $\eta = 0.49, 0.94, 2.11, 4.75$; $N_{\text{seg.}} = 180, 180, 180, 180, 180$.

gives the specific heat at constant volume, C_v:

$$C_v = \frac{\delta E(t \to \infty)}{\delta T(t \to \infty)}$$

$$= 1.5 \frac{\delta T(t = 0)}{\delta T(t \to \infty)} \tag{144}$$

Similarly,

$$\left(\frac{\delta P}{\delta T} \right)_v = \frac{\delta P(t \to \infty)}{\delta T(t \to \infty)} \tag{145}$$

Figure 15 gives the kinetic-energy and pressure relaxations for three LJ states

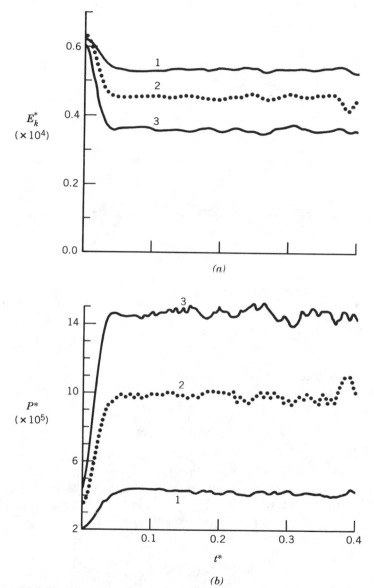

Figure 15.   Time-dependent effects of a temperature spike applied to 256 LJ molecules. The velocities were multipled by $1 + \chi$ at $t^* = 0$; $\chi = 6 \times 10^{-6}$. The states considered were (1) $\rho^* = 0.5$, $T^* = 3.46$; (2) $\rho^* = 0.8029$, $T^* = 3.54$; and (3) $\rho^* = 1.0397$, $T^* = 3.35$. Each calculation was run for 180 segments. (a) The average kinetic energy per molecule, E_k^*. (b) The pressure change, P^*. Equations (144) and (145) give C_v, $(\partial P/\partial T)_v$, respectively: (1) 1.76, 1.18; (2) 2.11, 3.30; and (3) 2.54, 6.25.

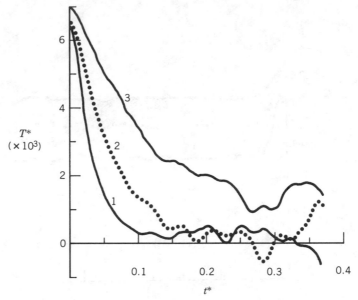

Figure 16. Temperature relaxation of one heated molecule in an MD cell of 256 LJ molecules. The single heated molecule is heated by scaling at $t^* = 0$ all velocity components by the factor $1 + \varepsilon$, where $\varepsilon = 10^{-3}$. (1) $\rho^* = 1.0397$, $T^* = 3.31$; (2) $\rho^* = 0.8029$, $T^* = 3.49$; (3) $\rho^* = 0.5$, $T^* = 3.54$. Calculations were done for 300 segments. The reduced thermal conductivities derived from these curves are 2.20, 5.08, and 11.55 $k_B(m/\varepsilon)^{-1/2}\sigma^{-2}$, respectively.

at $T^* = 3.5$. Because thermal equilibration is fast, there are no problems with long time tails. Excellent values for C_v and $(\delta P/\delta T)_v$ can be obtained after 100 such perturbations.

E. Thermal Conductivity

If just one of the molecules in the MD cell is heated by $\sim \chi$ above its surroundings, then the subsequent temperature decay of that particle is a direct measure of the thermal conductivity.[99] Three such temperature-decay curves for the LJ states at $T^* = 3.5$ are given in Fig. 16. Long time tails reappear in these relaxations. A simple model[92] gives the thermal conductivity as $\lambda = C_v/2\pi\sigma\tau$, where

$$T(t) = T(0)\exp\left(\frac{-t}{\tau}\right) \tag{146}$$

F. Self-Diffusion Coefficient

The self-diffusion coefficient D can be obtained from the direct response of a single molecule to an imposed force, F_{imp}^x. The resulting terminal x

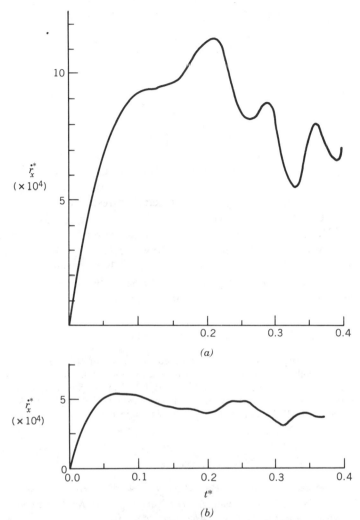

Figure 17. Plots of time-dependent x-direction velocity, $\dot{r}_x^*$, versus time in LJ reduced units for one molecule of a 256-molecule LJ MD cell. This velocity is induced by the application of a small force, $F_x^* = 2.4 \times 10^{-3} \times (N/\rho^*)^{1/3}$, to this arbitrarily chosen molecule. (a) $\rho^* = 0.8029$, $T^* = 3.64$, $D^* = 0.31$. (b) $\rho^* = 1.0397$, $T^* = 3.34$, $D^* = 0.083$. The diffusion coefficients were obtained using Eq. (148). Each calculation was run for 300 segments, each consisting of 150 time steps of $\Delta t^* = 0.0025$. The arrows denote the limiting velocities at which D^* was evaluated.

velocity $\dot{r}_t^x$ at long times directly gives the friction coefficient ξ:

$$\xi = \frac{F_{imp.}^x}{\dot{r}_t^x} \tag{147}$$

From this, we obtain

$$D = \frac{k_B T}{\xi} \tag{148}$$

using the Einstein expression.[41] Some examples of $\dot{r}_t^x$ are given in Fig. 17. The "overshoot" evident at $t^* \approx 0.1$ is a reflection of the negative region in the VACF, which $\dot{r}_x^*$ is simply an integrated manifestation of.

We see NEMD being applied to many practical problems associated with major rapid distortions of materials. There has recently been work on the modeling of shock waves[100] and plastic flow in crystals.[101] Equilibrium MD has also been applied to dislocations,[102] grain boundaries,[103] and super-cooled liquids.[104,105] Obviously, MD is increasingly being used to understand the molecular bases of many material properties.

Acknowledgments

In preparing this review we drew heavily on experience gained through association with many collaborators and colleagues. We take great pleasure in acknowledging those who have been so generous with their ideas and assistance: K. F. Carter, J. H. R. Clarke, N. Corbin, S. McQueen, N. Quirke, B. J. Ralston, K. Singer, W. Smith, S. Thompson, D. J. Tildesley, M. Whittle, and L. V. Woodcock. Mrs. K. Hales is thanked for typing the manuscript.

References

1. J. A. Barker and D. Henderson, *Rev. Mod. Phys.*, **48**, 587 (1976).

2. J. P. Hansen and I. R. McDonald, *Theory of Simple Liquids*, Academic, London, 1976.

3. W. E. Alley and B. J. Alder, *Phys. Rev. A*, **27**, 3158 (1983).

4. A. Rahman, *Rev. Nuovo Cimento I*, **1**, 315 (1969).

5. I. R. McDonald and L. V. Woodcock, *J. Phys. C, Solid State Phys.*, **3**, 722 (1970).

6. N. Quirke, J. W. Perram, and G. Jacucci, *Mol. Phys.*, **39**, 1311 (1980).

7. D. M. Heyes, C. J. Montrose, and T. A. Litovitz, *J. Chem. Soc. Faraday Trans. II*, **79**, 611 (1983).

8. A. M. Stoneham, *Handbook of Interatomic Potentials. I. Ionic Crystals* (AERE-R-9598), AERE, Harwell, U.K., 1981.

9. D. G. Bounds and A. Hinchcliffe, *Chem. Phys. Lett.*, **86**, 1 (1982).

10. M. K. Memon, R. W. Hockney, and S. K. Mitra, *J. Comp. Phys.*, **43**, 345 (1981).

11. W. F. Van Gunsteren, *Mol. Phys.*, **40**, 1015 (1980).

12. O. Edholm, H. J. C. Berendsen, and P. Van der Ploeg, *Mol. Phys.*, **48**, 379 (1983).

13. J. P. Ryckaert and G. Ciccotti, *J. Chem. Phys.*, **78**, 7368 (1983).

14. L. R. Pratt and S. W. Haan, *J. Chem. Phys.*, **74**, 1873 (1981).

15. L. R. Pratt and S. W. Haan, *J. Chem. Phys.*, **74**, 1864 (1981).

16. S. M. Thompson, K. E. Gubbins, D. E. Sullivan, and C. G. Gray, *Mol. Phys.*, **51**, 21 (1984).

17. D. J. Evans, *Phys. Lett. A*, **88**, 48 (1982).

18. S. W. De Leeuw and J. W. Perram, *Physica A*, **107**, 179, (1981).

19. M. Dixon, *Philos. Mag. B*, **47**, 531 (1983).

20. M. Dixon, *Philos. Mag. B*, **47**, 509 (1983).

21. D. J. Adams, *J. Chem. Phys.*, **78**, 2585 (1983).

22. S. H. Garofalini, *J. Non-Cryst. Solids*, **55**, 451 (1983).

23. D. M. Heyes, *Phys. Rev. B* **30**, 2182 (1984).

24. D. M. Heyes, *J. Chem. Phys.*, **74**, 1924 (1981).

25. D. M. Heyes, *Surf. Sci.*, **110**, L619 (1981).

26. G. N. Patey, D. Levesque, and J. J. Weis, *Mol. Phys.*, **45**, 733 (1982).

27. A. Giro, J. M. Gonzalez, J. A. Padro, and V. Torra, *An. Fis. Ser. A*, **77**, 57 (1981).

28. D. Fincham and D. M. Heyes, *Chem. Phys.*, **78**, 425 (1983).

29. L. Verlet, *Phys. Rev.*, **159**, 98 (1967).

30. C. W. Gear, *Numerical Initial Value Problems in Ordinary Differential Equations*, Prentice-Hall, Englewood Cliffs, New Jersey, 1971.

31. K. Singer, J. V. L. Singer, and A. J. Taylor, *Mol. Phys.*, **37**, 1239 (1979).

32. R. M. Lynden-Bell and I. R. McDonald, *Chem. Phys. Lett.*, **89**, 105 (1982).

33. C. S. Murthy, K. Singer, and R. Vallauri, *Mol. Phys.*, **49**, 803 (1983).

34. W. F. Van Gunsteren, H. J. C. Berendsen, and J. A. C. Rullmann, *Mol. Phys.*, **44**, 69 (1981).

35. D. Fincham, *Information Quarterly for MD & MC Simulations*, No. 10, Daresbury Laboratory, U.K., 1983, p. 43.

36. D. Fincham, *Information Quarterly for MD & MC Simulations*, No. 2, Daresbury Laboratory, U.K., 1981, p. 6.

37. A. Nordsieck, *Math. Comp.*, **16**, 22 (1962).

38. S. Toxvaerd, *J. Comp. Phys.*, **52**, 214 (1983).

39. S. Toxvaerd, *Phys. Rev. Lett.*, **51**, 1971 (1983).

40. D. M. Heyes and K. Singer, *Information Quarterly for MD & MC Simulations*, No. 6, Daresbury Laboratory, U.K., 1982, p. 11

41. D. M. Heyes, *J. Chem. Soc., Faraday Trans. II*, **79**, 1741 (1983).

42. D. M. Heyes, *Information Quarterly for MD & MC Simulations*, No. 9, Daresbury Laboratory, U.K., 1983, p. 32.

43. D. M. Heyes, *Information Quarterly for MD & MC Simulations*, No. 9, Daresbury Laboratory, U.K., 1983, p. 35.

44. N. Corbin and D. Fincham, *Information Quarterly for MD & MC Simulations*, No. 2, Daresbury Laboratory, U.K., 1981, p. 3.

45. D. J. Adams, *Information Quarterly for MD & MC Simulations*, No. 10, Daresbury Laboratory, U.K., 1983, p. 30.

46. W. Smith, *Information Quarterly for MD & MC Simulations*, No. 10, Daresbury Laboratory, U.K., 1983, p. 37.

47. D. Fincham and B. J. Ralston, *Comp. Phys. Commun.*, **23**, 127 (1981).

48. D. M. Ceperley, in *Supercomputers in Chemistry*, *ACS Symposium Series 173*, P. Lykos and I. Shavitt, eds., American Chemical Society, Washington, D.C., 1981, p. 125.

49. S. McQueen, unpublished work.

50. R. W. Hockney and J. W. Eastwood, *Computer Simulation Using Particles*, McGraw-Hill, New York, 1981, p. 278.

51. J. N. Cape and L. V. Woodcock, *J. Chem. Phys.*, **72**, 976 (1980).

52. G. R. Luckhurst and P. Simpson, *Chem. Phys. Lett.*, **95**, 149 (1983).

53. T. J. Sluckin, *Mol. Phys.*, **49**, 221 (1983).

54. J. R. Ray, *J. Chem. Phys.*, **79**, 5128 (1983).

55. M. P. D'Evelyn and S. A. Rice, *J. Chem. Phys.*, **78**, 5081 (1983).

56. S. H. Garofalini, *J. Chem. Phys.*, **78**, 2069 (1983).

57. F. F. Abraham, *Phys. Rev. Lett.*, **50**, 978 (1983).

58. P. S. Y. Cheung, *Mol. Phys.*, **33**, 519 (1977).

59. D. C. Wallace and G. K. Straub, *Phys. Rev. A*, **27**, 2201 (1983).

60. J. H. R. Clarke and J. Bruining, *Chem. Phys. Lett.*, **80**, 42 (1981).

61. K. S. Shing and K. E. Gubbins, *Mol. Phys.*, **46**, 1109 (1982).

62. J. G. Powles, *Chem. Phys. Lett.*, **86**, 335 (1982).

63. M. Fixman, *J. Chem. Phys.*, **78**, 4223 (1983).

64. P. L. Meyer, *Introductory Probability and Statistical Applications*, Addison-Wesley, Reading, Massachusetts, 1965, Theorem 13.1.

65. D. Fincham, N. Quirke, and D. J. Tildesley, to be published.

66. H. J. M. Hanley and D. J. Evans, *Mol. Phys.*, **39**, 1039 (1980).

67. J. G. Powles and W. Smith, *Information Quarterly for Computer Simulation of Condensed Phases*, No. 11, Daresbury Laboratory, U.K., 1984, p. 39.

68. R. J. Bearman and D. L. Jolly, *Mol. Phys.*, **44**, 665 (1981).

69. T. Tsang and W. D. Jenkins, *Mol. Phys.*, **41**, 797 (1980).

70. J. G. Powles, *Mol. Phys.*, **48**, 1083 (1983).

71. G. S. Grest, S. R. Nagel, A. Rahman, and T. A. Witten, Jr., *J. Chem. Phys.*, **74**, 3532 (1981).

72. J. J. Erpenbeck and W. W. Wood, *Phys. Rev. A*, **26**, 1648 (1982).

73. T. Gaskell, *Phys. Lett. A*, **90**, 51 (1982).

74. W. Smith, *Information Quarterly for MD & MC Simulations*, No. 7, Daresbury Laboratory, U.K., 1982, p. 12.

75. B. M. Ladanyi, *J. Chem. Phys.*, **78**, 2189 (1983).

76. F. F. Abraham, *J. Chem. Phys.*, **72**, 359 (1980).

77. S. Toxvaerd, *Phys. Rev. B*, **29**, 2821 (1984).

78. S. Nose and M. L. Klein, *J. Chem. Phys.*, **78**, 6928 (1983).

79. S. K. Mitra, *Philos. Mag. B*, **47**, L63 (1983).

80. L. V. Woodcock, *Chem. Phys. Lett.*, **10**, 257 (1971).

81. D. J. Evans and G. P. Morriss, *Chem. Phys.*, **77**, 63 (1983).

82. H. Tanaka, K. Nakanishi, and N. Watanabe, *J. Chem. Phys.*, **78**, 2626 (1983); D. Brown and J. H. R. Clarke, *Information Quarterly for Computer Simulation of Condensed Phases*, No. 11, Daresbury Laboratory, U.K., 1984, p. 11.

83. D. M. Heyes, unpublished results.

84. H. C. Andersen, *J. Chem. Phys.*, **72**, 2384 (1980).

85. J. M. Haile and H. W. Graben, *J. Chem. Phys.*, **73**, 2412 (1980).

86. D. Brown and J. H. R. Clarke, *Mol. Phys.*, **51**, 1243 (1984).

87. D. M. Heyes, *Chem. Phys.*, **82**, 285 (1983)

88. B. J. Alder and T. E. Wainwright, *J. Chem. Phys.*, **33**, 1439 (1960).

89. J. J. Erpenbeck and W. W. Wood, *J. Stat. Phys.*, **24**, 455 (1981).

90. G. Jacucci, *Physica A*, **118**, 157 (1983).

91. M. Whittle, *Information Quarterly for MD & MC Simulations*, No. 5, Daresbury Laboratory, U.K. 1982, p. 14.

92. D. M. Heyes, *J. Chem. Soc., Faraday Trans. II.* **80**, 1363 (1984).

93. M. P. Allen and D. Kivelson, *Mol. Phys.*, **44**, 945 (1981).

94. B. C. Eu, *Phys. Lett. A*, **96**, 29 (1983).

95. R. Bansal and W. Bruns, *J. Chem. Phys.*, **80**, 872 (1984).

96. J. J. Erpenbeck, *Physica A*, **118**, 144 (1983).

97. D. J. Evans, *Physica A*, **118**, 51 (1983).

98. B. L. Holian and D. J. Evans, *J. Chem. Phys.*, **78**, 5147 (1983).

99. M. J. Gillan and M. Dixon, *J. Phys. C*, **16**, 869 (1983).

100. D. H. Tsai and S. F. Trevino, *Phys. Rev. A*, **24**, 2743 (1981).

101. A. J. C. Ladd and W. G. Hoover, *Phys. Rev. B*, **28**, 1756 (1983).

102. A. J. C. Ladd and W. G. Hoover, *Phys. Rev. B*, **26**, 5469 (1982).

103. G. Ciccotti, M. Guillope, and V. Pontikis, *Phys. Rev. B*, **27**, 5576 (1983).

104. G. S. Grest, S. R. Nagel, and A. Rahman, *J. Phys. Colloq.*, **41**, No. C-8/293 (1980).

105. C. A. Angell and L. M. Torell, *J. Chem. Phys.*, **78**, 937 (1983).

NONADIABATIC SCATTERING PROBLEMS IN LIQUID-STATE VIBRATIONAL RELAXATION

MICHAEL F. HERMAN AND EDWARD KLUK[‡]

Department of Chemistry, Tulane University, New Orleans, Louisiana 70118

CONTENTS

I. INTRODUCTION

The great majority of condensed-phase work employs a classical description of the relevant motions.[1] This is particularly true of computer simulations. The use of a classical description for the translational and rotational motions of most molecules seems justified, at least by the apparent agreement between classical theories and experiment for many liquid-phase quantities. Quantum corrections to some equilibrium properties have been evaluated and found to be very small for most systems and temperatures of interest, as expected.[2]

On the other hand, the quantum nature of vibrational and electronic motion is important and must generally be accounted for. In many instances it

---

[‡]Present address: Division of Natural Science, Dickinson State College, Dickinson, North Dakota 58601.

is reasonable to assume that only the ground electronic and vibrational states are populated and essentially to "freeze" these motions out of the theory. The only reference to electronic motions is some model (usually pairwise) interaction potential between the atoms and molecules. The vibrational motion may be frozen out completely, with all bonds and bond angles treated as rigid; or these parameters may be treated classically; or only low-frequency bending motions may be treated classically while the high-frequency stretches are held rigidly fixed.

However, if one is interested in some property directly related to transitions between electronic or vibrational states, these options are not acceptable. New procedures need to be developed to account directly for electronic and vibrational relaxation in condensed media. In this article we will specifically have vibrational population relaxation in mind, although the formalism we will use here is also applicable to electronic problems. If such procedures can be developed and applied, it should be possible to study these processes in greater detail than has been possible heretofore. Questions concerning the efficiency of various types of solvent motions for promoting these phenomena and the subsequent relaxation toward equilibrium of the highly excited local solvent motions can be directly addressed.

In this article we will review some recent advances in the semiclassical theory of nonadiabatic interactions. These advances give reason for optimism that these methods will provide a mode of studying these condensed-phase relaxation problems in the near future. Since these methods are semiclassical in nature, they suggest how it may be possible to retain both a classical picture for the solvent motions and a quantum-mechanical description for the highly quantized degrees of freedom.

To apply the theory of nonadiabatic transitions to these relaxation problems, it is necessary to make a time-scale separation of the motions in the system. The "fast" motions, whether electronic or vibrational, are the highly quantized ones. We denote these coordinates by $\mathbf{r}$. The "slow" motions, such as center-of-mass translations, rotations, and so on may be treated quantum mechanically, classically, or semiclassically. These coordinates are collectively labeled $\mathbf{R}$. The total Hamiltonian of the system H, contains the kinetic-energy operator of the slow variables, T_s, the kinetic-energy operator of the fast motions, T_f, and the potential interaction, $V(\mathbf{r}, \mathbf{R})$:

$$H = T_s + T_f + V(\mathbf{r}, \mathbf{R})$$

$$= T_s + H_f \tag{1}$$

where $H_f = T_f + V(\mathbf{r}, \mathbf{R})$ is the Hamiltonian for the fast variables. Applying the usual adiabatic approximation,[3] the eigenvalues and eigenfunctions of H_f are found, treating $\mathbf{R}$ as a parameter:

$$H_f \psi_{f,n}(\mathbf{r}, \mathbf{R}) = E_{f,n}(\mathbf{R}) \psi_{f,n}(\mathbf{r}, \mathbf{R}) \tag{2}$$

The adiabatic energies $E_{f,n}(\mathbf{R})$ serve as the potential-energy surfaces for the slow motions. The time-independent Schrödinger equation for the slow degrees of freedom takes the form

$$\left[T_s + E_{f,n}(\mathbf{R}) \right] \psi_{s,n,m}(\mathbf{R}) = E_{s,n,m} \psi_{s,n,m}(\mathbf{R}) \tag{3}$$

and the wavefunction for the entire system in this approximation is merely $\psi_{n,m}(\mathbf{r}, \mathbf{R}) = \psi_{f,n}(\mathbf{r}, \mathbf{R}) \psi_{s,n,m}(\mathbf{R})$. This adiabatic approximation ignores the fact that T_s acts on $\psi_{f,n}$ as well as $\psi_{s,n,m}$, since both contain a $\mathbf{R}$ dependence. The approximation can be summarized as $T_s \psi_{f,n} \psi_{s,n,m} \approx \psi_{f,n} T_s \psi_{s,n,m}$. The nonadiabatic coupling, which is responsible for transitions between the quantum levels $E_{f,n}(\mathbf{R})$ in this picture, is given by

$$T_{n,m}^{NA} = T_s \psi_{f,n} \psi_{s,n,m} - \psi_{f,n} T_s \psi_{s,n,m} \tag{4}$$

These approximations are familiar as applied to the electronic–nuclear motion separation. When they are applied to the vibrational–solvent motion separation,[4,5] the vibrations are the fast motions, and all other motions, which we collectively call "solvent," are treated as slow. The $E_{f,n}(\mathbf{R})$ are quantized vibrational energies that depend on the solvent configuration, since this configuration alters the total interaction potential as a function of $\mathbf{r}$, the vibrational coordinates. In liquids the time-scale separation between vibrations and other motions is not as clear as the separation between electronic and nuclear motions. Consequently vibrational relaxation is an ubiquitous and often rapid process. Nevertheless, it still seems quite reasonable and useful to view the molecular vibrations as separate from, albeit perturbed by, the other motions.

It is worthwhile to point out at the start what this article is not. It is not in any sense a review of the vibrational-relaxation literature. Oxtoby has reviewed recent progress in this field.[6] It is also not intended to be an even-handed and thorough review of semiclassical theories of nonadiabatic processes, although we do briefly survey the major lines of work in this field. The bulk of the article is concerned with certain advances in the generalization of semiclassical procedures that appear to offer very promising possibilities for future application to these quantum relaxation problems.

II.   SEMICLASSICAL APPROACHES TO NONADIABATIC DYNAMIC SCATTERING

A.   Overview

The early work of Landau,[7] Zener,[8] and Stuckelberg[9] on nonadiabatic collisions and, in particular, atom–atom charge-transfer processes is semiclassical in nature. The motions of the nuclei are assumed to follow a classical trajectory. The relative positions and momenta of the atoms become functions of time. The nonadiabatic interaction, which depends on the nuclear position and momentum, becomes an implicit function of time as well, and the time-dependent Schrödinger equation is solved, yielding an expression for the transition probability. In the Landau–Zener model the nuclear motion is approximated as a constant-momentum trajectory, and the nonadiabatic Hamiltonian takes a very simple form. Despite its simplicity and the obvious deficiencies in the model from which it is derived, the Landau–Zener approximation has been extremely and widely useful in describing many nonadiabatic problems that involve a closely avoided crossing in the adiabatic surfaces. Extensions and generalizations of the model have been investigated.[10,11] One of these, for instance, accounts for changes in the nuclear velocity in the crossing region, arriving at an expression for the transition probability that is applicable at lower energies than is the original Landau–Zener result. Another considers the crossing of one state with a set of equally spaced states. In all these models a crucial approximation involves the treatment of the nuclear motion as a single trajectory. Methods of this sort, which we will call *effective-path* methods, have been analyzed in detail by Delos, Thorson, and co-workers.[12-14] They conclude that such methods can accurately describe the dynamics only if the differences between the momenta appropriate to the various adiabatic potentials are slight in the crossing region.

The classical electron model of Miller and co-workers[15-18] and the self-consistent eikonal approximation of Micha[19] are more recent attempts at devising a more general semiclassical nonadiabatic theory. They both contain prescriptions for the classical motion of the nuclei, and can thus be viewed as effective-path methods.

An alternative semiclassical procedure for nonadiabatic problems is the complex-trajectory method of Miller and George.[20,21] This method utilizes the analytical structure of the adiabatic Hamiltonian. In one-dimensional problems, the adiabatic electronic surfaces never cross for real values of the relative nuclear coordinate. However, they do have crossing points in the complex plane. If the trajectories corresponding to nuclear motion are allowed to become complex (i.e., to have complex values for t, r and p), some of the trajectories will wander around these crossing points of the adiabatic surfaces. When this happens, the trajectory may end on an adiabatic surface different from the one it originated on. The complex-trajectory method extracts transition amplitudes from such trajectories. One aspect of this method that seems paradoxical at first is that the nonadiabatic interaction never appears explicitly in the method. Pechukas and co-workers[22,23] have studied this and have shown that the nonadiabatic interaction has a singularity at the complex crossing point of the adiabatic surfaces, and that this singularity enables the transition amplitude to be evaluated from a trajectory that goes around the crossing point.

Although the complex-trajectory method yields valuable insights into nonadiabatic problems and has been used in the investigation of certain processes,[24-32] the needs to extend the adiabatic energy surfaces analytically throughout the complex plane and to consider complex as well as real trajectories severely limits its usefulness for many problems.

A third type of semiclassical technique for nonadiabatic dynamics is the surface-hopping technique.[33-43] The Preston–Tully surface-hopping method[33,34] assumes that the motions of the nuclei can be described to sufficient accuracy by classical trajectories on one of the adiabatic surfaces for all positions of the nuclei except near an avoided crossing seam. This seam represents the line of closest approach of the adiabatic surfaces in a two-dimensional problem and an $(N - 1)$-dimensional hypersurface in an N-dimensional problem. It is assumed that all nonadiabatic interactions are localized near this seam and can be accounted for by allowing the trajectory to hop between adiabatic surfaces at the seam. The probability of a hop is evaluated using the Landau–Zener model. The momentum for the trajectory after a transition is determined on the basis of conservation of energy and the requirement that only the component of momentum normal to the nonadiabatic interaction seam may change during a transition.

For reasons given above, the complex-trajectory method does not appear to offer a very promising route to the incorporation of nonadiabatic effects in simulation procedures. This leaves the effective-path and surface-hopping alternatives. Each has advantages and disadvantages. There is a certain degree of arbitrariness with regard to the nuclear dynamics in the effective-path methods. Various versions offer different prescriptions for the nuclear trajectory and the underlying average or effective potential. However, this is not necessarily a serious drawback. It might be compared to the use of perturbation theory in electronic-structure theory. Many perturbation expansions are possible; there is no correct one. However, there is a lack of numerical comparisons between different classical path methods with which to determine the relative utility of these techniques.

A more serious criticism with regard to vibrational-relaxation problems may be that by using a single average path, these methods do not allow for discrete hops. Consider the dynamics of a weakly bound van der Waals cluster of atoms about some diatomic. Suppose the diatomic is vibrationally excited, for instance, with a laser, and that this is followed by vibrational relaxation. In a fully quantum description the vibrational mode of the diatomic can be considered weakly coupled to the phonon modes of the cluster. Vibrational relaxation takes place by a conversion of the vibrational quantum into the cluster modes. The result could be that one or more atoms are kicked out of the cluster, or that no atoms are lost and the excess energy is distributed among the phonon modes. In a purely classical description, the hot diatomic would continuously lose energy to the surrounding cluster. If the coupling between the motions is weak enough, the diatomic-to-cluster energy flow will be slow and cluster modes will equilibrate among themselves. The results of these two descriptions may be quite different. The effective-path methods provide a description much like the classical picture except that they account for the diatomic vibrational motion by means of an average or effective vibrational energy that depends on the cluster configuration. A surface-hopping approach, on the other hand, is closer to the quantum description, since it allows for discrete vibrational transitions. Consequently, the transition produces a manifestly nonequilibrium cluster; this may account for interesting features of the subsequent evolution.

However, for surface-hopping methods to be generally useful for condensed-phase relaxation problems, they must be considerably generalized. Particularly in vibrational relaxation, it is not expected that there will be a closely avoided hopping seam as in the electronic-scattering problems. The transition probability must in some manner be smeared out into a continuous density. This may appear forbidding at first, since this means that transitions will be allowed at each point along the trajectories executed by the slow variables on an adiabatic vibrational potential surface, and thus each initial trajectory will lead to an infinity of paths. However, the interaction in these problems is generally weak, and terms containing multiple transitions should decline rapidly in importance as the number of transitions involved increases. Furthermore, the condensed-phase simulations are generally statistical in nature. The transition amplitude induced by the nonadiabatic interaction gives rise to the possibility that a transition takes place. At each time step the computer merely makes a decision to stay in the same adiabatic quantum state or to make a transition. Each option can be appropriately weighed. In this way the nonadiabatic aspects of the problem could be incorporated directly into the statistical nature of the calculation.

B.   Generalized Surface-Hopping Procedures

In this section we discuss in moderate detail recent advances in the semi-classical theory of nonadiabatic interactions. The goals of such work are the development and understanding of generalized surface-hopping techniques and their application to condensed-phase problems.

1.   Time-Independent Theory

a.   One-Dimensional Problems.   Before turning our attention to non-adiabatic problems, let us consider a single surface (adiabatic) problem in one dimension.[44] The adiabatic surface is given by $V(x)$. Now choose a partition of the x-axis $\{x_1, x_2, \ldots, x_N\}$ that defines $N+1$ regions. The jth region corresponds to $x_j < x < x_{j+1}$. The zeroth region is $x < x_1$ and the Nth is $x > x_N$. We approximate $V(x)$ in the jth region by $V_j(x)$, which is an average of $V(x)$ across this region. Thus we can envision a wavefunction corresponding to an incoming flux from the left (from $x = -\infty$) on this approximate potential. This incoming wavefunction has the form $e^{ip_0 x}$ where $p_j = \sqrt{2\mu(E - V_j)}$. When this incoming flux encounters the potential discontinuity at x_1, it splits into a forward-propagating wave $e^{ip_0 x_1} T_1 e^{ip_1(x - x_1)}$ in region 1 and a reflected wave $e^{ip_0 x_1} R_1 e^{-ip_0(x - x_1)}$ in region zero. The transmission coefficient T_1 and reflection coefficients R_1 are easily determined by the continuity of Ψ and its first derivative at x_1 to be $T_1 = 2p_0/(p_0 + p_1)$ and $R_1 = (p_0 - p_1)/(p_0 + p_1)$. The transmitted wave likewise gives rise to new transmitted and reflected waves when it encounters the discontinuity in the approximate potential at x_2. This transmitted wave propagates forward in region 2, whereas the reflected wave travels backward across region 1, encountering the potential discontinuity at x_1, once again. This results in the formation of two more outgoing waves at x_1. Such a sequence can be continued *ad infinitum*.

In this construction each wave satisfies the Schrödinger equation in its region, and the correct set of three waves, incoming, transmitted, and reflected taken together, are continuous at the boundary. Therefore, the wavefunction constructed by summing all waves generated by continuing this process indefinitely exactly satisfies the Schrödinger equation in each region and is continuous at every boundary. If the summation converges, this is an exact solution to the Schrödinger equation with the approximate potential.

This complicated construction is simplified by grouping terms by the number of reflection coefficients they contain. There is only one lead term corresponding to transmission at each boundary. Taking the limit as $N \to \infty$

and as the partition width approaches zero, this zeroth-order term has the limit

$$\Psi^{(0)} = e^{ip_0 x_1} \prod_j T_j e^{ip_j \Delta x_j}$$

$$\approx e^{ip_0 x_1} \exp\left\{ \sum_i \ln\left(1 - \frac{\Delta p_j}{2p_j}\right) + i \sum_j p_j \Delta x_j \right\}$$

$$\approx e^{ip_0 x_1} \exp\left\{ -\sum_j \frac{\Delta p_j}{2p_j} + i \sum_j p_j \Delta x_j \right\}$$

$$\rightarrow e^{ip_0 x_1} \exp\left\{ -\tfrac{1}{2}\ln\left(\frac{p_f}{p_0}\right) + i \int_{x_1}^{x_f} p\, dx \right\}$$

$$= e^{ip_0 x_1} \sqrt{\frac{p_0}{p_f}} \exp\left\{ i \int_{x_1}^{x_f} p\, dx \right\} \tag{5}$$

where $\Delta x_j = x_{j+1} - x_j$ and $\Delta p_j = p_{j+1} - p_j$. Equation (5) is just the WKB result. The summation of all terms with one reflection coefficient provides a first-order correction to this result, and so on.[44]

Essentially the same approach can be used to study one-dimensional nonadiabatic problems and to derive nonadiabatic corrections to adiabatic semiclassical propagation.[35] Let $\{\chi_1, \chi_2, \ldots\}$ be a complete adiabatic representation for the problem at hand. Once again the x-axis is partitioned. All adiabatic surfaces $W_j(x)$ are approximated by piecewise constant potentials, as before. The adiabatic states $\{\chi_j\}$ are also treated as constant throughout each region. At the region boundaries, each incoming flux produces transmitted and reflected outgoing fluxes on the same adiabatic surface and on all other surfaces. If we approximate the complete set $\{\chi_j\}$ by a set of N states, there are $2N$ outgoing fluxes for each incoming one. To determine the $2N$ reflection and transmission coefficients appropriate to a single specific incoming flux, it is necessary to express the adiabatic states on one side of the boundary in terms of those on the other side. If x_j is the boundary, then

$$\chi_m^{j+1} \approx \chi_m^j + \sum_n \chi_n^j \left\langle \chi_n^j \left| \frac{d}{dx} \chi_m^j \right.\right\rangle \Delta x_j \tag{6}$$

where $\langle \cdots \rangle$ denotes integration over the "fast" variables and the superscripts on the adiabatic wavefunctions denote the regions to which the wavefunctions correspond. If Eq. (6) is used to express the $\{\chi_m^{j+1}\}$ in terms of the $\{\chi_m^j\}$, and the wavefunctions corresponding to the incoming and the

$2N$ outgoing terms and their derivatives are compared on either side of the boundary x_j, then the coefficients of each χ_n^j in the resulting expression must be equal across the boundary. This follows from the mutual orthogonality of the γ^j. This provides $2N$ independent relations (N from the continuity of the wavefunction and N from the continuity of its derivative) between the amplitudes of fluxes on both sides of the boundary. From these relations the $2N$ reflection and transmission coefficients can be extracted.

If transmission coefficients T_{kk}^j corresponding to no change in adiabatic state are defined to be zeroth order, and all other reflection and transmission coefficients are defined as first order, then terms can be grouped as in the single-surface case. (This definition of order is reasonable because $T_{kk}^j \to 1$ as the partition width vanishes, whereas $T_{k\ell}^j \to 0$, $R_{k\ell}^j \to 0$, and $R_{k\ell}^j \to 0$, where $T_{k\ell}^j$ is defined to be the transmission coefficient for the surface k-to-surface ℓ process at boundary j and R_{kl}^j is the similarly defined reflection coefficient.) As before, there is only one zeroth-order term, and in the limit of vanishing partition width it becomes the WKB result for the initial adiabatic surface. Grouping all terms with a common set of first-order coefficients (e.g., one $k \to \ell$ transmission coefficient $T_{k\ell}^j$) and evaluating the limit for a vanishing width of the partition, the wavefunction is found to be

$$
\Psi = \chi_k(x)\Psi_k^{(0)}(x_0, x)
$$

$$
+ \sum_{m \neq k} \chi_m(x)\int_{x_0}^x dx_1\, \Psi_k^{(0)}(x_0, x_1)\tau_{km}^{(+)}(x_1)\Psi_m^{(0)}(x_1, x)
$$

$$
+ \sum_m \chi_m(x)\int_{\max[x_0, x]}^\infty dx_1\, \Psi_k^{(0)}(x_0, x_1)\rho_{km}^{(+)}(x_1)\Psi_m^{(0)}(x_1, x)
$$

$$
+ \sum_{m \neq k}\sum_{n \neq m} \chi_n(x)\int_{x_0}^x dx_1\int_{x_1}^x dx_2\, \Psi_k^{(0)}(x_0, x_1)\tau_{km}^{(+)}(x_1)\Psi_m^{(0)}(x_1, x_2)
$$

$$
\times \tau_{mn}^{(+)}(x_2)\Psi_n^{(0)}(x_2, x) + \cdots \tag{7}
$$

where $\Psi_m^{(0)}(x_0, x) = \sqrt{p_m(x_0)/p_m(x)}\,\exp\{i\int_{x_0}^x p_m(y)\,dy\}$ is the WKB wavefunction. The choice of the point x_0 determines the overall magnitude and phase of the wavefunction, but is arbitrary. The second and third terms are the $k \to m$ single-transition and single-reflection terms. The fourth term is the $k \to m \to n$ double-transition term. The wavefunction contains all possible sequences of transition (T-type) and reflection (R-type) events accompanied by single-surface semiclassical propagation between these nonclassical events. If this formal expression converges at a given value of x, it is exact. The reflection/transition (R/T) amplitudes (which are related to the previously mentioned reflection and transmission coefficients) are given by

$$\tau_{km}^{(\pm)} = \mp \frac{p_k + p_m}{2p_m} \left\langle \chi_k \left| \frac{d\chi_m}{dx} \right\rangle \right., \qquad k \neq m \qquad (8a)$$

$$\rho_{k,m}^{(\pm)} = \mp \frac{p_n - p_m}{2p_m} \left\langle \chi_k \left| \frac{d\chi_m}{dx} \right\rangle \right., \qquad k \neq m \qquad (8b)$$

and

$$\rho_{kk}^{(\pm)} = \mp \frac{1}{2p_k} \frac{dp_k}{dx} \qquad (8c)$$

The superscripts indicate whether the momentum just prior to the nonclassical event is positive $(+)$ or negative $(-)$.

Equations (7) and (8) are particularly noteworthy in that they constitute a generalized semiclassical hopping expression of the type we seek, and are formally exact if Eq. (7) converges. As such, they provide a guide for the development of many-dimensional and time-dependent formalisms and a framework for the testing of approximations.

Intuitively, one expects that the reflection terms should be less important than the transition corrections, and numerical evidence supports this.[35] The reflection without change in the adiabatic surface corrections have the exactly same form as the single-surface case.

The WKB wavefunctions and the R/T amplitudes become singular at the turning points of the classical motion, that is, where $p_m \rightarrow 0$. A uniform approximation can be derived[36] that has the same general form and interpretation as Eq. (7). In the derivation the potential $V(x)$ is replaced piecewise by its linear approximations rather than by average values. In this case the amplitudes remain well behaved at classical turning points.

   b. Many-Dimensional Case. Before discussing how the one-dimensional semiclassical nonadiabatic scattering formalism considered above might be generalized for many dimensions, we will find it useful to discuss many-dimensional generalizations of the one-dimensional WKB approximation for adiabatic problems.

The development of many-dimensional semiclassical dynamics proceeded rapidly in the early and middle 1970s, particularly in the work of Miller,[45] Marcus,[46-48] and Heller.[49] In this section we will be interested primarily in time-independent semiclassical wavefunctions, and the approach we will adopt is most closely related to the work of Marcus and co-workers.

Let us begin by expressing the wavefunction Ψ in terms of an amplitude A and phase $\phi/\hbar$, which are generally taken to be real functions of position, so that $\Psi = Ae^{i\phi/\hbar}$. Insertion of Ψ into the time-independent Schrödinger equation yields

$$\left[-\frac{\hbar^2}{2\mu} \left(\nabla^2 A + \frac{2i}{\hbar} \nabla A \cdot \nabla \phi + \frac{i}{\hbar} A \nabla^2 \phi - A(\nabla \phi)^2 / \hbar^2 \right) \right.$$
$$\left. + (V(\mathbf{r}) - E) A \right] e^{i\phi/\hbar} = 0 \qquad (9)$$

The two lowest-order terms of expansion in h lead to $\nabla^2 \phi = 2\mu(E - V)$ and $2\nabla A \cdot \nabla \phi + A \nabla^2 \phi = 0$. These equations can be solved numerically in the following fashion for scattering problems. Suppose the wavefunction is defined by some asymptotic incoming wave condition, and this asymptotic condition determines the value of the wavefunction amplitude everywhere on an asymptotic wavefront (surface of constant phase). A set of trajectories are initiated along this asymptotic wavefront with initial momenta determined by the system energy and the condition that the initial momenta be perpendicular to the wavefront. The condition $(\nabla \phi)^2 = 2\mu(E - V) = p^2$ can be integrated along each trajectory to obtain $\phi = \phi_0 + \int_{\mathbf{r}_0}^{\mathbf{r}} \mathbf{p} \, d\mathbf{r}$, where $\mathbf{r}_0$ is the initial point on the asymptotic wavefront and ϕ_0 is the value of ϕ on that wavefront. The second condition becomes $2\mathbf{p} \cdot \nabla A + A \nabla \cdot \mathbf{p} = 0$, using $\mathbf{p} = \nabla \phi$. This condition for A can be recast as the continuity equation $\nabla \cdot (A^2 \mathbf{p}) = 0$. Thus A^2 describes the density for a stationary flow, with the trajectories providing the streamlines of the flow, $A^2 \mathbf{p}$ being the flux, and the surfaces of constant ϕ constituting the wavefronts. The wavefronts and streamlines are everywhere perpendicular.

In developing a multidimensional semiclassical surface-hopping theory,[37] it seems reasonable to retain this semiclassical wave picture both before and after transitions occur. This requires that posttransition waves somehow be formed from a set of posttransition trajectories. There should be one posttransition trajectory for every initial or pretransition trajectory. Neighboring initial trajectories should make transitions at points near each other and result in neighboring posttransition trajectories. The set of transition points corresponding to a specific posttransition wave forms a surface, which we will call an R/T surface.

It is not difficult to show that for the posttransition wavefronts to be everywhere perpendicular to the corresponding trajectories, it is necessary that only the component of momentum perpendicular to the R/T surface change during the energy-conserving transition. This defines the posttransition momentum up to the sign of its component normal to the R/T surface.

Now picture space filled with a continuous set of R/T surfaces. A continuous set of first-order semiclassical waves are formed as the zeroth-order wave crosses these R/T surfaces. In terms of the trajectories, as each zeroth-order trajectory propagates, first-order trajectories continually arise, just as in the one-dimensional case discussed above. It is necessary to keep both the trajectory and the wave picture in mind. The trajectory picture describes how the equations can be solved numerically, but the wave picture is also needed to give meaning to the semiclassical equation for amplitude, $2\mathbf{p} \cdot \nabla A + A\nabla \cdot \mathbf{p} = 0$. An important remaining question, which we will address presently, is how to choose the R/T surfaces.

The one-dimensional nonadiabatic analysis and the discussion of single-surface semiclassical dynamics suggest that a semiclassical wavefunction for nonadiabatic problems can be cast in the form

$$\Psi(\mathbf{r}) = \chi_a(\mathbf{r}) \int d\sigma \int_0^\infty ds_1 \, Ae^{i\phi/\hbar} \delta(\mathbf{r} - \mathbf{r}_f)$$

$$+ \chi_b(\mathbf{r}) \int d\sigma \int_0^\infty ds_1 \int_0^\infty ds_2 \, Ae^{i\phi/\hbar} \tau_{ab}(\mathbf{r}_1) \delta(\mathbf{r} - \mathbf{r}_f)$$

$$+ \chi_b(\mathbf{r}) \int d\sigma \int_0^\infty ds_1 \int_0^\infty ds_2 \, Ae^{i\phi/\hbar} \rho_{ab}(\mathbf{r}_1) \delta(\mathbf{r} - \mathbf{r}_f)$$

$$+ \chi_a(\mathbf{r}) \int d\sigma \int_0^\infty ds_1 \int_0^\infty ds_2 \int_0^\infty ds_3 \, Ae^{i\phi/\hbar} \tau_{ab}(\mathbf{r}_1) \tau_{ba}(\mathbf{r}_2) \delta(\mathbf{r} - \mathbf{r}_f)$$

$$+ \cdots \tag{10}$$

It has been assumed only two adiabatic states, χ_a and χ_b, contribute significantly to the dynamics. The initial incoming wave is taken to be on adiabatic surface a. The σ integral is over the initial wavefront, and the δ function picks out only those trajectories that have their point $\mathbf{r}_f$ at $\mathbf{r}$. The integration variables $s_1, s_2, \ldots$ correspond to the arc lengths of the first, second, and so on classical segments of a given trajectory, respectively. These classical segments are separated by the intersurface jumps. The amplitude A and phase function ϕ are evaluated from the usual adiabatic semiclassical equations along each classical segment of a given trajectory and are taken to be continuous at the points of intersurface hops. The first term in Eq. (10) represents the zeroth-order wave and has no intersurface hops. Equation (10) allows for two types of nonclassical first-order events at each point along a trajectory. Transitions or T-type events preserve the sign of $\mathbf{p} \cdot \mathbf{n}$ when the trajectory hops between adiabatic surfaces, whereas reflections or R-type events change the sign of $\mathbf{p} \cdot \mathbf{n}$, where $\mathbf{n}$ is the unit vector normal to the R/T surface at the point of the nonclassical event. The second term in Eq. (10) accounts for the first-order trajectories with a single T-type event, while the third term is also first order with a single R-type event. The functions τ_{ab} and ρ_{ab} are the transition amplitudes associated with these two nonclassical events, respectively. The corresponding $b \rightarrow a$ amplitudes are denoted by τ_{ba} and ρ_{ba}. The vectors $\mathbf{r}_1, \mathbf{r}_2, \ldots$ are the positions of the first, second, and so on nonclassical events, and they depend on the starting point of the trajectory on the initial wavefront, on the arc length of each prior trajectory segment, and on whether the prior R/T events are R type or T type. There are 2^n nth-order terms, corresponding to the 2^n possible sequences of n R-type and T-type events.

Unlike the exact one-dimensional wavefunction expansion, this does not include reflections that are not accompanied by an intersurface hop. These terms function only as corrections to the single-surface semiclassical propagation and do not depend on the nonadiabatic nature of the problem. As a result, the wavefunction expansion (10) is not exact, but rather inherently semiclassical. If Ψ of the form (10) is inserted into the Schrödinger equation, a number of conditions are obtained that must be satisfied by the R/T amplitudes.[37] When the R/T surfaces are chosen to be everywhere perpendicular to the nonadiabatic coupling vector $\eta = \langle \chi_a | \nabla \chi_b \rangle$ (where ∇ is the gradient with respect to the "slow" variables and $\langle \cdots \rangle$ represents averaging over the "fast" variables), these conditions are satisfied if τ_{ab} and ρ_{ab} are given by

$$\tau_{ab} = - \frac{(\mathbf{p}_a + \mathbf{p}_b) \cdot \eta}{2\mathbf{p}_b \cdot \mathbf{n}} \tag{11a}$$

and

$$\rho_{ab} = - \frac{(\mathbf{p}_a - \mathbf{p}_0) \cdot \eta}{2\mathbf{p}_b \cdot \mathbf{n}} \tag{11b}$$

$\tau_{ba} = -\tau_{ab}$, and $\rho_{ba} = -\rho_{ab}$. The unit vector $\mathbf{n}$ is normal to the R/T surface and chosen such that $\mathbf{p}_i \cdot \mathbf{n} > 0$, where $\mathbf{p}_i$ is the incoming momentum.

For all the conditions from the Schrödinger equation to be satisfied, it is necessary to ignore terms of the forms $\nabla^2 A$ and $\mathbf{n} \cdot \nabla A$. The neglect of the $\nabla^2 A$ term is essentially consistent with the semiclassical approximation for single-surface problems. The neglect of the $\mathbf{n} \cdot \nabla A$ terms is also consistent with the semiclassical approximation.[37] In fact, if the exact wavefunction expansion (7) is inserted into the one-dimensional Schrödinger equation, both the $\nabla^2 A$ and $\mathbf{n} \cdot \nabla A$ terms are canceled by terms involving ρ_{aa}, the single-surface reflection correction.

It can be shown[37] that Eqs. (11) represent the only physically acceptable solutions to the conditions obtained from the Schrödinger equation and that there are no acceptable solutions unless the R/T surfaces are chosen to be everywhere perpendicular to η. This defines the R/T surfaces unequivocally.

An interesting feature of the transition amplitudes is that they become singular at points where the component of posttransition momentum perpendicular to the R/T surface, $\mathbf{p}_b \cdot \mathbf{n}$, vanishes. However, it is possible to show[38] that these points also correspond to infinite curvature for the wavefront of the posttransition trajectories. Singularities in single-surface semiclassical wavefunctions at points of infinite curvature and converging trajectories are well known. The singularity is a consequence of the continuity equation $\nabla \cdot (A^2 \mathbf{p}) = 0$ for the amplitude of a single-surface semiclassical wavefunction. The singularities in the transition amplitudes are actually necessary to introduce the correct semiclassical behavior in the posttransition wavefunction at these points.

The two-state approximation is employed in Eq. (10). If more adiabatic states contribute significantly to the dynamics, a different nonadiabatic coupling vector $\eta_{ab} = \langle \chi_a | \nabla \chi_b \rangle$ must be defined for each pair of states. This indicates that a different set of R/T surfaces is needed for each η_{ab}. If multisurface expansion, which is analogous to Eq. (10), is inserted into the Schrödinger equation, new terms arise due to the fact that the R/T surfaces and η_{ab} vectors for consecutive but different transitions will not in general be the same in the limit as the arc length between transitions tends to zero.[37] These new terms do not cancel. Thus, in a multisurface problem the semiclassical expansion analogous to Eq. (10) ignores these terms for the sake of simplicity. In one-dimensional problems, these geometrical difficulties do not arise; since there is only one dimension, all η vectors must be parallel. At present there is no numerical evidence to indicate how serious the interference between consecutive different transitions is.

A two-dimensional model for nonadiabatic scattering with two adiabatic surfaces has been solved in first-order approximation using this semiclassical expansion.[38] In this model the adiabatic surfaces are constant and the nonadiabatic coupling is radial. The agreement between the semiclassical and quantum results is good. It can be shown that in the $\hbar \to 0$ limit the quantum results reduce to the semiclassical solution. An interesting feature of this model is that the wavefronts for T-type and R-type posttransition semiclassical waves join at the points where the η component of the posttransition momentum vanishes, and are in effect two branches of one continuous wavefront.

2. Time-Dependent Theory

Many of the results of the time-independent analysis are applicable to time-dependent problems with only slight modification.[43] Beginning once again with single-surface semiclassical propagation, the equations obeyed by the time-dependent amplitude and phase functions are

$$\frac{\partial \phi}{\partial t} + \frac{\nabla^2 \phi}{2\mu} + V = 0 \tag{12}$$

and

$$2\mu \frac{\partial A}{\partial t} + 2\nabla A \cdot \nabla \phi + A \nabla^2 \phi = 0 \tag{13}$$

As before, $\nabla^2 A$ terms are higher order in $\hbar$ and are neglected. It is not too difficult to demonstrate that these equations are satisfied if ϕ is chosen to be Hamilton's principal function

$$\phi = \int L \, dt \tag{14}$$

where $L = T - V$ is the Langrangian, and A is given by

$$A = \left[\left(-\frac{1}{2\pi i \hbar} \right)^N \left| \frac{\partial^2 \phi}{\partial \mathbf{r}_i \partial \mathbf{r}_f} \right| \right]^{1/2} \tag{15}$$

In terms of these functions, the semiclassical solution for the time-dependent Schrödinger equation takes the form

$$U(\mathbf{r}_f, \mathbf{r}_i; t) = \left[\left(-\frac{1}{2\pi i \hbar} \right)^N \left| \frac{\partial^2 \phi}{\partial \mathbf{r}_i \partial \mathbf{r}_f} \right| \right]^{1/2} e^{i\phi/\hbar} \tag{16}$$

This well-known expression gives the value for the semiclassical propagator between points $\mathbf{r}_i$ and $\mathbf{r}_f$ by evaluating the action along a classical trajectory connecting the initial point $\mathbf{r}_i$ and final point $\mathbf{r}_f$ in time t. Knowledge of how this function varies with change in location of $\mathbf{r}_i$ and $\mathbf{r}_f$ yields the prefactor. If more than one classical trajectory connects $\mathbf{r}_i$ and $\mathbf{r}_f$ in the specified time, the contributions of this form from each trajectory are summed.

For multisurface problems, a surface-hopping expansion very similar in form to that for the time-independent case can be assumed:

$$\begin{aligned}
U_a(\mathbf{r}_1, \mathbf{r}_i; t) = {} & \chi_a A e^{i\phi/\hbar} + \chi_b \int_0^t dt_1 \left(\frac{p_a(t_1)}{\mu} \right) A e^{i\phi/\hbar} \tau_{ab}(t_1) \\
& + \chi_b \int_0^t dA_1 \left(\frac{p_a(t_1)}{\mu} \right) A e^{i\phi/\hbar} \rho_{ab}(t_1) \\
& + \chi_a \int_0^t dt_2 \int_0^{t_2} dt_1 \left(\frac{p_a(t_1)}{\mu} \right) \left(\frac{p_b(t_2)}{\mu} \right) A e^{i\phi/\hbar} \tau_{ab}(t_1) \tau_{ba}(t_2) \\
& + \cdots
\end{aligned} \tag{17}$$

The functions A and ϕ are again required to satisfy the single-surface conditions (12) and (13) for each single-surface segment of a trajectory, and to be continuous at the times of nonclassical intersurface hops. These hops occur at $t_1, t_2, \ldots$. There are two types of events, T type and R type, at each time. The velocities p_a/μ and p_b/μ are included in the first-, second-, and higher-order expressions, since the integration variables are taken to be times rather than arc lengths as in the time-independent case. These velocity factors could be incorporated in the amplitudes τ_{ab}, ρ_{ab}, τ_{ba}, and ρ_{ba}, but this would result in these amplitudes being defined differently in the time-dependent and time-independent cases.

As in the time-independent case, energy is conserved and only the component of momentum parallel to $\eta = \langle \chi_b | \nabla \chi_a \rangle$ is altered during intersurface hops. Trajectories that obey the classical equations of motion on a single surface at all times except for a finite number of transition times, and whose momentum obeys these conditions during the transitions, will be called pseudoclassical trajectories. Equation (17) represents the generalization of Eq. (16) where all pseudoclassical trajectories are allowed, as opposed to the single-surface case, where only strictly classical trajectories are allowed.

In applying the semiclassical expression for A, Eq. (15), the derivatives are interpreted as the rates of change of the appropriate functions as the initial or final point changes, subject to the constraints that the intersurface hops occur at the same R/T surfaces and that the trajectories remain pseudoclassical (only the η component of momentum changes during transitions). If the time-dependent propagator is inserted into the time-dependent Schrödinger equation, the same conditions on the transition amplitudes are obtained as in the time-independent case. Therefore, the transition amplitudes take the same form as before. The major difference is the use of Eqs. (12) and (13) as the equations determining A and ϕ instead of their time-independent counterparts.

The same remarks concerning problems when there are more than two important adiabatic states are pertinent here as in the time-independent case. Nonetheless, Eqs. (10) and (17), along with the expressions for the transition amplitudes, (11a) and (11b), significantly generalize the notion of semiclassical hopping and effectively remove the requirement of a strongly localized interaction. In fact, in their present form they appear to be more applicable to problems involving nonlocalized weak interactions.

 a. Rigid Gaussian Approximation. Each term in the time-dependent semiclassical nonadiabatic propagator expansion (17) has the form of intersurface hops at specific times, t_1, $t_2, \ldots$ and standard single-surface semiclassical propagation during the periods between these hopping times. If a specific initial wavefunction $\chi_a \Psi_a(\mathbf{r}, 0)$ is chosen, the corresponding wavefunction at time t is given by

$$\Psi(\mathbf{r}, t) = \int dr_i\, U_a(\mathbf{r}, \mathbf{r}_i; t)\Psi_a(\mathbf{r}_i, 0) \tag{18}$$

This expression holds for both quantum-mechanical and semiclassical propagation, depending on the approximation used for U_a. If the initial state does not correspond to the pure adiabatic states χ_a, then it can be expressed as a sum over adiabatic states of terms of this sort, and Eq. (18) must likewise be summed over the adiabatic states a.

Consider again a single-surface case. If $\Psi(\mathbf{r}_i, 0)$ is a localized wavepacket, it will propagate on a single adiabatic surface, in such a way that, in accord with the Ehrenfest theorem, its average position and momentum will execute a phase-space trajectory that closely approximates a classical trajectory over a time scale short enough that the packet remains highly localized. If in fact the spreading of the packet can be ignored, this suggests that a classical trajectory description of the dynamics should be applicable except at the points of intersurface hops.

However, consideration of some model problems, for instance the propagation of a free-particle Gaussian packet approximate for an argon atom, suggests that localized packets do spread significantly on the picosecond time scale relevant for most vibrational relaxation processes, at least for small molecules. On the other hand, Heller[50] has suggested that if $\Psi(\mathbf{r}, 0)$ is expanded in terms of a number of Gaussians, then the spreading of the classical trajectories that approximately describe the dynamics of the individual Gaussians should largely account for the spreading of $\Psi(\mathbf{r}, 0)$, even if the spreadings of the individual Gaussians are ignored. He has numerically applied this frozen-Gaussian approximation to several examples, with very good results.[50, 51]

To develop this a little more mathematically,[52] suppose we expand the initial wavefunction in an overcomplete set of Gaussian wavepackets[53–55]

$$\Psi(\mathbf{r}, 0) = (2\pi)^{-N} \int d\mathbf{a}\, \langle \mathbf{r} | \mathbf{a} \rangle \langle \mathbf{a} | \Psi(\mathbf{r}, 0) \rangle \tag{19}$$

where N is the dimensionality of $\mathbf{r}$, $\mathbf{a} = (\mathbf{r}_a, \mathbf{p}_a)$ is a phase-space point specifying the average position and momentum of the packet

$$\langle \mathbf{r} | \mathbf{a} \rangle = \left(\frac{2\gamma}{\pi} \right)^{N/4} \exp\left[-\gamma \frac{(\mathbf{r}-\mathbf{r}_a)^2}{\hbar} + i\mathbf{p}_a \cdot \frac{(\mathbf{r}-\mathbf{r}_2)}{\hbar} \right] \qquad (20)$$

and $d\mathbf{a} = d\mathbf{r}_a \, d\mathbf{p}_a$. The parameter γ is related to the packet width. It must be positive (we take it to be real) and independent of $\mathbf{a}$, but is otherwise arbitrary. It is not too difficult to show[52] that if the semiclassical approximation is used for the single-surface propagator and the stationary-phase approximation is applied to the integrations involved, then $\Psi(\mathbf{r}, t)$ can be written as

$$\Psi(\mathbf{r}, t) = \frac{1}{2\pi} \int d\mathbf{a} \, \langle \mathbf{r} | \mathbf{a}_t \rangle A(\mathbf{a}, t) e^{i\phi(\mathbf{a}, t)/\hbar} \langle \mathbf{a} | \Psi(\mathbf{r}, 0) \rangle \qquad (21)$$

In Eq. (21) $\langle \mathbf{r} | \mathbf{a}_t \rangle$ is the Gaussian wavepacket whose average position and momentum, $\mathbf{a}_t$, are obtained by integrating a classical path from the point $\mathbf{a}$ in phase space for time t. The shape of $\langle \mathbf{r} | \mathbf{a}_t \rangle$ is exactly the same as that of the original packet $\langle \mathbf{r} | \mathbf{a} \rangle$; that is, $\langle \mathbf{r} | \mathbf{a}_t \rangle$ is given by Eq. (20) with the same γ but with $(\mathbf{r}_a, \mathbf{p}_a)$ replaced by their time-propagated values $(\mathbf{r}_{at}, \mathbf{p}_{at})$. The additional phase $\phi(\mathbf{a}, t)$ is the classical action $\phi = \int L \, dt$ for the trajectory from $\mathbf{a}$ to $\mathbf{a}_t$ in time t. The prefactor A, which is given by a rather involved expression, accounts for amplitude changes due to features such as the spreading of the classical trajectories.[52] The important feature of this expression is that the quantum spreading of the individual Gaussian packets is ignored during the propagation, and yet the propagation of the entire wavefunction is still correct within the semiclassical approximation for propagator $U(\mathbf{r}, \mathbf{r}_1; t)$ and the stationary-phase approximation for integrations. This provides a possible justification for the use of classical trajectories in dynamics calculations for times much longer than the short times over which a single narrow wavepacket would remain localized. The localized-packet description of a single packet is a poor approximation, but it may be a reasonable description when many trajectories are used to describe the evolution of a general wavefunction or density, as is often done in numerical calculations.

The only thing required of the propagator for the proof is that it have the form $Ae^{i\phi}$, where ϕ is the classical action for the trajectory connecting the initial and final points in time t. If more than one trajectory satisfies these conditions, then the contributions from all such trajectories are summed. The semiclassical nonadiabatic propagator satisfies these conditions if the restriction to classical trajectories is weakened to include all pseudoclassical trajectories.[43] If the semiclassical nonadiabatic propagator is employed, as in Eq. (18), then the same rigid Gaussian result is obtained as in the single-surface case, except the packets must be allowed to hop between different adiabatic surfaces as they follow their trajectories. The transition amplitudes associated with these hops are the same as in the usual semiclassical propagator expression, Eq. (17).

The meaning of these equations is that the nonadiabatic propagation of a wavefunction or density can be obtained by expanding the initial wavefunction or density in terms of a set of gaussians of fixed shape, running pseudo-classical trajectories of duration t beginning at the phase-space points specified by the average position and momentum of these Gaussian packets, and finally constructing the propagated wavefunction or density by clothing the final point of each trajectory with a fixed-shape Gaussian multiplied by the appropriate A and $e^{i\phi}$ factors. Each final Gaussian is weighted by the expansion coefficient, $\langle a | \Psi(r, 0) \rangle$, for the Gaussian associated with the initial phase-space point of the trajectory.

III. VIBRATIONAL POPULATION RELAXATION AS A NONADIABATIC PHENOMENON

In this section we develop nonadiabatic expressions for condensed-phase processes. Vibrational population relaxation is the particular phenomenon with which this section is concerned, and although many of the equations are rather general, others contain approximations that may be less so.

A natural choice of quantity to investigate is the probability that a molecule in (adiabatic) vibrational state $|\alpha\rangle$ at time zero will be observed to be in (adiabatic) vibrational state $|\beta\rangle$ at time t. Let ρ represent the system density operator. In the canonical ensemble, $\rho = e^{-\beta H}/\mathrm{Tr}\, e^{-\beta H}$. This density is the density for both the vibrational and the nonvibrational degrees of freedom, and H is the nonadiabatic Hamiltonian for our system. The reduced density for a molecule in vibrational state $|\alpha\rangle$ is given by $\rho_\alpha = |\alpha\rangle \tilde{\rho}_\alpha \langle \alpha |$, where $\tilde{\rho}_\alpha = \langle \alpha | e^{-\beta H} | \alpha \rangle / \mathrm{Tr}_n \langle \alpha | e^{-\beta H} | \alpha \rangle$. Tr_n is the trace over nonvibrational degrees of freedom. Whereas ρ is stationary, ρ_α is not. The evolution of ρ_α is given by

$$\rho(t) = e^{-iHt} \rho_\alpha e^{iHt} \tag{22}$$

Since the adiabatic states form a complete set,

$$H = \sum_{\beta, \nu} |\beta\rangle \langle \beta | H | \nu \rangle \langle \nu | = \sum_{\beta, \nu} H_{\beta, \nu} \tag{23}$$

We can employ the perturbation expansion

$$
\begin{aligned}
e^{-iHt} = e^{-iH_0t} &- i\int_0^t dt_1\, e^{-iH_0(t-t_1)}H_1 e^{-iH_0 t_1} \\
&+ (-i)^2 \int_0^t dt_2 \int_0^{t_2} dt_1\, e^{-iH_0(t-t_2)}H_1 e^{-iH_0(t_2-t_1)}H_1 e^{-iH_0 t_2} \\
&+ \cdots
\end{aligned}
\tag{24}
$$

where $H = H_0 + H_1$ in Eq. (22) to obtain the second-order result

$$
\begin{aligned}
\rho_\alpha(t) = e^{-iH_\alpha t}\rho_\alpha e^{iH_\alpha t}
&- i\sum_\beta \int_0^t dt_1\, e^{-iH_\beta(t-t_1)}H_{\beta\alpha}e^{-iH_\alpha t_1}\rho_\alpha e^{iH_\alpha t} \\
&+ i\sum_\beta \int_0^t dt_1\, e^{-iH_\alpha t}\rho_\alpha e^{iH_\alpha t_1}H_{\alpha\beta}e^{iH_\beta(t-t_1)} \\
&+ \sum_{\alpha,\nu} \int_0^t dt_1 \int_0^t dt_2\, e^{-iH_\beta(t-t_1)}H_{\beta\alpha}e^{-iH_\alpha t_2}\rho_\alpha e^{iH_\alpha t_2}H_{\alpha\nu}e^{iH_\nu(t-t_2)} \\
&- \sum_{\beta,\nu} \int_0^t dt_2 \int_0^{t_2} dt_1\, e^{-iH_\nu(t-t_2)}H_{\nu\beta}e^{-iH_\beta(t_2-t_1)}H_{\beta\alpha}e^{-iH_\alpha t_1}\rho_\alpha e^{iH_\alpha t} \\
&- \sum_{\beta,\nu} \int_0^t dt_2 \int_0^{t_2} dt_1\, e^{-iH_\alpha t}\rho_\alpha e^{-H_\alpha t_1}H_{\alpha\beta}e^{iH_\beta(t_2-t_1)}H_{\beta\nu}e^{iH_\nu(t-t_2)} \\
&+ \cdots
\end{aligned}
\tag{25}
$$

where the definition $H_0 = \sum_\beta H_\beta = \sum_\beta |\beta\rangle\langle\beta|H|\beta\rangle\langle\beta|$ has been used. The probability of observing a molecule known to be in adiabatic state $|\alpha\rangle$ initially and in adiabatic state $|\beta\rangle \neq |\alpha\rangle$ at time t later is given by

$$
\begin{aligned}
P_{\alpha\beta}(t) &= \mathrm{Tr}\langle\beta|\rho_\alpha(t)|\beta\rangle \\
&= \mathrm{Tr}_n \int_0^t dt_1 \int_0^t dt_2\, \langle\beta|e^{-H_\beta(t-t_1)}H_{\beta\alpha}e^{-iH_\alpha t_1}\rho_\alpha^0 e^{iH_\alpha t_2}H_{\alpha\beta}e^{iH_\beta(t-t_2)}|\beta\rangle
\end{aligned}
\tag{26}
$$

through the second order. In obtaining our second-order result, we have replaced ρ_α^0 by its zeroth-order approximation $\rho_\alpha^0 = \exp(-\beta H_\alpha)/Q_\alpha^0$, and $Q_\alpha^0 = \mathrm{Tr}\rho_\alpha^0$. If we wish to extend Eq. (26) to higher orders, it is necessary to account for the fact that nonadiabatic effects alter ρ_α, and the first corrections to ρ_α^0 begin in the second order. Consequently, these corrections first appear in $P_{\alpha\beta}(t)$ in the fourth order.

If we employ the cyclic invariance of the trace and the fact that ρ_α^0 and H_α commute, the second-order expansion for $P_{\alpha\beta}(t)$ can be expressed as

$$
\begin{aligned}
P_{\alpha\beta}(t) &= \mathrm{Tr}_n \int_0^t dt_1 \int_0^t dt_2 \langle \beta | e^{-iH_\beta(t_2 - t_1)} H_{\beta\alpha} | \alpha \rangle \rho_\alpha^0 \langle \alpha | e^{iH_\alpha(t_2 - t_1)} H_{\alpha\beta} | \beta \rangle \\
&= \int_0^t dt_1 \int_{-t_1}^{t - t_1} dT \, \mathrm{Tr}_n \langle \beta | e^{-iH_\beta T} H_{\beta\alpha} | \alpha \rangle \rho_\alpha^0 \langle \alpha | e^{iH_\alpha T} H_{\alpha\beta} | \beta \rangle \\
&= \int_{-t/2}^{t/2} dt_1 \int_{-(t/2 + t_1)}^{t/2 - t_1} dT \, \mathrm{Tr}_n \langle \beta | e^{iH_\beta T} H_{\beta\alpha} | \alpha \rangle \rho_\alpha^0 \langle \alpha | e^{iH_\alpha T} H_{\alpha\beta} | \beta \rangle \\
&= \int_{-t/2}^{t/2} dt_1 \int_{-(t/2 + t_1)}^{t/2 - t_1} dT \, \mathrm{Tr}_n \rho_\alpha^0 \langle \alpha | e^{iH_\alpha T} H_{\alpha\beta} | \beta \rangle \langle \beta | e^{-iH_\beta T} H_{\beta\alpha} | \alpha \rangle \\
&= \int_{-t/2}^{t/2} dt_1 \int_{-(t/2 + t_1)}^{t/2 - t_1} dT \, \mathrm{Tr}_n \rho_\alpha^0 \langle \alpha | e^{iH_\alpha T} H_{\alpha\beta} e^{-iH_\beta T} H_{\beta\alpha} | \alpha \rangle \quad (27)
\end{aligned}
$$

Roughly speaking, the integral in Eq. (27) represents the average overlap of the quantum state of the system when the system propagates for time T on adiabatic energy surface α and then undergoes the $\alpha \to \beta$ transition with the resultant quantum state when the $\alpha \to \beta$ transition is followed by propagation for time T on adiabatic surface β. Assuming this average overlap decays to zero in a time scale τ, the limits in the T integration can be extended to $-\infty$ and ∞ without affecting the value of the integral for $\frac{1}{2}t - \tau > t_1 > -\frac{1}{2}t + \tau$. If $t \gg 2\tau$, $P_{\alpha\beta}(t)$ can be replaced by

$$
\begin{aligned}
P_{\alpha\beta}(t) &= \int_{-t/2}^{t/2} dt_1 \int_{-\infty}^{+\infty} dT \, \mathrm{Tr}_n \rho_\alpha^0 \langle \alpha | e^{iH_\alpha T} H_{\alpha\beta} e^{-iH_\beta T} H_{\beta\alpha} | \alpha \rangle \\
&= t \int_{-\infty}^{+\infty} dT \, \mathrm{Tr}_n \rho_\alpha^0 \langle \alpha | e^{iH_\alpha T} H_{\alpha\beta} e^{-iH_\beta T} H_{\beta\alpha} | \alpha \rangle \quad (28)
\end{aligned}
$$

In approximating Eq. (27) by Eq. (28), we assume t is large enough that the contributions from the edges of the t_1 integration range ($|t_1| > \frac{1}{2}t - \tau$), where the extension of the limits of the T integration is not justified, constitute such a small fraction of the entire range that they contribute negligibly to the integral. The ratio of $P_{\alpha\beta}(t)$ to t in this limit defines the $\alpha \to \beta$ vibrational relaxation rate via population, to the second order.

To evaluate the trace over nonvibrational degrees of freedom in Eq. (27) or (28), some representation must be chosen for these coordinates. If the Gaussian product representation (20) is employed, Eq. (27) yields

$$
\lim_{t \to \infty} \frac{P_{\alpha\beta}(t)}{t} = (2\pi)^{-2N} \int d\mathbf{a} \int d\mathbf{b} \int_{-\infty}^{\infty} dT \langle a | \rho_{\alpha}^{0} | b \rangle
$$

$$
\times \langle \alpha, \mathbf{b} | e^{iH_{\alpha}T} H_{\alpha\beta} e^{-iH_{\beta}T} H_{\beta\alpha} | \alpha, \mathbf{a} \rangle
$$

$$
= 2\,\mathrm{Re}(2\pi)^{-2N} \int d\mathbf{a} \int d\mathbf{b} \int_{0}^{\infty} dT \langle \mathbf{a} | \rho_{\alpha}^{0} | \mathbf{b} \rangle
$$

$$
\times \langle \alpha, \mathbf{b} | e^{iH_{\alpha}T} H_{\alpha\beta} e^{-iH_{\beta}T} H_{\beta\alpha} | \alpha, \mathbf{a} \rangle \qquad (29)
$$

where resolution of the identity

$$
\hat{1} = (2\pi)^{-N} \int d\mathbf{b} \, |\mathbf{b}\rangle\langle\mathbf{b}| \qquad (30)
$$

in terms of the Gaussian packets[53-55] has been used.

Equation (28) is readily interpreted in terms of the semiclassical rigid Gaussian nonadiabatic dynamics expressions provided in the previous section. [In fact, the non-adiabatic rigid gaussian expansion (17) is a semiclassical coordinate space representation of the operator expansion (24), which yields (28). The $H_{\alpha\beta}$ operators in (24) are replaced by transition amplitudes in (17) and the single surface semiclassical propagation in (17) corresponds to $e^{-iH_0 t}$ in (24).] To evaluate $P_{\alpha\beta}(t)$ initial phase-space points $\mathbf{a}$ and $\mathbf{b}$ are chosen from the distribution $\langle \mathbf{a} | \rho_{\alpha}^{0} | \mathbf{b} \rangle$. Pseudoclassical trajectories are integrated for each of these. The trajectory corresponding to $\mathbf{a}$ immediately hops to vibrational surface β and is integrated for time T, while the trajectory corresponding to $\mathbf{b}$ propagates for time T on surface α and then hops to surface β. These packets are held rigid and must be multiplied by the correct phase factor and preexponential factor and by the appropriate transition amplitude. The shape of the packet is unchanged by the propagation or intersurface hop, but the component of the average packet momentum parallel to the nonadiabatic coupling vector is altered during the hop to compensate for the change in the potential (at the average packet position) that results from the transition. The overlap of the propagated packets is then evaluated, and this is integrated over all values of $\mathbf{a}$ and $\mathbf{b}$ to yield the relaxation rate.

The density function has the form $\langle a|\rho_\alpha^0|b\rangle = \langle a|\exp(-\beta H_\alpha)/Q_\alpha^0|b\rangle$. In the nearly classical high-temperature limit, this can be approximated as

$$
\begin{aligned}
\langle a|\rho_\alpha^0|b\rangle &= \frac{1}{Q_\alpha^0} \int d\mathbf{r}_1 \int d\mathbf{r}_2 \langle a|\mathbf{r}_1\rangle\langle\mathbf{r}_1|e^{-\beta H_\alpha}|\mathbf{r}_2\rangle\langle\mathbf{r}_2|b\rangle \\
&= \frac{1}{Q_\alpha^0} \int d\mathbf{r}_1 \int d\mathbf{r}_2 \left(\frac{2\gamma}{\pi}\right)^{N/2}\left(\frac{\mu}{2\pi\beta}\right)^{N/2} \\
&\quad \times \exp\left\{-\gamma(\mathbf{r}_1-\mathbf{r}_a)^2 - i\mathbf{p}_a\cdot(\mathbf{r}_1-\mathbf{r}_a) - \frac{\mu(\mathbf{r}_1-\mathbf{r}_2)^2}{2\beta} - \beta V_\alpha(\mathbf{r}_{12})\right. \\
&\quad\quad \left. -\gamma(\mathbf{r}_2-\mathbf{r}_b)^2 + i\mathbf{p}_b\cdot(\mathbf{r}_2-\mathbf{r}_b)\right\}
\end{aligned}
\tag{31}
$$

In Eq. (31), the high-temperature approximation $\langle\mathbf{r}_1|e^{-\beta H_\alpha}|\mathbf{r}_2\rangle = [\mu/2\pi\beta]^{N/2}\exp\{-\mu(\mathbf{r}_1-\mathbf{r}_2)^2/2\beta - \beta V_\alpha(\mathbf{r}_{12})\}$ has been employed. The same mass μ has been used for all "slow" degrees of freedom. The generalization for problems involving more than one mass causes no essential difficulty. The adiabatic potential V_α, which corresponds to the system in vibrational state α, is evaluated at $\mathbf{r}_{12}$, the midpoint between $\mathbf{r}_1$ and $\mathbf{r}_2$. If γ is large and μ/β is large, the integrand in Eq. (31) is negligible unless $\mathbf{r}_1$ is near $\mathbf{r}_a$, $\mathbf{r}_2$ is near $\mathbf{r}_b$, and $\mathbf{r}_1$ is near $\mathbf{r}_2$. Under these conditions V_α can be treated as a constant over the important region and its value taken to be $V_\alpha(\mathbf{r}_{ab})$, where $\mathbf{r}_{ab} = (\mathbf{r}_a + \mathbf{r}_b)/2$. Since the two narrow packets are centered close to each other initially under these conditions, and because the packet momenta before and after transition differ, one can expect the overlap of the products embodied in Eq. (29) to decay rapidly with T. This indicates that the required trajectories have to be integrated only for short times. It might even be sufficiently accurate to treat the momenta of the trajectories as constant for these short times and to obtain an analytical expression for this overlap as a function of a, b, and T.

When V_α is treated as constant, or at most quadratic, $\langle a|\rho_\alpha^0|b\rangle$ can be evaluated in closed form. In the constant-V_α approximation, Eq. (31) becomes

$$
\begin{aligned}
\langle a|\rho_\alpha^0|b\rangle &= \frac{1}{Q_\alpha^0}\left(\frac{2\gamma}{\pi}\right)^{N/2}\left(\frac{\mu}{2\pi\beta}\right)^{N/2}\left[\frac{\pi}{2\gamma(\gamma+\mu/\beta)}\right]^{N/2} \\
&\quad \times \exp\left\{-\frac{\gamma\mu/\beta}{2(\gamma+\mu/\beta)}(\mathbf{r}_a-\mathbf{r}_b)^2 - \frac{(\mathbf{p}_a-\mathbf{p}_b)^2}{8\gamma} - \frac{\mathbf{p}_{ab}^2}{2(\gamma+\mu/\beta)}\right. \\
&\quad\quad \left. -\beta V_\alpha(\mathbf{r}_{ab}) + \frac{i\mu/\beta}{\gamma+\mu/\beta}\mathbf{p}_{ab}\cdot(\mathbf{r}_a-\mathbf{r}_b)\right\}
\end{aligned}
\tag{32}
$$

where $\mathbf{p}_{ab} = (\mathbf{p}_a + \mathbf{p}_b)/2$. Note that the partition function Q_α^0 is defined such that $\text{Tr}_n \rho_\alpha^0 = (2\pi)^{-N} \int d\mathbf{a} \langle \mathbf{a} | \rho_\alpha^0 | \mathbf{a} \rangle = 1$. When $\mathbf{b}$ is set to $\mathbf{a}$ in Eq. (32), all the terms in the exponent vanish except $-\beta[\mathbf{p}_a^2/2(\mu + \beta\gamma) + V_\alpha(\mathbf{r}_2)]$, which for high temperatures is approximately $-\beta$ times the classical Hamiltonian evaluated at the phase-space point $\mathbf{a}$. Therefore Q_α^0 is essentially the classical partition function, except for some unimportant constants related to the width of the Gaussians. It is also interesting that if both γ and μ/β are large, $(\mathbf{p}_{ab})^2$ and $(\mathbf{r} - \mathbf{r}_b)^2$ must be small for important packets, but $(\mathbf{p}_a - \mathbf{p}_b)^2$ can be large. However, if the initial momenta of the two packets differ too greatly, the overlap integral in Eq. (29) will decay over very short times, and this will result in a small contribution to the relaxation rate.

IV. CONCLUDING REMARKS

In this chapter we have attempted to review some recent developments in semiclassical theories of nonadiabatic transitions, and have indicated how these techniques might be fruitfully applied to the problem of vibrational population relaxation in liquids. Such liquid-phase calculations would necessarily be performed in a statistical fashion because of the large number of solvent degrees of freedom involved. There are many possible ways of organizing such computations. One way would be to employ a standard Monte Carlo algorithm to select numerous pairs of the phase-space points $\mathbf{a}$ and $\mathbf{b}$ to be associated with the two wavepackets according to Eq. (31). These points would be propagated on the two vibrational energy surfaces of interest and their overlaps evaluated at various times to obtain the integrals involved in the rate-constant expression (29). Alternatively, it might be more efficient to utilize an importance sampling scheme that takes into account the magnitude of the nonadiabatic coupling $H_{\alpha\beta}$. Whether this would lead to a significant increase in efficiency depends largely on the size of the fluctuations in the nonadiabatic coupling, about which little is known. These concerns are often best addressed by means of numerical models and preliminary calculations. Likewise, many of the assumptions and approximations introduced in the semiclassical nonadiabatic theory need to be tested numerically. Nonetheless, in the light of the recent advances in our description of nonadiabatic processes, there is certainly sufficient reason to expect that the direct simulation of these important liquid-phase processes will be feasible in the very near future.

Acknowledgments

This work was supported by NSF Grant CHE-8219380.

References

1. B. J. Berne, ed., *Modern Theoretical Chemistry*, Vols. V and VI, Plenum, New York, 1976.
2. J. O. Hirschfelder, C. F. Curtis and R. B. Bird, *Molecular Theory of Gases and Liquids*, Wiley, New York, 1954.
3. G. Baym, *Lectures on Quantum Mechanics*, Benjamin, New York, 1969.
4. S. H. Lin, *J. Chem. Phys.*, **65**, 1053 (1976).
5. M. F. Herman and B. J. Berne, *J. Chem. Phys.*, **78**, 4103 (1983).
6. D. W. Oxtoby, *Adv. Chem. Phys.*, **47**, 1981, p. 487.
7. L. D. Landau, *Phys. Z. Sowjetunion*, **2**, 46 (1932).
8. C. Zener, *Proc. R. Soc. London, Ser.* A, **137**, 696 (1932).
9. E. C. G. Stuckelberg, *Helv. Phys. Acta*, **5**, 369 (1932).
10. E. E. Nikitin, *Theory of Elementary Atomic and Molecular Processes in Gases*, Clarendon, Oxford, 1974.
11. M. S. Child, *Molecular Collision Theory*, Academic, New York, 1974.
12. W. R. Thorson, J. B. Delos, and S. A. Boorstein, *Phys. Rev.*, **A4**, 1052 (1971).
13. J. B. Delos, W. R. Thorson, and S. K. Knudson, *Phys. Rev.*, **A6**, 709 (1972).
14. J. B. Delos and W. R. Thorson, *Phys. Rev.*, **A6**, 720, 728 (1972).
15. W. H. Miller and C. W. McCurdy, *J. Chem. Phys.*, **69**, 5163 (1978).
16. C. W. McCurdy, H. D. Meyer, and W. H. Miller, *J. Chem. Phys.*, **70**, 3177 (1979).
17. H. D. Meyer and W. H. Miller, *J. Chem. Phys.*, **70**, 3214 (1979); *ibid.* **71**, 2156 (1979); *ibid.*, **72**, 2272 (1980).
18. C. W. McCurdy and W. H. Miller, *J. Chem. Phys.*, **73**, 3191 (1980).
19. D. A. Micha, *J. Chem. Phys.*, **78**, 7139 (1983).
20. W. H. Miller and T. F. George, *J. Chem. Phys.*, **56**, 5637 (1972).
21. R. K. Preston, C. Sloane, and W. H. Miller, *J. Chem. Phys.*, **60**, 4961 (1974).
22. J. P. Davis and P. Pechukas, *J. Chem. Phys.*, **64**, 3129 (1976).
23. J. T. Hwang and P. Pechukas, *J. Chem. Phys.*, **67**, 4640 (1977).
24. T. F. George and Y. W. Lin, *J. Chem. Phys.*, **60**, 2340 (1974).
25. Y. W. Lin, T. F. George, and K. Morokuma, *Chem. Phys. Lett.*, **30**, 49 (1975).
26. J. R. Laing, T. F. George, I. H. Zimmerman, and Y. W. Lin, *J. Chem. Phys.*, **63**, 842 (1976).
27. A. Komornicki, T. F. George, and K. Morokuma, *J. Chem. Phys.*, **65**, 48 (1976).
28. Y. M. Yuan, J. R. Laing, and T. F. George, *J. Chem. Phys.*, **66**, 1107 (1977).
29. J. R. Laing and T. F. George, *Phys. Rev.*, **A16**, 1082 (1977).
30. A. Komornicki, K. Morokuma, and T. F. George, *J. Chem. Phys.*, **67**, 5012 (1977).
31. J. M. Yuan and T. F. George, *J. Chem. Phys.*, **68**, 3040 (1978); *ibid.*, **70**, 990 (1979).
32. H. W. Lee, K. S. Lam, P. L. DeVries, and T. F. George, *J. Chem. Phys.*, **73**, 206 (1980).
33. R. K. Preston and J. C. Tully, *J. Chem. Phys.*, **54**, 4297 (1971).
34. J. C. Tully and R. K. Preston, *J. Chem. Phys.*, **55**, 562 (1971).
35. M. F. Herman, *J. Chem. Phys.*, **76**, 2949 (1982).
36. M. F. Herman, *J. Chem. Phys.*, **79**, 2771 (1983).
37. M. F. Herman, *J. Chem. Phys.*, **81**, 754 (1984).

38. M. F. Herman, *J. Chem. Phys.*, **81**, 764 (1984).
39. K. S. Lam and T. F. George, *J. Chem. Phys.*, **76**, 3396 (1982).
40. B. R. Johnson, *Chem. Phys.*, **2**, 381 (1973).
41. J. R. Laing and K. F. Freed, *Chem. Phys.*, **19**, 91 (1977).
42. M. F. Herman and K. F. Freed, *J. Chem. Phys.*, **78**, 6010 (1983).
43. M. F. Herman, *J. Chem. Phys.*, in press.
44. H. Bremmer, *Physica (Utrecht)*, **15**, 593 (1949); *ibid.*, *Commun. Pure Appl. Math.*, **4**, 105 (1951).
45. W. H. Miller, *Adv. Chem. Phys.*, **25**, 69 (1974).
46. R. A. Marcus, *J. Chem. Phys.*, **54**, 3965 (1971); *ibid.*, **56**, 311 (1972); *ibid.*, **59**, 5135 (1973).
47. J. N. L. Connor and R. A. Marcus, *J. Chem. Phys.*, **55**, 5636 (1971).
48. W. H. Wong and R. A. Marcus, *J. Chem. Phys.*, **55**, 5663 (1971).
49. E. J. Heller, *J. Chem. Phys.*, **62**, 1544 (1975); *ibid.*, *J. Chem. Phys.*, **65**, 4979 (1976); *ibid.*, *Chem. Phys. Lett.*, **34**, 321 (1975).
50. E. J. Heller, *J. Chem. Phys.*, **75**, 2923 (1981).
51. M. J. Davis and E. J. Heller, *J. Chem. Phys.*, **71**, 3383 (1979).
52. M. F. Herman and E. Kluk, *Chem. Phys.*, **91**, 27 (1984).
53. R. J. Glauber, *Phys. Rev.*, **131**, 2766 (1963).
54. J. R. Klauder, *J. Math. Phys.*, **4**, 1055 (1963); *ibid.*, **4**, 1058 (1963); *ibid.*, **5**, 177 (1964).
55. J. McKenna and J. R. Klauder, *J. Math. Phys.*, **5**, 878 (1964).

THE BREAKDOWN OF THE KRAMERS THEORY AS A PROBLEM OF CORRECT MODELING

FABIO MARCHESONI[‡]

Dublin Institute for Advanced Studies, 10 Burlington Road, Dublin 4, Ireland

CONTENTS

I. INTRODUCTION

Recently there has been a great revival of theoretical interest in the one-dimensional barrier-crossing problem and its applications to many physicochemical systems.[1-7] The problem has been modeled essentially as a "Brownian particle" moving into a double-well potential V. Since the original work of Kramers,[8] a number of investigators have improved and clarified several points. We mention in this context studies of multidimensional systems in the overdamped and underdamped limit,[9] of the effects of anharmonicities in the potential form,[7] of the role of non-Gaussian white thermal noise,[1,10] of the effect of a rate enhancement via parametric fluctuations,[11] and of the influence of the non-Markovian statistics of the heat bath.[2-6]

At the same time, some authors addressed the problem of the derivation of exact Langevin equations (LE), that is, the LE derived from a Liouville

‡Permanent address: Dipartimento di Fisica, Università di Perugia, I-06100 Perugia, Italy.

equation, and of their reduction to the mathematically more tractable phenomenological LE employed for modeling real physicochemical systems.[12-15] Mohanty et al.[15] studied in great detail the time dependence of the momenta of two Brownian particles of mass M interacting with a harmonic potential in a fluid of particles of mass m. Under the conditions $M \gg m$ and $\omega \tau_0 < 1$, where ω is the frequency of the Brownian oscillator and τ_0 is the relaxation time of the bath particles, a very general LE can be derived. Even though such conditions have been commonly assumed in the quoted literature,[1,11] the structure of this exact LE is still more complicated than that of the phenomenological LE actually treated. In particular, the friction coefficients are functions of $x(t)$, the separation of the oscillator particles, and the noise terms are generalized (i.e., not purely additive or purely multiplicative), Gaussian, and nonstationary.

The various approximations that must be made to reduce the LE derived from the Liouville equation to this simple one-dimensional LE are of three types: (1) The terms that describe the rotation of the oscillator in the fluid are neglected; (2) the $x(t)$ dependence of the friction coefficients, which arises from the interactions between the Brownian particles, is approximated *ad hoc*; and (3) the term involving the mean force exerted by the fluid on the oscillating Brownian particles is either neglected or approximated by a linear term in $x(t) - x_0$, where x_0 is the equilibrium interparticle separation of the oscillator.

With a few necessary restrictions, detailed in ref. 15, we can finally recover the phenomenological LE

$$\dot{x} = v, \qquad \dot{v} = -V'(x) - \lambda_0 v - 2\lambda_1 xv - \lambda_2 x^2 v + f(t) + x\eta(t) \quad (1.1)$$

where $f(t)$ and $\eta(t)$ are white Gaussian noises with

$$\langle f(t) \rangle \equiv 0, \qquad \langle f(t)f(0) \rangle = 2D_0\delta(t), \qquad \langle \eta(t) \rangle \equiv 0,$$
$$\langle \eta(t)\eta(0) \rangle = 2D_2\delta(t), \qquad \langle f(t)\eta(0) \rangle = 2D_1\delta(t) \quad (1.2)$$

where

$$D_i \equiv \lambda_i k_B T \qquad (k_B = \text{Boltzmann constant}) \quad (1.3)$$

Here $V(x)$ denotes the harmonic potential $\omega^2 x^2/2$. Lindenberg and Seshadri[14] were the first to obtain these LE, which they did by studying explicitly a specialized version of a *model* Hamiltonian introduced by Zwanzig[13] for a one-dimensional system interacting with a heat bath. Such a model admits as a peculiar feature an exact LE that can be derived by direct integration. The LE of Eq. (1.1) can then be recovered by employing the

Markovian approximation, in which the exact noise terms are assumed to be delta-correlated Gaussian stochastic processes.

Although criticism of ref. 15 is limited to the problem of the description of oscillating molecules in a fluid via one-dimensional LE with a simple noise structure, most arguments introduced by Mohanty et al.[15] and by Lindenberg and Seshadri[14] apply also to the problem of modeling the decay of a metastable state. This problem plays a central role in many areas of science, most notably in chemical kinetics, electron transport in semiconductors, and nonlinear optics. Detailed experimental work has recently been carried out by several groups[16,17] in order to answer the basic question: To what extent is a one-dimensional barrier-crossing picture applicable to actual physicochemical systems? Experimental discrepancies with the fundamental theory of Kramers[8] have been explained by recourse to one or more of the correcting mechanisms mentioned above. Memory effects due to the non-Markovian statistics of the heat bath coupled (phenomenologically) to the Brownian particle associated with the reaction coordinate $x(t)$ have been indicated as the most important cause of the remarkably increased activation rates of a number of chemical reactions in the high-friction limit.[1-6,17] The consequence of including such an additional mechanism is a "frequency-dependent friction,"[2-5] which is supposed to account for the unclear separation between the heat bath relaxation time scale τ_0 and the "mechanical" time scales related to the characteristic frequencies of the driving potential $V(x)$.

In contrast, nobody has heeded the suggestion, implicit in the exact approaches of ref. 12, that friction terms appearing in the LE modeling any single process under investigation may involve a dependence on the reaction coordinate itself, which generally will be nonfactorable. The present paper is aimed at extending Lindenberg and Seshadri's approach to the case in which the Brownian particle is driven by a double-well potential in the underdamped and overdamped limits. The x-dependent friction terms are shown to affect the rate of escape over the barrier (i.e., the relaxation process) distinctly in the two regimes. Our main conclusion is that the specific nature of the coupling between the Brownian particle and the heat bath cannot generally be neglected by substituting for the generalized friction term an effective one $(-\lambda_{\mathrm{eff}}\dot{x}(t))$ (ref. 14) that is proportional in some fashion to the solvent viscosity (that this can be done is referred to as the *hydrodynamic assumption*).[17]

The organization of this paper is as follows. In Section II we discuss, using projection operator techniques, the derivation of the LE (1.1)–(1.3) from Zwanzig's model Hamiltonian. Corrections due to the presence of anharmonicities in the Hamiltonian describing the heat bath and to the coupling with the system of interest are accounted for. In Section III we adapt

Lindenberg and Seshadri's model to the problem of the decay of a metastable state. The corresponding corrections to the Kramers activation rates are estimated in the case of small x-dependent friction terms in both the overdamped and the underdamped regime. In Section IV Lindenberg and Seshadri's derivation[14] of LE (1.1)–(1.3) is improved by taking into account the effects of the non-Markovian statistics of the heat bath. Finally, in Section V we summarize our findings and discuss their implications for applications to chemical–physical problems.

II.   THE ONE-DIMENSIONAL ZWANZIG MODEL

The most general Hamiltonian H for a one-dimensional system coupled to a heat bath can be decomposed as follows:

$$H = H_S + H_B + H_{SB} \tag{2.1}$$

Here H_S is the system Hamiltonian, H_B is the bath Hamiltonian, and H_{SB} is the term accounting for interaction between the system and the heat bath. Let us assume that our system is described in the phase space (x, p) by the classical Hamiltonian

$$H_S = \frac{p^2}{2M} + U(x) \tag{2.2}$$

H_S describes the dynamics of a particle of mass M bounded by a nonlinear potential $U(x)$. The heat bath consists of N independent harmonic oscillators of mass m_ν, canonical coordinates (q_ν, p_ν), and frequency ω_ν ($\nu = 1, \ldots, N$). The bath Hamiltonian is thus a diagonal bilinear form:

$$H_B = \sum_\nu \left(\frac{p_\nu^2}{2m_\nu} + \frac{\omega_\nu^2}{2} q_\nu^2 \right) \tag{2.3}$$

Following Zwanzig,[13] H_B may be written as

$$H_B = \tfrac{1}{2} X^T \cdot K \cdot X \tag{2.4}$$

where $X = (q_1, \ldots, q_N, p_1, \ldots, p_N)$, T denotes the transpose, and K is a symmetric nonsingular $2N \times 2N$ matrix. The choice of Eq. (2.3) corresponds to diagonalizing Zwanzig's bath Hamiltonian (without any loss of generality), and K reads accordingly as follows:

$$K_{ij} - \delta_{ij} \begin{cases} \omega_j & j \le N \\ m_{j-N}^{-1} & j > N \end{cases} \tag{2.5}$$

As pointed out by Mohanty et al.,[15] the assumptions we now make to give H_{SB} a more tractable form can affect dramatically both the potential and the dissipation terms in the final LE. Let us start by assuming the interaction to be linear in the bath coordinates:

$$H_{SB} = H_{SB}^{(0)} \equiv - \sum_{\nu} \omega_{\nu}^2 q_{\nu} a_{\nu}(x) \tag{2.6}$$

The total Hamiltonian can then be rewritten as

$$H = \frac{p^2}{2M} + V(x) + \frac{1}{2} \sum_{\nu} \left[\frac{p_{\nu}^2}{2m_{\nu}} + \omega_{\nu}^2 (q_{\nu} - a_{\nu}(x))^2 \right] \tag{2.7}$$

where

$$V(x) = U(x) - \frac{1}{2} \sum_{\nu} \omega_{\nu}^2 a_{\nu}^2(x) \tag{2.8}$$

The choice of Eq. (2.6) implies that the potential energy of the isolated system will be corrected by including a "static" portion of the interaction, as emerges from the following treatment.[‡]

One of the attractive features of this choice is that it enables us to establish explicitly the connection between the fluctuating force in the LE and the form of the interaction in the Hamiltonian. Following Lindenberg and Seshadri's restatement of Zwanzig's procedure,[14] we can integrate the Hamiltonian equations for the heat bath and eliminate the bath coordinates from the system equations. They then read

$$\dot{x} = \frac{p}{M}$$

$$\dot{p} = -V'(x) + \int_0^t d\tau \left\{ \sum_{\nu} a_{\nu}'(x(t)) a_{\nu}'(x(t-\tau)) \omega_{\nu}^2 \cos\left(\frac{\omega_{\nu}}{\sqrt{m_{\nu}}} \tau \right) \right\}$$

$$\times p(t-\tau) + [\bar{a}'(x(t))]^T F(t) \tag{2.9}$$

[‡] Discrepancies between our formulas (2.8), (2.10), and (2.18) and the corresponding ones of ref. 14 are due to some minor mistakes in that article.

where $\bar{a}(x)$ is the $2N$-dimensional column vector $(a_1(x), \ldots, a_N(x), 0, \ldots, 0)$. [A detailed derivation of the generalized LE, Eq. (2.9), can be found in refs. 13 and 14.] $F(t)$ is the initial condition-dependent portion, whose statistical properties can be likened to those of a Gaussian noise with zero mean value and autocorrelation functions

$$\langle F_\nu(t) F_\nu^T(t') \rangle = k_B T \omega_\nu^2 \cos\left(\frac{\omega_\nu}{\sqrt{m_\nu}} (t - t') \right) a_\nu'^2(x) \qquad (2.10)$$

To obtain the LE (1.1)–(1.3), Lindenberg and Seshadri choose a quadratic form for the coupling components $a_\nu(x)$ and, in addition, introduce the Markovian approximation so that the random forces $F(t)$ in Eq. (2.10) are delta correlated.

An alternative procedure consists of employing an equivalent Fokker–Planck formalism. The corresponding Fokker–Planck (FP) equation can be obtained by means of an adiabatic elimination procedure[18] that allows us to eliminate the bath variables, provided that $\omega \tau_0 < 1$, where $1/\omega$ now denotes a suitable mechanical time scale related to the effective potential $V(x)$. By changing the bath variables,

$$p_\nu \rightarrow P_\nu, \qquad q_\nu \rightarrow Q_\nu \equiv q_\nu - a_\nu(x) \qquad (2.11)$$

the Hamiltonian equations corresponding to the total Hamiltonian of Eq. (2.7) can be rewritten as

$$\dot{x} = \frac{p}{M}$$

$$\dot{p} = -V'(x) + \sum_\nu \omega_\nu^2 Q_\nu a_\nu'(x)$$

$$\dot{Q}_\nu = \frac{p_\nu}{m_\nu} - \frac{a_\nu'(x) p}{M} \qquad (2.12)$$

$$\dot{p}_\nu = -\omega_\nu^2 Q_\nu, \qquad \nu = 1, \ldots, N$$

The related Liouvillian operator $\mathbf{L}$, defined as $i[H, \ldots]$, where $[\ldots, \ldots]$ are the Poisson brackets, can be separated into an unperturbed part $(\mathbf{L}_0)$ and a perturbation $(\mathbf{L}_I)$:

$$\mathbf{L}_0 = \sum_\nu \left(-\frac{p_\nu}{m_\nu} \frac{\partial}{\partial Q_\nu} + \omega_\nu^2 Q_\nu \frac{\partial}{\partial p_\nu} \right) \qquad (2.13)$$

$$\mathbf{L}_I = -\frac{p}{M} \frac{\partial}{\partial x} + V'(x) \frac{\partial}{\partial p} - \sum_\nu \left(\omega_\nu^2 Q_\nu \frac{\partial}{\partial p} - \frac{p}{M} \frac{\partial}{\partial Q_\nu} \right) a_\nu'(x) \qquad (2.14)$$

For clarity we give further details of our perturbation technique in the Appendix. This applies in the presence of a clear-cut time-scale separation between the heat bath relaxation process and the mechanical driving by the potential $V(x)$ (i.e., when $\omega \tau_0 \ll 1$, in the notation of ref. 15). Our final result is a FP equation of the type

$$\frac{\partial}{\partial t} \rho(x, p; t) = \sum_{r=0}^{\infty} \Gamma_r \rho(x, p; t) \tag{2.15}$$

where $\rho(x, p; t)$ is the reduced distribution function in the relevant canonical coordinates of the system under study, and Γ_r are the perturbation terms of order r of the corresponding FP operator. In particular, we find

$$\Gamma_0 = -\frac{p}{M} \frac{\partial}{\partial x} + V'(x) \frac{\partial}{\partial p} \tag{2.16}$$

$$\Gamma_1 = \gamma(N, x) \left(k_B T \frac{\partial^2}{\partial p^2} + \frac{1}{M} \frac{\partial}{\partial p} p \right) \tag{2.17}$$

where

$$\gamma(N, x) \equiv \int_0^\infty d\tau \sum_\nu \omega_\nu^2 a_\nu'^2(x) \cos\left(\frac{\omega_\nu}{\sqrt{m_\nu}} \tau \right) \tag{2.18}$$

With the choice of ref. 14 for $a_\nu(x)$,

$$a_\nu(x) \equiv \Gamma_\nu x + \frac{\beta_\nu x^2}{2} \tag{2.19}$$

we readily obtain[‡]

$$\gamma(N, x) = \lambda_0 + 2\lambda_1 x + \lambda_2 x^2. \tag{2.20}$$

We remark that the FP equation (2.15)–(2.17) with the friction terms given by Eq. (2.20) corresponds exactly to the LE (1.1)–(1.3). The Markovian assumption is now implicit in the truncation of the series of Eq. (2.15) at $r = 1$.

Before going beyond such an approximation by calculating Γ_2, we briefly discuss the critical choice of Eq. (2.6) for the interaction Hamiltonian H_{SB}.

[‡] It is not our purpose here to establish the conditions under which the convergence of integral (2.18) can be proved. The assumption is appropriate if, for instance, N is large and $\omega_\nu / \sqrt{m_\nu}$ is an irrational number.

Although a very general choice of H_{SB} makes the model intractable, we can slightly improve our understanding of its role by assuming that the linear term $H_{SB}^{(0)}$ is *perturbed* by nonlinear corrections of the type

$$H_{SB} - H_{SB}^{(0)} = -\sum_{\nu} Q_{\nu}^n b_{\nu}(x) \tag{2.21}$$

for any $n > 1$. The additional interaction modifies the perturbation part $\mathbf{L}_I$ of the Liouvillian operator as follows:

$$\mathbf{L}_I = \mathbf{L}_I[\text{of Eq. (2.14)}] - \sum_{\nu} \left[Q_{\nu}^n b_{\nu}'(x) - n Q_{\nu}^{n-1} a_{\nu}'(x) b_{\nu}(x) \right] \frac{\partial}{\partial p}$$

$$- \sum_{\nu} n Q_{\nu}^{n-1} b_{\nu}(x) \frac{\partial}{\partial p_{\nu}} \tag{2.22}$$

The third term on the right-hand side of Eq. (2.22) does not contribute to our FP equation (see Appendix). Without a loss of generality, we may assume, for instance, that n is even, so that $\mathbf{L}_I$ can conveniently be rewritten as

$$\mathbf{L}_I = -\frac{p}{M} \frac{\partial}{\partial x} + \left(V'(x) - \sum_{\nu} Q_{\nu}^n b_{\nu}'(x) \right) \frac{\partial}{\partial p}$$

$$- \sum_{\nu} \left[Q_{\nu} \left(\omega_{\nu}^2 - n Q_{\nu}^{n-2} b_{\nu}(x) \right) \frac{\partial}{\partial p} - \frac{p}{M} \frac{\partial}{\partial Q_{\nu}} \right] a_{\nu}'(x) \tag{2.23}$$

On applying the perturbation technique outlined in the Appendix, we find for Γ_0 and Γ_1 the same formal expressions as in Eqs. (2.16) and (2.17), respectively, where $V(x)$ now reads

$$V(x) = U(x) - \frac{1}{2} \sum_{\nu} \omega_{\nu}^2 a_{\nu}^2(x) - \sum_{\nu} (n-1)!! \left(\frac{k_B T}{\omega_{\nu}^2} \right)^{n/2} b_{\nu}(x) \tag{2.24}$$

and $\gamma(N, x)$ exhibits an explicit dependence on $k_B T$. The explicit dependence on the temperature is due to the averages $\langle Q_{\nu}^m \rangle$ taken over the unperturbed equilibrium bath distribution (see Appendix).

The corrections to the isolated nonlinear potential $U(x)$ are the exact counterpart of the mean force exerted by the fluid on the molecular oscillator as it appears in the LE obtained by Mohanty et al.[15] If $a_{\nu}(x)$ and $b_{\nu}(x)$ are chosen to be polynomials in x, $\gamma(N, x)$ assumes a form still resembling that of Eq. (2.20):

$$\gamma(N, x) = \sum_{k=0} \lambda_k(T) x^k \qquad (2.25)$$

It is noteworthy that the same kind of corrections to Eqs. (2.16) and (2.17) can be determined also by assuming that the heat bath consists of nonlinear oscillators, provided that the nonlinearities can be treated perturbatively. If we add a nonlinear perturbation term to H_B in Eq. (2.3) and change variables as in Eq. (2.11), such a result follows immediately from our perturbation approach. We conclude that the T-dependence exhibited by both the phenomenological potential $V(x)$ and the friction terms $\gamma(N, x)$ [see Eqs. (2.24) and (2.25)] is general, and should be traced back to the intrinsic nonlinear features of the total system and in particular of the Hamiltonians H_B and H_{SB}. In Section V we discuss the physical relevance of such a dependence for applications to chemical–physical systems. For the purposes of Sections III and IV, however, nonlinear corrections to H_B and H_{SB} can be disregarded without loss of generality.

With the choices of Eqs. (2.3) and (2.6) for H_B and H_{SB}, we can easily compute Γ_2 of Eq. (2.15). On employing our adiabatic elimination technique, we readily find

$$\Gamma_2 = -\zeta_1(N, x) \frac{\partial}{\partial p}\left[k_B T \frac{\partial}{\partial p} + \frac{p}{M}\right]\left[-\frac{p}{M}\frac{\partial}{\partial x} + V'(x)\frac{\partial}{\partial p}\right]$$

$$+ \zeta_2(N, x) \frac{\partial}{\partial p}\left[-\frac{p}{M}\frac{\partial}{\partial x} + V'(x)\frac{\partial}{\partial p}\right]\left[k_B T \frac{\partial}{\partial p} + \frac{p}{M}\right]$$

$$- \zeta_3(N, x) \frac{\partial}{\partial p}\frac{p}{M}\left[k_B T \frac{\partial}{\partial p} + \frac{p}{M}\right] \qquad (2.26)$$

where

$$\zeta_1(N, x) = \frac{1}{2}\int_0^\infty ds_0 \int_0^{s_0} \sum_\nu a_\nu'^2(x)\omega_\nu^2 \cos\left(\frac{\omega_\nu}{\sqrt{m_\nu}}s_0\right) ds_1 \qquad (2.27a)$$

$$\zeta_2(N, x) = \frac{1}{2}\int_0^\infty ds_0 \int_0^{s_0} \sum_\nu a_\nu'^2(x)\omega_\nu^2 \cos\left[\frac{\omega_\nu}{\sqrt{m_\nu}}(s_0 - s_1)\right] ds_1 \qquad (2.27b)$$

$$\zeta_3(N, x) = \frac{1}{2}\int_0^\infty ds_0 \int_0^{s_0} \sum_\nu a_\nu'(x)a_\nu''(x)\omega_\nu^2 \cos\left[\frac{\omega_\nu}{\sqrt{m_\nu}}(s_0 - s_1)\right] ds_1$$

$$(2.27c)$$

The structure of Γ_2 is rather complicated. On following the procedure adopted for Γ_1, by choosing an explicit form for $a_\nu(x)$ and assuming the convergence of the integrals in Eqs. (2.27), ζ_1, ζ_2, and ζ_3 can be given the form of polynomials in x; eight new parameters (three from ζ_1 and ζ_2 each and two from ζ_3) control the non-Markovian corrections at the lowest per-turbation order. In Section IV we shall study numerically the role of the non-Markovian statistics of the heat bath under some stronger assumptions, in order to gain a deeper comprehension of the underlying dynamics.

III. ACTIVATION RATES IN THE MARKOVIAN LIMIT

In Section II we discussed the matter of under which assumptions the phenomenological LE (1.1) can be employed as a sensible description of a chemical reaction. Apart from the possible T-dependence of both the effective potential $V(x)$ and the friction terms arising from the inevitable non-linearities of $H_S + H_{SB}$, the Markovian statistics of the heat bath are understood in the system of Eqs. (1.1) as the main assumption. In this section we estimate the quantitative corrections that can be made to the rate of escape due to the multiplicative friction terms in λ_1 and λ_2 provided that these can be regarded as small in comparison with the usual dissipation term $-\lambda_0 v$ of the Kramers theory. Our treatment applies also to more general choices for $V(x)$ and $\gamma(N, x)$ than those of Eqs. (2.24) and (2.25), respectively.

A. The Overdamped Limit

We study first the limit most discussed in the literature,[2-8] that of high viscosity and large activation energy. For simplicity we assume our effective potential to be modeled as

$$V(x) = \frac{a^2}{4b} - \frac{ax^2}{2} + \frac{bx^4}{4} \tag{3.1}$$

This represents a symmetric double-well potential with two stable fixed points, $x_\pm = \pm(a/b)^{1/2}$, an unstable fixed point in $x = 0$, and an activation energy defined as

$$\Delta V = V(0) - V(x_\pm) = \frac{a^2}{4b} \tag{3.2}$$

The height of the barrier ΔV is assumed to be large compared with the thermal energy $k_B T$. Furthermore, the characteristic mechanical time scale mentioned in Section II is now given by $\sqrt{a}$, where $V''(0) = a$ and $V''(x_\pm) = 2a$.

Here, "high viscosity" will mean that $\lambda_0 \gg \sqrt{a}$, since we choose to consider the x-dependent friction terms as comparatively small. This is the well-known overdamped limit of our system.

We proceed further by applying the standard analysis,[1,3,7] which consists of eliminating the variable velocity perturbatively. We employ again the perturbation technique in the Appendix.

The FP operator corresponding to the LE (1.1)–(1.3) can be divided into a perturbation Γ_I and an unperturbed part Γ_0, given by

$$\Gamma_0 = \lambda_0 \left(\frac{\partial}{\partial v} v + k_B T \frac{\partial^2}{\partial v^2} \right) \tag{3.3}$$

$$\Gamma_I = -v \frac{\partial}{\partial x} + V'(x) \frac{\partial}{\partial v} + \left(\lambda_2 x^2 + 2\lambda_1 x \right) \left[\frac{\partial}{\partial v} v + k_B T \frac{\partial^2}{\partial v^2} \right] \tag{3.4}$$

The final result of our projection technique can be written as follows:

$$\frac{\partial}{\partial t} p(x;t) = \frac{1}{\lambda_0} \frac{\partial}{\partial x} D(x) j(x) p(x;t)$$

$$+ \frac{1}{\lambda_0^3} \left[\frac{\partial}{\partial x} j(x) \frac{\partial}{\partial x} j(x) - \frac{\partial'}{\partial x^2} j^2(x) \right] p(x;t) \tag{3.5}$$

where

$$j(x) = V'(x) + k_B T \frac{\partial}{\partial x} \tag{3.6}$$

and

$$D(x) = \left[1 - 2\frac{\lambda_1}{\lambda_0} x + \left(\frac{4\lambda_1^2}{\lambda_0^2} - \frac{\lambda_2}{\lambda_0} \right) x^2 + \frac{4\lambda_1\lambda_2}{\lambda_0^2} x^3 - \frac{\lambda_2^2}{\lambda_0^2} x^4 \right] \tag{3.6'}$$

Further details about this kind of calculation have been reported in refs. 18 and 19. We notice that generally the x-dependent friction coefficients ($2\lambda_1 x$ and $\lambda_2 x^2$) are to be taken as large in comparison with $\sqrt{a}$ (even if they are small with respect to λ_0). This implies that the second term on the right-hand side of Eq. (3.5) can be neglected. It is noteworthy that at any perturbation order the stationary distribution $\bar{p}(x)$ maintains its canonical form:[18]

$$\bar{p}(x) = \mathcal{N} \exp\left(-\frac{V(x)}{k_B T} \right) \tag{3.7}$$

where $\mathcal{N}$ is a normalization constant.

A useful estimate of the activation rate for this type of problem is introduced in refs. 5 and 20. If T_0 denotes the mean first-passage time (MFPT) to reach the barrier top, the activation rate μ is defined as

$$\mu = \frac{1}{T_0} \tag{3.8}$$

The first term on the right-hand side of Eq. (3.5) can be rewritten as

$$\frac{\partial}{\partial t} p(x;t) = \left[\frac{\partial}{\partial x} W(x) + k_B T \frac{\partial^2}{\partial x^2} D(x) \right] p(x;t) \tag{3.9}$$

where

$$W(x) \equiv V'(x)D(x) - k_B T D'(x) \tag{3.10}$$

With these assumptions made, the diffusion coefficient $k_B T D(x)$ is positive within the bistable region $[x_-, x_+]$, so the MFPT can be readily evaluated. If $x = \pm \infty$ are our natural reflecting boundaries and $x = 0$ an absorbing state, one finds for the MFPT[5,20]

$$T_0 = \frac{1}{k_B T} \int_{-\infty}^{+\infty} \frac{dx}{D(x)\bar{p}(x)} \int_{-\infty}^{x} \bar{p}(z)\, dz \tag{3.11}$$

In Eq. (3.11) the symmetry of both $V(x)$ and $\bar{p}(x)$ are taken into account.

In the case of a large barrier, that is, when $\Delta V / k_B T \gg 1$, we can evaluate Eq. (3.11) by using the method of steepest descent. We readily obtain the following expression for μ [Eq. (3.8)]:

$$\mu(\lambda_i) = \frac{\mu(\lambda_0)}{H(\lambda_1/\lambda_0, \lambda_2/\lambda_0)} \tag{3.12}$$

where $\mu(\lambda_0)$ is the well-known rate of escape calculated by Kramers[8] in the overdamped limit,

$$\mu(\lambda_0) = \frac{a}{\pi\sqrt{2}} \frac{1}{\lambda_0} \exp\left(-\frac{\Delta V}{k_B T} \right) \tag{3.13}$$

and $H(\lambda_1/\lambda_0, \lambda_2/\lambda_0)$ contains the corrections we are looking for:

$$H\left(\frac{\lambda_1}{\lambda_0}, \frac{\lambda_2}{\lambda_0} \right) = \left[1 + \left(\frac{4\lambda_1^2}{\lambda_0^2} - \frac{\lambda_2}{\lambda_0} \right) \frac{k_B T}{a} - \frac{\lambda_2^2}{\lambda_0^2} \left(\frac{k_B T}{a} \right)^2 \right] \tag{3.14}$$

We note that the contribution from the terms of the diffusion coefficient $k_B TD(x)$ that are odd in x vanish exactly for symmetric potentials.

We make now two remarks:

1. The restrictions under which our perturbation technique is valid can be determined from Eq. (3.14). The assumption of a definitive positive diffusion $k_B TD(x)$ within the bistable region is satisfied when

$$\frac{\lambda_2}{\lambda_0} \frac{k_B T}{a}, \quad \frac{\lambda_1}{\lambda_0} \left(\frac{k_B T}{a} \right)^{1/2} \ll 1 \qquad (3.15)$$

Such an inequality means that the x-dependent friction terms are small compared with $-\lambda_0 v$ (ref. 14).

2. The effects of the internal multiplicative noise $(\lambda_1, \lambda_2 \neq 0)$ on the activation rate are determined by the prefactor $H(\lambda_1/\lambda_0, \lambda_2/\lambda_0)^{-1}$. The dependence on the temperature is no longer controlled solely by the Arrhenius factor in Eq. (3.13), even assuming that λ_0, λ_1, and λ_2 are constant; the rate of escape increases or decreases depending on whether $4\lambda_1^2$ is smaller than $\lambda_0\lambda_2$ or not.

In Section V we shall discuss some consequences of the main results of the present section for applications to practical chemical–physical problems.

B. The Underdamped Limit

Let us now face the problem of "small" friction terms and large activation energies. Following Stratonovitch[21] and adopting the notation of ref. 14, we describe the system in terms of the displacement x and the energy envelope E:

$$E \equiv \frac{v^2}{2} + V(x) \qquad (3.16)$$

The energy-envelope technique is based on the assumption that the average energy envelope $\langle E(t) \rangle$ varies slowly compared with the average displacement $\langle x(t) \rangle$. This condition places two restrictions on the parameter values for which the technique is valid: (1) The damping must be weak in comparison with the characteristic mechanical frequencies; and (2) the variations in the average energy envelope must occur slowly in comparison with the average oscillation inside a single potential well. We shall justify the application of such a technique to our problem at the end.

On changing variables,

$$x \to x, \quad v \to E \qquad (3.17)$$

the FP equation corresponding to the LE (1.1)–(1.3) reads:[14]

$$\frac{\partial}{\partial t} P(x, E; t) = \left\{ -\frac{\partial}{\partial x} \{2[E - V(x)]\}^{1/2} + 2(\lambda_0 + 2\lambda_1 x + \lambda_2 x^2) \right.$$

$$\times \frac{\partial}{\partial E}[E - V(x)] - (D_0 + 2D_1 x + D_2 x^2)\frac{\partial}{\partial E}$$

$$\left. + 2(D_0 + 2D_1 x + D_2 x^2)\frac{\partial^2}{\partial E^2}[E - V(x)] \right\} P(x, E; t)$$

(3.18)

where the probability density $P(x, E; t)$ is related to $\rho(x, v; t)$ occurring in Eq. (2.15) by

$$P(x, E; t)\, dx\, dE = \rho(x, v; t)\, dx\, dv \qquad (3.19)$$

$P(x, E; t)$ can be written exactly as the product[21]

$$P(x, E; t) = w(x, t | E) p_E(E; t) \qquad (3.20)$$

where $w(x, t | E)$ is the probability density that the displacement at time t is x conditional on its energy envelope being E (and also conditional on the initial conditions). The method of Stratonovitch[21] is based on the assumption that independently of the initial condition $(x(0), E(0))$, $w(x, t | E)$ is proportional to the time that the system—with energy envelope E—spends at x. The time spent at x is, in turn, inversely proportional to the velocity at x, that is, to $v(t)$. Thus we obtain

$$w(x, t | E) = 2\phi'(E)[E - V(x)]^{1/2} \qquad (3.21)$$

where

$$\phi(E) \equiv \int_R [E - V(x)]^{1/2}\, dx \qquad (3.22)$$

and the prime denotes a derivative with respect to E. The region of integration R in Eq. (3.22) defines the domain of x for which $E \geq V(x)$. On substituting Eqs. (3.20) and (3.21) into Eq. (3.18) and integrating over x, we find an approximate FP equation for the reduced probability density $p_E(E; t)$:[14]

$$\frac{\partial}{\partial t} p_E(E; t) = \left\{ \frac{\partial}{\partial E}\left[\lambda_0 \frac{\phi(E)}{\phi'(E)} - D_0 + \frac{\lambda_2 \psi(E) - D_2 \psi'(E)}{\phi'(E)} \right] \right.$$

$$\left. + \frac{\partial^2}{\partial E^2}\left[\frac{D_0\phi(E) + D_2\psi(E)}{\phi'(E)} \right] \right\} p_E(E; t) \qquad (3.23)$$

where

$$\psi(E) \equiv \int_R x^2 [E - V(x)]^{1/2} dx \tag{3.24}$$

Note that in this approximation the contributions of the terms proportional to λ_1 vanish.

The equilibrium distribution $\bar{p}_E(E)$ of the FP equation (3.23) can be readily calculated:

$$\bar{p}_E(E) = \mathcal{N}\phi'(E)\exp\left(-\frac{E}{k_B T}\right) \tag{3.25}$$

where $\mathcal{N}$ is a normalization constant.

We propose the following definition of activation time T_E in the underdamped limit: T_E coincides with the average time needed for the energy envelope $\langle E(t) \rangle$ to reach the value of the activation energy ΔV starting from its mean value E_0; that is,

$$T_E(E_0) = \int_{E_0}^{\Delta V} \frac{dE}{\bar{p}_E(E)D(E)} \int_0^E p_E(E') dE' \tag{3.26}$$

where $D(E)$ is the diffusion coefficient in Eq. (3.23):

$$D(E) = \frac{D_0\phi(E) + D_2\psi(E)}{\phi'(E)} \tag{3.27}$$

Expression (3.26) is the counterpart of Eq. (3.11) and was obtained by solving the corresponding MFPT problem as outlined in ref. 21.

To estimate E_0 we must calculate explicitly $\phi(E)$ in Eq. (3.22). That integral involves complete elliptic integrals of the first and second kind. In the limit of high activation energies, however, we can suitably avoid this difficulty by linearizing $V(x)$ around the potential minima x_+. The height of the barrier is kept equal to ΔV and the frequencies of such parabolas are the same as those obtained by linearly expanding the potential $V(x)$ around x_+ and x_-, respectively. Thus we find an approximate expression for $\phi(E)$:

$$\phi(E) \approx \phi_0(E) \equiv 2 \int_{-\sqrt{E/a}}^{+\sqrt{E/a}} \sqrt{E - ax^2} \, dx = \frac{\pi}{\sqrt{a}} E \tag{3.28}$$

Analogously, for $\psi(E)$,

$$\psi_0(E) = \frac{\pi}{\sqrt{a}} \frac{E^2}{4a} \tag{3.29}$$

We note that such an estimate works fairly well in the limit $\Delta V/k_B T \gg 1$ and that the first corrections to $\phi(E)$ are proportional to $\phi_0(E)(E/\Delta V)^{1/2}$. On substituting $\phi_0(E)$ into Eq. (3.25), we determine that

$$E_0 = k_B T \tag{3.30}$$

We are now in a position to work out Eq. (3.26). Substituting Eqs. (3.25) and (3.27) with Eqs. (3.28) and (3.29) into Eq. (3.26) yields

$$T_E(k_B T) = \frac{(k_B T)^{-1}}{\lambda_0} \int_{k_B T}^{\Delta V} \frac{e^{E/k_B T}}{k_B T \phi_0(E) + (\lambda_2/\lambda_0)\psi_0(E)} dE$$
$$\times \int_0^E \phi_0'(E') e^{-E'/k_B T} dE' \tag{3.31}$$

On integrating by parts the right most integral, we obtain

$$T_E(k_B T) = \frac{(k_B T)^{-2}}{\lambda_0} \int_{k_B T}^{\Delta V} \frac{e^{E/k_B T}}{k_B T \phi_0(E) + (\lambda_2/\lambda_0)\psi_0(E)} dE$$
$$\times \int_0^E \phi_0(E') e^{-E'/k_B T} dE' \tag{3.32}$$

where terms $O(\Delta V/k_B T)$ are negligible compared to terms $O(e^{\Delta V/k_B T})$ from (3.32). We can now separate $T_E(k_B T)$ into two parts as follows:

$$T_E(k_B T) \equiv \langle \tau \rangle + \delta \langle \tau \rangle \tag{3.33}$$

where

$$\langle \tau \rangle \equiv \frac{(k_B T)^{-2}}{\lambda_0} \int_{k_B T}^{\Delta V} \frac{e^{E/k_B T}}{\phi_0(E)} dE \int_0^E \phi_0(E') e^{-E'/k_B T} dE' \tag{3.34}$$

and

$$\delta \langle \tau \rangle = -(k_B T)^{-2} \frac{\lambda_2}{\lambda_0} \int_{k_B T}^{\Delta V} \frac{\phi_0(E) e^{E/k_B T}}{\lambda_0 \phi_0(E) + \lambda_2 \psi_0(E)} dE \int_0^E \phi_0(E') e^{-E'/k_B T} dE' \tag{3.35}$$

Here $\langle \tau \rangle$ denotes the limit of $T_E(k_B T)$ for $\lambda_2 \to 0$, while $\delta \langle \tau \rangle$ is the correction due to the x-dependent friction terms.

The integrals in Eqs. (3.34) and (3.35) can be calculated explicitly by substituting Eqs. (3.28) and (3.29) into them:

$$\langle \tau \rangle = \frac{1}{\lambda_0}\left[\mathrm{Ei}\left(\frac{\Delta V}{k_B T}\right) - \mathrm{Ei}(1) \right] \qquad (3.36)$$

$$\delta\langle \tau \rangle = -\frac{1}{\lambda_0}\left[\mathrm{Ei}\left(\frac{\Delta V}{k_B T} + \frac{1}{\beta k_B T}\right) - \mathrm{Ei}\left(1 + \frac{1}{\beta k_B T}\right) \right] e^{-1/\beta k_B T} \qquad (3.37)$$

where $\beta \equiv \lambda_2 / 4a\lambda_0$ and $\mathrm{Ei}(x)$ denotes the exponential–integral function,[22] which can be expanded as

$$\mathrm{Ei}(x) = e^x \sum_{k=1}^{\infty} \frac{(k-1)!}{x^k} \qquad (3.38)$$

Employing Eq. (3.38), we can determine the leading term in Eq. (3.36):

$$\langle \tau \rangle = \frac{1}{\lambda_0} \frac{k_B T}{\Delta V} \exp\left(\frac{\Delta V}{k_B T}\right) \qquad (3.39)$$

In view of the approximations introduced in Eqs. (3.28) and (3.29), contributions proportional to $\langle \tau \rangle (k_B T/\Delta V)^k$, $k > 1$, are meaningless. The inverse of $\langle \tau \rangle$ in Eq. (3.39) coincides exactly with the well-known rate of escape found by Kramers[8] in the underdamped limit. This result makes us more confident about basing our approach on the energy-envelope technique and on definition (3.26) of activation time.

Analogously, expanding Eq. (3.37) to the first order in $k_B T/\Delta V$, we find

$$\delta\langle \tau \rangle = -\frac{k_B T}{\lambda_0}\left(\Delta V + \frac{1}{\beta}\right)^{-1} \exp\left(\frac{\Delta V}{k_B T}\right) \qquad (3.40)$$

Putting Eqs. (3.39) and (3.40) together, we conclude that

$$T_E(k_B T) \approx \langle \tau \rangle \left(1 - \frac{\beta \Delta V}{1 + \beta \Delta V}\right) \qquad (3.41)$$

We now make several remarks:

1. The activation rate in the underdamped limit,

$$\mu_E(\lambda_0, \lambda_2) \equiv T_E^{-1}(k_B T) \qquad (3.42)$$

is an increasing function of λ_2. In the frame of the Stratonovitch method,

Eqs. (3.26) and (3.27) prove this immediately. In the limit of high activation energies [see Eqs. (3.28) and (3.29)] we obtain, from Eq. (3.41),

$$\mu_E(\lambda_0, \lambda_2) = \langle \tau \rangle^{-1}(1 + \beta\Delta V) = \mu_E(\lambda_0)\left(1 + \frac{\lambda_2}{\lambda_0}\frac{a}{16b}\right) \qquad (3.43)$$

where $\mu_E(\lambda_0)$ is the inverse of the Kramers escape time for $\lambda_0/\sqrt{a} \to 0$ given in Eq. (3.39). If we compare this result to that of Eqs. (3.12)–(3.14) for the overdamped limit, we conclude that the x-dependent friction terms play different roles in the two viscosity regimes.

2. The restrictions under which Eq. (3.32) for T_E is valid can be summarized as follows:

$$\frac{\Delta V}{k_B T} \gg 1 \quad \text{(high activation energy)} \qquad (3.44)$$

$$\lambda_0\left(1 + \frac{\lambda_2}{\lambda_0}\frac{a}{16b}\right) \ll \sqrt{a} \quad \text{(low friction)} \qquad (3.45)$$

The second inequality can be justified by noting that its first term plays the role of an "effective" friction constant in $\mu_E(\lambda_0, \lambda_2)$ [see Eq. (3.43)] and that $\mu_E(\lambda_0)$ has been obtained in the limit $\lambda_0 \ll \sqrt{a}$. The same conclusion can be reached by supposing that the effective friction constant $\lambda_0 - \lambda_2\langle\psi_0(E)/\phi_0(E)\rangle_E$ in the denominator of the first integral in Eq. (3.32) is very small compared with $\sqrt{a}$ (ref. 14) [$\langle\ldots\rangle_E$ denotes the average with respect to the energy equilibrium distribution (3.25)]. When $\beta\Delta V$ (i.e., λ_2) is small, then Eq. (3.45) reduces simply to Stratonovitch's original weak damping condition $\lambda_0 \ll \sqrt{a}$. On the other hand, when $\beta\Delta V$ is large, relation (3.45) restricts the range of values of λ_0 and λ_2 to

$$\frac{\lambda_0}{\sqrt{a}} \ll \frac{\lambda_0}{\lambda_2}\frac{b}{16a} \ll 1 \qquad (3.46)$$

3. Equation (3.46) implies that there are ranges of parameter values within which the multiplicative fluctuations and the corresponding damping can have very strong dynamical effects. In such a range the condition that the energy-envelope variations are slow compared with the average oscillation inside a single potential well is certainly satisfied (note that $\Delta V/k_B T \gg 1$). To adopt Eq. (3.43) as a reliable estimate of the activation rate in the underdamped limit, we must impose the further restriction that contributions proportional to β must be larger than the inaccuracies implied by the approximations in Eqs. (3.28) and (3.29). Since the larger corrections are

proportional to $\mu_E(\lambda_0)(k_BT/\Delta V)^{1/2}$, we must require that the following inequality be satisfied (beside $\lambda_0 \ll \sqrt{a}$):

$$\frac{\lambda_0}{\lambda_2}\frac{b}{16a} \ll \left(\frac{\Delta V}{k_BT}\right)^{1/2} \tag{3.17}$$

In other words, our analytical expression for $\mu_E(\lambda_0, \lambda_2)$ is of practical use only if the value of λ_2 is not too small.

IV. ACTIVATION RATES IN THE PRESENCE OF MEMORY EFFECTS

The subject of this section has been treated by many authors[2-6] using many different approaches. Here we will study a particular case of the first-order correction to the Markovian limit analyzed in refs. 14 and 15. A completely general derivation has been studied in Section II; however, Eqs. (2.26) and (2.27) are of no use for practical purposes, because they contain too many unknown parameters. Since we are interested in a qualitative description of the effects of the non-Markovian statistics of the heat bath, we can simplify our problem as follows.

Let us assume that $a_\nu(x) \equiv a(x)$ for any $\nu = 1, \ldots, N$. In this case Eqs. (2.9) can be rewritten as

$$\dot{x} = v$$
$$\dot{v} = -V'(x) + a'(x)\int_0^\infty d\tau\varphi(t - \tau)\{a'(x(\tau))v(\tau)\} + a'(x)f(t) \tag{4.1}$$

where $M = 1$ in Eqs. (2.9) and

$$\varphi(t) \equiv \sum_\nu \omega_\nu^2\cos\left(\frac{\omega_\nu}{\sqrt{m_\nu}}t\right) \tag{4.2}$$

Equation (4.1) is a generalized LE and the function $\varphi(t)$ defined in Eq. (4.2) plays the role of a memory kernel. Generalizing the technique expounded in ref. 3, we make the problem (4.1)–(4.2) Markovian by introducing enough additional variables. In our case, the Laplace transform $\hat{\varphi}(z)$ of $\varphi(t)$ admits of a continued fraction expansion:[12c]

$$\hat{\varphi}(z) = \frac{\Delta_1^2}{z + \gamma_1}\frac{\Delta_2^2}{z + \gamma_2}\cdots\frac{\Delta_n^2}{z + \gamma_n} \tag{4.3}$$

Equations (4.1)–(4.3) are then equivalent to a set of $n+2$ Markovian equations:

$$\dot{x} = v$$

$$\dot{v} = -V'(x) + a'(x)\xi_1$$

$$\dot{\xi}_1 = -\gamma_1\xi_1 - \Delta_1^2 b(x)V + \xi_2$$

$$\dot{\xi}_2 = -\gamma_2\xi_2 - \Delta_2^2\xi_1 + \xi_3 \qquad\qquad (4.4)$$

$$\vdots$$

$$\dot{\xi}_n = -\gamma_n\xi_n - \Delta_n^2\xi_{n-1} + \eta(t)$$

where the random force $\eta(t)$ is a Gaussian white noise of zero mean and correlation

$$\langle \eta(t)\eta(0)\rangle = 2k_B T\gamma_n\Delta_1^2 \cdots \Delta_n^2\delta(t) \qquad\qquad (4.5)$$

and the function $b(x)$ is suitably related to $a(x)$:

$$b(x) = \frac{\gamma_1 a'(x)}{\Delta_1^2} \qquad\qquad (4.6)$$

We assume that $\varphi(t)$ is approximated by an exponential function, $\exp(-\gamma_1 t)$, that corrects the Markovian limit $\varphi(t) = \delta(t)$ in ref. 14; this implies that n is chosen to equal 1.

From now on we follow the perturbation approach described in Section II. The FP operator corresponding to the set of Eqs. (4.4) with $n=1$ must be separated into an unperturbed part,

$$\Gamma_0 = \gamma_1\left[\frac{\partial}{\partial\xi_1}\xi_1 + \langle\xi_1^2\rangle\frac{\partial^2}{\partial\xi_1^2}\right] \qquad\qquad (4.7)$$

and a perturbation

$$\Gamma_1 = -v\frac{\partial}{\partial x} + V'(x)\frac{\partial}{\partial v} - a'(x)\xi_1\frac{\partial}{\partial v} + \Delta_1^2 b(x)v\frac{\partial}{\partial\xi_1} \qquad (4.8)$$

where $\langle\xi_1^2\rangle = \gamma_1 k_B T$ [see Eq. (4.5)]. The perturbation expansion is supposed to converge for $\gamma_1 \gg \sqrt{a}$ (see ref. 19). In the notation of Mohanty et al.,[15] this corresponds to $\omega_0\tau \ll 1$ (see Section I).

The calculations are straightforward. The operator Γ_2 in Eq. (2.26) should now be replaced by ($M = 1$)

$$\Gamma_2 = \frac{u'(x)}{\gamma_1} \frac{\partial}{\partial v}\left(-v\frac{\partial}{\partial x} + V'(x)\frac{\partial}{\partial v}\right)u'(x)\left(k_BT\frac{\partial}{\partial v} + v\right)$$

$$-\frac{a'^2(x)}{\gamma_1}\frac{\partial}{\partial v}\left(k_BT\frac{\partial}{\partial v} + v\right)\left(-v\frac{\partial}{\partial x} + V'(x)\frac{\partial}{\partial v}\right) \qquad (4.9)$$

The FP equation for the reduced probability $\rho(x, v; t)$, corrected up to the first order in γ_1^{-1}, can be rewritten in a more compact manner as follows:

$$\frac{\partial}{\partial t}\rho(x, v; t) = \left\{-\frac{\partial}{\partial x}j(v) + \left(1 + \frac{a'^2(x)}{\gamma_1}\right)\frac{\partial}{\partial v}j(x)\right.$$

$$\left. + \frac{\partial}{\partial v}\left(a'^2(x) + a'(x)a''(x)\frac{v}{\gamma_1}\right)\right\}\rho(x, v; t) \qquad (4.10)$$

where $j(x)$ is defined in Eq. (3.6) and

$$j(v) = k_BT\frac{\partial}{\partial v} + v \qquad (4.11)$$

Adopting for $a(x)$ the choice of Eq. (2.19), that is, $a(x) = \Gamma x + \beta x^2/2$, we write down

$$a'(x)^2 = \lambda_2 x^2 + 2\lambda_1 x + \lambda_0, \qquad \lambda_1^2 = \lambda_0\lambda_2,$$

$$a'(x)a''(x) = \lambda_2 x + \lambda_1 \qquad (4.12)$$

Therefore the parameter controlling the relevance of the non-Markovian corrections is γ_1. For $\gamma_1 \to \infty$, the Markovian limit is recovered.

Following the prescription of refs. 3 and 18, we define the escape time from the reactant well, say x_-, to the product well, say x_+, as the area below the curve $\langle x(t)\rangle/\langle x(0)\rangle$. For fairly high values of the barrier ΔV, this curve is mostly exponential throughout all of the time domain except a narrow region close to $t = 0$. This fast relaxation depends significantly on the starting point distribution, $\rho(x, v; 0)$.[23] Let us assume $\rho(x, v; 0)$ to be given by $\delta(x - x_-)$. This choice may enhance the effect of the short time relaxation on our definition of rate of escape:

$$\mu' \equiv \hat{\Phi}^{-1}(0) \qquad (4.13)$$

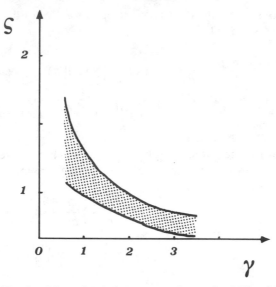

Figure 1.   Graphs of $\zeta = \mu(\lambda_i)/\mu(\lambda_0)$ versus γ for $a = b = 0.25$ and $k_BT = 0.025$. Upper curve: $\lambda_0 = 1$ and $\lambda_1 = \lambda_2 = 0$. Lower curve: $\lambda_0 = \lambda_1 = \lambda_2 = 1$.

where $\hat{\Phi}(0)$ is the Laplace transform of $\langle x(t) \rangle / \langle x(0) \rangle$ at zero frequency. However, for large enough values of $\Delta V / k_BT$, μ' can be relied on as a suitable estimate of the activation rate of the process. Definition (4.13) is especially well-suited to computational applications.

To apply the analytical approach of the foregoing section to the FP equation (4.10) would be cumbersome and of no practical use. For that reason we choose to employ a numerical algorithm that has been shown to give excellent performance in such computations.[23,24] This algorithm, called CFP, is based on a continued fraction expansion á la Mori,[12] and reviewed in ref. 24. Figure 1 displays our results for the overdamped regime. The most remarkable effect of the non-Markovian corrections is the increase of μ' as the heat bath relaxation time γ_1^{-1} increases. Curve 1 refers to the choice $a(x) = \Gamma x$, that is, $\lambda_1 = \lambda_2 = 0$. The small discrepancy between this curve and the Markovian limit $\mu(\lambda_0)$ of Eq. (3.13) is accounted for as an effect (of the interplay) of inertia and anharmonicities in the potential form.[7] These have been disregarded in working out Eqs. (3.9) and (3.11) using the steepest-descent method in the Smoluchowski approximation. The more accurate values of Larson and Kostin[7] are reproduced to a precision of several percent.

Curve 2 refers to the case $\lambda_0 = \lambda_1 = \lambda_2 = 1$ [see Eq. (4.12)]. In the Markovian limit, $\gamma_1 \to \infty$, the lower rate of escape confirms the predictions of Eqs. (3.12) and (3.14), provided that $\mu(\lambda_0)$ of Eq. (3.13) is replaced with Larson and Kostin's rate.[7] The dependence of the activation rate on the parameter γ_1 for λ_1, $\lambda_2 \neq 0$ is the main finding of the present section. Curve 1 closely reproduces results already obtained in ref. 3.

In the next section we will discuss the relevance of these results for applications to chemical–physical problems.

V. SUMMARY AND CONCLUSIONS

In this section we wish to draw some conclusions about the relevance of the phenomenological LE to applications to chemical–physical systems.

In Section II we reviewed the Lindenberg and Seshadri[14] derivation of LE (1.1) starting from Zwanzig's model Hamiltonian,[13] which describes a nonlinear one-dimensional system coupled with a heat bath of harmonic oscillators. If small nonlinearities are included in the interaction term—or in the heat bath Hamiltonian—a formally identical set of LE, (1.1)–(1.3), can be recovered in which both the effective potential $V(x)$ and the friction coefficients λ_i now depend on the system temperature T. In Section III we determined quantitatively the effects of x-dependent friction terms on the activation rate of a process modeled as the escape of a Brownian particle from one well (the reactant well) to another (the product well). Corrections to the Kramers theory in the overdamped limit are shown to depend on the relative magnitudes of λ_1^2 and $\lambda_0\lambda_2$. In Section IV the effects due to the non-Markovian statistics of the heat bath are accounted for in a simplified case in which the relevance of such a property is regulated by means of one new parameter only, $\gamma_1^{-1} \equiv \tau_0$. A finite heat bath correlation time τ_0 is proved to increase the rate of escape of the Brownian particle over the barrier. The main analytical tool employed throughout this paper is the perturbation technique of adiabatic elimination of fast-relaxing variables, described in the Appendix.

When in refs. 16 and 17 the experimenters claim that the Kramers theory fails in describing a number of chemical–physical processes, they usually refer to the phenomenological model of Eq. (1.1) with $\lambda_1 = \lambda_2 = 0$ (the Wang–Uhlenbeck LE[25]) and to the corresponding rate of escape, which for high friction constants coincides with Kramers's rate $\mu(\lambda_0)$ from Eq. (3.13). Various theorists[1-9] have improved this estimate by accounting for a variety of additional effects; all of them, however, take the Wang–Uhlenbeck model as a starting point or as the zero-order approximation of their perturbation approaches. The description obtained first by Lindenberg and Seshadri[14] and discussed in detail by Mohanty et al.[15] is to be regarded as a more realistic picture for actual chemical–physical systems. It reduces to the Wang–Uhlenbeck model under certain restrictions and approximations.[15]

We now compare the properties exhibited by the model of Eqs. (1.1)–(1.3) with the simpler Wang–Uhlenbeck picture.

A. Dependence on Temperature

If the viscosity is kept constant in the overdamped limit, the activation rate is supposed to depend on T by the Arrhenius law [see Eq. (3.13)]. In view of the findings of Sections II and III, however, we suggest that deviations from that fundamental rule could be revealed by means of detailed measurements. The physical origin of such corrections is twofold: First, when we approximated the Liouville description of the global system via a set of LE, we pointed out that the potential of the isolated Brownian particle, $U(x)$, was to be replaced by an "effective" potential, $V(x)$ [Eq. (2.8)]. This is the potential whose parameters (activation energy, characteristic frequencies, etc.) can be obtained from the experimental data of any single process. The inevitable anharmonicities of the real heat bath ($H_{SB} + H_B$) determine the T dependence of $V(x)$ [Eq. (2.24)] and of λ_i [Eq. (2.25)]. Second, even if we neglect this kind of dependence and refer to the "zero-order approximation" [Eqs. (1.1)–(1.3)], the x-dependent friction terms imply that a more reliable expression for the activation rate would now be $\mu(\lambda_i)$ of Eq. (3.12), where the temperature enters into the correction prefactor $H(\lambda_1/\lambda_0, \lambda_2/\lambda_0)$ as well. Slight deviations from the Arrhenius law have been measured recently (e.g., see ref. 17); a more detailed analysis would be of great interest.

B. Dependence on Viscosity

When experimental results for the dependence of the activation rate on dissipation have been compared with Kramers's predictions (i.e., with the Wang–Uhlenbeck model), it has been common[16,17] to assume a sort of hydrodynamical model for λ_0 in which

$$\lambda_0 \propto \eta \qquad (5.1)$$

where η is the solvent viscosity. If we adopt the LE (1.1)–(1.3) as an alternative phenomenological model, a new difficulty arises. Since we cannot fit a large number of parameters to the experimental data, one might think of taking η proportional to an "effective" or "average" damping λ.[14] Unfortunately, this choice is inconsistent with the results of Section III, where we showed that the x-dependent friction terms play different roles in correcting the activation time in the overdamped and underdamped limits: In Eq. (3.12), we would define λ as $\lambda_0 H(\lambda_1/\lambda_0, \lambda_2/\lambda_0)$, whereas in Eq. (3.43) λ would be read as $\lambda_0(1 + \beta \Delta V)$. Therefore it is no surprise that many experimental papers conclude that the Kramers theory breaks down because it incorrectly predicts the viscosity dependence of the activation rates.[16,17]

C. Dependence on Heat Bath Relaxation Time

This is an example of the additional mechanisms introduced[1-6] to account for the discrepancies in η dependence mentioned above. These improvements are no doubt well founded from a physical point of view, but are still to be regarded as perturbation corrections to the Wang–Uhlenbeck model. When we tried to apply one of those approaches[3] to the phenomenological LE (1.1)–(1.3), we found that the well-known increase of the activation rate with $\tau_0 = \gamma_1^{-1}$ depends dramatically on the choice of the friction parameters λ_i (see Fig. 1).

We conclude by remarking that the LE (1.1)–(1.3) are just an example of a generalized version of the Wang–Uhlenbeck model, and therefore one would be well advised, before using one-dimensional phenomenological LE of this type, to check under what assumptions these equations are valid descriptions of the dynamics of the specific chemical–physical system under investigation.

APPENDIX

The aim of this appendix is to give some technical rules for applying the adiabatic elimination procedure (AEP) of ref. 19 to the system of equations (2.12). We found it easier to carry out our projection procedure using a new set of heat bath variables:

$$\eta_{1\nu} = \frac{p_\nu}{\sqrt{m_\nu}} + i\omega_\nu Q_\nu, \qquad \nu = 1,\ldots,N \tag{A.1}$$

$$\eta_{2\nu} = \frac{p_\nu}{\sqrt{m_\nu}} - i\omega_\nu Q_\nu, \qquad \nu = 1,\ldots,N \tag{A.2}$$

The canonical equilibrium distribution ρ_{eq} is defined as

$$\mathbf{L}_0 \rho_{eq} \equiv 0 \tag{A.3}$$

In the (p_ν, Q_ν) frame ρ_{eq} reads

$$\rho_{eq}(p_\nu, Q_\nu) = \mathcal{N}\exp\left[-\frac{1}{k_B T}\sum_\nu\left(\frac{p_\nu^2}{2m_\nu} + \frac{\omega_\nu^2 Q_\nu^2}{2}\right)\right] \tag{A.4}$$

where $\mathcal{N}$ is a suitable normalization constant, while in the new one [Eqs. (A.1) and (A.2)],

$$\rho_{eq}(\eta_{1\nu}, \eta_{2\nu}) \equiv \mathcal{N}\exp\left(-\frac{1}{2k_B T}\sum_\nu \eta_{1\nu}\eta_{2\nu}\right) \tag{A.5}$$

In the new variables $(\eta_{1\nu}, \eta_{2\nu})$ the unperturbed part [Eq. (2.13)] and perturbation [Eq. (2.14)] of the FP operator can be rewritten as

$$\mathbf{L}_0 = -2\sum_\nu \frac{i\omega_\nu}{\sqrt{m_\nu}}\left(\eta_{1\nu}\frac{\partial}{\partial\eta_{1\nu}} - \eta_{2\nu}\frac{\partial}{\partial\eta_{2\nu}}\right) \tag{A.6}$$

and

$$\mathbf{L}_I = -\frac{p}{M}\frac{\partial}{\partial x} + V'(x)\frac{\partial}{\partial p}$$

$$-\sum_\nu i\omega_\nu a'_\nu(x)\left[\frac{p}{M}\left(\frac{\partial}{\partial\eta_{1\nu}} - \frac{\partial}{\partial\eta_{2\nu}}\right) + \frac{\partial}{\partial p}(\eta_{1\nu} - \eta_{2\nu})\right] \tag{A.7}$$

respectively.

Two basic rules of our AEP (see ref. 19, Section 3) are then to be recast
as follows:

$$P\frac{\partial}{\partial\eta_{j\nu}}e^{\mathbf{L}_0 t} \doteq Pe^{-\gamma_{j\nu}t}\frac{\partial}{\partial\eta_{j\nu}} \tag{A.8}$$

$$Pe^{\mathbf{L}_0 t}\eta_{j\nu}^n \doteq Pe^{-n\gamma_{j\nu}t}\eta_{j\nu}^n \tag{A.9}$$

where $j = 1, 2$ and

$$\gamma_{j\nu} = (-1)^{j-1}\frac{2i\omega_\nu}{\sqrt{m_\nu}} \tag{A.10}$$

Here we have used the notation of ref. 19. In particular, P is the projection
operator onto the subspace of the relevant variables (x, p). Finally, Eq. (A.5)
yields

$$\left\langle (\eta_{1\nu} - \eta_{2\nu})^2 \right\rangle = -k_B T \tag{A.11}$$

and $\langle \eta_{1\nu}^2 \rangle = \langle \eta_{2\nu}^2 \rangle = 0$. We are now in a position to apply straightforwardly
the perturbation technique described in detail in ref. 19; Eqs. (2.16)–(2.18),
(2.26), and (2.27) are readily recovered.

The treatment of the perturbation (nonlinear) corrections to $H_{SB}^{(0)}$ is based
on the counting rule expounded in ref. 19 (Section 3): Since $\rho_{eq}(p_\nu, Q_\nu)$ is a
Gaussian function in the variables Q_ν, the integral product of terms from $\mathbf{L}_I$
whose global power in Q_ν (for any ν) is odd vanishes. Employing this rule,
results such as Eqs. (2.24) and (2.25) are easily obtained.

References

1. J. L. Skinner and P. G. Wolynes, *J. Chem. Phys.*, **69**, 2143 (1978).

2. R. F. Grote and J. T. Hynes, *J. Chem. Phys.*, **74**, 4465 (1981); *ibid.* **73**, 2715, (1980).

3. F. Marchesoni, P. Grigolini, and P. Marin, *Chem. Phys. Lett.*, **87**, 451 (1982); F. Marchesoni and P. Grigolini, *J. Chem. Phys.*, **78**, 6287 (1983).

4. B. Carmeli and A. Nitzan, *Phys. Rev. Lett.*, **49**, 423 (1982); *ibid.*, *J. Chem. Phys.*, **79**, 393 (1983); *ibid.*, *Phys. Rev.*, **A29**, 1481 (1984).

5. P. Hänggi, *Phys. Rev.*, **A26**, 2996 (1982), *ibid.*, *J. Stat. Phys.*, **30**, 401 (1983); P. Hänggi and F. Mojtabai, *Phys. Rev.*, **A26**, 1168 (1982).

6. F. Guardia, F. Marchesoni, and M. San Miguel, *Phys. Lett.*, **100A**, 15 (1984).

7. R. S. Larson and M. D. Kostin, *J. Chem. Phys.*, **72**, 1392 (1980); *ibid.*, *J. Chem. Phys.*, **69**, 4821 (1978).

8. H. A. Kramers, *Physica*, **7**, 284 (1940).

9. R. Landauer and J. A. Swanson, *Phys. Rev.*, **121**, 1668 (1961); J. S. Langer, *Ann. Phys.*, **54**, 258 (1969); M. Büttiker, E. P. Harris, and R. Landauer, *Phys. Rev.*, **B28**, 1268 (1983).

10. N. G. van Kampen, *Progr. Theor. Phys.*, **64**, 389 (1978).

11. P. Hänggi, *Phys. Lett.*, **78A**, 304 (1980); S. Faetti, P. Grigolini, and F. Marchesoni, *Z. Phys.*, **B47**, 353 (1982).

12. (a) P. Mazur and I. Oppenheim, *Physica*, **50**, 241 (1970); (b) J. M. Deutch and I. Oppenheim, *J. Chem. Phys.*, **54**, 3547 (1971), (c) H. Mori, *Progr. Theor. Phys.*, **33**, 423 (1965); (d) N. G. van Kampen, *Phys. Rep.*, **24**, 171 (1976).

13. R. Zwanzig, *J. Stat. Phys.*, **9**, 215 (1973).

14. K. Lindenberg and V. Seshadri, *Physica*, **109A**, 483 (1981).

15. U. Mohanty, K. E. Shuler, and I. Oppenheim, *Physica*, **115A**, 1 (1982).

16. D. L. Hasha, T. Eguchi, and J. Jonas, *J. Chem. Phys.*, **75**, 1573 (1981); B. Bagchi and D. W. Oxtoby, *J. Chem. Phys.*, **78**, 2735 (1983); G. Rothenberger, D. K. Negus, and R. M. Hochstrasser, *J. Chem. Phys.*, **79**, 5360 (1983).

17. S. P. Velsko, D. H. Waldeck, and G. R. Fleming, *J. Chem. Phys.*, **78**, 249 (1983).

18. F. Marchesoni and P. Grigolini, *Physica*, **121A**, 269 (1983).

19. P. Grigolini and F. Marchesoni, in *Advances in Chemical Physics*, Vol. LXII, Wiley, New York, 1985.

20. P. Hänggi, F. Marchesoni, and P. Grigolini, *Z. Phys.*, **B56** 333 (1984).

21. R. L. Stratonovitch, *Topics in the Theory of Random Noise*, Vol. 1, Gordon & Breach, New York, 1967.

22. I. M. Ryzhik and I. S. Gradshteyn, *Tables of Integrals, Series and Products*, Academic, New York, 1980.

23. T. Fonseca, P. Grigolini, and P. Marin, *Phys. Lett.*, **88A**, 117 (1982); J. A. N. F. Gomes, T. Fonseca, P. Grigolini, and F. Marchesoni, *J. Chem. Phys.*, **79**, 3320 (1983); *ibid.* *J. Chem. Phys.*, **80**, 1826 (1984).

24. G. Grosso and G. Pastori-Parravicini, in *Advances in Chemical Physics*, Vol. LXII, Wiley, New York, 1985.

25. M. C. Wang and G. E. Uhlenbeck, *Rev. Mod. Phys.*, **17**, 323 (1945).

MOLECULAR DYNAMICS IN RIGID-ROD MACROMOLECULAR LYOTROPIC LIQUID CRYSTALS[‡]

J. K. MOSCICKI

Institute of Physics, Jagellonian University, ul. Reymonta 4, 30-059 Cracow, Poland

CONTENTS

[‡] The work reported here was carried out while the author was, in turn, with Laboratoire de Chimie Théorique, Université de Nancy I, Vandoeuvre-les-Nancy, and Laboratoire de Recherches sur les Interactions Gaz-Solides CNRS Maurice Letort, Villers-les-Nancy, France, and Baker Laboratory, Cornell University, Ithaca, New York, U.S.A.

I. INTRODUCTION

A large, and continuously increasing, number of elongated and rigid molecules are now known to form, under properly chosen conditions, a fluid state of matter that is intermediate (mesomorphic) between the isotropic liquid and crystalline solid states. This particular state of matter is called the liquid crystalline phase. Reinitzer[1] and Lehman[2] observed the first liquid crystalline phases nearly a century ago by varying the temperature of a single-component system of relatively short molecules. The cigarlike molecules of these *thermotropic* liquid crystals (t-LC) have axial (length/diameter) ratios of 3–6, are rigid and interact in a soft anisotropic manner. It is an essential property of the liquid crystalline phase that the molecules exhibit orientational order of their long axes. Thermotropic liquid crystals are generally divided into two distinctive classes, nematics and smectics, depending on the lack or presence of some degree of positional order of the molecules. In the nematics no long-range correlations exist between the centers of mass of elongated molecules; that is, molecules can translate freely. This feature is responsible for the high fluidity of these phases. In the smectics, positional correlations exist and the centers of the molecules are, on average, arranged in equidistant planes.

Nearly 50 years after the discovery of t-LC, it was discovered that extremely long, rigid, needlelike macromolecules can also form liquid crystalline phases, although not in response to varying temperature but rather on dissolution in appropriately chosen solvents. The macromolecule concentration is then the most important variable, and for that reason these LC are called *lyotropic* LC (l-LC). The first macromolecules observed to form l-LC were of biological origin, namely the rodlike molecules of tobacco mosaic virus.[3,4] Later l-LC were also discovered for a number of synthetic rigid-chain polymers including polyamino acids,[5-14] wholly aromatic polyamides,[15-18] aromatic polyheterocyclics,[19-21] polyribonucleotides,[22,23] polyisocyanides,[24,25] and polyisocyanates.[26-28]

It is characteristic of the lyotropic liquid crystalline behavior that on the concentration–temperature phase diagram the low-concentration pure isotropic phase and the high-concentration anisotropic phase are separated by a significant transition region in which different phases coexist.[12,18,29] The most complete experimental studies of the temperature–concentration phase diagram have been carried out on poly(γ-benzyl-L-glutamate) (PBLG) in the N,N'-dimethylformamide (DMF) system;[12] studies on other rodlike polymer chains have been limited to a narrower range of conditions.[12,18,29] Figure 1 shows the phase diagrams for the PBLG–DMF system and for the copolymer of n-hexylisocyanate and n-propylisocyanate (PHPIC) in the toluene system.[29] In both cases some critical temperature dividing the phase

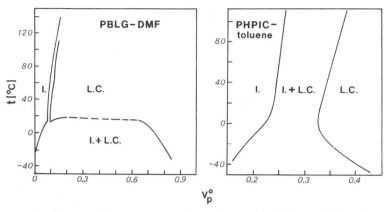

Figure 1.   Concentration–temperature phase diagrams for PBLG–DMF and PHPIC–toluene solutions. "I." and "L.C." indicate the isotropic and liquid crystalline phases, respectively.

diagram into two distinctive parts may be determined. For temperatures above this critical temperature the isotropic and nematic phases are separated by a relatively narrow biphasic range in which these two phases coexist. The low- and high-concentration boundaries of the biphasic range are essentially parallel to each other and nearly parallel to the temperature axis, deviating toward higher concentrations with increasing temperature. Below the critical temperature the transition range broadens significantly, predominantly on the high-concentration side. Miller and co-workers reported the existence of the reentrant nematic behavior[12] and presence of the nematic–nematic coexistence range[30] for the PBLG–DMF system; these, however, have not been observed for the PHPIC–toluene system.

Theoretical studies (refs. 31–39, see also refs. 8–15 of ref. 40) have shown that the l-LC behavior is due predominantly to the rigid-rod properties of macromolecules. That is, a solution of rodlike particles of sufficiently high axial ratio should separate spontaneously into two phases, one isotropic and another anisotropic–nematic, as a consequence solely of the rigidity and shape anisotropy of the solute particles[31–34] (athermal limit). To explain the character of the phase diagram, it was necessary to take into account the solvent–solvent and solute–solvent interactions[33,41] or solute–solute soft anisotropic (nematic) interactions.[37,39] Figure 2 shows a comparison among phase diagrams for three different axial ratios of monodisperse macromolecules (rods) in solution, calculated by Warner and Flory.[37] The statistical thermodynamics model worked out by Warner and Flory (WF) is based on the original lattice theory of Flory[33] and, in addition to accounting for re-

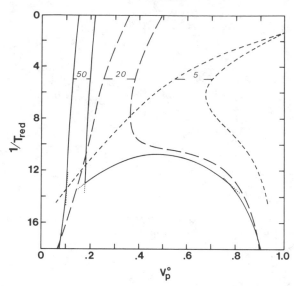

Figure 2.   Concentration–temperature phase diagrams calculated from the Warner–Flory phase equilibria model[37] for dispersions of rods with axial ratios $x = 5$, 20, and 50. T_{red} is the Warner–Flory reduced temperature.

pulsions on contact, it assumes the existence of the Maier–Saupe nematic potential, which each rodlike macromolecule experiences in the nematic phase. Although some have expressed reservations about use of the lattice model to describe properties of the liquid state, the results in Fig. 2 give us an excellent illustration of the subtle competition between steric repulsion and anisotropic attraction forces in the formation of the anisotropic–nematic phase. In the athermal limit, $1/T_{red} = 0$ (T_{red} is the WF reduced temperature,[37]) the existence of the nematic phase is possible only due to the steric repulsions. For the stable ordered phase to appear in this limit, the axial ratio of the molecules, x, has to be greater than the critical value of 6.4.[37,42,43] Thus for $x > 6.4$, and depending on the concentration, the solution may be in single isotropic or nematic phases or form biphasic material. For $x < 6.4$, even pure solute does not exhibit liquid crystalline behavior at high temperatures, and the only way to create conditions favorable for the formation of the liquid crystalline phase is to reduce the temperature (i.e., the feature characteristic of t-LC).

The lattice theory has recently been successfully applied by Flory and Ronca[42] and Warner[43] to study properties of the thermotropic nematic LC. The temperature–concentration phase diagrams predicted for l-LC reproduce main features of the experimentally observed ones for PBLG–DMF and

PHPIC–toluene (compare Figs. 1 and 2). In particular, for relatively short ($x \leq 20$) rodlike macromolecules, the theory anticipates existence of only one pair of coexisting phases: the isotropic and the nematic. This seems to be the case in the PHPIC–toluene system. For much longer rods, the phase diagram is significantly more complicated in its broad section and qualitatively resembles results for the PBLG–DMF system. Thus, although the results presented in Fig. 2 do not take into account such factors as flexibility or polydispersity of long stiff-chain polymeric molecules, they stress the importance that rigidity and shape anisotropy have for the formation of l-LC.

The needlelike character of l-LC macromolecules has a dramatic influence on their mobility in concentrated solutions. At concentrations typical for the lyotropic behavior, the rodlike macromolecules are so densely packed and entangled with one another that unrestricted translational motion of a particle and free rotational motion into the 4π solid angle are impossible. Any significant displacement or reorientation must then result from a cooperative process involving the neighbors. These restrictions on molecular motions have consequences for numerous different physical effects that have attracted experimental and theoretical interest in both their molecular dynamics and their relation to the phase behavior of the macromolecular solutions. The earliest studies in this field concerned steady-state viscosity. One of the most striking effects observed in viscometric studies of l-LC is that the apparent viscosity does not increase monotonically with increasing macromolecular concentration. Initially it was found for PBLG in m-cresol[44] and DMF,[45] as well as poly($para$-benzamide) (PBA) in dimethyl acetamide,[17,18] that viscosity goes through a sharp maximum associated with the first appearance of small amounts of the anisotropic phase in solution. However, more recent work[21,26-29,45-47] has indicated that the viscosity–concentration curve is much more complicated. First, the onset of liquid crystallinity may be related not to the concentration at the maximum viscosity, but rather to a shoulder on the low-concentration side of this maximum. Second, the maximum always appears within the biphasic range of concentration. Third, for concentrations higher than the one characteristic of the maximum viscosity, the viscosity decreases sharply toward a minimum, which usually occurs in the concentration range of the pure nematic phase, and increases again with further rises in concentration.

Initial attempts to describe the entire course of the viscosity–concentration curve, from the isotropic fluid through the biphasic zone and into the nematic phase, were purely phenomenological in nature.[48,49] The first attempt to provide a molecular theory of viscosity for a solution of rodlike molecules was undertaken by Doi[50] and reconsidered later by Doi and Edwards,[51] although these works were limited to the isotropic phase. Quite recently Doi[39] extended the theory to cover the nematic phase as well. The theory approximates a solution of macromolecules by a solution of mono-disperse, infinitely thin, and very long rods that interact with each other via (1) repulsive forces (i.e., the rods are impenetrable) and (2) soft anisotropic attractive forces. The basic assumption concerning their dynamics is that due to entanglement the only unrestricted motion of a rod is the translational one in the direction of the long axis. The network formed by neighbors effec-tively prevents translational diffusion perpendicular to the long axis and limits "free" reorientation of a rod in a solvent to "wobbling" within some small solid angle defined by the neighbors. Any significant change in the di-rection of the long axis of the rod is due only to changes in the environment of the rod. As a result, the rotational diffusion of the rods and, in turn, the zero-shear viscosity of the solution are strongly dependent on the rod con-centration, axial ratio, and the orientational order of the rods. Theoretical viscosity–concentration curves explained[39,52] the appearance of the sharp viscosity maximum inside the biphasic zone, although they did not predict the shoulder or the renewed increase of the viscosity in the nematic phase. Significantly improved predictions, especially for the viscosity in the nematic phase, may be obtained by taking into account the finite thickness of rods at high concentrations and also the restrictions on their translational mobility (see refs. 39 and 53, especially Fig. 5 of ref. 39 and Fig. 4 of ref. 53).

Apart from the viscosity studies, the molecular-dynamics theory of Doi[39,54] and Doi and Edwards,[51] (hereafter referred to as DE) has recently been applied to interpret results of experimental studies of the dynamic light-scattering,[55-57] dynamic Kerr effect,[58] and the dynamic Cotton–Mouton effect[59] in the isotropic phase of semidiluted solutions of rigid-rod mac-romolecules. The latter studies also showed that in the isotropic phase, espe-cially when the solution concentration is close to the isotropic–biphasic transition concentration, it is necessary to take into account the finite thick-ness of the rods[50,53] and concentration-dependent restriction on the trans-lational motion[53] in order to obtain better agreement between theory and experiment. Although the dynamic light-scattering, Cotton–Mouton, and Kerr effects are very sensitive to the molecular dynamics of macromolecules, their use is limited to the transparent isotropic phase.

To our knowledge, the only investigations complementary to the viscosity studies probing the molecular dynamics of entire macromolecules in a whole interesting range of concentration have been performed by studying the dielectric relaxation effect for poly(n-alkylisocyanate) (PAIC) in toluene sys-tems.[60-62] The most characteristic and striking results are significant differences between values of dielectric parameters in the isotropic and nematic phases. The values of the maximum dielectric loss, ε''_m, and the di-electric increment (or "magnitude"), $\Delta\varepsilon$, in the nematic phase were more than 5 times smaller than those in the isotropic one. At the same time the relaxation process became more than an order of magnitude faster in the nematic than in the isotropic phase. Due to these differences all parameters underwent dramatic changes across the whole concentration range of interest. In the isotropic phases ε''_m and $\Delta\varepsilon$ were linearly increasing functions of concentration, while the logarithm of the frequency of maximum dielectric loss, f_m, decreased nearly linearly at the same time. The distinct departure of ε''_m and $\Delta\varepsilon$ below the linear relation and substantially stronger nonlinearity of $\log f_m$ that occurred with increasing concentration marked the first appearance of the anisotropic material in the solution. These tendencies persisted through a significant part of the biphasic zone, but at higher concentrations all three quantities underwent drastic variations, with ε''_m and $\Delta\varepsilon$ falling and $\log f_m$ growing rapidly in a narrow concentration range. In the remaining part of the biphasic range, and in the nematic phase, ε''_m and $\Delta\varepsilon$ leveled off or even increased slightly. Concurrently, the $\log f_m$ curve went through a maximum and slightly decreased at the highest concentrations studied.

To explain the latter results, we proposed[62,63] a simple tentative model of the dielectric relaxation in l-LC. A main feature of the model was the assumption that in the nematic phase rods cannot perform an end-over-end reorientation, even on a very long time scale; instead, the rod can undergo spatially restricted rotations around the nematic director. To rationalize this situation, the theory of small-step rotational diffusion in a cone worked out by Warchol and Vaughan[64] and Wang and Pecora[65] (we henceforth refer to these four authors collectively as WVWP) was applied. The theory, applicable to the nematic as well as the isotropic phases, predicts that as the cone angle is reduced, $\Delta\varepsilon$ and ε''_m both decrease, and f_m increases in complementary manner, a main feature of the experimental results. Although qualitative agreement between calculated and experimentally observed dielectric relaxation behavior was surprisingly good, so many assumptions and approximations were made in developing the final expressions that the numerical results could not have been expected to have predictive value. With the DE theory available, we found it worthwhile to extend the theory to polydisperse rods and apply it to both viscosity and dielectric relaxation. In the first instance,[53] the theory has been redeveloped to include explicitly effects arising from the finite diameter of the rods and frictional interaction between translationally diffusing rods. Subsequently, the modified DE theory was extended to a polydisperse system of rods and the results were applied to the study of the zero-shear viscosity and the complex dielectric permittivity over the entire isotropic–biphasic–nematic range of concentration and temperature. The present paper summarizes the results of these studies.

The molecular-dynamics theory for rods in semidiluted solutions is reviewed and discussed in terms of restrictions on the rotational and translational motions of rods in the next section. This includes a formulation of the general diffusion equation and an evaluation of the rotational diffusion coefficient and/or rotational relaxation time for a polydisperse system. Application of these results to the bulk complex dielectric permittivities and viscosities of the isotropic and nematic phases is considered in Sections III and IV, respectively.

In Section V, theoretical results are compared with the existing dielectric relaxation[60-62] and viscosity[26-29] data for poly(n-hexylisocyanate) (PHIC) in toluene systems. The concentration-dependent behavior of the numerically simulated viscosity as well as of the complex permittivity are shown to be in excellent qualitative agreement with experimental observations. The same observation applies to the results of the extended simulation of the complex dielectric permittivity over not only the concentration but also the temperature range.

II. DYNAMICS OF RODLIKE PARTICLES IN CONCENTRATED SOLUTION

The first tentative theoretical approaches to the problem of the molecular dynamics of the stiff-chain macromolecules in concentrated solution were made by Doi,[50] who attempted to explain the concentration dependence of the steady-state zero-shear viscosity of the isotropic solution. His basic idea was derived directly from the theory of the molecular dynamics of linear flexible polymers in concentrated solutions or melts worked out by de Gennes.[66] The essential aspect of de Gennes's theory is the idea of entanglement between chains. As a result of this entanglement, each chain is confined inside some hypothetical tubelike region defined by the restrictions imposed by other chains. Any mobility of the chain inside the tube is due to the "reptation" mechanism.[66] However, the situation in a concentrated solution of rodlike macromolecules is quite different. Due to their inherent rigidity, their mobility is far more restricted than is that of flexible chains. This particularly concerns the rotational Brownian motion, which is strongly limited by interaction with other molecules. Although the intermolecular potential between these macromolecules is very complicated, the most important and characteristic contribution comes from steric repulsions between molecules. To stress the importance of the rigidity and shape anisotropy of macromolecules for their dynamics, Doi simplified the system to a system of hard, thin, monodisperse rods. For very small concentrations of rods in solution (see Fig. 3a), each rod can diffuse unrestrictedly, the diffusion coefficients being defined by the viscous properties of the solvent.[67,68] When the concentration of the rods increases, the rods begin first to collide with each other, but with further increases of concentration collisions are so

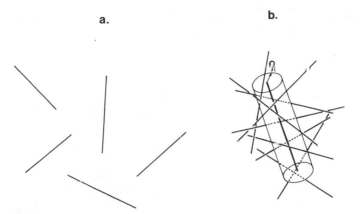

Figure 3. Schematic representation of rods in (a) diluted and (b) semidiluted solutions. ρ_t is the radius of the tubelike cage in which the rod is trapped by its neighbors.

frequent that rod mobility becomes dramatically restricted (Fig. 3b). This especially affects the reorientational motion and translational motion in the direction perpendicular to the rod long axis. Despite these limitations, the rods can undergo end-over-end reorientation, (albeit on a much longer time scale), thanks to the relatively unrestricted mobility of each rod along the long axis. This may be explained as follows. On the short time scale, a rod (hereafter we call it the test rod) is trapped by the neighbors in a sort of tubelike cage. The distance ρ_t between the test rod and the nearest neighboring rod may be considered as a good measure of the size of this trap. The cage defines space available to the rod for perpendicular translational diffusion, which is of the order of ρ_t, and for rotational diffusion, which is of the order of $\Delta\Omega \approx (\rho_t/L)^2$, where L is the rod length. Because each rod has freedom to move parallel to its long axis, it is adequate to consider the cage as a segment of some imaginary tube (which we may consider as a hypothetical path) meandering through a crowd of other rods. Due to this meandering, each rod is able to change its orientation significantly, although on a much, much longer time scale.

The original concept of Doi[50] was later reconsidered by Doi and Edwards,[51] who put it in a much more rigorous mathematical form. Quite recently Doi extended the theory to the cases of an anisotropic (nematic) solution[39] and of a solution in the presence of an external ordering field.[54] The work of Doi and Edwards[39,50,51,54] was originally restricted to a solution that was concentrated enough to make the entanglement between rods the predominant factor in the determination of rod dynamics, but dilute enough to keep the average distance between rods sufficiently large to enable them to diffuse freely along their axes. Thus, if n denotes the number concentration of rods, the concentration range adequate for application of the theory is $L^{-3} \ll n \ll (dL^2)^{-1}$, where d is the rod diameter. However, in the concentration range close to the isotropic–biphasic transition, $n \approx (dL^2)^{-1}$, the distance between rods is comparable to their diameter and the rods can no longer be considered as infinitely thin. Doi[50] attempted to take this into account by redefining the solid angle available to the rod for free rotational diffusion within the cage, $\Delta\Omega$, but still considering the translational motion undisturbed. Despite this problem he obtained significantly better agreement between the experimental and theoretical viscosity–concentration curves in the isotropic solution in the vicinity of the transition concentration. Quite recently,[53] we proposed an improvement on the model to take into account restrictions on the translational mobility of the rods. The redeveloped model explains not only the critical behavior of the viscosity in the vicinity of the isotropic–biphasic transition concentration but also the behavior over the entire isotropic–biphasic–nematic range of concentration.

Although DE theory helped significantly to explain the main features of the viscosity behavior of stiff-chain macromolecular solutions, it was developed for a model system of perfect monodisperse rods. For better understanding of the molecular dynamics in real systems it is necessary to consider additional effects arising from such factors as some inherent flexibility of the chains and their significant polydispersity. Efforts have been made to incorporate these effects into the theory, although for the isotropic solution only. Odijk[69] considered the molecular dynamics of entangled stiff chains, and Marrucci and Grizzuti[70] outlined how polydispersity has to be accounted for. In the following sections we apply the DE theory to a polydisperse system of rods in either the isotropic or nematic phase. The approach will also take into account our recent correction of DE theory to account for the decrease of the translational diffusivity at high concentrations.[53]

A. The Kinetic Equation

Let us consider a solution of rods that are polydisperse in length but equal in diameter. Let $F_\alpha(\mathbf{r}, \mathbf{u}; t)$ denote the distribution function for α-type species (i.e., rods L_α in length), where $\mathbf{r}$ is the position vector of the center of mass and $\mathbf{u}$ is the unit vector in the direction of the rod long axis. The distribution function is normalized such that

$$\int d^3r \int d^2u\, F_\alpha(\mathbf{r}, \mathbf{u}; t) = 1 \tag{1}$$

and it satisfies a diffusion equation of the general form

$$\frac{\partial F_\alpha}{\partial t} = \mathcal{T}_r[F_\alpha] + \mathcal{T}_u[F_\alpha] \tag{2}$$

where $\mathcal{T}_r$ and $\mathcal{T}_u$ are terms describing the translational and rotational motion of the α-rod, respectively. In a very dilute solution (i.e., one in which $n_p^0 \ll (L_p^0)^{-3}$, where n_p^0 and L_p^0 are the number concentrations of all rods in the solution and the number-average rod length, respectively), the Brownian motion of a test rod is independent of other rods and determined by viscous properties of the solvent. Thus the diffusion may be considered as the "free space" one. The translational and rotational terms in Eq. (2) have in this case the form[51]

$$\mathcal{T}_r[F_\alpha] = \left\{ D_{\alpha\|}^{t0}\left(\mathbf{u}\cdot\frac{\partial}{\partial\mathbf{r}}\right)^2 + D_{\alpha\perp}^{t0}\left[\frac{\partial^2}{\partial\mathbf{r}^2} - \left(\mathbf{u}\cdot\frac{\partial}{\partial\mathbf{r}}\right)^2\right] \right\} F_\alpha \tag{3a}$$

$$\mathcal{T}_u[F_\alpha] = D_\alpha^{r0}\nabla_u\cdot\nabla_u F_\alpha \tag{3b}$$

where $D_{\alpha\|}^{t0}$ and $D_{\alpha\perp}^{t0}$ are the translational diffusion constants for the motion of the rod parallel and perpendicular to its long axis, respectively; D_α^{r0} denotes the rotational diffusion constant, and ∇_u is the gradient operator on the sphere $|\mathbf{u}| = 1$.[39] The diffusion coefficients of a single rod are determined by the hydrodynamics and are given by Kirkwood and co-workers[67,68] as

$$D_{\alpha\|}^{t0} = k_BT\frac{\ln(L_\alpha/d)}{2\pi\eta_s L_\alpha}, \qquad D_{\alpha\perp}^{t0} = \frac{D_{\alpha\|}^{t0}}{2} \tag{4a}$$

$$D_\alpha^{r0} = k_BT\frac{\ln(L_\alpha/d)}{3\pi\eta_s L_\alpha^3} \tag{4a}$$

where η_s, k_B, and T are the solvent viscosity, the Boltzmann constant, and the absolute temperature, respectively.

For the concentration range of interest (i.e. $(L_p^0)^{-3} \ll n_p^0 \ll [d(L_p^0)^2]^{-1}$), the rod may be considered trapped inside the "cage" or "trap" formed by the network of entangled neighboring rods. Let ρ_t denote the diameter of the cage tube. The diffusion equation, Eq. (2), has to take account of the restrictions on rod mobility in a concentrated solution. In particular, the expressions for $\mathcal{T}_r$ and $\mathcal{T}_u$ given in Eqs. (3) are not valid in this case. As long as one is interested in the long-time-scale translational motion, that is, motion on a time scale longer than ρ_t^2/D_α^{t0}, the perpendicular diffusivity of the rod is enormously smaller[71,72] than that parallel to the rod long axis. This mobility is so small that it is probably unmeasurable experimentally and may be neglected; that is, $D_{\alpha\parallel}^{t0} = 0$. Thus, in the limit, $\mathscr{T}_r$ reduces to

$$\mathscr{T}_r\,[F_\alpha] \simeq D_\alpha^{t0}\left(\mathbf{u}\cdot\frac{\partial}{\partial\mathbf{r}}\right)^2 F_\alpha \tag{5}$$

where $D_\alpha^{t0} \equiv D_{\alpha\parallel}^{t0}$.

To take into account the particular mechanism of the rod reorientation in concentrated solution, Doi[54] used a two-vector orientational distribution function,

$$\Phi_\alpha(\mathbf{w},\mathbf{u};t) = \int d^3r\,F_\alpha'(\mathbf{r},\mathbf{w},\mathbf{u};t) \tag{6}$$

where the auxiliary unit vector $\mathbf{w}$ represents the direction of the cage axis. The vector $\mathbf{u}$ is then assumed to fluctuate around $\mathbf{w}$ due to the existence of some coupling potential, $V_c(\mathbf{u}-\mathbf{w})$, originating from repulsion forces between neighboring rods. Diffusivity of $\mathbf{u}$ is still considered to be characterized by the free-space diffusion coefficient, D_α^{r0}, but that of $\mathbf{w}$ by a distinctly smaller diffusion coefficient $\tilde{D}_\alpha^r$. What is more important, $\tilde{D}_\alpha^r$ in general is not a constant but rather is dependent on the orientational order existing in solution, that is, on Φ_α. Thus, the exact kinetic equation for Φ_α inevitably becomes a very complicated one. To make the problem soluble, Doi[39,54] replaced $\tilde{D}_\alpha^r$ with its average value, D_α^r. In this approximation the kinetic equation reduces to

$$\frac{\partial\Phi_\alpha}{\partial t} = D_\alpha^r\nabla_\mathbf{w}\cdot\left\{\nabla_\mathbf{w}\Phi_\alpha + (k_BT)^{-1}\Phi_\alpha\nabla_\mathbf{w}V_c(\mathbf{u}-\mathbf{w})\right\}$$
$$+ D_\alpha^{r0}\nabla_\mathbf{u}\cdot\left\{\nabla_\mathbf{u}\Phi_\alpha + (k_BT)^{-1}\Phi_\alpha\nabla_\mathbf{u}[V_c(\mathbf{u}-\mathbf{w})+V(\mathbf{u})]\right\} \tag{7}$$

where the first term describes diffusion of $\mathbf{w}$ and the second that of $\mathbf{u}$. A detailed account of how the average diffusion coefficient D_α^r can be estimated is given in the next section. The second term in Eq. (7) contains an additional potential $V(\mathbf{u})$, which is included to account for the existence of the "nematic potential" acting on each rod in the ordered, anisotropic–nematic state of solution, although in general $V(\mathbf{u})$ may be considered[65] as originating from, for example, externally applied fields (such as electric or magnetic ones). The coupled motion of the two vectors, $\mathbf{u}$ and $\mathbf{w}$, may be visualized as that of a pair of chopsticks. In the absence of the potential $V(\mathbf{u})$, the slowly moving vector $\mathbf{w}$ changes the equilibrium position of the quickly fluctuating vector $\mathbf{u}$. In the presence of $V(\mathbf{u})$ acting on $\mathbf{u}$, the motion of $\mathbf{w}$ also becomes affected, thanks to the existence of the coupling potential $V_c(\mathbf{u}-\mathbf{w})$.

Because the time scales characteristic for the fast motion of $\mathbf{u}$ and slow motion of $\mathbf{w}$ differ by a few orders of magnitude, the distribution function $\Psi_\alpha(\mathbf{w}, \mathbf{u}, t)$ may be separated into a slowly varying part $f_\alpha(\mathbf{w}; t)$ and a rapidly varying part $\phi_\alpha(\mathbf{u}; \mathbf{w}, t)$:

$$\Phi_\alpha(\mathbf{w}, \mathbf{u}; t) = f_\alpha(\mathbf{w}; t)\phi_\alpha(\mathbf{u}; \mathbf{w}, t) \tag{8}$$

Consequently, the kinetic equations for f_α and ϕ_α may be written independently as

$$\frac{\partial \phi_\alpha}{\partial t} = D_\alpha^{r0}\nabla_\mathbf{u} \cdot \left\{ \nabla_\mathbf{u}\phi_\alpha + (k_B T)^{-1}\phi_\alpha\left[V_c(\mathbf{u} - \mathbf{w}) + V(\mathbf{u})\right] \right\} \tag{9}$$

and

$$\frac{\partial f_\alpha}{\partial t} = D_\alpha^r\nabla_\mathbf{w} \cdot \left\{ \nabla_\mathbf{w}f_\alpha + (k_B T)^{-1}f_\alpha U(\mathbf{w}) \right\} \tag{10}$$

$U(\mathbf{w})$, which acts on the cage axis, results from the presence of the potential $V(\mathbf{u})$ and the coupling $V_c(\mathbf{u} - \mathbf{w})$ between the vectors $\mathbf{u}$ and $\mathbf{w}$:[54,73]

$$U(\mathbf{w}) = V(\mathbf{w}) + \Delta\Omega_\alpha\delta(V; \mathbf{w}) + O\left(\Delta\Omega_\alpha^2\right) \tag{11}$$

where $\delta(V; \mathbf{w})$ is a differential function of $V(\mathbf{w})$.[54] It is important to note that the potential for the cage axis, $U(\mathbf{w})$, is different from the potential for the rod, $V(\mathbf{u})$. Although the correction term in $\Delta\Omega_\alpha$, the solid angle characterizing the size of the trap, is small and has a negligible influence on such long time effects as dielectric relaxation or the steady-flow viscosity, it becomes important for short-time-scale effects such as the Kerr or Cotton–Mouton effect.

Because the dielectric relaxation and the viscosity are of primary interest for this paper, we restrict our further considerations to the long-time-scale effects. From Eqs. (9) and (10) it is clear that any large-scale reorientation of the rod is due solely to the reorientation of the cage axis; that is, the rod may be considered as merely a probe sensing the latter motion. Let us then calculate the rotational diffusion coefficient of the cage, D_α^r.

B. The Diffusion Coefficient D_α^r

To calculate the diffusion coefficient of the unit vector $\mathbf{w}$ (i.e. of the cage axis), we have to consider in detail all possible mechanisms by which the test rod may change the trap or change $\mathbf{w}$. For simplicity of further calculation, we set $d = 1$, so that the length, L_α, and the axial ratio, $x_\alpha = L_\alpha/d$, become numerically equivalent. Additionally, let the polydispersity of the solute be represented by the ratio of the number concentration of rods of axial ratio x_α, n_α^0, to the total number concentration of particles, n_p^0, in solution:

$$p^0(x_\alpha) = \frac{n_\alpha^0}{n_p^0} \tag{12}$$

or, alternatively, by the volume fraction ratio

$$P^0(x_\alpha) = \frac{v_\alpha^0}{v_p^0} = \left(\frac{x_\alpha^0}{x_p^0}\right) p^0(x_\alpha) \tag{13}$$

where v_α^0 and v_p^0 denote the volume fractions of the α-species and of the whole solute in the solution, respectively. The superscript "0" indicates quantities characteristic of the whole solute.

We focus our attention on highly concentrated solutions of polydisperse rods, shown schematically in Fig. 3b. The test rod is chosen from the β-species. Because of entanglement, the rotational mobility of the test rod is restricted to the cage formed by neighbors. For simplicity, we approximate this cage as a tube closed on both sides by circular lids, with the tube axis pointing in the direction of $\mathbf{w}$.[53] The rod axis remains within the solid angle around $\mathbf{w}$ defined by the trap,

$$\Delta\Omega_\beta \approx \left(\frac{\rho_t}{x_\beta}\right)^2 \tag{14}$$

where the dimensionless quantity ρ_t represents the cage radius in units of d (the rod diameter), until a change of the cage takes place. This may happen if the test rod diffuses translationally along its axis a sufficient distance to escape, or if the restricting neighbor, say one of the α-type, does the same, creating a "defect" in the trap. Thus the time of residence in the trap, that is, the time between jumps of $\mathbf{w}$ from one orientation to another, may be estimated as the time necessary for the shorter of the two rods to diffuse a distance equal to its own length, namely,

$$\Delta t \propto \frac{x_\alpha^2}{D_\alpha^{t0}} \qquad \text{for } x_\alpha < x_\beta \tag{15a}$$

or

$$\Delta t \propto \frac{x_\beta^2}{D_\beta^{t0}} \qquad \text{for } x_\alpha \geq x_\beta \tag{15b}$$

where D_i^{t0}, $i = \alpha, \beta$, denotes the longitudinal translational diffusion constant of the i-type rod [see Eqs. (4)]. The probability that the neighbor is of α-type is $P^{\alpha}(x_{\alpha})$, so the average time between jumps of $\mathbf{w}$ may be written as[53,70]

$$\overline{\Delta t} \propto \sum_{\alpha < \beta} \Delta t_{\alpha} P^0(x_{\alpha}) + \Delta t_{\beta} \sum_{\alpha \geq \beta} P^0(x_{\alpha}) \tag{16}$$

A small change of orientation gained by $\mathbf{w}$ via an escape-from-trap mechanism is only of the order of $\Delta \Omega_{\beta}$, so any significant reorientation of the trap axis must be realized by repetition of the process. The effective diffusion coefficient for the reorientation of $\mathbf{w}$ [see Eq. (10)] may be thus estimated as

$$D_{\beta}^r \propto \Delta \Omega_{\beta} (\overline{\Delta t})^{-1} \tag{17}$$

To evaluate an expression for D_{β}^r, we estimate the radius ρ_t in the same manner as in ref. 53. Consider an imaginary tube with arbitrary radius R that is coaxial with the test rod and of the test rod length. The number of α-type rods in direct contact with the tube is

$$N_{\alpha}(R) = \left(\frac{\ell}{\pi}\right) v_{\alpha}^v R x_{\beta} \langle \sin(u_{\alpha} u_{\beta}) \rangle + 4 n_u^0 R^2 \langle \cos(u_u u_{\beta}) \rangle \tag{18}$$

where $\mathbf{u}_i$, $i = \alpha, \beta$, denotes the unit vector along an i-type rod, so $(u_{\alpha} u_{\beta})$ is the angle between the test rod and a penetrating one; and the sine and cosine are averaged over all possible orientations of both rods:

$$\langle \cdots \rangle = \int d^2 u_{\alpha} \int d^2 u_{\beta} f(\mathbf{u}_{\alpha}) f(\mathbf{u}_{\beta}) \ldots; \quad d^2 u_i = \sin \theta_i \, d\theta_i \, d\phi_i; \quad i = \alpha, \beta \tag{19}$$

Here $f(\mathbf{u}_i)$ is the orientational distribution function of i-type species. If the solution is in the isotropic state, $f(\mathbf{u}_{\alpha}) = f(\mathbf{u}_{\beta}) = 1/4\pi$, $\langle \sin(u_{\alpha} u_{\beta}) \rangle = \pi/4$, and $\langle \cos(u_{\alpha} u_{\beta}) \rangle = \frac{1}{2}$. However, if the ordering potential $V(\mathbf{u})$ is present, $f(\mathbf{u}_{\alpha})$ and $f(\mathbf{u}_{\beta})$ are no longer constant, and in the limit of strong alignment, the averages $\langle \sin(u_{\alpha} u_{\beta}) \rangle$ and $\langle \cos(u_{\alpha} u_{\beta}) \rangle$ may be written in the quadrupolar approximation[53] as

$$\langle \sin(u_{\alpha} u_{\beta}) \rangle \simeq \frac{\pi}{4}(1 - s_{\alpha} s_{\beta}) \tag{20a}$$

$$\langle \cos(u_{\alpha} u_{\beta}) \rangle \simeq \frac{1}{2}(1 + 2 s_{\alpha} s_{\beta}) \tag{20b}$$

where

$$s_i = 1 - \frac{3}{2} \int d^2 u_i \, f(\mathbf{u}_i) \sin^2(\theta_i); \quad i = \alpha, \beta \tag{20c}$$

is the order parameter of i-type rods in the nematic phase. Note that Eqs. (20) are universal; that is, they hold also in the isotropic phase, where $s_\alpha = s_\beta = 0$.

Averaging Eq. (18) over different types of species in the system, we obtain the total number of rods in contact with the tube:

$$N(R) = 2v_p^0 \left\{ Rx_\beta \sum_\alpha (1 - s_\alpha s_\beta) P^0(x_\alpha) + R^2 \sum_\alpha (1 + 2s_\alpha s_\beta) P^0(x_\alpha) \right\}$$

$$= 2v_p^0 \left\{ Rx_\beta (1 - s_\beta S) + R^2 (1 + 2s_\beta S) \right\} \qquad (21)$$

where $S = \sum_\alpha s_\alpha P^0(x_\alpha)$ is the average order parameter of the system.

To estimate ρ_t, we now reduce R until $R \simeq \rho_t$, that is, until $N(R \simeq \rho_t) = 1$ (refs. 51, 53):

$$2v_p^0 (1 + 2s_\beta S) \rho_t^2 + 2v_p^0 (1 - s_\beta S) x_\beta \rho_t = 1 \qquad (22)$$

Solution of this equation with respect to $(1/\rho_t)$ gives

$$\frac{1}{\rho_t} = 2v_p^0 x_\beta (1 - s_\beta S) F \qquad (23)$$

where

$$F = \frac{1}{2} \left[\sqrt{\frac{1 + (2/v_p^0 x_\beta^2)(1 + 2s_\beta S)}{(1 - s_\beta S)^2}} + 1 \right] \qquad (24)$$

Equation (23) may be now substituted into Eq. (14) to calculate $\Delta \Omega_\beta$.

The remaining problem concerns the behavior of the longitudinal translational diffusion constant, D_β^{t0}, at high concentrations. As long as the average distance between rods is larger than the rod diameter ($\rho_t \gg 1$), we may assume that the diffusion coefficient is equal to that of the free space; that is, it is given by the Riseman–Kirkwood relation[67]

$$D_\beta^{t0} \propto \left(\frac{k_B T}{2\pi \eta_s} \right) \frac{\ln(x_\beta)}{x_\beta} \qquad (25)$$

This estimation is valid for the concentration range $(L_p^0)^{-3} \ll n_p^0 \ll [d(L_p^0)^2]^{-1}$. However, with further concentration increases, as soon as both dimensions become comparable, as in the isotropic phase close to the transition concentration ($\approx [d(L_p^0)^2]^{-1}$) or in the nematic phase, the translational motion of rods becomes restricted in two ways. First, the friction forces are no longer dependent solely on the viscous properties of the solvent alone, because rods are now in frequent direct contact and slip over one another. Second, a diffusing rod finds in its path impenetrable obstacles (i.e., other rods) and is also trapped translationally. These factors lead to a significant decrease in the diffusion constant:[53]

$$D_\beta^{t0} \to D_\beta^{t\text{eff}} = \frac{D_\beta^{t0}}{R_\beta^r} \tag{26}$$

where we will call R_β^r the retardation factor.

To evaluate R_α^r we consider both origins of the retardation in turn. We start with the retardation caused by the steric restriction, that is, by impenetrable obstacles in the rod path. Because of the finite thickness of the test rod, jammed and entangled neighbors begin to block the path from time to time at sufficiently high concentrations. The rod thus becomes both rotationally and translationally trapped. It may, however, wobble within the trap and after some time, find a free "window" for translation and escape. On finding a new path the rod diffuses until meeting the next barrier. Consequently, the time of residence in the trap is augmented:

$$\Delta t_\beta' = \Delta t_\beta + \delta t_\beta \tag{27}$$

δt_β is related to the frequency, ν_β, with which the rod has an opportunity to penetrate the barrier and to the average distance between barriers, γ:

$$\delta t_\beta \approx \frac{\gamma x_\beta}{\nu_\beta} \tag{28}$$

The frequency ν_β we calculate as follows. The barrier is created by rods penetrating the end lid of the tubelike cage. Let $\pi(d^*/2)^2$ be the effective fraction of the lid surface penetrated by the single neighbor; d^* denotes the "effective" diameter of the rod. The frequency may be then estimated[53] as proportional to the free surface left on the lid by the obstructing neighbors:

$$\nu_\beta \propto \frac{\pi r_\beta^2 - m_\beta'\pi(d^*/2)^2}{\pi r_\beta^2}$$

$$\propto 1 - \frac{m_\beta'(d^*/2)^2}{r_\beta^2} \tag{29}$$

where r_β is the lid radius, and the number of obstructing neighbors m'_β, is easily estimated taking into account only the second term on the right-hand side of Eq. (21):

$$m'_\beta \simeq 2v_p^0(1+2s_\beta S)r_\beta^2 \tag{30}$$

Combining Eqs. (29) and (30) we have

$$v_\beta \approx 1 - d^{*2}\left(\frac{v_p^0}{2}\right)(1+2s_\beta S) \tag{31}$$

and finally, from Eqs. (15), (31), (28), and (27),

$$\Delta t'_\beta = \frac{x_\beta^2}{D_\beta^{t0}} + \frac{\gamma x_\beta}{v_\beta}$$

$$= \frac{x_\beta^2}{D_\beta^{t0}/R'_\beta} = \Delta t_\beta R'_\beta \tag{32}$$

where

$$R'_\beta - 1 = \frac{\gamma D_\beta^{t0}}{x_\beta\left[1 - d^{*2}\left(v_p^0/2\right)(1+2s_\beta S)\right]} \tag{33}$$

We note immediately the importance of R'_α for the dynamics of rods at high concentrations. The denominator in Eq. (33) tends to 0 as the concentration approaches a critical value of

$$v_g = \frac{2(d^*)^{-2}}{1+2s_\beta S} \tag{34a}$$

At this concentration the translational mobility becomes "frozen," because $R'_\beta \to \infty$ and thus $D_\beta^{teff} \to 0$. One may then identify v_g as the critical concentration for solidification or gelation of the solution. Equation (34a) links the effective rod diameter, d^*, to the solidification concentration:

$$d^* = \sqrt{2(1+2s_\beta S)^{-1}v_g^{-1}} \tag{34b}$$

This suggests a simple interpretation of d^* as the critical distance between rods in solution in the solid or gel state.

The diffusion coefficient D_β^{t0} is still considered unaffected by the increased concentration [Eq. (4a)]. Expressed via the friction coefficient Ξ_β^s, it can be written as

$$D_\beta^{t0} = \frac{k_B T}{\Xi_\beta^s} \tag{35}$$

where

$$\Xi_\beta^s = x_\beta \zeta_s \Psi_\beta \tag{36}$$

and $\zeta_s = 3\pi\eta_s$ is the friction constant per unit rod length arising from the viscous properties of a solvent, and $\psi_\beta = 1/\ln x_\beta$ may be considered as a shape factor associated with the rod. Let us now assume that the rod also experiences friction resulting from cohesive interactions with surrounding rods, and that the fraction of the rod exposed to these interactions is ϕ_β. Then to the first approximation

$$\Xi_\mu^{eff} \sim \left\{ x_\beta (1 - \phi_\beta) \zeta_s + x_\beta \phi_\beta \zeta_r \right\} \Psi_\beta$$
$$\simeq \Xi_\beta^s \Delta_\beta \tag{37}$$

with

$$\Delta_\beta = 1 + \phi_\beta \delta\zeta; \qquad \delta\zeta = \frac{\zeta_r - \zeta_s}{\zeta_s} \tag{38}$$

where ζ_r denotes the friction coefficient arising from rod–rod interactions. We note here that as long as collisions between diffusing rods are rare, that is, as long as $\phi_\beta \approx 0$ and thus $\Delta_\beta \approx 1$, Eq. (35) remains valid. In a solution densely packed with rods, the rod is in contact with more than one neighbor at the same time and ϕ_β may be written as

$$\phi_\beta = m_\beta'' \langle \Delta\phi_\beta \rangle \tag{39}$$

where m_β'' is the average number of neighbors interacting with the test rod and $\langle \phi_\beta \rangle$ is the average contribution to ϕ_β arising from single collision. The interactions giving rise to the friction are active over very short distances. It is thus reasonable to consider the rod–rod frictional interaction as a step function of the distance between interacting parts of the two rods. The fraction of the rod length exposed to these interactions is, roughly speaking, proportional to the sine of the angle between the two rods, $(u_\alpha u_\beta)$; thus, in a very crude approximation, we write[53]

$$\langle \Delta\phi_\beta \rangle \propto \frac{1}{\left\langle \sin(u_\alpha u_\beta) \right\rangle}$$

$$\propto \frac{1}{1 - s_\beta S} \tag{40}$$

where $\langle \dots \rangle$ denotes a double average over the orientational distribution and polydispersity of rods. It seems reasonable to assume that the critical distance for the interactions is of the order of d^*; m_β'' is then estimated from Eq. (21) as

$$m_\beta'' = 2v_p^0 d^* \left\{ x_\beta(1 - s_\beta S) + d^*(1 + 2s_\beta S) \right\} \tag{41}$$

Substitution of Eqs. (39)–(41) into (38) gives

$$\Delta_\beta = 1 + x_\beta d^* \delta \zeta v_p^0 \left\{ 1 + \frac{d^*}{x_\beta} \left(\frac{1 + 2s_\beta S}{1 - s_\beta S} \right) \right\} \tag{42}$$

Replacing Ξ_β^s in Eq. (35) by the effective value, Ξ_β^{eff}, and substituting the result into Eq. (33), we have, finally,

$$R_\beta^r - 1 \simeq \Delta_\beta + \frac{\gamma D_\beta^{t0}}{x_\beta \left[1 - d^{*2}\left(v_p^0/2\right)(1 + 2s_\beta S) \right]} - 1$$

$$\simeq \delta \zeta d^* x_\beta v_p^0 \left\{ 1 + \frac{d^*}{x_\beta} \left(\frac{1 + 2s_\beta S}{1 - s_\beta S} \right) \right\}$$

$$+ \frac{\gamma D_\beta^{t0}}{x_\beta \left[1 - d^{*2}\left(v_p^0/2\right)(1 + 2s_\beta S) \right]} \tag{43}$$

where D_β^{t0} denotes, as before, the diffusion coefficient in free space.

Equation (43) completes the set of equations necessary for calculation of the effective diffusion coefficient of the cage axis **w** [Eq. (17)]. Note that although the temperature was incorporated into equations wherever possible, the dependence of rod dynamics on temperature is very complicated, because nearly all parameters appearing in the set of equations are temperature dependent. We will come back to this problem prior to performing numerical calculations. The extended DE theory will now be applied to evaluate, in turn, the complex dielectric permittivity and the steady-flow, small shear viscosity of the solution in the isotropic and nematic phases.

III. DIELECTRIC RELAXATION

The rigid-rod macromolecules studied experimentally are in most cases polymers built up from repetition units, each of which contributes a small permanent electric dipole moment to the net, effective electric dipole moment of the molecule, μ_β. This dipole moment is, because of symmetry, parallel to the long molecular axis. In consequence, μ_β is proportional to the molecular axial ratio:

$$\mu_\beta = N_\beta \mu_1 = x_\beta \mu_0 \tag{44}$$

where μ_1 is the repetition-unit contribution to the dipole moment in the direction parallel to the axis, N_β is the number of repetition units in the β-type molecule; and μ_0 is the contribution per unitary segment (for which $x_1 = 1$). The dielectric relaxation effect gives information on the dipole-moment correlation function, $C(t) = \langle \mu(0) \cdot \mu(t) \rangle$:[74]

$$\frac{\hat{\varepsilon}(\omega) - \varepsilon_\infty}{\varepsilon_s - \varepsilon_\infty} = \mathscr{L}_{j\omega}\left(-\frac{dC(t)}{dt} \right) \tag{45}$$

where $\hat{\varepsilon}(\omega) = \varepsilon'(\omega) - j\varepsilon''(\omega)$ denotes the complex dielectric permittivity, and ε_∞ and ε_s are the dielectric constant of induced polarization and the static dielectric constant, respectively. $\varepsilon'(\omega)$ and $\varepsilon''(\omega)$ are called the frequency-dependent dielectric constant and the dielectric loss factor. $\mathscr{L}_{j\omega}$ is the Laplace operator. Hence, dielectric relaxation is a very sensitive means of studying the rotational dynamics of rodlike particles in semidilute solutions.

A. Isotropic Phase

As was shown in the preceding sections, the dipole moment of each molecule takes part in two types of rotational motion: rapid fluctuations around the direction of the trap, and slow, small-step rotational diffusion coupled with the reorientation of the trap. To describe the situation more quantitatively the coupling between vectors $\mathbf{u}$ and $\mathbf{w}$ must be defined. Doi[54] considered the coupling potential in the form of

$$V_c(\mathbf{u} - \mathbf{w}) = \frac{k_B T}{\Delta \Omega_\beta}(1 - \mathbf{u} \cdot \mathbf{w}) \tag{46}$$

although, as was pointed out by Williams,[73] very similar results may be obtained if one regards fluctuations of **u** as free diffusion in a cone coaxial with **w**. We follow here the approximation of Doi. With the electric field applied in the Z-direction, the electric field potential acting on the β-type rod is

$$V(\mathbf{u}) = -\mathbf{\mu}_\beta \cdot \mathbf{E} = -\mu_\beta E \cos(u_\beta u_E) \qquad (47)$$

and the contribution from β-rods to the macroscopic electric polarization may be written as

$$\Pi_\beta = n_\beta^0 \mu_\beta \langle\langle \cos(u_\beta u_E) \rangle\rangle \qquad (48)$$

where

$$\langle\langle \ldots \rangle\rangle = \int d^2u \int d^2w\, \Phi(\mathbf{w}, \mathbf{u}; t) \ldots \qquad (49)$$

If the applied field alternates with frequency f and circular frequency $\omega = 2\pi f$, (i.e., if $E(t) = E_0 \mathrm{Re}\{\exp(j\omega t)\}$), the polarization may be written in terms of the complex polarizability, $\hat{\alpha}_\beta(\omega) = \alpha'_\beta(\omega) + j\alpha''_\beta(\omega)$, as[74]

$$\Pi_\beta(t) = E_0 \mathrm{Re}\{\hat{\alpha}(\omega)\exp(j\omega t)\} \qquad (50)$$

After elaborate calculations Doi showed that for the isotropic solution

$$\hat{\alpha}_\beta(\omega) = \frac{n_\beta^0 \mu_\beta^2}{3k_B T} \left\{ \frac{1 - 2\Delta\Omega_\beta}{1 + j\omega\tau_\beta^r} + \frac{2\Delta\Omega_\beta}{1 + j\omega\tau_\beta^{r0}} \right\} \qquad (51a)$$

where

$$\tau_\beta^r = \left(2D_\beta^r\right)^{-1} \quad \text{and} \quad \tau_\beta^{r0} = \frac{\Delta\Omega_\beta}{D_\beta^{r0}} \qquad (52)$$

Neglecting the local field problem, in the dilute-gas approximation, Eq. (51a) may be rewritten in terms of the complex dielectric permittivity as

$$\frac{\hat{\varepsilon}_\beta(\omega) - \varepsilon_\infty}{\varepsilon_s - \varepsilon_\infty} = \frac{n_\beta^0 \mu_\beta^2}{3k_B T} \left\{ \frac{1 - 2\Delta\Omega_\beta}{1 + j\omega\tau_\beta^r} + \frac{2\Delta\Omega_\beta}{1 + j\omega\tau_\beta^{r0}} \right\} \qquad (51b)$$

and the autocorrelation function, $C_\beta(t)$, is [Eq. (45)]

$$C_\beta(t) = \left\langle \mu_\beta(0) \cdot \mu_\beta(t) \right\rangle$$

$$\approx \mu_\beta^2 \left\{ (1 - 2\Delta\Omega_\beta) \exp\left(\frac{-t}{\tau_\beta^r} \right) + 2\Delta\Omega_\beta \exp\left(\frac{-t}{\tau_\beta^{r0}} \right) \right\} \qquad (53)$$

From Eqs. (51)–(53) it is clear that each reorientation mode should contribute to the dielectric relaxation process, although it must be noted that the relaxation due to the fluctuation mode could be very difficult to observe; the strength of this relaxation, $2\Delta\Omega_\beta$, is very small (e.g., for a typical axial ratio of about 30–40, in the semidilute regime, $2\Delta\Omega_\beta$ is of the order of 10^{-3}). In fact, the high-frequency domain characteristic of τ_β^{r0} has not yet been observed by means of dielectric relaxation spectroscopy. [It must be noted that the slow (τ_β^r) and fast (τ_β^{r0}) processes have been observed using electrically induced optical birefringence (the Kerr effect) measurements of PBLG solutions.[75] However, because the use of this method is limited to the isotropic phase, it is of no interest for the present paper. A detailed discussion of the Kerr effect in the system of monodisperse rods in an isotropic solution can be found in a paper by Doi.[54]]

To a very good approximation, then, we may consider only one dielectric relaxation process as present in the solution, and the total complex dielectric permittivity of the isotropic phase is given to within a proportionality factor by

$$\frac{\hat{\varepsilon}(\omega) - \varepsilon_\infty}{\varepsilon_s - \varepsilon_\infty} \approx \frac{v_p^0}{3k_B T} \sum_\alpha \frac{x_\alpha P^0(x_\alpha)}{1 + j\omega\tau_\alpha^r} \qquad (54)$$

B. Nematic Phase

The straightforward generalization of Doi's approach[54] to the case of the nematic phase required solution of Eq. (7) with the modified potential $V(\mathbf{u})$ accounting for the presence of the nematic potential, $V_N(u_\beta n)$:

$$V(\mu) = -\mu_\beta \cdot \mathbf{E} + V_N(u_\beta n) \qquad (55)$$

where $(u_\beta n)$ is the angle between vectors $\mathbf{u}_\beta$ and $\mathbf{n}$. However, before we proceed any further, let us first recall briefly the dielectric relaxation process in nematic t-LC.

It is well known from experimental[76] and theoretical[77,78] studies of dielectric relaxation in short, rodlike-molecule nematic t-LC that the rotational molecular dynamics in these systems is affected by the presence of the nematic potential. In comparison with the isotropic phase, each rod is still able to reorient into the whole of 4π-space, but the nematic potential acts as a strong perturbation on the motion of the rod, and two different types of dielectric behavior are observed: (1) the short-time-scale behavior related to fluctuation within the nematic potential around the nematic director, $\mathbf{n}$; and (2) the long-time-scale behavior related to end-over-end reorientation. The first type of dielectric relaxation effect appears at frequencies slightly higher than those characteristic of dielectric relaxation in the isotropic phase. Because this motion is spatially restricted, $\langle \mu_\beta^2 \rangle$ is only partially relaxed and the strength of the dielectric process is relatively small, that is, $\langle \mu_\beta^2 \rangle / \mu_\beta^2 \ll 1$. The remainder of $\langle \mu_\beta^2 \rangle$ is relaxed in the second dielectric process, governed by the end-over-end reorientation, but because of the nematic potential, this motion is significantly retarded and the process appears at much lower frequencies. For molecules with a high axial ratio, the difference between fast and slow relaxation should be even more dramatic, because steric effects come into play. The nematic potential modifies the situation in such a way that although the molecule can, on average, point "up" or "down" and fluctuate around the director, the end-over-end motion becomes less and less likely as the axial ratio increases. We may expect this type of situation to become even more pronounced in rigid-rod macromolecules in solution. Because the ratio at which a rod changes its direction from "up" to "down" and vice versa is very small, the dielectric relaxation associated with this motion should be pushed down to such low frequencies that it will be practically unobservable in the frequency domain.[79] One may consider only short-time-scale behavior to be detectable in the lyotropic nematic phase under consideration (i.e. fluctuations around $\mathbf{n}$). Instead of solving Eq. (7) for $\Phi_\beta(\mathbf{w}, \mathbf{u}; t)$, we propose a much simpler description, justified as above, of the dielectric relaxation in the lyotropic nematic phase. The model is based on the assumption that although the rod cage axis, vector $\mathbf{w}$, undergoes small-step diffusion similar to that in the isotropic phase, the free space available for this diffusion is restricted to some cone coaxial with $\mathbf{n}$ and characterized by the cone angle $\theta_{0\beta}$, which is a function of the degree of orientational order of the β-rods in solution. In other words, the tubelike hypothetical path along which the rod diffuses translationally is allowed by the highly ordered crowd of other rods to fluctuate only around the direction of $\mathbf{n}$. Thus, in the nematic phase both vectors, $\mathbf{u}$ and $\mathbf{w}$, are trapped—the first one by the entanglement between rods, the second by the coupled action of the entanglement and nematic potential.

The cone-restricted diffusion of the elongated molecule has been considered by many authors.[64,65,80,81] Warchol and Vaughan[64] and later Wang and Pecora[65] discussed a consequence of such a restriction on the dielectric relaxation. They showed that if the dipole moment, μ, diffuses freely within a cone characterized by the spherical polar angles (θ, ϕ), $0 \leq \theta \leq \theta_0$ and $0 \leq \phi \leq 2\pi$, the dipole correlation function is

$$C(t) = \mu^2 \left\{ \delta_1^0 + \delta_2^0 \exp\left[-\nu_2^0(\nu_2^0 +1)D't \right] + \delta_1^1 \exp\left[-\nu_1^1(\nu_1^1 +1)D't \right] \right\}$$

(56)

where

$$\delta_1^0 = \tfrac{1}{4}(1 + \cos\theta_0)^2$$
(57a)

$$\delta_2^0 = \tfrac{1}{12}(1 - \cos\theta_0)^2$$
(57b)

$$\delta_1^1 = \tfrac{1}{3}(2 - \cos\theta_0 - \cos^2\theta_0)$$
(57c)

and D^r is the diffusion constant characteristic of unrestricted, isotropic reorientation of μ. The dependence of ν_n^m on θ_0 is defined by the boundary conditions and can be found, in the form of $\nu_n^m = \nu_n^m(\theta_0)$ plots in ref. 65. It follows from Eqs. (56) and (57) that a fraction $\mu^2 \delta_1^0$ of the total $\langle \mu^2 \rangle$ that remains unrelaxed by a motion in the cone depends strongly on the cone angle. In the limit of unrestricted free diffusion (i.e., $\theta_0 = \pi$), $\delta_1^0 = 0$ and whole $\langle \mu^2 \rangle$ is relaxed. In the opposite case, as θ_0 decreases, less and less $\langle \mu^2 \rangle$ would be relaxed. Furthermore, for sufficiently small θ_0 the second term in Eq. (56) is negligible and the relaxation process becomes described by only δ_1^1 and ν_1^1. For θ_0 decreasing below $\pi/3$, $\nu_1^1 > 3$ and increases rapidly (see Fig. 2 of ref. 65). Thus, as the space available for random motion decreases, both the relaxation strength, $\mu^2 \delta_1^1$, and relaxation time, $\tau_1^1 = (\nu_1^1(\nu_1^1 +1)D')^{-1}$, decrease markedly.

Before applying the cone model to the highly ordered lyotropic nematic phase, it will be instructive to compare the model results with predictions of existing theories of dielectric relaxation in thermotropic nematics.[77,78] Due to the fact that the thermotropic nematic can easily be ordered, dielectric relaxation studies are made on oriented samples of the material. For nematogen molecules with the dipole moment parallel to the long molecular axis, a relaxation process observed in the direction parallel to the nematic director, $\mathbf{E} \| \mathbf{n}$, may be attributed to the end-over-end reorientation of molecules:

$$\frac{\hat{\varepsilon}_\|(\omega) - \varepsilon_{\|\infty}}{\varepsilon_{\|s} - \varepsilon_{\|\infty}} \simeq \frac{(\mu^2/3k_B T)\delta_\|}{1 + j\omega\tau_\|}$$

(58a)

while the fluctuations around $\mathbf{n}$ govern a process observed in the perpendicular direction, $\mathbf{E} \perp \mathbf{n}$:

$$\frac{\hat{\varepsilon}_\perp(\omega) - \varepsilon_{\perp\infty}}{\varepsilon_{\perp s} - \varepsilon_{\perp\infty}} \simeq \frac{(\mu^2/3k_B T)\delta_\perp}{1 + j\omega\tau_\perp} \tag{58b}$$

where δ_i, $i = \|, \perp$, is the fraction of $\langle\mu^2\rangle$ relaxed in the "i" process, and τ_i is the characteristic relaxation time. In what follows, the strength of the process in the direction parallel to $\mathbf{n}$ is always significantly larger than in the perpendicular direction, but its relaxation is almost two orders of magnitude slower than that of the latter process. For example, for $S = 0.6$, Nordio et al.[78] predict $\delta_\| = 0.724$, $\delta_\perp = 0.12$ and $\tau_\|/\tau_\perp = 28.2$, in agreement with experimental data.[76] Both quantities, δ_i and τ_i, are strongly dependent on the order parameter S, and tend to values characteristic of the isotropic phase as the orientational order of the phase diminishes [i.e., $\delta_i(S = 0) = 1$ and $\tau_i(S = 0) = \tau_0$]. It is convenient to scale the relaxation times in the nematic phase in units of τ_0:

$$\tau_i = g_i(S)\tau_0; \qquad i = \|, \perp \tag{58c}$$

so $g_\|(S) \geq 1$ and $g_\perp(S) \leq 1$.[77,78]

The only essentially important correlation time in Eq. (56) can also be rescaled in the same manner:

$$\tau_1^1 = \left\{ \nu_1^1(\theta_0)\left[\nu_1^1(\theta_0) + 1\right]D'\right\}^{-1}$$
$$= g_1^1(\theta_0)\tau_0 \tag{59}$$

where $\tau_0 = (2D')^{-1}$ and

$$g_1^1(\theta_0) = 2\left\{ \nu_1^1(\theta_0)\left[\nu_1^1(\theta_0) + 1\right]\right\}^{-1} \tag{60}$$

Because free diffusion in a cone implies homogeneous distribution of orientations within the cone, one can relate θ_0 to the order parameter S:

$$S = 1 - \frac{3}{2}\frac{\int_0^{\theta_0} d\theta \sin^3\theta}{\int_0^{\theta_0} d\theta \sin\theta} = \frac{\cos\theta_0(1 + \cos\theta_0)}{2} \tag{61}$$

and $g_1^1(S)$ and $g_\perp(S)$, both of which characterize fluctuations of μ around the nematic director, can be compared. Comparison (see Fig. 4) shows that for high values of S ($S > 0.8$), as occur in l-LC, the cone model may be considered a fairly good approximation of the molecular dynamics around $\mathbf{n}$.

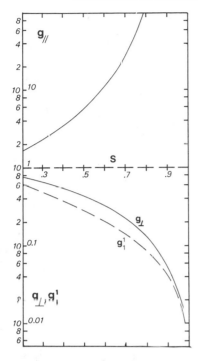

Figure 4. Comparison between the normalized dipole rotational correlation times for cone-restricted diffusion, g_1^1, and for diffusion in the presence of the nematic potential, $g_\perp$ and $g_\parallel$, as a function of the order parameter S. Correlation times are normalized to unity in the limit of isotropic diffusion.

Applying the cone model to the diffusion of the dipole moment of the cage in the nematic phase, $[\mu_\beta(1-2\Delta_\beta)w]$, the dipole autocorrelation function may be written as

$$
\begin{aligned}
C_\beta(t) &= \langle \mathbf{\mu}_\beta(0) \cdot \mathbf{\mu}_\beta(t) \rangle \\
&\simeq \mu_\beta^2(1 - 2\Delta\Omega_\beta)\delta_1^0(\theta_{0\beta}) \\
&\quad + \mu_\beta^2(1 - 2\Delta\Omega_\beta)\delta_1^1(\theta_{0\beta})\exp\left[\frac{-t}{\tau_1^1(\theta_{0\beta})}\right] \\
&\quad + \mu_\beta^2 2\Delta\Omega_\beta\exp\left(\frac{-t}{\tau_\beta^{r0}}\right)
\end{aligned}
$$

$$(62)$$

where δ_1^0 and δ_1^1 are defined in Eqs. (57), τ^{r0} in Eq. (52), and

$$
\tau_1^1(\theta_{0\beta}) = g_1^1(\theta_{0\beta})(2D_\beta^r)^{-1} \tag{63}
$$

$\theta_{0\beta} = \theta_{0\beta}(s_\beta)$ denotes the effective cone angle for a β-type rod.

The first term on the right side of Eq. (62) denotes a fraction of $\langle \mu_\beta^2 \rangle$ that remains unrelaxed in the nematic phase. The remainder is relaxed by the cone-restricted diffusion of the trap axis, $\mu_\beta^2(1 - 2\Delta\Omega_\beta)\delta_1^1$, or by the fluctuations of the dipole inside the trap, $\mu_\beta^2 2\Delta\Omega_\beta$. Although the latter contribution is very small and, as in the isotropic phase, can be neglected, it is written explicitly in Eq. (62) to avoid possible confusion between the trap-axis ($\mathbf{w}$) fluctuations around the nematic director $\mathbf{n}$ and the rod-fluctuation mode, that is, the trap-restricted random fluctuations of the rod ($\mathbf{u}$) around $\mathbf{w}$.

Neglecting the contribution from the rod-fluctuation mode, substitution of Eq. (62) into Eq. (45) and appropriate averaging over polydispersity yields the complex dielectric permittivity of the bulk, macroscopically unoriented nematic phase:

$$\frac{\hat{\varepsilon}(\omega) - \varepsilon_\infty}{\varepsilon_s - \varepsilon_\infty} \simeq \frac{v_p^0}{3k_B T} \sum_\alpha \frac{x_\alpha P^0(x_\alpha)\delta_{1\alpha}^1}{1 + j\omega g_{1\alpha}^1 \tau_\alpha^r} \tag{64}$$

where $\delta_{1\alpha}^1 \equiv \delta_1^1(\theta_{0\alpha})$, $g_{1\alpha}^1 \equiv g_1^1(\theta_{0\alpha})$, and $\tau_\alpha^r = (2D_\alpha^r)^{-1}$.

IV. STEADY-FLOW VISCOSITY

Although the rotational dynamics of rodlike macromolecules can be studied over the entire concentration–temperature range of interest (i.e., the isotropic, biphasic, and nematic regions) most directly by means of the dielectric relaxation effect, steady-flow viscometry of the system can offer similar information.[39,50,51,53,54]

The general Newton's law of viscosity for an incompressible fluid gives[82]

$$\underset{\sim}{\sigma} = -\eta\underset{\sim}{\kappa} \tag{65a}$$

where $\underset{\sim}{\sigma}$ and $\underset{\sim}{\kappa}$ are the stress and rate of deformation tensors, respectively, and η denotes the shear viscosity of the fluid. If a fluid is non-Newtonian, the relation between $\underset{\sim}{\sigma}$ and $\underset{\sim}{\kappa}$ becomes complicated. We can, however, still use Eq. (65a) for the solution of rodlike particles, although η is then a function of $\underset{\sim}{\kappa}$:[51]

$$\underset{\sim}{\sigma} = -\eta(\underset{\sim}{\kappa})\underset{\sim}{\kappa} \tag{65b}$$

Note that in the case of simple homogeneous flow (e.g., Couette flow), $\underset{\sim}{\kappa}$ reduces to the velocity gradient tensor. To calculate the viscosity from Eqs. (65), the stress tensor has to be evaluated for a given flow condition, namely, $\underset{\sim}{\kappa}$. This problem has been solved for steady flow of the isotropic solution of monodisperse rods by Doi[50] and Doi and Edwards,[51] and for the nematic phase by Doi[39] and Marrucci.[83] In the present section the main results of these authors are adopted to the case of polydisperse rods.

A.   Isotropic Phase

Doi and Edwards[51] have shown that for the isotropic solution of monodisperse rods the stress-tensor components can be written as

$$\sigma_{ik} = -3n^0 k_B T u_i u_k - \frac{n^0 k_B T}{2D^{r0}} \sum_{l,m} \kappa_{lm} u_l u_m u_i u_k - \eta_s(\kappa_{ik} + \kappa_{ki}) + P\delta_{ik} \quad (66)$$

where P is the hydrodynamic pressure and κ_{ik} denotes components of the tensor $\underset{\sim}{\kappa}$. Equation (66) was derived with some simplifying assumptions. First, it is obtained in the dilute-solution approximation; that is, the intermolecular forces are neglected. From a dimensional analysis, the contribution to σ_{ik} from these forces is of the magnitude of $n^0 k_B T \cdot n^0 L^2$, and as long as the concentration is relatively small ($n^0 \ll L^{-2}$), this contribution is negligible.[51] Although the latter contribution may become substantial as n^0 becomes comparable to L^{-2}, Doi[39] neglected it also when extending the viscosity theory to the nematic phase. On the other hand, we believe that these interactions are sufficiently accounted for by introduction of the retardation factor, R^r [Eqs. (26) and (43)], into the expression for the diffusion coefficient of the trap, D^r. Second, Eq. (66), derived for the "shish kebab" model of a rod introduced by Riseman and Kirkwood,[67] neglects the hydrodynamic interactions, which again is justified[51] in the limit of zero-shear flow.

In practice, Eq. (66) can be simplified even further. Because the viscosity of a semidilute solution is a few orders of magnitude greater than η_s, the third term on the right side of Eq. (66) may be neglected. Further, in the limit of zero-shear flow the second term is much smaller than the first one and can also be skipped.

Hence, averaging over all possible orientations of a rod, Doi and Edwards[51] finally arrived at a very simple expression for the stress tensor:

$$\sigma_{ik} = -3n^0 k_B T \langle\langle u_i u_k - \delta_{ik}/3 \rangle\rangle \quad (67)$$

where in terms of the two-vector distribution function, Eq. (6), the average $\langle\langle u_i u_k - \delta_{ik}/3 \rangle\rangle$ is calculated according to Eq. (49). To calculate this average, Doi[54] considered the behavior of rods in the presence of the velocity gradient. As a result of this gradient, in the diffusion equations [Eqs. (9) and

(10)] additional terms due to the flow have to be added:

$$\frac{\partial \phi}{\partial t} = D^{r0} \nabla_{\mathbf{u}} \cdot \left\{ \nabla_{\mathbf{u}} \phi + (k_B T)^{-1} \phi \nabla_{\mathbf{u}} (V_c + V) \right\} - \nabla_{\mathbf{u}} \cdot (\mathbf{u} \times \underline{\kappa} \cdot \mathbf{u} \phi) \quad (68a)$$

$$\frac{\partial f}{\partial t} = D^r \nabla_{\mathbf{w}} \cdot \left\{ \nabla_{\mathbf{w}} f + (k_B T)^{-1} f \nabla_{\mathbf{w}} U) \right\} - \nabla_{\mathbf{w}} \cdot (\mathbf{w} \times \underline{\kappa} \cdot \mathbf{w} f) \quad (68b)$$

(The subscript "α," used before to distinguish between different species, is skipped here, as we are considering for the moment only monodisperse rods.) If no external field except the velocity gradient is present, then the potential U, arising now only from the flow, is negligibly small; furthermore, for a very small flow rate, ϕ is close to the equilibrium distribution. Hence we have[54]

$$\sigma_{ik} = -3n^0 k_B T \langle\langle u_i u_k - \delta_{ik}/3 \rangle\rangle$$
$$= -3n^0 k_B T (1 - 3\Delta\Omega) \langle w_i w_k - \delta_{ik}/3 \rangle \quad (69)$$

where

$$\langle w_i w_k - \delta_{ik}/3 \rangle = \int d^2 w f(\mathbf{w})(w_i w_k - \delta_{ik}/3)$$

If the considered flow is the steady Couette flow when $\underline{\kappa}$ is given by

$$\underline{\kappa} = \begin{pmatrix} 0 & \kappa & 0 \\ 0 & 0 & 0 \\ 0 & 0 & 0 \end{pmatrix} \quad (70)$$

then the steady-state solution of Eq. (68b), f, includes only one dimensionless parameter, (κ/D^r), and consequently any average over the distribution function f depends on some combination of (κ/D^r). From Eqs. (65) and (68)–(70) it follows that

$$\eta(\kappa) = -\frac{\sigma_{xy}}{\kappa}$$
$$= 3n^0 k_B T \frac{\Lambda(\kappa/D^r)}{\kappa} \quad (71a)$$

with $\Lambda(\kappa/D^r)$ being some function of (κ/D^r). In the limit of very small ("zero") shear, the viscosity has a finite, constant value, which implies that Λ/κ approaches a constant as $\kappa \to 0$, and

$$\eta(0) \propto \frac{n^0 k_B T}{D^r}$$
$$\propto n^0 k_B T \tau^r \quad (71b)$$

that is, the zero-shear viscosity is proportional to the rotational relaxation time.

To generalize the latter to polydisperse systems, it may be assumed that the bulk viscosity arises from contributions of different species, with each contribution given by Eq. (71b):

$$\eta_{\mathrm{iso}} \propto k_B T \sum_\alpha n^0_\alpha \tau^r_\alpha$$

$$\propto k_B T v^0_p \sum_\alpha \frac{P^0(x_\alpha)\tau^r_\alpha}{x_\alpha} \tag{72}$$

In other words, the bulk viscosity of the isotropic phase is proportional to the number average of the rotational relaxation time of all rods in the solution.

B. Nematic Phase

In the nematic phase the potential $U(\mathbf{w})$ in Eq. (68b) is the total potential experienced by the trap axis and contains contributions from the soft "nematic" anisotropic interactions between rods, from the flow, and, generally, from an external orienting field such as a magnetic field. In this case the stress tensor can be written as[39]

$$\sigma_{ik} = -3n^0 k_B T \langle w_i w_k - \delta_{ik}/3 \rangle - n^0 \langle (\nabla_{\mathbf{w}} U)_i u_k \rangle \tag{73}$$

Doi[39] calculated the stress tensor in the absence of an orienting field for simple Couette flow of the nematic fluid [Eq. (70)]. His result for monodisperse rods in the limit of very slow flow is

$$\sigma_{xy} \simeq -n^0 k_B T (2D^r)^{-1} \kappa \frac{(1-s)^2(1+2s)(1+3s/2)/(1+s/2)^2}{3} \tag{74}$$

where $s(=S)$ denotes, as before the order parameter. Hence, by definition, the zero-shear viscosity becomes

$$\eta = -\frac{\sigma_{xy}}{\kappa}$$

$$\simeq \frac{n^0 k_B T \tau^r}{3} \left\{ \frac{(1-s)^2(1+2s)(1+3s/2)}{(1+s/2)^2} \right\} \tag{75}$$

For t-LC it is customary to study the viscosity of the liquid crystalline phase not only in flow-oriented samples but also in samples oriented by an external (e.g., magnetic) field, using three different configurations of the nematic-director orientation with respect to the flow direction:[84,85]

$$\text{(a) } \mathbf{n} \| (\mathbf{v} \times \nabla v); \quad \text{(b) } \mathbf{n} \| \mathbf{v}; \quad \text{and (c) } \mathbf{n} \| \nabla \mathbf{v}, \tag{76}$$

where $\mathbf{v}$ is the velocity of the flow. Viscosities associated with these geometries, η_a^M, η_b^M, and η_c^M, are traditionally referred to as Miesowicz viscosities, after the author who performed pioneering experimental viscosity studies of the thermotropic nematics.[85]

Miesowicz viscosities for the solution of monodisperse rods have been evaluated by Marrucci,[83] who extended calculations of Doi[39] to account for the presence of the external ordering magnetic field:

$$\eta_i^M \simeq \left\{ n^0 k_B T \frac{\tau^r}{3} \right\} \begin{cases} (1-s) & \text{for } i = a \\ \dfrac{(1-s)^2}{1+s/2} & \text{for } i = b \\ \dfrac{(1+2s)^2}{1+s/2} & \text{for } i = c \end{cases} \tag{77}$$

It can easily be shown that extension of Eqs. (75) and (77) to the case of polydisperse rods requires the same approach as for the isotropic-phase averaging over all different species in the system; that is,

$$\eta_{\text{nem}} \propto v_p^0 k_B T \sum_\alpha \frac{P^0(x_\alpha)\tau_\alpha^r}{x_\alpha} \left\{ \frac{(1-s_\alpha)^2(1+2s_\alpha)(1+3s_\alpha/2)}{(1+s_\alpha/2)^2} \right\} \tag{78a}$$

$$\eta_i^M \propto v_p^0 k_B T \sum_\alpha \frac{P^0(x_\alpha)\tau_\alpha^r}{x_\alpha} \begin{cases} (1-s_\alpha) & \text{for } i = a \\ \dfrac{(1-s_\alpha)^2}{1+s_\alpha/2} & \text{for } i = b \\ \dfrac{(1+2s_\alpha)^2}{1+s_\alpha/2} & \text{for } i = c \end{cases} \tag{78b}$$

V. COMPARISON WITH EXPERIMENTAL RESULTS

In the preceding sections we have developed the expressions for the complex dielectric permittivity, $\hat{\varepsilon}(\omega)$, and for the steady-flow viscosity, η, with the intent of confronting the molecular-dynamics model with the experimental results. Unfortunately, very few experimental dielectric and viscometric studies have been made over a sufficiently large concentration–temperature range to permit in-depth assessment of the present model. Adequate dielectric data are available only for poly(n-hexylisocyanate) in toluene solution,[62] and similar viscosity measurements were carried out for PBA in sulfuric acid[18] and PHPIC in toluene.[29] In the remaining cases, the dielectric[61] and viscosity[17,18,26-29,45-47] studies were limited to the isothermal (room temperature) concentration-dependent measurements. Because we would like to verify the theory in the most comprehensive way, the only choice for an example system is represented by the poly(n-alkylisocyanate)s in toluene solutions for which sufficient data are available. Thus, before outlining the numerical simulation procedure and discussing calculations, we find it necessary to review the most important results of experimental studies of poly(n-alkylisocyanate)s in solution.

A. The Example System: Poly(n-Alkylisocyanate)s in Solution

Diluted solutions of poly(n-alkylisocyanate)s (PAIC) in different solvents have been studied extensively for many years; an excellent survey of results can be found in a paper by Bur and Fetters.[86] It was established that the spatial configuration of PAIC chains consists of helices that can be pictured as close to the *cis–trans* conformation (see Fig. 5). This explains the very good solubility of PAIC in nonpolar solvents, since in this configuration the alkyl side chains are in positions to shield the highly polar backbone from the solvent medium. In fact, it was found that the conformation of the PAIC chain is practically unaffected by these solvents. From dielectric and Kerr-effect measurements, it was established that each chain possesses a permanent dipole moment parallel to the helix axis. The magnitude of this dipole moment, which is proportional to the degree of polymerization or, in other words, to the length of the macromolecule, and the lack of a dipole moment in the direction perpendicular to the molecule long axis also confirm the helical nature of the chain. In both dilute[86,87] and concentrated[88] solution,

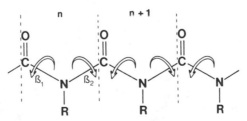

Figure 5.   Poly(n-alkylisocyanate) chain in all-*trans* conformation, $(\beta_1, \beta_2) = (180°, 180°)$. The stable helix conformation is close to a *cis–trans* structure, $(\beta_1, \beta_2) = (\pm 140°, \mp 20°)$.[86]

PAIC helices turn out to behave as rigid rods up to a weight-average degree of polymerization $\langle n_w \rangle$ of at least a few hundreds. For values of $\langle n_w \rangle$ greater than 10^3, they deviate from rodlike behavior and gradually adopt a worm-like conformation.

The inherent rigidity of PAIC macromolecules stimulated Bur and co-workers to search for lyotropic liquid crystallinity, although without success. This was discovered only recently, by Aharoni and co-workers.[26-29,88] A large number of members of a homogeneous series of PAIC have been investigated in solvents such as toluene, tetrachloroethane, and bromoform. It was found that the middle members of the series, with alkyl-side-chain lengths in the range $4 \leq m \leq 13$ (where m is the number of carbon atoms in the chain) exhibit lyotropic mesomorphic behavior. Polarized-light optical microscopy studies revealed that the lyotropic mesophase is nematic in character.

The concentration–temperature phase diagram of PAIC solution resembles in its main details the phase diagrams observed for the other rodlike l-LC[12,18] (see Fig. 1). The most characteristic feature is the coexistence of the isotropic and nematic phases over a wide range of concentrations and temperatures. On the phase diagram this biphasic range separates regions of solely isotropic (low concentration) and nematic (high concentration) phases. The biphasic material has the nature of a heterogeneous mixture (see Fig. 6). At low volume fractions of the minor component (i.e., close to the iso-tropic–biphasic transition concentration, v_p^*, or to the biphasic–nematic transition concentration, v_p^{**}), the "guest" phase is suspended in the "host" one in the form of more or less defined spherulites (Fig. 6a and e). As concentration increases in the biphasic range, the size of the spherulites grows at the expense of the host phase, and eventually "phase inversion" occurs at some critical concentration v_p^i. It is characteristic of PAIC solutions that at this concentration, the volume fraction of the isotropic phase is smaller than the volume fraction of the nematic phase.

Due to the heterogeneous nature of the biphasic material, the isotropic and anisotropic components could be physically separated,[27,28] and, more importantly, it was found that the high- and low-molecular-weight species partition preferentially into the isotropic and nematic phase, respectively. Unfortunately, there is very little experimental information available concerning the way in which the polydisperse system is partitioned between the isotropic and nematic phases, that is, concerning the polydispersity of macromolecules in either phase, and how this partitioning changes over the bi-phasic range.

The viscous properties of concentrated solutions of numerous PAIC have been extensively investigated across a broad range of concentrations, from isotropic through biphasic to nematic fluids,[26-29] and temperatures.[29] Typical results are presented in Fig. 7.

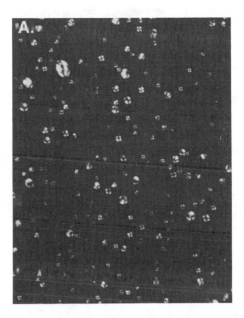

Figure 6.   Micrographs of PHIC solutions in tetrachloroethylene in biphasic range of concentration, magnification 200×. The micrographs are labeled in the order of increasing solution concentration; for example, *a* corresponds to a concentration close to v_p^*, and *e* to a concentration close to v_p^{**}. (Courtesy of Dr. S. M. Aharoni.)

As far as isothermal viscosity versus concentration behavior, $\eta(v_p^0)$, is concerned, the following general observations were drawn from these measurements:

1. Within the isotropic phase, the viscosity is a rapidly increasing function of concentration. This increase becomes critical as concentration approaches the isotropic–biphasic transition concentration, v_p^*.

2. In most cases, passage through the transition concentration v_p^* is manifested by the appearance of a *shoulder* on the viscosity–concentration curve (see Fig. 7). This feature is solvent independent.

3. Within the biphasic range the viscosity curve passes through a maximum at a certain critical concentration of polymer, v_p^m. At this concentration, the volume fractions of the isotropic phase and the nematic phase in the biphasic material are approximately equal.

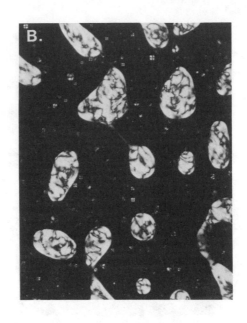

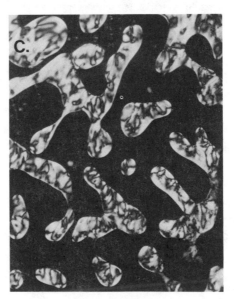

Figure 6. (*Continued*)

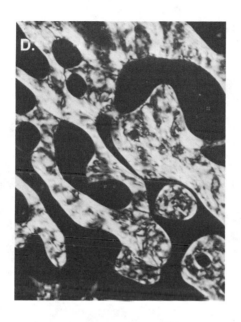

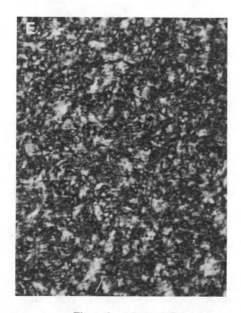

Figure 6. (*Continued*)

667

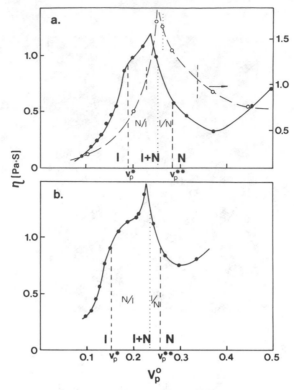

Figure 7. Viscosity versus concentration curves for (a) PHPIC and (b) PHIC (M_w = 75,000) in toluene solutions.[29] Full curve in a corresponds to M_w = 43,000; the broken curve to M_w = 41,000. Vertical broken lines indicate the position of either v_p^* or v_p^{**}; the dotted lines, the position of the phase inversion v_p^i. "I" and "N" stand for isotropic and nematic, respectively; "N/I" and "I/N" indicate the guest–host character of the heterogeneous suspension.

4.   The viscosity decreases rapidly from the maximum with further concentration increases. The curve bottoms out for concentrations well above the biphasic–nematic transition concentration, v_p^{**}.

5.   Within the nematic phase, after reaching a minimum, the concentration increases again.

From the concentration–temperature-dependent viscosity studies, shown in Fig. 8, it was additionally noted that within each single phase, isotropic or nematic, the viscosity was a decreasing function of temperature, and that the viscosity maximum in the biphasic range showed a tendency to move toward v_p^{**} as temperature increased.

Dielectric relaxation studies have been carried out for two samples, the copolymer of n-butyl- and n-nonylisocyanate, PBNIC (134,000),[61] and PHIC

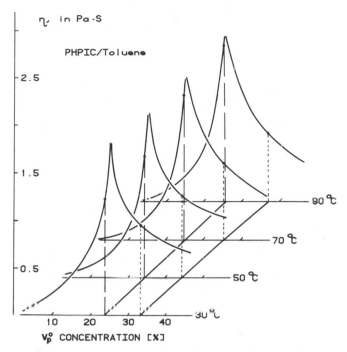

Figure 8. Viscosity of PHPIC (41,000)–toluene solution as a function of concentration and temperature.[29] The broken vertical lines indicate v_p^* (— —) and v_p^{**} (- - -). Solid lines in (v_p^0, t) plane are the boundaries of the biphasic range. (Henceforth, numbers in parentheses after polymer names will denote M_w.)

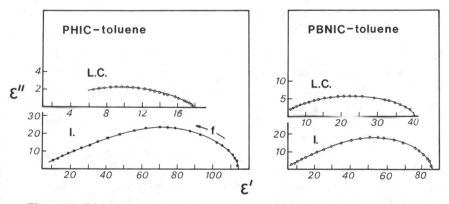

Figure 9. Dielectric relaxation in PHIC (76,000) and PBNIC (134,000) solutions in toluene.[61,62] Typical Argand diagrams of the complex dielectric permittivity, $\hat{\varepsilon} = \varepsilon' - j\varepsilon''$, in the isotropic (I.) and nematic (L.C.) phases.

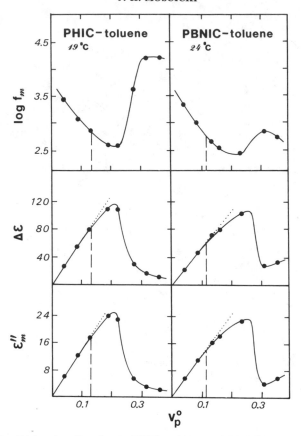

Figure 10.   Dielectric relaxation in PHIC (76,000) and PBNIC (134,000) solutions in toluene.[61,62] $\log f_m$, $\Delta\varepsilon$, and ε_m'' are shown as functions of concentration at room temperature. Vertical broken lines indicate v_p^*.

(76,000),[60,62] with toluene as a solvent.[‡] The relaxation process was studied as a function of both concentration and temperature over the isotropic, biphasic, and nematic states of solutions. Only one relaxation process, with a broad distribution of relaxation times, was observed in the isotropic and nematic phases (Fig. 9), and it showed up at relatively low frequencies (10^{-2}–10^5 Hz). The most important results are summarized in Fig. 10. All three characteristic dielectric parameters, the dielectric increment, $\Delta\varepsilon = \varepsilon_s - \varepsilon_\infty$; the maximum of the dielectric loss, ε_m''; and the frequency at the dielectric loss maximum, f_m ($= (2\pi\tau_d)^{-1}$, where τ_d is the average dielectric re-

_____

[‡]Numbers in parentheses denote M_w.

laxation time) underwent significant changes across the isotropic–biphasic–nematic concentration range. Summing up the results, we note that:

1. Since the dielectric increment and the dielectric loss maximum are closely related to each other,[74] they demonstrated very similar changes across the isotropic–biphasic–nematic range of concentration. Within the isotropic phase up to the transition concentration v_p^*, both were linear in v_p^0. Above v_p^*, they initially increased further, although visibly more slowly and no longer linearly in v_p^0. Well within the biphasic range, $\Delta\varepsilon$ and ε_m'' went through maxima and dropped rapidly afterwards. In the nematic phase, $\Delta\varepsilon$ and ε_m'' apparently continued to decrease with increasing concentration, at least for PHIC.

2. The logarithm of the frequency of maximum dielectric loss monotonically decreased across the isotropic range and through the initial part of the biphasic range, reaching a minimum at approximately the same concentration, v_p^c, at which the maxima in $\Delta\varepsilon$ and ε_m'' occurred. Further increases of concentration produced a rapid increase of $\log f_m(v_p^0)$ toward a maximum, which appeared around or just above v_p^{**}. In the nematic phase, $\log f_m$ again decreased. [Note that since $\tau_d = (2\pi f_m)^{-1}$, $\log f_m$ ($= -\log \tau_d + $ const) directly reflects the behavior of the relaxation time.]

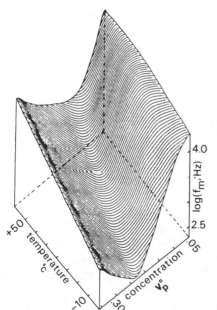

Figure 11. Dielectric relaxation in PHIC (76,000)–toluene.[62] $\log f_m$ as a function of concentration and temperature.

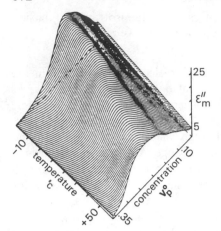

Figure 12. Dielectric relaxation in PHIC (76,000)–toluene.[62] ε_m'' as a function of concentration and temperature.

3. Although the general character of the $\Delta\varepsilon(v_p^0)$, $\varepsilon_m''(v_p^0)$, and $\log f_m(v_p^0)$ curves remained qualitatively the same with changing temperature, the influence of temperature was pronounced. With increasing temperature, the positions of the maxima of the $\Delta\varepsilon$ and ε_m'' curves moved toward higher concentrations, while the maximum height decreased. At the same time the minimum of $\log f_m$ shifted toward higher concentrations and its depth decreased. (See Figs. 11 and 12.)

It must be noted at this point that, fortunately, viscosity studies have been carried out for the same ($M_w = 76,000$)[28] or compatible ($M_w = 75,000$)[29] samples of PHIC as for the dielectric studies. It is thus possible to compare predictions of the theoretical model with these viscosity and dielectric relaxation data simultaneously.

B. Simulation of the Dielectric and Viscosity Behaviors

The viscous and dielectric properties of the PAIC solutions underwent the most pronounced changes as the solutions traversed the biphasic range. The dramatic character of these changes resulted first of all from the heterogeneous nature of the biphasic material (the volume fraction of the biphasic material occupied by either of the component isotropic and nematic phases varied with concentration and temperature), and secondly from the variable and significantly different viscous and dielectric properties of the component phases. Although the former origin of these changes is understandable, the latter is less obvious and is due to the preferential partitioning of the longest species into the nematic phase at the expense of the isotropic one. Consequently, on variation of the concentration or temperature, the polydispersity of the species in either phase varies continuously, influencing each phase property. The simulation of dielectric relaxation and viscosity thus requires detailed knowledge about the phase behavior of the considered system, in particular a quantitative description of each component phase at any concentration–temperature point of the biphasic range. Unfortunately, neither the polydispersity nor the partitioning of PAIC macromolecules in the biphasic material is known. Therefore, prior to performing the simulation, we chose a model function for the polydispersity of rods and calculated the phase behavior of this system in solution.

1.   Model System of Polydisperse Rods in Solution

Although we have great freedom of choice in defining the model system of rods, in selecting the polydispersity function we expect the system to meet all possible requirements imposed by the experimental evidence:

1.   The model values of v_p^* and v_p^{**} should be comparable to those experimentally observed.

2.   The average rod length should be close to the average macromolecule length calculated from M_w [29,86]

Our earlier numerical studies of the dielectric relaxation[61,89] indicated that the distribution function $P^0(x_\alpha)$ could be in the form of a "core" Gaussian-like distribution with a small high-rod-length "tail", contributing no more than few hundredths to the total volume fraction of rods:

$$P^0(x_\alpha) = \left(\frac{x_\alpha}{x_p^0}\right)\exp\left[-4\ln 2\left(\frac{x_\alpha - x_0}{\Delta}\right)^2\right] + \mathrm{tail}(x_1, x_h) \qquad (79)$$

where x_0 is the position of the center of the core distribution, Δ is the half-width, and x_1 and x_h define, respectively, the lower and upper limits of the rod length in the tail. We investigated[89] the phase equilibria in the athermal limit of numerous systems in solution, with polydispersity given by Eq. (79) with different forms of tails, and found that although v_p^* and, to a lesser extent, v_p^{**} are sensitive to the presence or absence of the tail, the tail position, and the tail magnitude, they are virtually insensitive to the tail shape. The presence of the tail in the distribution was found[61] to be essential for explanation of the isothermal dielectric relaxation behavior across the biphasic range. We believe that the persistent increase of $\Delta\varepsilon$ and ε_m'' in the initial part of the biphasic range, $v_p^* \lesssim v_p^0 \ll v_p^{**}$ (see Fig. 10), reflects the presence of a small amount of very long chains in solution. For this reason the polydispersity function given in Eq. (79) is also used in the present studies.

Adjustment of the distribution function parameters, $\{x_0, \Delta, x_1, x_h\}$, was done by trial-and-error calculations of the phase diagram and comparison with the experimental results in Fig. 1. The phase diagram calculations were made with an aid of the statistical thermodynamics theory worked out by Warner and Flory[37] (WF). Of various existing theories of phase equilibrium in solutions of rigid-rod macromolecules,[31-38,41] we found the WF theory particularly suitable for application in the present studies. Based on the original lattice theory of Flory,[33] the theory accounts for both steric repulsions and the existence of soft anisotropic interactions between rods via pairs of segments in contact with each other, resulting, in the nematic phase, in the Maier–Saupe potential[37]

$$U(u_\alpha n) = -k_B T^* v_p^0 S \left\{ 1 - \tfrac{3}{2} \sin^2(u_\alpha n) \right\} \tag{80}$$

where T^* defines the scale of the segment interaction. Therefore, the basic assumptions of the phase-equilibrium theory and of the molecular-dynamics model are compatible. An additional good aspect of the WF theory is that it is already formulated for the case of polydisperse rods. Because the use of the WF theory is only auxiliary, for clarity in this paper we have moved a short outline of the theory, together with an algorithm of the FORTRAN program developed for the calculations, to Appendix A.

As mentioned in the preceding subsection, because it provides the most complete dielectric and viscosity results, the PHIC (76,000)–toluene system is chosen for consideration. From the known M_w and the monomeric molecular weight of $M_0 = 127$, we calculate the average degree of polymerization, $\langle n_w \rangle \approx 600$. Dev et al.[90] found the contribution from the monomer to the total length of a helix to be about 0.9–1.1 Å and the helix diameter to be about 22–25 Å. Combining this information, we obtained $x_p^0 \simeq 25 \pm 5$. Because the presence of the tail shifts the average rod length toward higher values, the maximum of the core distribution, Eq. (79), should be positioned at slightly lower values, $x_0 \leq x_p^0$. Polarized light microscopy of the dielectrically studied solutions revealed[91] that the isotropic–biphasic transition at room temperature (r.t.) occurred at approximately $v_p^*(\text{r.t.}) \approx 0.14$. The value of $v_p^{**}(\text{r.t.})$ is much more difficult to estimate although it was probably in the range 0.27–0.30.[62,91] These values are in fairly good agreement with the results for the viscometrically studied solutions[28,29] and were used as matching points in the subsequent phase diagram trial-and-error calculations.

The model polydispersity function chosen for numerical calculations defines the tail as an isosceles triangle stretching from a base formed by x_1 and x_h. Because the longest species effectively determines the value of v_p^*, and the shortest species that of v_p^{**},[91,92] the tail position has a direct influence on the values of v_p^*, while the position of the center of the core distribution has an influence on the values of v_p^{**}. A few trial calculations show that a satisfactory phase diagram (see Appendix A, Fig. A2) is obtained for $x_0 = 25$, $\Delta = 10$, and a tail stretching from 65 to 80 and contributing 2% to $\sum_\alpha P^0(x_\alpha)$ ($=1$).

The phase-equilibrium calculation provides us not only with information about the phase diagram but also with a quantitative description of the partitioning of rods between coexisting phases in the biphasic range and a detailed characterization of the biphasic material at any concentration–temperature point of the range. This includes the volume fractions of the sample occupied by the nematic and isotropic components, ϕ^N and $(1 - \phi^N)$, respectively; the volume concentrations of rods in both components, v_p^N and v_p^I, respectively; and the average rod length of species in either phase, x_p^N and x_p^I, respectively. Additionally, the order parameter s_α of each α-type rod in the nematic phase is also given over the whole concentration–temperature range of interest.

Because of the limited space and scope of this paper, only some of these results are shown in the Appendix A. Detailed discussion of the results will be presented elsewhere.

2. Dielectric and Viscous Properties of the Biphasic Material

Calculation of the bulk dielectric or viscous properties of the model solution in the single isotropic or nematic phase is simple, once v_p^0, $P^0(x_\alpha)$ and, for the nematic phase, s_α for all α-type rods present in the distribution are known (from now on $\{s_\alpha\}$ will denote the set of all s_α and S). This calculation is not straightforward, however, for the biphasic range. Because of the heterogeneity of the biphasic material, additional effects arise. The manners in which heterogeneity influences the effective complex dielectric permittivity and the steady-state viscosity are substantially different, and each case has to be considered separately.

a. *Complex Dielectric Permittivity.* The bulk dielectric relaxation behavior of the heterogeneous mixture of two systems, each characterized by a complex dielectric permittivity originating in the pure dipole relaxation, has not yet been treated theoretically. The complexity of the problem is well illustrated by the very simple example of a bricklike sandwich made up of two parallel slabs of material of thicknesses d_1 and d_2, the first having a complex permittivity $\hat{\varepsilon}_1(\omega)$, and the second a complex permittivity $\hat{\varepsilon}_2(\omega)$ (see Fig. 13). If the brick is placed in an external electric field $\mathbf{E}$ in such a way that $\mathbf{E}$

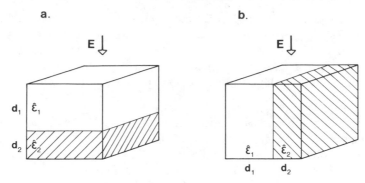

Figure 13.   Schematic representation of two possible orientations of a two-slab bricklike sample with respect to a measuring electric field.

is perpendicular to the interface between slabs (fig. 13a), the apparent complex permittivity is given by

$$\frac{1}{\hat{\varepsilon}_{app}(\omega)} = \frac{\Phi_1}{\hat{\varepsilon}_1(\omega)} + \frac{(1-\Phi_1)}{\hat{\varepsilon}_2(\omega)} \tag{81a}$$

where $\Phi_1 = d_1/(d_1 + d_2)$. If the geometry of measurement is changed and **E** is now parallel to the interface (Fig. 13b), we have

$$\hat{\varepsilon}_{app}(\omega) = \Phi_1 \hat{\varepsilon}_1(\omega) + (1 - \Phi_1)\hat{\varepsilon}_2(\omega) \tag{81b}$$

It is thus clear that the $\hat{\varepsilon}_{app}$ value of a heterogeneous mixture consisting of a suspension of one phase in another is extremely difficult to estimate. However, because we intend to compare the present theory with experimental results for PHIC–toluene systems, we may benefit from the observed behavior of the studied biphasic solutions and the particular geometry of the dielectric experiments by making significant simplifications. First, due to the difference between the effective densities of the conjugated nematic and isotropic phases, in the presence of gravitation or centrifugal forces the phases tend to separate into two layers over the course of a few hours, with the lighter material floating on top of the heavier.[61] Second, in the experiments, the dielectric cell electrodes were placed vertically.[61] Next, because of the very high viscosity of the solutions, the cell was filled with sample with the aid of a centrifuge. Since we always waited at least a few hours before initiating measurements, there is a good chance that the studied samples were nearly at equilibrium and made up of two well-separated slabs. Since in this geometry the measuring electric field was perpendicular to the gravitational one, the situation was similar to that in Fig. 13b. Therefore, we assume that Eq. (81b) is directly applicable in the biphasic range:

$$\hat{\varepsilon}_{bi}(\omega) = (1 - \Phi^N)\hat{\varepsilon}_{iso}(\omega) + \Phi^N\hat{\varepsilon}_{nem}(\omega) \tag{82}$$

where $\hat{\varepsilon}_{iso}$ and $\hat{\varepsilon}_{nem}$ are given by Eqs. (54) and (64). It is very important to note that by using a two-slab model of the biphasic sample we also avoid the discontinuity in dielectric properties at the phase inversion.

   b. *Viscosity.* A simple but intuitively justified idealization of the biphasic material, especially when the volume fraction of the "guest" phase is small, was proposed for calculation of the bulk viscosity by Matheson[48] and later considered also by Aharoni[49] and Samulski.[52] This idea of approximating the biphase as a uniform suspension of identical colloidal spheres of the "guest" phase in the "host" one in order to calculate the apparent viscosity, η_{bi}, was based on the original work of Taylor,[94] which, for a sufficiently low volume fraction Φ_g of the guest phase, predicts that

$$\eta_{bi} \rightarrow \eta_h(1 + \lambda_0\Phi_g) \tag{83}$$

where

$$\lambda_0 = \frac{2.5(\eta_g + 2\eta_h/5)}{\eta_g + \eta_h} \tag{84}$$

and the subscripts "g" and "h" denote the guest and host phases, respectively. Matheson[48] has modified this equation to extend its validity to higher volume fractions of the guest phase and thus make it applicable well within the biphasic range:

$$\eta_{bi} = \eta_h\exp\left[\lambda_0\Phi_g\left(\frac{1 - \Phi_g}{\Phi_c}\right)\right] \tag{85}$$

where Φ_c is a numerical constant equal to the maximum volume that can be filled by close-packed identical spheres, 0.74.

   On the low-concentration side of the biphasic material ($v_p^* \le v_p^0 \ll v_p^{**}$), Eq. (85) concerns the nematic phase suspended in the isotropic medium; on the high-concentration side ($v_p^* \ll v_p^0 \le v_p^{**}$) (see Fig. 6d), this is reversed. Although this approximation is a very crude one, it gave quite reasonable behavior for the bulk viscosity in the biphasic range,[48,49,52,53] especially for concentrations not far from either v_p^* or v_p^{**}.

Within the suspension model, the maximum in the viscosity curve is a consequence solely of a "switch" from nematic-in-isotropic (N/I) to isotropic-in-nematic (I/N) type suspension behavior. An unsolved problem remains concerning the concentration at which this change happens. For a suspension of identical spherulites, the phase inversion happens for $\Phi^N \approx 0.5$, and we may expect the maximum to appear at a concentration associated with this value of Φ^N. For the real solutions, however, the suspended domains become less and less spheroidal as concentration departs from either transition concentration, v_p^* or v_p^{**} (see Fig. 6). Furthermore, domains are easily shear deformable. Under these circumstances the maximum position v_p^m, and the phase inversion concentration, v_p^i, are uncorrelated; Aharoni[29] noted that although the viscosity maximum appeared at approximately $\Phi^N \approx 0.5$, the phase inversion in the investigated PAIC–toluene samples was observed at much higher volume fractions ($\Phi^N \approx 0.7$) of the nematic component. In the absence of more precise information on where the viscosity maximum should appear, for simplicity we will define the maximum as the intersection between the N/I and I/N viscosity curves, which roughly corresponds to the point where $\Phi^N = 0.5$. Having established all necessary relations of the bulk $\hat{\varepsilon}(\omega)$ and η with concentration, we may now briefly discuss effects related to temperature.

3. Temperature Effects

Apart from the effects of temperature as a parameter, explicitly incorporated into equations for the diffusion coefficient, D_α^r; complex permittivity; and steady-flow viscosity [see, respectively, Eqs. (17) and adjoiningly, Eqs. (54), (64), (72), and (78)], we anticipate additional significant temperature effects arising from the strongly temperature-dependent translational diffusivity of rods at high concentration, $D_\alpha^{r\text{eff}}$ [Eq. (26)] [more precisely, from $R_\alpha^r = R_\alpha^r(T)$, Eq. (43)] and from the phase behavior of the solution.

Developing Eq. (43) we introduced three parameters, $\delta\zeta$, d^*, and γ, the first two of which are, in principle, temperature dependent. The constant $\delta\zeta$, defined in Eq. (38), can also be expressed in terms of the viscosity of a solvent, η_s, and the viscosity resulting from the cohesive forces between rods, η_r: $\delta\zeta_r = \eta_r/\eta_s - 1 \approx \eta_r/\eta_s$ (because $\eta_r \gg \eta_s$). Since the dependence of viscosity on temperature is usually explained as an activational process, we may expect $\delta\zeta_r$ to behave roughly in the same manner:

$$\delta\zeta \simeq A_\eta \exp\left(\frac{\delta E_\eta}{k_B T}\right) \qquad (86)$$

where δE_η measures the difference between activation energies characteristic for η_r and η_s, and A_η is some numerical constant.

Even more important changes of R'_α originate from the temperature-sensitive quantity $d*$. As we mentioned before, the retardation factor R'_α tends to infinity as concentration approaches the critical value of $2(1+2s_\alpha S)^{-1}/d*^2$ [see Eq. (34a)]. This critical concentration, which we identified with gelation or solidification of the solution, is $v_g^I = 2/d*^2$ for the isotropic and $v_g^N = 2/3d*^2$ for the perfectly ordered nematic phase. Thus, knowing the temperature dependence of v_g^I and v_g^N, one may estimate $d* = d*(T)$.

Unfortunately, $v_g^I(T)$ and $v_g^N(T)$ have not been studied for PAIC in solution. Some qualitative notions of what these temperature dependences look like can be obtained by applying the lattice theory of Flory[33,95] to the problem of the fluid–crystal phase equilibrium. Let us consider the crystalline polymer in equilibrium with its solution. Given the usual condition that the chemical potentials of a polymer in the crystalline state and in solution must be equal, the lattice theory predicts[95-97] for the isotropic solution of monodisperse rods that

$$\frac{1}{T_M} - \frac{1}{T_M^I} = \left(Rx/\Delta H_f^I \right) \left[(1 - 1/x)(1 - v_p^0) - \left(\ln v_p^0/x \right) - \chi_1 \left(1 - v_p^0 \right)^2 \right]$$

(87)

where T_M is the melting point for the crystalline polymer in equilibrium with the isotropic solution, T_M^I is the melting point in the absence of a solvent, $(\Delta H_f^I/x)$ is the heat of fusion per mole of unitary segments, and χ_1 is the polymer–solvent interaction parameter. R denotes the gas constant. For the crystalline polymer–nematic solution equilibrium, application of the Flory–Ronca model[42] yields

$$\frac{1}{T_M} - \frac{1}{T_M^N} = \left(Rx/\Delta H_f^N \times \right)$$

$$\left[\left\{ -\left[\ln \left(v_p^0/x \right) + (y - 1)v_p^0 + 2 - C - \ln \left(\sigma/x^2 \right) - \ln \left(y^2 \right) \right]/x \right\} - \chi_1 \left(1 - v_p^0 \right)^2 \right]$$

(88)

where $C = 2\ln(\pi c/8)$, $\sigma = (4x/\pi)^2[1 - (\pi/4x)^2]^{1/2}$, and y is the Flory disorder index describing departure of the system from perfect alignment.

The phase equilibrium between the crystalline and fluid states for the polydisperse system has yet to be considered. One may, however, speculate to some extent on what would happen on the basis of results for monodisperse rods. For demonstrative purposes we have plotted together in Fig. 14 the fluid–fluid phase diagram for polydisperse rods with a narrow Gaussian-like distribution centered around $x_p^0 = x_0 = 35$ (ref. 93) and the isotropic-crystal and nematic-crystal phase diagrams calculated for mono-

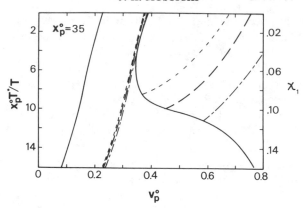

Figure 14.   Comparison between the fluid–fluid phase diagram for polydisperse rods with a Gaussian-like distribution of rod length and an average rod length of 35 (—) and the isotropic-crystal and nematic-crystal phase diagrams calculated for monodisperse rods with rod lengths of 30 (— · —), 35 (— —), and 40 (---).

disperse rods from Eqs. (87) and (88) with $x_0 = 30$, 35, or 40. We took $y(v_p^0)$ to be the same as the average one calculated for the polydisperse system[93] and $T_M^I = 700$ K, $T_M^N = 350$ K, $(\Delta H_f^I/x) = 600$ cal mol^{-1}, and $(\Delta H_f^I/x) = 250$ cal mol^{-1}. The interaction parameter was assumed to be temperature dependent:[96,97]

$$\chi_1 = 1/2 - \delta_1 + \delta_1 \vartheta / T$$

where δ_1 is an entropy parameter and ϑ is the Flory theta temperature. Because WF theory does not take into account the solvent–solute interactions, that is, it assumes $\chi_1 = 0$, we had to match the phase diagrams. We noted that the broadening of the fluid–fluid phase diagram calculated for the case in which only solute–solvent interactions are present[33,41,93] occurs when $\chi_1 \approx -1$. For the WF phase diagram the same happens for $x_p^0 T^*/T \approx 10$ (where T^* defines the scale of rod–rod interaction) and we found that for $\vartheta = 250$ K and $\delta_1 = 4.0$, $\chi_1 = 0$ corresponds approximately to $x_p^0 T^*/T = 0$, and $\chi_1 = -1$ to $x_p^0 T^*/T = 10$ (see Fig. 14). This procedure is very crude, which makes the entropy parameter very high; nevertheless, we believe that the results may serve as an intuitive demonstration of the situation in a solution of polydisperse rods. First of all, note that in Fig. 14 fluid–crystal equilibrium curves refer only to the single isotropic and nematic phases and not to the biphasic material, for which the fluid–crystal equilibria are far more difficult to calculate and outside the scope of this paper. The fluid–crystal equilibrium curves calculated for $x_0 = 35$ will be considered now as representative for the polydisperse system with this value of the average axial ratio, while the curves calculated for $x_0 = 30$ and $x_0 = 40$ indicate the anticipated behavior of the equilibrium curves for, respectively, the shorter and longer species in the distribution. Because the solubility of the considered system is extremely good, the isotropic-crystal equilibrium curve runs at much higher concentrations than the isotropic–biphasic one. It is not surprising that the curves calculated for $x_0 = 30$, 35, and 40 are very close together, indicating that solidification from the isotropic phase should happen to all species at the same time. In the nematic phase, the situation is very different. The nematic-crystal equilibrium curve intersects the biphasic–nematic one at sufficiently low temperatures, as the melting point decreases rapidly with decreasing concentration. The $x_0 = 30$ and $x_0 = 40$ curves are now separated from that for $x_0 = 35$. This indicates, reflecting the preferential partitioning of the long species into the nematic intrusion for concentrations just above v_p^*, that the longest species may initiate a successive fluid–solid transition in the nematic phase, occurring either on a reduction in temperature or an increase in concentration, with the shortest species solidifying last.

The behavior of the isotropic-crystal and nematic-crystal equilibrium curves indicates in turn the temperature dependence of the rod effective diameter. In the isotropic phase $d*$ is a decreasing function of temperature, decreasing for the considered example from about 2.3 in the athermal limit to about 2.9 for $x_p^0 T*/T = 16$. In the nematic phase $d*$ varies in a much more subtle way, from 0.95 (≈ 1) in the athermal limit (from ref. 92, $s_\alpha S \approx$ 0.97 for $x_\alpha = 35$) to 1.3 ($s_\alpha S \approx 0.82$) at $x_p^0 T*/T \approx 10$, although this corresponds to a very significant change of v_g^N, from about 0.76 to 0.45, respectively. That $d*$ values in the isotropic phase are twice as large as in the nematic phase reflects the problem of packing rods to a high concentration in either of the phases. In the isotropic phase, randomly oriented rods become jammed and immobilized at concentrations at which there is still a lot of free space between them, whereas in the nematic phase the rods can pack as close to the perfect packing limit as intermolecular interactions allow.

Because extensive additional studies are required to account properly for the temperature dependence of $d*$ in the polydisperse system of rods, effects related to $d*$ are not accounted for in the present calculations. As we show later, however, their inclusion would further improve agreement between the theoretical and experimental dielectric relaxation and viscosity data.

Due to the concentration–temperature dependent phase situation, within the biphasic range of concentration all parameters describing the system are functions of temperature; v_p^I and v_p^N, x_p^I and x_p^N, $P^I(x_\alpha)$ and $P^N(x_\alpha)$, and $\{s_\alpha\}$; $\{s_\alpha\}$ are temperature dependent also in the pure nematic phase above v_p^{**}. This detailed information is supplied by the WF phase-equilibrium calculations; hence the related temperature effects will be accounted for automatically in the numerical simulations.

Finally, we note that the solvent viscosity itself is temperature dependent and, we assume, is governed by a law similar to Eq. (86).

4. Numerical Results

Numerical simulations of dielectric relaxation and viscosity were performed with the following assumptions made concerning the undefined numerical constants present in the equations:

1.   To make the phase-separation calculations and the molecular-dynamics model compatible, the absolute temperature in all equations was replaced by the WF reduced temperature, $\theta = T/T^*$, with $T^* = 60$ K, so that $\theta^{-1} = 0.2$ corresponds to 300 K.

2.   The activation energies for $\delta\zeta$ and η_s were set equal to each other, with their value equal to the average activation energy of dielectric relaxation in PHIC–toluene:[60] $\delta E_\eta/k_B = 3000$ K ($\delta E_\eta/k_B T^* \approx 50$). [Trial calculation shows that with $\delta E_\eta/k_B$ equal to the value for pure toluene (≈ 1043 K[98]), the theoretical dielectric relaxation time varies with temperature more slowly than does the experimental time; thus we decided to use the activation energy of the relaxation process instead.]

3.   The effective rod diameter d^* was set equal to 1.

4.   Constant factors appearing in Eq. (43) after substitution of Eq. (86) for $\delta\zeta$ and η_s (in D_β^{t0}), were $A_\eta^I = 1 \times 10^{-6}$ and $k_B\gamma^I = 5 \times 10^{-3}$ for the isotropic phase, and $A_\eta^N = 1 \times 10^{-4}$ and $k_B\gamma^N = 1$ for the nematic phase. Both A_η and $k_B\gamma$ scale effects arising from restrictions on the translational diffusion of rods at high concentrations. As was argued in ref. 53, these restrictions are expected to be relatively small in the isotropic phase. This leads to

$$\delta\zeta^I d^* x_p^0 v_p^0 \ll 1$$

and

$$\frac{\gamma^I D_\alpha^{t0}}{x_p^0 \left(1 - v_p^0/2\right)} \ll 1$$

which for the present system at room temperature gives $A_\eta^I < 2 \times 10^{-5}$ and $k_B\gamma^I < 1/8$. In the nematic phase the motion of rods is severely restricted, requiring that $A_\eta^N > 2 \times 10^{-5}$ and $k_B\gamma^N > 1/8$.

5.   The proportionality constants not explicitly written down in equations leading to D_α^r [Eq. (17)] were eliminated by choosing some reference

(R) state and defining the reduced dielectric relaxation time as

$$\tau_{\text{red}}^d = \frac{\tau\left(v_p^0, \theta^{-1}\right)}{\tau^R\left(v_p^R, (\theta^R)^{-1}\right)}$$

$$= \frac{f_m^R\left(v_p^R, (\theta^R)^{-1}\right)}{f_m\left(v_p^0, \theta^{-1}\right)}, \quad \text{for } 2\pi f_m \tau = 1 \tag{89}$$

and viscosity as

$$\eta_{\text{red}} = \frac{\eta\left(v_p^0, \theta^{-1}\right)}{\eta^R\left(v_p^R, (\theta^R)^{-1}\right)} \tag{90}$$

An additional assumption was made as to what very long rods, that is, those from the "tail" of the distribution, do, in view of their inherent flexibility in real systems. In a previous numerical study of dielectric relaxation,[63] we found that satisfactory results are obtained if one assumes the rotational diffusion of these tail rods to be like that of the wormlike chains governed by the reptation mechanism proposed by de Gennes.[66] Although we believe that it would be more appropriate to consider their dynamics as being that of stiff semiflexible chains, for example, in the manner proposed recently by Odijk,[69] rather than that of wormlike ones, the problem was simplified by accepting the wormlike model for the tail rods in both the isotropic and nematic phases, that is, by assuming $\tau_\alpha \propto x_\alpha^3$.

Two types of numerical studies were carried out. First, we noted that for a monodisperse system of rods the dielectric relaxation time and normalized viscosity, η/v_p^0, are equal to the rotational relaxation time, $(2D')^{-1}$. For polydisperse systems the effective values of these parameters may be considered as averages of the rotational relaxation time, although the averaging is done differently in each case. We compared the results of different averaging procedures with each other and with experimental data. Numerical calculations were carried out for five different values of the inverse WF reduced temperature (referred to for simplicity as the inverse temperature), $\theta^{-1} =$ 0.08, 0.14, 0.20, 0.26, and 0.32. For each temperature the state at $v_p^0 = v_p^*$ and $\{s_\alpha = 0\}$ was chosen as the reference state, and the reduced dielectric relaxation time and reduced normalized viscosity were calculated with respect to this state. Results for the two extreme temperatures and a comparison between results for $\theta^{-1} = 0.26$ and experimental data for PHIC solutions are shown in Figs. 15 and 16, respectively. Experimental results selected for comparison include those for PHIC (76,000)–toluene,[62] studied by means of

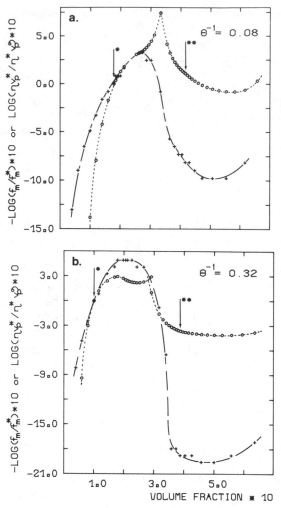

Figure 15. The concentration dependence of the reduced rotational relaxation time as calculated from the numerical dielectric relaxation time ($+$) and viscosity ($\bigcirc$) data for two extreme values of θ^{-1}: (a) 0.08 and (b) 0.32. Arrows indicate the lower ($*$) and upper ($**$) boundaries of the biphasic range. Note the log scale of the reduced rotational relaxation time.

dielectric relaxation, and those for PHIC (76,000)–tetrachloroethylene (TCE)[28] and PHIC (75,000)–toluene,[29] studied by means of steady-flow viscometry. [For purposes of comparison, the viscosity results for the PHIC–TCE have been plotted in Fig. 15 on a reduced concentration scale, where $v_p^0 = v_p^0(\text{TCE}) \cdot \{ v_p^*(\text{toluene})/v_p^*(\text{TCE}) \}$. Such rescaling of concentration seems justified in light of the athermal characters of both toluene and

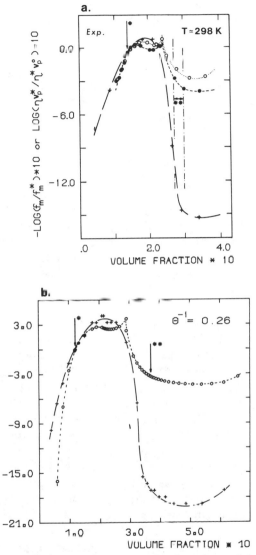

Figure 16. Comparison between the concentration dependence of the reduced rotational relaxation time as calculated from (a) experimental (T = 298 K) and (b) numerical ($\theta^{-1} = 0.26$) dielectric relaxation time (+) and viscosity (O,●) data. Arrows indicate the lower (*) and upper (**) boundaries of the biphasic range. Note the log scale of the reduced rotational relaxation time.

685

TCE, on the one hand, and of the solvent-invariant behavior of the viscosity–concentration curve (in different solvents there are only changes in the values of v_p^* and v_p^*), on the other.]

Further numerical studies were done separately for the complex dielectric permittivity and for the viscosity. We simulated both the isothermal and the temperature-dependent concentration behavior in each case. All calculations were carried out with state at $\theta^{-1} = 0.08$, $v_p^0 = v_p^*$ and $\{s_\alpha\} = 0$ as the reference state. Results are summarized in Figs. 17–24.

C. Discussion

1. Rotational Relaxation Time

Comparison between the numerical and experimental data for the rotational relaxation time shows excellent qualitative agreement (see Fig. 16). Noting this, we recall that the position, v_p^m, of the viscosity maximum in the numerical curves is determined solely as the intersection of N/I suspension and I/N suspension curves and corresponds approximately to the model-system phase-inversion concentration, v_p^i. In real systems, the position of v_p^m within the biphasic range, $v_p^* < v_p^m < v_p^{**}$, depends on the nature of the heterotropic mixture. If the guest phase is finely dispersed as spherical domains in the host phase anywhere in the biphasic range of concentrations, the viscosity maximum may be associated with the inversion phase concentration. On the other hand, as noted by Matheson,[48] "if some or all of the minor phase is present in domains that are continuous over distances that are of the scale of the experimental apparatus, then the maximum will occur at some ill-defined point in the range of biphasic stability." This happens also if the domains are not well defined as spherulites or are deformable in flow. Observation of the experimentally studied biphasic material under a cross-polarized microscope revealed[26,27,91] [see Fig. 6a and b], that as long as the guest nematic phase forms a definitely minor constituent of the heterogeneous mixture, $v_p^* < v_p^0 \ll v_p^{**}$, it takes the form of well-defined spherulites negligibly deformed by the shear, and Eq. (85) may be considered a good approximation.

This picture changes decisively as soon as the amounts of the isotropic and nematic phases are comparable. First, the guest-phase domains depart from a spherical shape and are easily deformed, even at a very small shear rate. Second, the guest phase is no longer well defined: We have observed in this concentration range regions where the N/I suspension transforms gradually into the I/N suspension and vice versa (e.g., see Fig. 6d). We would guess that such a sample of the biphasic material forms when the flow structures resemble more or less a multilayer "sandwich" with layers parallel to the direction of the Couette flow. If the viscosities of the nematic and isotropic phases are similar, the maximum should appear in this case at a concentration at which the volume fractions of both components are comparable, even if the phase inversion for this sample in the absence of flow and at equilibrium occurs at a much higher concentration and value of Φ^N. This type of behavior has been observed for PAIC in solutions; in most cases the maximum position corresponded to $\Phi^N \approx 0.5$, whereas the phase inversion occurred at a concentration corresponding to $\Phi^N \approx 0.7$.[29]

In conclusion, the apparent agreement between the numerically calculated and observed positions of the viscosity maximum is rather accidental, and does not attest to the validity of the spherulite suspension model over the entire biphasic ranges of PAIC solutions.

Differences in the behaviors of the calculated and observed average rotational relaxation times in the nematic phase require additional explanation. This behavior is a result of two competing effects. The first is an increase of the rod rotational mobility due to the high orientational order of the phase (i.e., high values of $\{s_\alpha\}$). The second is a decrease of the translational mobility of rods as a result of increasing contacts between rods. In the low-concentration range of the nematic phase, that is, just above v_p^*, the former effect prevails and the bulk η and τ are decreasing functions of concentration. With a further increase in concentration the orientational order of the phase approaches the perfect alignment limit; therefore the rate at which $\{s_\alpha\}$ varies decreases and the second factor takes over. Each curve passes through a minimum and increases, at an increasing rate, as the concentration comes closer and closer to the gelation concentration, v_g^N. The difference between the positions of the minima on the experimental and numerical curves has been anticipated, as we have assumed in calculations that the cohesive forces are active only when rods are in direct contact with each other, that is, when $d^* = 1$, which imposes the perfect packing limit on v_g^N. As we argued in Section V.B.3, the experimental values of v_g^N are much lower than this limit, and knowledge of them would help markedly to improve agreement between the theoretical and experimental curves in the nematic phase.

2.   Dielectric Relaxation

Simulated dielectric relaxation behavior is in excellent qualitative agreement with almost all details of the experimental data for PHIC–toluene solutions. Direct comparison between Figs. 17 and 10 shows that in the whole isotropic–biphasic–nematic range of concentration the respective calculated and observed isothermal curves for all dielectric parameters, namely log f_m, $\Delta\varepsilon$, and ε_m'', follow the same patterns. Both ε_m'' and $\Delta\varepsilon$ increase linearly with concentration until the low-concentration boundary of the biphasic range is reached. Further increases in v_p^0 lead to an increasing downward deviation from linearity; a dramatic decrease occurs at a critical concentration, v_p^c. This

decrease persists through the rest of the biphasic range and continues in the
nematic phase, although less steeply. Log f_m varies with concentration in a
complementary manner (see Figs. 10 and 17). Across the isotropic phase
log f_m values decrease monotonically with increasing concentration. Above
v_p^*, log f_m initially decrease further, at a gradually increasing rate, until the
same critical concentration, v_p^c, as for $\Delta\varepsilon$ and ε_m'' is reached. With further in-

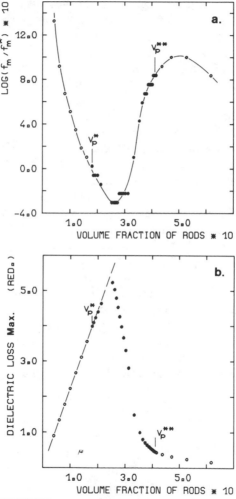

Figure 17. Simulated complex dielectric permittivity of polydisperse rods in solution. Concentration dependences of (a) log f_m, (b) the dielectric loss maximum, and (c) the dielectric increment for $\theta^{-1} = 0.08$. Note the reduced scale of f_m. The scales for the dielectric loss maximum and the dielectric increment are arbitrary.

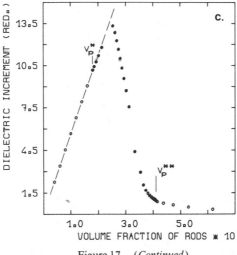

Figure 17. (*Continued*)

creases in v_p^0 the $\log f_m$ values begin to increase rapidly in a manner complementary to the changes in $\Delta\varepsilon$ and ε_m'' until it reaches a maximum within the nematic phase. Unfortunately, the experimental studies of the PIIIC toluene system did not extend beyond this maximum, although the existence of the maximum and subsequent decrease of $\log f_m$ with concentration was observed for the PBNIC–toluene system (see Fig. 10).

Quantitative comparison between numerical and experimental results shows that the extents of the range over which $\log f_m$ varies on going from the isotropic through the biphasic to the nematic solution are very similar in both cases. In Fig. 10, the difference between the $\log f_m$ minimum in the biphasic range and the maximum in the nematic phase is about 1.6, while in Fig. 17a we find this difference to be about 1.3–1.4. There is, however, much less quantitative agreement between the respective results for ε_m'' and $\Delta\varepsilon$. The experimental results show a relative drop in value across the biphasic range, from the maximum within the range to the value at the transition concentration, v_p^{**}, of a factor of about 5, the same change in the numerical data is by a factor of about 15. Because *this* difference is of a factor of 3, it suggests an easy explanation if one assumes that the nematic phase of the PAIC–toluene solution studied was oriented by the measuring electric field. Such an assumption is not unreasonable; in the course of the preliminary static Kerr-effect measurements in the isotropic solution in the vicinity of v_p^* we observed[79] a tendency for macromolecules to orient parallel to the electric field, even in very weak (~ 10 V cm^{-2}) fields. The induced orientational polarization decayed over the course of at least several tens of minutes, evidence of quasipermanent orientation of the polymer chains. Because the electric field in the dielectric experiment[60-62] was of the same order, we may expect that the nematic phase was more or less homogeneously ordered in the direction of the field. In such a case, the observed magnitude of the complex dielectric permittivity would have been 3 times larger than that expected for the unoriented nematic sample. Average complex permittivity of the unoriented nematic may be written as[74]

$$\langle \hat{\varepsilon}(\omega) \rangle = \frac{\hat{\varepsilon}_{\parallel}(\omega) + 2\hat{\varepsilon}_{\perp}(\omega)}{3} \tag{91}$$

where $\hat{\varepsilon}_{\parallel}$ and $\hat{\varepsilon}_{\perp}$ are permittivities measured parallel and perpendicular to the nematic director. For PAIC solutions, $\hat{\varepsilon}_{\perp} \approx 0$, hence $\langle \hat{\varepsilon}(\omega) \rangle \approx \hat{\varepsilon}_{\parallel}(\omega)/3$. The theoretical expression for the complex dielectric permittivity of the nematic phase, Eq. (64), was developed on the assumption of a random distribution of the nematic director within the sample. Therefore, it is equivalent to the average $\langle \hat{\varepsilon}(\omega) \rangle$, and values of ε_m'' and $\Delta\varepsilon$ should be one-third the values expected for the homeotropic nematic phase. Hence, by multiplying the calculated values of ε_m'' and $\Delta\varepsilon$ by 3, very good quantitative agreement between experimental and theoretical curves can be reached. [Being unaware of the possibility that the nematic phase might be ordered, we found (refs. 61, 62; see also ref. 73) a pronounced discrepancy between the nematic cone angles, θ_0, calculated from the dielectric increment values in the isotropic and nematic phases, $\theta_0 \approx 22°$, and from the values of the relaxation time, $\theta_0 \approx 12°$. This discrepancy disappears once the homeotropic order of the nematic phase is accounted for: θ_0 calculated using the corrected value for $\Delta\varepsilon$ in the nematic phase becomes about 12.5°.]

The Argand diagram representation of the calculated complex dielectric permittivities for a number of different concentrations in the biphasic range is shown in Fig. 18. As long as the dielectric properties of the solution are dominated by the isotropic material (i.e., when $v_p^* < v_p^0 < v_p^c$), $\varepsilon'' - \Delta\varepsilon$ curves have the character of the Davidson–Cole skewed arc.[74] Above v_p^c, the arcs become more and more symmetric and closer to semicircles, which is characteristic of single-relaxation time behavior. Although not shown here, the shape of the Argand diagram does not undergo any significant change within the isotropic or nematic phases, and resembles in these phases those curves characteristic of v_p^* ($= 0.16$) and v_p^{**} ($= 0.41$), respectively.

The particular shape of the Argand plot is, in the present case, a consequence of the rod-length distribution and, in turn, of the relaxation time, dipole moment, and rod-volume-fraction distributions. In the isotropic phase the Davidson–Cole asymmetry of the arc is due to the short species in solu-

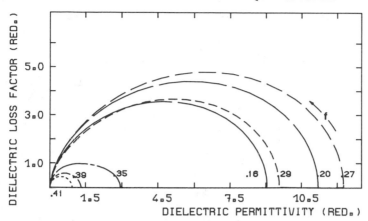

Figure 18. Simulated complex dielectric permittivities of polydisperse rods in solution. Variation of Argand diagram across the biphasic range of concentration, for $\theta^{-1} = 0.08$. Numbers on curves indicate different concentrations; $v_p^* = 0.16$ and $v_p^{**} = 0.41$.

tion making only a small contribution to the dielectric process strength ($\sim x_\alpha^2$), but relaxing in significantly shorter time (x_α^7) than do the long rods, thus leading to the high-frequency tail of the Argand plot. In the nematic phase the strength and relaxation time for each type of α-rod are reduced by factors of $1/\delta_{1\alpha}^1$ and $1/g_{1\alpha}^1$, respectively, both of which are directly related to the order parameter s_α. These reductions are far more drastic for long rods than for short ones, compensating for the effect of the rod length on the strength and relaxation time, leading, in turn, to the more symmetric Cole–Cole arcs.

The shape asymmetry of the calculated Argand plots is in good qualitative agreement with experimental results (see Fig. 9), although the experimental arcs are much more depressed. This discrepancy is anticipated due to the presence of additional relaxation mechanisms broadening the distribution of relaxation times, such as those originating from the limited flexibility of macromolecules, but not taken into account by the present model.

In the isotropic phase, rising temperature provokes a decrease of ε_m'' and an increase of $\log f_m$, changing the slopes of $\log f_m(v_p^0)$ and $\varepsilon_m''(v_p^0)$ curves at the same time. Most characteristic of the biphasic range is the temperature dependence of the position v_p^c of the minimum in $\log f_m(v_p^0)$ and of the maximum in $\varepsilon_m''(v_p^0)$ curves, which is an increasing function of temperature. This increase is accompanied by a decline in the height of the maximum and the depth of the minimum. These features are very similar to those experimentally observed [compare Fig. 11 with Fig. 19, and 12 with Fig. 20]. At the biphasic–nematic transition concentration, v_p^{**}, $\log f_m$ is a decreasing func-

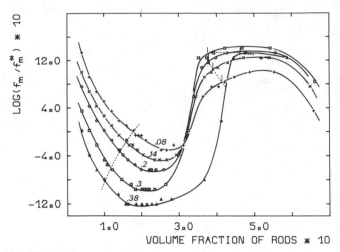

Figure 19. Simulated complex dielectric permittivity of polydisperse rods in solution. Log f_m is shown as a function of concentration and θ^{-1}. Boundaries of the biphasic range are marked by dotted lines; numbers on curves indicate different θ^{-1} values. Note the reduced scale of f_m.

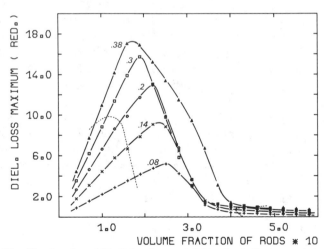

Figure 20. Simulated complex dielectric permittivity of polydisperse rods in solution. The dielectric loss maximum is shown as a function of concentration and θ^{-1}. Boundaries of the biphasic range are marked by dotted lines; numbers on curves indicate different θ^{-1} values.

692

tion of temperature, which is the opposite of the situation at v_p^* (Fig. 19). This is complemented by a decrease in ε_m'', although this is not as clearly visible as for $v_p^0 = v_p^*$ (Fig. 20). The behaviors of both $\log f_m$ and ε_m'' are due to the sensitivity of $\hat{\varepsilon}(\omega)$ to the orientational order in the nematic phase and are influenced by the temperature dependence of $\{s_\alpha\}$, which is a decreasing function of temperature. For decreasing values of $\{s_\alpha\}$, $\{g_{1\alpha}^1\}$ values increase significantly, leading to smaller values of the effective f_m. At the same time, $\{\delta_{1\alpha}^1\}$ values increase, but because the distribution of relaxation times is now significantly broader, the effective ε_m'' appears to be smaller (the Davidson–Cole arc is flattened; see the discussion of the results in Fig. 17).

There are obvious discrepancies between the experimental and theoretical $\log f_m(v_p^0)$ curves in the nematic phase. Contrary to the experimentally observed behavior, the calculated values of $\log f_m$ increase with increasing temperature. This reverse behavior has its origin again in the assumed temperature independence of d^*. As a result of this assumption, the downturn of $\log f_m(v_p^0)$ appears at a very high concentration and its position is temperature insensitive. Therefore, over a significant portion of the nematic-phase range, the calculated values of $\log f_m$ are controlled solely by $\{s_\alpha\}$, and thus are higher at low temperatures. The opposite behavior of the experimental data, shown in Fig. 11, is a direct result of the temperature dependent onset of solidification of the solution. With decreasing temperature, the gelation or precipitation of the solid polymer from solution occurs at much lower concentrations. Therefore the translational mobility becomes severely restricted at much lower concentrations, leading in turn to smaller values of $\log f_m$. Effects arising from v_g^N (and hence from d^*) dominate effects arising from $\{s_\alpha\}$. The maximum of $\log f_m$ decreases and moves toward lower concentrations as temperature decreases.

The present model enables us also to simulate the temperature dependence of the dielectric loss curve, $\varepsilon''(f)$, at fixed concentration. Two concentration regions are of particular interest. The first is concentrations in the vicinity of v_p^c, that is, the region in which $\Delta\varepsilon(v_p^0)$, $\varepsilon_m''(v_p^0)$, and $\log f_m(v_p^0)$ undergo the most dramatic changes. The second is concentrations close to v_p^{**}, at which even small amounts of the isotropic material influence significantly the dielectric properties of the biphasic material. If the temperature dependence of d^* is unaccounted for, comparison between theory and experiment is limited to the first region, in which changes in $\hat{\varepsilon}(\omega)$ result primarily from the heterogeneous nature of the biphasic material and, furthermore, the influence of v_g^N on molecular dynamics is not significant. Numerical results characteristic of the dielectric relaxation behavior for this region are shown in Fig. 21a. The relative position of the particular concentration chosen for calculations in the model phase diagram is such that the numerical results may be compared with the experimental data presented in Fig. 21b. Al-

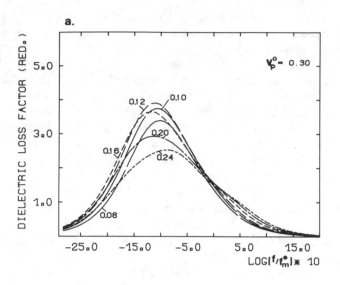

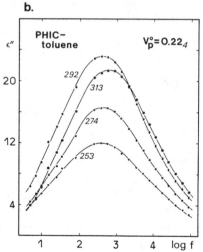

Figure 21.   Comparison between (a) simulated and (b) experimental[62] dielectric loss versus log frequency curves for the concentration in the biphasic range at which the most dramatic changes in $\hat{\varepsilon}$ occur. Numbers on curves indicate different θ^{-1} values and temperatures, respectively.

though this comparison is only qualitative, the numerical and experimental curves show very similar trends with respect to the dependence of the dielectric loss maximum magnitude and position on temperature. As long as the temperature is high (i.e., $T > 292$ K for Fig. 21b or $\theta^{-1} \leq 0.12$ for Fig. 21a), the variation of $\varepsilon''(f)$ with temperature is dominated by the isotropic material. With decreasing temperature f_m moves to *lower* frequencies while ε''_m *increases* at the same time. Below a certain temperature, the situation essentially reverses: f_m shifts to *higher* frequencies and ε''_m *decreases* as the temperature decreases. This is more pronounced in the numerical data, since the loss curve is much narrower and the rod rotational motion is not influenced by the lowered critical concentration for gelation or precipitation.

In light of all these remarkable agreements between the theoretical and experimental results, we may argue that the dielectric relaxation behavior of PHIC–toluene solutions has to be considered due predominantly to the rigid-rod nature of PHIC chains.

3. Steady-Flow Viscosity.

Typical isothermal viscosity–concentration curves for four different rod orientations with respect to the flow velocity in the nematic phase are shown in Fig. 22. We note that calculated Miesowicz viscosities show the same ordering in values as that already well known[84,85] for ordinary t-LC, namely $\eta^M_c > \eta^M_b > \eta^M_a$, although the quantitative differences among the calculated viscosities are far larger than those among the experimental values. It is doubtful whether it is possible to observe such large differences experimentally. An extremely stiff entangling network of rods should make it very difficult to maintain the desired orientation of molecules during the flow. It is much more probable that the flow ordering prevails over or is comparable with externally induced ordering, allowing determination of some average viscosity values. For the same reason the experimental Miesowicz viscosity curves in the biphasic range may differ significantly from those presented in Fig. 22. In both N/I and I/N suspensions, the viscosity curves are calculated for the rod orientation in the nematic phase undisturbed by the flow. In a real experimental situation the tumbling motion of the nematic spherulites would probably even out all preorientations of spherulites and lead to one universal viscosity behavior for N/I suspensions. For the I/N material, the above arguments concerning the viscosity measurements in the nematic phase hold and one may expect differences among the four curves to be less pronounced.

It is a characteristic of the effective viscosity in the biphasic range to continue to behave like a pure isotropic phase over quite a large range of concentrations above v^*_p until the Miesowicz and flow-induced orientation viscosities split away at a certain concentration, say v^a_p. This behavior is a consequence of the nature of the model-system phase transition. As can be

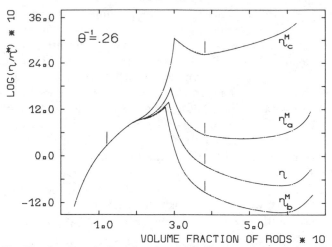

Figure 22. Simulated steady-flow viscosity of polydisperse rods in solution. The concentration dependence of the viscosity is shown over the isotropic–biphasic–nematic range of concentration for four different cases: the flow-oriented nematic phase (η) and three different Miesowicz geometries of the nematic phase orientation with respect to the direction of flow. $\theta^{-1} = 0.08$. Boundaries of the biphasic range are indicated by vertical bars.

deduced from Fig. A3 in Appendix A, in the (v_p^*, v_p^0) concentration range, the first nematic droplets are not only formed from but also populated by predominantly very long rods. However, the volume fraction of these droplets is very small and rises only to $\Phi^N \approx 0.01$ at v_p^a. Hence, the influence of the guest phase on the apparent viscosity of the biphasic material is not very significant over the (v_p^*, v_p^a) range, and the viscosity is due predominantly to the viscous properties of the host isotropic phase. The eventual small differences among η_a^M, η_b^M, η_c^M, and η are effectively suppressed by the model assumption of wormlike-chain rotational dynamics[66] for the very long rods. Above v_p^a, Φ^N becomes significantly dependent on concentration, leading to the observed separation of the viscosity curves.

The precise determination of the Miesowicz viscosities would require the use of special viscometers, and to my knowledge this type of study has not yet been made for rigid-rod macromolecules in solution. Most of the data in the literature on viscosity measurements in the nematic–lyotropic phase are for the flow-oriented sample, and agreement between the theory and experiment in this case is very good (see Fig. 16). It is interesting to note that in addition to conventional viscometry methods,[100] the Stokes falling ball (FB) method has been used to measure the viscosity of PHIC–toluene. The viscosity–concentration curve in the nematic phase showed a deeper minimum, with much lower minimum values than those in corresponding curves determined with standard viscometers (compare Fig. 1 of ref. 88 with Fig. 7). These discrepancies probably result from a specific flow geometry in the FB experiment. As Diogo[101] has shown for t-LC, the apparent viscosity of the nematic phase measured by the Stokes method, η_{FB}, is lower than that measured under Couette flow and close to the Miesowicz viscosity η_b^M. Combining results of Diogo[101] and Marrucci,[83] the expression for η_{FB} for the case of rods in solution may be written as

$$\eta_{FB} = \eta_b^M - \delta\eta \qquad (92)$$

where

$$\delta\eta \propto v_p^0 k_B T \sum_\alpha \frac{P^0(x_\alpha)\tau_\alpha^r}{x_\alpha}\left\{(s_\alpha/2)^2 - \frac{s_\alpha(1-s_\alpha)}{1+s_\alpha/2}\right\}$$

$\delta\eta$ is always positive as long as there is a high degree of orientational order in the nematic phase (i.e., when $\{s_{\hat{a}}\} > 0.75$). Despite the fact that for a very high orientational order, then η_{FB} calculated from Eq. (92) may eventually become negative (it is negative in the limit of perfect order, $\{s_{\hat{a}}\} = 1$), Eq. (92) offers at least an indicative explanation for the relatively low viscosity observed when the nematic phase is studied by the Stokes method. It seems also to support an idea of Aharoni,[88] who used the viscosity minimum value to investigate the rigidity of PHIC chains in solution as a function of the chain average molecular weight. According to the Diogo–Marucci equation, any departure from rigidity, which would result in decreasing order of the nematic phase, would lead to an increase in viscosity. Decreasing $\{s_{\hat{a}}\}$ leads to a decrease in $\delta\eta$, even to values below 0 (for $\{s_{\hat{a}}\} < 0.75$), and consequently to higher η_{FB} values. Aharoni[88] has observed, for PHIC in toluene solution, that as long as $M_w < 45{,}000$, the viscosity minimum remained small and constant, but it increased linearly for higher M_w. This increase may indeed be due to a decrease in chain rigidity. However, that a flow behavior of a solution of extremely long rods flowing around a ball more complicated than that considered by Diogo[101] occurs cannot be excluded.

From here on we restrict our attention to the case in which the nematic phase is flow oriented. To investigate the influence of the rod polydispersity, the viscosity curve obtained for the model system, $\mathscr{G}t$ (for "Gaussian with tail"), is compared with the curve calculated for a system of rods with the asymmetric Gaussian distribution of the rod length, $a\mathscr{G}$ (for "asymmetric Gaussian")[93] (see Fig. 23). Within the isotropic phase, the viscosity behavior is essentially the same in both cases, despite pronounced differences in polydispersity. This result confirms the universality of the prediction of Doi,[50] who showed, for the case of monodisperse rods, that η/η^* has to be a uni-

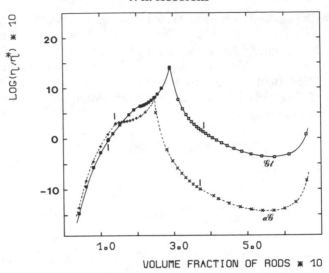

Figure 23.   Simulated steady-flow viscosity of polydisperse rods in solution. Comparison between the viscosity curves calculated for two different polydisperse systems with the rod length polydispersities in the forms of the asymmetric Gaussian[93] ($a\mathcal{G}$) and of the Gaussian with the very long rod tail ($\mathcal{G}t$). Boundaries of the biphasic range are indicated by vertical bars.

versal function of the similarly reduced concentration, v_p^0/v_p^*. As concentration increases above v_p^*, visible differences show up in the initial part of the biphasic range. In contrast to the viscosity for $\mathcal{G}t$, which continues to behave in the manner characteristic of the isotropic phase, the viscosity curve for $a\mathcal{G}$ shows a sudden change at v_p^*. This results from the strong concentration dependence of Φ^N, which starts at the very beginning of the biphasic range. The behavior of both curves in the rest of the biphasic and nematic ranges is very similar.

An important result in Fig. 23 is the appearance in both cases of a shoulder on the viscosity curve due to the isotropic–biphasic transition. Recalling our earlier results for the monodisperse-rod solution (see Fig. 4 of ref. 53), in which no shoulder was observed, it becomes clear that this shoulder is a strong indication of the rod polydispersity. For a reasonably narrow polydispersity function, the shoulder should be sharp and located at v_p^*. If, however, a sample contains residual amounts of long rods, the shoulder is smeared over the initial (N/I) part of the biphasic range. In fact, both types of shoulder have been observed for PAIC solutions[26-29] (see Fig. 7), although for other rodlike macromolecules in solution only the sharp shoulder was noticed.[21,47]

In two different samples, PHPIC (43,000)–toluene and PHIC (70,000)–toluene, Aharoni[29] effectively did not observe the shoulder. The first case is shown in Figs. 7a and 8. Particularly interesting is a comparison between the viscosity curves for two PHPIC samples with close average molecular weight values, $M_w = 41,000$ and 43,000 (Fig. 7a). With respect to the results for the PHPIC (43,000)–toluene system, despite very similar M_w, the viscosity curve for PHIC (41,000)–toluene has a pronounced shoulder at v_p^* and significantly shifted positions of the characteristic concentrations, v_p^* and v_p^{**}, and the phase-inversion concentration, v_p^i. We believe that these differences originate in the different polydispersities of the samples. This seems to be manifested not only by the shoulder but also by the positions of the characteristic concentrations. The presence of the very long rods depresses v_p^*, while that of the very short ones augments v_p^{**}.[92] As can be seen from Fig. 7a, v_p^* for PHPIC (41,000) was lower than for PHPIC (43,000), but the reverse was true of v_p^{**}. Because both polydispersity functions are centered around approximately the same value of the molecular weight, they should have reverse asymmetry. For the PHPIC (41,000) sample, one would expect a relatively small distribution half-width on the low-molecular-weight side and a large half-width on the other side (i.e. a "tail" of the very long chains). The PHPIC (43,000) sample should feature a narrow high-molecular-weight half of the distribution and a significant amount of the shortest species. An argument in favor of this hypothesis is supplied by the position of the phase-inversion concentration v_p^i for each sample. In PHPIC (41,000), v_p^i appeared close to v_p^{**}, about seven-tenths of the way through the biphasic range (see Fig. 7a); that is, Φ^N increased essentially linearly across the range, in the manner characteristic of the $a\mathcal{G}$ distribution.[93] For the second sample, v_p^i was closer to v_p^* than to v_p^{**}; that is, most of the sample was transformed into the nematic phase in the initial part of the biphasic range, so the further significant increase of concentration was necessary only to incorporate into the nematic phase the shortest rods.

The narrow distribution of molecular weight or length should also affect the phase behavior of such a system. For monodisperse rods in solution, Warner and Flory[37] have shown a significant difference between the phase diagrams for rods with axial ratios below and above 20. The phase diagram calculated for very long rods ($x > 20$) is complicated, showing critical and triple points. In contrast, for low x values the phase behavior complexity disappears and the phase diagram shows a narrow biphasic range that grows wider and wider as the temperature decreases below some critical value. This type of phase diagram was observed for PHPIC (43,000) (see Fig. 1). This is not a surprising result, since the average axial ratio for this sample is about 15–20, so even if the sample contained species longer than $x = 20$, polydispersity would have "smeared out" any tendency to show the complex phase behavior.[93] On the other hand one cannot rule out the possibility that the complex phase behavior manifested its presence in the viscosity measurements for the PHIC (70,000)–toluene system (see Fig. 2 of ref. 29). The similarity between the viscosity curves for PHIC (70,000) and PHPIC (43,000) solutions in the isotropic and biphasic ranges of concentration in toluene speaks in favor of a similarly shaped polydispersity function in both cases. First, the PHIC (70,000)–toluene viscosity curve showed only a very small step at v_p^*; second, the values of all characteristic concentrations (v_p^*, v_p^{**}, and v_p^i) were high; and third, the phase inversion occurred at a concentration visibly closer to v_p^* than to v_p^{**}. In the nematic phase, however, Aharoni[29] repeatedly observed rather large irregularities in an otherwise smooth viscosity curve. Although these irregularities may be due to hydrodynamic effects,[102] the average axial ratio in the sample, $21 \leq x_p^0 \leq 28$, was sufficiently large for the complex phase behavior to have appeared and influenced the steady-flow viscosity behavior.

Numerical calculations show also that the viscosity–concentration behavior is sensitive to variations in temperature. Viscosity curves calculated for a few different inverse temperatures, θ^{-1}, are compared in Fig. 24. The most characteristic and pronounced features are an increase of the maximum and a shift in its position toward lower concentrations with decreasing temperature, on the one hand, and an increase in the magnitude of a shoulder on the viscosity curve in the low-concentration part of the biphasic range, on the other. It must also be noted that whereas in the isotropic phase viscosity in-

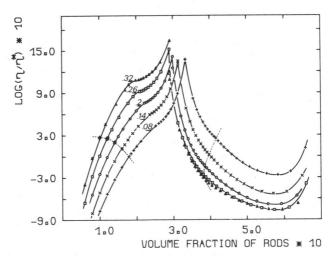

Figure 24. Simulated steady-flow viscosity of polydisperse rods in solution. Reduced viscosity is shown as a function of concentration and θ^{-1}. Boundaries of the biphasic range are marked with dotted lines; numbers on curves indicate different θ^{-1} values.

creases with decreasing temperature, the inverse happens in the nematic phase. The shift of the viscosity maximum position is a consequence of the phase behavior; that is, when temperature is varied the maximum position changes approximately in the same manner as does the phase-inversion concentration v_p^i ($\Phi^N = 0.5$).[93] The broad shoulder associated with a transformation in the viscosity behavior, from the isotropic phase alike to the biphasic suspension alike, is especially visible at low temperatures (high θ^{-1} values), mainly due to the significant narrowing of the concentration range over which this transition takes place (see Fig. A3 in Appendix A) and an increase in the viscosity value in both conjugated phases.

The lack of extensive viscosity studies of solutions of rigid-rod macromolecules over a broad temperature range limits the extent to which the present results can be interpreted. The only available viscosity data for a large part of the phase diagram are for PHPIC–toluene[29] and PBA–sulfuric acid.[18] For the reasons discussed above, the viscosity curves for PHPIC (43,000)–toluene did not show the shoulder at v_p^* (see Fig. 8). If we neglect this difference in shape between the experimental and theoretical viscosity curves, they show very similar behaviors up to the viscosity maximum within the biphasic range. Visible differences show up on the high-concentration side of the viscosity maximum. First, the calculated viscosity shows a clear tendency to increase with increasing temperature, in contrast to the decrease observed in PHPIC–toluene solutions (Fig. 8). Second, the minima in the calculated viscosity curves appear at very high concentrations and their locations and magnitudes are practically temperature independent. The viscosity measurements in the PHIC–toluene system were terminated before reaching the minimum, but a minimum is visible in the viscosity curves for PBA–sulfuric acid.[18] For this system, the decreasing temperature causes the minimum position to move to lower concentrations and the minimum viscosity value to increase at the same time. As for the dielectric relaxation time, these discrepancies result from the "athermal" approximation applied when accounting for the frictional interactions between diffusing rods, that is, the temperature-independent limit value of d^* ($= 1$). As a consequence of this approximation, the calculated minimum appears at very high concentrations and is relatively temperature independent. Once the temperature dependence of v_g^N is known, we may expect far better qualitative agreement between the numerical and experimental viscosity data over the entire isotropic–biphasic–nematic concentration–temperature range.

VI. CONCLUDING REMARKS

We have presented in this paper the extended Doi–Edwards (DE) molecular-dynamics theory for polydisperse rods in concentrated solution. Its application to the problems of dielectric relaxation and steady-flow viscosity produced numerical results that are in far-reaching qualitative agreement with experimental results for stiff-chain poly(n-alkylisocyanate) (PAIC) macromolecules in semidiluted solutions. On the one hand, this remarkable agreement emphasizes that the dynamics of polymeric macromolecules in solution is that of rigid, rodlike objects. On the other hand, it speaks in favor of the basic concepts proposed in the Doi[50] model of macromolecular dynamics, in which the molecular rod in semidiluted solution is allowed to diffuse in the direction of its long axis in a more or less unrestricted fashion, but the freedom of any other motion is dramatically limited.

Considering the well-known fact that stiff-chain macromolecules are always flexible to some extent, the degree of this flexibility being a function primarily of the degree of polymerization,[57,86-88,90,103] but also of the solvent character[103] and concentration,[86,88] the rodlike approximation of the chain is a very crude one. The surprisingly good explanation of the dielectric relaxation and viscosity behavior provided by the rod model is due to the fact that both effects sense the reorientation of the molecule end-to-end vector, or the hydrodynamic length, which is smaller than its contour length but nevertheless comparable to it. Under these circumstances, the limited but inherent rigidity and the entanglement between chains conform at least partially to the basic clever assumptions made by Doi concerning the reptation–rotation mechanism of the chain end-over-end reorientation.

The reptation–rotation model has received considerable and steadily growing support,[55-63] especially from the light-scattering Cotton–Mouton- or Kerr-effect studies of concentrated isotropic solutions in the vicinity of the isotropic–biphasic transition. Better understanding of these experimental data required consideration of the inherent flexibility of chains,[55,57-59] but very little attention has been paid until now to the problem of the translational mobility of rigid chains in semidiluted solutions.[53,57] Our earlier[53] and present studies show, however, that the increasing difficulty of translational diffusion at high macromolecule concentrations gives rise to significant effects observed experimentally in the dielectric relaxation and viscosity behaviors of the lyotropic liquid crystalline phases. Accounting for the enhancement of friction in the isotropic phase would help also to explain the experimental observation[55-59] of a much more dramatic increase in the rotational relaxation time as concentration approaches v_p^* than is predicted by the original DE theory.[50,51]

It is necessary to note here that in addition to the DE theory, based on purely geometrical arguments, recent efforts at understanding the macromolecular dynamics[104,105] and phase behavior[38] of semidiluted solutions have made use of the theory of Onsager.[31] Very recently, Fesjian and Frisch[104] used the Onsager theory to obtain an expression for the rod rotational relaxation time in the presence of an external flow for a concentrated solution of monodisperse rods in either the isotropic and nematic phases; their results are qualitatively similar to those of Doi.[39]

Finally, we wish to point out that despite polydispersity, the PAIC make up an excellent model system with which to study not only molecular dynamics but also phase transitions, and especially for verifying existing phase-equilibrium theories.[37,41,106–108]

Thanks to the presence of flexible side chains, PAIC show very good solubility in a variety of common solvents. Furthermore, in most cases these solvents behave as athermal solvents,[109] although recent studies show[110] that the lateral alkyl chains interact to some extent with the solvent molecules. However, the latter interactions are limited to the side chains, and due to existing side-chain disorder in both the isotropic and lyotropic–nematic solutions, the rigid polyisocyanate backbone chain may be considered to be enveloped by an additional "film" of alkyl "solvent"; this makes the improved solubility characteristic of PAIC readily understandable.

Acknowledgments

The final version of this paper was written at Baker Laboratory of Cornell University, and I acknowledge with gratitude partial support by NSF Solid State Chemistry Grant DMR81-02047. I wish to express my special thanks to Professors A. Cassuto, J. H. Freed, and J.-L. Rivail, for their hospitality during my visits to Nancy (France) and Ithaca, New York (U.S.A.), which made possible the completion of this work. I am greatly indebted to Dr. S. M. Aharoni for supplying me with micrographs of the PHIC–TCE biphasic material and numerical data concerning the viscosity of PHIC–toluene solutions. I am also very thankful to Dr. K. Karnicka-Moscicka and Dr. V. Rosato for many helpful discussions during the course of this work. The author appreciated the continuous assistance of Dr. M. Alnot at the MITRA-15 computer. Special appreciation goes to Ms. M. Babb for her many helpful suggestions concerning improving the readability of the paper.

APPENDIX A: THE WARNER–FLORY PHASE EQUILIBRIUM THEORY OF LYOTROPIC LIQUID CRYSTALLINE SYSTEMS

The theory of phase transition used in this paper is exactly that developed by Warner and Flory.[37] We recall here briefly the main results of their work; for details, in particular those concerning the Flory lattice theory, the reader should consult the original and related papers.[33–37,41,42,106]

The theory concerns a system of rods, identical in diameter but polydisperse in length, dispersed in a solvent. For simplicity, the diameter is set to unity so that the length and the axial ratio will be numerically equivalent. Solvent molecules are considered isodiametric, and $x_s = 1$. The polydispersity of the system is given by function $p^0(x_\alpha) = n_\alpha^0/n_p^0$ or, alternatively, by the volume-fraction ratio, $P^0(x_\alpha)$, as defined in Eq. (13).

Let the volume fraction of rods v_p^0 (in other words, the volume concentration) be within the biphasic range. Then for any concentration from that range, the solute mass, the composition of the biphasic material, and the total number of solute rods must be conserved:

$$\Phi^N v_p^N + (1 - \Phi^N) v_p^I = v_p^0 \tag{A-1}$$

$$\Phi^N v_\alpha^N + (1 - \Phi^N) v_\alpha^I = v_\alpha^0 \tag{A-2}$$

$$\Phi^N \frac{v_p^N}{x_p^N} + (1 - \Phi^N) \frac{v_p^I}{x_p^I} = \frac{v_p^0}{x_p^0} \tag{A-3}$$

The phase equilibrium between the isotropic and nematic phases requires the chemical potentials of the same component of the solution in both phases to be equal:

$$\mu_1^I = \mu_1^N \quad \text{and} \quad \mu_{x_\alpha}^I = \mu_{x_\alpha}^N, \quad \text{for all } x_\alpha \tag{A-4}$$

By definition, the chemical potentials are

$$\frac{\mu_i^I - \mu_i^0}{RT} = \left(\frac{\partial G^I}{\partial n_i} \right)_{T,V,\{n_i\}} = -\left(\frac{\partial \ln Z_M^I}{\partial n_i} \right)_{T,V,\{n_i\}} \tag{A-5a}$$

$$\frac{\mu_i^N - \mu_i^0}{RT} = \left(\frac{\partial G^N}{\partial n_i} \right)_{T,V,\{n_i\},\text{eq}} = -\left(\frac{\partial \ln Z_M^N}{\partial n_i} \right)_{T,V,\{n_i\},\text{eq}} \tag{A-5b}$$

where G is the Gibbs free energy, Z_M is the partition function, and T, V, and $\{n_i\}$ denote, respectively, the absolute temperature, sample volume, and the set of mole numbers of all components (i.e., solute and solvent particles) except n_i. The subscript "eq" signifies the orientational equilibrium of rods in the nematic phase.

In general, the partition function Z_M is considered to be composed of four factors: the combinatory or "steric" factor Z_{comb}, the orientational factor Z_{orient}, and two factors introducing the exchange free energies of interaction between the solvent and the rods, Z_{s-r}, and between the rods, Z_{r-r}:

$$Z_M = Z_{comb} \times Z_{orient} \times Z_{s-r} \times Z_{r-r}$$

The Warner–Flory (WF) treatment neglects the solvent–solute interactions, Z_{s-r}, although they may be easily incorporated in the same way as in ref. 41. Accounting for these interactions is also not essential for the present work; in the experimental systems we deal with (i.e., PAIC in simple solvents), most of the solvents behave as the "athermal" ones;[109] therefore these interactions are of minor importance for the phase equilibria. The WF theory takes into account two predominant factors causing rodlike molecules to align and form the anisotropic-lyotropic phase: (1) the steric difficulties associated with packing at high concentrations, which are investigated using Flory's lattice model approach; and (2) soft anisotropic forces similar to those that are responsible for the thermotropic nematic behavior of short, cigarlike molecules, with the nematic potential in the form proposed by Flory and Ronca:[42]

$$U(u_\alpha n) = -k_B T^* v_p^0 S \{ 1 - \tfrac{3}{2} \sin^2(u_\alpha n) \}$$

where T^* scales the interaction between unitary segments of rods.

The WF expression for the free energy of the nematic mixture in its equilibrium state of orientational order is

$$\frac{G^N}{k_B T} = n_s^N u + n_s^N \ln n_s^N - (n_s^N \mid n_p^N) \ln n_0 + n_p^N v - n_p^N + \sum_\alpha n_\alpha^N \ln n_\alpha^N$$

$$- \sum_\alpha n_\alpha^N \ln(\sigma f_1^\alpha) - \frac{v_p^N x_p^N n_p^N S(1 - S/2)}{\theta} \qquad \text{(A-6)}$$

where

$$f_m^\alpha = \int d\beta \cdot \sin^m \beta \cdot \exp \left\{ - x_\alpha \left[\left(\frac{4}{\pi} \right) a + \tfrac{3}{2} v_p^N S \theta^{-1} \cdot \sin \beta \right] \cdot \sin \beta \right\} \qquad \text{(A-7)}$$

$$a = -\ln \left[1 - v_p^N \left(1 - \frac{y}{x_p^N} \right) \right] \qquad \text{(A-8)}$$

θ is the reduced temperature, $\theta = T/T^*$, referred to the unit rod length,[37,42] and β, for simplicity of notation, has replaced $(u_\alpha n)$. Parameters S and y are, respectively, the average order parameter

$$S = \sum_\alpha s_\alpha \cdot P^N(x_\alpha) = 1 - \frac{3}{2} \sum_\alpha \left(\frac{f_3^\alpha}{f_1^\alpha} \right) P^N(x_\alpha) \qquad \text{(A-9)}$$

where s_α is the order parameter of the α-species; and the average Flory disorder index

$$y = \sum_\alpha y_\alpha P^N(x_\alpha) = \frac{4}{\pi} \sum_\alpha \left(\frac{f_2^\alpha}{f_1^\alpha}\right) P^N(x_\alpha) \tag{A-10}$$

where y_α is the disorder index of the α-species. Both parameters characterize the orientational order of the nematic phase (for the isotropic phase, $s_\alpha = S = 0$ and $y = x_p^I$). n_s^N and n_p^N denote, respectively, the number concentrations of the solvent and solute particles, n_0 is the total number of lattice sites available for the solution, and σ is some arbitrary constant associated with the orientational part of the partition function, Z_{orient}. For simplicity σ is set equal to 1.

Note that the equilibrium orientational distribution function of the α-species, $f^\alpha(\beta)$, is given by

$$f^\alpha(\beta) = \sin\beta \cdot \frac{\exp\left\{-x_\alpha\left[\left(\frac{4}{\pi}\right)a + \frac{3}{2}v_p^N \cdot S \cdot \theta^{-1}\sin\beta\right]\sin\beta\right\}}{f_1^\alpha} \tag{A-11}$$

Adaptation of Eq. (A-6) to the case of the isotropic phase is straightforward. Taking the derivatives defined in Eqs. (A-5) one may calculate the chemical potentials and, following the equilibrium conditions, Eq. (A-4), obtain the *partition rules* for the system of rods partitioned between the isotropic and nematic components of the biphasic material:

$$\frac{v_\alpha^I}{v_\alpha^N} = \frac{\exp(-\xi x_\alpha)}{f_1^\alpha} \tag{A-12}$$

$$\ln\left(\frac{1-v_p^N}{1-v_p^I}\right) = \xi - S \cdot v_p^N \cdot \theta^{-1} - a \tag{A-13}$$

where

$$\xi = \frac{v_p^I}{x_p^I}(x_p^I - 1) - \frac{v_p^N}{x_p^N}(y - 1) \tag{A-14}$$

To ensure the equilibrium of the orientational distribution of rods in the nematic phase, Eqs. (A-7)–(A-10) must be satisfied simultaneously. To ensure the phase equilibria in biphasic material, Eqs. (A-1)–(A-3) also have to be fulfilled simultaneously. With the aid of Eq. (13), combination and rearrangement of Eqs. (A-2) and (A-12) gives

$$\frac{v_\alpha^N}{v_p^0} = \frac{(1/\Phi^N)P^0(x_\alpha)}{1+\left[(1-\Phi^N)/\Phi^N\right]\exp(-\xi x_\alpha)/f_1^\alpha} \tag{A-15}$$

Defining

$$I_m = \sum_\alpha \frac{(x_\alpha)^{-m} \cdot P^0(x_\alpha)}{1+\left[(1-\Phi^N)/\Phi^N\right]\exp(-\xi x_\alpha)/f_1^\alpha}; \quad m=0,1 \tag{A-16}$$

and

$$I_y = \sum_\alpha \frac{(f_2^\alpha/f_1^\alpha)P^0(x_\alpha)}{1+\left((1+\Phi^N)/\Phi^N\right)\exp(-\xi x_\alpha)/f_1^\alpha} \tag{A-17}$$

we have

$$\frac{v_p^N}{v_p^0} = \frac{I_0}{\Phi^N} \tag{A-18}$$

$$x_p^N = \frac{I_0}{I_1} \tag{A-19}$$

$$y = \frac{(4/\pi)I_y}{I_1} \tag{A-20}$$

Substitution of Eq. (A-18) into Eq. (A-1) leads to

$$\frac{v_p^I}{v_p^0} = \frac{\Phi^N}{1-\Phi^N}\left(\frac{1}{\Phi^N} - \frac{v_p^N}{v_p^0}\right) \tag{A-21}$$

Now Eq. (A-13) may be rewritten as

$$v_p^0 = \frac{1-\exp\left(\xi - a - S\cdot v_p^0\cdot\theta^{-1}\cdot A\right)}{A - B\cdot\exp\left(\xi - a - S\cdot v_p^0\cdot\theta^{-1}\cdot A\right)} \tag{A-22}$$

where

$$A = \frac{v_p^N}{v_p^0} \tag{A-23}$$

$$B = \frac{v_p^I}{v_p^0} \tag{A-24}$$

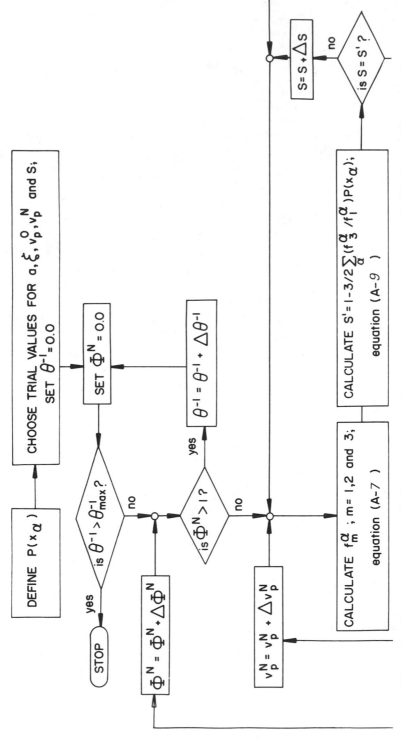

Figure A1. Flow chart of the numerical procedure developed to find the self-consistent solution of the WF phase-equilibrium conditions for any arbitrary polydispersity function at any given (θ^{-1}, Φ^N) point.

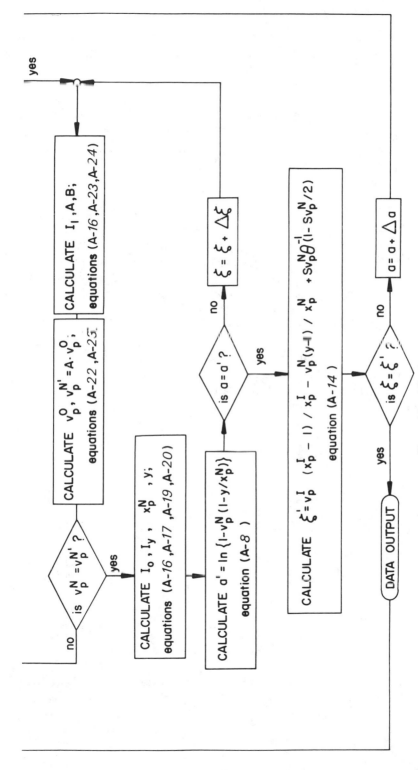

Figure A1. (*Continued*)

This completes the set of equations required to solve the problem of phase equilibrium. For a given distribution function $P^0(x_\alpha)$, the above set of equations must be satisfied simultaneously for any value of $\theta^{-1} \geq 0$ and for any value of Φ^N such that $0 \leq \Phi^N \leq 1$. This may be ensured numerically in a manner similar to that already described in refs. 35 and 89 (see Fig. A1).

An extensive study of the results of the phase-equilibrium calculations for the model system of rods discussed in this paper [Eq. (79)] will be given elsewhere. We present here only the phase diagram of the solution and curves for the typical behavior of Φ^N, v_p^N, and v_p^N across the biphasic range for three selected temperatures (see Figs. A2 and A3) (we defined the boundary of the biphasic range as $v_p^* = v_p^0$ for $\Phi^N = 1 \times 10^{-5}$ and $v_p^{**} = v_p^0$ for $\Phi^N = 1 - 1 \times 10^{-5}$). It is characteristic of the phase diagram of the system (Fig. A2), that it resembles in its main features the one for monodisperse rods with $x = 20$ calculated by Warner and Flory.[37] As we argued in ref. 93 and in Section V, despite the presence of the very long rods in the distribution, polydispersity leads to "smearing out" of any tendency of the system to demonstrate the complex phase behavior characteristic of monodisperse species with $x > 20$ (see ref. 37).

The three temperatures were chosen to be typical for different temperature ranges on the phase diagram: the region with the narrow biphasic range ($\theta^{-1} = 0.08$), the region with the broad biphasic range ($\theta^{-1} = 0.4$), and the region where the change from the narrow to the broad biphasic range occurs ($\theta^{-1} = 0.2$) (see Fig. A3). Results are qualitatively the same as those obtained earlier[89] for a similar distribution function, but for the solution in the

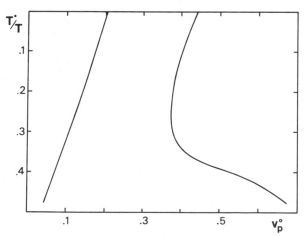

Figure A2.   Calculated WF phase diagram for the model system of polydisperse rods {20,10,65,80}.

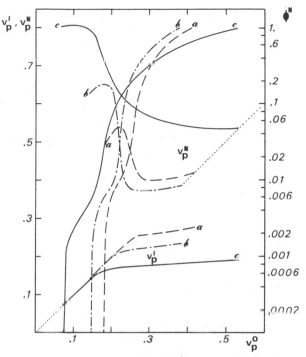

Figure A3.   Calculated concentration dependences of Φ^N, v_p^I, and v_p^N for three different inverse WF reduced temperatures θ^{-1}: (a) 0.08; (b) 0.2; (c) 0.4. Dotted lines indicate the isotropic ($v_p^I = v_p^0$) and nematic ($v_p^N = v_p^0$) phases. Note the log scale of Φ^N.

athermal limit. The presence of the very long rods effectively suppresses v_p^* and is responsible for the system phase behavior in the initial part of the biphasic range. We note that practically all the very long rods are transferred to the nematic component of the biphasic material, immediately after the concentration exceeds v_p^*. At each temperature this is demonstrated by a rapid, nearly instant increase of Φ^N. At progressively higher concentrations, the Φ^N curve first forms a shoulder and later asymptotically approaches the curve characteristic of the basic, Gaussian part of the distribution.[89] The transition across the biphasic range may be considered a two-step process, in the first part of which the "tail" rods are transferred to the nematic phases and in the second part of which the rest of the rods follow.

Apart from causing the shifts in value of v_p^* and v_p^{**}, temperature also has an influence on the partitioning of rods in the biphasic range, which is again pronounced in the vicinity of v_p^*. This shows up in the behavior of v_p^N curves. For very low temperatures, $\theta^{-1} = 0.4$, v_p^N remains large and constant as long as the longest species are being transferred into the nematic phase. At higher temperatures, $\theta^{-1} = 0.2$ and 0.08, v_p^N increases in the initial part of the biphasic range. This increase is more pronounced the higher the temperature is. Note that the transfer of the very long rods is extended over a larger part of the concentration scale at low temperatures; that is, the initial rapid increase of Φ^N is smaller and the shoulder broader for $\theta^{-1} = 0.4$ than for $\theta^{-1} = 0.2$ or 0.08.

APPENDIX B: FREQUENTLY USED ABBREVIATIONS AND SYMBOLS

We present here a glossary of those abbreviations and symbols used most frequently in this article. Some less frequently used symbols are not included.

Abbreviations

DMF	N, N'-dimethylformamide
PAIC	Poly(n-alkylisocyanate)
PBA	Poly($para$-benzamide)
PBLG	Poly(γ-benzyl-L-glutamate)
PBNIC	Copolymer of n-butylisocyanate and n-nonylisocyanate
PHIC	Poly(n-hexylisocyanate)
PHPIC	Copolymer of n-hexylisocyanate and n-propylisocyanate
TCE	Tetrachloroethylene
DE	Doi–Edwards
WF	Warner–Flory
WVWP	Warchol–Vaughan–Wang–Pecora
LC	Liquid crystals
l-LC	Lyotropic liquid crystals
t-LC	Thermotropic liquid crystals

Symbols

$C_\alpha(t)$	Dipole correlation function for the electric dipole moment of the α-type rod
d^*	Effective diameter of rod
D_α^{t0}	Translational diffusion coefficient for the diffusion of the α-type rod in the direction parallel to its long axis
$D_\alpha^{r0}, D_\alpha^r$	Rotational diffusion coefficient in free space and in semi-dilute solution for the α-type rod
f_m	Frequency at the maximum of dielectric loss
k_B	Boltzmann constant

$\mathbf{n}$	Nematic director
n_p^0	Total number concentration of rods in solution
$P^0(x_\alpha)$	Rod-length distribution function in an undiluted system of rods
$P^I(x_\alpha)$	Rod-length distribution function of species partitioned into the isotropic phase
$P^N(x_\alpha)$	Rod-length distribution function of species partitioned into the nematic phase
R_α^r	Retardation factor for the retardation of rotational motion of the α-type rod due to restricted translational mobility
S	Average order parameter for the nematic phase
s_α	Order parameter for the α-type rods
$\{s_\alpha\}$	Set of order parameters, including S, for all different types of rods present in the system.
t	Time
T	Absolute temperature
$\mathbf{u}$	Unit vector in the direction of the rod long axis
v_p^0	Total volume fraction or the concentration of rods in solution
v_p^I	Total volume fraction or the concentration of rods in the lyotropic isotropic phase
v_p^N	Total volume fraction or the concentration of rods in the lyotropic nematic phase
v_p^m	Concentration at the viscosity maximum in the biphasic range
v_p^i	Concentration at the phase inversion
v_p^c	Concentration at the maximum of ε_m'' and $\Delta\varepsilon$ in the biphasic range
v_p^*, v_p^{**}	Isotropic–biphasic and biphasic–nematic transition concentrations
v_g	Gelation or solidification concentration
$\mathbf{w}$	Unit vector in the direction of the rod cage axis
x	Axial ratio of a particle
x_α	Axial ratio of the α-type rods
y	Flory disorder index
ε'	Frequency-dependent dielectric constant
ε''	Dielectric loss factor
$\hat{\varepsilon}$	Complex dielectric permittivity
ε_m''	Maximum of the dielectric loss
$\Delta\varepsilon = \varepsilon_s - \varepsilon_\infty$	Dielectric increment or strength of the dielectric process; ε_s is the static dielectric constant, and ε_∞ is the dielectric constant of induced polarization

η_s	Solvent viscosity
η	Solution viscosity
η_{bi}	Solution viscosity in the biphasic range
η_{red}	Reduced viscosity of solution
$\eta_a^M, \eta_b^M, \eta_c^M$	Miesowicz viscosities
$\underset{\sim}{\kappa}, \kappa$	Rate of deformation tensor and its only nonzero element in the case of the Couette flow
μ_α	Electric dipole moment of the α-type rod
ρ_t	Average radius of the rod tubelike cage
$\underset{\sim}{\sigma}$	Stress tensor
$\tau_\alpha^{r0}, \tau_\alpha^r$	Rotational relaxation time in free space and in semidilute solution for the α-type rods
Φ^N	Volume fraction of the nematic component of biphasic material
$\Delta\Omega$	Solid angle of the rod cage

References

1. F. Reinitzer, *Monatsschrift*, **9**, 421 (1888).
2. O. Lehman, *Z. Phys. Chem.*, **4**, 462 (1889).
3. F. C. Bawden and N. W. Pirce, *Proc. R. Soc. Lond. Ser. B*, **123**, 274 (1937).
4. H. Freudlich, *J. Phys. Chem.*, **41**, 1151 (1937).
5. A. E. Elliot and E. J. Ambrose, *Discuss. Faraday Soc.*, **9**, 246 (1950).
6. C. Robinson, *Trans. Faraday Soc.*, **52**, 571 (1956).
7. C. Robinson, J. C. Ward, and R. B. Beevers, *Discuss. Faraday Soc.*, **25**, 29 (1958).
8. C. Robinson, *Tetrahedron*, **13**, 219 (1961).
9. C. Robinson, *Mol. Cryst.*, **1**, 467 (1966).
10. E. T. Samulski and A. V. Tobolsky, in *Liquid Crystals and Ordered Fluids*, Vol. 1, J. F. Johnson and R. S. Porter, eds., Plenum, New York, 1970, p. 494.
11. E. T. Samulski and A. V. Tobolsky, *Biopolymers*, **10**, 1013 (1971).
12. W. Miller, C. C. Wu, E. L. Wee, G. L. Santee, J. H. Rai, and K. G. Goebel, *Pure Appl. Chem.*, **38**, 37 (1974).
13. R. W. Filas and H. Stefanou, *J. Phys. Chem.*, **79**, 941 (1975).
14. D. B. Dupre, *Polym. Eng. Sci.*, **21**, 717 (1981).
15. S. L. Kwolek, P. W. Morgan, and W. R. Sorenson, *Macromolecules*, **10**, 1390 (1977).
16. T. I. Bair, P. W. Morgan, and F. L. Killian, *Macromolecules*, **10**, 1396 (1977).
17. S. P. Papkov, V. G. Kulichikhin, V. D. Kalmykova, and A. Y. Malkin, *J. Polym. Sci. Polym. Phys. Ed.*, **12**, 1753 (1974).
18. V. G. Kulichikhin, G. I. Kudryavtsev, and S. P. Papkov, *Int. J. Polym. Mater.*, **9**, 239 (1982).
19. G. C. Berry, in *Contemporary Topics in Polymer Science*, Vol. 2, E. M. Pearce and J. R. Schaefgen, eds., Plenum, New York, 1977, p. 55.
20. T. E. Helminiak, *Prepr. Am. Chem. Soc. Div. Org. Coat. Plast.*, **40**, 475 (1979).
21. E. W. Choe and S. N. Kim, *Macromolecules*, **14**, 920 (1981).

22. E. Iizuka and J. T. Yang, in *Liquid Crystals and Ordered Fluids*, Vol. 3, J. F. Johnson and R. S. Porter, eds., Plenum, New York, 1978, p. 197.

23. E. Iizuka, *Polym. J.*, **15**, 525 (1983).

24. F. Millich, *Adv. Polym. Sci.*, **19**, 117 (1975).

25. S. M. Aharoni, *J. Polym. Sci. Polym. Phys. Ed.*, **17**, 603 (1970)

26. S. M. Aharoni, *Macromolecules*, **12**, 94 (1979).

27. S. M. Aharoni and K. Walsh, *J. Polym. Sci. Polym. Lett. Ed.*, **17**, 321 (1979).

28. S. M. Aharoni and K. Walsh, *Macromolecules*, **12**, 271 (1979).

29. S. M. Aharoni, *J. Polym. Sci. Polym. Phys. Ed.*, **18**, 1439 (1980).

30. P. S. Russo and W. G. Miller, *Macromolecules*, **16**, 1960 (1983).

31. L. Onsager, *Ann. N. Y. Acad. Sci.*, **51**, 627 (1949).

32. A. Ishikara, *J. Chem. Phys.*, **18**, 1446 (1950); *ibid.*, **19**, 1142 (1951).

33. P. J. Flory, *Proc. R. Soc. Lond. Ser. A*, **234**, 73 (1956).

34. P. J. Flory and A. Abe, *Macromolecules*, **11**, 1119 (1978).

35. P. J. Flory and R. S. Frost, *Macromolecules*, **11**, 1126 (1978).

36. R. F. Frost and P. J. Flory, *Macromolecules*, **11**, 1134 (1980).

37. M. Warner and P. J. Flory, *J. Chem. Phys.*, **73**, 6327 (1980).

38. R. Diebleck and H. N. W. Lekkerkerker, *J. Phys. Lett. (Paris)*, **41**, 351 (1980).

39. M. Doi, *J. Polym. Sci. Polym. Phys. Ed.*, **19**, 229 (1981).

40. W. G. Miller, *Annu. Rev. Phys. Chem.*, **29**, 519 (1978)

41. R. R. Matheson and P. J. Flory, *Macromolecules*, **14**, 54 (1981).

42. P. J. Flory and G. Ronca, *Mol. Cryst. Liq. Cryst.*, **54**, 289 (1979); *ibid.*, 311 (1979).

43. M. Warner, *Mol. Cryst. Liq. Cryst.*, **80**, 67 (1982); *ibid.*, 79 (1982). [Warner's critical rod axial ratio for the formation of the stable nematic phase, $x = 8.75$, is higher than the value of 6.4 calculated by Flory and Ronca (ref. 42).]

44. J. Hermans, *J. Colloid Sci.*, **17**, 638 (1962).

45. G. Kiss and R. S. Porter, *J. Polym. Sci. Polym. Symp.*, **65**, 193 (1978).

46. E. Iizuka, *Mol. Cryst. Liq. Cryst.*, **25**, 287 (1974).

47. C. Balbi, E. Bianchi, A. Ciferri, A. Tealdi, and W. R. Kirgbaum, *J. Polym. Sci. Polym. Phys. Ed.*, **18**, 2037 (1980).

48. R. R. Matheson, *Macromolecules*, **13**, 643 (1980).

49. S. M. Aharoni, *Polymer*, **21**, 1413 (1980).

50. M. Doi, *J. Phys. (Paris)*, **36**, 607 (1975).

51. M. Doi and E. F. Edwards, *J. Chem. Soc., Faraday II*, **74**, 560 (1978); *ibid.*, 918 (1978).

52. E. T. Samulski, personal communication.

53. J. K. Moscicki, *Mol. Phys.*, **51**, 919 (1984).

54. M. Doi, *J. Polym. Sci. Polym. Phys. Ed.*, **20**, 1963 (1982).

55. J. F. Maguire, *J. Chem. Soc., Faraday II*, **77**, 513 (1981).

56. K. M. Zero and R. Pecora, *Macromolecules*, **15**, 87 (1982).

57. P. S. Russo, F. E. Karasz, and K. H. Langley, *J. Chem. Phys.*, **80**, 5312 (1984).

58. J. F. Maguire, J. P. McTague, and F. Rondelez, *Phys. Rev. Lett.*, **45**, 1891 (1980).

59. H. Nakamura and K. Okano, *Phys. Rev. Lett.*, **50**, 186 (1983).

60. J. K. Moscicki, G. Williams, and S. M. Aharoni, *Polymer*, **22**, 571 (1981).

61. J. K. Moscicki, G. Williams, and S. M. Aharoni, *Polymer*, **22**, 1361 (1981).

62. J. K. Moscicki, G. Williams, and S. M. Aharoni, *Macromolecules*, **15**, 642 (1982).

63. J. K. Moscicki and G. Williams, *J. Polym. Sci. Polym. Phys. Ed.*, **21**, 213 (1983).

64. M. P. Warchol and W. E. Vaughan, *Adv. Mol. Relax. Proc.*, **13**, 317 (1978); W. E. Vaughan, *Ann. Rev. Phys. Chem.* **30**, 103 (1979).

65. C. C. Wang and R. Pecora, *J. Chem. Phys.*, **72**, 5333 (1980).

66. P. G. de Gennes, *J. Chem. Phys.*, **55**, 572 (1971).

67. J. Riseman and J. G. Kirkwood, *J. Chem. Phys.*, **18**, 512 (1950).

68. J. G. Kirkwood and P. L. Auer, *J. Chem. Phys.*, **19**, 281 (1951).

69. T. Odijk, *Macromolecules*, **16**, 1340 (1983); *ibid.*, **17**, 502 (1984).

70. G. Marrucci and N. Grizzuti, *J. Polym. Sci. Polym. Lett. Ed.*, **21**, 83 (1983).

71. P. G. de Gennes, *J. Phys. (Paris)*, **42**, 472 (1981).

72. D. Frenkel and J. F. Maguire, *Phys. Rev. Lett.*, **47**, 1025 (1981); *ibid.*, *Mol. Phys.*, **49**, 503 (1983).

73. G. Williams, *J. Polym. Sci. Polym. Phys. Ed.*, **21**, 2037 (1983).

74. C. J. F. Bottcher and P. Bordewijk, *Theory of Dielectric Polarization*, Vol. 2, Elsevier, Amsterdam, 1978.

75. Y. Mori, N. Ookubo, R. Hayakawa, and Y. Wada, *J. Polym. Sci. Polym. Phys. Ed.*, **20**, 2111 (1982).

76. An excellent review of experimental studies of dielectric relaxation in thermotropic liquid crystals may be found in J. P. Parneix, D. Sc. Thesis, Université de Sciences et Techniques de Lille, Lille, France, 1982.

77. A. J. Martin, G. Meier, and A. Saupe, *Symp. Faraday Soc.*, **5**, 119 (1971).

78. P. L. Nordio, G. Riggatti, and U. Segre, *Mol. Phys.*, **25**, 129 (1973).

79. J. K. Moscicki, V. Rosato and G. Williams, unpublished results. In the course of static and dynamic Kerr-effect studies of PHIC–toluene solutions we observed a build-up of quasipermanent birefringence, even in very weak electric fields. This birefringence decayed over the course of a few tens of minutes. Both the rise and the decay of the birefringence result from the end-over-end reorientation and ordering of the macromolecules in the direction of the electric field.

80. O. H. Griffith and P. C. Jost, in *Spin Labeling*, L. J. Berlinger, ed., Academic, London, 1976, Chapter 12.

81. K. Kinosita, S. Kawato, and A. Ikegami, *Biophys. J.*, **20**, 289 (1977).

82. R. B. Bird, W. E. Steward, and E. N. Lightfoot, *Transport Phenomena*, Wiley, New York, 1960.

83. G. Marrucci, *Mol. Cryst. Liq. Cryst.*, **72**, 153 (1982).

84. M. Miesowicz, *Mol. Cryst. Liq. Cryst.*, **97**, 1 (1983).

85. M. Miesowicz, *Nature*, **136**, 261 (1935); *ibid.*, **158**, 27 (1946).

86. A. J. Bur and L. J. Fetters, *Chem. Rev.*, **76**, 727 (1976).

87. H. Block, E. M. Gregson, A. Ritchie, and S. M. Walker, *Polymer*, **24**, 859 (1983).

88. S. M. Aharoni, *Polym. Bull.*, **5**, 95 (1981).

89. J. K. Moscicki and G. Williams, *J. Polym. Sci. Polym. Phys. Ed.*, **21**, 197 (1983).

90. S. B. Dev, R. Y. Lochhead, and A. M. North, *Disc. Faraday Soc.*, **49**, 244 (1970).

91. J. K. Moscicki, unpublished polarized-light microscopy data.

92. J. K. Moscicki and G. Williams, *Polymer*, **24**, 85 (1983).

93. J. K. Moscicki, *J. Polym. Sci. Polym. Phys. Ed.*, in press.

94. G. I. Taylor, *Proc. R. Soc. Lond. A*, **138**, 41 (1932).

95. P. J. Flory, *J. Chem. Phys.*, **17**, 223 (1949).

96. A. Chiferri and W. R. Krigbaum, *Mol. Cryst. Liq. Cryst.*, **69**, 273 (1981).

97. W. R. Krigbaum and A. Chiferri, *J. Polym. Sci. Polym. Lett. Ed.*, **18**, 253 (1980).

98. *Handbook of Chemistry and Physics*, 63th Ed., CRC Press, Cleveland, 1982, p. F-46.

99. H. Kresse, *Fortschr. Phys. (Germany)*, **30**, 508 (1982); *ibid. Adv. Liq. Cryst.* **6**, 109 (1983).

100. J. R. Van Wazer, J. W. Lyons, K. Y. Kim, and R. E. Colwell, *Viscosity and Flow Measurement*, Wiley-Interscience, New York, 1963.

101. A. C. Diogo, *Mol. Cryst. Liq. Cryst.*, **100**, 153 (1983).

102. K. F. Wissbrun, *J. Rheol.*, **25**, 619 (1981).

103. G. Conio, E. Bianchi, A. Ciferri, and W. R. Krigbaum, *Macromolecules*, **17**, 856 (1984).

104. S. Fesjian and H. L. Frisch, *J. Chem. Phys.*, **80**, 4410 (1984).

105. T. Odijk, personal communication.

106. P. J. Flory, *Adv Polym. Sci.*, **59** (in press).

107. E. L. Wee and W. G. Miller, in *Liquid Crystals and Ordered Fluids*, Vol. 3, J. F. Johnson and R. S. Porter, eds., Plenum, New York, 1978, p. 371.

108. A. R. Khokhlov and A. N. Semenov, *Physica*, **112A**, 605 (1982).

109. S. M. Aharoni, personal communication

110. J. K. Moscicki, B. Robin-Lherbier, and D. Canet, *J. Phys. Lett. (Paris)*, **45**, L-379 (1984).

THE LOCAL FIELD IN THE STATISTICAL-MECHANICAL THEORY OF DIELECTRIC POLARIZATION

WOLFFRAM SCHRÖER

Department of Chemistry, University of Bremen, Bibliothekstrasse, Postfach 330440, 2800 Bremen 33, Federal Republic of Germany

CONTENTS

I. MACROSCOPIC CONCEPTS IN THE THEORY OF DIELECTRIC POLARIZATION

A. Introduction

The equations of motion of moving particles are determined by the forces and torques acting on the particles. The various dynamical models discussed in the other chapters of this volume distinguish between the stochastic forces and the systematic forces, caused by external fields and in some cases also by intermolecular interactions. The field that on average acts on a particle, termed the local field, is only in the case of dilute gases given by the applied field. In dense media the average field caused by the surrounding molecules has to be taken into account.

The most important part of the classical electrostatic models of dielectric polarization due to Debye,[1] Fröhlich,[2] Onsager,[3] and Böttcher and Scholte[4,5] is the modeling of that local field. In this paper we will investigate how this classical work relates to the modern statistical-mechanical theory of dielectric polarization. The early work on dielectrics started from an electrostatic model that considered the behavior of a molecule in a macroscopic electrostatic field. In the theories of Debye the local field was at first approximated by the Maxwell field, and later by the Lorentz field. Onsager assumed the cavity field to be the local field in his first approximation. The cavity field is the field present in a real cavity embedded in a polarized medium. The Maxwell field is the field experienced by a point charge in a polarized medium. It is the sum of the applied field and the field due to the surface charges in a uniformly polarized system. The Lorentz field is the field experienced by a charge in a virtual cavity. (By "virtual cavity" we mean that the polarization remains uniform and is not perturbed by the surface charges of the cavity.) In Onsager's electrostatic model it is assumed that the local field is given by the cavity field and the reaction field. This is the field due to the polarization of the medium caused by the moment of the molecule being investigated. In their generalization of the Onsager model for nonspherical molecules, Böttcher and Scholte[4,5] replaced the spherical molecular cavity with an ellipsoidal cavity. The most recent review of these classical models is given in a book by Böttcher.[6]

Modern statistical-mechanical theories of dielectric polarization, which can be said to have started with those of Kirkwood[7] and Yvon,[8] attempt to calculate the dielectric permittivity from the dipole fluctuations of a macroscopic sample using statistical mechanics. The more recent advances are reviewed in articles by Adelman and Deutch;[9] Stell, Patey, and Høye;[10] and Wertheim.[11] In order to gain insight, but also in order to apply the results of these theories in dynamical models, we wish to extract formally exact and numerically accurate expressions of the local field from these theories and thus obtain well-founded estimates of the systematic forces.

B. Dipole Fluctuations of a Sample *In Vacuo*

Following Kirkwood,[7] we consider a canonical ensemble of samples comprising N polarizable dipolar molecules interacting *in vacuo* with a fixed uniform field E_0. The sample shape is regular, but not necessarily spherical, in order to ensure uniform polarization. This restriction is not applied in most of the related recent work.[12-15] Uniform polarization is required to ensure invariance against the choice of the molecular point of reference, since contributions of higher multipole densities to the dielectric polarization[16-19] are not taken into account. No contributions to the polarization due to short-range distortion,[20,21] higher moments,[22] and hyperpolarizabilities[23] will be considered, so that the potential energy H of the sample is the sum of the dipolar contributions and H_0, which represents the two-body short-range interactions:

$$H = H_0 + \frac{1}{2} \sum_{i,k} \mu_i \cdot T_{ik} \cdot m_k - \sum_i m_i \cdot E^0 - \frac{1}{2} \sum_i E^0 \cdot A_i \cdot E^0 \qquad (1)$$

This Hamiltonian was given by Mandel and Mazur[24] and was also applied in the recent work of Wertheim.[15,25] $T(r)$ is the dipole–dipole interaction tensor

$$T(r) = -\nabla\nabla \frac{1}{r} \qquad (2)$$

A_i is the generalized polarizability of a molecule and is determined from the molecular polarizabilities α_j of all molecules in the sample.

$$A_1 = \alpha_1 \cdot \left[1 - \sum_k T_{1k} \cdot \alpha_k + \sum_{k,k'} T_{1k} \cdot \alpha_k \cdot T_{kk'} \cdot \alpha_{k'} - \cdots \right] \qquad (3)$$

Similarly, the moment m_i is given by the permanent moment μ_i corrected for the induced contributions.

$$m_1 = \mu_1 - \alpha_1 \left[\sum_k T_{1k} \cdot \mu_k - \sum_{k,k'} T_{1k} \cdot \alpha_k \cdot T_{kk'} \cdot \mu_{k'} + \cdots \right] \qquad (4)$$

The dipole-dipole interaction term has been the subject of dispute.[26-29] There is no question now that the expression $\sum m_i T_{ik} m_k$ is incorrect and erroneously replaces the sum of a geometric series with the product of two terms of that series.

The polarization $\mathbf{P}$ is determined by the average molecular moment $\langle\mathbf{p}_1\rangle$ and the number density ρ:

$$\mathbf{P} = \rho\langle\mathbf{p}_1\rangle \tag{5}$$

The average moment $\langle\mathbf{p}_1\rangle$ may be separated into a term $\langle\mathbf{m}_1\rangle$ depending on the permanent molecular dipoles and another term $\langle\boldsymbol{\pi}_1\rangle$, which is the moment induced by the external field and the moments induced by induced moments:

$$\langle\mathbf{p}_1\rangle = \langle\mathbf{m}_1\rangle + \langle\boldsymbol{\pi}_1\rangle \tag{6}$$

To the first order of the applied field $\mathbf{E}^0$, Eq. (6) reads

$$\langle\mathbf{p}_1\rangle = \left[\beta\langle\mathbf{m}_1\textstyle\sum\mathbf{m}_i\rangle_0 + \langle\mathbf{A}_1\rangle_0\right]\cdot\mathbf{E}^0 \tag{7}$$

The averages $\langle\dots\rangle_0$ involve the distribution functions of the isotropic system. Terms $\langle\mathbf{m}_1\sum\mathbf{m}_i\rangle_0$ and $\langle\mathbf{A}_1\rangle_0$ are tensors of rank 2 and should be distinguished from the scalar properties $\langle\mathbf{m}_1\cdot\sum\mathbf{m}_i\rangle$ and A_1.

From the phenomenological relation between the polarization $\mathbf{P}$ and the Maxwell field $\mathbf{E}$,

$$\mathbf{P} = \frac{\varepsilon-1}{4\pi}\mathbf{E} \tag{8}$$

and Eq. (7), it follows that

$$\frac{\varepsilon-1}{4\pi}\mathbf{E} = \rho\left[\beta\langle\mathbf{m}_1\textstyle\sum\mathbf{m}_i\rangle_0 + \langle\mathbf{A}_1\rangle_0\right]\cdot\mathbf{E}^0 \tag{9}$$

According to the solution of the boundary-value problem of classical electrostatics, the applied field is related to the Maxwell field by the shape-dependent function

$$\mathbf{E} = \mathbf{E}^0 - 4\pi\mathbf{S}\cdot\mathbf{P} \tag{10}$$

The electric field in the medium is given by the applied field counteracted by a correction due to the surface polarization. In the case of the ellipsoid, the shape function $\mathbf{S}$ is a diagonal second-rank tensor with the components[30,31]

$$S_{zz} = \frac{abc}{2}\int_0^\infty \frac{d\lambda}{(c^2+\lambda)D(\lambda)}$$
$$D(\lambda) = \left[(a^2+\lambda)(b^2+\lambda)(c^2+\lambda)\right]^{1/2} \tag{11}$$

where a, b, and c are the half-axes in the x-, y-, and z-directions of the ellipsoid. In the more symmetric special cases of the needle, the sphere, and the disc, the components of $\mathbf{S}$ take the values 0, $\frac{1}{3}$, and 1.

From Eqs. (9) and (10) the general relation between ε and the molecular properties for an ellipsoidal sample *in vacuo* is obtained:

$$\frac{\varepsilon - 1}{3[1 + (\varepsilon - 1)\mathbf{S}]} = \frac{4\pi}{3}\rho\left[\beta\langle\mathbf{m}_1\sum\mathbf{m}_i\rangle_0 + \langle A_1\rangle_0\right] \qquad (12)$$

Because the intermolecular dipole correlations are our main interest, $\langle A_1\rangle_0$ is expressed by ε_∞, the high-frequency limit of ε. In the case of the sphere, the result of Harris and Alder[26] is obtained:

$$\frac{3(\varepsilon - \varepsilon_\infty)}{(\varepsilon + 2)(\varepsilon_\infty + 2)} = \frac{4\pi}{3}\rho\beta\frac{1}{3}\left\langle\mathbf{m}_1\cdot\sum_k\mathbf{m}_k\right\rangle_0^{\text{sphere}} \qquad (13)$$

Obviously, the averages in Eqs. (12) and (13) depend on the shape of the sample. If the sample is surrounded by a dielectric, different macroscopic expressions are obtained; these are given in the following section.

C. Dipole Fluctuations of an Ellipsoidal Sample in Any Medium

The dielectric constant inside the ellipsoidal sample is ε and that outside it, ε'. The solution of the boundary-value problem gives for the Maxwell field $\mathbf{E}$ inside the sample in this general case[30]

$$\mathbf{E} = \left(1 + \frac{\varepsilon - \varepsilon'}{\varepsilon'}\mathbf{S}\right)^{-1}\cdot\mathbf{E}_a \qquad (14)$$

$\mathbf{E}_a$ is the Maxwell field at a large distance from the small sample inside a large polarized body.

It is instructive to separate the various contributions to $\mathbf{E}$. The field $\mathbf{E}$ can be decomposed into the field $\mathbf{E}_a$ and corrections due to the fields of the surface polarization inside and outside the boundary of the sample and to the reaction fields caused by the surface polarization:

$$\mathbf{E} = \mathbf{E}_a + 4\pi\mathbf{S}\cdot[\mathbf{P}_a - \mathbf{P}_i] + \frac{4\pi}{3}abc\mathbf{R}\cdot[\mathbf{P}_a - \mathbf{P}_i] \qquad (15)$$

The reaction field is, in the case of the ellipsoid,[5]

$$\mathbf{R}\cdot\mathbf{M} = \frac{3}{abc}\mathbf{S}\frac{(\varepsilon' - 1)(1 - \mathbf{S})}{1 + (\varepsilon' - 1)(1 - \mathbf{S})}\cdot\mathbf{M} \qquad (16)$$

$\mathbf{R}$ is the diagonal reaction-field tensor. $\mathbf{M}$ takes the values $\mathbf{M}_i$ and $\mathbf{M}_a$, which are the dipole moments of a uniformly polarized ellipsoidal sample that give the same field outside the sample as the surface polarization:

$$\mathbf{M}_i = \frac{4\pi}{3} abc\mathbf{P}_i \quad \text{and} \quad \mathbf{M}_a = \frac{4\pi}{3} abc\mathbf{P}_a \tag{17}$$

The cavity field is obtained if the polarization $\mathbf{P}_i$ equals 0 inside the sample.

The field acting on the sample that is independent of the molecules inside the sample is the cavity field. The contributions of $\mathbf{P}_i$ in Eq. (15) are part of the intermolecular interactions and are therefore the subject of statistical calculations. Because of the reaction field, the terms in Eqs. (6) and (7) depend now on the surrounding medium, so that the average moment is

$$\langle \mathbf{p}_1 \rangle_{\varepsilon'} = \left[\beta \Big\langle \mathbf{m}_1 \sum_k \mathbf{m}_k \Big\rangle_{\varepsilon'} + \langle \mathbf{A}_1 \rangle_{\varepsilon'} \right] \cdot \frac{\varepsilon'}{\varepsilon' \cdot 1 + (1 - \varepsilon')S} \cdot \mathbf{E}_a \tag{18}$$

Using the macroscopic expression of the polarization $\mathbf{P}_i$ after expressing $\mathbf{E}_a$ by $\mathbf{E}$, the general fluctuation expression is

$$\frac{\varepsilon - 1}{3} \frac{\varepsilon' 1 + (1 - \varepsilon')S}{\varepsilon' 1 + (\varepsilon - \varepsilon')S} = \frac{4\pi}{3} \rho \left[\beta \Big\langle \mathbf{m}_1 \sum \mathbf{m}_i \Big\rangle_{\varepsilon'} + \langle \mathbf{A}_1 \rangle_{\varepsilon'} \right] \tag{19}$$

For a spherical sample, Eq. (19) reads

$$\frac{\varepsilon - 1}{3} \frac{2\varepsilon' + 1}{2\varepsilon' + \varepsilon} = \frac{4\pi}{3} \rho \left[\frac{1}{3} \beta \Big\langle \mathbf{m}_1 \cdot \sum \mathbf{m}_i \Big\rangle_{\varepsilon'} + \langle \mathbf{A}_1 \rangle_{\varepsilon'} \right] \tag{20}$$

For the special case of rigid dipoles, Eq. (20) was recently given by Perram[32] and by Pollock and Alder.[33] Equation (19) includes the cases of the sphere *in vacuo* ($\varepsilon' = 1$) and the sphere in its own medium ($\varepsilon' = \varepsilon$). If the sphere is surrounded by a metal ($\varepsilon' = \infty$) the same expression is obtained as for the needle before. The common feature in both cases is the lack of a surface polarization.

To express the average $\langle \mathbf{m}_1 \sum \mathbf{m}_k \rangle_{\varepsilon'}$ in terms of macroscopic quantities, $\langle \mathbf{A}_1 \rangle_{\varepsilon'}$ is expressed in terms of the high-frequency limit ε_∞ of the dielectric constant and those components $\mathbf{E}^\infty$ of the Maxwell field $\mathbf{E}$ that are independent of the permanent moments of the molecules inside the ellipsoid. The field $\mathbf{E}^\infty$ is obtained from Eq. (15):

$$\mathbf{E}^{\infty} = \mathbf{E}_a + 4\pi \mathbf{S} \cdot \left[\mathbf{P}_a - \mathbf{P}_i^\infty \right] - 4\pi \frac{abc}{3} \mathbf{R} \cdot \left[\mathbf{P}_a - \mathbf{P}_i^\infty \right]$$

$$= \frac{\varepsilon'}{\varepsilon'\mathbf{1} + (\varepsilon_\infty - \varepsilon')\mathbf{S}} \cdot \mathbf{E}_a \tag{21}$$

which gives the macroscopic exact expression for $\langle \mathbf{A}_1 \rangle_{\varepsilon'}$.[34] The general form of the dipole fluctuations is

$$\frac{4\pi}{3} \rho\beta \left\langle \mathbf{m}_1 \sum \mathbf{m}_i \right\rangle_{\varepsilon'} = \frac{(\varepsilon - \varepsilon_\infty)\left[\varepsilon'\mathbf{1} + (1 - \varepsilon')\mathbf{S} \right]^2}{3\left[\varepsilon'\mathbf{1} + (\varepsilon - \varepsilon')\mathbf{S} \right] \cdot \left[\varepsilon'\mathbf{1} + (\varepsilon_\infty - \varepsilon')\mathbf{S} \right]} \tag{22}$$

For a spherical sample, Eq. (22) simplifies to

$$\frac{(2\varepsilon' + 1)^2 (\varepsilon - \varepsilon_\infty)}{3(2\varepsilon' + \varepsilon)(2\varepsilon' + \varepsilon_\infty)} = \frac{4\pi}{3} \rho\beta \frac{1}{3} \left\langle \mathbf{m}_1 \cdot \sum \mathbf{m}_i \right\rangle_{\varepsilon'} \tag{23}$$

Equation (22) generalizes the Fröhlich formula for the dipole fluctuations of a dielectric sphere in a medium of the same dielectric constant to the very general case of an ellipsoid in any medium.

The macroscopic expressions for the dipole fluctuations in terms of the dielectric constant are derived in a straightforward but rigorous manner using continuum electrostatics. In the derivation the dipole-independent contribution $\langle \mathbf{A}_1 \rangle_{\varepsilon'}$ is estimated exactly from macroscopic electrostatics and not, as in Fröhlich's original derivation, by nonappropriate classical averaging of the electronic fluctuations,[35] or by a model of point dipoles in a continuum with permittivity ε_∞.[36] Fröhlich's formula for a spherical sample in a medium with the same dielectric constant is in contradiction to the formula recently given by Wertheim,[25]

$$\frac{(\varepsilon - 1)(2\varepsilon + 1)}{9\varepsilon} - \frac{(\varepsilon_\infty - 1)(2\varepsilon_\infty + 1)}{9\varepsilon_\infty} = \frac{4\pi}{3} \rho\beta \frac{1}{3} \left\langle \mathbf{m}_1 \cdot \sum \mathbf{m}_i \right\rangle_{\varepsilon} \tag{24}$$

Wertheim's expression is obtained if in Eq. (20) $\langle \mathbf{A}_1 \rangle_{\varepsilon_\infty}$ is subtracted instead of $\langle \mathbf{A}_1 \rangle_{\varepsilon}$. Wertheim's result is founded on the inability of the dipoles in the surrounding medium to follow the electronic fluctuations. Because we investigate the dielectric polarization using a static field, electronic fluctuations are irrelevant and Wertheim's formula is not the correct fluctuation expression required in the calculation of $\langle \mathbf{m} \rangle_\varepsilon$. As pointed out by Felderhof,[37] Wertheim and Fröhlich consider the fluctuations of different dipole moments. Wertheim's formula includes moments induced by the fluctuating fields due to rigid dipoles in the medium outside the sample. Because those fluctuations contribute to $\langle \mathbf{A} \rangle_\varepsilon$, Fröhlich's fluctuation formula represents the dipole fluctuations caused by the permanent dipoles inside the sample. In the classical theories the concept of a sphere embedded in the medium was introduced to account for the effect of long-range interactions on the fluctuations. However, a great deal of care must be taken in using this concept, since, because of the reaction field, the dipole fluctuations of the embedded sphere are not independent of the surrounding dielectric.

D. The Onsager Model for Many Particles

As in the Onsager model,[3] the reaction field will now be taken into account explicitly. Modified expressions for $\langle \mathbf{m}_1 \Sigma \mathbf{m}_i \rangle_{\varepsilon'}$ and $\langle \mathbf{A}_1 \rangle_{\varepsilon'}$ are thus required. The polarizability $\langle \mathbf{A}_1 \rangle_{\varepsilon'}'$ of the sample with respect to all external charges, including the charges induced by the polarization of the sample, is obtained by investigating $\langle \pi_1 \rangle_{\varepsilon'}$:

$$\langle \pi_1 \rangle_{\varepsilon'} = \left\langle \mathbf{A}_1 \cdot \left[\frac{\varepsilon'}{\varepsilon'\mathbf{1} + (1-\varepsilon')\mathbf{S}} \cdot \mathbf{E}_a + \sum_j \mathbf{R} \cdot \pi_j \right] \right\rangle \tag{25}$$

$\langle \pi_1 \rangle_{\varepsilon'}$ is given by a term induced by the cavity field and a reaction-field contribution. The average $\langle \mathbf{A}_1 \cdot \Sigma \pi_j \rangle$ may be separated into terms in which $\mathbf{A}_1$ and π_j are not correlated and others, termed *irreducible terms* in which $\mathbf{A}_1$ and π_j are correlated:

$$\langle \pi_1 \rangle_{\varepsilon'} = \langle \mathbf{A}_1 \rangle_{\varepsilon'}' \cdot \left[\frac{\varepsilon'}{\varepsilon'\mathbf{1} + (1-\varepsilon')\mathbf{S}} \mathbf{E}_a + N\mathbf{R} \cdot \langle \pi_1 \rangle_{\varepsilon'} \right] + \left\langle \mathbf{A}_1 \cdot \mathbf{R} \sum \pi_j \right\rangle_{\text{irr}} \tag{26}$$

In the limit of a macroscopic sample, the last term can be neglected. Expressing the moment $\langle \pi_1 \rangle_{\varepsilon'}$ by means of the macroscopic expression Eq. (26) gives, using the explicit formula Eq. (16) for the reaction field $\mathbf{R}$,[34]

$$\frac{\varepsilon_\infty - 1}{3} \frac{1}{1 + (\varepsilon_\infty - 1)\mathbf{S}} = \frac{4\pi}{3} \rho \langle \mathbf{A}_1 \rangle_{\varepsilon'}' \tag{27}$$

which is the expression for an ellipsoidal sample *in vacuo*. We conclude that quite generally for a macroscopic sample the polarizability $\langle \mathbf{A}_1 \rangle_{\varepsilon'}'$ agrees with the polarizability of the sample *in vacuo*, $\langle A_1 \rangle_0$, independent of the surrounding medium.

In polar fluids the reaction field modifies the size of the molecular dipole moment. The part of the moment that is independent of external charges is denoted m'_1. The moments m_1 that depend on the reaction field are given by

$$\sum_i m_i = \sum_i \left[m'_i + \sum_j A'_i \cdot R \cdot m_j \right] \tag{28}$$

In a macroscopic sample the contributions in which m_j and A_i are correlated may be neglected, so that

$$m_i = \frac{1}{1 - \frac{4\pi}{3} \rho abc R \cdot \langle A_1 \rangle_0} m'_i \tag{29}$$

It remains to consider the contributions of the reaction field to the intermolecular correlations. The Hamiltonian of the sample in a dielectric medium contains additional reaction-field terms, given by

$$U_R = - \sum_{i \le k} m'_i \cdot R \cdot m_k \tag{30}$$

The self term in Eq. (30) is an isotropic energy term and has no influence on the average orientation. The distinct terms give a contribution to the angular correlation function:

$$\langle m_1 \rangle_{\varepsilon'} = \beta \langle m_1 \sum m_k \rangle'_{\varepsilon'} \frac{\varepsilon'}{\varepsilon' 1 + (1 - \varepsilon') S} \cdot E_a + \sum_{k \ne k'} \beta \langle m_1 m'_k \cdot R \cdot m_{k'} \rangle \tag{31}$$

The reaction-field term may again be decomposed into irreducible terms, in which $m_{k'}$ is correlated with m_1 or m'_k, and reducible terms:

$$\sum_{k \ne k'} \langle m_1 m'_k \cdot R \cdot m_{k'} \rangle = \sum_k \langle m_1 m'_k \rangle_{\varepsilon'} \cdot R \cdot (N - 1) \langle m_1 \rangle$$
$$+ \sum_{k \ne k'} \langle m_1 m'_k \cdot R \cdot m_{k'} \rangle_{irr} \tag{32}$$

The irreducible term is of the order $O(1/V)$ and can be neglected for a macroscopic sample. Using the explicit expression for the reaction field, Eq. (16), and expressing **m** in terms of **m'** according to Eq. (29), we obtain from Eq. (31) the reaction-field-corrected fluctuation formula

$$\frac{\varepsilon - \varepsilon_\infty}{3} \frac{1}{[1+(\varepsilon-1)S]\cdot[1+(\varepsilon_\infty-1)S]} = \frac{4\pi}{3}\rho\beta\left\langle \mathbf{m}'_1 \sum_k \mathbf{m}'_k \right\rangle'_{\varepsilon'} \qquad (33)$$

which agrees with Eq. (22) in the special case of an ellipsoidal sample *in vacuo*. So we obtain the general result that for macroscopic samples of ellipsoidal shape explicit consideration of the reaction field leads to the same expression obtained for a sample *in vacuo*. This result is independent of the permittivity of the surrounding medium.

The field due to all external charges, which includes the reaction field, is the same as the field acting on the sample *in vacuo*. We conclude that it is a matter of convenience whether we investigate a sample *in vacuo* or a sample embedded in a dielectric. For an analytical theory the system *in vacuo* is simpler to treat, particularly since it is not necessary to consider the shape-dependent contributions explicitly. The theory of the dipole-moment fluctuations for an embedded sample is, in principle, more complicated than that for a sample *in vacuo*.

This conclusion holds also for computer simulations. The situation is shown quite clearly if a large spherical sample is simulated.[38,39] The application of periodic boundary conditions removes the discontinuity at the boundary and promises reliable results with a few hundred particles, which corresponds to a sample size of less than 10 average next-neighbour distances. To take into account the long-range dipolar interactions various techniques have been used, such as the reaction-field technique,[40-42] the spherical cutoff,[41] the mirror image,[41] and the Ewald–Kornfeld summation.[32,33] Use of this last technique, however, introduced an uncertainty about which fluctuation expression had to be used to calculate the dielectric permittivity. This problem seems now to be understood.[32,33,43,44] If the Ewald summation is applied with a reaction field depending on ε', a spherical sample surrounded by a medium with dielectric permittivity ε' is simulated.[32] By applying techniques such as the Ewald summation, one does not circumvent the requirement that the sample be large enough to ensure the convergence of short-range correlation integrals within the simulated sample, so that macroscopic relations such as Eq. (19) can be applied.[45]

E. The Onsager Model

We now consider an ellipsoidal sample containing only one particle, taking into account that this ellipsoid may assume any orientation with respect to the applied field. This requires that the macroscopic expression of the average moment be angle averaged:

$$\langle \mathbf{p} \rangle_{\varepsilon'} = \frac{\varepsilon - 1}{4\pi\rho} \frac{1}{3} \sum \frac{\varepsilon'}{\varepsilon' + (\varepsilon - \varepsilon')S_{\lambda\lambda}} \mathbf{E}_a \tag{34}$$

The macroscopic angular-averaged expressions of $\langle \boldsymbol{\pi} \rangle_{\varepsilon'}$ and $\langle \mathbf{m} \rangle_{\varepsilon'}$ are obtained analogously. The microscopic expression of $\langle \boldsymbol{\pi} \rangle_{\varepsilon'}$ is obtained from Eq. (26) by taking $N = 1$ and replacing $\mathbf{A}_1$ with the polarizability tensor $\boldsymbol{\alpha}$.

$$\langle \boldsymbol{\pi} \rangle_{\varepsilon'} = \frac{1}{3} \sum_{\lambda} \frac{\alpha_{\lambda\lambda}}{1 - R_{\lambda\lambda}\alpha_{\lambda\lambda}} \frac{\varepsilon'}{\varepsilon' + (1 - \varepsilon')S_{\lambda\lambda}} \mathbf{E}_a \tag{35}$$

The index λ denotes the components of the polarizability tensor with respect to the main axes. The microscopic expression of $\langle \mathbf{m} \rangle_{\varepsilon'}$ is obtained from Eqs. (31) and (32) using a similar method: Taking $N = 1$ and replacing $\mathbf{m}'$ with the permanent moment $\boldsymbol{\mu}$,

$$\langle \mathbf{m} \rangle_{\varepsilon'} = \frac{1}{3}\beta \sum \left(\frac{\mu_\lambda}{1 - R_{\lambda\lambda}\alpha_{\lambda\lambda}} \right)^2 \frac{\varepsilon'}{\varepsilon' + (1 - \varepsilon')S_{\lambda\lambda}} \mathbf{E}_a \tag{36}$$

In this model the microscopic expression for the molecular moment $\langle \mathbf{p} \rangle_{\varepsilon'}$ is

$$\langle \mathbf{p} \rangle_{\varepsilon'} = \frac{1}{3} \sum_{\lambda} \left[\frac{\alpha_{\lambda\lambda}}{1 - R_{\lambda\lambda}\alpha_{\lambda\lambda}} + \left(\frac{\mu_\lambda}{1 - R_{\lambda\lambda}\alpha_{\lambda\lambda}} \right)^2 \right] \frac{\varepsilon'}{\varepsilon' + (1 - \varepsilon')S_{\lambda\lambda}} \mathbf{E}_a \tag{37}$$

For simplicity it is assumed that the polarizability tensor is symmetric and that its main axes are the main axes of the ellipsoid. The Onsager approximations are in this case

$$\alpha_{\lambda\lambda} = \frac{\varepsilon_\infty - 1}{1 + (\varepsilon_\infty - 1)S_{\lambda\lambda}} \frac{abc}{3} \quad \text{and} \quad \frac{abc}{3} = \frac{1}{4\pi\rho} \tag{38}$$

With these approximations, Eq. (36) gives

$$\langle \mathbf{m} \rangle_{\varepsilon'} = \frac{1}{3}\beta \sum_{\lambda} \left[\frac{\mu_\lambda[1 + (\varepsilon_\infty - 1)S_{\lambda\lambda}]}{\varepsilon' + (\varepsilon_\infty - \varepsilon')S_{\lambda\lambda}} \right]^2 \varepsilon'[\varepsilon' + (1 - \varepsilon')S_{\lambda\lambda}]\mathbf{E}_a \tag{39}$$

so that the relation between the molecular dipole moment and ε, ε', and ε_∞

is

$$\frac{\varepsilon - \varepsilon_\infty}{3} \frac{1}{3} \sum_\lambda \frac{\varepsilon' + (1 - \varepsilon') S_{\lambda\lambda}}{[\varepsilon' + (\varepsilon - \varepsilon') S_{\lambda\lambda}][\varepsilon' + (\varepsilon_\infty - \varepsilon') S_{\lambda\lambda}]}$$

$$= \frac{4\pi}{3} \rho\beta \frac{1}{3} \sum \left(\frac{\mu_\lambda [1 + (\varepsilon_\infty - 1) S_{\lambda\lambda}]}{\varepsilon' + (\varepsilon_\infty - \varepsilon') S_{\lambda\lambda}} \right)^2 [\varepsilon' + (1 - \varepsilon') S_{\lambda\lambda}] \qquad (40)$$

If the permanent dipole is lying in the direction of one of the main axes, the sum in Eq. (39) contains only one term. We now assume that the dielectric constant takes the same values inside and outside the molecule. Equation (40) then simplifies to

$$\frac{\varepsilon - \varepsilon_\infty}{3} \frac{1}{3} \sum_\lambda \frac{\varepsilon + (1 - \varepsilon) S_{\lambda\lambda}}{\varepsilon[\varepsilon + (\varepsilon_\infty - \varepsilon) S_{\lambda\lambda}]} = \frac{4\pi}{3} \rho\beta \frac{\mu^2}{3} \left(\frac{1 + (\varepsilon_\infty - 1) S_{\mu\mu}}{\varepsilon + (\varepsilon_\infty - \varepsilon) S_{\mu\mu}} \right)^2 [\varepsilon + (1 - \varepsilon) S_{\mu\mu}]$$

$$(41)$$

which is a generalization of the Onsager model for nonspherical molecules. For spherical molecules, Eq. (41) leads to the Onsager equation:

$$\frac{\varepsilon - \varepsilon_\infty}{(\varepsilon_\infty + 2)^2} \frac{2\varepsilon + \varepsilon_\infty}{\varepsilon} = \frac{4\pi}{3} \rho\beta \frac{\mu^2}{3} \qquad (42)$$

The Böttcher–Scholte model is obtained if on the left-hand side of Eq. (41) S is approximated by the value of the sphere:

$$\frac{\varepsilon - \varepsilon_\infty}{3} \frac{2\varepsilon + 1}{2\varepsilon + \varepsilon_\infty} = \frac{4\pi}{3} \beta \frac{\mu^2}{3} \rho \left(\frac{1 + (\varepsilon_\infty - 1) S_{\mu\mu}}{\varepsilon + (\varepsilon_\infty - \varepsilon) S_{\mu\mu}} \right)^2 [\varepsilon + (1 - \varepsilon) S_{\mu\mu}] \qquad (43)$$

In the Onsager model, the cavity contains only one particle, so contributions to the intermolecular correlations due to the reaction field are not present [see Eq. (32)]. This is why the Onsager model does not give the same fluctuation expression obtained for a sample *in vacuo*. If the sample immersed in a medium contains a small number of particles, the situation becomes very complicated. The neglected irreducible components of the fluctuations [see Eqs. (26), (29), and (32)] then need to be taken into account. The reaction field and the cavity field become dependent on the size of the sample[45] and on the positions of the particles,[46] leading to multipole contributions from the reaction field.[47] This should also be taken into account in models involving one particle, since only for the Onsager model of a sphere with a central point dipole and point polarizability does the applied reaction-field expression hold exactly.[48] In the generalization of this model to ellipsoidal molecules, the orientation of the molecules with respect to the field has been taken into account. The Böttcher–Scholte model agrees with our result in the case where there are small deviations from the spherical shape.

E.   Universality of the Clausius–Mossotti Equation

In the preceding sections it was shown that the polarizability of a macroscopic sample in a medium with respect to all external charges, including the reaction-field contributions, is the same as that of a sample interacting *in vacuo* with the applied field. Now we will show that an expression of the form of the Clausius–Mossotti equation holds quite generally, independently of sample shape, for a sample interacting *in vacuo* with a uniform field.

We take advantage of the cluster structure of the distribution functions in the evaluation of the averages $\langle A_1 \rangle_0$ and $\langle m_1 \Sigma m_k \rangle_0$ in the fundamental equation (7). From the analysis of the interaction patterns and of the structure of the distribution functions[15,18] it follows that the two averages may be written as sums of products of partial averages. The most long-range partial average is

$$\langle T \rangle = \rho \int g(r) T(t)\, d\mathbf{r} \tag{44}$$

where $g(r)$ is the radial distribution function.

Equation (7) can be expanded in terms of $\langle T \rangle$ and partial averages $\langle A_1 \rangle^*$ and $\langle m_1 \Sigma m_k \rangle^*$ that are free of averages in which the only connection between two partial averages is a $\langle T \rangle$ link. The averages are therefore short range and independent of sample shape. The average moment $\langle p_1 \rangle$ can now be written

$$\langle p_1 \rangle = [\mathbf{x} - \mathbf{x} \cdot \langle T \rangle \cdot \mathbf{x} + \mathbf{x} \cdot \langle T \rangle \cdot \mathbf{x} \cdot \langle T \rangle \mathbf{x} - \cdots] \cdot \mathbf{E}^0 \tag{45}$$

where $\mathbf{x} = \beta \langle m_1 \Sigma m_k \rangle^* + \langle A_1 \rangle^*$, and summed to

$$\langle p_1 \rangle = \left(\beta \left\langle m_1 \sum m_k \right\rangle^* + \langle A_1 \rangle^* \right) \cdot \left[\mathbf{E}^0 - \langle T \rangle \cdot \langle p_1 \rangle \right] \tag{46}$$

The summation of Eq. (45) is possible if the system is uniformly polarized and the series is convergent. This implies that the number of particles determining the average $\langle \ldots \rangle^*$ is small compared with the number of particles in the system, so that the finite series can be approximated by an infinite geometric series.

To evaluate the average $\langle \mathbf{T} \rangle$, we write the radial distribution function as a sum of a step function $g_0(r)$ and a function $g_{00}^{000}(r)$ that vanishes at the origin and at large separations. The contribution to $\langle \mathbf{T} \rangle$ determined by $g_0(r)$ is

$$\rho \int g_0(r) \mathbf{T}(\mathbf{r}) \, d\mathbf{r} = -\rho \int_{V - (4\pi\sigma^3/3)} \nabla\nabla \frac{1}{\mathbf{r}} \, dr$$

$$= 4\pi\rho \left(\mathbf{S}^* - \tfrac{1}{3}\mathbf{1} \right) \qquad (47)$$

where $\mathbf{S}^*$ is a tensor depending on the shape and is identical with the tensor $\mathbf{S}$ given in Eq. (11). For a spherical sample, Eq. (47) gives no contribution. The integral determined by g_{00}^{000} gives no contribution either, since one may show that all terms cancel using a stepwise evaluation of the integral. Because the value of the integral in Eq. (44) is independent of the value of the cutoff diameter σ this is in principle an arbitrary quantity. Returning to Eq. (46), the field $\mathbf{E}^0$ is expressed in terms of the Maxwell field $\mathbf{E}$ by using Eq. (10) and applying the result (47). We find that the shape factors cancel and that the square bracket is Eq. (46) represents the Lorentz field, which is the first approximation to the local field independent of the boundary conditions:

$$\mathbf{E}^0 - \langle \mathbf{T} \rangle \cdot \langle \mathbf{p}_1 \rangle = \mathbf{E} + \frac{4\pi}{3} \rho \langle \mathbf{p}_1 \rangle \qquad (48)$$

The Lorentz field results from the long-range contributions to the polarization. It is not obtained from a model for the local field as in the Debye theory. We recall that the Maxwell field is given by the applied field and a shape-dependent contribution due to the surface polarization. All shape dependence is taken into account in the Maxwell field, so that a formulation in terms of the Maxwell field is shape independent.

The dielectric constant is determined by short-range averages in an equation of the form of the Clausius–Mossotti equation:

$$\frac{\varepsilon - 1}{\varepsilon + 2} = \frac{4\pi}{3} \rho \left[\beta \frac{1}{3} \left\langle \mathbf{m}_1 \cdot \sum \mathbf{m}_k \right\rangle^* + \langle A_1 \rangle^* \right] < 1 \qquad (49)$$

The inequality in Eq. (49) is a necessary condition for the convergence of the series in Eq. (45). It restricts the validity of the theory to ordinary dielectrics and excludes materials with nonlinear dielectric responses, such as ferroelectrics and conductors. Equation (49) was given by Nienhuis and Deutch[12] and Ramshaw[14] for rigid dipoles, and by Wertheim[15] for this general case. It holds rigorously only for uniformly polarized systems, whereas in general fields, multipole contributions may contribute to the free energy of a capacitance and therefore to the experimentally determined value of ε.[19]

The averages $\langle \ldots \rangle^*$ are partial averages that are free of the long-range $\langle \mathbf{T} \rangle$ contributions and agree only in the case of a spherical sample *in vacuo* with the average $\langle \ldots \rangle_0$, since then $\langle \mathbf{T} \rangle = 0$. The density expansion of $(\varepsilon - 1)/(\varepsilon + 2)$, known as the dielectric virial expansion, may therefore be regarded not only as the density expansion of the averages $\langle \ldots \rangle_0$ of a spherical sample *in vacuo*, but also as the density expansion of the averages $\langle \ldots \rangle^*$, which are independent of the macroscopic boundary conditions. It is worth noting that the accuracy of any calculation of ε is limited by the accuracy of the Clausius–Mossotti function, since it contains in the average $\langle \ldots \rangle^*$ the specific features of the system.

II. RIGID DIPOLES

A. General Theory

The most important model of a polar molecule is the rigid dipole. Because of the absence of the polarizability, many-body interactions are not present and the properties of this model fluid can be calculated using standard methods of statistical mechanics based on two-body interactions. Since the influence of the boundary conditions on the dipole fluctuations is now well understood also in the case of periodic boundary conditions,[32,33,43,44] the simulation methods allow reliable estimates to be made of the dielectric permittivity. There still exist, however, some technical problems. In particular, the reliability of these calculations is determined by the number of particles considered in the sample.[45]

The analytical theories allow the calculation of fluctuation expressions that are freed of the long-range components $\langle \mathbf{T} \rangle$, as has been outlined in Section I.F. The reliability of these calculations is determined by the approximations involved and has to be judged by comparison with the results of simulations and with rigorous results. It is obvious that the analytical theories for hard-sphere dipoles, such as the linearized hypernetted chain (LHNC) theory[49] and the mean spherical model (MSM)[50] fail at low densities, since they are based on a high-temperature approximation of the Meyer f-functions to the first order in β. They neglect entirely the second dielectric virial coefficient B, which makes necessary an expansion at least to the third order in β.[51] Since B is positive for hard-sphere dipoles, both theories must predict small values of ε at low densities. By comparison with simulations[13,33] it is found that at high densities the LHNC theory nevertheless predicts values of ε that are too large, whereas the values predicted by the MSM remain too small. Both theories predict much larger values of ε than does Onsager's macroscopic model, which also neglects the second dielectric virial coefficient but seems to model quite well the behavior of real polar fluids, as has been pointed out by Cole.[52]

Padé approximants based on the first virial coefficients have been successfully applied to describe the equation of state of hard-sphere fluids,[53] and have proved useful for the estimation of thermodynamic properties of hard-sphere dipole fluids.[54,55] In this work the cluster expansion of the dipole correlation factor is approximated by a self-consistent Padé approximant in order to gain physical insight and to trace which terms determine the dielectric response. This approximant can be regarded as a generator of the graphs in the various theories, and can be corrected in a systematic manner for the terms not considered in the hard-sphere dipole theories mentioned above and also for contributions due to the nonspherical molecular shape.[18]

From Kirkwood's theory[4] the average molecular moment $\langle \mu \rangle$ in a sample of regular shape containing rigid dipoles interacting *in vacuo* with a uniform field $\mathbf{E}^0$ is, to the first order in the field, given by

$$\langle \mu \rangle = \frac{\beta}{3} \left[\mu^2 + \frac{\rho}{(8\pi^2)^2} \int \mu(\Omega_1) \cdot \mu(\Omega_2) h^{(2)}(\Omega_1, \Omega_2, \mathbf{r}_{12}) \, d\Omega_1 \, d\Omega_2, d\mathbf{r}_{12} \right] \mathbf{E}^0 \tag{50}$$

where $\mathbf{r}_{12} = \mathbf{r}_2 - \mathbf{r}_1$, ρ is the number density, and $h^{(2)}$ is the specific, generally angular-dependent part of the pair correlation function. Using the Ornstein–Zernicke equation, Eq. (50) can be reformulated into the self-consistent equation[14,56]

$$\langle \mu \rangle = \frac{\beta \mu^2}{3} \left[\mathbf{E}^0 + \frac{4\pi}{3} \rho \delta \cdot \langle \mu \rangle \right] \tag{51}$$

The tensor δ is defined by the correlation integral (52) involving the direct correlation function C[14,56]

$$y \frac{\mu^2}{3} \delta = \beta \frac{\rho}{(8\pi^2)^2} \int \mu_1(\Omega_1) \mu_2(\Omega_2) C(\Omega_1, \Omega_2, \mathbf{r}_{12}) \, d\Omega_1 \, d\Omega_2 \, d\mathbf{r}_{12} \tag{52}$$

where

$$y = \frac{4\pi}{3} \rho \beta \frac{\mu^2}{3}$$

Equation (52) holds rigorously provided the dipole axis is a symmetry axis.

For less symmetrical molecules, δ is an ordered product[18,57] of similar integrals involving the components $C_{\nu,\mu}^{110}$ of the rotational invariant expansion[58,59] of the direct correlation function, where μ and ν take the values 0 or 1. Working out the most long-range part of the direct correlation function, which is

$$-\beta\mu(\Omega_1)\mathbf{T}(\mathbf{r}_{12})\mu(\Omega_2)g_0(r) \tag{53}$$

Eq. (51) may be rewritten in terms of the Lorentz field and a correlation tensor δ^*. This factor is determined by the C^{110} contributions of the direct correlation function. The corresponding integral in Eq. (52) is convergent and therefore independent of the shape of the sample. The Lorentz field is the first systematic approximant to the local field.

$$\langle\mu\rangle = \frac{\beta\mu^2}{3}\left[\mathbf{E}_L + \frac{4\pi}{3}\rho\delta^*\cdot\langle\mu\rangle\right] \tag{54}$$

In isotropic systems all the components of the diagonal tensor δ^* have the same value, so that the tensor notation may be relaxed. The relation between the macroscopic value of ε and the molecular properties is[14]

$$\frac{\varepsilon-1}{\varepsilon+2} = y\frac{1}{1-y\delta^*} \tag{55}$$

To calculate ε, an accurate estimate of δ^* is required. Since the correlation factor δ^* is determined by the direct correlation function C and not by its convolution h, as the Kirkwood g factor in Eq. (50) is, δ^* is more closely related to gas-phase properties than g is, its density dependence is more rapidly convergent, and the evaluation of the cluster series is simpler.

B. Structure of the Direct Correlation Function

To evaluate δ^*, as well as to gain insight into the mechanism of the intermolecular correlations, it is necessary to discuss the cluster expansion of the direct correlation function.[60-62] For convenience we recall some basic definitions. Denoting the particles by dots and the Meyer f-functions by lines, the two-particle direct correlation function can be represented by all unlabeled connected and cut-point-free graphs, with two white points or terminals denoting the fixed particles "one" and "two" and with any number of field points denoted by black dots. Integration is carried out over all positions and orientations of the field particles which have the weight of the density. A cut point is the only connection between two subgraphs. Cut-point-free graphs are termed irreducible. Among the cut points, we distinguish bridge points and articulation points. A bridge point connects two subgraphs both of which contain a terminal. An articulation point connects two subgraphs one of which contains no terminal. The graphs of the correlation function h are obtained if bridge points are admitted. The first graphs of the cluster expansion of the direct correlation function are:

$$C = \circ\!\!-\!\!\circ + \triangle + \square + \boxslash + \boxtimes +$$

$$+ \tfrac{1}{2} \diamond\!\!\!\!\diamond + \tfrac{1}{2} \diamond\!\!\!\!\diamond + \tfrac{1}{2} \diamond\!\!\!\!\diamond + \tfrac{1}{2} \diamond\!\!\!\!\diamond \qquad (56)$$

The numerical factors are the reciprocals of the symmetry number, which is the number of possible permutations of black dots in the labeled graph that do not change the connections between the particles. Use of the symmetry number ensures that a physical cluster is counted only once.

The graphs of the n-particle distribution function[62-64] are closely related to the graphs of the direct correlation function. They are defined by all irreducible graphs with n fixed points and any number of field points. With the exception of the terminals, which are connected by e-bonds, all connections are f-bonds. The e-bonds and the f-bonds are defined by the Boltzmann factor of the two-particle interaction energy:

$$f = e - 1, \qquad e = \exp[-\beta u] \qquad (57)$$

Since the liquid structure is determined to a large degree by repulsive interactions,[65] it is convenient to separate the interaction potential into two parts: the repulsive potential u_0 and the attractive potential u_1. Then the f-functions may also be separated. In the case of hard-sphere dipoles, this separation reads

$$f = f_0 + f_1(1 + f_0),$$
$$f_1 = \exp[-\beta \mu_1 \cdot \mathbf{T} \cdot \mu_2] - 1 \qquad (58)$$

From the definition of the direct correlation function and the n-particle distribution function, it follows that the integral of the direct correlation function consists of a pure hard-sphere term and irreducible averages of graphs consisting of f_1-bonds[13,18,66] ("irreducible" denotes the fact that the averages cannot be reduced to products of independent averages). The averages are determined by distribution functions of the hard-sphere reference system.

A complete evaluation of the direct correlation function is not possible, even in the simplest case, that of hard spheres. In analytical theories, therefore, only certain groups of graphs are evaluated. In the cases of orientation-dependent potentials, the problem is even more complicated, since in a rotationally invariant expansion in terms of Wigner matrices,[58,59] f_1 is represented by an infinite set of different symmetry components. Analytical theories such as the MSM theory[50,66] and the LHNC theory[49] are based on the high-temperature approximation of f_1 in which only the linear term in β of the expansion of f_1 is considered. Symmetry requires that only terms containing chains or bundles of chains of dipole interactions ending at terminals and closed loops contribute to δ^*. Neglecting the loops that modify the radial distribution function, the direct correlation function can be approximated by a Meyer f-function involving a shielded dipole interaction $\mu_1\mathscr{T}_{12}\mu_2$;

$$C(1,2) = C_{HS}(r) + \langle\langle \exp[-\beta\mu_1\mathscr{T}_{12}\mu_2]\rangle\rangle_{12} - 1 \tag{59}$$

Another useful representation is

$$C(1,2) = C_{HS}(r) - \beta\mu_1\langle\langle \mathscr{T}\rangle\rangle_{12}\mu_2 + C'(1,2) \tag{60}$$

The exponential takes into account exactly the symmetry number representing the interchangeability of the dipole chains. The double bracket indicates that the average $\langle\langle \rangle\rangle$ is irreducible:

$$\langle\langle \mathscr{T}\rangle\rangle = \rho \int \langle\langle \mathscr{T}\rangle\rangle_{12} \, d\mathbf{r}_{12} \tag{61}$$

The average contains no terms that are products of independent averages. The averaging involves distribution functions of the hard-sphere reference system. The shielded dipole–dipole interaction represents the series

$$\mu_1\mathscr{T}_{12}\mu_2 = \mu_1 T_{12}\mu_2 - \mu_1 T_{13}\beta\overline{\mu_3\mu_{3'}} T_{3'2}\mu_2 + \cdots \tag{62}$$

where

$$\overline{\mu_1\mu_{1'}} = \beta\frac{\mu^2}{3} + \mu_1 h'(1,1')\cdot\mu_{1'} \tag{63}$$

Here h' is determined by the Ornstein–Zernicke equation involving only $C'(1,2)$. Equation (60) enables us to reformulate Eq. (50) as

$$\langle\mu\rangle = \frac{\beta\mu^2}{3} \frac{1}{1 - y\delta'} [\mathbf{E}_L + \langle\langle \mathscr{T}\rangle\rangle * \langle\mu\rangle] \tag{64}$$

The square bracket can be regarded as a microscopic expression of the local field. δ' is a short-range correlation factor defined by Eq. (52) but involving only C'. The asterisk indicates that the shape-dependent contribution $\langle T \rangle$ to $\langle\langle \mathscr{T} \rangle\rangle$ is taken into account in the Lorentz field $\mathbf{E}_L$.

The renormalized Boltzmann factor in Eq. (59) is a fairly complete representation of the direct correlation function. It includes the MSM and the LHNC, which are linear in $\mathscr{T}$ and provide approximate values of $\langle\langle \mathscr{T}\rangle\rangle^*$. Equation (64) includes the second dielectric virial coefficient, which is neglected in the MSM and the LHNC theory.[51] It provides a tool for systematic improvement of these analytical theories. It is free of artificial terms contained in the QHNC theory.[68,69] Finally, it is a nice physical interpretation that the part of the direct correlation function determining the dipole–dipole correlation is an averaged Boltzmann factor involving a shielded dipole–dipole interaction.

C. The Mean Spherical Model and the Linearized Hypernetted Chain Theory

In what follows we will show that the MSM and the LHNC theories are indeed merely different approximations of $\langle\langle \mathscr{T} \rangle\rangle^*$ and neglect C' of Eq. (60) entirely. The pair correlation function $g(1,2)$ may generally be written in terms of the potential of the average force $\psi(1,2)$:

$$g(1,2) = \exp[-\beta\psi(1,2)] \tag{65}$$

where

$$\psi(1,2) = u(1,2) + \frac{1}{\beta}(b(1,2) + d(1,2))$$

Here b represents all two-terminal graphs of the correlation function h with at least one bridge point. It is the integral in the Ornstein–Zernicke (OZ) equation. The term d represents the $(1,2)$ irreducible graphs that have no bridge points and do not decompose into independent chains if the graphs are cut at the terminals. In the analytical theories the d graphs are generally not taken into account. This is the only approximation in the hypernetted chain theory; in the Percus–Yevick (PY) approximation only the first-order contribution of b to g is considered. In the LHNC theory u and b are separated into spherical and orientation-dependent parts (u_1, b_1). The pair correlation function is expanded into the nonspherical components of the potential of the average force, keeping the linear contributions only. The pair correlation function is then

$$g(1,2) = g(r) - \beta\mu_1 \cdot \mathbf{T}_{12} \cdot \mu_2 g(r) + b_1 g(r) \tag{66}$$

The nonspherical part C_1 of the direct correlation function is

$$C_1(1,2) = -\beta\mu_1 \cdot T_{12} \cdot \mu_2 g(r) + b_1 h(r) \tag{67}$$

where $g(r)$ and $h(r)$ are the radial distribution functions of the hard-sphere reference system. The first term vanishes if $r < \sigma$. Denoting h by a line and $-\beta\mu T\mu \cdot g$ by a wavy line, the iterative solution of the OZ equation generates the following graphs representing C_1:

$$\tag{68}$$

These are chains of dipole–dipole interactions where the first and the last particle are correlated by an h-bond. Nonadjacent particles may be correlated by noncrossing h-bonds. The graphs represent a shielded dipolar interaction averaged using distribution functions of the hard-sphere reference system based on the Kirkwood superposition approximation generalized for more than three particles. The h-bond between the terminals ensures that the averages are irreducible. Beyond the superposition approximation, terms such as

$$\tag{69}$$

should be considered. They are contained in the formulation of Eq. (59). The graphs of the order T^2 have been calculated by Rushbrooke.[70] In a PY theory, the direct correlation function contains only terms linear in b, so that

$$C_1(1,2) = -\beta\mu_1 T_{12}\mu_2 g(r) + b_1(1,2) f_0(r) \tag{70}$$

The second term vanishes for $r > \sigma$. The iteration of the OZ equation generates the same graphs as are given in Eq. (68); the lines, however, now represent an f_0-bonds.

The mean spherical model is defined by the closure

$$C_1(1,2) = -\beta\mu_1 T_{12}\mu_2 \quad \text{if } r > \sigma \tag{71}$$

so that Eq. (67) now reads

$$C_1(1,2) = -\beta\mu_1 T_{12}\mu_2 g_0(r) + b_1(1,2) \cdot f_0(r) \tag{72}$$

where g_0 is the zero-density limit of the hard-sphere radial distribution function. Again the iteration of the OZ equation leads to the same graphs, but the wavy lines now represent $-\beta\mu T\mu g_0$. In other words, in the mean spherical model the average $\langle\langle \mathscr{T} \rangle\rangle$ is evaluated by approximating the higher distribution functions by a generalized superposition approximation based on the zero-density limit of $g(r)$.

The calculation of the dielectric polarization requires the estimation of the density expansion of certain correlation integrals, Eqs. (50)–(52). The calculation of the correlation functions h or C is not necessary. It has been proved for thermodynamic calculations[53-55] that Padé approximants allow quite accurate extrapolation on the basis of few exactly known coefficients. This method is applied here in order to reproduce the results of the MSM and the LHNC theory with the aim of tracing the determining terms and investigating the influence of exactly known contributions not considered in these theories [see Eq. (67), (70), and (72)].

The first contribution to the shielded dipolar interaction in the mean spherical model is

$$\frac{4\pi}{9}\beta\frac{\rho^2}{(8\pi^2)^2}\int\mu_1\left(\triangle\right)\mu_2\,d\Omega_1\,d\Omega_2\,d\mathbf{r}_2\,d\mathbf{r}_1 = -\tfrac{15}{16}y^3 \tag{73}$$

The calculation of this expression using the Fourier technique given by Jepsen[71,72] shows that the numerical factor depends on the size of the field particle. If the field particle is a point particle, the numerical factor is -2, whereas doubling or tripling the size of the field particle reduces this value to -0.736 and -0.183, respectively. An increase in the size of the field particle reduces the shielding of the dipolar interaction.

The irreducible four-body term of the MSM theory is

$$\frac{4\pi}{9}\rho^2\frac{\beta}{(8\pi^2)^2} = \int\mu_1\left\{\square + 2\boxslash\right\}\mu_2\,d\Omega_1\,d\Omega_2\,d\mathbf{r}_1\,d\mathbf{r}_2 = \tfrac{7}{12}y^3 \tag{74}$$

which is obtained by integrating the contraction of two convolutions that are both proportional to the dipole–dipole interaction.

If all field particles in the interaction chain are treated as point particles, if correlations between nonadjacent particles are neglected, and if only second-rank contractions are considered then, we find that the following approximation holds:

$$\delta^* = -2y\frac{1}{1+y} \tag{75}$$

This suggests that δ^* may be approximated also in the case of hard-sphere dipoles by a geometric series based on the first two terms given in Eq. (73)

and (74). To take into account correlations between nonadjacent field particles, subdiagrams such as $\triangle$ and $\square$ are treated as effective particles with diameters larger than the hard-sphere diameter. This reduces the value of the cluster integrals. For simplicity it is assumed that in all clusters the size of the effective particle reduces the values of the cluster integrals by the same factor. The correlation factor δ is now determined by the self-consistent set of equations

$$\delta* = -\frac{15}{16}x\frac{1}{1+(28/45)x}, \qquad x = y\frac{1}{1-\gamma\delta*} \tag{76}$$

Taking the size factor γ as equal to $\frac{1}{3}$, which corresponds to an effective-particle size slightly more than twice the size of the real particle, the calculated values of ε agree well with those obtained from Wertheim's solution for the MSM.[50]

If the effective field particles are treated as point particles, which is the continuum approximation, Eq. (76) reduces to

$$\delta* = -\frac{2}{2}x\frac{1}{1+x}, \qquad x = y\frac{1}{1-\delta*} = \frac{\varepsilon-1}{\varepsilon+2} \tag{77}$$

which, when solved for y, gives the Onsager–Kirkwood equation:

$$y = \frac{(\varepsilon-1)(2\varepsilon+1)}{9\varepsilon} \tag{78}$$

The cavity field, which is applied in the Onsager model as an approximation to the local field, is obtained from our self-consistent Padé approximant to the MSM by treating groups of short-range correlated particles as point dipoles. This is the continuum approximation.

In Patey's LHNC theory[49] the f_0-bonds are replaced by h-bonds. Using the linear term of the density expansion of the hard-sphere radial distribution function, Rushbrooke[70] showed that

$$\frac{4\pi}{3}\left(\frac{\rho}{8\pi^2}\right)^2\int d\Omega_1\,d\Omega_2\,d\mathbf{r}_1\,d\mathbf{r}_2|\mu_1\{\triangle+2\;\triangle\!\!\!\!\slash+\;\diamondsuit+\diamondsuit$$

$$+\;\diamondsuit+\diamondsuit\}\cdot\mu_2$$

$$= -\frac{15}{16}y^3(1+2\times0.4014\rho*-0.7836\rho*-0.05272\rho*) \tag{79}$$

where $\rho* = \rho\cdot\sigma^3$. The contribution of the last two graphs is very small. They may be regarded as averages involving those parts of the three-particle distribution function that are not considered in the superposition approximation, and therefore not in the LHNC theory either. In any case, the correction terms in Eq. (79) essentially cancel. The correction to the result of the MSM theory is therefore very small. Equation (79) shows, however, that the density-dependent corrections to the various graphs may nevertheless be considerable. The contribution due to a $\mathbf{T} \cdot g(r)$ bond is enhanced if the density dependence of the radial distribution function is taken into account by means of a factor $c = 1 + 0.4014\rho^*$. Correcting the higher terms in the expan-

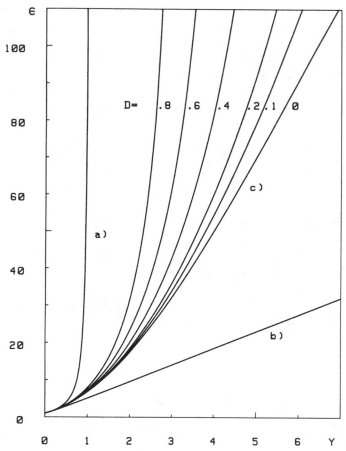

Figure 1.   Calculated values of ε of a dipolar hard-sphere fluid from (a) the Debye theory, (b) the Onsager model, and (c) the renormalized cluster expansions that approximate the MSM theory ($D = 0$) and the LHNC theory ($D = \rho\sigma^3 = 0.1, 0.2, 0.4, 0.6, 0.8$).

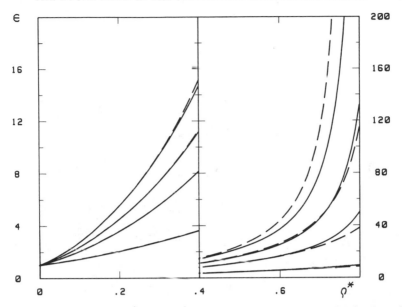

Figure 2. The value of ε of a dipolar hard sphere fluid as a function of a density, calculated using the cluster approximant of the LHNC theory (- - -) and Patey's[69] empirical approximant (——); $\beta\mu^2/\sigma^3 = 1, 2, 2.5, 3$; $\rho^x = \rho\sigma^3$.

sion of δ^* by multiplying the factor $28/45$ in Eq. (76) by the factor c, we obtain results that agree well with the predictions of the LHNC theory.

Figure 1 shows values of ε calculated as a function of y by using the density expansions that approximate the MSM and LHNC theory. Like the LHNC theory, our approximant reduces to the MSM result at low densities. The figure shows also that the Onsager theory predicts much smaller values of ε than do the microscopic models, since it overestimates the shielding of a dipolar interaction by a third particle. All three theories agree at small values of y with predictions of the Debye model, which is the exact low-density limit. Fig 2 shows that the cluster approximant of the LHNC theory agrees well with Patey's exact solution, calculated here from a power series given by Patey as an empirical approximation.[69] Only at large values of y and at very high densities do the deviations become noticeable. In the representation of the Clausius–Mossotti function $(\varepsilon - 1)/(\varepsilon + 2)y$, shown in Fig. 3 the difference between the exact and the approximate LHNC theory is negligible. It is remarkable that in this representation the MSM and the LHNC theory also give very similar plots. This figure shows clearly that both models assume a zero value of the second dielectric virial coefficient, since all the curves start with initial slopes of zero.

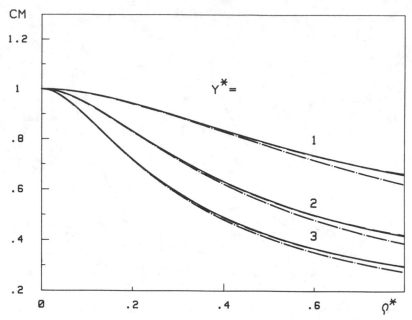

Figure 3. Clausius–Mossotti function of a dipolar hard-sphere fluid as calculated using the cluster approximant of the LHNC theory [Eqs. (76), (79)] (- - -), the MSM model (·····), and Patey's[69] empirical power series (——); $\beta\mu^2/\sigma^3 = 1, 2, 3$; $\rho^* = \rho\sigma^3$.

D. The Renormalized Second Dielectric Virial Coefficient of Hard-Sphere Dipoles

Since the MSM and the LHNC theory are based on the high-temperature approximations of the f_1-function, they cannot include the second dielectric virial coefficient, which is of a least third order in the dipolar interactions. The exact hard-sphere result was recently given by Buckingham and Joslin:[51]

$$B\rho^2 = \frac{4\pi}{9}\beta\left(\frac{\rho}{8\pi^2}\right)^2\int\mu_1\left[\exp(-\beta\mu_1 T_{12}\mu_2) + \beta\mu_1 T_{12}\mu_2 - 1\right]$$

$$\times\mu_2(1 + f_0)\,d\Omega_1\,d\Omega_2\,d\mathbf{r}_1\,d\mathbf{r}_2$$

$$= y^2\sum_{n=1}^{\infty}\frac{12}{n}(y^*)^{2n}\left(\frac{2^n(n+1)}{(2n+3)!}\right)^2\sum_{t=0}^{n}\frac{(3t-n)(2t)!}{t!2}$$

$$= y^2 y^{*2} 12\left(\frac{1}{300} + \frac{1}{3675}y^{*2} + \frac{1}{79380}y^{*3} + \cdots\right) \tag{80}$$

where

$$y^* = \frac{\beta\mu^2}{\sigma^3}$$

This is always positive, enhancing the value of the Clausius–Mossotti function. The importance of this contribution to δ^* is modified at higher densities, where f_0 should be replaced by h_0, the dipole interactions by shielded interactions, and y^* by some renormalized values, taking into account that shielded dipole interactions connected at the terminals may involve other short-range-correlated particles. Those corrections can be derived on the basis of the calculated third dielectric virial coefficient.[73] The leading term of the density expansion of the hard-sphere pair correlation function enhances the first term in Eq. (80) by a factor $F = 1 + 0.9289 \, \rho^*$. With the shielding factor of the reaction field, given in Section III, the dipolar interactions are reduced by a factor f:

$$f = \frac{1}{1 + \frac{15}{32}x} \tag{81}$$

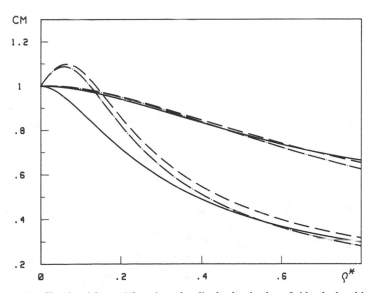

Figure 4.   Clausius–Mossotti function of a dipolar hard-sphere fluid calculated by taking into account the renormalized second dielectric virial coefficient. In the dashed curve, the density dependence of the radial distribution function is taken into account; this is neglected in the dotted and dashed curve. The unbroken curve is calculated from Patey's[69] empirical power series. $\beta\mu^2/\sigma^3 = 1, 3; \; \rho^* = \rho\sigma^3$.

The renormalized second dielectric virial coefficient now reads

$$B\rho^2 = y^2 a_2 = y^2 Ff(fx^*)^2 12\left[\frac{1}{300} + \frac{1}{3675}(fx^*)^2\right], \qquad x^* = \frac{x}{4\pi\rho^*/3}$$

(82)

where the x^* term takes into account that the terminals may be short-range-correlated to other particles.

The dielectric permittivity is determined using Eq. (55), where

$$\delta^* = a_2 - \frac{15}{16}x\left(1 + \frac{28}{45}xc\right)^{-1}, \qquad x = \frac{y}{1 - \frac{1}{3}\delta^*}$$

(83)

If the density dependence of the hard-sphere pair correlation function is neglected, as in the MSM, F and c take values of 1.

Figure 4 shows that consideration of a_2 leads to a pronounced maximum in the Clausius–Mossotti function at low densities. This is quite different

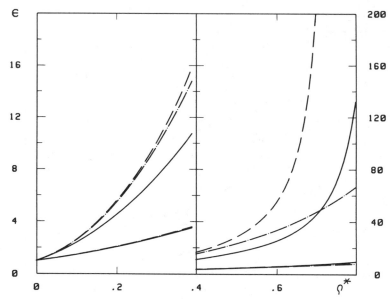

Figure 5. Values of ε for a dipolar hard-sphere fluid calculated by taking into account the renormalized second dielectric virial coefficient. In the dashed curve, the density dependence of the radial distribution function is taken into account; this is neglected in the dotted and dashed curve. The unbroken curve is calculated from Patey's[69] empirical power series. $\beta\mu^2/\sigma^3 = 1, 2.5$; $\rho^* = \rho\sigma^3$.

from the behavior predicted by the MSM or the LHNC theory. Again, the difference caused by the density dependence of the pair correlation function is fairly small in this representation. The enhancement of the dielectric permittivity by a_2 at low densities is reduced at higher densities. The behavior agrees then essentially with that predicted by the theories that entirely neglect this term. It is remarkable that the MSM corrected with the renormalized second dielectric virial coefficient predicts values of ε that are in excellent agreement with those from the molecular-dynamics calculations of Pollock and Adler.[33] This modification enlarges the deviations of the LHNC predictions from the molecular-dynamics results. As in the case of the equation of state of hard spheres, the simpler theory provides better results because of cancellation of errors (Fig. 5).

E. Nonspherical Dipoles

If the molecules have a nonspherical shape, the direct correlation function of the nonpolar reference fluid may have components that contribute to the correlation factor. There are further components that, contracted with the dipole–dipole interaction, contribute to the correlation factor.

Since such terms can exist even if the dipolar axis is not a polar axis with respect to the molecular shape and therefore enable the consideration of small deviations from sphericity, these contributions are the ones to start with in a perturbation theory.[74] In a rotationally invariant expansion[58,59] of the pair correlation function, the components of interest are g^{202}, g^{222}, and g^{022}. The correlation integral of Eq. (52) gives, in terms of S-functions,

$$-\frac{\rho}{8\pi^2}\beta^2\int\mu_1^2 T_{12}\mu_2^2\left[g_{00}^{202}S_{00}^{202} + g_{00}^{022}S_{00}^{022} + g_{00}^{222}S^{222}\right]d\Omega_1\,d\Omega_2\,dr_{12}$$

$$= \frac{4\pi}{3}\rho\beta\frac{\mu^2}{3}\beta\frac{\mu^2}{3}\left[6\times 5^{-3/2}\int\frac{1}{r}2g^{202}(r)\,dr - \frac{70^{1/2}}{125}\int\frac{1}{r}g^{222}(r)\,dr\right]$$

$$= y\frac{\beta\mu^2}{3}a_2' \tag{84}$$

In principle, a_2' is density dependent, because of the density dependence of the pair correlation function. At higher densities the shielding of the dipole–dipole interactions by the factor f must also be taken into account. To treat nonspherical molecules we need only to replace the hard-sphere contribution a_2 in Eq. (83) by the sum of a_2 and $a_2'f$:

$$a_2 \Rightarrow a_2 + a_2'f$$

In the low-density limit the pair correlation functions in Eq. (84) are just

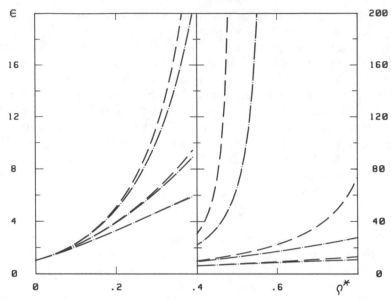

Figure 6.   Shape dependence of ε [see Eqs. (83), (84)]. For a disc shape ($a'_2 = -\frac{1}{3}$), ε is enhanced compared with a spherical shape ($a'_2 = 0$); it is reduced for rod-shaped molecules ($a'_2 = \frac{1}{3}$). In the dashed curves, the density dependence of the radial distribution function is taken into account; this is neglected in the dotted and dashed curves. $\beta\mu^2/\sigma^3 = 2$.

Meyer f-functions. For hard rods, g^{222} and g^{202} are both negative.[74-76] Because the integrals are of the same order of magnitude, it follows that in this case a'_2 is negative. For hard discs, g^{202} is positive, so a'_2 may become positive too. The attractive dispersion interactions give components in which all the signs are reversed. However, the pair correlation functions are essentially determined by the repulsive interactions and modified only slightly by attractive interactions. We conclude that a rod shape will reduce the dielectric constant in comparison with that of the hard-sphere dipole, whereas a disc shape causes an enhancement. This is well known for real gases.[75]

Because the components g^{202} and g^{222} are essentially determined by the molecular shape, a'_2 is expected to be temperature independent to a good approximation. Figure 6 gives the ε values for different molecular shapes of a rigid dipole fluid as a function of the density; it is apparent how strongly molecular shape influences ε. The figure shows that the influence of the density dependence of the radial distribution function of the reference system is relatively unimportant. This influence is almost negligible in the representation of the Clausius-Mossotti function in Fig. 7. The quadrupole interactions give contributions to the components g^{202} and g^{222} of the pair corre-

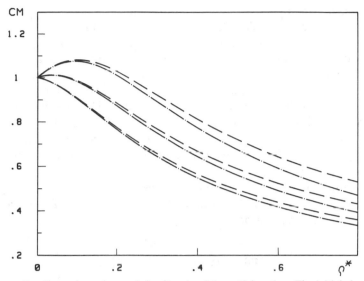

Figure 7.   Shape dependence of the Clausius–Mossotti function: The initial slope is positive for disc-shaped and spherical molecules, but negative for rodlike molecules. The parameters and symbols are given in Fig. 6

lation function that are of second order in β. Such terms enhance ε and counteract the reduction of ε predicted by the LHNC theory when quadrupole–quadrupole and dipole–quadrupole interactions are included in first order only.[69] The molecular shape, which is rodlike in most polar molecules, and the quadrupole interactions are probably the reasons why real dielectric fluids are quite well described in terms of the Onsager model,[52] which for the model of hard-sphere dipole fluids predicts values of ε that are too small.

III.   DIELECTRIC POLARIZATION OF NON-POLAR FLUIDS

A.   Graph-Theoretical and Algebraic Analysis of Dielectric Polarization

In the Kirkwood–Yvon theory,[8,77] the average moment $\langle \pi \rangle$ of a molecule induced by an external field $\mathbf{E}^0$ acting on a sample *in vacuo* is

$$\langle \pi \rangle = \langle \mathbf{A} \rangle \cdot \mathbf{E}^0 \tag{85}$$

where $\mathbf{A}$ is a generalized polarizability, defined in Eq. (3). In Eq. (85) it is assumed that the dielectric polarization of nonpolar fluids can be described

by a classical dipole–dipole interaction mechanism involving the gas-phase
molecular polarizability. The contributions of higher multipole interactions,
the hyperpolarizability, and short-range repulsion to the molecular polarization are not taken into account. Such effects modify the polarizability of
molecular pairs, causing at large separations an increase and at short separations a reduction of the pair polarizability.[78] These effects might be taken
into account in a theory based on Eq. (85) by introducing a density-dependent polarizability.[79] In a more systematic approach, these effects should be
considered explicitly, as in the work on collision-induced effects in light
scattering,[21,80,81] where mechanisms other than the DID mechanism have
been considered also.

As outlined in Section I the graphs representing $\langle A \rangle$ can be arranged as
the sum of chains of partial averages $\langle A \rangle^*$ connected by $\langle T \rangle$ bridges. The
average moment $\langle \pi \rangle$ can be written

$$\langle \pi \rangle = \langle A \rangle^* \left[E^0 - \langle T \rangle \cdot \langle \pi \rangle \right] \tag{86}$$

so that

$$\frac{\varepsilon - 1}{\varepsilon + 2} = \frac{4\pi}{3} \rho \langle A \rangle^* \tag{87}$$

If we denote T by a line and the polarizability α by a dot, all interaction
patterns contributing to A can be represented by graphs that can be drawn
without lifting pen from paper. Such graphs are called Eulerian graphs.[82]
Interaction patterns that start and end at the same particle are shown as the
one-terminal graphs. These terms represent self-induction or reaction-field
terms. The two-terminal graphs, in which the start and the end of the interaction concern different particles, represent shielded dipolar interactions. The
terminals are denoted by open dots. In the interaction patterns, two particles may be connected by various T-bonds; in other words, the interaction
patterns are Eulerian multigraphs.

The graphs of the n-particle averages contributing to $\langle A \rangle^*$ are obtained
by multiplying the interaction patterns by the graphs of the corresponding
n-particle distribution functions. It is therefore necessary to introduce different lines for the $T(1 + h)$-, T-, and h-bonds (wavy, dotted, and straight lines,
respectively).

The average $\langle A \rangle^*$ again is a sum of ordered products of partial averages
that form a chain and are connected by 1- and 2-cut points. (An n-cut point
is a bridge or articulation point where an interaction path is cut n times.[15])
The partial averages, which still contain articulation points and higher bridge

points, are the following:[18]

$$A_1 = \left\langle\!\left\langle \frac{\alpha_1}{1 \mid \Sigma' T_{1k} \alpha_k} \right\rangle\!\right\rangle$$

$$A_2 = \left\langle\!\left\langle \frac{\alpha_1 \Sigma T_{1k}}{1 + \Sigma' \alpha_k T_{kk'}} \right\rangle\!\right\rangle$$

$$A_3 = \left\langle\!\left\langle \frac{\Sigma T_{1k} \alpha_k}{1 + \Sigma' T_{kk'} \alpha_{k'}} \right\rangle\!\right\rangle$$

$$\langle\langle \mathscr{T} \rangle\rangle^* = \left\langle\!\left\langle \frac{\Sigma T_{1k} \alpha_k T_{kk'}}{1 + \Sigma' \alpha_{k'} T_{k'k''}} \right\rangle\!\right\rangle \tag{88}$$

The summations in the denominators are restricted by the requirement that the interaction patterns form Eulerian graphs, which is indicated by a prime. Among these averages one-terminal and two-terminal terms have to be distinguished.

$\langle A \rangle E^0$ can be summed to

$$\langle \pi \rangle = \frac{1}{1 + A_2} A_1 \frac{1}{1 + A_3} \left[E_L + \langle\langle \mathscr{T} \rangle\rangle^* \langle \pi \rangle \right] \tag{89}$$

The term in square brackets can be regarded as the local field acting on a generalized polarizability represented by the other terms. Equation (89) is a rigorous reformulation of Eq. (85). The validity of Eq. (89) is not restricted to the electrostatic case; it holds quite generally in the high-frequency region also. In this last case, it should be taken into account in Eq. (85) that E^0 varies in space and time and that the **T** tensor should be replaced by a retarded propagator. Following the procedure given in the literature,[83,84] it can be shown that in the dipole approximation the nonuniformity is taken into account in the Maxwell field and that in the irreducible averages $\langle\langle \mathscr{T} \rangle\rangle^*$ and $\langle A \rangle$ the propagator can be replaced by the **T** tensor, since these averages are determined by short-range correlation everywhere except in the vicinity of the critical region.

The one-terminal $\langle\langle \mathscr{T} \rangle\rangle^*$ factors that are connected by a 2-articulation point to the main graph represent the reaction-field tensor and are denoted $\langle\langle \overset{\circ}{\mathscr{T}} \rangle\rangle$. The Onsager reaction field is an approximation to this partial average taken from continuum electrostatics.

Since any particle may be the source of a set (blossom) of one-terminal graphs (decorations), it is convenient to introduce a renormalized polarizability α', which is a reaction-field-corrected effective polarizability:

$$\alpha' = \frac{\alpha}{1 - \langle\langle \mathscr{F} \rangle\rangle \alpha} \tag{90}$$

The averages given in Eq. (88) may be formulated in terms of α'. In this form the averages are free of 2-articulation points. With

$$\langle\langle \mathscr{T} \rangle\rangle^* = \langle\langle \bar{\mathscr{T}} \rangle\rangle + \langle\langle \mathring{\mathscr{F}} \rangle\rangle \tag{91}$$

and

$$A = \frac{1}{1 + A_2} A_1 \frac{1}{1 + A_2} \tag{92}$$

the dielectric permittivity is determined by

$$\frac{\varepsilon - 1}{\varepsilon + 2} = \frac{4\pi}{3} \rho \frac{A}{1 - \langle\langle \bar{\mathscr{T}} \rangle\rangle A} \tag{93}$$

For spherical molecules the polarizability has only zero-rank components, so that

$$A_2 = A_3 = 0 \quad \text{and} \quad A_1 = \alpha' = \tfrac{1}{3}\mathrm{Tr}\,\alpha' \tag{94}$$

Since $\langle\langle \mathscr{T} \rangle\rangle^*$ is at least of the order $\bar{\alpha}$, it follows from Eq. (93) that the density expansion of $(\varepsilon - 1)/(\varepsilon + 2)$ for spherical molecules contains no term of the order α^2, as has been shown already by Kirkwood.[77]

B. The Microscopic Fields

In terms of the renormalized polarizability α', the first graphs of the reaction-field tensor $\langle\langle \mathring{\mathscr{F}} \rangle\rangle$ and the shielded dipolar interaction $\langle\langle \bar{\mathscr{T}} \rangle\rangle$ are

$$\tag{95}$$

In these graphs all higher cut points are excluded [the single superchain (SSC) approximation].[15,85] This links the theory of dielectric polarization of nonpolar molecules to the analytical theories of hard-sphere dipoles. In the SSC approximation the graphs representing $\langle\langle \bar{\mathscr{T}} \rangle\rangle$ are the same as those taken into account in the approximations of the direct correlation function in the MSM and the LHNC theory.[49,50] The only differences are that $\beta\mu^2/3$ is replaced by α' and that in the rigid-dipole case distribution functions of the reference system are used, whereas in the case of polarizable molecules the distribution functions of the system considered are applied.

The approximation that allows the representation of $\langle\langle \mathring{\mathscr{T}} \rangle\rangle$ and $\langle\langle \bar{\mathscr{T}} \rangle\rangle$ by the graphs given above is reasonable if α/σ^3 is relatively small. In a density expansion of $\langle\langle \mathscr{T} \rangle\rangle^*$ the leading terms are of low order in α, which makes terms involving higher cut points and a higher connectivity small. Interaction patterns where two particles are connected by more than two shielded interaction chains are therefore neglected. By using this simplification, one avoids a difficult counting problem of graphs[86,87] that has not yet been solved. Whereas in the usual graphs in terms of two-body Meyer f-functions a graph is uniquely related to a certain cluster, this is not so in general for the graphs representing the induced interactions. A particular graph may represent different physical situations that are characterized by the number of ways a certain interaction pattern can be drawn. In fact, the number of directed graphs is required.[82,87] Only graphs containing n-cut points with $n \leq 2$ and where particles are connected by not more than two shielded dipolar interactions can be counted in the same way as the Meyer graphs, which are simple graphs.

Evaluating the first terms of the reaction-field series given in Eq. (95) on the basis of the low-density limit of the hard sphere radial distribution function, and using the Kirkwood superposition approximation, we find that[UU]

$$\mathring{\big\}} = \frac{4\pi}{3}\rho 2 \frac{\alpha}{\sigma^3} \quad \text{and} \quad \bigtriangledown = -\left(\frac{4\pi}{3}\rho\alpha\right)^2 \cdot \frac{1}{\sigma^3}\frac{15}{16} \qquad (96)$$

Replacing the field particles by effective field particles and assuming that the two terms are the first terms of a geometric series that gives the reaction field, we find that

$$\langle\langle \mathring{\mathscr{T}} \rangle\rangle = 2\frac{1}{\sigma^3}\frac{x}{1 + \frac{15}{32}x} \qquad (97)$$

The dielectric permittivity is now determined by the self-consistent set of equations

$$\frac{\varepsilon - 1}{\varepsilon + 2} = y' \frac{1}{1 - \alpha'\langle\langle \mathscr{T} \rangle\rangle}$$

$$\alpha'\langle\langle \bar{\mathscr{T}} \rangle\rangle = -\frac{15}{16}y'x\frac{1}{1 + (28/45)xc}$$

$$x = \frac{y'}{1 - \frac{1}{3}\alpha'\langle\langle \bar{\mathscr{T}} \rangle\rangle} \qquad (98)$$

$$y' = \frac{4\pi}{3}\alpha\frac{1}{1 - \alpha\langle\langle \mathring{\mathscr{T}} \rangle\rangle}$$

The density-dependent factor c is the same as that used in our approximant to the LHNC theory [see Eq. (79)]. The classical expression for the reaction field is obtained if the size of the third particle in the three-particle graph is neglected:

$$\text{[graph]} = -2\left(\frac{4\pi}{3}\alpha\rho\right)^2\frac{1}{\sigma^3} \tag{99}$$

The two-particle term is the same in the microscopic and in the macroscopic approach. Replacing the field particles again by effective particles, which means in the continuum approximation that

$$x = \frac{\varepsilon - 1}{\varepsilon + 2} \tag{100}$$

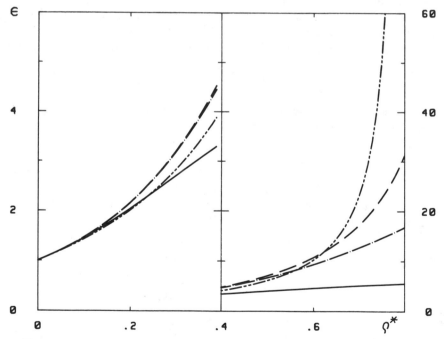

Figure 8. Graphs of ε for polarizable spheres as calculated by the Clausius–Mossotti theory (-·····-), the Onsager model (—), and the renormalized cluster expansion [Eq. (98)] with the density dependence of the radial distribution function taken into account (---) or neglected (-··-). $\alpha/\sigma^3 = 0.3$.

we regain Onsager's reaction field expression:

$$\alpha\langle\langle\mathscr{F}\rangle\rangle = 2\frac{\alpha}{\sigma^3}x\frac{1}{1+x}$$

$$= \frac{\alpha}{\sigma^3}\frac{2(\varepsilon-1)}{2\varepsilon+1} \tag{101}$$

Since, as shown in Section II, $E_L + \langle\langle\mathscr{F}\rangle\rangle\langle\mathbf{p}\rangle$ reduces to the cavity field, the microscopic theory [Eq. (98)] reduces to the Onsager model if the size of the field particles is neglected. The microscopic reaction-field factor is in general larger than in the Onsager model. In the macroscopic model the Lorentz field is also much more shielded than in the microscopic theory. For a given molecular model polarizability and hard-core diameter, the classical model is therefore expected to yield smaller values of the dielectric permittivity than is the microscopic theory. This is a consequence of the finite size of the particles determining the average field acting on a particle.

Figure 8 plots the dielectric permittivity values calculated for $\alpha/\sigma^3 = 0.3$ by the Clausius–Mossotti (CM) equation, the Onsager model and the microscopic theory in the two modifications that model the LHNC and the MSM theories. The values of ε predicted by the models decrease in the order CM

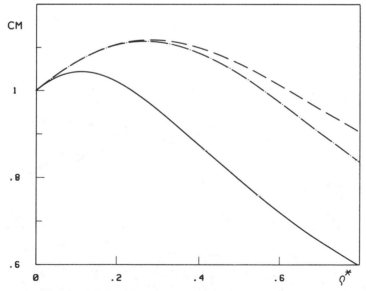

Figure 9. Clausius–Mossotti functions of polarizable spheres. The parameters and symbols are given in Fig. 8.

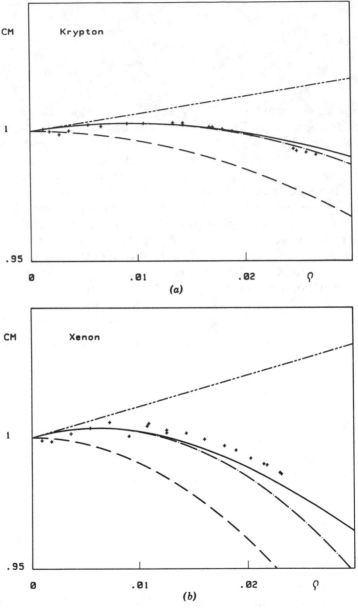

Figure 10. Clausius–Mossotti functions of (a) krypton, (b) xenon, and (c) carbon disulfide. The calculated curves are obtained by taking into account the second dielectric virial coefficient (- · ·-), the third dielectric virial coefficient (– –), the sum of the two (- · · ·), and the cluster approximant [Eq. (98)] (—). Plus marks represent experimental values.

756

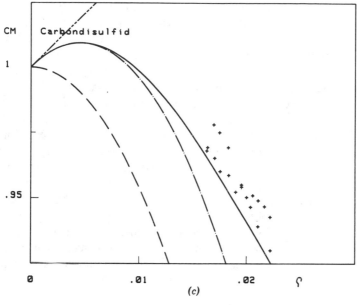

Figure 10. (*Continued*)

LHNC-MSM-Onsager. This holds true for $\alpha/\sigma^3 \leq 0.3$. For larger values of α/σ^3, the microscopic theories predict larger values for ε than the CM equation does, and the Onsager model predicts even smaller values. Consequently the microscopic theories predict a dielectric catastrophe at smaller densities than the CM theory does.

The graphs of the CM function (Fig. 9)

$$\frac{\varepsilon-1}{\varepsilon+2}\frac{1}{4\pi\alpha\rho/3} \qquad (102)$$

give some more insight into the processes determining the dielectric permittivity. The Onsager model, the MSM, and the LHNC theory all predict a positive deviation from the CM equation at low densities, caused by the reaction field, and a negative deviation, caused by the shielding due to a third particle. As in the case of rigid dipoles, the difference between CM functions for the MSM and the LHNC theory is quite small, whereas the predicted values of ε are very different. In noble gases and organic substances, the values of α/σ^3 are very small, so that the deviations of the CM function are small too. Figure 10 shows the CM function as a function of density for krypton, xenon,[89,90] and carbon disulfide.[91] The experimental values are compared with estimates that consider only the second dielectric virial coefficient [Eq. (96)], the third dielectric virial coefficient, and the sum of the two, and with predictions of the microscopic theories [Eq. (98)]. For the noble gases, the three dielectric virial coefficients give a quite reasonable description of the behavior, in spite of the fact that the experimental data include high-pressure measurements taken at up to 10,000 atm. The improvement achieved by use of the renormalized cluster expansion is fairly small, but significant in the cases of Xe and CS_2. The positive deviations in the latter cases are founded in the neglect of the influence of attractive interactions on the pair correlation function. Results for the noble gases He, Ne, and Ar are omitted here because by neglecting the collision-induced contributions due to short-range distortions, the DID model overestimates the second dielectric virial coefficient. Such collision-induced contributions become less important for the heavier noble gases.[92]

The consideration of higher terms in α causes a decrease in the Clausius–Mossotti function that is much more pronounced than would be predicted by a theory in which only the α^2 term is estimated as accurately as possible.[93]

C.   The Dielectric Catastrophe

The value of ε becomes very large if the value of the function $(\varepsilon - 1)/(\varepsilon + 2)$ approaches 1. This dielectric catastrophe was regarded as an artifact of the simple Debye theory, since in the case of polar systems the Debye theory predicts this divergence, followed by negative values of ε at densities where the observed behavior is absolutely normal.[6] Removing this divergence was one of the aims of the Onsager model.

The divergence of ε is, however, a quite common effect if highly polarizable particles such as free atoms of metals are considered.[94] As early as 1927 Herzfeld[95] proposed taking the divergence of the dielectric permittivity as a criterion of the transition from the nonmetallic to the metallic state. The corresponding critical density can be calculated from a theory of dielectric polarization, which in the simplest case is the CM theory. This concept has been applied recently by Ross[96] and Edwards and Sienko.[96] It turns out, however, that the CM theory predicts the dielectric catastrophe at densities that are too large by an order of magnitude[97,98] (see Table I). Using the value of the distance of closest approach taken from the pair correlation function[98] for σ and experimental values of α,[99] much better agreement with experiment is obtained if the microscopic reaction field is considered. Since the values of α/σ^3 are on the order of 1 for metals, it is sufficient to consider only the first term in the expansion of $\langle\langle \mathscr{T} \rangle\rangle^*$. The dielectric permittivity

TABLE I

Estimation of the Critical Density (ρ_c) of the Dielectric Catastrophe

	α ($\mathring{A}^3$)	σ ($\mathring{A}$)	ρ_c ($\times 10^{21}$ molecules cm^{-3})		
			CM^a	Eq. (103)	Exp.
H_2	0.79	2.9	606	572	614
J_2	2×4.96	4.98	48.0	41.5	39.5
Xe	4.15	4.3	57.6	51.4	—
Hg	5.5	2.6	43.5	26.5	27.0
Na	20.0	3.0	12.0	4.9	5.5
K	36.0	3.6	6.6	2.6	2.9
Rb	40.0	4.4	6.0	3.1	2.4
Cs	52.5	4.7	4.6	2.3	1.9

^{a} Value predicted by Clausius–Mossotti function.

is, in this approximation, predicted to diverge if

$$\frac{4\pi}{3}\alpha\mu - \frac{1}{1 + 2\alpha/\sigma^3} \tag{103}$$

The distance of closest approach σ, which is the appropriate distance to choose as the hard-sphere diameter, is different from the atomic diameter of the free atom, which is determined by the unpaired electron. The value of σ is only slightly smaller than the average distance in the gas-phase dimers, as one would expect.[100] In the double logarithmic plot of the susceptibility $\varepsilon - 1$ versus the order parameter ($\rho_c/\rho - 1$), we find almost straight lines with the slope varying between -1 and -1.4 depending on the value of α/σ^3. The order parameter varies over five orders of magnitude. The susceptibility is calculated from Eq. (98). The slope in Fig. 11 can be interpreted as a critical exponent that deviates from the value of 1 predicted by the CM theory. This is a consequence of an elaborate mean-field model depending on molecular properties and not of the universality of critical long-range fluctuations.[101] From the experiments done so far, it is not yet possible to distinguish clearly between the two possibilities. Another open question is whether the divergence of the dielectric permittivity is identical with the Mott transition[102] from the nonmetallic to the metallic state, or whether it is an optical anomaly occurring in the region of this transition. In mercury, optical anomalies that are probably due to the presence of charged droplets are observed at much lower densities than those assumed for the nonmetal–metal transition.[103]

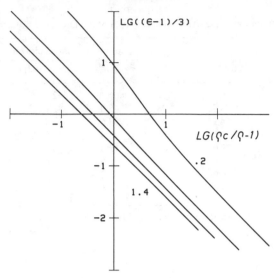

Figure 11.   Susceptibility $(\varepsilon - 1)$ as a function of the order parameter $(\rho_c/\rho - 1)$ as calculated from Eq. (98).

The details of the nonmetal–metal transition cannot be expected to be described by a classical theory assuming rigid spheres with a polarizable center. The divergence of ε in such a classical theory indicates a transition to a state that cannot be described by this theory. It provides an improved estimate of the critical density of the nonmetal–metal transition (in fluids) compared with that obtained by application of the Herzfeld criterion and the CM theory on the basis of the parameters of "polarizability" and "collision diameter," which can be determined by independent experiments. The theory can be generalized to fluid mixtures,[103] which allows its application to metallic solutions.[104] Whether it can be applied to the metal–insulator transition in solids[105] and surface films[106] requires further investigation.

IV. POLAR FLUIDS

A.   Graph-Theoretical Analysis

The theory of the dielectric polarization of nonpolar fluids requires the evaluation of irreducible averages of many-body interaction patterns by means of distribution functions of the system. The theory of dielectric polarization of rigid dipoles requires the estimation of the direct correlation function of the fluid. It is not obvious how the two theories may be combined to treat the more general case of the dielectric polarization of fluids consisting of polar molecules.

The major problem in any attempt to apply the formalism adequate for rigid dipole fluids to systems of polarizable dipoles is that no function corresponding to the direct correlation function is clearly defined in systems with many-body interactions. It is therefore not straightforward to extend the theory of rigid dipoles to the general case. The common feature of the theories for nonpolar fluids and rigid-dipole fluids is that the original average is decomposed, because of its nodal structure, into a power series of irreducible components.

In a series of papers,[11,15] Wertheim presented a graph-theoretical analysis of the pair correlation function in fluids comprising polarizable dipoles. He pointed out that the direct correlation function is not the proper two-terminal subgraph in the convolution giving the pair correlation function. The units that can be arranged in a chain are one- and two-terminal subgraphs; these have to be distinguished by the nature of the cut points which may have weights of α or $\beta\mu^2$. The cut points also have to be distinguished in terms of the number of interaction lines cut. The subunits may still contain α-articulation points, which are not removed in the usual renormalization procedure, in which the graphs of the one-particle distribution function in the fugacity expansion of the distribution function are summed in order to get a density expansion.[60-62] The articulation points are retained because at an α-cut point, a many-body interaction is cut in such a way that the obtained subgraphs are incomplete in a physical sense. They do not exist on their own as additive contributions to the partition function. Blocks containing one or more interrupted interaction paths (extra chains) can be interpreted as an average of those cut interaction paths determined by the distribution functions of the system. A block containing an extra chain has at least one α-cut point.

Since any particle may be the source of a blossom of one-terminal diagrams, it is convenient to formulate the dielectric polarization in terms of α' and μ', which are the polarizability and the dipole moments renormalized for the reaction field:

$$\alpha' = \frac{\alpha}{1 - \langle\langle \mathscr{F} \rangle\rangle \alpha} \quad \text{and} \quad \mu' = \frac{\mu}{1 - \langle\langle \mathscr{F} \rangle\rangle \alpha} \tag{104}$$

The renormalized graphs of the interaction patterns, which are now freed of 2-α-articulation points, can be separated into a chain of subdiagrams connected by $\beta\mu'^2$- and $1 - \alpha'$-bridge points. As in the case of nonpolar molecules those subdiagrams can be interpreted as partial averages, so that the average $\langle \mathbf{m}_1 \Sigma \mathbf{m}_k \rangle^*$ is a sum of such terms.

B. Irreducible Averages

The fact that the bridge particle may or may not be included in the partial average requires us to introduce further terms in addition to those given in Eq. (88). A partial average $\langle \mu'_1 \Sigma \mu'_k \rangle$ which contains no $1 - \alpha$-bridge point factors, quite analogously to the rigid dipole case, into subgraphs connected by $\beta\mu'^2$ bridge points, which we denote δ' and $\langle\langle \mathcal{T} \rangle\rangle$. $\langle\langle \mathcal{T} \rangle\rangle$, however, is determined by the distribution functions of the nonpolar reference system, whereas an average-shielded interaction between terminals of which at least one is an α-cut point is determined by the distribution functions of the system. If this difference is neglected, the sum of the ordered products determining the dielectric polarization can be worked out. The factors are A_1, A_2, A_3, and $\langle\langle \mathcal{T} \rangle\rangle$, as given in Eq. (88), and the additional terms

$$
\left.
\begin{aligned}
B_1 &= \beta \frac{\mu'^2}{3} \frac{1}{1 - y'\delta'} \\[2mm]
B_2 &= \frac{1}{1 - y'\delta'} \beta \langle\langle \mu'\mu' \mathcal{T} \rangle\rangle \\[2mm]
B_3 &= \beta \langle\langle \mathcal{T} \mu'\mu' \rangle\rangle \frac{1}{1 - y'\delta'} \\[2mm]
C_2 &= \frac{1}{1 - y'\delta'} \beta \langle\langle \mu'\mu' \mathcal{T} \alpha \rangle\rangle \\[2mm]
C_3 &= \beta \langle\langle \alpha \mathcal{T} \mu'\mu' \rangle\rangle \frac{1}{1 - y'\delta'}
\end{aligned}
\right\} \tag{105}
$$

where

$$
y' = \frac{4\pi}{3} \rho \frac{\beta\mu'^2}{3} \tag{106}
$$

The leading term in the expansion of the averages B_2, B_3, C_2, and C_3 is of first order in $\mathbf{T}$, while in the expansion of B_1 it is of second order in $\mathbf{T}$. The notation $\langle\langle \ldots \rangle\rangle$ indicates that the partial average is free of $\beta\mu'^2$- and $1 - \alpha'$-bridge points. The undetermined bridge point in B_2 and B_3 is a $1 - \alpha'$-bridge point.

The dielectric polarization is determined as a sum of ordered products of such terms. The general expression of the average molecular moment is a complicated continued fraction to involved to be useful in practical applications.[18] Great simplification can be achieved if the polarizability anisotropy can be neglected. In this case, since

$$A_1 = \alpha \quad \text{and} \quad C_2 = C_3 = A_2 = A_3 = 0,$$

$$\langle \mathbf{p} \rangle = \left[(1 - B_2) \alpha'(1 - B_3) + B_1 \right] \left[\mathbf{E}_L + \langle\langle \mathscr{F} \rangle\rangle \langle \mathbf{p} \rangle \right] \tag{107}$$

In a perturbation treatment in which the averages are worked out using distribution functions of the reference system containing spherical nonpolar molecules, B_2 and B_3 are also equal to zero, so that

$$\langle \mathbf{p} \rangle = \left[\alpha' + B_1 \right] \left[\mathbf{E}_L + \langle\langle \mathscr{F} \rangle\rangle \cdot \langle \mathbf{p} \rangle \right] \tag{108}$$

This expression is of the same form as Böttcher's generalization of the Onsager model, since B_1 contains the short-range orientational correlations. $\langle\langle \mathscr{F} \rangle\rangle$ and the reaction field are given by molecular averages, and not by their macroscopic approximants as in the classical model:

$$\langle \mathbf{p} \rangle = \frac{1}{1 - \alpha \langle\langle \mathring{\mathscr{F}} \rangle\rangle} \left[\alpha + \beta \frac{\mu^2}{3} \frac{1}{1 - \alpha \langle\langle \mathring{\mathscr{F}} \rangle\rangle} \frac{1}{1 - y'\delta'} \right] \left[\mathbf{E}_L + \langle\langle \mathscr{F} \rangle\rangle \langle \mathbf{p} \rangle \right] \tag{109}$$

C.   Self-Consistent Padé Approximant

Starting from Eq. (109) it is a simple matter to adapt the methods derived for rigid dipoles and nonpolar molecules to the more general case of polarizable dipoles. As in the case of nonpolar molecules, $n - \alpha'$-bridge points and subgraphs in which more than two interaction lines connect two α-points have to be discarded. This approximation is a generalization of Wertheim's SSC approximation[15] including the renormalized second dielectric virial coefficient. To calculate the dielectric permittivity, the following set of equations is solved iteratively:

$$\frac{\varepsilon - 1}{\varepsilon + 2} = \frac{z}{1 - \langle\langle \mathscr{F} \rangle\rangle z}$$

$$z = \frac{4\pi}{3} \rho \frac{1}{1 - \alpha \langle\langle \mathring{\mathscr{F}} \rangle\rangle} \left[\alpha + \frac{1}{1 - \alpha \langle \mathring{\mathscr{F}} \rangle} \frac{\beta \mu^2}{3} \frac{1}{1 - y'\delta'} \right]$$

$$\alpha \langle\langle \mathring{\mathscr{F}} \rangle\rangle = 2 \frac{\alpha}{\sigma^3} \frac{x}{1 + (15/32)x} \tag{110}$$

$$\langle\langle \mathscr{F} \rangle\rangle = -\frac{15}{16} x \frac{1}{1 + (28/45)xc}$$

$$x = \frac{z}{1 - \frac{1}{3} \langle\langle \mathscr{F} \rangle\rangle z}$$

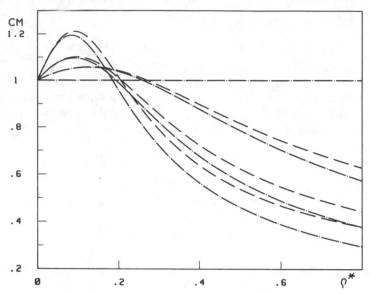

Figure 12. Clausius–Mossotti functions of polarizable and polar spheres as calculated from Eqs. (110). In the dashed curves, the density dependence of the radial distribution function is taken into account; this is neglected in the dotted and dashed curves. $\alpha/\sigma^3 = \frac{1}{8}$, $\beta\mu^2/\sigma^3 = 0, 1, 2, 3$.

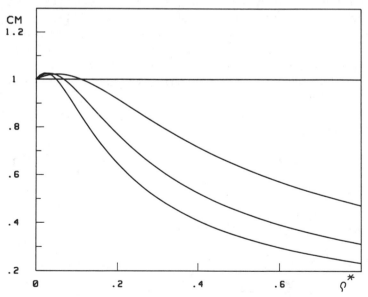

Figure 13. Clausius–Mossotti functions of polarizable polar spheres as calculated from the Onsager model for the same parameters as in Fig. 12.

764

In the expression of δ' [see Eqs. (64) and (81)], the dipole moment is replaced by its reaction-field-corrected value and x is taken from Eq. (110). The graphical representation is the same as in the theory of nonpolar fluids [Eq. (95)]. The field points represent $\alpha' + \beta\mu'^2$, however.

As outlined in Section III.B, the restrictions on the connectivities of the interaction patterns are reasonable if the polarizability is small and causes only minor corrections to the rigid-dipole result. This approximation ensures that the interaction path is uniquely determined by the graphical representation (this is not so in the general case in which the application of directed graphs would be appropriate).[87] The results of the numerical calculations are quite similar to those obtained for rigid dipoles (Fig. 12). δ' causes a much more pronounced maximum in the CM function at low densities than that in the Onsager model (Fig. 13), where it is caused by the reaction field only. Depending on the value of α/σ^3 the dielectric permittivity may be increased compared with that of the hard-sphere dipole system.

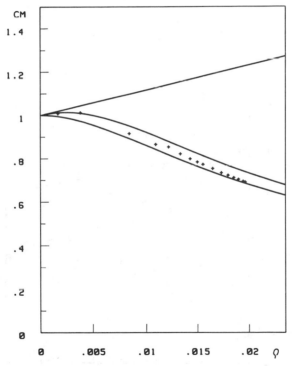

Figure 14.   Clausius–Mossotti functions of CF_3H for $T = 403$ K. The slope of the straight line is determined by the second dielectric virial coefficient. The upper curve is calculated from Eqs. (110). In the lower curve, reaction-field contributions are neglected. Plus signs represent experimental data.

D.   Dielectric Polarization of CF_3H, CH_3F, and CH_3CN

The dielectric permittivities of simple polar substances such as CH_3F, CF_3H, and CH_3CN have been measured over a large temperature range and up to pressures as high as 3000 atm by Franck and collaborators.[106,107] For these substances the second dielectric virial coefficient is also known from the measurements of Sutter and Cole[108] and Buckingham[109] at extremely low densities. To compare our theory with the experimental results, we have calculated the CM function as a function of density. Unfortunately, the pVT data are not available for the full temperature–density range treated in the dielectric measurements. Only in the case of CH_3CN are pVT data known.[110]

For CH_3F and CF_3H, pVT data are available only up to 150 atm.[111,112] It was therefore necessary to estimate the density data in the high-pressure region on the basis of the available data. Taking advantage of the fact that isochores are almost straight lines[113] and using the densities from the known saturation curve, the slopes of the isochores were obtained using a perturbation method. The slope is given by the slope of the isochore of methane[114] at

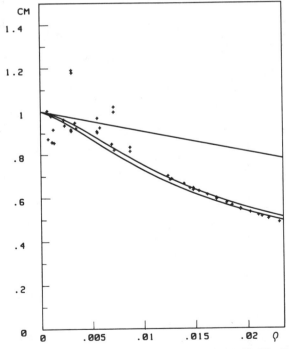

Figure 15.   Clausius–Mossotti functions of CH_3F for $T = 361$ K. The lines and symbols are explained in Fig. 14.

the same reduced density corrected by addition of a factor that is quadratic in the density and adjusted to the slope in the region where the pVT data of methane and the fluorinated compounds are known. This method has been checked using the data for acetonitrile.

CF_3H is an almost spherical molecule. The molecular parameters used in the calculation are $\mu = 1.649$ D, $\alpha = 3.57$ Å^3, and $\sigma = 3.94$ Å. The values of μ and α are taken from Sutter and Cole;[108] σ is the Lennard–Jones value in the Stockmeyer potential used by Barnes and Sutton.[145] Figure 14 shows the experimental values of the CM function and of predictions for that function from Eq. (110). The lower curve is obtained by neglecting all reaction-field contributions. The straight line gives the value of the CM function based on the second dielectric virial coefficient. The second dielectric virial coefficient

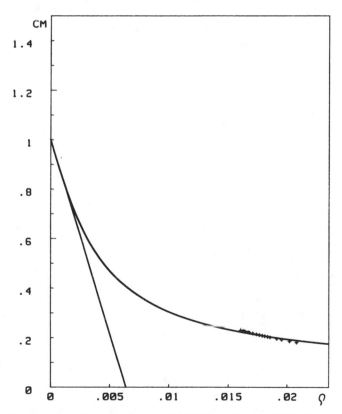

Figure 16.   Clausius–Mossotti functions of CH_3CN for $T = 423$ K. The lines and symbols are explained in Fig. 14. The influence of the reaction-field contributions on the CM function is very small.

obtained from the low-density limit of Eq. (110) agrees nicely with the experimental findings based on measurements at extremely low densities (2×10^{-5} mol cm^{-3}).[108] Sutter and Cole find 1125 cm^3 mol^{-2} for 50°C, whereas we calculate 1000 cm^3 mol^{-2} for 37°C. At high temperatures (143°C) they find 704 cm^3 mol^{-2}; we calculate 590 cm^3 mol^{-2} for 130°C. No correction for the molecular shape was made.

In the application of the theory to molecules that show more pronounced deviations from a spherical shape, a'_2 is treated as a density- and temperature-independent parameter. CH_3F is a slightly rod-shaped molecule. Besides the high- and low-pressure measurements quoted above, some data are available also for the intermediate region.[115] The CM function (Fig. 15) is calculated with $\mu = 1.851$ D, $\alpha = 2.97$ Å^3, and $\sigma = 4.75$ Å. With $a'_2 = \frac{1}{3}$, the calculated curves and the second dielectric virial coefficients agree quite well with the experimental data over the entire region investigated. The same is true for acetonitrile (Fig. 16), for which $\mu = 3.96$ D, $\alpha = 1.3$ Å^3, and $\sigma = 6.1$ Å; a'_2 is adjusted to 0.8 in order to obtain agreement with all experimental data. This value allows for the large negative values of the second dielectric virial coefficient, which increase in magnitude at lower temperatures. The value $a'_2 = 0.8$ is also obtained by evaluating Eq. (84) using Bertagnolli's[76] pair distribution function calculated for a molecule of the shape of acetonitrile.

Experimental and calculated values of the second dielectric virial coefficients for CH_3CN, CH_3F, and CF_3H are listed in Table II.

TABLE II
Second Dielectric Virial Coefficients

Compound	Experimental		Calculated	
	Temperature (K)	B (cm^6 mol^{-2})	Temperature (K)	B (cm^6 mol^{-2})
CH_3CN	354	-10×10^5	277	-7.4×10^4
	373	-6×10^5	323	-5.7×10^4
	405	-3×10^5	373	-4.5×10^4
	433	-2×10^5	423	-3.6×10^4
	—	—	523	-2.0×10^4
CH_3F	323	-1307	272	-1057
	369	-606	323	-739
	416	-331	359	-642
	—	—	383	-527
	—	—	433	-445
	—	—	472	-364
CF_3H	323	1125	310	1000
	416	704	403	590

V. CONCLUSIONS

Reinvestigation of the macroscopic concepts of the classical theory of dielectric polarization shows that the CM function $(\varepsilon - 1)/(\varepsilon + 2)\rho\mathscr{A}$ is the ultimate expression describing the contributions of intermolecular correlations to the dielectric polarization. $\mathscr{A}$ is the first dielectric virial coefficient. All other fluctuation expressions are related to this function by purely macroscopic relations.

Self-consistent Padé approximants based on the first terms of an expansion in irreducible clusters have been applied in this paper as a very useful tool in the estimation of the CM function and of the dielectric constant. The method is easy to apply and is helpful in explaining the results of simulations and analytical theories. The direct correlation function of the hard-sphere dipole system may be approximated by a Boltzmann factor involving a shielded dipole–dipole interaction. The MSM and the LHNC theory consider the linear term of this Boltzmann factor only. In the continuum approximation where groups of short-range correlated particles are treated as point particles, the two theories simplify to the Kirkwood–Onsager model.

In the case of polarizable molecules the CM function can be written as an ordered product of partial averages of cut interaction paths. The graphical representation contains articulation points. The corresponding one-terminal graphs, which have the form of a cactus, represent the reaction field. The molecular expressions reduce to the classical reaction field if the continuum approximation is used. The statistical-mechanical theory leads to expressions of the local field of the same form as in the Onsager model but supplemented with terms determined by specific short-ranged correlations, which also determine the averaged forces acting on a molecule.

With the method described in this article it is possible to reproduce the results of much more involved theories. Since our formalism is capable of including the second dielectric virial coefficient and the polarizabilities of molecules (a task that is very difficult in analytical theories that calculate the pair distribution function by solving the OZ equation), the theory could successfully be applied to describe experimental data for real fluids.

The renormalized cluster expansion requires only the calculation of small cluster integrals, and it should not be too difficult to treat more complicated molecular models and realistic intermolecular potentials involving multipole moments and higher polarizabilities, which are known to give important contributions to the second dielectric virial coefficient. The treatment of more realistic models seems necessary if we are to gain deeper understanding of the dielectric properties of dense fluids based on knowledge of the low-density behavior. Concerning the comparison with experimental results, it must be noted that very few substances have been investigated over a wide temperature and density range. But even for these few examples, data in the intermediate-density region near the critical density are very scarce. More dielectric measurements of simple molecules accompanied by pVT measurements in this region are required if we are to be able to judge theories of dielectric polarization in real polar fluids.

Acknowledgments

The author is much indebted to Professor A. D. Buckingham for stimulating him to do this research and for many helpful discussions. Discussions with Professor F. Harary are also acknowledged. A continuous help was the assistance of Mrs. C. Rybarsch in doing the computer work.

This research was carried out with the aid of a fellowship granted by the European Research Programme sponsored by the Royal Society. Grants from the Deutsche Forschungsgemeinschaft and from the Fonds der Chemischen Industrie are also acknowledged.

References

1. P. Debye, *Polare Molekeln*, Hirzel, Leipzig, 1929.

2. H. Frölich, *Theory of Dielectrics*, 2nd Ed., Oxford University Press, Oxford, 1968.

3. L. Onsager, *J. Am. Chem. Soc.*, **58**, 1468 (1936).

4. C. J. F. Böttcher, *Theory of Electric Polarization*, Elsevier, Amsterdam, 1952.

5. T. G. Scholte, *Physica*, **15**, 437 (1949).

6. C. J. F. Böttcher, *Theory of Electric Polarization*, 2nd Ed., Elsevier, Amsterdam, 1973.

7. J. G. Kirkwood, *J. Chem. Phys.*, **7**, 911 (1939).

8. J. Yvon, *Actualités Scientifiques et Industrielles*, Vol. 543, Hermann & Cie., *La Propagation et la Diffusion de la Lumière*, Paris, 1937.

9. S. A. Adelman and J. M. Deutch, *Adv. Chem. Phys.*, **31** 103 (1975).

10. G. Stell, G. N. Patey, and J. S. Høye, *Adv. Chem. Phys.*, **48**, 183 (1981).

11. M. S. Wertheim, *Annu. Rev. Phys. Chem.*, **30**, 471 (1979).

12. G. Nienhuis and J. M. Deutch, *J. Chem. Phys.*, **55**, 4213 (1971).

13. J. S. Høye and G. Stell, *J. Chem. Phys.*, **61**, 562 (1974); *ibid.*, **64**, 1952 (1976).

14. J. D. Ramshaw, *J. Chem. Phys.*, **57**, 2684 (1972); *ibid.*, **66**, 3134 (1977); *ibid.*, **68**, 4149 (1978).

15. M. S. Wertheim, *Mol. Phys.*, **26**, 1425 (1973); *ibid.*, **33**, 95 (1977).

16. L. Rosenfeld, *Theory of Electrons*, North Holland, Amsterdam, 1951, p. 19.

17. S. R. de Groot, *Studies in Statistical Mechanics IV: The Maxwell Equations*, North Holland, Amsterdam, 1969, p. 55.

18. W. Schröer, Thesis, Cambridge University, Cambridge, 1981.

19. W. Schröer, *J. Mol. Struct.*, **84**, 329 (1982).

20. C. A. ten Seldam and S. R. de Groot, *Physica*, **18**, 905 (1952).

21. P. Madden, *Annu. Rev. Phys. Chem.*, **31**, 523 (1980).

22. A. D. Buckingham, *Adv. Chem. Phys.*, **12**, 107 (1967).

23. A. D. Buckingham and J. A. Pople, *Proc. Phys. Soc.*, **68A**, 403 (1955).

24. M. Mandel and P. Mazur, *Physica*, **24**, 116 (1958).

25. M. S. Wertheim, *Mol. Phys.*, **36**, 1217 (1978).

26. F. E. Harris and B. J. Alder, *J. Chem. Phys.*, **21**, 1031 (1953); *ibid.*, **22**, 1806 (1954).

27. H. Fröhlich, *J. Chem. Phys.*, **22**, 1804 (1954); *ibid.*, *Physica*, **22**, 889 (1956).

28. A. D. Buckingham, *J. Chem. Phys.*, **23**, 2370 (1955); *ibid.*, *Proc. R. Soc. Lond.*, **A238**, 235 (1956).

29. B. K. P. Scaife, *Proc. Phys. Soc.*, **70B**, 314 (1957).

30. J. A. Stratton, *Electromagnetic Theory*, McGraw-Hill, New York, 1941, p. 207.

31. R. Becker and F. Sauter, *Theorie der Elektrizität*, Vol. 1, Teubner, Stuttgart, 1969.

32. S. de Leew, J. S Perram, and E. R. Smith, *Proc. R. Soc. Lond.*, **A373**, 27, 57 (1980).

33. E. L. Pollock and B. G. Alder, *Physica*, **102A**, 1 (1980); B. G. Alder, E. Alley, and E. L. Pollock, *Ber. Bunsenges. Phys. Chem.*, **85**, 944 (1981).

34. W. Schröer, *Ber Bunsenges. Phys. Chem.*, **86**, 916 (1982).

35. Ref. 2, p. 181, Eq. (B1.28).

36. Ref. 2, p. 46.

37. B. U. Felderhof, *J. Phys. C, Solid State Phys.*, **12**, 2423 (1979).

38. C. Brot, G. Bossis, and C. Hesse-Bezot, *Mol. Phys.*, **40**, 1053 (1980).

39. G. Bossis and C. Brot, *Mol. Phys.*, **43**, 1095 (1981).

40. J. A. Barker and R. O. Watts, *Mol. Phys.*, **26**, 789 (1973).

41. D Levesque, G. N. Patey, and J. J. Weis, *Mol. Phys.*, **34**, 1077 (1977).

42. D. J. Adams and L M Adams, *Mol, Phys.*, **42**, 907 (1981).

43. G. N. Patey, D. Levesque, and J. J. Weis, *Mol Phys*, **45**, 733 (1982).

44. M. Neumann, *Mol. Phys.*, **50**, 841 (1983).

45. M. Neumann and O. Steinhauser, *Mol. Phys.*, **39**, 437 (1981).

46. A. J. Dekher, *Physica*, **12** 209 (1946).

47. J.-L Rivail and B. Tenryn, *J. Chim. Phys.*, **79**, 1 (1982); D. Rinaldi, *Computers and Chemistry*, **6**, 155 (1982).

48. J. A. Abbot and H. C. Bolton, *Trans. Faraday Soc.*, **48**, 422 (1952).

49. G. N. Patey, *Mol. Phys.*, **34**, 427 (1977).

50. M. S. Wertheim, *J. Chem. Phys.*, **55**, 4291 (1971).

51. A. D. Buckingham and C. G. Joslin, *Mol. Phys.*, **40**, 1513 (1980).

52. R. Cole, in *Molecular Liquids Dynamics and Interactions*, A. J. Barnes, W. J. Orville-Thomas, and J. Yarwood, eds., D. Reidel, Dordrecht, 1984, p. 59.

53. F. H. Ree and W. G. Hoover, *J. Chem. Phys.*, **42**, 939 (1964).

54. G. S. Rushbrooke, G. Stell, and J. S. Høye, *Mol. Phys.*, **26**, 1119 (1973).

55. G. Stell, J. S. Rasaiah, and H. Narang, *Mol. Phys.*, **27**, 1393 (1974).

56. W. Schröer, *J. Mol. Phys.*, **41**, 239 (1980).

57. J. D. Ramshaw, *J. Chem. Phys.*, **68**, 5199 (1978).

58. L. Blum and A. J. Torruella, *J. Chem. Phys.*, **56**, 303 (1972).

59. A. Stone, *Mol. Phys.*, **36**, 241 (1978).

60. G. E. Uhlenbeck and G. W. Ford, *Studies of Statistical Mechanics*, **1**, 123 (1962).

61. G. S. Rushbrooke, in *Physics of Simple Liquids*, H. N. V. Temperley, G. S. Rushbrooke, and J. S. Rowlinson, eds., Elsevier, Amsterdam, 1968, p. 25.

62. G. Stell, in *The Equilibrium Theory of Classical Fluids*, H. L. Frisch and J. L. Lebowitz, eds., Benjamin, New York, 1964, p. II–171.

63. M. S. Wertheim, *J. Math. Phys.*, **8**, 927 (1976).

64. H. J. Raveché and R. D. Mountain, in *Progress in Liquid Physics*, C. Coxton, ed., Wiley, New York, 1978, p. 469.

65. J. D. Weeks, D. Chandler, and H. C. Anderson, *J. Chem. Phys.*, **54**, 5237 (1971).

66. J. L. Lebowitz, G. Stell, and S. Baer, *J. Math. Phys.*, **6**, 1282 (1965).

67. J. L. Lebowitz and J. K. Percus, *J. Math. Phys.*, **4**, 248 (1961).

68. G. N. Patey, *J. Mol. Phys.*, **35**, 1413 (1978).

69. G. N. Patey, D. Levesque, and J. J. Weis, *Mol. Phys.*, **38**, 219, 1635 (1979).

70. G. S. Rushbrooke, *Mol. Phys.*, **37**, 761 (1979).

71. D. W. Jepsen and H. Friedman, *J. Chem. Phys.*, **38**, 846 (1963).

72. D. W. Jepsen, *J. Chem. Phys.*, **44**, 774 (1966).

73. C. G. Joslin, *Mol. Phys.*, **42**, 1507 (1981).

74. J. A. Pople, *Proc. R. Soc. Lond.*, **A221**, 498 and 508 (1954).

75. A. D. Buckingham and J. A. Pople, *Trans. Faraday Soc.*, **51**, 1173 (1955).

76. H. Bertagnolli, *Ber. Bunsenges. Phys. Chem.*, **81**, 739 (1977).

77. J. G. Kirkwood, *J. Chem. Phys.*, **4**, 592 (1936).

78. P. R. Certain and P. Fortune, *J. Chem. Phys.*, **61**, 2620 (1974).

79. J. De Boer, F. van der Maesen, C. A. ten Seldam, *Physica*, **19**, 265 (1953).

80. H. Posch, *Mol. Phys.*, **34**, 1059 (1979); *ibid.*, **40**, 1137 (1980).

81. A. D. Buckingham and G. Tabisz, *Mol. Phys.*, **36**, 583 (1978).

82. F. Harary, *Graph Theory*, Addison-Wesley, London, 1969.

83. M. Born and E. Wolf, *Principles of Optics*, 5th Ed., Pergamon, Oxford, 1975, p. 99.

84. P. Mazur and M. Mandel, *Physica*, **22**, 299 (1956).

85. M. S. Wertheim, *Mol. Phys.*, **25**, 211 (1973).

86. F. Harary and E. M. Palmer, *Graphical Enumeration*, Academic, New York, 1973.

87. F. Harary and R. W. Robinson, personal communications.

88. B. Linder and D. Hoernschemeyer, *J. Chem. Phys.*, **46**, 784 (1967).

89. D. Vidal and M. Lallemand, *J. Chem. Phys.*, **64**, 4293 (1967).

90. M. Lallemand and D. Vidal, *J. Chem. Phys.*, **66**, 4776 (1977).

91. W. E. Danforth, Jr., *Phys. Rev.*, **38**, 1224 (1936) [through W. Fuller-Brown, Jr., *J. Chem. Phys.*, **18**, 1200 (1950).

92. R. H. Orcutt and R. H. Cole, *J. Chem. Phys.*, **46**, 697 (1967).

93. H. W. Graben, G. S. Rushbrooke, and G. Stell, *Mol. Phys.*, **30**, 373 (1975).

94. P. P. Edwards and M. J. Sienko, *J. Chem. Ed.*, **60**, 691 (1983).

95. K. F. Herzfeld, *Phys. Rev.*, **29**, 701 (1927).

96. M. Ross, *J. Chem. Phys.*, **56**, 4651 (1972).

97. M. Shimoji, *Liquid Metals*, Academic, New York, 1977, p. 309.

98. T. E. Faber, *Theory of Liquid Metals*, Cambridge University Press, Cambridge, 1972, p. 131.

99. E. Pollack and E. Bederson, *Phys. Rev.*, **124**, 1431 (1961).

100. G. Herzberg, *Molecular Spectra and Molecular Structure*, 2nd Ed., Vol. I, Van Nostrand, New York, 1950, p. 502.

101. S. K. Ma, *Modern Theory of Critical Phenomena*, Benjamin, Reading, Massachusetts, 1976.

102. N. F. Mott, *Metal–Insulator Transition*, Taylor and Francis, London, 1974.

103. W. Ilcfner, R. Sonnehorn-Schmick, and F. Hensel, *Ber. Bunsenges. Phys. Chem.*, **86**, 844 (1982).

104. W. Schröer, *Bunsentagung Kaiserslautern*, 1984.

105. N. F. Hess, K. D. Conde, T. F. Rosenbaum, and G. A. Thomas, *Phys. Rev.*, **B25**, 5578 (1982).

106. C. Bartels, B. Danzfuss, and J. W. Schulze, in *Passivity of Metals and Semiconductors*, M. Froment, ed., Elsevier, Amsterdam, 1984, p. 503.

107. E. U. Franck, in *Organic Liquids*, A. D. Buckingham, E. Lippert, and S. Bratos, eds., Wiley, London, 1978, p. 181.

108. H. Sutter and R. H. Cole, *J. Chem. Phys.*, **52**, 132 (1970).

109. A. D. Buckingham and R. E. Raab, *J. Chem. Soc.*, 5511 (1961).

110. A. Z. Francesconi, E. U. Franck, and H. Lentz, *Ber. Bunsenges. Phys. Chem.*, **79**, 897 (1975).

111. Y. C. Hou and J. J. Martin, *A.I.Ch.E.*, Jr., **5**, 125 (1975).

112. A. Michels, A. Vissar, R. J. Lumbeck, and G. J. Wohlers, *Physica*, **18**, 144 (1952).

113. J. S. Rowlinson and F. L. Swinton, *Liquids and Liquid Mixtures*, Butterworth, London, 1982, p. 78.

114. B. Armstrong, K. M. de Reuch, eds., *International Thermodynamic Tables of the Fluid State*, Vol. 5, *Methane*, Pergamon, Oxford, 1978.

115. A. N. M. Barnes and L. E. Sutton, *Trans. Faraday Soc.*, **67**, 2915 (1971).

116. H. G. David, S. D. Hamann, and J. F. Pearse, *J. Chem. Phys.*, **20**, 969 (1952).

ADVANCES IN MICROWAVE AND SUBMILLIMETER-WAVE DIELECTRIC SPECTROSCOPIC TECHNIQUES AND THEIR APPLICATIONS

J. K. VIJ

Department of Microelectronics and Electrical Engineering, School of Engineering, Trinity College, Dublin 2, Ireland

and

F. HUFNAGEL

Microwave Laboratory, Institute of Physics, Johannes Gutenberg University, Mainz, Federal Republic of Germany

CONTENTS

I. THEORY OF MEASUREMENTS IN THE MICROWAVE FREQUENCY RANGE

For a uniform isotropic nonmagnetic dielectric material, Maxwell's equations are

$$\nabla \times \mathbf{H} = \mathbf{J} + \frac{\partial \mathbf{D}}{\partial t} \tag{1}$$

$$\nabla \times \mathbf{E} = - i\omega\mu_0 \mathbf{H} \tag{2}$$

$$\nabla \cdot \mathbf{D} = \rho; \qquad \mathbf{D} = \varepsilon_0 \varepsilon_r \mathbf{E} \tag{3}$$

$$\nabla \cdot \mathbf{B} = 0 \tag{4}$$

where $\mathbf{J}$ is the current density due to charges in motion, and the other symbols have their usual significances. The electronic conduction due to charges in the dielectric will be assumed to be negligible; thus $\mathbf{J} = 0$.

For sinusoidal variation of the electric field (i.e., $\mathbf{E} = \mathbf{E}_0 \exp(i\omega t)$), Eq. (1) can be written as

$$\nabla \times \mathbf{H} = i\omega \varepsilon_0 \varepsilon_r^*(\omega) \mathbf{E} \tag{5}$$

where ε_0 is the absolute permittivity of free space and $\varepsilon_r^*(\omega)$ is the complex relative permittivity at frequency f ($= \omega/2\pi$ Hz).

Taking the curl of Eq. (2), we obtain

$$\nabla \times \nabla \times \mathbf{E} = -i\omega \mu_0 (\nabla \times \mathbf{H}) \tag{6}$$

and substituting for $\nabla \times \mathbf{H}$ from Eq. (5), we get

$$\nabla(\nabla \cdot \mathbf{E}) - \nabla^2 \mathbf{E} - \omega^2 \mu_0 \varepsilon_0 \varepsilon_r^*(\omega) \mathbf{E} \tag{7}$$

For a dielectric without free charges, Eq. (3) requires that $\nabla \cdot \mathbf{E} = 0$, and hence Eq. (7) becomes

$$\nabla^2 \mathbf{E} + \omega^2 \mu_0 \varepsilon_0 \varepsilon_r^*(\omega) \mathbf{E} = 0 \tag{8}$$

For plane-wave propagation in the x-direction and with the electric field in the z-direction, Eq. (8) reduces to the partial differential equation

$$\frac{\partial^2 E_z}{\partial x^2} + \omega^2 \mu_0 \varepsilon_0 \varepsilon_r^*(\omega) E_z = 0 \tag{9}$$

with a solution

$$E_z(x) = E_{0z}(0) \exp(-\gamma x) \exp(i\omega t) \tag{10}$$

Here, $E_{0z}(0)$ is the electric field in the z-direction at $x = 0$, and γ, the propagation factor, is given by

$$\gamma = i\omega \left(\mu_0 \varepsilon_0 \varepsilon_r^*(\omega) \right)^{1/2} \tag{11}$$

$$= i\frac{\omega}{c} \left(\varepsilon_r^*(\omega) \right)^{1/2} \tag{12}$$

where $c = (\varepsilon_0 \mu_0)^{-1/2}$ is the velocity of light *in vacuo*.

In general, γ is complex and is usually written as

$$\gamma = \alpha + i\beta \qquad (13)$$

where α is the attenuation constant and β is the phase constant.

The complex relative permittivity can be written as

$$\varepsilon_r^*(\omega) = \varepsilon'(\omega) - i\varepsilon''(\omega) \qquad (14)$$

where $\varepsilon'(\omega)$ is the real part of the complex relative permittivity and will be referred to from now on as the permittivity, and $\varepsilon''(\omega)$ is the imaginary part and will be called the dielectric loss. Equation (12) now becomes

$$\alpha + i\beta = i(\omega/c)\left(\varepsilon'(\omega) - i\varepsilon''(\omega)\right)^{1/2} \qquad (15)$$

Squaring both sides and equating real and imaginary parts, we find that

$$\alpha^2 - \beta^2 = -(\omega/c)^2 \varepsilon'(\omega) \qquad (16)$$

$$2\alpha\beta = (\omega/c)^2 \varepsilon''(\omega) \qquad (17)$$

Therefore, the experimental determination of α and β at a particular frequency will yield the real and imaginary parts of the complex relative permittivity at any frequency, as long as the conditions for the solution of Eq. (8) are satisfied.

For dilute solutions of polar liquids in nonpolar solvents, Eq. (16) can be simplified considerably, because in many cases $\alpha^2 \ll (\omega/c)^2 \varepsilon'(\omega)$, and thus

$$\beta \simeq (\omega/c)\left(\varepsilon'(\omega)\right)^{1/2} \qquad (18)$$

On using this expression for β in Eq. (17), we get

$$\varepsilon''(\omega) \simeq \left(\frac{\lambda_0}{\pi}\right)\alpha(\omega)\left[\varepsilon'(\omega)\right]^{1/2} \qquad (19)$$

Again, for dilute solution $\varepsilon'(\omega) \simeq \varepsilon'(0)$; Eq. (19) thus becomes

$$\varepsilon''(\omega) \simeq \left(\frac{\lambda_0}{\pi}\right)\alpha(\omega)\left[\varepsilon'(0)\right]^{1/2} \qquad (20)$$

Consequently, $\varepsilon''(\omega)$ can be obtained from an experimental determination of the attenuation suffered by an electromagnetic wave traversing the dielectric medium.

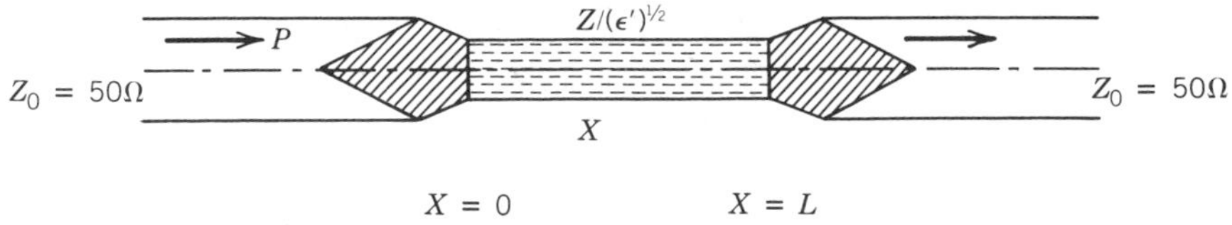

$$X = 0 \qquad\qquad X = L$$

Figure 1.   Schematic of a coaxial line dielectric cell.

It must be emphasized that the wave equation (9) holds for waves in which both the electric and magnetic vectors are transverse to the direction of the propagation (TEM waves). It must be modified for TE- or TM-wave propagation modes in waveguides.

The mode of propagation in a coaxial waveguide is known to be of the TEM type;[1] hence the equations derived above are directly applicable to the determination of $\varepsilon'(\omega)$ and $\varepsilon''(\omega)$. Nevertheless, the cell containing the dielectric must be designed in such a way that its impedance matches that of the coaxial waveguide, so that reflections at the interfaces will be as small as possible.

We assume that the reflections of the wave at the interfaces ($X = 0$ and $X = L$) can be neglected (Fig. 1). Let the electric power incident at $X = 0$ be P_0 watts (Fig. 1) and let the power transmitted at $X = L$ be P_L watts. Then, according to the definition of the attenuation constant,

$$P_L = P_0 \exp[-2\alpha(\omega)L] \tag{21}$$

$$\alpha(\omega) = \frac{1}{2L}\log_e\left(\frac{P_0}{P_L}\right) = \frac{2.3025}{2L}\log_{10}\left(\frac{P_0}{P_L}\right) \tag{22}$$

It is usual to express the logarithmic ratio of two powers in bels (B). Thus, we say the ratio has the value $10\log_{10}(P_0/P_L)$ decibels (dB).

$$\alpha(\omega) = \frac{0.1151}{L\ (\mathrm{cm})}D \quad (\mathrm{dB}) \tag{23}$$

Significant advances in the techniques for measuring dielectric properties at high frequency have taken place during the last decade. The developments include swept-frequency microwave transmission and reflection spectrometers, spot-frequency interferometric techniques at 2 and 1 mm, Kremer's method using an untuned cavity resonator. These are reviewed, and advances in their designs to date discussed, below.

II. MICROWAVE SWEPT-FREQUENCY TRANSMISSION
SPECTROMETER (2–18 GHz)

A. Introduction

The spectrometer discussed in this section is based on the principle of re-cording the power transmitted by a dielectric as a function of frequency. The main advantage of this device lies in its applicability as a conventional spec-trometer in the microwave region. It is much faster than the conventional spot-frequency methods and can achieve an accuracy of $\pm 3\%$ for $\varepsilon''(\omega)$ in dilute solutions. The method is slightly inferior in accuracy to the spot-frequency ones. However, it is not only superior in accuracy to the time-domain method[2] in the microwave range, but also has the added advantage of providing an immediate representation of the frequency dependence of the attenuation constant without requiring Fourier transformations to the frequency domain.

The method was developed by Price and Wegdam[3] (1977) in Aberyst-wyth, Wales, and by Hufnagel and Hammel[4] (1978) in Mainz, Germany. General principles of the method are reviewed briefly below, and the tech-nique as developed and used in Mainz is given.

B. General Principles of the Method

In Eq. (22), $\alpha(\omega)$ is directly related to the attenuation in power suffered by a TEM wave in passing through the dielectric medium. However, when the sample thickness L is of the order of the wavelength, it is extremely dif-ficult to eliminate reflections at the interfaces and multiple reflections within the sample. This situation is illustrated in Fig. 2, where t_{jk} is the transmis-sion coefficient and r_{jk} is the reflection coefficient at the $j-k$ boundary. We

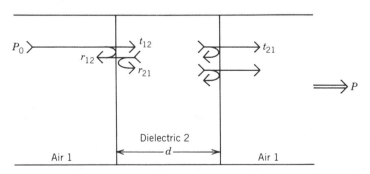

Figure 2.   Schematic of reflection and transmission by a dielectric. (Reproduced by permission from ref. 3.)

assume that the dielectric cell (Fig. 4) is designed in such a manner that the reflections between the cell and the source on one hand and the cell and detector on the other are reduced significantly. We can then apply the conventional theory of transmission lines (cf. ref. 5).

It may be shown that

$$t_{jk} = \frac{2Z_k}{Z_k + Z_j} \qquad (24)$$

and

$$r_{jk} = \frac{Z_k - Z_j}{Z_k + Z_j} \qquad (25)$$

where Z_k is the characteristic impedance of the coaxial line containing the dielectric, and Z_j is that of the line with air.

The ratio of the transmitted power, P_t, to the incident power, P_0, can be expressed as

$$T = \frac{P_t}{P_0} = \frac{|t_{12}|^2 |t_{21}|^2 \exp(-2\alpha L)}{1 - 2|r_{21}|^2 \cos(2\beta L)\exp(-2\alpha L) + |r_{21}|^4 \exp(-4\alpha L)} \qquad (26)$$

where $|r|$ and $|t|$ denote the magnitudes of the complex reflection and transmission coefficients. For large attenuations such that $\alpha L > 1$, the denominator of Eq. (26) approaches unity, with the result that $\ln(P_t/P_0)$ is a linear function of the sample thickness. The slope of the curve $\ln(P_t/P_0)$ versus L can be used to determine α. If $\alpha L < 1$, periodic variations may be superimposed on such a linear plot owing to departures from unity of the denominator in Eq. (26). In addition, variations are introduced into the plot of $\ln(P_t/P_0)$ versus frequency, f, due to optical interference with the waves in the coaxial line arising from reflections at the air–dielectric interfaces and from multiple reflections within the cell. The design of the dielectric cell and of the overall system must ensure significant reduction of such optical interference between the waves.

C. System Design

The block diagram of the spectrometer is given in Fig. 3. The source is amplitude modulated at 1 kHz. The frequency of the oscillator is swept from 2.0 to 18.5 GHz on manual triggering; the X-input of the X–Y recorder is synchronized with this trigger. The detector is a broad-band Schottky diode (Wiltron Co., Palo Alto, California) designed to detect radio-frequency power

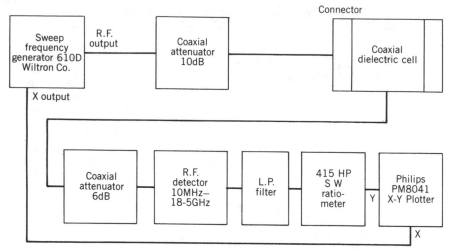

Figure 3.   Block diagram of 2–18 GHz transmission spectrometer.

from 10 MHz to 18.5 GHz. The power level of the signal to be measured is adjusted with the coaxial attenuators so that the detector will be run in its linear range of operation.

The most important part of the system is the dielectric cell. It is designed to have a characteristic impedance, Z_0, with a dielectric of 50 Ω, and is connected on both sides to commercial N-type connectors. The diameter of the inner conductor of the cell increases gradually from an initial size of 1 mm to 3 mm (Fig. 4), whereas the inner diameter of the outer conductor is held fixed at 5 mm. The liquid compartment is 64 cm long and is closed on both sides with specially designed windows (Fig. 4). These windows are made of PTFE (Teflon) of permittivity approximately 2.0, so their refractive index matches that of the dielectric sample. The windows are tapered as shown in Fig. 4 so as to ensure a significant reduction in reflections at the interfaces between the cell and the source on one side and the detector and the cell on the other. The total length of the cell is 91 cm.

It is often desirable to calculate the small increment in attenuation $\Delta\alpha(\omega)$ between an extremely dilute solution and a nonpolar solvent. If the power transmitted by the pure solvent, P_1, can be used as a reference, the transmitted power, P_t, is then independent of the transmission coefficients $|t_{12}|$ and $|t_{21}|$, since these are almost the same for the solutions and their solvents. Therefore, we obtain, from Eq. (26),

$$P_t = P_1 \frac{\exp(-2\Delta\alpha L)}{1 - 2|r_{21}|^2\cos(2\beta L)\exp(-2\Delta\alpha L) + |r_{21}|^4\exp(-4\Delta\alpha L)} \quad (27)$$

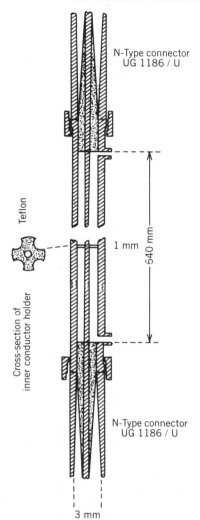

N-Type connector
UG 1186 / U

Teflon

Cross-section of
inner conductor holder

1 mm

540 mm

N-Type connector
UG 1186 / U

3 mm

Figure 4.   A coaxial line dielectric cell.

The maximum departure of the denominator of Eq. (27) from unity for the dilute solutions under discussion is estimated to be 1.5%. Such departures give rise to small variations in $\ln(P_t/P_1)$ versus L plots. The dependence of $\Delta\alpha(\omega)$ on frequency introduces variations in $\ln(P_t/P_1)$ versus f plots, especially at lower frequencies (i.e., 2–10 GHz); these variations gradually decrease as frequency increases (Fig. 5). These signal variations are further reduced using a double T low-pass Tschebychev filter with a cutoff frequency of 12 dB per octave.

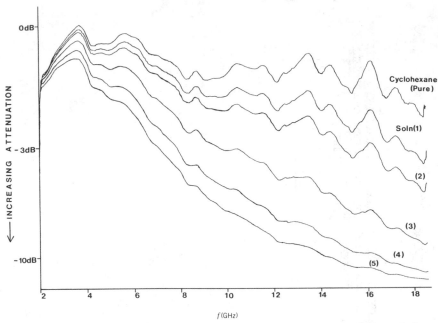

Figure 5.   Plots of transmitted power versus frequency (2–18 GHz) for some dilute solutions.

Differential attenuation, $\Delta\alpha(\omega)$, can therefore be determined from the ratio of the power transmitted by the solvent, P_{t_1}, to that transmitted by the solution, P_{t_2}.

$$\Delta\alpha(\omega) = \frac{1}{2L}\ln\left(\frac{P_{t_1}}{P_{t_2}}\right) \tag{28}$$

$$= \frac{0.1151}{L \text{ (cm)}} D \text{ (dB)} \tag{29}$$

where D (dB) is the difference in transmitted power between the solvent and the solution in decibels. Knowing $\Delta\alpha(\omega)$, $\Delta\varepsilon''$ can be determined using Eq. (19).

The spectrometer displays P_t versus f; curves for a few solutions are given in Fig. 5. It must be emphasized that to determine $\Delta\alpha$ from these plots, one must adjust the reference positions for zero attenuation and infinitely large attenuation accurately. D (dB) can then be determined simply from the curves using

$$D \text{ (dB)} = 10\log_{10}\left(\frac{d_0}{d_1}\right) \tag{30}$$

where d_0 and d_1 are the distances (in arbitrary units) of the curves from the x-axis for solvent and solution, respectively.

D.   Limitations of the Method

This method has two main limitations:

1. It is suitable only for low- to medium-dielectric-loss liquids or solutions (up to $\tan\delta \approx 5 \times 10^{-2}$).

2. The temperature range of measurement is limited because of the danger of the inner conductor of the coaxial cell buckling. The cell, however, is thermostated and the temperature can be varied over a range of 10–60°C.

In a new reflection spectrometer designed and built in Mainz, the frequency range has been extended to the K-band (26–40 GHz). The cell is designed so that a number of samples can be studied one after the other and the temperature can be varied down to that of liquid nitrogen. This is described in the next section.

III.   K-BAND (26–40 GHz) SWEPT-FREQUENCY REFLECTOMETER

A.   Introduction

This method involves the measurement of the attenuation suffered by a K-band signal in traversing a dielectric-filled waveguide. The remote end of the cell is short-circuited by a plunger and the reflected beam is further attenuated by the dielectric. The advantages of this method are:

1. The bottom window of the cell is dispensed with (Fig. 6,) and therefore the problem of sealing is avoided:

2. Measurement of $\varepsilon''(\omega)$ for supercooled liquids is possible:

3. The frequency dependence of $\varepsilon''(\omega)$ can be determined in a few minutes.

B.   Experimental Setup

The block diagram of the reflectometer is shown in Fig. 6. The sweep oscillator employs a traveling-wave tube modulated by a square wave at 1 kHz. The frequency of the oscillator is swept over its full range on manual triggering. The power output of the oscillator (5 mW) is almost independent of frequency. The bidirectional coupler makes possible the use of the oscillator for other experiments. A variable attenuator adjusts the power level so that the output of the detector is linearly related to the input power. A flexi-

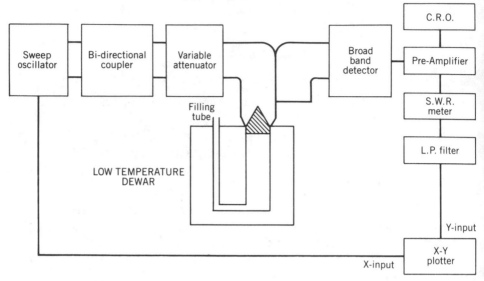

Figure 6.   Block diagram of a K-band swept-frequency reflectometer.

ble waveguide of length 10 cm (attenuation 67 dB m^{-1}) is used to isolate the waveguide cell thermally and mechanically from the oscillator. The waveguide dielectric cell is designed to operate in the TE_{10} mode. When filled with dielectric, the impedance is matched to that of the system. For this purpose, the dimensions of the waveguide cell are those appropriate to the Q-band (5.7 mm $\times$ 2.8 mm) rather than the K-band (7.11 mm $\times$ 3.56 mm). To avoid reflections at the dielectric–air interface, the conical Teflon window has an apex angle of 30°. The reflected wave is separated from the incident wave by means of the directional coupler.

The output of the detector on an X–Y recorder as a function of frequency is shown in Fig. 7 for (1) cell filled with cyclohexane and (2) a cell filled with a dilute solution of acetone. The output is riddled with noise, which indicates the presence of undesired reflections. However, for a fixed length, L, of dielectric, the differential attenuation for a solution, $\Delta\alpha(\omega)$, can be determined, with limited accuracy, using Eqs. (29) and (30).

The value of $\Delta\varepsilon''(\omega)$ is calculated using Eq. (34) with $\lambda_c = 2a = 1.14$ cm. (see next section). The accuracy of $\Delta\varepsilon''(\omega)$ at the moment is estimated to be 5%, but in the near future will be decreased by increasing the signal-to-noise ratio by means of on-line signal processing. The permittivity $\varepsilon'(\omega)$ can be determined using a bridge configuration of the reflectometer in which the phase of the wave reflected from the cell system is compared with that of the reference arm.

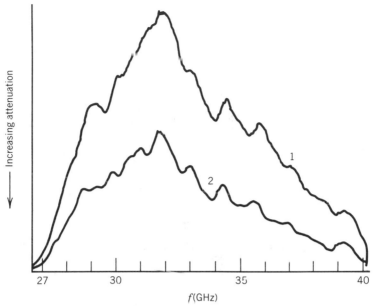

Figure 7.   Plots of reflected power versus frequency f (26.5 40 GHz) for cyclohexane (1) and a dilute solution of acetone (2).

The dielectric cell (Fig. 8) is designed in such a way that measurements can be made on a number of samples one after another at extremely low temperatures. A group of four waveguides is arranged to lie on the circumference of a cylinder so that a particular cell can be placed in the correct position in the spectrometer.

IV.   SPOT-FREQUENCY METHODS

A.   Introduction

The methods for measuring dielectric loss up to a frequency of approximately 75 GHz are based on the transmission of a TE_{10} wave through a waveguide. For this mode, the wave equation (9) is modified such that the propagation factor is given by

$$\gamma = \alpha + i\beta = \frac{i\omega}{c}\left[\varepsilon_r^*(\omega) - \left(\frac{\lambda_0}{\lambda_c}\right)^2\right]^{1/2} \tag{31}$$

where λ_c is the cutoff wavelength for the particular mode of propagation. For the TE_{10} mode, $\lambda_c = 2a$ (where a is the width of the guide). The wave-

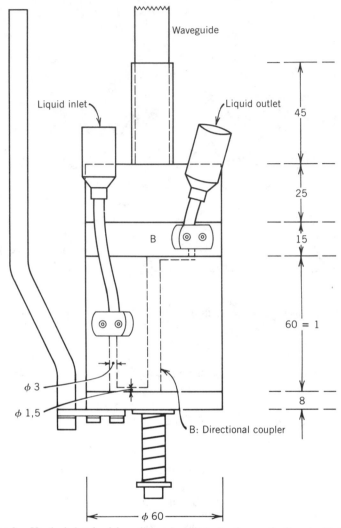

Figure 8.   Vertical sketch of four dielectric cells (reflection method) embedded in a cylinder; shown is one of the four cells that can be inserted in the system on turning it with the lever; dimensions in millimeters. Dimensions in mm.

length in an air-filled guide, λ_g, can be determined from Eq. (31), with $\varepsilon'(\omega)=1$ and $\varepsilon''(\omega)=0$ and $\beta=2\pi/\lambda_g$.

Experimentally, λ_g can be measured easily.

$$\frac{1}{\lambda_g^2}=\frac{1}{\lambda_0^2}-\frac{1}{\lambda_c^2} \tag{32}$$

Equation (31) can be solved for β ($= 2\pi/\lambda_d$) in terms of α, ϵ', λ_0 and λ_c, here λ_d is the wavelength of the wave in the dielectric; $\epsilon''(\omega)$ can also be expressed in terms of α. Explicit expressions for β and $\epsilon''(\omega)$ are as follows:

$$\beta = \frac{2\pi}{\lambda_d} = \frac{2\pi}{\lambda_0}\left[\epsilon'(\omega) - \left(\frac{\lambda_0}{\lambda_c}\right)^2 + \left(\frac{\lambda_0\alpha}{2\pi}\right)^2\right]^{1/2} \tag{33}$$

$$\epsilon''(\omega) = \frac{\lambda_0}{\pi}\alpha(\omega)\left[\epsilon'(\omega) - \left(\frac{\lambda_0}{\lambda_c}\right)^2 + \left(\frac{\lambda_0\alpha}{2\pi}\right)^2\right]^{1/2} \tag{34}$$

Using Eq. (23), Eq. (34) becomes

$$\epsilon''(\omega) = A \cdot D\left[\epsilon'(\omega) - \left(\frac{\lambda_0}{\lambda_c}\right)^2 + \left(\frac{AD}{2}\right)^2\right]^{1/2} \tag{35}$$

where

$$A = \frac{\lambda_0}{20\pi L \log_{10}e} \tag{36}$$

and the free-space wavelength, λ_0, can be calculated either from λ_g using Eq. (32) or from the frequency monitored by a calibrated wavemeter. L is the length of the guide filled with dielectric, as before.

Based on the determination of the attenuation D in decibels for various lengths of the waveguide cell, methods have been developed for operating at spot frequencies of 10, 18, and 75 GHz. However, at 34 GHz, a different approach has been used that does not involve the explicit measurement of L. With it, $\epsilon''(\omega)$ can be measured to an accuracy of within 1%. This method is also capable of determining ϵ' and will be reviewed briefly next.

B. 34-GHz Transmission Method

This method is based on the general ideas about the transmission of microwave power through a dielectric first given by Lane and Saxton[6]. The technique was developed in Mainz by Klages, Hufnagel, and Kramer[7] in 1960 and involves measuring the increase in attenuation with length of the sample in the guide. The value of ϵ' can also be determined for dilute solutions. A cylindrical reservoir (Fig. 9) is filled with the liquid to be tested; a motor-driven piston in the cylinder forces the liquid to rise in the guide. The system, which is fully automated, is shown schematically in Fig. 9. The rate of increase of the attenuation with the depth of the liquid can be calculated from the cross-sectional area of the piston, the inner cross-sectional area of the guide, and the speeds of the motor and the chart recorder.

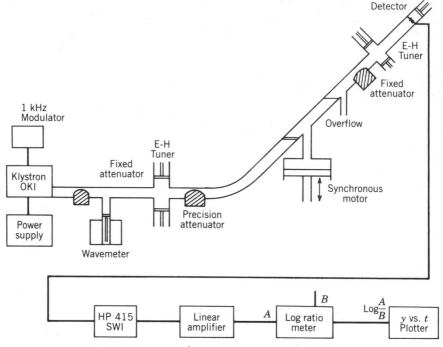

Figure 9.   Block diagram of 34-GHz transmission method.

The detector is a bolometer biased by means of a source of constant current and its demodulated signal is fed into a selective amplifier. The output of the amplifier is directly proportional to the transmitted power. The output of the logarithmic ratio meter is displayed on the $y-t$ chart recorder as a function of time.

The main advantage of this method is that it makes it possible to estimate the magnitude of wall losses, to measure ε'' for nonpolar solvents, and to eliminate completely the effect of these losses for dilute solutions. The dependence of power attenuation on the depth of the dielectric is estimated from the parameters of the experimental system. An expression based on Eq. (35) for $\Delta\varepsilon''$ (the increment in the dielectric loss of a solution from its solvent) is

$$\Delta\varepsilon'' = mA_1 A_2 \left(1 + \frac{\Delta\varepsilon'}{B}\right) \tag{37}$$

where m is the slope of the curve on the plotter (y/t),

$$A_1 = \left(\frac{1}{20\pi \log_{10}e}\right)\lambda_0 \,(\text{cm})\left[\varepsilon' - \left(\frac{\lambda_0}{\lambda_c}\right)^2\right]^{1/2} \tag{38}$$

and A_2 is a parameter of the experimental system given by

$$A_2 = \frac{S_R S_A A_W}{A_R S_M} \tag{39}$$

where S_A is the attenuation in decibels per centimeter on the recorder, S_R is the speed of the recorder in centimeters per second, A_R and A_W are the inner-cross sectional areas of the reservoir and the waveguide cell, and S_M is the speed of the synchronous motor in centimeters per second.

The permittivity ε' is estimated by determining the half-wavelength of the wave in the dielectric cell from the plots on the recorder; then, using a relation derived from Eq. (33),

$$\varepsilon' = \left(\frac{\lambda_0}{\lambda_d}\right)^2 + \left(\frac{\lambda_0}{\lambda_c}\right)^2 - \left(\frac{\lambda_0 \alpha}{2\pi}\right)^2 \tag{40}$$

$$\simeq \left(\frac{\lambda_0}{\lambda_d}\right)^2 + \left(\frac{\lambda_0}{\lambda_c}\right)^2 \tag{41}$$

for the very dilute solutions under discussion.

V. MEASUREMENT OF DIELECTRIC PROPERTIES AT FREQUENCIES OF 145.6 AND 285 GHz— GENERAL PRINCIPLES

For frequencies greater than 75 GHz, the waveguide measuring techniques are unsuitable, mainly because of the increase in the guide wall losses with frequency. At such frequencies, the power attenuation in the dielectric is comparable to that of the waveguide cell; for example, at 140 GHz, the guide wall loss is 11 dB m^{-1}. In addition, the dimensions of the waveguides are small. As a consequence, matching the impedance of the guide cell with that of the system is extremely difficult to achieve. At these frequencies, therefore, free-space propagation through the dielectric has to be attained.

A quasioptical technique at 150 GHz was given by Garg, Kilp, and Smyth[8] and also Kilp[9] involving the microwave analogue of the Michelson interferometer. A technique different from this one, for use at a frequency of 285 GHz, was given by Kilp.[10] This method could be operated in the following modes, depending on the type of dielectric: (1) the usual transmission mode, (2) as a Mach–Zehnder interferometer; and (3) as a Michelson interferometer. The common feature in these modes of operation is that the incident beam is split in two; the modes differ with respect to the manner in which the reference beam is combined with that transmitted or reflected after traversing the dielectric. The measurement techniques have now been

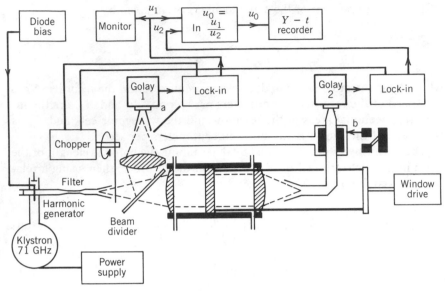

Figure 10.   Block diagram of the 285-GHz (~1 mm) dielectric setup. For dispersion measurements, the setup can readily be changed into a Mach–Zehnder interferometer by inserting a reflector (a) and a second beam splitter (b). (Reproduced by permission from ref. 10.)

modified to make use of the most recent developments in millimeter-wave devices and waveguide components, with a view toward improving the reproducibility and accuracy of the measurements. The techniques are reviewed briefly and the salient features of the newer designs are discussed in Sections VI and VII.

Open-space propagation of a plane TEM wave through the dielectric is the essential requirement. The main problems lie therefore in launching the wave in free space and in the design of a cell in which the reflections at the air–window and dielectric–window interfaces and at the edges are minimized. The diffraction of the plane wave by the cell must be significantly reduced. ε' and ε'' are calculated from the measurements of α, λ_d, and λ_0 using Eqs. (18) ($\beta = 2\pi/\lambda_d$) and (19).

VI.   DIELECTRIC MEASURING TECHNIQUE AT 285 GHz

A.   Introduction

The block diagram of the method of Kilp[10] is shown in Fig. 10. The power source is a klystron (YK 1010), mechanically tunable between 67 and 73

GHz, with a maximum power output of approximately 150 mW. The fourth harmonic frequency is produced by the nonlinearity in the characteristics of a In-doped Si point-contact diode.[11] The harmonics lower than the fourth are cut off by an adjustable waveguide high-pass filter, whereas the output did not show any measurable level of power for a signal higher than the fourth harmonic. The power output of the fourth harmonic at 285 GHz is estimated to be between 1 and 10 μW.

Kilp's design, though extremely imaginative, suffers from three drawbacks that may give rise to serious errors in ε' and ε'' in some instances, namely:

1. The power stability of the harmonic generator over long periods of time is poor.
2. The reference signal is too weak, due to the low reflection coefficient of the beam divider.
3. The collimation of the beam incident on the cell is inadequate; it is necessary to make the beam as near to a perfect plane wave as possible and also to eliminate reflections from the edges of the cell.

The first two defects are remedied by replacing the klystron with a 151-GHz 47138H Impatt diode oscillator with a power output of 15 mW (discussed in Section VII.B). The point-contact diode (In-doped Si) will be producing fundamental and higher harmonics. The fundamental wave is cut off using a high-pass filter; the power level of the third harmonic is measured to be 15 dB less than that of the second. The power level of the second harmonic is further increased by inserting a solid conical polyethylene (PE) connector into the sending horn (Fig. 11). The beam is then fed into an oversized cylindrical copper tube of inner diameter 15 mm and collimated by a specially designed plano-convex lens of variable focal length. A PE lens is screwed into a copper tube holding a 5-mm-thick polystyrene foam sheet (Fig. 11). The density of the sheet, and hence its refractive index, can be varied from 1.05 to 1.17 by varying the extent to which the lens is screwed into the tube. The imaging of the beam is therefore adjusted by optimal adjustment of the lens.

The beam is then amplitude modulated by a 10.8-Hz square-wave signal, which also supplies a reference voltage for the Brookdeal lock-in amplifier. For low to medium dielectric losses (up to $\varepsilon'' \approx 10^{-2}$), the novel part of this technique is the design of the two-chamber cell.[10] These chambers contain liquids, one of known and the second of unknown ε''; the liquids have approximately the same refractive index in order to reduce the effects of reflections at interfaces between the air and the windows. This is discussed in the next section.

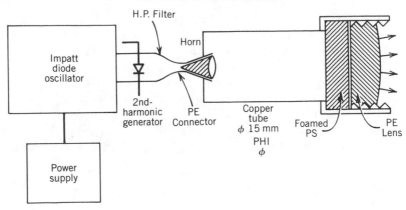

Figure 11.   Collimation of the beam at 302 GHz.

B.   Differential-Absorption Millimeter-Wave Dielectric Cell

The differential-absorption millimeter-wave cell, shown as a part of the 285-GHz setup, in Fig. 10, consists of two chambers separated by a plane window (called the septum) driven by a synchronous motor through a gearbox. The two plano-convex lenses and the septum are made of Teflon with a refractive index of approximately 1.45. The cell chambers are filled with liquids differing in ε'' but having approximately the same ε'. The refractive indices of organic liquids (between 1.3 and 1.6) are within 10% of that of Teflon. As a consequence, less than 0.3% of the power is reflected from the interface between the septum and the two liquids. A part of the incident power is scattered from the convex surfaces of the two plano-convex lenses. Since the scattered waves do not enter either the source or the detector, they can be ignored for purposes of analysis.

When one of the chambers is empty, however, the records of transmitted power show interference from multiple reflections. The partial beams causing the optical interference are shown in Fig. 12. We assume that E_1 is the amplitude of the primary beam transmitted without reflection. E_2 to E_4 are the amplitudes of beams generated from the reflection coefficients (r) of the air–Teflon interfaces and the generator mismatch (ρ). To a first approximation, one can show that the detector power signal, P, is proportional to

$$\left| \sum_{i=1}^{N} E_i \right|^2 \tag{42}$$

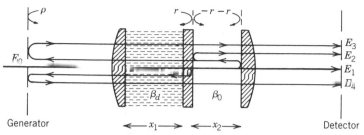

Figure 12.   Schematic of the optical interference between beams when one chamber is filled. (Reproduced by permission from ref. 10.)

and is due to the superposition of three wave periods:

$$x_a = \frac{\pi}{\beta_0} = \frac{\lambda_0}{2}$$

$$x_b = \frac{\pi}{\Delta\beta} = \frac{1}{2}\frac{\lambda_0\lambda_d}{\lambda_0 - \lambda_d}; \qquad \Delta\beta = \beta_d - \beta_0 \tag{43}$$

$$x_c = \frac{\pi}{\beta_d} = \frac{\lambda_d}{2}$$

Here β_0 and β_d are phase constants related to the wavelengths λ_0 and λ_d in air and liquid, respectively. The wave periods correspond to an optical interference of E_2, E_3, and E_4 with the predominant beam E_1.

For purposes of mathematical analysis, we make the following assumptions:

1. The beam in the cell is parallel.
2. There are no reflections at the interface between the two chambers of the cell.
3. The chambers are filled, one with the solution and the other with pure solvent.

On the basis of these assumptions, the power transmitted through the dielectric is given by

$$P_t = P_0|t|\exp(-2\alpha_1 x_1)\exp(-2\alpha_2 x_2) \tag{44}$$

where x_1 and x_2 are the thicknesses of the two liquids corresponding to their absorption coefficients, α_1 and α_2, and $|t|$ is the magnitude of the overall transmission coefficient of the cell.

Since the total length of the cell, $L = x_1 + x_2$, is fixed, Eq. (44) can be written as

$$P_t = P_0 |t| \exp(-2\alpha_2 L) \exp(-2\Delta\alpha x_1) \qquad (45)$$

with

$$\Delta\alpha = \alpha_1 - \alpha_2$$

P_t is an exponential decay function of x_1. Since the output signal is proportional to $\ln(P_t/P_0)$, its slope ($= 2\Delta\alpha$) is used to calculate $\Delta\varepsilon''$ using Eq. (19).

The cell is especially suitable for differential-absorption measurements; the accuracy of α_1 is dependent on that of α_2. From these and other independent experiments,[12] we know that cyclohexane has one of the lowest absorption coefficients in the millimeter-wave range, and it can therefore be used as a reference liquid.

C. Transmission Mode

The salient feature of the transmission mode (as in Kilp's design), particularly for a source with a power level of a few microwatts, is that Golay detector 1 (Fig. 10) receives a fraction of the radiation, enabling monitoring of the output of the harmonic generator. This signal is used as a reference in a logarithmic ratio meter, the output of which is then independent of some small fluctuations and drifts in the power of the harmonic generator. In the present design, Golay detector 1 is dispensed with due to a significant (at least 10-fold) increase in the power level of the source. The reference for the logarithmic ratio meter is derived from a dc power supply.

This mode of operation is suitable for low-loss liquid solutions. The linear dependence of the logarithmic ratio of the transmitted to the incident power on the sample thickness is used to determine $2\Delta\alpha$ and hence $\Delta\varepsilon''$ using Eq. (19).

D. Mach–Zehnder Interferometer Mode

In the Mach–Zehnder mode, a 45° mirror ("a" in Fig. 10) is switched in front of the monitor detector. This mirror reflects the reference beam into a second waveguide joining the first at an angle of 90°. A second 45° beam divider ("b" in Fig. 10) is placed in the T-junction and combines both beams to produce an interference signal to be recorded in detector 2. In this case, the resulting signal has a wave period $x_d = 2\pi/\Delta\beta$, where $\Delta\beta = \beta_1 - \beta_2$; β_1 and β_2 are phase constants for liquids in chambers 1 (dilute solution) and 2 (solvent), respectively.

It can easily be shown that

$$\frac{\lambda_0}{\lambda_d} \simeq 1 + \frac{\lambda_0}{x_d} \qquad (46)$$

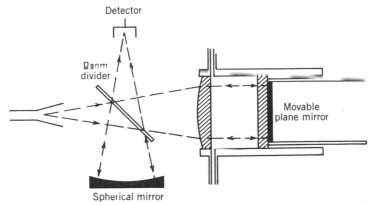

Figure 13.   Michelson interferometer for dispersion measurements in high-loss liquids. (Reproduced by permission from ref. 10.)

$\Delta\varepsilon''$ is calculated from the slope ($= 2\Delta\alpha$) as in the previous section, whereas ε' is calculated from an average λ_d using an equation derived from Eq. (18):

$$\varepsilon'(\omega) \simeq \left(\frac{\lambda_0}{\lambda_d}\right)^7 \tag{47}$$

E.   Michelson Interferometer Mode

For high-loss liquids, the amplitude of the transmitted signal in the transmission method is reduced considerably. The reference signal is also too weak, since it is being reflected twice from the beam divider. This makes the fringe contrast between its maxima and minima too weak for an accurate measurement of the wavelength in the dielectric. The Michelson interferometer mode has been used to measure ε' and ε'' in chlorobenzene. The reference beam and the beam reflected from a plane mirror fixed to the movable window (Fig. 13) are approximately of the same amplitude α; λ_0 and λ_d are determined from the plots of $\ln(P/P_0)$ versus displacement of the septum displayed on the chart recorder.

An expression for ε'' for the present setup based on Eq. (19) can be easily derived:

$$\varepsilon'' = \log_e 10 \frac{c\sqrt{\varepsilon'}}{f\pi} \frac{S}{v_m v_s} M \tag{48}$$

where v_m is the speed with which the septum is moved (cm s^{-1}) and is controlled by the motor, v_s is the speed of the chart recorder (s cm^{-1}), S is the sensitivity of the recorder (V cm^{-1}), and M is the slope of the curve on the plotter.

VII. DIELECTRIC MEASUREMENT AT 145 GHz

A. Introduction

Garg, Kilp, and Smyth[8] in Princeton were the first to set up the microwave analogue of the Michelson interferometer for ε' and ε'' measurements of nonpolar and slightly polar liquids. They used the second harmonic component of a 71-GHz klystron (Amperex DX151). The recent developments in semiconductor high-frequency oscillators, in particular, the Impatt diode and several millimeter-wave components, have led to refinements of the basic technique. These have led in turn to improvements in the frequency stability and reproducibility, and consequently in the accuracy, of dielectric measurements of liquids over a range of ε''. The equipment is now designed to be operable both in the transmission mode and the Michelson interferometer mode with a specially designed beam splitter.

B. Transmission Method

The experimental setup is shown in fig. 14. The Impatt diode (47138H-1101) oscillator generates a 145.6-GHz signal with a power level of approximately 5 mW. The diode is a sandwich of four diffused layers of $n^+ p n_i p^+$ Read[13]–Misawa[14] type (n_i is the number density of carriers in intrinsic silicon). The diode generates millimeter-wave radiation, the frequency of which depends on the doping depth when it is operated on its reverse-bias avalanche mode. The signal is pulse modulated at a frequency of 1 kHz. A combination of a radiating horn and a lens converts the signal into a near-perfect plane wave propagating through the dielectric.

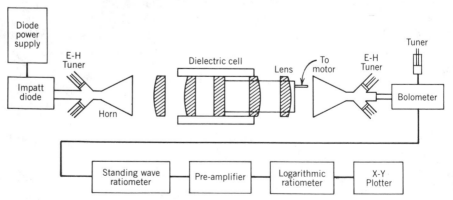

Figure 14.   Block diagram of 146-GHz transmission setup.

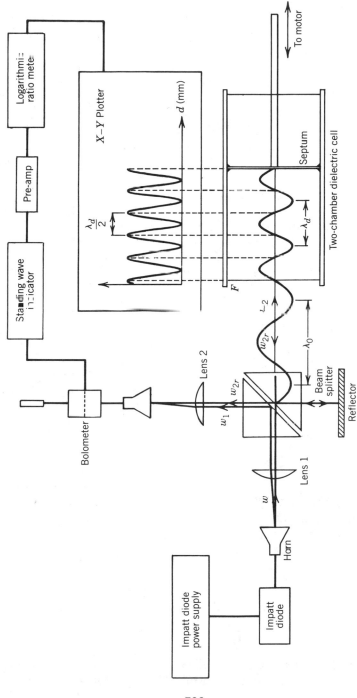

Figure 15. A 146-GHz Michelson interferometer.

The method makes use of the two-chamber cell already described in Section VI.B. The septum of the cell is driven at constant velocity by a synchronous motor through a variable-ratio gear box. The signal is received by the horn and fed into an E-H tuner, which is used to match the wave impedance with the line. The $X-Y$ plotter produces a logarithmic plot of the transmitted signal, from which the differential-absorption coefficient, $\Delta\alpha$, is calculated.

The independence of $\Delta\alpha$ for a solution from its position in a chamber, whether in a chamber to the left or right of the solvent, proves that the incident beam is well collimated and furthermore that it behaves as an almost perfect plane wave. In an earlier design,[9] systematic errors in permittivity and dielectric loss measurements were detected for the solution on interchanging its chamber with that for the solvent. The success of the design for this method is dependent on (1) the high stability of the frequency and power output of the Impatt diode oscillator and (2) the improvement in the collimation of the incident beam.

C. Michelson Interferometer Mode

For high-loss liquids, the transmitted signal through the cell is too weak to enable an accurate measurement of $\Delta\alpha$ to be made. Hence, a different mode of the interferometer must be used.

The apparatus shown in Fig. 14 is readily converted into the Michelson interferometer mode (Fig. 15). A special beam divider has been designed using a PE cube cut along its diagonal. The two halves of the cube are mounted so as to maintain a distance of 0.6 mm between them, with the result that the incident beam is split in two equal amplitudes. An earlier design[9] used a Mylar beam divider; its drawback was that the reference signal was too weak to produce a reasonable fringe contrast between its maxima and minima.

Another important feature of the present design is that a metal reflector used earlier[9] to reflect the signal that had passed through the dielectric has been dispensed with, which was possible because of the stability in power and frequency of the oscillator output in the new design. A Teflon–air interface is sufficient to produce the good fringes. The values of ε' and ε'' are calculated from the determined values of α, λ_d, and λ_0 using Eqs. (47) and (48).

VIII. REFRACTIVE INDICES OF SOLIDS AT MILLIMETER WAVELENGTHS

A. Introduction

Chamberlain and Gebbie[15] developed an interferometric technique for determining the refractive index of a solid. The path difference between the

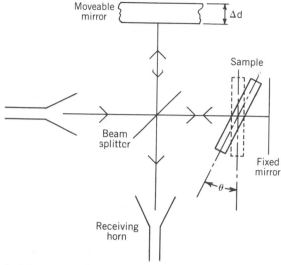

Figure 16.   Chamberlain–Gebbie interferometer at 146 GHz for measuring refractive index of a solid.

reference and the unknown arm is altered by changing the angle of incidence of the sample (Fig. 16). Their system was illuminated with monochromatic radiation of 0.337 mm wavelength from a C–N maser. This method has now been extended in Mainz to a 2-mm-wavelength apparatus driven by an Impatt diode oscillator and modified to yield values of n accurate to 1 part in 10^3. The principle of the method will now be reviewed and the setup described. Results of measurements on three polymers used as material for windows and for optical lenses in millimeter- and submillimeter-wave applications are given.

B.   Principle of the Method

If the mirrors of a Michelson interferometer (Fig. 16) illuminated with monochromatic radiation of wavelength λ are set in the zero-path-difference position, the insertion into one arm of a plane specimen of refractive index n and thickness t, placed so that the angle of incidence is θ, produces a path difference $x(\theta)$, given by

$$x(\theta) = 2t\left[(n^2 - \sin^2\theta)^{1/2} - \cos\theta\right] \tag{49}$$

If the specimen is rotated so that the angle of incidence changes from 0 to θ, the path difference increases by the amount

$$x(\theta) - x(0) = 2t\left[(n^2 - \sin^2\theta)^{1/2} - \cos\theta + 1 - n\right] \tag{50}$$

As a result, m fringes pass the detector, where

$$x(\theta) - x(0) = m\lambda \tag{51}$$

or

$$2t\left\{(n^2 - \sin^2\theta)^{1/2} - \cos\theta + 1 - n\right\} = m\lambda \qquad (52)$$

or

$$(n^2 - \sin^2\theta)^{1/2} = \frac{m\lambda}{2t} + \cos\theta + n - 1 \qquad (53)$$

On squaring both sides and rearranging the terms, we obtain:

$$y = 2nx - 1 \qquad (54)$$

where

$$y = \left(1 - \frac{m\lambda}{2t}\right)\left\{\left(1 - \frac{m\lambda}{2t}\right) - 2\cos\theta\right\} \qquad (55)$$

and

$$x = \left(1 - \frac{m\lambda}{2t}\right) - \cos\theta \qquad (56)$$

Chamberlain[16] determined n from a plot of y versus θ, with θ as an independent variable and m determined experimentally. From Eq. (51) we note that integral values of m are obtained for discrete values of θ only. For any other θ, m may not be an integer and values of n for materials may therefore be subject to error.

C. Experimental Technique and Results

The problem just mentioned is solved with a new technique whereby the path difference $x(\theta) - x(0)$ is compensated for by moving the mirror (Fig. 16) to a new position. This mirror is attached to a precision micrometer and can be displaced forward or backward at right angles to its plane. The path difference, $\Delta d = x(\theta) - x(0)$, is measured repeatedly and averaged for a particular θ value, for both clockwise and anticlockwise rotation of the sample, to take into account the fact that the incident beam may not be a perfect plane wave.

Plots of Δd versus θ are observed to be nonlinear for the polymers. To determine n, y values are plotted against x, as done by Chamberlain[15] on

TABLE I
Refractive Indices for Polymers at 2.2 mm

Material	n ($\lambda = 2.2$ mm)
Polystyrene	1.545 ± 0.01
Polymethylacrylate	1.625 ± 0.01
Polytetrafluorethylene (Teflon)	1.415 ± 0.01^a

^{a}Literature value of n ($\lambda = 0.337$ mm) = 1.391 ± 0.017 (from Chamberlain[15]).

replacing $m\lambda$ with Δd in Eq. (55) and (56). Measurements of n for three polymers are given in Table I.

IX. MILLIMETER-WAVE SPECTROSCOPY—KREMER'S UNTUNED CAVITY RESONATOR METHOD

A. Introduction

Kremer[17,18] has reported a new spectroscopic method for measuring ε'' values of solid or liquid samples, of any well-defined shape, as a function of frequency in the millimeter-wave range. The method is suitable for studies down to the temperature of liquid helium. The method is based on the use of untuned cavity resonators, first suggested by Lamb[19] for the measurement of microwave absorption spectra of gases. Gebbie and Bohlander[20] adapted these ideas to the measurement of power absorption in gases at submillimeter wavelengths (12–45 cm^{-1}).

This method offers great potential for the future development of a millimeter-wave dielectric spectrometer for studies of samples of arbitrary shape at temperatures as low as that of liquid helium. A spectrometer is being developed in Mainz for high-precision $\varepsilon''(\omega)$ measurements on dilute solutions, details of which will be published later.[21]

The theory of the method and the development of Kremer's technique are now reviewed briefly.

B. Theory

If a millimeter-wave source supplies radiation to a greatly oversized resonator, then an almost isotropic and homogeneous electric field is created inside a high-Q cavity. The quality factor, Q, of such a multimode resonator is only slightly frequency dependent, provided that the wavelength remains very small relative to the cavity dimensions and that the isotropy and homogeneity of the field are also maintained through the use of a mechanical mode stirrer.

The sample to be measured is introduced into the cavity, and the resultant change in Q can easily be measured with good precision. Since any radiation scattered or reflected by the sample is returned to the cavity, the measured loss is due entirely to absorption by the sample.

Following Lamb,[19] the quality factor of a cavity is defined as

$$Q = \frac{\omega \times \text{number of photons stored in cavity}}{\text{rate of loss of photons from cavity}} \qquad (57)$$

where ω is the angular frequency of the radiation in the cavity in rads per second.

The theory of the measurement technique can be developed in terms of Q factors, as was done by Gebbie and Bohlander,[20] or it can be formulated in terms of rate equations.

Let N_0 represent the number of photons in the cavity when no sample is present, and N_s, the number after the sample is introduced. If the radiation is fed into the cavity at a constant rate M, an equilibrium population will then be established that satisfies the following two equations: (1) With the cavity empty,

$$\frac{dN_0}{dt} = 0 = M - l_R N_0 \tag{58}$$

and (2) with the sample in the cavity,

$$\frac{dN_s}{dt} = 0 = M - l_R N_s - l_s N_s \tag{59}$$

where l_R and l_s represent the loss coefficient (due to the wall or scattering loss at the source and the detector) in the untuned resonator and the absorption in the sample, respectively. The detector signal, within its linear range, is directly proportional to the photon flux. The latter is taken to be homogeneous and isotropic throughout the cavity.

Let the ratio of the detector signal with the sample present to that measured with the cavity empty to be R_s. Then

$$R_s = \frac{N_s}{N_0} \tag{60}$$

Combining Eqs. (59) and (60), we get

$$l_s = l_R \left(\frac{1 - R_s}{R_s} \right) \tag{61}$$

A more convenient method is to compare two samples with different loss coefficients (l_{s1} and l_{s2}). Equation (61) then becomes

$$\frac{l_{s2}}{l_{s1}} = \frac{R_{s1}}{R_{s2}} \cdot \frac{1 - R_{s2}}{1 - R_{s1}} \tag{62}$$

It can be shown quite easily that the loss coefficient is proportional to the corresponding integral representing the total loss by absorption in the sample:

$$l_{s1} \propto I_1 = \int_0^{\pi/2} A_T(\theta_0, \varepsilon', \varepsilon'', d, \lambda) \sin\theta_0 \cos\theta_0 \, d\theta_0 \tag{63}$$

where A_T is an absorption factor explicitly dependent on the real and imaginary parts of the complex permittivity of the sample, the angle of incidence θ_0, the thickness of the sample d, and the wavelength λ_0 of the radiation. Equation (62) can be rewritten in terms of a ratio of integrals as

$$\frac{I_2(\varepsilon', \varepsilon'', d_1, \lambda_0)}{I_1(\varepsilon', \varepsilon'', d_2, \lambda_0)} = \frac{R_{s1}}{R_{s2}} \cdot \frac{1 - R_{s2}}{1 - R_{s1}} \tag{64}$$

For a given λ_0 value, measurement of R_{s1} and R_{s2} fixes the ratio of the two integrals. Numerical methods in conjunction with calibration techniques have been devised to determine ε' and ε'' for a sample from Eq. (64). Details of how ε'' is related to the ratio (I_2/I_1) will be published later.[21]

The value of ε'' for a sample is found to be almost independent of the sample geometry, provided a homogeneous electric field is created inside the resonator.

C. Experimental System

The experimental system of Kremer and Genzel[18] is shown in Fig. 17a. The oversized cavity (resonator) is shown in more detail in Fig. 17b. The resonator is made out of two aluminum hemispheres, one of which is fixed and the other of which (the "mode stirrer") rotates at a frequency of 25 Hz. The split between the two hemispheres is less than 0.15 mm wide, and the Q value of the cavity is of the order of 10^5. The inner surface of the resonator is randomly pitted with a spherical bit. This inner structure and the rotating mode-stirrer are required in this method to minimize the degree of degeneracy of the modes in the resonator. The detector signal is amplified and averaged with a time constant that is long compared with the period of the mode-stirrer in order to integrate over the modes of the resonator.

The frequency of the input radiation is varied within the frequency range 26.5–40 GHz using a Hewlett-Packard traveling-wave tube oscillator, whereas a signal of discrete frequency at 75 GHz comes from a Gunn diode oscillator. The diameter of the resonator is 120 cm, at least 100 times the longest wavelength of the input radiation.

X. ABSORPTION COEFFICIENT AND REFRACTIVE INDEX MEASUREMENTS USING SUBMILLIMETER LASERS

A. C–N Laser

FIR stimulated emission at 0.337 ± 0.001 mm in HCN and related molecules subjected to pulsed electrical discharges was first observed by Gebbie, Stone, and Findlay[22] in 1964. A peak power of 10 W was estimated for these

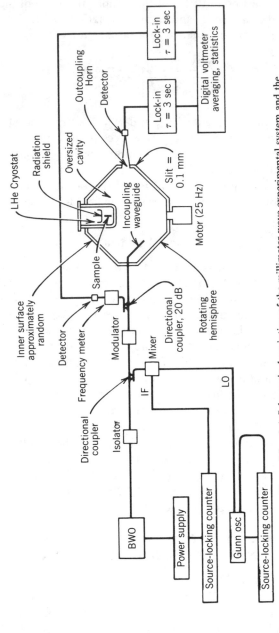

Figure 17. (a) Schematic description of the millimeter-wave experimental system and the oversized cavity (with cryostat). (b) Schematic description of the oversized cavity. From F. Kremer, A. Poglitsch, D. Böhme, and L. Genzel, in K. Button, ed., *Infrared and Millimeter Waves*, Vol. 11, (Academic Press, New York), 1979, p. 141. Reproduced by permission.

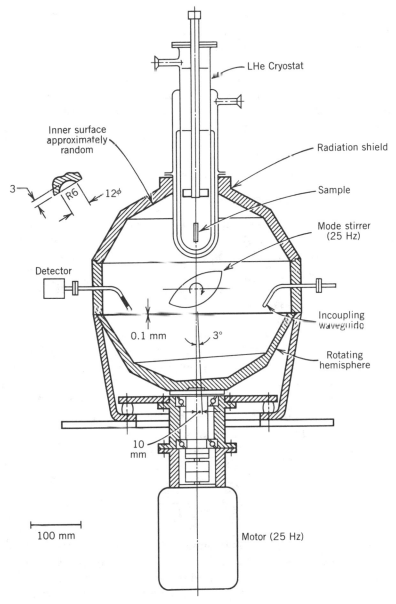

Figure 17. (*Continued*)

807

pulses. However, the source could not be exploited for measurement of the power absorption coefficient, $A(\omega)$, and refractive index, $n(\omega)$, due to the lack of fast detectors. In subsequent work by Gebbie et al.,[23] an estimated 10 mW of continuous power was obtained in CH_3CN, which they used as a source in measurements of A and n for halogenobenzenes. Using this source, Chamberlain et al.[15] designed a new technique for measuring n for solids.

Optical constants (A and n) at this wavelength were established for a number of liquids and solids; these are listed by Chantry.[24] A and n values determined using a laser as the source were in agreement with those obtained using broad-band interferometric techniques. This made the laser technique complementary rather than competitive.

B. Optically Pumped Methanol Laser

Optically pumped submillimeter-wave continuous lasing in methanol was first reported by Chang et al.[25] in the U.S.A. in 1970. The study was followed by the work of Dyubko et al.[26,27] in Russia. By 1980 a total of 83 lines at wavelengths ranging from 37.9 to 1223.7 μm $\bar{\nu} = 264$–8.2 cm^{-1} had been discovered by Henningsen[28] in Copenhagen, and Petersen et al.[29] at the National Bureau of Standards in Colorado.

Reid[30] was the first to set up an optically pumped methanol laser (Apollo Instruments Inc., U.S.A.) and to measure $A(\omega)$ for methylene chloride, which he did at 15 wavelengths ranging from 47 to 300 μm ($\bar{\nu} = 33$–213 cm^{-1}). This was followed soon by work on a number of liquids by Vij et al.[31,32] and Reid et al.[33] A number of new lines[30,31] were also observed, and some of these have now been independently confirmed by Hufnagel and Helker,[34] who discovered a few more new lines. These are listed in Tables II and III, along with CO_2 pump lines.

The plane of polarization is measured with reference to a CO_2 laser. The level of polarization, P_L, is defined as

$$P_L = \frac{P_{max} - P_{min}}{P_{max} + P_{min}} \tag{65}$$

where P_{max} and P_{min} are the maximum and minimum powers (in the direction of polarization and at right angles to it, respectively). These power levels are measured using a parallel wire-grid polarizer interposed between the exit from the resonator and the detector. The lower and upper pressure limits in the resonator are for methanol pressure at half of the maximum power output.

The wavelengths of the submillimeter-wave laser were measured using a Fabry Perot interferometer (Fig. 18). This consists of two mirrors separated by a distance of 1 m; one of the mirrors is fixed, whereas the other is moved

TABLE II
Optically Pumped CH_3OH Wavelengths

λ (μm)	$\bar{\nu}$ (cm^{-1})	Relative power	Polarization	Pump line	λ Pump line
42.2	237.0	Strong	⊥	9P32	—
46.9[a]	213.2			9P6/8	(9.45)
70.5	141.8	Strong	⊥	9P34	(9.68)
77 9	128.4			10R16	(10.27)
96.5	103.6	Strong	‖	9R10	(9.33)
108.0[a]	92.6			9R4	(9.37)
118.0	84.7			9P14	(9.48)
118.8	84.2	Strong	⊥	9P36	(9.69)
142.0[a]	70.4			9P22	(9.57)
151.3	66.1			9R26	(9.26)
163.0	61.3	Strong	‖	10R38	(10.14)
174.0[a]	57.5			9P4/6	(9.43)
191.3[b]	52.3	Strong	‖	9R16	(9.3)
204.0[a]	49.0			10R8/10	(10.33)
300.0[a]	33.3			10P26/28	(10.61)

[a] Weak new lines.
[b] Strong new line (also confirmed in Mainz).

TABLE III
Optically Pumped CH_3OH New Lines on 9R16 CO_2 Pump Line

				Wavelength of Pump Line = 9.293 μm (9R16)			
λ (μm)	Power output (mW)	Limits of pressure in resonator (mbar)	Pressure for optimum power (mbar)	Plane of polarization	Level of polarization	$^{12}C^{16}O_2$ Pump line λ_0 (μm)	Power in CO_2 laser (W)
(191.6 ± 0.20)	6	0.28–0.53	0.4	‖	0.5	9.293 (9R16)	7
(66.35 ± 0.05)	7	0.12–0.53	0.3	‖	0.3	9.293 (9R16)	
(56.84 ± 0.04)	7	0.11–0.43	0.25	⊥ (?)	0.1	9.293 (9R16)	
(36.70 ± 0.05)	2	0.3–0.5	0.4	‖	0.4	9.293 (9R16)	
(47.80 ± 0.04)	7	0.09–0.42	0.25	⊥	0.6	9.569 (9P22)	8.5
(41.91 ± 0.04)	7		0.25			9.519 (9P16)	7.5

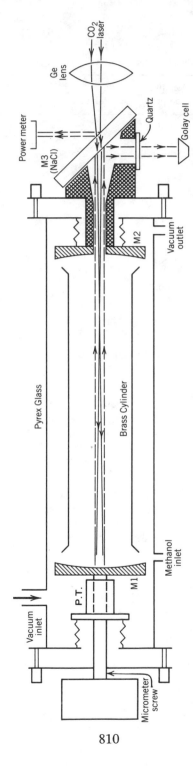

Figure 18. Schematic diagram of a FIR Laser. M_1 and M_2 are gold-coated mirrors of focal length 5 m each; M_2 has a hole of diameter 2 mm inclined at an angle of 2.2° to the laser axis. M_3 is a NaCl window of thickness 5 mm, and PT denotes piezoelectric translator.

through a distance of 1 mm on being coupled to a precision micrometer driven by a synchronous motor. The submillimeter radiation is reflected by a fixed NaCl mirror and is monitored with a Golay detector and a lock-in amplifier. In a separate exercise, the output was monitored using a Grubb–Parsons interferometer. The signal was observed to be purely sinusoidal, which is indicative of the monochrome nature of the line. The power level of the submillimeter radiation was found to lie between 1 and 7 mW, depending on the line.

The wavelengths of the CO_2 pump lines were measured using a double-grid monochromatic spectrum analyzer (manufactured by Oriel Co., W. Germany).

C.   Measurement of the Absorption Coefficient, $A(\omega)$, Using a Submillimeter-Wave Laser

A block diagram of a submillimeter-wave laser system is shown in Fig. 19. The signal for the submillimeter-wave laser for a particular line (Tables II and III) is stabilized by adjusting the CO_2 pump lines and the length of the resonator, which contains methanol. The pressure of methanol in the resonator cavity is adjusted for optimum output.

The laser system can be operated in a mode such that the CO_2 power can be chopped on external application of a signal at frequencies between 10 and 15 Hz. In an alternate mode the CO_2 laser is operated continuously. The submillimeter wave is chopped externally using a Brookdeal chopper at a

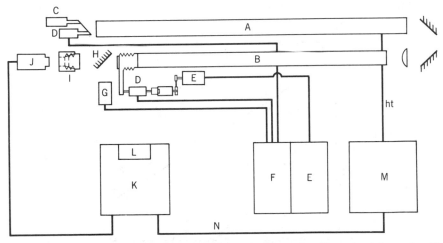

Figure 19.   Block diagram of an Apollo laser. A, CO_2 cavity; C, grating micrometer; D, piezoelectric units; E, step motor; F, L.C.S. 1 and L.C.S. 2 feedback amplifiers; G, pyroelectric detector; H, beam divider; I, variable-path-length cell; J, Golay detector; K, lock-in amplifier; L, voltmeter; M, CO_2 console; N, reference signal; ht, high tension. (Reproduced by permission from ref. 30.)

frequency of 10.8 Hz. The detector system consists of a Golay detector (Unicam IR50) in conjunction with a lock-in amplifier; the reference to the lock-in amplifier was fed from the chopper. Wavelengths of the CH_3OH lines have been measured using this laser, as reported in the previous section.

The absorption coefficient is calculated using the Beer–Lambert law:

$$I(d) = I_0(d_0)\exp\left[-A(d-d_0)\right] \tag{66}$$

where I_0 and I are the beam intensities for sample lengths (in centimeters) d_0 and d, respectively. To measure A, the sample is placed in a variable-path-length cell (VCOI, made by Analytical Accessories Ltd., Kent) between the exit and the detector. The initial path length was chosen as a compromise between obtaining a linear detector response and a high signal-to-noise ratio. The output of the detector, which is proportional to the beam intensity, is monitored as a function of the variable sample path length d. A plot of $\ln I$ versus d is then made. The linearity of the plot[32] (Fig. 20) shows the stability of the laser system, with respect to both amplitude and frequency.

The major advantage of the laser system lies in the extra power available at a single wavelength. This is used to penetrate an accurately measurable thickness of intensely absorbing liquid. The second advantage is that the difference in the thickness, $(d-d_0)$, rather than the actual thickness, is used in Eq. (66). The difference in the thickness is more accurately measurable in a variable-pathlength cell. An accuracy of 3–5% is estimated for $A(\bar{\nu} = \omega/2\pi c)$ in intensely absorbing liquids such as water and methanol. For

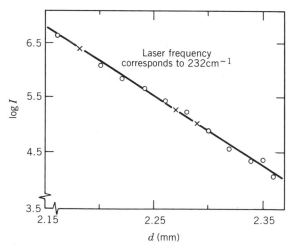

Figure 20.   Plot of $\log_{10} I$ versus d (sample thickness): $\bigcirc$, d increasing; $\times$, d decreasing. (Reproduced by permission from ref. 32.)

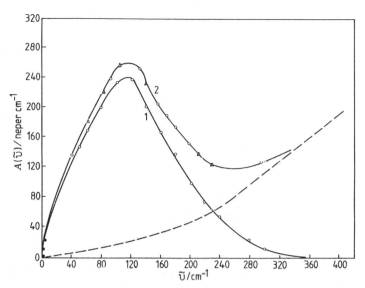

Figure 21. Power absorption coefficient $A(\bar{\nu})$ of pure methanol: interferometric (○) and microwave (●) ε'' measurements; △, laser points. Curve 1, corrected from higher-frequency dispersion (shown by dotted lines); curve 2, experimental (Reproduced by permission from ref. 32.)

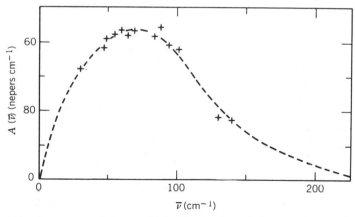

Figure 22. Interferometric (---) and submillimeter-laser (+) measurements of power absorption coefficient $A(\bar{\nu})$ as a function of wavenumber $\bar{\nu}$ for CH_2Cl_2. (Reproduced by permission from ref. 33.)

813

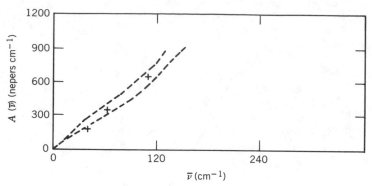

Figure 23. Power absorption coefficient $A(\bar{\nu})$ as a function of wavenumber $\bar{\nu}$ for water. Symbols as in Fig. 22. (Reproduced by permission from ref. 33.)

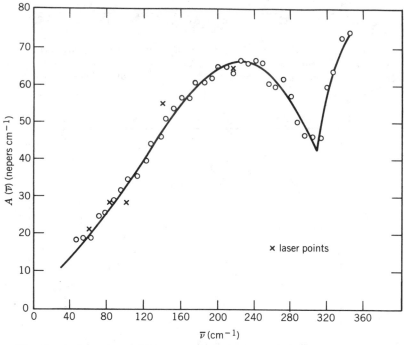

Figure 24. Power absorption coefficient $A(\bar{\nu})$ for heptanol-4: interferometric ($\bigcirc$) and laser ($\times$) measurements. (Reproduced by permission from ref. 31.)

814

weakly to moderately absorbing liquids, such as the anilines and chlorobenzene, the results of laser spectroscopy are complementary to those obtained from interferometry. Curves of $A(\bar{\nu})$ (neper cm^{-1}) versus $\bar{\nu}$ (cm^{-1}) are shown in Figs. 21–24 for some of the intensely absorbing liquids.

XI.   FREQUENCY DEPENDENCE OF DIELECTRIC LOSS FOR KETONES

A.   Introduction

In this section we present results of dielectric measurements on ketones in the microwave and submillimeter-wave regions. These results are analyzed in terms of the comparative behavior of ketone molecules, as well as with molecules of identical shape, when subjected to an alternating field. It has been known for some time that aliphatic ketones show, in general, Debye-like dielectric dispersion in the microwave frequency range, in contrast to that found for alkyl bromides, alkyl ethers, and alcohols. The Debye times, τ_D, are also known to be much lower compared than those of the corresponding alkyl bromides, with a ε'' peak lying in the millimeter-wave region. With this in mind, and in order to use the sophisticated high-precision millimeter-wave dielectric measuring techniques and Bruker submillimeter-wave spectrometer, we have made a systematic study of a number of ketones in cyclohexane and decalin solutions.

Figures 25–28 show $\varepsilon''(f)$ versus f for solutions of some ketones in cyclohexane and decalin.

B.   Bézot–Quentrec–Mori Formalism

Molecular parameters are deduced by fitting the observed dependence of the dielectric loss on frequency to the Bézot–Quentrec[35] equation. This equation is based on the semiempirical formalism of Mori,[36,37] with second-order truncation in the memory kernel.

In the form expressed by Delker and Klages,[38] the Bézot–Quentrec equation[35] becomes

$$\frac{\Delta\varepsilon''}{\varepsilon_s - \varepsilon_\infty} = \frac{C_1\omega\tau_{D1}}{1+\omega^2\tau_{D1}^2} + \frac{C_2\omega\tau_{D2}}{1+\omega^2\tau_{D2}^2} + \frac{C_3\omega T_P}{\left(1-\omega^2 T_P T_J\right)^2 + \omega^2\left(T_P - \omega^2 T_P T_J T_M\right)^2}$$

with

$$\tag{67}$$

$$C_1 + C_2 + C_3 = 1 \tag{68}$$

Here C_1, C_2, and C_3 are the weight factors for the three processes: two Debye type with dielectric relaxation times τ_{D1} and τ_{D2}, and the third, the Poley resonance absorption with a time constant of T_P. The molecular significance of the constants T_P, T_J and T_M will be discussed later.

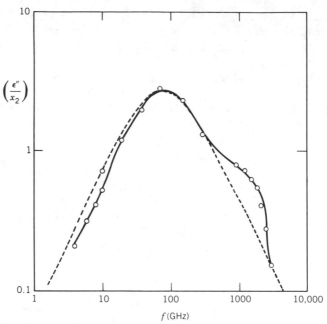

Figure 25. Plots of (ε''/x_2) versus f (GHz) on log–log scale for acetone in cyclohexane; where x_2 is the mole fraction: unbroken line ($\odot$), experimental; dashed line, Debye function with a single τ_D.

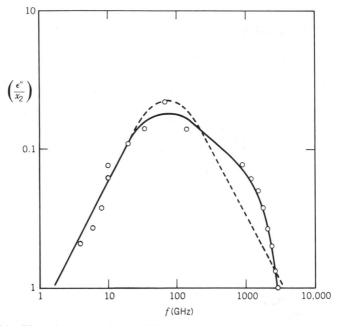

Figure 26. Plots of (ε''/x_2) versus f (GHz) for acetone in decalin; symbols as in Fig. 25.

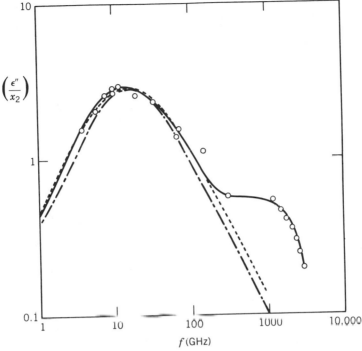

Figure 27. Plots of (ε''/x_2) versus f (GHz) for undecanone-6 in cyclohexane; unbroken line, ($\odot$), experimental; —·—·—, Debye function with a single τ_D; - - -, Fröhlich $P_2 = 1.8$.

Equation (67) is based on the knowledge, gained dealing with rigid molecules, than almost 90% of the contribution to the dispersion step, $\varepsilon_s - \varepsilon_\infty$, is due to Debye-type absorption (with one or more than one macroscopic relaxation time). The remaining 10% contribution comes from the resonance-type (Poley) absorption. Furthermore, the time constants for the two processes differ by a factor of at least 20. We are therefore justified in separating the two processes, at least qualitatively, to arrive at some time constants. These can be used to study the comparative behavior of different molecules of a given type in the presence of an alternating electric field. The last term in Eq. (67) is similar to the Evans[39]–Reid[40]–Mori formalism, except that in that equation,

$$K_0(0) = \frac{1}{\tau_D T_J} \quad \text{and} \quad K_1(0) = \frac{1}{T_J T_M} \tag{69}$$

The frequency for which $A(\omega) = \omega \varepsilon''(\omega)/[n(\omega)c]$ is a maximum is determined from Eq. (67) to be

$$f_{max}\left(= \frac{\omega_{max}}{2\pi} \right) = \frac{1}{2\pi\sqrt{T_P T_J}} \quad \text{Hz} \tag{70}$$

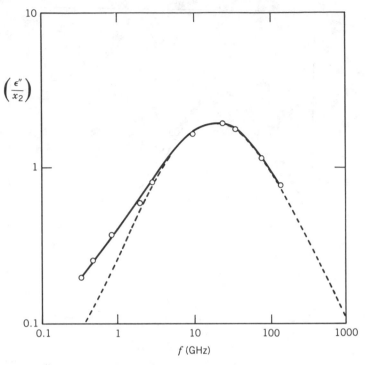

Figure 28.   Plots of (ε''/x_2) versus f (GHz) for heptanone-4 in decalin; straight line ($\odot$), experimental (up to 140 GHz); dashed line, Fröhlich $P_2 = 2.0$.

C.   Debye Rotational Diffusion

For $\tau_{D2}/\tau_{D1} \leq 5$, it is not possible to discern with certainty, from an experimental $\varepsilon''(f)$ versus f curve with an overall accuracy of 3% in ε'', whether or not these times are discrete. A distribution of relaxation times between the limits τ_1 and τ_2 proposed by Fröhlich[41] is a physically acceptable picture for the case in which the experimental curves depart slightly from the perfect Debye $\varepsilon''(f)$ versus f curve. In Fröhlich's model,[41] a polar molecule encounters a distribution of heights of the potential barriers (H) in the condensed phase. These heights are equally distributed over a range between H_0 and $H_0 + v$ such that

$$H = H_0 + v, \qquad 0 \leq v \leq v_0 \tag{71}$$

Thus v_0 is the maximum perturbation in the height of a potential barrier from an unperturbed height of H_0.

The dielectric relaxation time lies between the limits

$$\tau_1 \leq \tau \leq \tau_2 = \tau_1 \exp\!\left(\frac{v_0}{kT}\right) \tag{72}$$

with

$$\tau_1 = \left(\frac{h}{kT}\right)\exp\left(\frac{\Delta H}{kT}\right) \tag{73}$$

The geometric mean of the two limiting times τ_2 and τ_1 is therefore

$$\tau_D = \tau_1\exp\left(\frac{v_0}{2kT}\right) = \sqrt{\tau_1\tau_2} \tag{74}$$

In our own analysis based on the shape of the observed dielectric spectra under discussion, we assume that τ_{D1} and τ_{D2}, if they do exist, differ by a factor of at least 10. The least of the two times (i.e., τ_{D2}) is distributed according to the model proposed by Fröhlich. In essence, we neglect a distribution in τ_{D1} due to the extreme difficulty of extracting this information from the observed dielectric spectrum.

The second term in Eq. (67) may be replaced using Fröhlich's[41] equation, with the result that

$$\frac{\Delta\varepsilon''}{\varepsilon_s - \varepsilon_\infty} - C_1\frac{\omega\tau_{D1}}{1 + \omega^2\tau_{D1}^2} + \frac{C_2}{P_2}\tan^{-1}\left(\frac{\sinh P_2/2}{\cosh\ln\omega\tau_D}\right)$$

$$+ \frac{C_3\omega T_P}{\left(1 - \omega^2 T_P T_J\right)^2 + \omega^2\left(T_P - \omega^2 T_J T_M T_P\right)^2} \tag{75}$$

where $P_2 \equiv v_0/kT$, and τ_1 and τ_2 are the lowest and highest limiting relaxation times of τ_{D2}. In the limit $P_2 \to 0$, the second term of Eq. (75) reduces to the Debye equation with a single relaxation time τ_{D2}. The results of fitting Eq. (75) are listed in Table IV. The analysis of the dielectric spectrum for most of the solutions in cyclohexane yields C_1, in which case with $P_2 = 1$, Eq. (75) produces a curve hardly distinguishable from the Debye curve with a single τ_D after excluding the submillimeter-wave absorption.

τ_D is determined from the condition

$$\frac{\partial(\Delta\varepsilon'')}{\partial f} = 0$$

which yields

$$\omega_0\tau_D = 1 \tag{76}$$

where $f_0 = (\omega_0/2\pi)$ Hz is the frequency at which $\Delta\varepsilon''$ is a maximum.

TABLE IV

Material	$\dfrac{\varepsilon_s - \varepsilon_\infty}{x_2}$	C_1	τ_{D1} (ps)	C_2	τ_{D2} (ps)	P_2	C_3	T_P (ps)	T_J (ps)	T_M (ps)
			Solvent: Cyclohexane							
Acetone	7.10	—	—	0.93	2.0	0	0.07	0.23	0.06	0.04
Hexanone-2	7.65	—	—	0.93	4.0	0	0.07	0.28	0.06	0.04
Heptanone-2	7.46	—	—	0.93	6.6	1.2	0.07	0.24	0.06	0.04
Heptanone-4	7.21	—	—	0.95	4.4	1.2	0.05	0.28	0.05	0.04
Nonanone-5	7.35	—	—	0.95	6.9	1.6	0.05	0.28	0.05	0.04
Undecanone-6	7.28	—	—	0.90	10.0	1.8	0.10	0.26	0.055	0.04
Acetophenone	9.04	—	—	0.91	9.4	1.6	0.09	0.24	0.08	0.04
			Solvent: Decalin							
Acetone	4.36	—	—	0.85	2.7	0	0.15	0.25	0.08	0.04
Heptanone-2	5.81	0.05	183	0.77	9.3	1.5	0.17	—	—	—
Heptanone-4	5.20	0.06	193	0.84	8.2	1.6	0.10	—	—	—
Undecanone-6	5.64	0.07	235	0.84	17.0	4.7	0.09	—	—	—

D. Relationship between τ_D and the Effective Molecular Radius

Following Kauzmann's,[42] application of the Eyring's theory of rate processes to dielectric relaxation, Hufnagel[43] has established for rigid molecules a relation between dielectric relaxation times and effective molecular radius, as follows:

$$\tau_D = \tau_0 \exp(\sigma r_{\text{eff}}) \tag{77}$$

where r_{eff} is given by:

$$r_{\text{eff}} = \left[\left(\frac{\cos\theta}{b} \right)^2 + \left(\frac{\sin\theta}{a} \right)^2 \right]^{-1/2} \tag{78}$$

The molecule is assumed to be ellipsoidal, with $2a$ and $2b$ the lengths of the molecule along the major and minor axis, respectively. The angle θ is that between the molecular dipole moment and the major axis. For a spherical molecule, $\theta = 0$ and $b = a = r_{\text{eff}}$. τ_0 is a time constant dependent on solvent and temperature; σ is the structural factor for the solvent and is also dependent on temperature.

Equation (77) has been worked out, on the basis of measurements of τ_D, for 24 molecules in different solvents and at different temperatures and values of r_{eff} calculated using Pauling radii of the molecules. From the values of σ and τ_0 in Table V, the molecular radii presented in Table VI were calculated.

TABLE V

Values of σ and τ_0 for Solvents

Solvent	σ ($\mathring{A}^{-1}$)	τ_0 (ps)
Cyclohexane	1.27	0.087
Decalin	1.51	0.091

TABLE VI

Molecular Radii Calculated from Relaxation Times

Solute	r_{eff} ($\mathring{A}$) in cyclohexane	r_{eff} ($\mathring{A}$) in decalin
Acetone	2.35	2.74
Hexanone-2	3.04	—
Heptanone-2	3.54	3.81
Heptanone-4	3.14	3.76
Nonanone-5	3.59	—
Undecanone-6	3.96	4.55
Acetophenone	3.90	—

E.   Dielectric Relaxation Mechanisms for Ketones

1.   Acetone and Hexanone-2

The calculated value of τ_{D2} for acetone in cyclohexane is 2.0 ps, which compared with a literature value[4] of 2.6 ps, and τ_{D2} for acetone in decalin is 2.7 ps. In both cases, an almost zero distribution in the relaxation times is observed and the weight factor for τ_{D1} is zero. The same picture is presented by hexanone-2, which has a similar dielectric spectrum.

The determined value of $\langle r_{eff} \rangle$ is in accord with the molecules being rigid and spherical. As a consequence of these characteristics, τ_{D2} is not appreciably dependent on the viscosity of the solvent, as is experimentally observed. The overall molecular rotation is suggested as the likely dielectric relaxation mechanism.

2.   Heptanones and the Higher Aliphatic-Series Ketones

Perusal of Table IV shows that even for the higher aliphatic-series ketones, the weight factor for τ_{D1} in cyclohexane is zero. Nevertheless, a weight factor of 5–7% for τ_{D2} is observed in decalin solutions. An explanation of the mechanism of τ_{D1} will be given later.

The τ_{D2} values for these ketones correspond to $\langle r_{\text{eff}} \rangle$ values slightly bigger than those of spherical, rigid molecules. This is especially so for undecanone-6. There is also a corresponding increase in the distribution of the relaxation time, reflected in an increase in Fröhlich's parameter P_2 from 1.2 for heptanone-2 to 1.8 for undecanone-6. However, the increase in P_2 is not so significant as to allow τ_{D2} to be resolved further.

The results seems to suggest that the various segments of the molecules, at least up to the heptanones in cyclohexane, coil up together to form a more or less ball-like structure. The space-filling models of these molecules show such a structure to be feasible. However, the flexibility of the various molecular segments increases with an increase in the number of methyl groups on each side of the ketone group, making such molecules ellipsoidal. The biggest increase observed in τ_{D2} is from nonanone-5 (6.9 ps) to undecanone-6 (10.0 ps) (both in cyclohexane). A corresponding increase in P_2 is also observed. These results point toward the possibility of a range of shapes existing for the bigger molecules, which would arise naturally from the molecular flexibility proposed in our model.

It follows from these results that the higher aliphatic-series ketone molecules relax predominantly via a tumbling motion. This implies that rotational diffusion takes place along the major axis of the ellipse, perpendicular to the direction of the dipole moment, involving a displacement of less volume. The small probability of these molecules relaxing along the minor axis of the ellipsoidal structure is reflected in a weight factor of 5–7% for τ_{D1} in a viscous solvent, due to an easy resolution of the spectrum.

The Debye times for hexanone-2 and nonanone-5 in cyclohexane are in agreement with those of Crossley.[45] These are the only two systems commonly studied. The τ_{D2} values for hexanone-2 are exactly the same, whereas our value of 6.9 ps for nonanone-5 compares well to the 7.2 ps calculated by Crossley.[45] However, our proposed model for the relaxation mechanism differs from that proposed by him. He suggests that for molecules such as hexanone-2, the main relaxation mechanism is through the intramolecular rotation of the terminal acetyl group. We feel that such a rotation will be severely hindered, if it occurs at all. This rotation is likely to lead to a wider distribution in the relaxation times than is observed experimentally. Furthermore, the relaxation due to this type of motion would be a contributing factor to the Poley absorption.[46] It will be extremely difficult to extract any information about an intramolecular relaxation mechanism from the resonance absorption, due to the similar orders of magnitude of the time scales for this mechanism and for the resonance absorption.

Our results are in agreement with the proposed model, which assumes an almost rigid structure for the smaller ketone molecules and an increase in the flexibility of the molecular segments with an increase in their size. A tumbling type of motion is proposed as the mechanism of dielectric relaxation for the larger ketone molecules.

F. Poley Absorption

It is now widely accepted that Poley dispersion is caused by the molecular librations[39] of a dipolar molecule in a potential well. These librations arise from the molecular inelastic collisions. In a dilute solution, the molecular dipole is continually being hit by the neighboring molecules.

Delker and Klages[38] have considered Poley absorption as a damped libration of molecules in a potential minimum. The damping is caused mainly by a random fluctuating force due to the neighbors. They identify T_M (Mori's time constant) as the correlation time of the torque moment acting on the dipolar molecule. The transfer of torque to the dipolar molecule as a result of inelastic collisions is assumed not to take place instantaneously. T_M is therefore a measure of how long a molecular collision lasts.

T_J has been identified as the correlation time of the angular momentum and is interpreted as the time during which the state of free angular rotation of a molecule is preserved. It is obviously inversely proportional to the damping constant. The Poley time constant, T_P, determines the angular frequency of maximum power absorption, ω_{max}.

These time constants are related through the following relations:

$$T_P = \frac{\Gamma}{\omega_{max}} \tag{79}$$

$$T_J = \frac{1}{\omega_{max}\Gamma} \tag{80}$$

where Γ is a dimensionless damping constant for the mechanism of libration, and

$$\frac{1}{\omega_{max}^2} = T_P T_J \tag{81}$$

which is the same as Eq. (70).

Values of T_M, T_P, and T_J calculated from experimental data and Eq. (75) are given in Table IV. T_M is dependent on the solvent, whereas T_P and T_J are implicitly dependent on the moment of inertia of the molecule through the damping constant. These are also dependent, to a lesser extent, on the solvent.

A detailed interpretation of these constants awaits fuller development of an analytical theory. A comparatively simple model of the Brownian motion of a particle in a periodic cosine potential is outlined in Section XIII.

XII. ITINERANT-OSCILLATOR MODEL OF BROWNIAN MOTION

An itinerant-oscillator (I.O.) model[47,48] assumes the dipolar molecule to be a planar rotator and its neighbors to be represented by an annulus. This annulus (cage) is assumed to undergo Brownian rotation, providing a separable mechanism for Debye-type relaxation at low frequencies, especially for a large friction coefficient ($\beta \geq 20$ ps^{-1}). The libration of the disk, which is coupled to the annulus, gives rise to a narrow resonance absorption. A major disadvantage of this model is that a simple analytical solution is possible only when the potential of the coupling between the dipole and its neighbor is harmonic. As a consequence, the model cannot produce loss curves broader than the Debye half-width value of 1.14 decades, nor can it generate rather broad power absorption curves such as are observed experimentally. Furthermore, the model does not account for relaxation behavior in strong electric fields.

The I.O. model may be extended to include the possibility of a dipole rotating relative to the inner disk. This means the model has three components: the outer annulus, the disk to which the annulus is coupled, and within the disk a dipolar molecule coupled to the disk. If it is assumed that the disk, as well as the annulus, is subject to damping, then in principle the model is capable of accounting for broad relaxation spectra with resonant absorption peaks. However, the I.O. model is conceptually unrealistic, especially for low-viscosity (ambient) fluids consisting of almost spherical molecules, because the harmonic coupling prevents the central molecule from undergoing large angular reorientations. A more realistic I.O. model would be one in which the coupling potential is spatially periodic, with several possible equilibrium positions. However, the numerical solution of the equations for such a model is an extremely difficult task, though some theoretical and computational studies have been published by Evans et al.[49]

XIII. BROWNIAN MOTION IN A PERIODIC COSINE POTENTIAL

A. Introduction

Praestgaard and van Kampen[50] have developed a new model of Brownian motion somewhat different from that proposed earlier by Lassier and Brot.[51] According to this model, a molecule has several stable positions of equilibrium, or potential minima. These are separated by potential barriers whose heights are larger than kT. The molecule librates about one of these minima; this libration is subject to damping and to a random force due to the surroundings. Because of the random energy fluctuations, the molecule may occasionally reach the height of the potential barrier and thus escape to a neighboring equilibrium site.

B. Solution of the Kramers Equation for the Planar Rotator with Several Equilibrium Positions

The periodic potential is assumed to be of the form

$$V(\theta) = -V_0\cos M\theta, \qquad -\pi \le \theta \le \pi \tag{82}$$

where M is the number of equilibrium positions in a period of 2π. We assume that the equation of motion is the Langevin equation,

$$I\ddot{\theta}(t) + I\beta\dot{\theta}(t) + V'(\theta) = \dot{W}(t) \tag{83}$$

where $V'(\theta)$ is the derivative of V with respect to θ, $\dot{W}(t)$ is the random torque, β is the friction coefficient, and I is the moment of inertia. The equation cannot be solved for the cosine potential due to its nonlinearity. However, the transition probability density, $\rho(\dot{\theta}, \theta, t | \dot{\theta}_0, \theta_0, 0)$ associated with a single molecule moving in one spatial dimension under the influence of a potential $V(\theta)$, evolves in time according to the Kramers equation:[52]

$$\frac{\partial \rho}{\partial t} + \dot{\theta}\frac{\partial \rho}{\partial \theta} - \frac{V'}{I}\frac{\partial \rho}{\partial \dot{\theta}} = \beta\frac{\partial}{\partial \dot{\theta}}\left(\dot{\theta}\rho + \frac{kT}{I}\frac{\partial \rho}{\partial \dot{\theta}}\right) \tag{84}$$

Equation 84 has been solved numerically by Reid[53] for (1) the δ-function initial probability distribution for angular velocity and position, and (2) the Maxwell–Boltzmann distribution in both velocity and position. The procedure is briefly reviewed and an account of its application is given below for the following cases:

1. Rate of escape of a particle from the potential minimum.
2. Frequency dependence of the mobility of ions in superionic conductors.
3. Dielectric relaxation of dipolar molecules.

The solution of the equation is of the form (with $\alpha^2 = kT/I$) (α has a different meaning than first introduced)

$$\rho = \exp\left(-\frac{\dot{\theta}^2}{4\alpha^2}\right)\sum_{n=0}^{\infty} D_n\left(\frac{\dot{\theta}}{\alpha}\right)\sum_{p=-\infty}^{\infty} A_p^n(t)\exp(ip\theta) \tag{85}$$

where

$$D_n(x) = \exp\left(\frac{x^2}{4}\right)(-1)^n\frac{d^n}{dx^n}\left[\exp\left(\frac{-x^2}{2}\right)\right] \tag{86}$$

are orthogonal Hermite polynomials. On substituting Eq. (85) into Eq. (84), we obtain a set of differential-difference equations as follows:

$$\dot{A}_p^n(t) + n\beta A_p^n(t) - i\gamma M\left[A_{p-M}^{n-1}(t) - A_{p+M}^{n-1}(t)\right]$$
$$+ ip\alpha\left[A_p^{n-1}(t) + (n+1)A_p^{n+1}(t)\right] \qquad (87)$$

where

$$\gamma = \frac{V_0}{2I\alpha} \qquad (88)$$

Coffey et al.[54] have described the formulation of Eq. (87) into an infinite matrix of matrices for the special case $M = 1$, and a similar method is used by Risken and Vollmer[55] in connection with calculation of the eigenvalues. Any $A_p^n(t)$ can be solved in terms of their initial values, depending on the initial conditions.

To solve Eq. (87), we restrict the maximum values of n and p such that $A_p^n(t) = 0$ for $n \geq N$ or $p \geq P$. The solution of Eq. (87) would require a computer with a large memory. This difficulty is avoided[53] by Laplace-transforming Eq. (87), and thus obtaining

$$(s + n\beta)\tilde{A}_p^n(s) - i\gamma M\left[\tilde{A}_{p-M}^{n-1}(s) - \tilde{A}_{p+M}^{n-1}(s)\right]$$
$$+ ip\alpha\left[\tilde{A}_p^{n-1}(s) + (n+1)\tilde{A}_p^{n+1}(s)\right] = A_p^n(0) \qquad (89)$$

where

$$\tilde{A}_p^n(s) = \mathscr{L}\left[A_p^n(t)\right] = \int_0^\infty A_p^n(t)e^{-st}\,dt \qquad (90)$$

Here $\mathscr{L}$ denotes the Laplace transform.

On defining the normalized sum and difference of coefficients,

$$\tilde{S}_p^n(s) = \frac{\tilde{A}_{-p}^n(s) + \tilde{A}_p^n(s)}{2A_0^0(0)} \qquad (91)$$

$$\tilde{D}_p^n(s) = \frac{\tilde{A}_{-p}^n(s) - \tilde{A}_p^n(s)}{2A_0^0(0)} \qquad (92)$$

we see that the terms for $+p$ and $-p$ in Eq. (89) can be combined; we also find that the difference terms are eliminated, leaving the following reduced recurrence relations:

$$
\begin{aligned}
&\left(a_n + p^2 b_n\right)\tilde{S}_p^n(s) + d_n\tilde{S}_{p-2M}^{n-2}(s) + (M - 2p)e_n\tilde{S}_{p-M}^{n-2}(s) \\
&- \left(2d_n - p^2 f_n\right)\tilde{S}_p^{n-2}(s) + (2p + M)e_n\tilde{S}_{p+M}^{n-2}(s) \\
&+ d_n\tilde{S}_{p+2M}^{n-2}(s) + (Mne_n - pg_n)\tilde{S}_{p-M}^n(s) + (Mne_n + pg_n)\tilde{S}_{p+m}^n(s) \\
&+ p^2 c_n\tilde{S}_p^{n+2}(s) = x_n y_n S_p^n(0)
\end{aligned}
\tag{93}
$$

with

$$
S_p^n(0) = \frac{A_{-p}^n(0) + A_p^n(0)}{2A_0^0(0)}
\tag{94}
$$

and

$$
\begin{aligned}
&a_n = x_n(x_n - \beta); \quad b_n = \alpha^2\left(nx_n + (n+1)y_n\right); \\
&c_n = (n+1)(n+2)\alpha^2 y_n; \quad d_n = \gamma^2 M^2 x_n; \quad e_n = \alpha\gamma M x_n \\
&f_n = \alpha^2 x_n; \quad g_n = \frac{b_n\gamma M}{\alpha}; \quad x_n = s + (n+1)\beta; \\
&y_n = x_n - 2\beta
\end{aligned}
\tag{95}
$$

In the matrix form

$$
[F][\tilde{S}_p^n(s)] = [S_p^n(0)], \qquad n = n_0, n_0 + 2, n_0 + 4, \ldots,
$$
$$
p = 0, 1, 2 \cdots
\tag{96}
$$

where the brackets indicate matrices. Note that n_0 may be either 0 or 1, and p now ranges over positive values only. The effect of Eqs. (91) and (92) is to reduce the sizes of the square matrices to be dealt with from order $[(2P + 1)N]$ to $[(P + 1)N/2]$.

C. Angular-Velocity Autocorrelation Function for Delta-Function Initial Conditions

The probability density at $t = 0$ for delta-function initial conditions for angular velocity and position is

$$
\rho(0) = \delta\left(\dot{\theta}(0) - \dot{\theta}_0\right)\delta\left(\theta(0) - \theta_0\right)
\tag{97}
$$

The angular-velocity autocorrelation function (acf) is expressed as

$$
\left\langle \dot{\theta}(t)\dot{\theta}(0)\right\rangle = \int \dot{\theta}(t)\dot{\theta}(0)\rho\delta\left(\dot{\theta}(0) - \dot{\theta}_0\right)\delta\left(\theta(0) - \theta_0\right) d\dot{\theta}_0\, d\theta_0\, d\dot{\theta}(t)\, d\theta(t)
\tag{98}
$$

where $\dot{\theta}(0)$ and $\theta(0)$ are the initial values (i.e., at $t = 0$) of angular velocity and position, respectively. From Eq. (85), it follows that

$$\langle \dot{\theta}(t)\dot{\theta}(0) \rangle = \Bigg\langle \dot{\theta}(0) \int_{-\infty}^{\infty} \int_{-\pi}^{\pi} \exp\left(-\frac{\dot{\theta}(t)^2}{4\alpha^2}\right) \sum_n D_n\left(\frac{\dot{\theta}(t)}{\alpha}\right)$$

$$\times \sum_{p=-\infty}^{\infty} A_p^n(t)\exp(ip\theta(t))\dot{\theta}(t)\, d\dot{\theta}(t)\, d\theta(t) \Bigg\rangle \quad (99)$$

Integrating with respect to $\theta(t)$, and using the orthogonality property

$$\int_{-\pi}^{\pi} \exp(ip\theta(t))\exp(-ir\theta(t)) = 2\pi\delta_{p,r} \quad (100)$$

we get

$$\langle \dot{\theta}(t)\dot{\theta}(0) \rangle_{\text{del}} = (2\pi)^{3/2}\alpha^2 A_0^1(t)\dot{\theta}(0) \quad (101)$$

Using the inverse Laplace transform of Eq. (91), Eq. (101) becomes

$$\langle \dot{\theta}(t)\dot{\theta}(0) \rangle_{\text{del}} = \mathcal{N} S_0^1(t)\dot{\theta}(0) \quad (102)$$

where $\mathcal{N} = (2\pi)^{3/2}\alpha^2$ is a normalization constant.

On taking the Laplace transform of Eq. (102), we obtain

$$\mathscr{L}\langle \dot{\theta}(t)\dot{\theta}(0) \rangle_{\text{del}} = \mathcal{N}\tilde{S}_0^1(s)\dot{\theta}(0) \quad (103)$$

where $\dot{\theta}(0) = \dot{\theta}_0$ is the velocity at $t = 0$, which is a constant for delta-function initial conditions.

The Laplace transform of the angular-velocity acf can thus be determined on solving the matrix equation (96), for $\tilde{S}_0^1(s)$, provided $S_p^n(0)\dot{\theta}(0)$ is known for all values of n and p.

The values $S_p^n(0)$ depend on the initial probability condition. We can work out $A_p^n(0)$ by assuming that the particle starts at $t = 0$ with a position θ_0 and a velocity $\dot{\theta}_0$ such that the initial probability density is given by

$$\rho(0) = \exp\left(-\frac{\dot{\theta}(0)^2}{4\alpha^2}\right) \sum_n \sum_p D_n\left(\frac{\dot{\theta}(0)}{\alpha}\right) A_p^n(0)\exp(ip\theta(0))$$

$$= \delta(\dot{\theta}(0) - \dot{\theta}_0)\delta(\theta(0) - \theta_0) \quad (104)$$

From Eq. (104) $A_p^n(0)$, and hence $S_p^n(0)$, can now be determined. We multiply both sides of this equation by $e^{-ir\theta(0)}$ integrate with respect to $\theta(0)$, and use the orthogonality property, obtaining

$$e^{-\dot{\theta}(0)^2/4\alpha^2}\sum_n D_n\left(\frac{\dot{\theta}(0)}{\alpha}\right)2\pi A_r^n(0) = e^{-ir\theta_0}\delta\left(\theta(0)-\theta_0\right) \tag{105}$$

We multiply both sides of this equation by $De_m(\dot{\theta}_0/\alpha)[= D_m(\dot{\theta}_0/\alpha)$ $\exp\frac{1}{4}(\dot{\theta}_0/\alpha)^2]$ and integrate with respect to $\dot{\theta}(0)$ to obtain

$$A_p^n(0) = \frac{De_n(\dot{\theta}_0/\alpha)e^{-ip\theta_0}}{n!(2\pi)^{3/2}\alpha} \tag{106}$$

and

$$\dot{\theta}_0 S_p^n(0) = \dot{\theta}_0\frac{De_n(\dot{\theta}_0/\alpha)}{n!}\cos p\theta_0 \tag{107}$$

If we assume further that the particle starts from the bottom of the potential minimum, that is, that $\theta_0 = 0$, Eq. (107) becomes

$$\dot{\theta}_0 S_p^n(0) = \dot{\theta}_0\frac{De_n(\dot{\theta}_0/\alpha)}{n!} \tag{108}$$

This is independent of p, and thus the computing time is much reduced. Knowing the column matrix $[\![\dot{\theta}_0 S_p^n(0)]\!]$, we can compute $\tilde{S}_0^1(s)$ numerically, using Eq. (96). Hence the Laplace transform of the angular-velocity acf is determined from Eq. (103).

D.   Angular-Velocity Autocorrelation Function for Equilibrium Initial Conditions

We assume that the initial velocity and position is Maxwell–Boltzmann (MB) distributed such that the initial probability is

$$\rho_{MB}(0) = \frac{C\exp\left(-I\dot{\theta}(0)^2/2 - V_0\cos M\theta(0)\right)}{kT} \tag{109}$$

where C is a constant. One can show that the equilibrium acf is given by

$$\mathcal{L}\langle\langle\dot{\theta}(t)\dot{\theta}(0)\rangle_{del}\rangle_{eq} = \mathcal{N}\langle\dot{\theta}_0\tilde{S}_0^1(s)\rangle_{eq} \tag{110}$$

where $\langle\ldots\rangle_{eq}$ denotes the average under the equilibrium initial conditions. The initial values are averaged over the MB distribution in velocity and position, with the result

$$\left\langle S_p^n(0)\dot{\theta}(0) \right\rangle_{eq} = \mathcal{N}C \int\int \exp\left(-\frac{\dot{\theta}_0^2}{2\alpha^2} \right) \dot{\theta}_0 De_m\left(\frac{\dot{\theta}_0}{\alpha} \right) \cos p\theta_0$$

$$\times \exp\left[-2(\gamma/\alpha)\cos M\theta_0 \right] d\dot{\theta}_0 \, d\theta_0 \qquad (111)$$

These averages vanish for $n > 1$, and for $n = 1$, we conclude that

$$\left\langle S_p^1(0)\dot{\theta}(0) \right\rangle_{eq} = \mathcal{N}C(2\pi)^{3/2}\alpha^2 i^p J_p\left(\frac{-i2\gamma}{\alpha} \right) \qquad (112)$$

From the theory of Bessel Functions,[56]

$$J_p(iz) = (i)^p I_p(z)$$

Equation (112) then becomes

$$\left\langle S_p^1(0)\dot{\theta}(0) \right\rangle_{eq} = \mathcal{N}C(2\pi)^{3/2}\alpha^2(-1)^p I_p\left(-\frac{2\gamma}{\alpha} \right) \qquad (113)$$

where $I_p(z)$ is a modified Bessel function of first order with a real argument. For small values of z, Eq. (113) becomes

$$\left\langle S_p^1(0)\dot{\theta}(0) \right\rangle_{eq} = NC(2\pi)^{3/2}\alpha^2 \sum_{r=0}^{\infty} \frac{1}{r!(p+r)!}\left(\frac{\gamma}{\alpha} \right)^{p+2r} \qquad (114)$$

For $n > 1$, $\left\langle S_p^n(0)\dot{\theta}(0) \right\rangle_{eq} = 0$. Equation (114) is found to converge for $\gamma/\alpha \leq 3$.

The equilibrium angular velocity can thus be computed using Eqs. (96) and (110).

E. Frequency-Dependent Mobility

The frequency-dependent mobility is given by the following equation:[57]

$$\mu(\omega) = \text{Real}\left[I(kT)^{-1} \int_0^{\infty} \exp(-i\omega t)\langle \dot{\theta}(0), \dot{\theta}(0) - \dot{\theta}(t) \rangle_{eq} \right] \qquad (115)$$

which becomes, on using Eq. (110)

$$\mu(\omega) = \mathcal{N} \, \text{Real}\left[I(kT)^{-1}\langle \dot{\theta}(0)\tilde{S}_0^1(s) \rangle_{eq} \right]_{s=i\omega} \qquad (116)$$

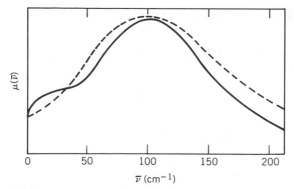

Figure 29.   Mobility spectrum $\mu(\omega)$ for AgI at 453 K: unbroken line, experimental; dashed line, numerical with $\alpha = 187$, $\beta = 374$, $\gamma = 187$ ($\times 10^{12}$ Hz). (Reproduced by permission from ref. 53.)

The normalized mobility spectrum is therefore just the angular-velocity acf evaluated under the MB initial distribution of velocity.

Acceptably close agreement is found in Fig. 29 between the experimental and numerical $\mu(\omega)$ spectra[58] for Ag I (a superionic superconductor).

F.   Applications to Static Mobility and Rate of Escape from a Well

The static or zero-frequency ionic/charge mobility of a particle is related to the $s = 0$ value of $\tilde{C}_{del}(s)$, as shown below in the discussion of the translational acf

$$\mathscr{L}\langle \theta(0), \theta(0) - \theta(t) \rangle_{del} = \frac{1}{s} \langle \theta(0), \theta(0) - s\tilde{\theta}(s) \rangle_{del}$$

which on using the property of the Laplace transform

$$\mathscr{L}\{\dot{\tilde{f}}(t)\} = s\tilde{f}(s) - f(0) \tag{117}$$

becomes

$$-\frac{1}{s}\langle \theta(0), \mathscr{L}\dot{\tilde{\theta}}(t) \rangle_{del} = \frac{1}{s}\langle \theta(0), \tilde{\theta}(s) \rangle_{del} \tag{118}$$

since

$$\langle \theta(0)\mathscr{L}\dot{\tilde{\theta}}(t) \rangle = -\langle \dot{\theta}(0)\theta(s) \rangle \tag{119}$$

Using Eq. (117), Eq. (118) can be written as

$$\mathscr{L}(\theta(0), \theta(0) - \theta(t))_{\text{del}} = \frac{1}{s^2}\left\langle \dot{\theta}(0), \mathscr{L}\tilde{\dot{\theta}}(t) + \theta(0)\right\rangle_{\text{del}}$$

$$= \frac{1}{s^2}\left\langle \dot{\theta}(0), \mathscr{L}\tilde{\dot{\theta}}(t)\right\rangle_{\text{del}} = \frac{1}{s^2}\tilde{C}_{\text{del}}(s) \quad (120)$$

because it is assumed that $\dot{\theta}(0) = 0$; hence $\langle\dot{\theta}(0), \theta(0)\rangle = 0$. Now we have

$$\tilde{C}_{\text{del}}(s) = s^2\mathscr{L}\langle\theta(0), \theta(0) - \theta(t)\rangle_{\text{del}} \quad (121)$$

Equation (121) implies that the value at $s = 0$ of $\tilde{C}_{\text{del}}(s)$ is the coefficient (with a change in sign) of the linear term in the time expansion of $\langle\theta(0), \theta(0) - \theta(t)\rangle_{\text{del}}$, known as the diffusion constant, or D.

From Eq. (115) with $s = 0$,

$$\mu(0) = \frac{I}{kT}D = \frac{D}{\alpha^2} \quad (122)$$

This is a well-known Einstein equation. The value of $\mu(0)$ can be derived from the algorithm involving δ-function initial conditions, since these produce a result almost identical to that based on the MB distribution in velocity and the position in the limit $s = 0$. This has been confirmed using the algorithms; the former initial conditions are preferred over the latter for the static case, because they reduce the computing time.

Marchesoni and Vij[59] have shown that $D = \kappa$, where κ is the rate of escape from a potential minimum for a periodic potential with $M = 1$. We show that $\kappa = 8\eta$, where η is the rate of escape calculated from Kramers's[52,60] predictions. Calculation of η is possible only in the limits of high β (overdamped limit) or $\beta \to 0$ (underdamped limit). The numerical technique therefore has the advantage of yielding values for the diffusion coefficient (or the rate of escape from a potential minimum) over a wide range of friction coefficients. In the two limiting cases, there is a close agreement between the refined version[60] of Kramers's results and our results (Fig. 30).

G.   Dielectric Loss and the Power Absorption Coefficient

The dielectric spectrum obtained numerically from this model for $M > 1$ produces a relaxation and libration spectrum; the weight factors for the two processes are dependent of γ, α, β, and M.

The complex permittivity $\varepsilon^*(\omega)$ is related to the one-sided Fourier transform of $K_{\text{eq}}(t)$, given by the Kubo–Scaife equation[57,61] (in the absence of

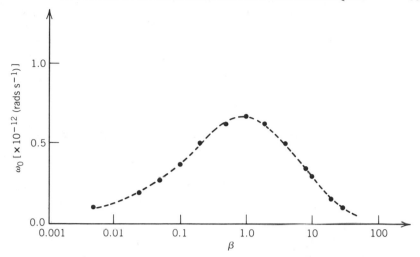

Figure 30.   The peak position ω_0 versus β for $M = 2$, $\alpha = 1$, $\gamma = 4$. Numerical results (●) compared with theoretical predictions of Praestgaard and van Kampen[50] (dashed line). (Reproduced by permission from ref. 59.)

internal-field corrections):

$$\varepsilon^*(\omega) = \varepsilon'(\omega) - i\varepsilon''(\omega) = \varepsilon_\infty + (\varepsilon_s - \varepsilon_\infty)\left[1 - i\omega\int_0^\infty K_{eq}(t)\exp(-i\omega t)\,dt\right]$$

$$(123)$$

Equation (123) can be written for $\varepsilon''(\omega)$ as

$$\varepsilon''(\omega) = \omega(\varepsilon_s - \varepsilon_\infty)\mathscr{L}\left[K_{eq}(t)\right] \tag{124}$$

$K_{eq}(t)$ is the normalized orientational acf, defined as

$$K_{eq}(t) = \frac{\langle\cos\theta(t)\cos\theta(0)\rangle_{eq}}{\langle\cos^2\theta(0)\rangle_{eq}} \tag{125}$$

where $\langle\ldots\rangle_{eq}$ denotes the average evaluated under the equilibrium initial conditions.

On adopting a procedure similar to that used in deriving Eq. (102), we find that

$$\mathscr{L}\left[K_{eq}(t)\right] = \mathscr{L}\left[\frac{\left\langle\cos\theta(t)\cos\theta(0)\right\rangle_{eq}}{\left\langle\cos^2\theta(0)\right\rangle_{eq}}\right]$$

$$= \left\langle\tilde{S}_1^0(s)\cos\theta(0)\right\rangle_{eq} \tag{126}$$

For equilibrium initial conditions, Eq. (96) can be written as

$$[\![F]\!][\![\left\langle\cos\theta_0\tilde{S}_p^n(s)\right\rangle]\!] = [\![\left\langle\cos\theta_0 S_p^n(0)\right\rangle_{eq}]\!] \tag{127}$$

where the elements of the matrix on the right hand side of Eq. (127) are evaluated as follows:

$$\left\langle\cos\theta_0 S_p^n(0)\right\rangle_{eq} = \int\cos\theta_0 S_p^n(0)\rho_{MB}(0)\,d\theta_0\,d\dot\theta_0 \tag{128}$$

where $\rho_{MB}(0)$ is the initial Maxwell–Boltzmann distribution given by Eq. (109).

Simple algebra shows that averages under equilibrium conditions vanish for $n > 0$, and those for $n_0 = 0$ are as follows: For $M = 1$,

$$\left\langle\cos\theta_0 S_p^0(0)\right\rangle = \tfrac{1}{2}\left(C_{1p-11} + C_{p+1}\right) \tag{129a}$$

For $M = 2$,

$$\left\langle\cos\theta_0 S_{2p+1}^0(0)\right\rangle_{eq} = \tfrac{1}{2}\left(C_p + C_{p+1}\right) \tag{129b}$$

and for $M = 3$,

$$\left\langle\cos\theta_0 S_{3p-1}^0(0)\right\rangle_{eq} = \left\langle\cos\theta_0 S_{3p+1}^0(0)\right\rangle_{eq} = \frac{C_p}{2} \tag{129c}$$

where

$$C_p = (-1)^p I_p\left(-\frac{2\gamma}{\alpha}\right) \tag{130}$$

If we know the right-hand side of Eq. (127), we can solve for $\left\langle\tilde{S}_1^0(s)\cos\theta_0\right\rangle_{eq}$.

The dielectric spectrum is obtained from Eq. (124) and is found to be in reasonable agreement with experiment (Fig. 31). Marchesoni and Vij[59] have found that the numerical dielectric spectrum produces three time constants: τ_e, corresponding to the rate of escape from one well into another; $\tau_d = \beta^{-1}$, corresponding to damping; and τ_p, corresponding to the libration of a dipolar molecule within a well.

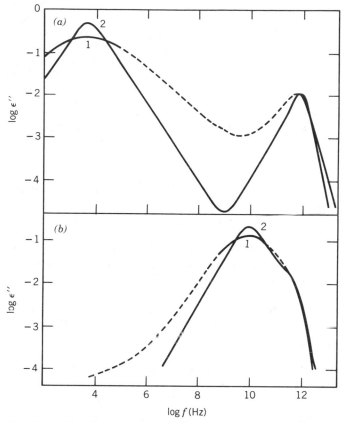

Figure 31.   Plots of log ε'' versus log f for 10% (volume/volume) bromobenzene in decalin. (a) At 144 K: 1, experimental; 2, numerical with $\alpha = 0.7$, $\beta = 12$, $\gamma = 2.85$ THz, $M = 4$. (b) At 293 K: 1, experimental; 2, numerical with $\alpha = 1$, $\beta = 6$, $\gamma = 0.5$, $M = 4$. (Reproduced by permission from ref. 62.)

The power absorption coefficient $A(\omega)$ is related to the dielectric loss through the equation (see ref. 61)

$$A(\omega) = \frac{\omega \varepsilon''(\omega)}{n(\omega)c} \qquad (131)$$

where $n(\omega)$ is the refractive index.

The weight factors of the various mechanisms contributing to $\varepsilon''(\omega)$ and hence $A(\omega)$ is dependent on the ratio of γ/α and β. The analysis of the relevant results is to be published separately.

Acknowledgments

The authors are grateful to Professor B. K. P. Scaife for very useful discussions concerning the contents and their presentation. One of the authors (J. K. V.) is especially indebted to him for helpful criticism and encouragement. This author is also thankful to Dr C. Reid, Dr. M. W. Evans, Dr F. Marchesoni, Dr W. T. Coffey, Mr. M. Helker, and Mr. Séan White for useful discussions.

The Apollo laser system used for some of the experiments was purchased by Dr. M. W. Evans through an S.E.R.C research grant.

The Trinity College Dublin Trust is thanked for financial assistance toward a visit to Mainz by J. K. V. in 1984, and the D.A.A.D., for a Visiting Professorship in 1982.

References

1. A. B. Bronwell and R. E. Beam, *Theory and Applications of Microwaves*, McGraw-Hill, New York, 1947, p. 337.

2. A. Suggett, *Dielectric and Related Molecular Processes*, Vol. 1, The Chemical Society, London, 1972, p. 100.

3. A. H. Price and G. H. Wegdam, *J. Phys. E*, **10**, 478 (1977).

4. F. Hufnagel and H. Hammel, Microwave Laboratory Instruction Sheets, University of Mainz, Mainz (unpublished); H. Hammel, Diplomarbeit, University of Mainz, Mainz, 1978.

5. W. C. Johnson, *Transmission Lines and Networks*, McGraw-Hill, New York, 1950.

6. J. A. Lane and J. A. Saxton, *Proc. R. Soc. Lond.*, **A213**, 400 (1952).

7. G. Klages, F. Hufnagel, and H. Kramer, unpublished work; H. Kramer, Ph.D dissertation, University of Mainz, Mainz, 1959.

8. S. K. Garg, H. Kilp, and C. P. Smyth, *J. Chem. Phys.*, **43**, 2341 (1965).

9. H. Kilp, *Z. Angew. Physik*, **30**, 288 (1970).

10. H. Kilp, *J. Phys. E*, **10**, 985 (1977).

11. W. Haydl, Fraunhöfer Institute for Applied Solid State Physics, Freiburg, West Germany, The diode is fabricated by Dr. W. Haydl.

12. U. Stumper, *Rev. Sci. Instruments*, **44**, 165 (1973).

13. W. T. Read, Jr., *Bell Syst. Tech. J.*, **37**, 401 (1958).

14. T. Misawa, in *Proceedings of a Symposium on Submillimeter Waves*, Polytechnic Press of Brooklyn, New York, 1970, pp. 53–67.

15. J. E. Chamberlain and H. A. Gebbie, *Nature*, **206**, 602 (1965).

16. J. Chamberlain, in *Proceedings of a Conference on Measurement of High Frequency Dielectron Properties of Materials*, J. Chamberlain and G. W. Chantry, eds., IPC Science and Technology Press, Surrey, 1973, p. 104.

17. F. Kremer and J. R. Izatt, *Int. J. Infrared Millimeter Waves*, **2**, 675 (1981).

18. F. Kremer and L. Genzel, *Sixth International Conference on Infrared and Millimeter Waves*, Miami, December 1981.

19. W. E. Lamb, Jr., *Phys. Rev.*, **70**, 308 (1946).

20. H. A. Gebbie and R. A. Bohlander, *Appl. Optics*, **11**, 723 (1972).

21. F. Hufnagel, to be published.

22. H. A. Gebbie, N. W. B. Stone, and F. D. Findlay, *Nature*, **202**, 685 (1964).

23. H. A. Gebbie, N. W. B. Stone, F. D. Findlay, and E. C. Pyatt, *Nature*, **205**, 377 (1965).

24. G. W. Chantry, *Submillimetre Wave Spectroscopy*, Academic, London, 1971, p. 341.

25. T. Y. Chang, T. J. Bridge, and E. G. Burkhardt, *Appl. Phys. Lett.*, **17**, 249 (1970).

26. S. F. Dyubko, V. A. Svich, and L. D. Fesenko, *Sov. Phys. Tech. Phys.*, **16**, 592 (1972).

27. S. F. Dyubko, V. A. Svich, and L. D. Fesenko, *Sov. Phys. Tech. Phys.*, **18**, 1121 (1974).

28. J. O. Henningsen, *IEEE J. Quant. Electronics*, **QE-14**, 958 (1978).

29. F. R. Petersen, K. M. Evenson, D. A. Jennings, and A. Scalabrin, *IEEE J. Quant. Electronics*, **QE-16**, 319 (1980).

30. C. J. Reid, *Spectrochim. Acta*, **38A**, 697 (1982).

31. J. K. Vij, C. J. Reid, and M. W. Evans, *Chem. Phys. Lett.*, **92**, 528 (1982).

32. J. K. Vij, C. J. Reid, and M. W. Evans, *Mol. Phys.*, **50**, 935 (1983).

33. C. J. Reid, J. K. Vij, P. L. Rosselli, and M. W. Evans, *J. Mol. Liq.*, **29**, 37 (1984).

34. F. Hufnagel, M. Helker, and J. K. Vij, to be published.

35. B. Quentrec and P. Bézot, *Mol. Phys.*, **27**, 879 (1974).

36. H. Mori, *Progr. Theor. Phys.*, **33**, 423 (1965).

37. H. Mori, *Progr. Theor. Phys.*, **34**, 399 (1966).

38. R. Delker and G. Klages, *Z. Naturforsch.*, **36A**, 611 (1981).

39. M. W. Evans and C. J. Reid, in *Molecular Dynamics*, M. W. Evans, G. J. Evans, W. T. Coffey, and P. Grigolini, Wiley-Interscience, New York, 1982, Chapter IV.

40. C. J. Reid and J. K. Vij, *J. Chem. Soc., Faraday Trans. II*, **78**, 1649 (1982).

41. H. Fröhlich, *Theory of Dielectrics*, Oxford University Press, Oxford, 1958, p. 95.

42. W. Kauzmann, *Rev. Mod. Phys.*, **14**, 12 (1942).

43. F. Hufnagel, *Z. Naturforsch.*, **25A**, 1143 (1970).

44. M. D. Magee and S. Walker, *J. Chem. Phys.*, **50**, 1019 (1969).

45. J. Crossley, *J. Chem. Phys.*, **56**, 2549 (1972).

46. J. Ph. Poley, *J. Appl. Sci. Res. (B)*, **4**, 337 (1955).

47. N. E. Hill, *Proc. Phys. Soc.*, **82**, 723 (1963).

48. J. H. Calderwood and W. T. Coffey, *Proc. R. Soc. Lond.*, **A356**, 269 (1977).

49. M. W. Evans, M. Ferrario, and W. T. Coffey, *Adv. Mol. Relax. Int. Processes*, **20**, 1 (1981).

50. E. Praestgaard and N. G. van Kampen, *Mol. Phys.*, **43**, 33 (1981).

51. B. Lassier and C. Brot, *Chem. Phys. Lett.*, **1**, 591 (1968); *ibid.*, *J. Chem. Phys.*, **65**, 1723 (1968); *ibid.*, *Disc. Faraday Soc.*, **48**, 39 (1969).

52. H. A. Kramers, *Physica*, **7**, 284 (1940).

53. C. J. Reid, *Mol. Phys.*, **49**, 331 (1983).

54. W. T. Coffey, P. Grigolini, and P. Marin, *Proc. R. Irish Acad.*, **82A**, 167 (1982).

55. H. Risken and H. D. Vollmer, *Z. Phys.*, **B31**, 209 (1978).

56. I. S. Gradshteyn and I. M. Ryzhik, *Table of Integrals, Series and Products*, Academic, New York, 1965.

57. R. Kubo, *J. Phys. Soc. Jpn.*, **12**, 570 (1957).

58. P. Fulde, L. Pietronera, W. R. Schneider, and S. Strassler, *Phys. Rev., Letters* **35**, 1776 (1975).

59. F. Marchesoni and J. K. Vij, *Z. phys.* **B58**, 187 (1985).

60. M. Büttiker, E. P. Harris, and R. Landauer, *Phys. Rev.*, **B28**, 1268 (1983).

61. B. K. P. Scaife, *Complex Permittivity*, The English Universities Press, London, 1971.

62. C. J. Reid and J. K. Vij, *J. Chem. Phys.*, **79**, 4624 (1983).

AUTHOR INDEX

Numbers in parentheses are reference numbers and indicate that the author's work is referred to although his name is not mentioned in the text. Numbers in *italics* show the pages on which the complete references are listed.

SUBJECT INDEX